Chemistry in Context

A Project of the American Chemical Society

Chemistry in Context

Applying Chemistry to Society

Sixth Edition

Lucy Pryde Eubanks
Clemson University

Catherine H. Middlecamp
University of Wisconsin–Madison

Carl E. Heltzel
Environmental Risk Management (ErMC^2)

Steven W. Keller
University of Missouri–Columbia

A Project of the American Chemical Society

McGraw-Hill
Higher Education

Boston   Burr Ridge, IL   Dubuque, IA   New York   San Francisco   St. Louis
Bangkok   Bogotá   Caracas   Kuala Lumpur   Lisbon   London   Madrid   Mexico City
Milan   Montreal   New Delhi   Santiago   Seoul   Singapore   Sydney   Taipei   Toronto

CHEMISTRY IN CONTEXT: APPLYING CHEMISTRY TO SOCIETY, SIXTH EDITION

Published by McGraw-Hill, a business unit of The McGraw-Hill Companies, Inc., 1221 Avenue of the Americas, New York, NY 10020. Copyright © 2009 by the American Chemical Society. All rights reserved. Previous editions © 2006, 2003, 2000, 1997 and 1994. No part of this publication may be reproduced or distributed in any form or by any means, or stored in a database or retrieval system, without the prior written consent of The McGraw-Hill Companies, Inc., including, but not limited to, in any network or other electronic storage or transmission, or broadcast for distance learning.

Some ancillaries, including electronic and print components, may not be available to customers outside the United States.

This book is printed on acid-free paper.

1 2 3 4 5 6 7 8 9 0 DOW/DOW 0 9 8

ISBN 978–0–07–304876–5
MHID 0–07–304876–3

Publisher: *Thomas D. Timp*
Senior Sponsoring Editor: *Tamara Good-Hodge*
Managing Developmental Editor: *Shirley R.. Oberbroeckling*
Senior Marketing Manager: *Todd Turner*
Senior Project Manager: *Gloria G. Schiesl*
Senior Production Supervisor: *Kara Kudronowicz*
Senior Media Project Manager: *Sandra Schnee*
Senior Coordinator of Freelance Design: *Michelle D. Whitaker*
Cover/Interior Designer: *Rick Noel*
(USE) Cover Image: *© Jeremy Woodhouse/Masterfile*
Lead Photo Research Coordinator: *Carrie Burger*
Photo Research: *Pam Carley*
Supplement Producer: *Mary Jane Lampe*
Compositor: *Aptara*
Typeface: *10/12 Times Roman*
Printer: *R. R. Donnelley Willard, OH*

The credits section for this book begins on page 569 and is considered an extension of the copyright page.

Library of Congress Cataloging-in-Publication Data

Chemistry in context : applying chemistry to society — 6th ed. / Lucy Pryde Eubanks ... [et al.].
  p. cm.
 Includes index.
  ISBN 978–0–07–304876–5 — ISBN 0–07–304876–3 (hard copy : alk. paper) 1. Biochemistry. 2.
Environmental chemistry. 3. Geochemistry. I. Eubanks, Lucy T.
QD415.C482 2009
540 — dc22

2007040348

Brief Contents

Contents

Contents

Preface

Following in the tradition of its first five editions, the goal of *Chemistry in Context*, Sixth Edition, is to establish chemical principles on a need-to-know basis within a contextual framework of significant social, political, economic, and ethical issues. We believe that by using this approach, students not majoring in a science develop critical thinking ability, the chemical knowledge and competence to better assess risks and benefits, and the skills that can enable them to make informed and reasonable decisions about technology-based issues. The word *context* derives from the Latin word meaning "to weave." Thus, the spiderweb motif on the cover continues with this edition because a web exemplifies the complex connections between chemistry and society.

Chemistry in Context is not a traditional chemistry book for nonscience majors. In this book, chemistry is woven into the web of life. The chapter titles of *Chemistry in Context* reflect today's technological issues and the chemistry principles imbedded within them. Global warming, acid rain, alternative fuels, nutrition, and genetic engineering are examples of such issues. To understand and respond thoughtfully in an informed manner to these vitally important issues, students must know the chemical principles that underlie the sociotechnological issues. This book presents those principles as needed, in a manner intended to better prepare students to be well-informed citizens.

Organization

The basic organization and premise remain the same as in previous editions. The focal point of each chapter is a real-world societal issue with significant chemical context. The first six chapters are core chapters in which basic chemical principles are introduced and expanded upon on the need-to-know basis. These six chapters provide a coherent strand of issues focusing on a single theme—the environment. Within them, a foundation of necessary chemical concepts is developed from which other chemical principles are derived in subsequent chapters. Chapters 7 and 8 consider alternative (nonfossil-fuel) energy sources: nuclear power, batteries, fuel cells, and the hydrogen economy. The emphases in the remaining chapters are carbon-based issues and chemical principles related to polymers, drugs, nutrition, and genetic engineering. Thus, a third of the text has an organic/biochemistry flavor. These latter chapters provide students with the opportunity to focus on additional interests beyond the core topics, as time permits. Most instructors teach seven to nine chapters in a typical one-semester course. However, others find that *Chemistry in Context* contains ample material for a two-semester course.

What's New and Improved

Art Program

The art program for the sixth edition has been updated for consistency and accuracy, with new art added where needed. Chemical structures emphasize the important details of bonding and reactive sites. Details of chemical processes are emphasized in many figures, and real-world data have been updated and their presentation clarified to help students understand the information.

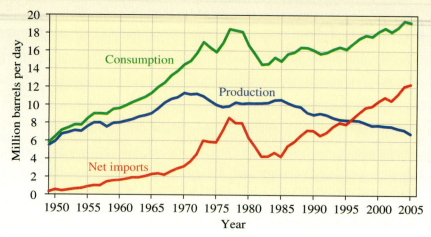

Figure 4.11

U.S. petroleum product use, domestic production, and imports. At present, more than 60% of the total oil used in the United States is imported, and projections show oil imports will continue to increase.

Source: Department of Energy, Energy Information Administration, *Annual Energy Review 2005*.

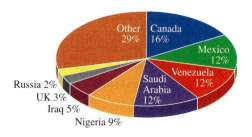

Figure 4.12

Sources of crude oil and petroleum products imported by the United States in 2004.

Source: Department of Energy/EIA.

Molecular Representations

Many types of representations are available to help the student understand molecular architecture. Lewis structures give essential information about bonding and can be interpreted to predict bond angles shown in structural formulas. Space-filling models provide another representation. This sixth edition uses the newest version of Spartan to produce charge-density diagrams. This type of representation shows charge distribution within molecules and is particularly helpful when explaining solubility, acidic and basic properties, and the reasons certain reactions take place.

More complex molecular structures are shown using structural formulas, but the sixth edition makes increased use of line-angle representations as well.

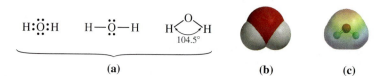

Figure 3.10

Representations of H_2O.
(a) Lewis structures and structural formula; (b) Space-filling model; (c) Charge-density model.

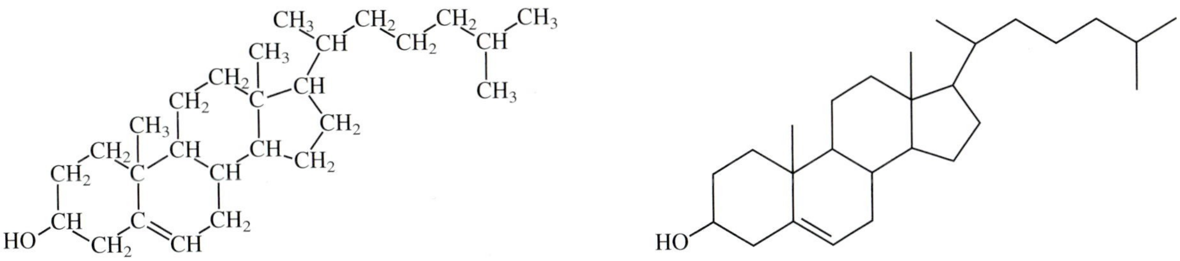

Figure 10.22
Representations of cholesterol.

Chapter 0, "Why the Spiderweb?"

Chapter 0, first introduced as a feature of the fifth edition, again walks the student through how to use all of the resources available to them in *Chemistry in Context*. Much of the information found in this preface in previous editions is now placed in the new introductory chapter. Although written for students, Chapter 0 also serves to explain to instructors the pedagogy, problem-solving opportunities, and many of the media resources of the sixth edition.

New and Updated Content

The major focus of a new edition is to update topical content. All information is as up-to-date as possible using a printed format. The resources on the Internet allow students to acquire real-time data, seek out current information, and make their own risk–benefit analysis about topics at the interface between science and society.

This edition introduces many new or expanded topics while keeping the same chapter organization. The discussion of air quality focuses on production of air pollutants and how they interact to affect both outdoor and indoor air quality. The role of particulates is more fully explored, as is the reality that air pollutants do not respect international borders. Looking at a more global perspective, ozone depletion is an issue that cannot be relegated to the past, despite successes in identifying major causes and obtaining the cooperation of much of the global community. Certainly global warming and its effects on climate change are now at the forefront of international attention. This edition provides more background for understanding Earth's energy processes and the molecular mechanisms producing warming, presents an accumulating base of scientific knowledge, and discusses the global implications of our actions (or lack thereof). The sixth edition has a sharper focus on biofuels, discussing alternatives to help us move away from dependence on fossil fuels. Measuring acidic precipitation has been given increased attention, as has the role of reactive nitrogen in understanding the problems of acid rain. Discussion of nuclear energy considers issues of the past, explores international practices, and assesses opportunities for resurgence in this industry. Other energy-related topics in the sixth edition include expanded coverage of newer generations of fuel cells, hybrid cars, and the hydrogen economy. Both biodegradable plastics and recycling receive more attention in this edition. There is new coverage of nanomedicine, the union of nanoscale technology and medical treatment, and an expanded discussion of drug discovery through combinatorial synthesis. Nutrition topics have again been reorganized to enable students to better understand and make decisions about popular diets. Current information on stem cell research is introduced in the final chapter of the text, along with cloning, transgenic foods, and the Human Genome Project.

Running through the text are the themes of green chemistry, energy, global connections, and applications of nanotechnology. Often these issues are connected. For example, how can the demand for increased energy use, necessary for equitable economic development around the world, be met in a globally responsible manner? How is energy produced in different parts of the world? Can nanotechnology help with challenges ranging from the safe storage of hydrogen for cleaner burning fuels to the development of

new sunscreens? What is the role of green chemistry in developing alternative products and processes? A broader view of how the themes mentioned here are carried out in the sixth edition can be found in the content grid on the Instructor Center of the *Online Learning Center*.

New and Updated Resources on the *Online Learning Center* (www.mhhe.com/cic)

The *Online Learning Center* (OLC) is a comprehensive, book-specific Web site offering excellent tools for both the instructor and the student. Instructors can create an interactive course with the integration of this site, and a secured Instructor Center stores your essential course materials to save you preparation time before class. This Instructor Center offers the Instructor's Resource Guide, additional labs, and a Presentation Center. The Student Center offers Web Exercises, Figures Alive Interactives, and quiz questions for each chapter. The *Online Learning Center* content has been created for use in most course management systems.

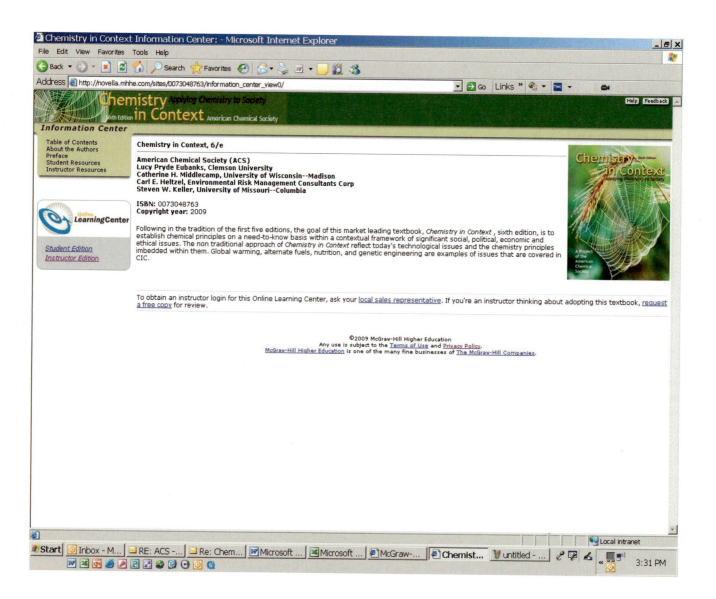

Figures Alive Interactives, marked by this icon ![icon] near the figure in the text, lead the student through the discovery of various layers of knowledge inherent in the figure and enables them to develop their own understanding. Each chapter has an interactive learning experience tied to a specific figure in the chapter. The self-testing segments built into *Figures Alive!* are based on the same categories as the chapter-end problems—*Emphasizing Essentials, Concentrating on Concepts,* and in many cases, *Exploring Extensions.*

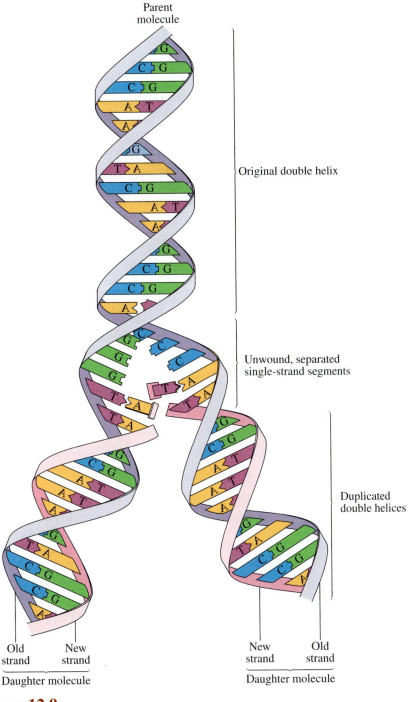

Figure 12.9

Diagram of DNA replication. The original DNA double helix (*top portion of figure*) partially unwinds, and the two complementary portions separate (*middle*). Each of the strands serves as a template for the synthesis of a complementary strand (*bottom*). The result is two complete and identical DNA molecules.

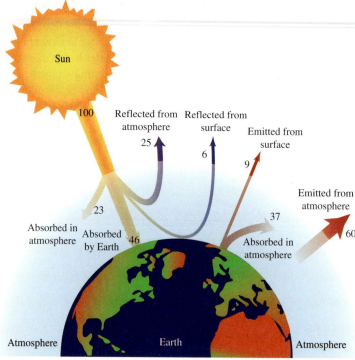

 **Figure 3.2**

Earth's energy balance by percent. Yellow represents a mixture of wavelengths. Shorter wavelengths of radiation are shown in blue, longer in red.

The *Instructor's Resource Guide*, edited by Anne K. Bentley (Lewis & Clark College), can be found on the *Online Learning Center* under Instructor Resources. The guide contains:

- A chemical topic matrix that lists chemical principles commonly covered in a general chemistry course.
- Answers for suggested responses to many of the open-ended questions in the Consider This and the solutions to the in-chapter and chapter-end exercises and questions.
- The instructors guide for the laboratory experiments.

The *McGraw-Hill Presentation Center* is a multimedia collection of visual resources allowing instructors to utilize artwork from the text in multiple formats to create customized classroom presentations, visually based tests and quizzes, dynamic course Web site content, or attractive printer support materials. The *McGraw-Hill Presentation Center* is found in the instructor center of the *Online Learning Center* and contains the images, photos, and tables from the text. To access the Instructor materials, request registration information from your McGraw-Hill sales representative.

Instructor's Testing and Resources Online contains the Test Bank written by the author team of Julie M. Smist (Springfield College), Marcia L. Gillette (Indiana University-Kokomo), Mark B. Freilich (University of Memphis), Thomas Zona (Illinois State University), Amy J. Phelps (Middle Tennessee State University), and Eric Bosch (Southwest Missouri State University). This resource contains approximately 65 multiple-choice questions for every chapter. The questions are comparable to the problems in the text in content coverage. The Test Bank is formatted for easy integration into the following course management systems: WebCT, and Blackboard. You may also choose to use these questions as models for writing your own classroom-specific test questions.

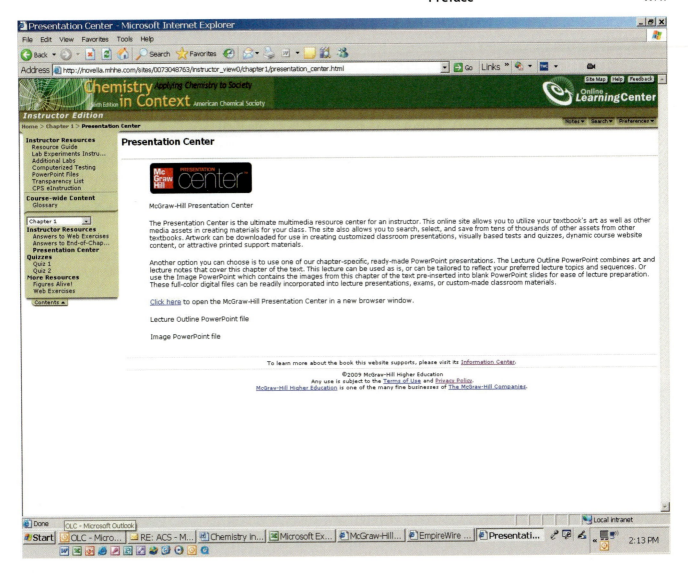

Other New and Updated Resources

For those whose course includes a laboratory component, a *Laboratory Manual*, compiled and edited by Gail A. Steehler (Roanoke College), is available for the sixth edition. The experiments use microscale equipment (wellplates and Beral-type pipets) and common materials. Project-type and cooperative–collaborative laboratory experiments are included. *New* experiments are included on ozone and biodiesel. Additional experiments are available on the *Online Learning Center*, as is the Instructor's Resource Guide.

Special Acknowledgments

It is always a pleasure to bring a new textbook or new edition to fruition. But the work is not done by just one individual. It is a team effort, one that comprises the work of many talented individuals. The sixth edition builds on the proud tradition of prior author teams, led by A. Truman Schwartz of Macalester College for the first and second editions, and by Conrad L. Stanitski from the University of Central Arkansas for the third and fourth editions. We have been fortunate to have the unstinting support and encouragement of the ACS Division of Education, led during much of the preparation of this edition by the now-retired Sylvia A. Ware. The new director, Mary M. Kirchoff, continues this legacy of enthusiasm and understanding of our mutual goals. We also recognize the able assistance of Jerry A. Bell and Corrie Y. Kuniyoshi of the ACS Division of Education office during preparation of the sixth edition.

The McGraw-Hill team has been superb in all aspects of this project. Marty Lange (Director of Editorial), Thomas Timp (Publisher), Tamara Hodge (Senior Sponsoring Editor), and Shirley Oberbroeckling (Senior Developmental Editor) led this outstanding team. Todd Turner serves as the Marketing Manager. The Senior Project Manager is Gloria Schiesl, who coordinates the production team of Carrie Burger (Lead Photo Researcher), Kara Kudronowicz (Production Supervisor), and Melissa Leick (Projects Coordinator). The Lead Media Producer is Daryl Bruflodt and Sandra Schnee serves as Senior Media Project Manager. The team also benefited from the knowledgeable editing of Linda Davoli and from the persistent work of Pam Carley in tracking down elusive images. Dwaine Eubanks of LATEst IDEas, Inc., brought both his chemical knowledge and computer-based artistic skills together to continue the high standard for the art in this edition. His ability to respond quickly and expertly to the needs of the author team was integral to our success.

The sixth edition is the product of a collaborative effort among writing team members—Lucy Pryde Eubanks, Catherine H. Middlecamp, Carl E. Heltzel, and Steven W. Keller. This is the maiden voyage in this realm for Steve Keller as a new coauthor and colleague. We welcome him to the team and have benefited from his diverse expertise.

We are very excited by the new features of this sixth edition, which exemplify how we continue to "press the envelope" to bring chemistry in creative, appropriate ways to nonscience majors, while being honest to the science. We look forward to your comments.

Lucy Pryde Eubanks
Senior Author and Editor-in-Chief
January 2008

Further Acknowledgments

Reviewers for *Chemistry in Context*, Sixth Edition

A sincere thank you to the following individuals for their comments:

John R. Allen	*Southeastern Louisiana University*
Edward J. Baum	*Grand Valley State University*
Eric Bosch	*Missouri State University*
Donna Budzynski	*San Diego Mesa College*
Cynthia H. Coleman	*State University of New York–Potsdam*
Rebecca W. Corbin	*Ashland University*
Sheree J. Finley	*Alabama State University*
Lawrence A. Fuller	*State University of New York–Oswego*
Anne Gaquere	*University of West Georgia*
Amy Grant	*El Camino College*
Rick D. Huff	*Genesee Community College*
Milt Johnson	*University of South Florida*

Margaret G. Kimble	*Indiana University-Purdue University–Fort Wayne*
Kimball S. Loomis	*Century College*
Elizabeth Maschewske	*Grand Valley State University*
S. Walter Orchard	*Tacoma Community College*
Somnath Sarkar	*Central Missouri State University*
Stacy Sparks	*University of Texas at Austin*
Heeyoung Tai	*Miami University*
Joseph C. Tausta	*Oneonta State College*
Victor H. Vilchiz	*Virginia State University*
Jerry Walsh	*University of North Carolina at Greensboro*
Lou Wojecinski	*Kansas State University*
Thomas A. Zona	*Illinois State University*
Martin G. Zysmilich	*George Washington University*

Reviewers for *Chemistry in Context Laboratory Manual*, Sixth Edition survey:

Frank Carey	*Wharton County Junior College*
Donald W. Carpenetti	*Marietta College*
Marguerite Crowell	*Plymouth State University*
Al Gotch	*Mount Union College*
Peter Hamlet	*Pittsburg State University*
Tara L. S. Kishbaugh	*Eastern Mennonite University*
Scott Mason	*Mount Union College*
Sheldon L. Miller	*Chestnut Hill College*
Keith E. Peterman	*York College*
Pamela C. Turpin	*Roanoke College*

Chemistry in Context

A Project of the American Chemical Society

Chapter

0

Why the Spiderweb?

Dear Students,

Have you wondered why the image of a spiderweb appears on the cover of your chemistry text? Perhaps you connected it with the incredible influence that the Web has wielded on all aspects of our lives. However, when the first edition of *Chemistry in Context* was published in 1994, few college courses used the resources of the Web. Therefore, this was not the origin of the spiderweb motif. Rather, our title, *Chemistry in Context,* provides the clue to the choice of the spiderweb. The word *context* derives from the Latin word meaning "to weave." The spiderweb reminds you that this text emphasizes the strong and complex connections that exist among chemistry, societal needs, and personal concerns. Therefore, we continue the tradition of using a spiderweb in this sixth edition.

The spiderweb motif carries another and more subtle message, however. Spiders are industrious little creatures, often rebuilding their webs each day. Their most difficult task always is to establish the first anchoring thread for the web. Spiders do this by releasing a long sticky silken thread that blows with the wind until it attaches and becomes securely fastened. In *Chemistry in Context,* the authors have established anchoring threads for you by choosing several of today's real-world issues that have significant chemical context. You will need to work every day to build your own network of connecting strands, forming a strong, resilient web of knowledge, attitudes, and skills that will help you in *Applying Chemistry to Society,* the subtitle of this text.

Because we want your web to be well formed, we have designed ways to enhance your weaving. You will soon discover that the chemistry is presented when you need to apply it. This approach will help you to become a well-informed citizen no matter what career path you choose. For example, if you are following the thread of learning about air quality, you need to know which substances are found in air and why even very low concentrations of them can affect your health. If you are considering the issues surrounding global warming, you need to understand Earth's energy balance and how scientific evidence is gathered and evaluated. Production and use of energy are threads that are woven throughout the text, discussed in contexts including combustion of fossil fuels, nuclear power, fuel cells, and solar energy technologies. In selecting the water you drink, chemical principles can help you to make informed choices that have both health and financial consequences. Many of the later chapters deal with the chemistry involved in personal issues such as nutrition, drugs, and genetic engineering. In every case, chemistry can help you make important societal and personal choices.

As is the case with our industrious spiders, consistent practice will help you to refine your web of knowledge. Do not just "read" this text, much as you might read a novel. Rather, stop and do the activities embedded in the text. Also spend time exploring the *Online Learning Center.* Become an active learner, for this is one of the best ways for you to increase your understanding. Here is a sampling of the learning opportunities that you will find within *Chemistry in Context.*

In-Chapter Features

Your Turn exercises give you a chance to practice new skills or calculations. They may relate to a figure, table, or other information just introduced in the text. Answers are often given following the exercise or in Appendix 4. Plan to complete all the Your Turn activities as you proceed through a chapter in *Chemistry in Context.* Here is an example.

Your Turn 2.2 — Finding the Ozone Layer

Use Figure 2.1 and values given in the text to answer these questions.

a. What is the altitude of maximum ozone concentration?
b. What is the range of altitudes in which ozone molecules are more concentrated than in the troposphere?

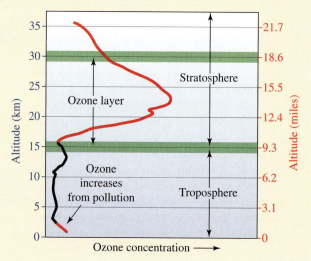

Figure 2.1

Ozone concentrations at different altitudes.

c. What is the maximum number of ozone molecules per billion molecules and atoms of all types found in the stratosphere?
d. What is the maximum number of ozone molecules per billion molecules and atoms of all types found in ambient air just meeting the EPA limit for an 8-hour average?

Consider This activities give you a chance to use what you are learning to make informed decisions. They may require you to consider opposing viewpoints, to do a risk–benefit analysis, to predict the consequences of a particular action, or to make and defend a personal decision. These activities may require additional research, often from Web sources, as is the case with this example.

Consider This 1.32 — Radon Testing

As a public service, local and national agencies provide information about radon on the Web.

a. Find two Web sites about radon provided by government agencies. Cite the source and the URL for each. You might find it helpful to use the keywords *radon detection, air quality,* and *EPA* in your search.
b. Find a company on the Web that sells radon test kits. Describe the kit, including its price.
c. Compare the dangers of radon described on your Web sites from parts **a.** and **b.** Is commercial information about radon different from that provided as a public service? If so, report the differences and suggest reasons why.

Sceptical Chymist activities require you to marshal your analytical skills to respond to various statements and assertions. The unusual spelling comes from an influential book written in 1661 by Robert Boyle, an early investigator studying the properties of air. His experimentation challenged some of the "conventional wisdom" of the time, which is what you will do in these activities. Here is an example.

Sceptical Chymist 5.8 — Bottled Water and Claims of Purity

The Web site for Penta Ultra Premium Purified Drinking Water states:

"Penta ultra-premium purified drinking water is the cleanest-known bottled water."

Consider the claims made by the company about the product of their 13-step purification process:

- Studies on human cells (in vitro) show that Penta water increases cell survivability by 266%.
- Penta water differs from water in having a higher boiling point, a higher surface tension, and a lower viscosity.
- Studies on human cells (in vitro) show that DNA chromosomal mutation rates were 271% greater in lab distilled water than in Penta water.
- Penta water is a new composition of matter.

As a Sceptical Chymist, evaluate each of these claims. Be sure to explain your reasoning.

Figures Alive! animations and activities are on the Web at the McGraw-Hill *Online Learning Center* (www.mhhe.com/cic). Figures Alive! bring textbook figures "alive" through animations and interactive questions that guide you to practice chapter essentials, better develop chapter concepts, and explore extensions of chapter material. We encourage you to visit this resource again and again!

Marked with this icon, ![icon], you will find a set of animations for one figure in each chapter. For example, this figure from Chapter 2 "comes alive" at the *Online Learning Center*.

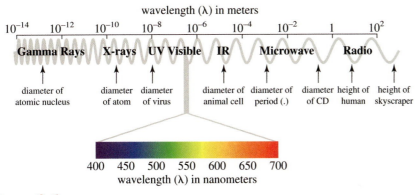

Figure 2.6

The electromagnetic spectrum. The wavelength variation from gamma rays to radio waves is not drawn to scale.

Figures Alive! Visit the *Online Learning Center* to learn more about relationships in the electromagnetic spectrum. Practice, using the interactive exercises. Look for the **Figures Alive!** icon elsewhere in this chapter.

Figure 8.17
Hydrogen can be produced via photosynthesis by various kinds of algae.

The **Green Chemistry** icon 🌍 alerts you to topics in the text related to green chemistry; that is, the designing of chemical products and processes that reduce or eliminate the use or generation of hazardous substances. Here is an example from Chapter 8, Energy from Electron Transfer.

In both a figurative and literal example of green chemistry, some scientists are looking to biological organisms for hydrogen production. Certain species of unicellular green algae form hydrogen gas as a product of photosynthesis (Figure 8.17). Again, the efficiency of the process currently is much too low for commercial applications. New types of algae, created though genetic modifications, are more effective at using the light as well as being more resistant to chemical degradation.

End-of-Chapter Features

A **Conclusion** brings together the general themes of each chapter, often relating these to earlier chapters or to those that lie ahead. After the conclusion comes a **Chapter Summary** with a list of items keyed to specific sections in the chapter. The summary provides an efficient way to review topics and to better understand the progression of ideas within the chapter. Do not simply read over the summary items, but use them to guide further study. For example, here is a partial list of items from Chapter 7, The Fires of Nuclear Fission.

Chapter Summary

Having studied this chapter, you should be able to:

- Rank the sources that contribute to your annual dose of radiation, both natural and human-made (7.7)
- Apply the concept of half-life to radiocarbon dating and the storage of nuclear waste (7.8)
- Describe the issues associated with the production and storage of high-level radioactive waste, including spent nuclear fuel (7.9)
- Take an informed stand on how high-level radioactive wastes should be handled and stored (7.9)
- Evaluate news articles on nuclear power and nuclear waste with confidence in your ability to understand the scientific principles involved (7.9–7.11)

End-of-chapter questions are grouped into three categories:

- **Emphasizing Essentials** These questions give you the opportunity to practice fundamental skills. They most closely relate to the Your Turn exercises in the chapter.
- **Concentrating on Concepts** These questions ask you to integrate and apply the chemical concepts developed in the chapter and to relate them to societal issues. These questions most closely resemble the Consider This activities in the chapter.
- **Exploring Extensions** These questions challenge you to go beyond the information presented in the text. They provide an opportunity for you to extend and integrate the facts, concepts, and communication skills from the chapter. Some questions closely relate to the type of analysis practiced in the Sceptical Chymist activities in the chapter.

See Appendix 5 for the answers to questions with numbers in blue. Questions marked with this icon 🖥️ require the resources of the Internet.

End-of-Text Material

Yes, there is more! You will want to check out these resources as well. At the back of the book are several **Appendices.** One contains conversion factors and another will help you review operations with exponents. You will find the appendix on logarithms

particularly useful for understanding the concept of pH in Chapter 6. The appendices include answers to Your Turns that have not already been given in the text as well as answers to selected end-of-chapter questions. A very useful resource is the **Glossary,** a handy place to do a quick vocabulary check. Each term is keyed to the page where the term is first explained. The **Index** will lead you efficiently to information in the text, figures, and tables.

The Online Learning Center

Several resources are found at McGraw-Hill's *Online Learning Center.*

- **Quiz questions** Each chapter has two sets of multiple-choice quiz questions. Check your understanding and receive immediate feedback.

- **Web links** The *Online Learning Center* provides quick access to many of the Web sites useful for those Consider This and Sceptical Chymist activities marked with this Web icon.

- You will also find chapter overviews, information about *Chemistry in Context* and its authors, and other useful features. This site is frequently updated, so check for new resources not listed here.

As you journey through *Chemistry in Context,* remember the image of that industrious spider. Its beautiful web illustrates that many interconnections are necessary to create a harmonious whole, and so it is with the topics in this text. Remember to think about how the particular thread under discussion connects to others. Said another way, any individual risk-benefit analysis must be considered in light of its potential connections with other societal and personal concerns. Looking at the "big picture" of how strands are connected is what some call a life cycle analysis.

And now, dear students, it is time for you to start weaving your own webs. We hope this orientation will prove useful. Our best wishes for a successful and most enjoyable experience with *Chemistry in Context,* Sixth Edition.

Most sincerely yours,

The Author Team

Chemistry in Context, Sixth Edition

Left to right, first row.
Gail A. Steehler, *Roanoke College, Author (Lab Manual)*
Lucy Pryde Eubanks, *Clemson University, Author and Editor-in-Chief*
Catherine H. Middlecamp, *University of Wisconsin–Madison, Author*

Left to right, second row.
I. Dwaine Eubanks, *LATEst IDEas, Inc., Technical Illustrator*
Steven W. Keller, *University of Missouri–Columbia, Author*
Carl E. Heltzel, *Environmental Consultant, Author*

Chapter

1

The Air We Breathe

The "blue marble," our Earth, as seen from outer space.

"The first day or so, we all pointed to our countries. The third or fourth day, we were pointing to our continents. By the fifth day, we were aware of only one Earth."

Prince Sultan Bin Salmon Al-Saud, Saudi Arabian astronaut

Individually and collectively, we take the air we breathe for granted. Yet our atmosphere is a fragile, thin veil of essential gases interspersed with pollutants in differing amounts. It surrounds the third planet from the Sun, helping to make habitable the place we call home. The striking words of astronaut James Irwin compel us to consider the awesome spectacle of our home planet: "Finally it shrank to the size of a marble, the most beautiful marble anyone can imagine." Only a few men and women have actually observed what James Irwin saw in July, 1971, but most of us have seen the spectacular photographs of the Earth taken from outer space. From that vantage point, our planet looks magnificent—a blue and white ball composed of water, earth, air, and fire. It is where thousands upon thousands of species of plants and animals live in a global community. More than 6 billion of us belong to one particular species with special responsibilities for the protection of our beautiful "blue marble."

As we move in from outer space, an aerial view of Earth from a satellite reveals more detail about the blue marble. The landforms visible in the computer-enhanced photograph of Figure 1.1 include rivers, lakes, islands, mountains, forests, and prairies. Truly, our planet has great geological diversity, and many biological species inhabit these varied environments. Although the gases of our atmosphere are invisible in this photo, we can see white areas of condensed water vapor, better known as clouds. These clouds that both shade us and give us rain, as well as the invisible air that surrounds them, are resources beyond price.

At ground level we arrive at the communities that we know best: the cities, towns, ranches, and farms where we live, study, play, work, and sleep. Our families, friends, and neighbors can be found here. We are shaped by the people, customs, and laws, as well as by the climate, natural resources, and air quality in these regional environments.

As individuals, we simultaneously inhabit these concentric communities. Our personal lives are embedded not only in our immediate surroundings, but also in our countries and the entire globe. Changes in any of these environments affect us, and we, in turn, have obligations at each level of community, from personal to global. This book is about some of those responsibilities and the ways in which a knowledge of chemistry can help us meet them with intelligence, understanding, and wisdom.

Wherever you live, to be an informed (and healthy) member of your community, you should know about the air you breathe. In air are chemical substances that are essential for your existence, as well as a few that can endanger it. To understand the chemical complexities of air, you will need to become familiar with certain chemical facts and concepts. Therefore, this chapter begins by considering the composition of air, its major and minor constituents (including pollutants), and how the concentration of each can be expressed.

Figure 1.1

The Great Lakes, imaged by SeaWiFs (Sea-viewing Wide Field-of-view Sensor) aboard a satellite launched in 1997.

1.1 Everyday Breathing

We begin by asking you to do something you do automatically and unconsciously thousands of times each day—to take a breath. You certainly do not need textbook authors to tell you to breathe! A doctor or nurse may have encouraged your first breath, but from then on nature took over. Even when you hold your breath in a moment of fear or suspense, you soon involuntarily gasp a lungful of that invisible stuff we call air. Indeed, you could not survive long without a fresh supply of air.

> ### Consider This 1.1   Take a Breath
>
> What total volume of air do you inhale (or exhale) in a typical day? Although you could simply guess, a simple experiment can enable you to come up with a reasonably accurate answer. First determine how much air you exhale in a single "normal" breath and how many breaths you "normally" take per minute. Once you establish this information, calculate how much air you exhale per day (24 hours). Describe the experiment you performed, provide the data you obtained, and list any factors you believe may have affected the accuracy of your answer.

The previous activity addresses how much air you breathe, but not the equally important topic of *what* you breathe. No matter where you live, the lungful of air you just inhaled contains some substances that, depending on their amounts, can be harmful. In some places, the threat to your health can be so great that laws curtail certain activities in order to lessen the pollution. It is impossible to remove all pollutants from the air, because most have natural sources. Nonetheless, with few exceptions, air quality in the United States has shown a general trend of improvement over the past three decades. The improvements have occurred through a combination of governmental actions, chemical ingenuity, and allowing the atmosphere to regenerate naturally.

In 1970, the Clean Air Act was passed. This federal mandate established national air quality standards to reduce air pollution. Table 1.1 shows the dramatic decreases in six air pollutants since the 1980s. Four of the pollutants are atmospheric gases: carbon monoxide, nitrogen dioxide, ozone, and sulfur dioxide. The other two, particulate matter (PM) and lead, are minuscule suspended particles on the order of a millionth of a meter in diameter. PM includes dust, soot, dirt, and even microscopic droplets of liquid, bacteria, or viruses. As you can see in Table 1.1, particulate matter is classified by size. **PM$_{10}$** has an average diameter of 10 μm or less, which is on the order of 0.0004 inches. **PM$_{2.5}$**, also called fine particles, has an average diameter less than 2.5 μm and thus is even tinier. From these definitions, you can see that PM$_{2.5}$ is a subset of PM$_{10}$. A **micrometer** (μm) is 10^{-6} of a meter (m) and is sometimes simply referred to as a micron.

The EPA was formed in 1970 by President Richard Nixon. Senators from earlier years also played a role.

These six pollutants are labeled by the U.S. Environmental Protection Agency (the EPA) as **criteria air pollutants,** or more simply, **criteria pollutants.** For each, the EPA has set permissible levels in the air based on their effects on human health and on the environment. In Section 1.3, we will revisit air quality and the effects of these different pollutants, but first we will examine the details of what you breathe.

> ### Consider This 1.2   Visit the EPA
>
> The U.S. Environmental Protection Agency maintains an extensive Web site. In particular, the EPA's Office of Air and Radiation contains many consumer-friendly documents on air quality. Select a document and report its title, URL, and several interesting things that you learned. A direct link is provided at the *Online Learning Center.*

Table 1.1	Changes in Air Quality	
Criteria Air Pollutant	**1983–2002 (%)**	**1993–2002 (%)**
Carbon monoxide	−65	−42
Nitrogen dioxide	−21	−11
Ozone		
1 hr	−22	−2
8 hr	−14	+4
Sulfur dioxide	−54	−39
Particulate matter		
PM_{10}	…	−13
$PM_{2.5}$	…	−8
Lead	−94	−57

Source: EPA, *National Air Quality and Emission Trends Report, 2003 Special Studies Edition.* http://www.epa.gov/air/airtrends/aqtrnd03/

… Trend data not available

Note: These percentages represent overall changes. Changes for a particular urban area may differ.

1.2   What's in a Breath? The Composition of Air

The air we breathe is a **mixture,** that is, a physical combination of two or more substances present in variable amounts. For the moment we will focus on only five components of air: oxygen, nitrogen, argon, carbon dioxide, and water. The first four normally exist as gases. Although we usually think of water as a liquid, it can also be a gas, in which case we may refer to it as "water vapor." Just like oxygen and nitrogen, water vapor is a gas that you cannot see. Although you *can* see steam and clouds, these are not water vapor as such. Rather, they are condensed water vapor, that is, tiny droplets of liquid water (Figure 1.2).

The concentration of water vapor in air varies by location. It can be close to 0% in dry desert air or as much as 5% in a tropical rain forest. Because of this variability, reference tables typically list the composition of air with no humidity. The normal composition of dry air is 78% nitrogen, 21% oxygen, and 1% other gases by volume. **Percent means "parts per hundred"** and is sometimes abbreviated as pph. In this case, the parts are molecules (or, in a few cases, atoms).

Figure 1.3 displays the composition of air in the form of a pie chart and a bar graph. Both of these are important, widely used methods for displaying numerical information, and we will use both in this text. The pie chart emphasizes the fractions of the total, whereas the bar graph uses height to emphasize the relative amounts of each. Regardless of how we present the data, notice that 99% of dry air is made up of only two substances: nitrogen and oxygen.

Life on Earth bears the stamp of oxygen. Indeed, it is difficult to conceive of life on any planet without this remarkable chemical. Oxygen is absorbed into our blood via the lungs and reacts with the foods we eat to release the energy needed for all life processes within our bodies (see Chapter 11). Oxygen is also a participant in burning, rusting, and

At ground level, ozone is an air pollutant. At high altitudes, ozone is beneficial, as you will find out in Chapter 2.

Mixtures do not need to be gases. Soil is a mixture of solids and liquids. We will revisit mixtures in Section 1.6.

Figure 1.2

Clouds consist of condensed water vapor. As we will see in Chapter 3, clouds are one of several factors that influence the Earth's energy balance.

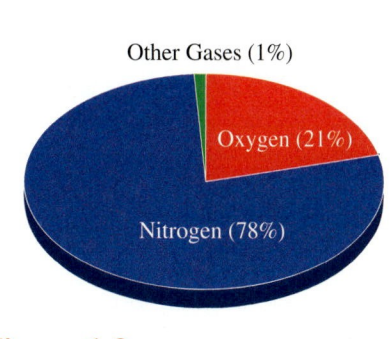
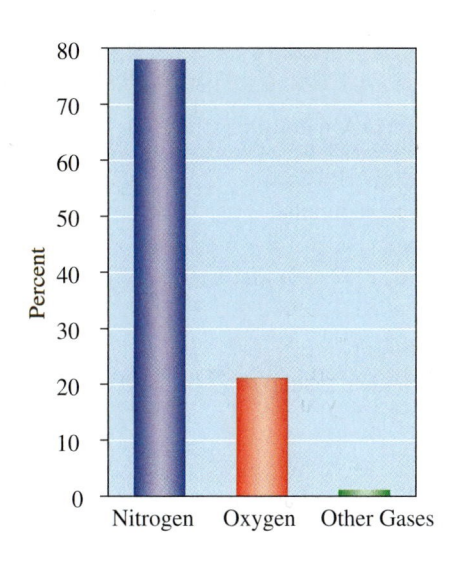

Figure 1.3

The composition of dry air, by volume.

Figures Alive! Visit the *Online Learning Center* to learn more about the molecules and atoms that make up air. Look for the Figures Alive! icon throughout this text.

other corrosion reactions. As a constituent of water and of many rocks, oxygen is the most abundant element by mass in the Earth's crust and in the human body. Given this broad distribution, it is somewhat surprising that oxygen was not isolated as a pure substance until the 1770s. But once isolated, oxygen proved to be of great significance in establishing the principles of the young science of chemistry.

Consider This 1.3   More Oxygen?

Humans are accustomed to living in an atmosphere of 21% oxygen where a paper match burns completely in less than a minute, a fireplace consumes a small pine log in about 20 minutes, and you exhale a certain number of times a minute. Burning, rusting, and the rate of most metabolic processes in plants and animals depend on the concentration of oxygen. How would life on Earth be different if the oxygen content in the atmosphere were doubled? List at least four effects.

Section 6.12 describes the route by which atmospheric nitrogen becomes part of living plants and animals.

Nitrogen is the most abundant substance in the air and constitutes over three fourths of the air we inhale. However, it is much less reactive than oxygen and is exhaled from our lungs unchanged (Table 1.2). Although nitrogen is essential for life and is a part of all living things, most plants and animals obtain the nitrogen they require from other sources, not directly from the atmosphere.

Table 1.2	**Typical Composition of Inhaled and Exhaled Air**	
Substance	**Inhaled Air (%)**	**Exhaled Air (%)**
Nitrogen	78.0	75.0
Oxygen	21.0	16.0
Argon	0.9	0.9
Carbon dioxide	0.04	4.0
Water vapor	0.0	4.0

Every time we exhale, we add carbon dioxide to the atmosphere. Table 1.2 indicates the difference between inhaled dry air and exhaled air. Clearly some changes have taken place that use up oxygen and give off both carbon dioxide and water. Not surprisingly, chemistry is involved. In the biological process of metabolism, oxygen reacts with foods to yield carbon dioxide and water. However, most of the water in exhaled air is simply the result of evaporation from the moist surfaces within the lungs. Note that even exhaled air still contains 16% oxygen. Some people mistakenly think that in respiration most of the oxygen is replaced with carbon dioxide. But if this were true, mouth-to-mouth resuscitation would not work.

Just under 1% of the air you breathe is argon, a gas so unreactive that it is considered chemically inert. This inertness is recognized in the name *argon*, which means "lazy" in Greek. As you can see from Table 1.2, you simply exhale the argon that you inhale, chemically unchanged.

The percentages we have been using to describe the composition of the atmosphere are based on volume. Thus, we could closely approximate 100 liters (L) of dry air by mixing 78 L of nitrogen, 21 L oxygen, and 1 L argon (78% nitrogen, 21% oxygen, and 1% argon). Because the volume of a gas sample increases with temperature and decreases with pressure, all gas volumes must be compared at the same temperature and pressure.

An alternative way to represent the composition of air is in terms of the molecules and atoms present in the mixture. This works because equal volumes of gases contain equal numbers of molecules, providing the gases are at the same temperature and pressure. Thus, if you took a sample of 100 of the molecules and atoms in air (an unrealistically small amount of air), 78 would be nitrogen molecules, 21 would be oxygen molecules, and 1 would be an argon atom. In other words, when we say that air is 21% oxygen, we mean that there are 21 molecules of oxygen per 100 molecules and atoms in the air. We soon will explain why nitrogen and oxygen are found as molecules, and argon, in contrast, is found as an atom.

Some atmospheric components are present at less than one part per hundred, or 1%. Such is the case with carbon dioxide, which has a concentration of about 0.0385%. Although we could express this as 0.0385 molecules of carbon dioxide per 100 molecules and atoms in the air, it does not make sense to talk about a fraction of a molecule. Accordingly, we scale the measurement from parts per hundred to **parts per million (ppm)**, which means one part out of a million and is 10,000 times less concentrated than 1 part per hundred (pph). Having 0.0385 pph is equivalent to having 385 ppm. Through a series of relationships, we can show that the difference between pph (= percent) and ppm is a factor of 10,000 or moving the decimal four places:

0.0385%	means	0.0385 parts per hundred
	means	0.385 parts per thousand
	means	3.85 parts per ten thousand
	means	38.5 parts per hundred thousand
	means	385 parts per million

Thus, out of a sample of air consisting of 1,000,000 molecules/atoms, 385 of them will be carbon dioxide molecules. Hence, the carbon dioxide concentration is 385 ppm, or 0.0385%.

Look for more about carbon dioxide in Chapter 3.

1 liter = 1.06 quart. Appendix 1 contains this and many other conversion factors.

Changing between % and ppm involves moving the decimal point four places. Practice using the exercises in Figures Alive!

Sceptical Chymist 1.4   Really One Part Per Million?

It has been said that a part per million is the same as one second in nearly 12 days. Is this a correct analogy? How about one step in a 568-mile journey? A pinch of salt on 20 pounds of potato chips? Check the validity of these three analogies and explain your reasoning. Then come up with a new one of your own.

Your Turn 1.5   **Using ppm**

a. The EPA sets the permissible limit for the average concentration of carbon monoxide in an 8-hour period at 9 ppm. Express this concentration as a percentage.

b. Exhaled air typically contains about 75% nitrogen. Express this concentration in parts per million.

Answers

a. 0.0009%   b. 750,000 ppm

In the next section, we will see the health effects of substances present in the air at the miniscule level of parts per million—or even lower.

1.3   What Else Is in a Breath?

Our noses tell us that the air is different in a pine forest, a bakery, an Italian restaurant, a locker room, and a barnyard. Even blindfolded, we can *smell* where we are. Pine needles, fresh bread, garlic, sweat, and manure all have distinctive odors that are carried by molecules. Hence, air must contain trace quantities of substances not included among the five substances listed in Table 1.2. Although the major components of air are odorless (at least to our noses), many of the other airborne substances have pronounced odors. In fact, the human nose is an extremely sensitive odor detector. In some cases, only a minute trace of a substance is needed to trigger the olfactory receptors responsible for detecting odors. Thus, tiny amounts of substances can have a powerful effect on our noses, as well as on our emotions.

Our noses also warn us to avoid certain places. But some of the more dangerous air pollutants have no odor. As a result, it may be necessary to rely on specialized scientific equipment to monitor the presence of such substances in the air. It is rather surprising that the gases that cause serious air pollution are present in relatively small amounts, generally in the range of parts per million to parts per billion. Yet even at such low concentrations, they can do significant harm.

In this chapter, we focus on four gases that contribute to air pollution at the surface of the Earth. One of these gases, carbon monoxide, is odorless; the other three—ozone, sulfur dioxide, and nitrogen dioxide—have characteristic (and unpleasant) odors. With sufficient exposure, each of these is hazardous to health, even at concentrations well below 1 ppm. Together with particulate matter (PM), they represent the most serious air pollutants at the Earth's surface. Let's now examine the health effects of each.

Carbon monoxide earned a nickname as "the silent killer" because you cannot detect it with your senses. Once in the lungs, carbon monoxide enters the bloodstream and disrupts the delivery of oxygen throughout the body. In an extreme case, such as breathing auto exhaust or furnace emissions in a confined space, carbon monoxide can be fatal. Charcoal grills, kerosene heaters, and propane stoves also generate carbon monoxide (Figure 1.4) and should be used outdoors or vented if used indoors. With low-level exposure to carbon monoxide, victims first may experience dizziness, headache, and nausea, symptoms all easily mistaken for the onset of a respiratory infection. To add to the difficulties in diagnosing carbon monoxide poisoning, people exposed at the same time may show different symptoms. Carbon monoxide exposure can be serious for individuals with cardiovascular disease, and emergency medical care may be necessary.

Ozone is closely related to oxygen, as we soon will see. Unlike carbon monoxide, it has a sharp odor, one that you may have detected around photocopiers, electric motors, transformers, or welding apparatus. Even at very low concentrations ozone is toxic, and it can reduce lung function in normal, healthy people during periods of exercise. Symptoms include chest pain, coughing, sneezing, and pulmonary congestion. At the Earth's surface,

Figure 1.4
Propane camping stoves can be used outdoors with no venting.

ozone is definitely a bad actor, but as you will see in Chapter 2, it plays an essential role at higher altitudes.

Sulfur oxides and *nitrogen oxides* are respiratory irritants that can affect your breathing. The people most susceptible to these two pollutants include the elderly, young children, and individuals with diseases such as emphysema or asthma. A particularly severe example of these effects occurred during the killer London fog of 1952 that lasted five days and resulted in approximately 4000 deaths. The oxides of sulfur and nitrogen also contribute to acid precipitation, a subject explored in Chapter 6.

Particulate matter, or PM, is the least well understood of the pollutants that we are considering. PM is not just one chemical; rather it includes a mixture of tiny solid particles and microscopic liquid droplets that either are emitted into the air, form in the air from other pollutants, or are blown up into the air by the wind. Sometimes particulate matter is visible, and you may recognize it as soot or smoke. Of most concern, however, are the particles too tiny to see. Once airborne, these invisible particles can get deep into the lungs and cause all kinds of mischief. As a health hazard, particles smaller than 2.5 μm are the most worrisome, and the standards for these are stricter, as we soon will see in Table 1.5.

Particulate matter can irritate your eyes, nose, throat, and lungs even if you are a healthy adult. For those with lung diseases the consequences can be more serious. Equally important, exposure can aggravate heart disease. As reported in *Circulation,* a journal of the American Heart Association, $PM_{2.5}$ particles in urban air are linked with an increased risk of heart attacks. In the study, the risk for heart attack in Boston peaked both 2 hours and 24 hours after patients were exposed to increased levels of the particles. A 2003 article in *Circulation* examines pollution data from over 150 cities and reveals that the risk of heart disease increases as the amount of $PM_{2.5}$ increases. Thus particulate matter affects *both* the cardiovascular system and the lungs.

To better understand the effects of *what* you are breathing *when* you are likely to be breathing it, the EPA has developed the color-coded Air Quality Index (AQI) shown in Table 1.3 and a distinctive logo (Figure 1.5). If you live in a metropolitan area, you may find the daily AQI forecast listed in your daily newspaper. National newspapers carry the information as well. For example, *USA Today* currently reports the daily AQI for 36 cities. If you had checked their listing on a hot summer day (July 11, 2006, to be exact), you would have found that 12 of these cities were listed as good and 19 were moderate. Five cities were unhealthy for sensitive groups: Baltimore, Charlotte, Columbus, Philadelphia, and Washington, D.C. The "sensitive group" varies with the pollutant and includes those with respiratory diseases such as asthma, the elderly, and children.

To see how air quality can vary over time, examine Table 1.4. For the more than 2 million people who lived in Houston in 2005, the air was of good or moderate quality 87% of the time. Alternatively, the air was unhealthy for over 40 days that year, with the ozone and $PM_{2.5}$ being the pollutants with highest concentrations. In some years, Houston had more

PM comes from many sources, including trucks, cars, coal-burning power plants, fires, and blowing dust.

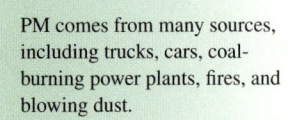

Figure 1.5
EPA air quality logo.

Table 1.3	Levels for the Air Quality Index	
Air Quality Index (AQI) Values	**Levels of Health Concern**	**Colors**
When the AQI is in this range:	*...air quality conditions are:*	*...as symbolized by this color.*
0–50	Good	Green
51–100	Moderate	Yellow
101–150	Unhealthy for sensitive groups	Orange
151–200	Unhealthy	Red
201–300	Very unhealthy	Purple
301–500	Hazardous	Maroon

Source: EPA, http://www.epa.gov/airnow/aqibroch/aqi.html

Table 1.4	Air Quality Index Values for Houston			
	Good	Moderate	Unhealthy for Sensitive Groups	Unhealthy
Year	(0–50)	(51–100)	(101–150)	(>150)
1997	258	54	33	20
1998	253	70	23	19
1999	223	82	34	26
2000	166	147	37	16
2001	180	144	25	16
2002	196	136	24	9
2003	169	154	26	16
2004	175	152	29	10
2005	138	180	39	7
2006	77	88	13	4

unhealthy days; in other years less. The variations largely reflect those in the local weather patterns. Regional events such as forest fires and volcanic eruptions also can influence the air quality.

Values reported are the number of days per year. These values do not always add up to 365 (=1 year), as no data were reported on some days. The 2006 data are only reported through June. "Unhealthy" includes unhealthy, very unhealthy, and hazardous.

Consider This 1.6   Los Angeles, Boston, or Houston?

You can use the AirData Web site of the EPA to look up air quality data similar to that shown in Table 1.4. Select a year, a locale, and generate the report. How does the air quality in the city you selected compare with Houston? The *Online Learning Center* has stepwise directions to help you obtain the data.

Air quality may also be listed by individual pollutant. For example, Figure 1.6 shows the air quality forecast for carbon monoxide, ozone, and particulates on a hot summer day in Phoenix, AZ. Although the color coding is not employed, the criteria are the same.

Consider This 1.7   Red Alert

Figure 1.6 shows that the *Arizona Republic* uses numerical values in its air quality forecasts. In contrast, this same information is conveyed by some other newspapers using color codes.

a. What are the advantages of using colors (green, yellow, orange, red) to represent air quality? The disadvantages?
b. A "red alert" is forecast today for ozone. List five actions that you could take to help reduce the pollution, particularly if everyone were to follow your suggestions.

Pollution

Good: 0-50 Moderate: 51-100
Unhealthful: More than 100

Carbon monoxide	9
Ozone	56
Particulates	62

Figure 1.6

Air quality forecast from the *Arizona Republic* newspaper for central Phoenix on July 16, 2006. These values are scaled based on 100 as unhealthy.

Air pollution is primarily an urban problem, and more than 50% of all Americans live in cities with populations over 500,000. Many of these cities (as we saw with Houston) fail to meet the national air quality standards at certain times. In spite of recent improvement in air quality, we still have difficulties, especially with nitrogen oxides, particulate

matter, and ozone. Furthermore, our present air quality standards may provide only a small margin of safety in protecting public health. The American Lung Association has estimated that $50 billion in health benefits could be realized annually in the United States if air quality standards were met throughout the country. Even more striking are the figures from a study in 2006 in the *American Journal of Respiratory and Critical Care Medicine.* The authors report that "for each decrease of 1 microgram of soot per cubic meter of air, death rates from cardiovascular disease, respiratory illness and lung cancer decrease by 3 percent—extending the lives of 75,000 people a year in the United States." These figures are nothing to sneeze (or wheeze) at.

We face difficult political and economic choices. Are we willing to spend the money that would be needed to clean up the air we breathe? If we want to stimulate the economy, what regulations can we afford to drop or relax? Would the economic gains compensate for the hidden health costs? In weighing the risks against the benefits, tighter regulations could mean a boon to health and a significant reduction in health care costs. The improved air quality that we now enjoy could be short-lived. How do we assess risks? We now turn to this question.

1.4 Taking and Assessing Risks

Air quality provides an opportunity for our first look at the subject of risk, one to which we will return repeatedly throughout this text. Indeed, it is an issue central to life itself, because everything we do carries a certain level of risk. Some activities that carry high risks are labeled as such. For example, by law cigarette packages are required to carry a warning. Similarly, a bottle of wine carries the words "GOVERNMENT WARNING" followed by a statement about the risks of birth defects for pregnant women and about the risks of driving a car or operating machinery under the influence of alcohol. Some practices have been declared illegal because the level of risk is judged unacceptable to society. However, many activities carry no warning. In these cases, presumably the risk is quite low, the risk is obvious or unavoidable, or the benefits of the activity far outweigh the risk.

One feature of such warnings is a characteristic of risk itself. The warnings do not say that a specific individual *will* be affected by a particular activity. They only indicate the statistical probability, or chance, that an individual will be affected. For example, if the odds of dying from an accident while traveling 300 miles in a car are approximately one in a million, this means that, on average, one person out of every million people traveling 300 miles by car would be killed in an accident. Such predictions are not simply guesses, but are the result of **risk assessment:** evaluating scientific data and making predictions in an organized manner about the probabilities of an occurrence.

> **Consider This 1.8**   **Cell Phones**
>
> Driving down the highway? If you are using a cell phone, your ability to handle your car may be compromised. State the risks and the benefits of answering a call while driving. Are the risks obvious or unavoidable? Do the benefits of the activity far outweigh the risk? Do other factors apply?

For air pollutants, the assessment of risk involves two factors: the **toxicity,** the intrinsic health hazard of a substance, and the **exposure,** the amount of the substance encountered. Exposure is the easier factor to evaluate because it depends simply on the concentration of the substance in the air, the length of time a person is exposed, and the amount of air inhaled into the lungs in a given period. As you saw earlier in this chapter, the last factor depends on lung capacity and breathing rate.

Concentrations of pollutants in air are usually expressed either as parts per million (ppm) or as micrograms per cubic meter ($\mu g/m^3$). Earlier, when talking about particulate

1 cubic meter (m³) is a volume
1 meter × 1 meter × 1 meter.

1 m³ = 1000 liters (about
250 gallons)

1 μg is approximately the mass of
a period printed on a page.

matter, we encountered the prefix *micro-* in micrometers (μm), meaning 10^{-6}, or a millionth of a meter. Thus, one **microgram** (μg) is 10^{-6}, or a millionth of a gram (g).

Let us use carbon monoxide (CO) as an example. Millions of tons of carbon monoxide are spread throughout the atmosphere. Yet in and of itself, this prodigious amount is not indicative of the risk. In some places the concentration of CO is so low that no health concerns arise. In others, such as near a gas appliance with faulty ventilation, the concentration can be dangerously high. To assess the risk, then, we need to consider *both* the exposure *and* the toxicity of the substance.

Consider, for example, an air sample containing 5000 μg CO per cubic meter of air. Is breathing this concentration of CO harmful? The most straightforward way to evaluate toxicity of an air pollutant is to compare the exposure level to the National Ambient Air Quality Standards (NAAQS). Here, the term **ambient** refers to the outside air, that is, the air surrounding or encircling us. Although a pollutant may be present, it is not considered hazardous unless it exceeds the amount that causes harmful effects.

Consult Table 1.5 to see that two standards are reported for carbon monoxide, one for a 1-hour exposure and another for an 8-hour exposure. The value for the 1-hour exposure, 4×10^4 μg CO/m³ (or 35 ppm) is higher because a higher concentration can be tolerated for a short period of time. Notice that the value 4×10^4 μg CO/m³ is expressed in **scientific notation,** that is, a system for writing numbers as the product of a number and 10 raised to the appropriate power. Scientific notation avoids writing strings of zeros either before or after the decimal point. This particular value, 4×10^4, is equivalent to 40,000. Here, the easy way to understand this conversion is to simply count the number of zeros to the right of the initial 4. There are four of them. The number 4 is then multiplied by 10^4 to obtain 4×10^4 μg CO/m³. We realize that this may be confusing because our number has two 4s, so let us express the 8-hour standard in scientific notation as well. Here, 1×10^4 μg CO/m³ is equivalent to 10,000 μg CO/m³. To see this, again simply count to see that there are four zeros to the right of the initial 1.

Table 1.5	National Ambient Air Quality Standards (NAAQS), 1999	
Pollutant	**Standard (ppm)**	**Approximate Equivalent Concentration (μg/m³)**
Carbon monoxide		
8-hr average	9	1×10^4
1-hr average	35	4×10^4
Nitrogen dioxide		
Annual average	0.053	100
Ozone		
8-hr average	0.08	157
1-hr average	0.12	235
Lead		
Quarterly average	…	1.5
Particulates[*]		
PM_{10}, annual average	…	50
PM_{10}, 24-hr average	…	150
$PM_{2.5}$, annual average	…	15
$PM_{2.5}$, 24-hr average	…	65
Sulfur dioxide		
Annual average	0.03	80
24-hr average	0.14	365
3-hr average	0.50	1300

… Data not available

*PM_{10} refers to airborne particles 10 μm in diameter or less. $PM_{2.5}$ refers to particles less than 2.5 μm in diameter.

Scientific notation becomes even more useful when considering much larger numbers, such as the number of molecules in a typical breath. The value is more than 20,000,000,000,000,000,000,000 molecules, a number large enough to take your breath away! In scientific notation, this particular number is written as 2×10^{22} molecules. If you need help with exponents, consult Appendix 2.

Getting back to breathing air containing 5000 µg CO/m^3 of air, we now can express this in scientific notation as 5×10^3 µg CO/m^3. Clearly, this value is *less* than the concentrations for either period. In the case of an 8-hour exposure, 5×10^3, or 5000, is less than 1×10^4, or 10,000. Similarly, for a 1-hour period, 5×10^3 is also less than 4×10^4, or 40,000.

Your Turn 1.9   Time Matters

For carbon monoxide, we just saw that the limits of exposure were higher for 1 hour than for 8 hours. Do the limits set for other pollutants follow this pattern as well? Consult Table 1.5 to find out.

Let's examine the other pollutants found in Table 1.5. Again, based on scientific studies, these values are the maximum concentrations considered to be safe for the general population. These values give us a basis on which to evaluate the relative amounts that are hazardous. For example, for an 8-hour average exposure, compare 9 ppm for carbon monoxide with 0.08 ppm for ozone. Doing the math, ozone is about 100 times more hazardous to breathe than carbon monoxide! Nonetheless, carbon monoxide still is exceedingly dangerous. In the case of breathing air polluted with ozone, your senses can detect it and you are likely to move to less polluted air (perhaps indoors) if you can. In contrast, you may not know that you are inhaling carbon monoxide because it is odorless (and after breathing it for a while your judgment may be impaired).

Your Turn 1.10   Particle Size Matters Too

We stated earlier that "fine" particulate matter (<2.5 µm) have more serious health consequences than "coarse" particulate matter (<10 µm). Does Table 1.5 bear this out?

Although the standards for the pollutant gases are expressed as parts per million, the concentrations of sulfur dioxide and nitrogen dioxide are sufficiently low that they also could conveniently be reported in **parts per billion (ppb),** meaning one part out of one billion, or 1000 times less concentrated than 1 part per million.

sulfur dioxide 0.030 ppm = 30 ppb

nitrogen dioxide 0.053 ppm = 53 ppb

As you can see from these values, to convert from parts per million to parts per billion, you need to move the decimal point three places to the right.

Whether in parts per million or parts per billion, toxicities are difficult to accurately assess because it is unethical to experiment on humans with pollutants such as carbon monoxide or sulfur dioxide. Even if data were available to calculate the risks from a given pollutant, we still would have to ask what level of risk was acceptable and for what groups of people. Various government agencies are charged with establishing safe limits of exposure for the major air pollutants. Table 1.5 gives current outdoor air quality standards established by the EPA for the pollutants discussed in this chapter. Some states, including California and Oregon, have their own stricter standards.

Learn more about the ozone layer in Figure 2.1 in the next chapter.

Air has a lower density at higher elevations. The concept of density is introduced in Section 5.6.

Your Turn 1.11　Living Downwind

Copper metal can be recovered from copper ore by smelting, a refining process that may release sulfur dioxide (SO_2) into the atmosphere. A woman living downwind of a smelter typically inhales 1050 µg of SO_2 in a day.

a. If her lungs handled about 15,000 L (15 m³) of air per day, would she exceed the 24-hour average for the National Ambient Air Quality Standards for SO_2? Support your answer with a calculation.

b. If it were the same amount each day, would she exceed the annual average?

Finally, an important factor in dealing with risks is not only the actual risk, but people's perception of a particular risk. For example, the risks of driving far exceed those of flying. Each day in the United States, more than 100 people die in automobile accidents. Yet some people avoid taking a flight because of their fear of falling out of the sky. Similarly, some fear living near a nuclear power plant. Yet as recent hurricanes have demonstrated, living in a coastal area can be a far riskier proposition. In both of these cases, other factors may be at work. For example, media coverage of an airline crash may heighten people's fears about flying. After any accident, look for a large ripple effect of public concern.

Consider This 1.12　Risk Analysis

A publication of the American Chemical Society, *Chemical Risk: A Primer,* states: "The general public is uncomfortable with uncertainties. Too often we think in terms of absolutes and demand that scientists and decision makers be held accountable for their risk decisions." Do you agree or disagree with these statements? Support your opinion with reasonable arguments, giving a specific example from your personal experience in considering a risk of importance to you.

1.5　The Atmosphere: Our Blanket of Air

The most familiar kinds of air pollution occur in the **troposphere,** the region of the atmosphere that lies directly above the surface of the Earth. Figure 1.7 shows the regions of the atmosphere with reference points in relation to altitude. As one rises in the troposphere, the temperature decreases until it reaches about −40 °C (also −40 °F). That temperature roughly marks the beginning of the **stratosphere,** the region of the atmosphere above the troposphere that includes the ozone layer. The temperature of the stratosphere increases from about 240 °C at 20 kilometers (km) to 0 °C (32 °F) at 50 km. Above that altitude, the temperature of the atmosphere again begins to decrease on passing through the **mesosphere,** the region of the atmosphere above an altitude of 50 km. The issues we will study in the first three chapters of this book will take us to these various regions, which differ in atmospheric properties and phenomena (see Figure 1.7). Bear in mind that no sharp physical boundaries separate these layers. The atmosphere is a continuum with gradually changing composition, concentrations, pressure, and temperature. In fact, temperature changes account for the organization of the atmosphere.

The relative concentrations of the major components of the atmosphere are nearly constant at all altitudes. For example, the concentration of oxygen remains about 21% and nitrogen about 78%. However, you might know from the experience of hiking in high mountains or from flying in a plane (Figure 1.8) that the air gets "thinner" with increasing altitude. As you climb higher up into the atmosphere, there is less air, that is, fewer molecules in a given volume. Moreover, as you climb higher, the mass of air above you decreases. Somewhere above 100 km, the atmosphere simply fades into the almost perfect vacuum of outer space.

Pollutants are minor components of the atmosphere. Their concentrations usually are not constant at different altitudes. In general, do you think the concentrations are higher or lower near the surface of the Earth? In the special case of pollutants emitted from a tall smokestack, what would you expect?

1.6 Classifying Matter: Mixtures, Elements, and Compounds

As we described the atmosphere and air quality, we used a bit of chemical terminology. Before proceeding, some clarification is in order. First, we will examine the way chemists describe the composition of different types of matter. Matter can be classified either as a single pure substance or as a mixture of two or more pure substances, as seen in Figure 1.9.

Much of the matter we encounter in everyday life is in the form of mixtures. A breath of air is a mixture of gases. Polluted air also is a mixture that, depending on the pollutants, has different compositions. Exhaled air is a different mixture from inhaled air. Earlier (see Section 1.2), we defined a mixture as a physical combination of two or more substances present in variable amounts. As composition of a mixture changes, so do many of its properties. For example, gasoline is a mixture of many compounds. As the composition of gasoline is changed, its properties change as well.

Elements and compounds are the two pure substances of most interest to us (see Figure 1.9). The two most plentiful components of air are nitrogen and oxygen. These both are examples of **elements,** substances that cannot be broken down into simpler ones by any *chemical* means. There are over 110 elements, and all common forms of matter are composed of one or more of these elements. About 90 occur naturally on planet Earth and, as far as we know, in the universe. The others have been created from existing elements through artificially induced nuclear reactions. Plutonium is probably the best known of the artificially produced elements, although it does occur in trace concentrations in nature. In some cases, the total amount of a newly created form of matter is so small that there is some uncertainty (and even some controversy) over just how many elements have been identified.

An alphabetical list of the elements and their **chemical symbols, one- or two-letter abbreviations for the elements,** appears in the inside back cover of the text. These symbols, established by international agreement, are used throughout the world. Some of the symbols are quite obvious to those who speak English. For example, oxygen is O, nitrogen is N, carbon is C, and sulfur is S. Most symbols, however, consist of two letters: Ni for nickel, Cl for chlorine, Ca for calcium, and so on. Other symbols appear to have little relationship to their English names. Thus, Fe is iron, Pb is lead, Au is gold, Ag is silver, Sn is tin, Cu is copper, and Hg is mercury. All these metals were known to the ancients and hence were given names long ago. Their symbols reflect their Latin names, for example, *ferrum* for iron, *plumbum* for lead, and *hydrargyrum* for mercury.

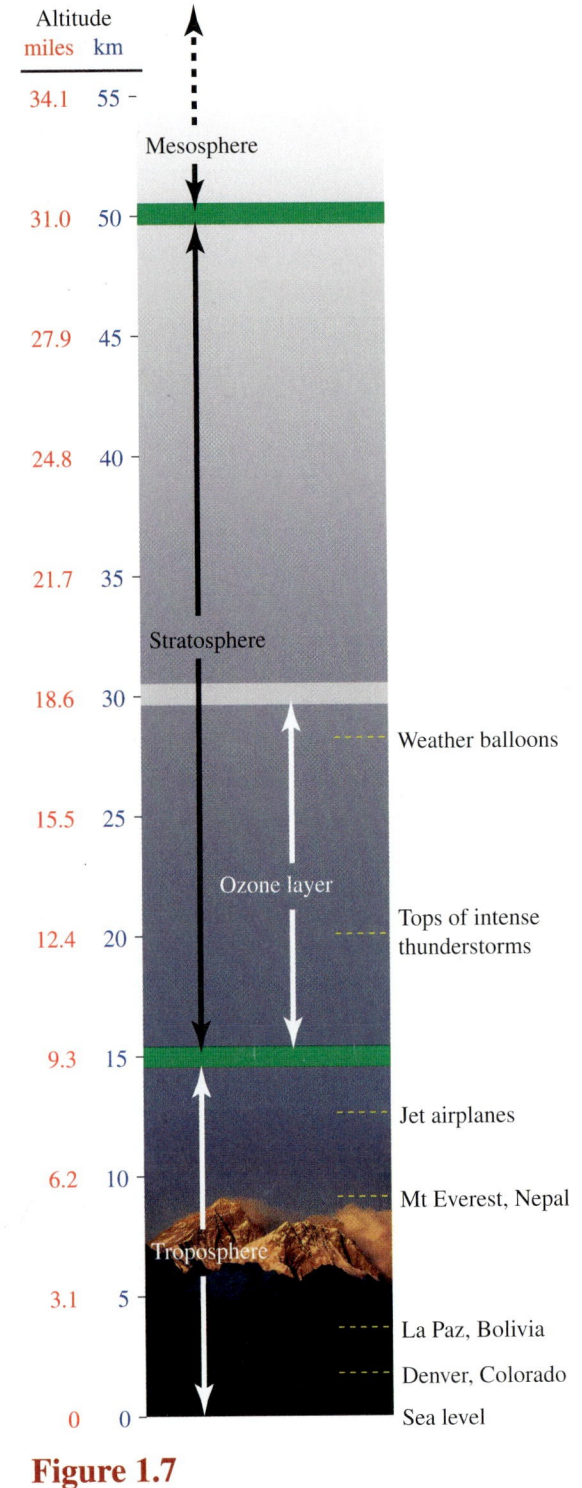

Figure 1.7

The regions of the lower atmosphere.

Chemical symbols also may be referred to as atomic symbols.

Figure 1.8
Flight across the Rockies. A view out of the aircraft window shows the upper troposphere at about 35,000 feet.

Elements have been named for properties, planets, places, and people. Hydrogen (H) means "water former," a name that reflects the fact that this flammable gas burns in oxygen to form water. Neptunium (Np) and plutonium (Pu) were named after the two most recently discovered members of our solar system. Berkelium (Bk) and californium (Cf) honor the Berkeley lab where a team of researchers first produced these two elements. Albert Einstein, Dmitri Mendeleev, and Lise Meitner (codiscoverer of nuclear fission) have attained elementary immortality in einsteinium (Es), mendelevium (Md), and meitnerium (Mt), respectively. The most recently discovered elements are darmstadtium (Ds) and roentgenium (Rg). The former was named after Darmstadt, the city in Germany in which it was discovered. The latter was named after Wilhelm Roentgen and is the heaviest element currently known. Only a few atoms of each have been produced.

It is particularly appropriate that Mendeleev should have his own element, because the most common way of arranging the elements reflects the periodic system developed by this 19th-century Russian chemist. Figure 1.10 is the **periodic table,** an orderly arrangement of all the elements based on similarities in their properties. We will explain how it is ordered and the significance of all the numbers in Chapter 2. For

Lothar Meyer, a German chemist, also developed a periodic table at the same time as Mendeleev.

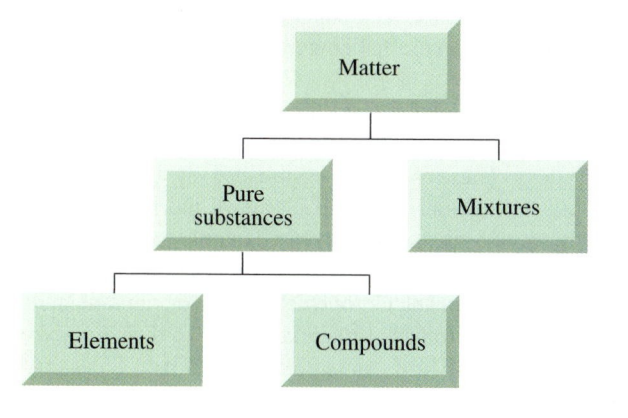

Figure 1.9
The classification of matter.

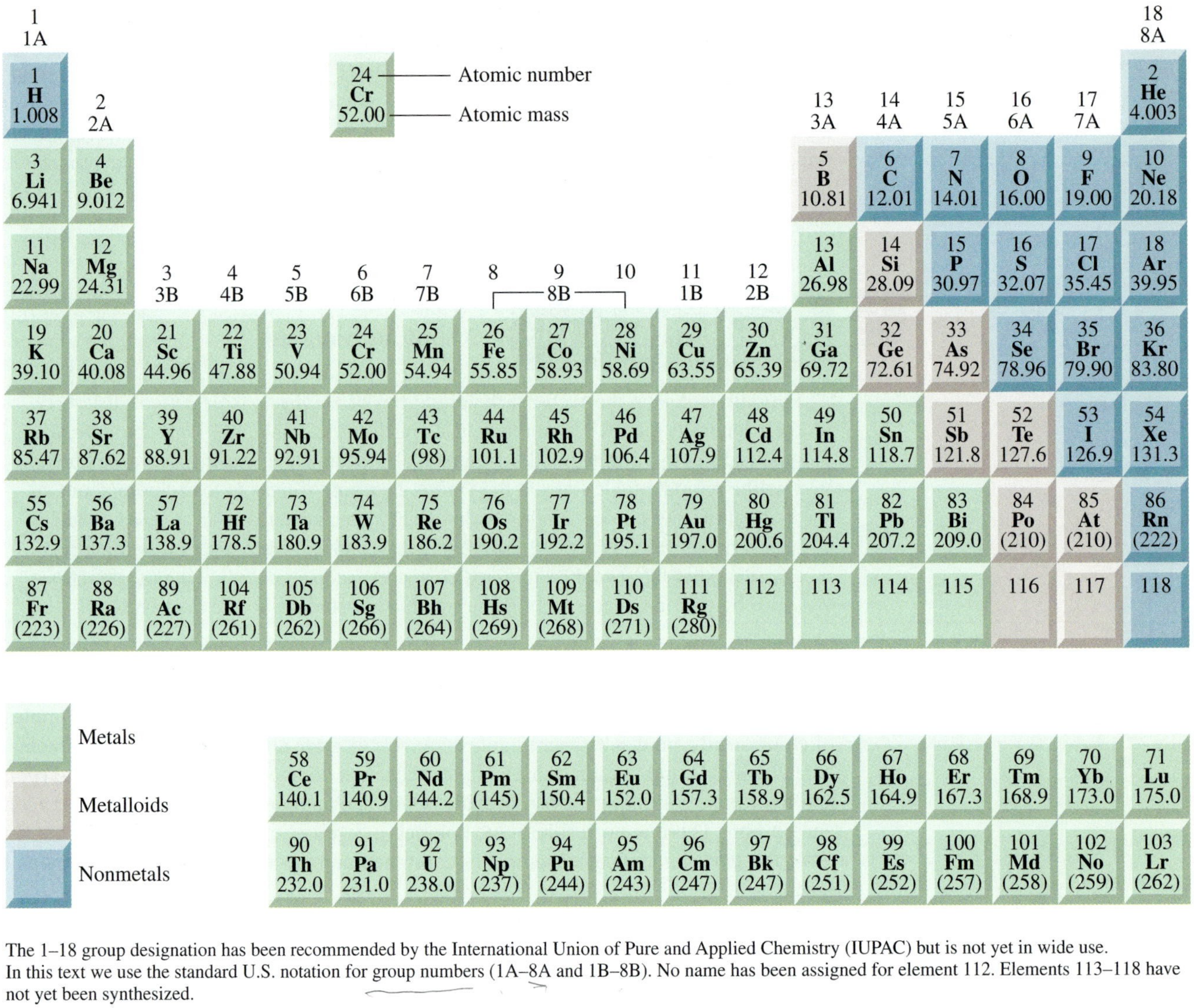

Atomic number — 24
Atomic mass — Cr 52.00

1 1A	2 2A	3 3B	4 4B	5 5B	6 6B	7 7B	8	9 8B	10	11 1B	12 2B	13 3A	14 4A	15 5A	16 6A	17 7A	18 8A
1 H 1.008																	2 He 4.003
3 Li 6.941	4 Be 9.012											5 B 10.81	6 C 12.01	7 N 14.01	8 O 16.00	9 F 19.00	10 Ne 20.18
11 Na 22.99	12 Mg 24.31											13 Al 26.98	14 Si 28.09	15 P 30.97	16 S 32.07	17 Cl 35.45	18 Ar 39.95
19 K 39.10	20 Ca 40.08	21 Sc 44.96	22 Ti 47.88	23 V 50.94	24 Cr 52.00	25 Mn 54.94	26 Fe 55.85	27 Co 58.93	28 Ni 58.69	29 Cu 63.55	30 Zn 65.39	31 Ga 69.72	32 Ge 72.61	33 As 74.92	34 Se 78.96	35 Br 79.90	36 Kr 83.80
37 Rb 85.47	38 Sr 87.62	39 Y 88.91	40 Zr 91.22	41 Nb 92.91	42 Mo 95.94	43 Tc (98)	44 Ru 101.1	45 Rh 102.9	46 Pd 106.4	47 Ag 107.9	48 Cd 112.4	49 In 114.8	50 Sn 118.7	51 Sb 121.8	52 Te 127.6	53 I 126.9	54 Xe 131.3
55 Cs 132.9	56 Ba 137.3	57 La 138.9	72 Hf 178.5	73 Ta 180.9	74 W 183.9	75 Re 186.2	76 Os 190.2	77 Ir 192.2	78 Pt 195.1	79 Au 197.0	80 Hg 200.6	81 Tl 204.4	82 Pb 207.2	83 Bi 209.0	84 Po (210)	85 At (210)	86 Rn (222)
87 Fr (223)	88 Ra (226)	89 Ac (227)	104 Rf (261)	105 Db (262)	106 Sg (266)	107 Bh (264)	108 Hs (269)	109 Mt (268)	110 Ds (271)	111 Rg (280)	112	113	114	115	116	117	118

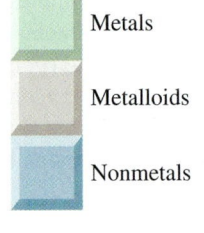

Metals

Metalloids

Nonmetals

58 Ce 140.1	59 Pr 140.9	60 Nd 144.2	61 Pm (145)	62 Sm 150.4	63 Eu 152.0	64 Gd 157.3	65 Tb 158.9	66 Dy 162.5	67 Ho 164.9	68 Er 167.3	69 Tm 168.9	70 Yb 173.0	71 Lu 175.0
90 Th 232.0	91 Pa 231.0	92 U 238.0	93 Np (237)	94 Pu (244)	95 Am (243)	96 Cm (247)	97 Bk (247)	98 Cf (251)	99 Es (252)	100 Fm (257)	101 Md (258)	102 No (259)	103 Lr (262)

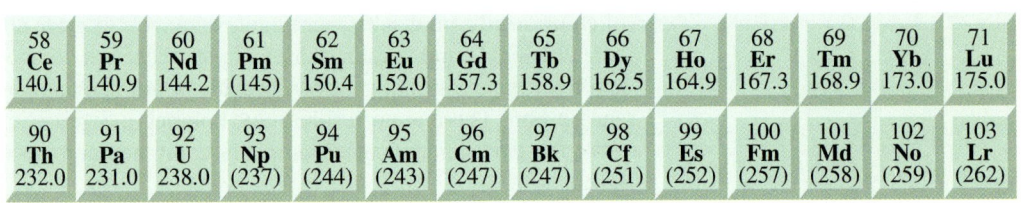

The 1–18 group designation has been recommended by the International Union of Pure and Applied Chemistry (IUPAC) but is not yet in wide use. In this text we use the standard U.S. notation for group numbers (1A–8A and 1B–8B). No name has been assigned for element 112. Elements 113–118 have not yet been synthesized.

Figure 1.10
The periodic table of the elements.

the moment, it is sufficient to note that about the time of the American Civil War, Mendeleev arranged the 66 elements that were then known into vertical columns according to the properties they had in common. These vertical columns are called **groups** and are given numbers. The members of Group 1A include lithium (Li), sodium (Na), potassium (K), and three other very reactive metals. Similarly, Group 7A consists of very reactive nonmetals, including fluorine (F), chlorine (Cl), bromine (Br), and iodine (I). Nitrogen and oxygen, the two most common elements in the atmosphere, are side by side in Groups 5A and 6A.

Some helpful generalizations come from examination of the elements that make up the periodic table. For example, the vast majority of elements are solids; some are gases; and only two, bromine and mercury, are liquids at room temperature and pressure. Most elements are metals, as indicated by the light green shading on the periodic table in Figure 1.10. **Metals** are elements that are shiny and conduct electricity and heat well; they include familiar substances such as iron, gold, and copper. Far fewer are **nonmetals**, elements that have varied appearances and don't conduct well, such as sulfur, chlorine, and oxygen. A mere eight elements fall into a category known as **metalloids**, sometimes also called semimetals. These elements fall on the line between metals and nonmetals on the periodic table and do not fall cleanly into either group.

Metals, nonmetals, and the ions they form are discussed in Section 5.7

Semiconductors are explained in Section 8.9.

Radioactivity is explained in Chapter 7.

The semiconductors silicon and germanium are examples of metalloids. The Group 8A elements are known as the **noble gases,** elements that are inert and do not readily undergo chemical reactions. In fact, some of the noble gases (helium and neon) do not combine chemically with *any* other elements. Radon is a noble gas that is radioactive, as are over a dozen other naturally occurring elements. As we will see in Section 1.13, radon affects the quality of indoor air. Thus, the periodic table is a very handy database, an amazingly useful way of organizing the building blocks of the universe. We will refer to it often throughout the text.

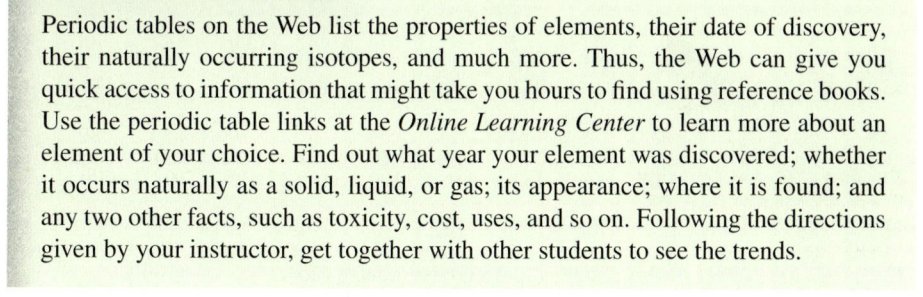

Consider This 1.14   Adopt an Element

Periodic tables on the Web list the properties of elements, their date of discovery, their naturally occurring isotopes, and much more. Thus, the Web can give you quick access to information that might take you hours to find using reference books. Use the periodic table links at the *Online Learning Center* to learn more about an element of your choice. Find out what year your element was discovered; whether it occurs naturally as a solid, liquid, or gas; its appearance; where it is found; and any two other facts, such as toxicity, cost, uses, and so on. Following the directions given by your instructor, get together with other students to see the trends.

Two other components of the atmosphere, water and carbon dioxide, are examples of **compounds,** pure substances made up of two or more elements in a fixed, characteristic chemical combination. For example, water is a compound of the elements oxygen and hydrogen. Similarly, carbon dioxide (CO_2) is a compound of the elements oxygen and carbon. There is no residual uncombined oxygen or carbon in carbon dioxide. In CO_2, the two elements are chemically combined and are no longer in their elemental forms.

As its name implies, carbon dioxide is made up of chemically combined carbon and oxygen in a fixed composition. All pure samples of carbon dioxide contain 27% carbon and 73% oxygen by weight (or mass). Thus, a 100-g sample of carbon dioxide will always consist of 27 g of carbon and 73 g of oxygen, chemically combined to form this particular compound. These values never vary, no matter the source of the carbon dioxide. This illustrates the fact that every compound exhibits a constant characteristic chemical composition.

A small paper clip weighs about a gram.

In contrast, carbon monoxide (CO) is also a compound of carbon and oxygen. However, pure samples of carbon monoxide contain 43% carbon and 57% oxygen by weight. Thus, 100 g of carbon monoxide contain 43 g of carbon and 57 g of oxygen, a much different composition from that of carbon dioxide. This is not surprising, because carbon monoxide and carbon dioxide are two different compounds.

The physical properties (such as boiling point) and chemical reactivity of compounds are also constant. Consider water (H_2O), a compound for which a 100-g sample consists of 11 g of hydrogen and 89 g of oxygen, (11% hydrogen and 89% oxygen by weight). At room temperature, pure water is a colorless, tasteless liquid. At sea level, it boils at 100 °C and it freezes at 0 °C. All samples of pure water have these same properties.

Although roughly only 100 elements exist, over 20 million compounds have been isolated, identified, and characterized. Some are very familiar naturally occurring substances such as water, salt, and sugar. But most known compounds do not exist in nature; rather, they were chemically synthesized by men and women across our planet. The motivation for making new compounds is almost as varied as the compounds themselves. The reasons may be to make synthetic fibers and plastics, to find drugs to cure AIDS or cancer, to create materials with a memory, or just for the creativity and intellectual fun of it. In Chapters 9 and 10, we will see examples of new compounds synthesized by chemists.

1.7 Atoms and Molecules

The definitions we just gave for elements and compounds made no assumptions about the nature of matter. We now know that elements are made up of **atoms,** the smallest unit of an element that can exist as a stable, independent entity. The word *atom* comes from the Greek for "uncuttable." Although today we know that it is possible to "cut" atoms using specialized processes, atoms remain indivisible by ordinary chemical or mechanical means.

Atoms are extremely small, many billions of times smaller than anything we can directly see. Because of this small size, huge numbers of atoms must be in any sample of matter that we can see or touch or weigh by conventional means. For example, in a single drop of water you might find 5×10^{21} atoms. This is about a trillion times greater than the approximately 6 billion people on Earth, enough to give each person a trillion atoms from that water drop.

As Figure 1.11 reveals, the invisible have recently become visible. Using a scanning tunneling microscope, scientists at the IBM Almaden Research Center lined up 112 carbon monoxide molecules on a copper surface to spell "NANO USA." **Nanotechnology** refers to work at the atomic and molecular (nanometer) scale: 1 nanometer (nm) = 1×10^{-9} m. Each letter is 4 nm high by 3 nm wide. At this size, about 250 million nanoletters could fit on a cross section of a human hair, corresponding to about three hundred 300-page books.

The existence of atoms provides a means of refining our earlier definitions of elements and compounds. Elements are made up of atoms of one type. For example, the element carbon is made up of carbon atoms only. By contrast, compounds are made up of the atoms of two or more elements. In the compound carbon dioxide, two types of atoms are present: carbon and oxygen. Similarly, water is made up of both hydrogen and oxygen atoms.

See Section 2.2 for more on atomic structure. See Section 7.2 for more about "atom cutting" (nuclear fission).

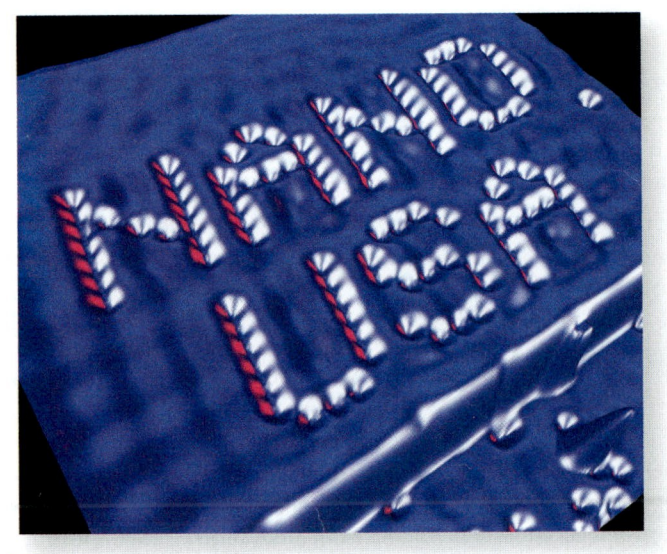

Figure 1.11
Carbon monoxide molecules lined up on a copper surface, as seen with a scanning tunneling microscope.

But we need to be careful with our language. The carbon and oxygen atoms in carbon dioxide are not present as such. Rather, the carbon and oxygen atoms are chemically combined to form a carbon dioxide **molecule,** a fixed number of atoms held together by chemical bonds in a certain spatial arrangement. More specifically, two oxygen atoms are combined with one carbon atom to form a carbon dioxide molecule. We represent this molecule with its chemical formula, CO_2. Similarly, in the water molecule, H_2O, three atoms are bonded together: two hydrogens and one oxygen. If we use black for carbon atoms, red for oxygen atoms, and white for hydrogen atoms, we can represent these molecules as follows.

water molecule   carbon dioxide molecule

Section 3.3 explains the shapes of these two molecules.

Millions of compounds other than carbon dioxide and water exist as molecules.

A **chemical formula** is a symbolic way to represent the elementary composition of a substance. It reveals both the elements present (by chemical symbols) and the atom ratio of those elements (by the subscripts). For example, in CO_2 the elements C and O are present in a ratio of one carbon atom for every two oxygen atoms. Similarly, H_2O indicates two hydrogen atoms for each oxygen atom. Note that when an atom is used once in a formula, such as the O in H_2O or the C in CO_2, the subscript of "1" is omitted. Look for more about the chemical formulas in the section that follows.

Elements have chemical formulas as well. Some elements exist as single atoms, such as helium or radon. We represent these as He and Rn, respectively. Other elements exist as molecules. For example, nitrogen and oxygen are found in our atmosphere as N_2 and O_2 molecules. We call these **diatomic molecules,** meaning that they contain two atoms per molecule. These representations clearly show the difference.

oxygen molecule   nitrogen molecule   helium atom

Table 1.6 summarizes our discussion of elements, compounds, and mixtures. It lists both what scientists can observe experimentally and the theory that scientists use to explain the atomic level that we cannot see. These descriptions complement each other.

Your Turn 1.16   **Elements and Compounds**

Name the element(s) present in each substance and identify each as an element or compound.

a. sulfur dioxide, SO_2 b. carbon tetrachloride, CCl_4
c. hydrogen peroxide, H_2O_2 d. sucrose, $C_{12}H_{22}O_{11}$
e. chlorine, Cl_2 f. nitrogen monoxide, NO

Answers
a. sulfur, oxygen (compound)   e. chlorine (element)

We now can apply these concepts to Earth's atmosphere. Air is a mixture, and its composition varies with time of day, location, and altitude. Some of its components, such as nitrogen, oxygen, and argon, are elements; others, notably carbon dioxide and water, are compounds. All of the compounds are present as molecules (i.e., CO_2 and H_2O). Some of the elements also exist as molecules (i.e., N_2 and O_2), whereas others are uncombined atoms (i.e., Ar and He).

Table 1.6	Classification of Matter	
Substance	**Observable Properties**	**Atomic Level**
Element	Cannot be broken down into simpler substances	One type of atom
Compound	Fixed composition, but capable of being broken down into elements	Two or more different atoms in a fixed combination
Mixture	Variable composition of elements, compounds, or both	Variable assortment of atoms, molecules, or both

Dry air is composed mainly of the elements nitrogen and oxygen, that is, N_2 molecules and O_2 molecules. If the air is humid, add in some water vapor in the form of H_2O molecules. Remember also the 385 ppm of carbon dioxide, meaning that there are 385 CO_2 molecules per 1×10^6 molecules and atoms in the air. Which atoms are these? Air contains just under 1% Ar (argon) atoms, as well as tiny amounts of He (helium) and Xe (xenon) atoms and extremely small amounts of Rn (radon) atoms.

Sceptical Chymist 1.17   The Chemistry of Lawn Care

News reports and advertisements should be viewed with a critical eye for their scientific accuracy. For example, a lawn care service advertisement reports that its fertilizers are "a balanced blend of nitrogen, phosphorus, and potassium. They have an organic nature, made up of carbon molecules. These fertilizers are biodegradable and turn into water." Suggest changes to make this ad more chemically correct.

1.8 Names and Formulas: The Vocabulary of Chemistry

If chemical symbols are the alphabet of chemistry, then chemical formulas are the words. The language of chemistry, like any other language, has rules of spelling and syntax. In this section we will help you to "speak chemistry" using both chemical formulas and names. In addition, you will find some duplication: a chemical formula can be known by more than one name (consider the alternatives as nicknames), but each name corresponds uniquely to one chemical formula.

In selecting the rules, we will follow the "need-to-know" philosophy. In essence, we will help you learn what you *need to know* to understand the topic at hand and leave the details of other naming rules for later sections. In this chapter, you need to know the chemical names and formulas of compounds that relate to the air you breathe. Accordingly, we will work on these now.

Earlier, we named several chemicals found in the atmosphere: carbon monoxide, carbon dioxide, sulfur dioxide, ozone, water vapor, and nitrogen dioxide. Although it may not be obvious by looking at the list of names, some are "systematic names," whereas the others are "common names."

Systematic names more or less follow a set of rules and are intended to lighten the burden on your memory. To see how systematic names work, let's start with carbon dioxide and carbon monoxide for the compounds CO_2 and CO, respectively. These names, like many (but not all) systematic names, contain prefixes that indicate how many atoms are present. *Di-* means "two," and thus the name carbon *di*oxide means two oxygen atoms are present

Check Section 5.8 for the rules of naming ionic compounds.

Remember that ozone, O_3, is an element, not a compound.

for each carbon atom. The chemical formula of CO_2 indicates these two oxygen atoms with a subscript 2. There is no subscript of 1 here because by convention, the 1 is assumed. In contrast, the prefix *mono-* means one, and carbon monoxide is simply CO. Again the subscript of 1 is assumed for both atoms. Consult Table 1.7 for a list of the prefixes.

Table 1.7	Prefixes Used in Naming Compounds		
Prefix	**Meaning**	**Prefix**	**Meaning**
mono-	1	hexa-	6
di- or bi-	2	hepta-	7
tri-	3	octa-	8
tetra-	4	nona-	9
penta-	5	deca-	10

To be successful in your use of names, you need to pay attention to three other details as well. First, the elements in CO and CO_2 are in a particular order. The name of the more metallic element comes first, followed by the name of the less metallic one. The periodic table is your guide. The metallic elements are on the left side (*light green* in Figure 1.10), and the nonmetallic elements on the right (*light blue*). Although carbon is a nonmetal, it still is closer to the metals than oxygen and hence is written first.

Second, notice that the name of the second element is modified to end in the suffix *-ide*. Oxygen becomes oxide, and similarly sulfur becomes sulfide and chlorine becomes chloride.

Third, the prefix *mono-* is omitted for the first element. For example, sulfur trioxide is not named *mono*sulfur trioxide, and CO_2 is not *mono*carbon dioxide. In contrast, the prefix *mono-* must be used if it applies to the second element, as in carbon monoxide (CO).

Your Turn 1.18   Oxides of Sulfur and Nitrogen

a. Write chemical formulas for nitrogen monoxide, nitrogen dioxide, dinitrogen monoxide, and dinitrogen tetraoxide.
b. Give chemical names for SO_2 and SO_3.

Answers
a. NO, NO_2, N_2O, and N_2O_4. *Note:* NO and N_2O also go by the common names nitric oxide and nitrous oxide.

Water is an example of a common name. From the rules just described, you might have expected H_2O to be called dihydrogen monoxide. Makes sense! Remember, though, that water was given its name long before anybody knew anything about hydrogen and oxygen. So chemists, being reasonable folks, call the stuff they swim in and drink by the common name water, just like everybody else does. Ozone (O_3) is another common name, as is ammonia (NH_3). Common names cannot be figured out. If you don't already know them, then you simply have to memorize them. Fortunately, the list is short, and we will introduce each name as needed.

See Section 4.7 for more about hydrocarbons.

In the next two sections, we will explore combustion reactions. Following our need-to-know philosophy, for combustion you will need to know the names of several common **hydrocarbons,** that is, compounds of hydrogen and carbon. Methane (CH_4) is the simplest hydrocarbon. Other small hydrocarbons include ethane, propane, and butane. Although methane may not appear to be a systematic name, it indeed is one if you are willing to accept *meth-* as meaning one carbon atom. Similarly, *eth-* means two carbons *prop-* means three, and *but-* means four. As you will further explore in Chapter 4, the suffix *-ane* is used with

hydrocarbons that contain no carbon-to-carbon double or triple bonds. With two carbons, ethane is C_2H_6. Similarly, propane is C_3H_8, and butane is C_4H_{10}. Just as *mono-, di-, tri-* and *tetra-* are used to count, so are *meth-, eth-, prop-,* and *but-*. The latter, however, are more versatile and used not only at the beginning of chemical names, but also embedded within them as the next exercise shows.

Your Turn 1.19   **Mother Eats Peanut Butter**

What does each name indicate about the number of carbon atoms?

   a. ethanol (found in some gasolines)
   b. methylene chloride (an indoor air pollutant)
   c. propane (the major component in LPG, liquid petroleum gas)
   d. methyl tertiary-butyl ether (MTBE, a gasoline additive)

Hint: Many generations of students have used the memory aid "<u>m</u>other <u>e</u>ats <u>p</u>eanut <u>b</u>utter" to remember the sequence *meth-, eth-, prop-, but-*.

Answers

   b. The *meth-* in methylene indicates one carbon in the chemical formula.
   d. This one is tricky. The *methyl-* in the name denotes one carbon atom, and the *butyl* denotes four carbon atoms, for a total of five carbon atoms. The chemical formula for MTBE is $C_5H_{12}O$, as you will see in Section 4.9.

Of course hydrocarbon molecules can have more than four carbon atoms. With larger molecules, use the counting scheme given in Table 1.7. For example, the octane molecule has eight carbon atoms.

1.9 Chemical Change: Oxygen's Role in Burning

The first pollutant listed in Table 1.5 is carbon monoxide, CO; however, all air, polluted or not, contains carbon dioxide, CO_2. Carbon monoxide and carbon dioxide can both arise from the same source: combustion. **Combustion is the rapid combination of oxygen with a substance.** When elemental carbon or carbon-containing compounds burn in air, oxygen combines with the carbon to form CO_2 or CO (or both). Similarly, combustion reactions produce water and sulfur dioxide by the burning of hydrogen and sulfur, respectively.

See Section 4.3 for more about combustion.

   Combustion is a major type of **chemical reaction,** a process whereby substances described as **reactants** are transformed into different substances called **products.** Chemical reactions can be represented by a **chemical equation,** a representation of a chemical reaction using chemical formulas. To students, chemical equations are probably better known as "the thing with the arrow in it." Chemical equations are the sentences in the language of chemistry. They are made up of chemical symbols (corresponding to letters) that often are combined in the formulas of compounds (the "words" of chemistry). Like a sentence, a chemical equation conveys information, in this case about the chemical change taking place. But, as we now will see, a chemical equation must also obey some of the same constraints that apply to a mathematical equation.

   At its most fundamental level, a chemical equation is very simple indeed. It is a qualitative description of the reaction:

$$\text{Reactant(s)} \longrightarrow \text{Product(s)}$$

By convention, the reactants are always written on the left and the products on the right. The arrow represents a chemical transformation and is read as "is converted to" or "yields." Thus, reactants are converted to products in the sense that the reaction gives products whose properties are different from those of the reactants.

Figure 1.12
The burning of charcoal in air.

The color coding we use for atoms reflects the standard used in many model kits and in molecular modeling software.

The combustion of carbon to produce carbon dioxide (e.g., the burning of charcoal in air, Figure 1.12) can be represented in several ways. One way is by a "word equation":

$$\text{carbon} + \text{oxygen} \longrightarrow \text{carbon dioxide}$$

It is more common to use chemical formulas to represent the elements and compounds involved:

$$C + O_2 \longrightarrow CO_2 \qquad [1.1]$$

This compact symbolic statement conveys a good deal of information. A translation of equation 1.1 might sound something like this: "One atom of the element carbon reacts with one molecule of the element oxygen to yield one molecule of the compound carbon dioxide."

Using black for carbon and red for oxygen, we can represent the rearrangement of atoms in this reaction as follows.

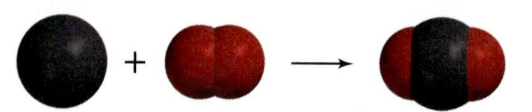

Similarly, using yellow for sulfur, we can represent the burning of sulfur to produce the air pollutant sulfur dioxide.

$$S + O_2 \longrightarrow SO_2 \qquad [1.2]$$

Note that this equation is balanced because an equal number of both sulfur atoms and oxygen atoms are in the reactants and products.

It is possible to pack even more information into a chemical equation by specifying the physical states of the reactants and products. A solid is designated by (*s*), a liquid by (*l*), and a gas by (*g*). Because carbon and sulfur are solids, and oxygen, carbon dioxide, and sulfur dioxide are gases at ordinary temperatures and pressures, equations 1.1 and 1.2 become:

$$C(s) + O_2(g) \longrightarrow CO_2(g)$$
$$S(s) + O_2(g) \longrightarrow SO_2(g)$$

We will designate the physical states when this information is particularly important, but usually omit it for simplicity.

Note that equation 1.1 has some of the characteristics of a mathematical equation; in this case, the number and kinds of atoms on the left equal those on the right:

Left side: 1 C, 2 O   Right side: 1 C, 2 O

This is the test for a correctly balanced equation. Atoms are neither created nor destroyed in a chemical reaction, and the elements present do not change when converted from reactants to products. This relationship is called the **law of conservation of matter and mass: in a chemical reaction, matter and mass are conserved.** The mass of the reactants consumed equals the mass of the products formed. The total mass does not change, as matter is neither created nor destroyed.

Atoms are rearranged during a chemical reaction. That is what a chemical change is all about: The atoms in the products are in a different arrangement than they were as reactants. Therefore, there is no requirement that the number of *molecules* must be the same on both sides of the arrow. In equation 1.1, one atom of carbon plus one molecule of oxygen yields one molecule of carbon dioxide. This looks suspiciously like 1 + 1 = 1. This is no cause for alarm; a chemical equation is not exactly the same as a mathematical equation. Remember, a chemical equation represents a transformation, not a simple equality. In a correctly balanced chemical equation, some things must be equal, others need not be. Table 1.8 summarizes.

Here is an analogy. The building materials used to construct a warehouse (reactants) can be disassembled and rearranged to build three houses and a garage (products).

Table 1.8	Characteristics of Chemical Equations

Always Conserved
Identity of atoms in reactants = Identity of atoms in products
Number of atoms in reactants = Number of atoms in products
Mass of all reactants = Mass of all products

May Change
Number of molecules in reactants vs. Number of molecules in products
Physical states (*s*, *l*, or *g*) of reactants vs. physical states of products

Equation 1.1 describes the combustion of pure carbon in an ample supply of oxygen. However, if the oxygen supply were limited, the products would include CO. For the sake of argument, let us say that pure carbon monoxide is formed. First we write the chemical formulas for the reactants and product:

$$C + O_2 \longrightarrow CO \text{ (unbalanced equation)}$$

Is this chemical equation balanced? No. There are two oxygen atoms on the left but only one on the right. We cannot balance the equation by simply adding an additional oxygen atom to the product side. Once we write the *correct* chemical formulas for the reactants and products, we cannot change them. To do so would imply a different reaction. All we can do is to use whole-number coefficients (or occasionally fractional ones) in front of the various chemical formulas. In simple cases like this, the coefficients can be found quite easily by simple trial and error. If we place a 2 to the left of the symbol CO, it signifies two molecules of carbon monoxide. This corresponds to a total of two carbon atoms and two oxygen atoms. Two oxygen atoms are also on the left side of the arrow, so the oxygen atoms have been balanced.

> A **subscript** follows a chemical symbol, as in O_2 or CO_2.
> A **coefficient** precedes a symbol or a formula, as in 2 C or 2 CO.

$$C + O_2 \longrightarrow 2\,CO \text{ (still not balanced)}$$

But now the carbon atoms do not balance. Fortunately, this is easily corrected by placing a 2 in front of the C.

$$2\,C + O_2 \longrightarrow 2\,CO \text{ (balanced equation)} \qquad [1.3]$$

The balanced equation can be represented with models, again using black for carbon atoms and red for oxygen atoms.

It is evident from comparing equations 1.1 and 1.3 that, relatively speaking, more O_2 is required to form CO_2 from carbon than is needed to form CO. This matches the conditions we stated for the formation of carbon monoxide, namely, that the supply of oxygen was limited.

Another air pollutant, nitrogen monoxide (also called nitric oxide), is produced from nitrogen and oxygen. In the presence of something hot, such as an automobile engine or a forest fire, these two atmospheric gases will combine. Again, we begin with chemical formulas.

$$N_2 + O_2 \xrightarrow{\text{high temperature}} NO \text{ (unbalanced equation)}$$

The equation is not balanced: two oxygen atoms are on the left, but only one is on the right. The same is true for nitrogen atoms. Placing a 2 in front of NO supplies two nitrogen *and* two oxygen atoms, and the equation is now balanced.

> Nitrogen and oxygen both are diatomic molecules.

$$N_2 + O_2 \xrightarrow{\text{high temperature}} 2\,NO \qquad [1.4]$$

Balance these chemical equations. Also draw representations of all reactants and products, analogous to equation 1.4. *Note:* In your drawings, both H_2O and NO_2 are bent molecules, with O and N as the central atom, respectively.

a. $H_2 + O_2 \longrightarrow H_2O$

b. $N_2 + O_2 \longrightarrow NO_2$

Answer

a. $2\,H_2 + O_2 \longrightarrow 2\,H_2O$

Consider This 1.21   **Advice from Grandma**

A grandmother offered this advice to rid the garden of pesky caterpillars. "Hammer some iron nails about a foot up from the base of your trees, spacing them every 3 to 5 inches." According to this grandmother, the iron nails convert the tree sap (a sugary substance containing carbon, hydrogen, and oxygen atoms) into ammonia (NH_3), a substance the caterpillars cannot stand. Comment on the accuracy of grandma's chemistry (allowing that the nails may still work, regardless of her explanation).

1.10   Fire and Fuel: Air Quality and Burning Hydrocarbons

Look for other examples of burning fuels throughout Chapter 4.

As we mentioned earlier, hydrocarbons are compounds of hydrogen and carbon. Hydrocarbons have many natural sources, but we typically obtain them from petroleum. Methane (CH_4), the simplest hydrocarbon, is the primary component of natural gas. Gasoline and kerosene are complex mixtures of many different larger hydrocarbon molecules.

Given an ample supply of oxygen, a hydrocarbon will burn completely, sometimes referred to as "complete combustion." All the carbon will combine with oxygen to form carbon dioxide, and all the hydrogen will combine with oxygen to form water. For example, here is the complete combustion of methane.

$$CH_4 + O_2 \longrightarrow CO_2 + H_2O \text{ (unbalanced equation)}$$

This equation is not as easy to balance as the previous ones because O appears in *both* of the products: CO_2 and H_2O. With equations like this one, it is easiest to start with an element that is present only in *one substance* on each side of the arrow. Here carbon is present only in CH_4 and CO_2, so we may begin with it (but don't need to change anything because for carbon, the expression is already balanced). Next, hydrogen also is present only in CH_4 and H_2O, so balance the H next by placing a 2 in front of H_2O.

When balancing chemical equations, it is fine to use fractional coefficients.

$$CH_4 + O_2 \longrightarrow CO_2 + 2\,H_2O \text{ (still not balanced)}$$

For combustion equations, do the oxygen last. A CO_2 molecule contains two O atoms, so there are four O atoms on the right of the equation and two O atoms on the left. Balance the number of O atoms by placing a 2 before O_2.

$$CH_4 + 2\,O_2 \longrightarrow CO_2 + 2\,H_2O \text{ (balanced equation)} \qquad [1.5]$$

A nice feature of chemical equations is that you can tell if they are balanced by counting atoms on both sides of the arrow. Carbon and hydrogen are easy to count because they appear on both sides of the equation. Here is the bookkeeping for oxygen:

Left: $2 \text{ O}_2 \text{ molecules} \times \dfrac{2 \text{ O atoms}}{1 \text{ O}_2 \text{ molecule}} = 4 \text{ O atoms}$

Right: $\left(1 \text{ CO}_2 \text{ molecule} \times \dfrac{2 \text{ O atoms}}{1 \text{ CO}_2 \text{ molecule}} \right)$

$+ \left(2 \text{ H}_2\text{O molecules} \times \dfrac{1 \text{ O atom}}{1 \text{ H}_2\text{O molecule}} \right) = 4 \text{ O atoms}$

Most automobiles run on the complex mixture of hydrocarbons that we call gasoline. One component is octane, C_8H_{18}. If sufficient supply of oxygen reaches the engine, octane burns to form carbon dioxide and water.

Look for more about gasoline in Sections 4.7 and 4.8.

$$2 \text{ C}_8\text{H}_{18} + 25 \text{ O}_2 \longrightarrow 16 \text{ CO}_2 + 18 \text{ H}_2\text{O} \qquad [1.6]$$

With less oxygen, the hydrocarbon will burn incompletely, sometimes referred to as "incomplete combustion." In this case, equation 1.6 will not occur as written. Instead, CO will be one of the products. An extreme situation is represented by equation 1.7, in which all of the carbon in the octane is converted to carbon monoxide.

$$2 \text{ C}_8\text{H}_{18} + 17 \text{ O}_2 \longrightarrow 16 \text{ CO} + 18 \text{ H}_2\text{O} \qquad [1.7]$$

Note that the coefficient of O_2 in equation 1.6 is 25, whereas the corresponding coefficient in equation 1.7 is 17. Less oxygen is needed to form CO.

Your Turn 1.22   Balancing Equations

Demonstrate that equations 1.6 and 1.7 are balanced by counting the number of atoms of each element on either side of the arrow.

Answer
Equation 1.6 contains 16 C, 36 H, and 50 O on each side.

What really happens in a car's engine is a combination of these and other chemical reactions. Most of the carbon released in automobile exhaust is in the form of CO_2, although some CO and soot (unburned carbon) are produced. The relative amounts of CO and CO_2 indicate how efficiently the car burns the fuel, which in turn indicates how well the engine is tuned. States that monitor auto emissions sample exhaust with a probe that detects CO (Figure 1.13). The measured CO concentrations are compared with established standards, for example, 1.20% in the state of Minnesota. If the vehicle fails the emissions test, it must be serviced.

Consider This 1.23   Auto Emissions Report

a. For which four substances are emissions being reported in Figure 1.13?
b. Which of these are criteria pollutants? *Hint:* See Table 1.1.
c. How is NO produced in an engine? *Hint:* See equation 1.4.
d. Why is the green line missing on the CO_2 graph?

Answer
d. Because CO_2 is not classified as an air pollutant, it has no unacceptable range.

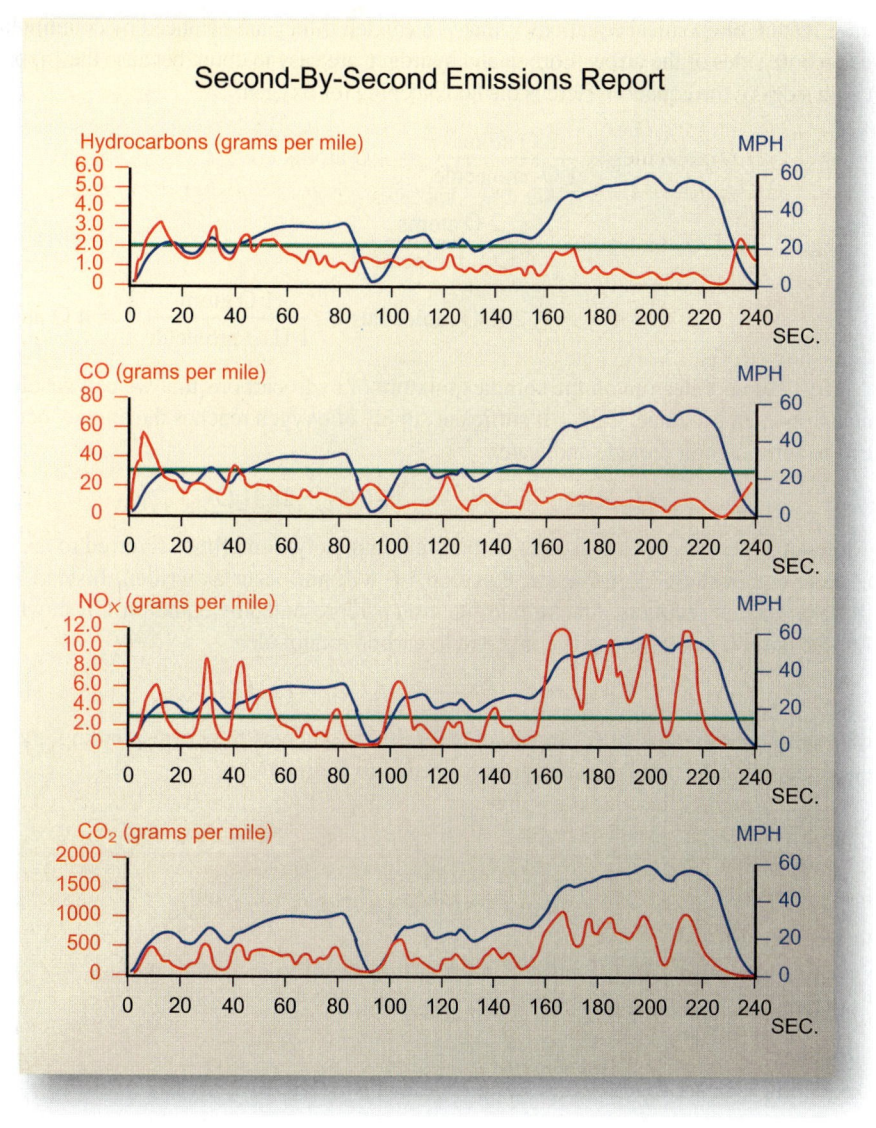

Figure 1.13

An auto emission report. All the emissions below the green line are in the acceptable range.

1.11   Air Pollutants: Direct Sources

By now you should realize that each breath you inhale contains mainly nitrogen and oxygen. In contrast, any pollutants are present in minuscule amounts. But even at the level of parts per million (or billion), these pollutants can compromise the quality of the air. How are these pollutants produced? In this section, we will examine two major sources: coal-fired plants that generate electricity and motor vehicles such as cars and trucks.

CO$_2$ emissions are measured, but not yet regulated. Look for more about CO$_2$ emissions throughout Chapter 3.

See Section 4.6 for more about coal and its chemical composition.

Your Turn 1.24   Tailpipe Gases

What comes out of the tailpipe of an automobile? Start your list now and build on it as you work through this section.

Hint: Some of the air that enters the engine also comes out in the exhaust.

In the United States, burning coal is the major source of electric power. Burning coal also is the major source of SO$_2$ emissions. Consisting mostly of carbon and hydrogen,

coal burns to form carbon dioxide (or carbon monoxide) and water. But coal also contains other elements. For example, most coals contain 1–3% sulfur and rock-like minerals. When coal is burned, the sulfur forms SO_2, and the minerals are converted into fine ash particles. If not removed, the SO_2 and particles go right up the smokestack. The hundreds of millions of tons of coal burned in this country translate into millions of tons of SO_2 and ash.

Once emitted, sulfur dioxide can react with oxygen to form sulfur trioxide, SO_3.

$$2\,SO_2 + O_2 \longrightarrow 2\,SO_3 \qquad\qquad [1.8]$$

Although normally quite slow, this reaction is much faster in the presence of small ash particles. The ash particles also aid another process. If the humidity is high enough, they promote the conversion of water vapor into an aerosol of tiny water droplets that we call fog. **Aerosols** consist of particles, both liquid and solid, that remain suspended in the air rather than settling out. Smoke is a familiar aerosol made up of tiny particles of solids and liquids.

Section 6.11 describes how sulfuric acid aerosols contribute to haze.

The aerosol of concern is one made up of tiny droplets of sulfuric acid, H_2SO_4. It forms because sulfur trioxide dissolves readily in water droplets to produce sulfuric acid.

$$H_2O + SO_3 \longrightarrow H_2SO_4 \qquad\qquad [1.9]$$

If inhaled, the sulfuric acid aerosol droplets are small enough to become trapped in the lung tissue and cause severe damage.

The good news? Sulfur dioxide emissions in the United States are declining. For example, in 1985, approximately 20 million tons of SO_2 was emitted from the burning of coal. Today the value is closer to 16 million tons. This decrease can be credited to the Clean Air Act of 1970 that mandated reductions in emissions from coal-fired electric power plants. More stringent regulations were established in the Clean Air Act Amendments of 1990. But continued progress will not come cheaply. You will find more information about the strategies and technologies for reducing atmospheric SO_2 and the economic and political costs in Chapter 6.

Your Turn 1.25 SO_2 from the Mining Industry

Burning coal is not the only source of sulfur dioxide. As you saw in Your Turn 1.11, smelting is another source. For example, silver and copper metal can be produced from their sulfide ores. Write the balanced chemical equations.

a. Silver sulfide (Ag_2S) is heated with oxygen to produce silver and sulfur dioxide.
b. Copper sulfide (CuS) is heated with oxygen to produce copper and sulfur dioxide.

Answer
a. $Ag_2S + O_2 \longrightarrow 2\,Ag + SO_2$

With more than 220 million vehicles (more than one for every two Americans), the United States has more vehicles per capita than any other nation. Would you expect sulfur dioxide to be coming out of all these tailpipes? Happily, the answer is no, because most cars have internal combustion engines fueled by gasoline. We already discussed the combustion of octane in gasoline to form carbon dioxide and water vapor (equation 1.6). Because gasoline contains little or no measurable sulfur, burning it is not a significant source of sulfur dioxide. Nonetheless, each tailpipe puffs out its share of air pollutants. The ubiquitous motor car adds to the atmospheric concentrations of carbon monoxide, nitrogen oxides, and particulate matter. So do, of course, SUVs and trucks.

Cars account for about 60% of carbon monoxide emissions nationwide. But remember to think in terms of *all* the tailpipes out there, not just those attached to cars. Nonroad vehicles such as farm and construction equipment, boats, snowmobiles, and gasoline-powered lawn mowers also emit carbon monoxide. So do heavy trucks, SUVs, and the three m's: motorcycles, minibikes, and mopeds.

Your Turn 1.26   **Other Tailpipes**

What about the exhaust from engines on gasoline-powered lawn mowers and on boats? At the EPA Web page entitled *Nonroad Vehicles and Equipment*, look up a source of interest. Summarize your findings and be sure to cite your sources.

> Wildfires contribute additional CO, increasing emissions as much as 10% each year.

A dramatic reduction in CO emissions has occurred even though the number of cars has risen. Based on measurements by the EPA at over 250 sites, the average CO concentration has dropped 60% from 1990 to 2005. Assuming that wildfires are not included, today's levels are the lowest reported in three decades. The decrease is due to several factors, including improved engine design, computerized sensors that better adjust the fuel–oxygen mixture, and most importantly, that all new cars since the mid-1970s must have **catalytic converters** (Figure 1.14), devices installed in the exhaust stream to reduce emissions. In general, a **catalyst** is a chemical substance that participates in a chemical reaction and influences its rate without undergoing permanent change. Catalytic converters in vehicles have two functions. The first is to lower carbon monoxide emissions using metals such as platinum and rhodium to catalyze the combustion of CO to CO_2. Other catalysts convert nitrogen oxides back to N_2 and O_2, the two atmospheric gases that formed them.

> The air in a pine forest contains VOCs. The wonderful smell comes from volatile compounds emitted by the trees.

A modern high-performance automobile capable of operating at high speeds and with fast acceleration not only emits carbon in the form of carbon monoxide, but also in the form of unburned hydrocarbons. These are often referred to as VOCs, or volatile organic compounds. This term requires some explaining. A substance is **volatile** if it readily passes into the vapor phase. Gasoline and nail polish remover are both volatile; when you spill a few drops, these drops readily evaporate. A substance is an **organic compound** if it contains mainly carbon and hydrogen. For example, organic compounds include the hydrocarbons methane and octane mentioned earlier, as well as compounds containing O in addition to C and H, such as alcohol and sugar. We will discuss organic compounds more fully beginning in Chapter 4.

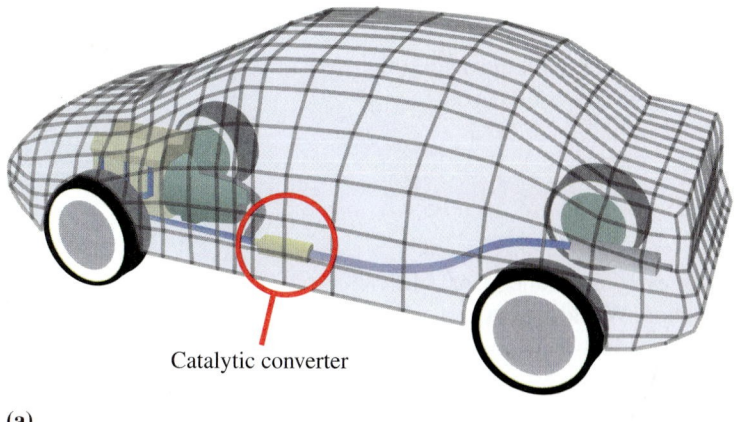

Catalytic converter

(a)

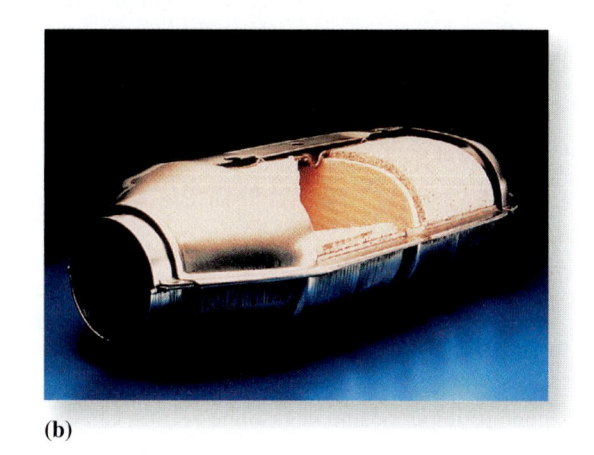

(b)

Figure 1.14

(a) Location of catalytic converter in car. (b) Cutaway view of a catalytic converter. Metals such as platinum and rhodium coat the surface of ceramic beads.

In the case of tailpipe emissions, **volatile organic compounds (VOCs)** are vapors of incompletely burned gasoline molecules or fragments of these molecules. This incomplete combustion is a result of either insufficient oxygen or insufficient time in the engine cylinders for all the hydrocarbons to be burned. The exhaust gas still contains oxygen, as not all of it is consumed in the engine. Catalytic converters utilize this oxygen to lower the amounts of VOCs emitted by burning them to form carbon dioxide and water. Look for more in Section 1.12 about the connection between VOCs and ozone formation.

Another success, although much harder won, accompanied the advent of the catalytic converter. For more than 50 years, a lead-containing compound, tetraethyl lead (TEL), was added to gasoline to make it burn more smoothly and eliminate "knocking." About a teaspoon of TEL was added to every gallon of gasoline. TEL worked beautifully in "knocking out the knock," but unfortunately released lead through the tailpipe onto the roadsides and city streets. In many of its chemical forms, lead is highly toxic and acts as a cumulative poison that can cause a wide variety of neurological problems, especially in children. Although its toxicity was well known and documented, the struggles to remove lead from gasoline lasted over 60 years.

The catalytic converter and TEL are connected in that the latter destroys the effectiveness of the former. Therefore, the cars and trucks built with catalytic converters since 1976 were formulated to run on unleaded gasoline (gasoline without TEL). After over 20 years of phasing in unleaded fuel, in 1997 leaded fuel finally was banned by law in the United States. Accordingly, today at the gas pump you see all fuel labeled as unleaded (Figure 1.15). The result has been a dramatic decrease (95%) in lead emissions from vehicles, from 1980 to 1999. Unfortunately, lead has yet to be banned globally, and several dozen countries still allow almost a gram of lead per liter of fuel. High levels of lead are found in cities such as Bangkok, Cairo, Jakarta, and Mexico City.

The United States has had less success curbing the emission of nitrogen oxides. Nitrogen and oxygen are present wherever there is air. And, whenever air is subjected to high temperatures, as in an internal combustion engine or in a coal-fired power plant, N_2 and O_2 combine to form two molecules of NO as we saw earlier in equation 1.4.

Many countries in the Middle East and Africa still use leaded gasoline.

(a)

(b)

Figure 1.15

Pumps for unleaded gasoline from **(a)** the United States and **(b)** United Kingdom.

Unlike N_2, NO is very reactive. It reacts with oxygen to form NO_2.

$$2\,NO + O_2 \longrightarrow 2\,NO_2 \qquad [1.10]$$

However, this reaction does not occur in a short time (e.g., while driving your car to work) because it requires high concentrations of NO to proceed quickly. The concentration of NO in polluted air is on the order of 100 ppb, a concentration not high enough for the NO to quickly react with O_2. Given this, how is NO_2 formed from NO? To answer this question, we need to bring in two other players: VOCs (mentioned earlier) and the hydroxyl radical, $\cdot OH$. The latter is a reactive species containing an unpaired electron indicated by the dot. In Chapter 2, you will meet other reactive species with unpaired electrons.

Although both $\cdot OH$ and VOCs occur naturally, VOC concentrations are considerably higher in polluted air. The following complex chain of events converts NO to NO_2.

$$VOC + \cdot OH \longrightarrow A$$
$$A + O_2 \longrightarrow A'$$
$$A' + NO \longrightarrow A'' + NO_2 \qquad [1.11]$$

Here, A, A′, and A″ represent reactive molecules that are synthesized from the VOCs. As promised, this reaction is complex! But then again so is the chemistry of air pollution. We emphasize the bottom line: If the air contains sufficient concentrations of NO, O_2, VOCs, and $\cdot OH$, you have the right ingredients to form NO_2. In the process, other reactive molecules (A, A′, and A″) form as well. NO_2 is toxic and a player in the formation of tropospheric ozone, as we will see in the next section. Nitrogen dioxide also contributes to acid rain, the subject of Chapter 6.

The quantity of nitrogen oxides emitted into the atmosphere has increased about 9% since 1980, though they have shown some slight decreases in the last few years. Given the significantly increased number of vehicles and miles driven, any decrease is impressive. Despite early claims from the auto industry that it would be impossible, or too costly, to meet the emissions standards, the industry has, in fact, achieved these goals by improving catalytic converters, engine designs, and gasoline formulations.

Consider This 1.27   How You Drive

Burning less gasoline translates to fewer emissions out the tailpipe. Which driving practices (other than not getting into the car) conserve fuel? Which practices expend it more quickly than necessary? Think about your behavior on highways, city streets, and even in parking lots. Consider when you accelerate, coast, and brake. List at least four ways that drivers could burn less gasoline while still getting to their destination.

Chapter 8 discusses fuel cells and other alternatives to gasoline-powered vehicles.

An obvious way to reduce pollutants is not to have them form in the first place. Over the past decade, an important initiative known as "green chemistry," the use of chemistry to prevent pollution, has taken place. **Green chemistry** is the designing of chemical products and processes that reduce or eliminate the use or generation of hazardous substances. Begun under the EPA's Design for the Environment Program, green chemistry reduces pollution through fundamental chemical breakthroughs in designing and redesigning processes that make chemical products, with an eye toward making them environmentally friendly, that is, "benign by design." In this regard, Dr. Barry Trost, a Stanford University chemist, advocates an "atom economy" approach to the synthesis of commercial chemical products such as pharmaceuticals, plastics, or pesticides. Such syntheses would be designed so that all reactant atoms end up as desired products, not as wasteful by-products. This approach will save money, as well as materials; undesired products would not be produced as waste, which requires disposal.

Dr. Lynn R. Goldman, who served from 1993 to 1998 in the Office of Prevention, Pesticides, and Toxic Substances at the EPA, remarked "Green chemistry is preventative medicine for the environment." Innovative "green" chemical methods already have had an impact on a wide variety of chemical manufacturing processes by decreasing or eliminating the use or creation of toxic substances. For example, the use of green chemical principles has led to cheaper, less wasteful, and less toxic production of ibuprofen, pesticides, new materials for disposable diapers and contact lenses, new dry-cleaning methods, and recyclable silicon wafers for integrated circuits. The research chemists and chemical engineers who developed these and other green chemistry approaches have received the Presidential Green Chemistry Challenge Award. Begun in 1995, it is the only presidential-level award recognizing chemists and the chemical industry for their innovations for a less polluted world; its theme is "Chemistry is not the problem, it's the solution." Throughout this book, we will discuss applications of green chemistry and designate these with the Green Chemistry icon.

1.12 Ozone: A Secondary Pollutant

Ozone definitely is a bad actor in the troposphere. As we mentioned earlier, ozone affects your respiratory system; even at very low concentrations it will reduce lung function in normal, healthy people who are exercising outdoors. Low concentrations of ozone also take a toll on vegetation, damaging the leaves and needles of trees.

In the previous section, we made no mention of ozone coming out of a tailpipe or being produced when coal is burned to generate energy. How then is ozone produced? Before we answer this question, examine Figure 1.16 and do the activity that accompanies it (Consider This 1.28).

6 AM

10 AM

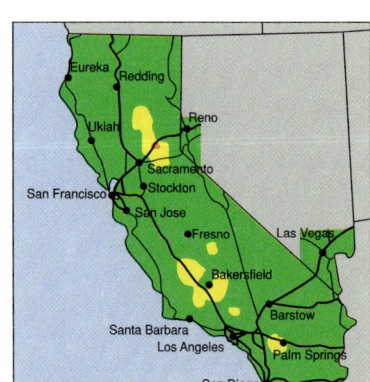

Noon

4 PM

10 PM

Figure 1.16

Ozone level maps for a summer day in California, July 2006.

Consider This 1.28 **Ozone Around the Clock**

Ozone concentrations vary over the course of a day, as shown in Figure 1.16.

a. Near which city is the air hazardous to one or more groups? *Hint:* Refer back to the color-coded AQI (see Figure 1.5).
b. Around what time does the level of ozone level peak?
c. Can moderate levels of ozone exist in the absence of sunlight? Assume that sunrise was at about 6 AM and sunset at about 8 PM.

Answer

c. Yes, there can be, but not for long. After sundown, the ozone levels drop.

The previous activity raises several interrelated questions. How is ozone produced? Why is it more prevalent in some areas than others? And what role does sunlight play in ozone production? We now address these.

Unlike the pollutants described in the previous section, ozone is not directly emitted into the atmosphere. Rather, it is a **secondary pollutant,** that is, it is produced from chemical reactions among two or more other pollutants, in this case, VOCs and NO_2. Recall from Section 1.11 that NO_2 does not come directly out of the tailpipe either. Rather, automobile engines produce NO. But over time and in the presence of VOCs and $\cdot OH$, NO in the atmosphere is converted to NO_2 as you saw in equation 1.11.

Nitrogen dioxide meets several fates in the atmosphere, and the one of interest to us occurs when the Sun gets higher in the sky. The energy provided by sunlight splits one of the bonds in the NO_2 molecule:

> Recall from the previous section that $\cdot OH$ is the hydroxyl radical.

$$NO_2 \xrightarrow{\text{sunlight}} NO + O \qquad [1.12]$$

Focus on the oxygen atom produced in equation 1.12. It can in turn react with an oxygen molecule to produce ozone.

$$O + O_2 \longrightarrow O_3 \qquad [1.13]$$

> In Section 2.1, we will define allotropes and use O_3 and O_2 as examples.

We now have an explanation for why ozone is linked to sunlight. Ozone formation requires O, which in turn is produced when sunlight splits NO_2. No sunlight, no ozone. Thus once the Sun goes down, the ozone concentrations drop off sharply, as you saw in Figure 1.16. What happened to the ozone? The O_3 molecule is consumed in just a matter of hours. It reacts with many things, including animal and plant tissue.

Note that equation 1.13 contains three different forms of the element oxygen: O, O_2, and O_3. All three are found in nature, but O_2 is by far the most abundant as it constitutes about one fifth of the air we breathe. Our atmosphere naturally contains tiny amounts of protective ozone up in the stratosphere as we shall see in Chapter 2, and even tinier amounts of oxygen atoms that are too reactive to persist very long.

> "Good" ozone is in the stratosphere. "Bad" ozone is in the troposphere.

Consider This 1.29 **O_3 Summary**

Summarize what you have learned about ozone formation by developing your own way to arrange these chemicals sequentially and in relation to one another: O, O_2, O_3, VOCs, NO, NO_2. Chemicals may appear as many times as you would like, and also you may wish to include sunlight.

Because sunlight is involved in ozone formation, you might suspect that the concentration of ozone varies by the weather, the season, and by latitude. Your suspicion would be correct. High levels of tropospheric ozone are much more likely to occur on long sunny summer days, especially in a congested urban area. Stagnant air can increase the buildup of air pollution; clouds, wind, and rain can mitigate it. For example, let's look back at the information for Houston presented in Table 1.4. From the data, we see

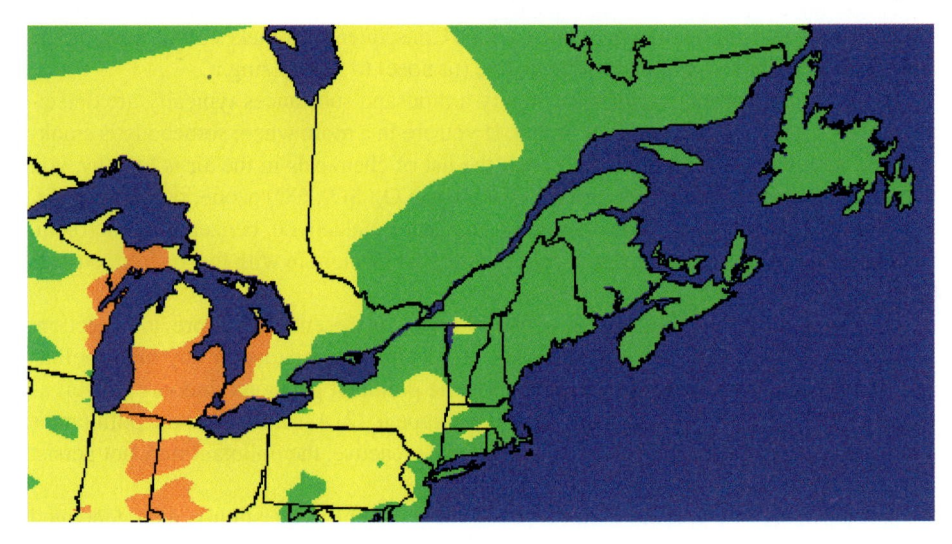

Figure 1.17

Tropospheric ozone map for May 23, 2007.

Source: Environment Canada.

that the air was unhealthy for somebody (and in some cases for everybody) on 42 days in 2003 and on 39 days in 2004. Closer examination of the data indicates that two pollutants were mainly responsible for the unhealthy air: ozone and $PM_{2.5}$. Ozone was more likely to be the culprit on hot sunny summer days.

Consider This 1.30　　**Ozone Maps**

Ozone levels are reported for almost all parts of the United States. To see the data, go to the AIRNOW Web site, courtesy of the EPA. A direct link is provided at the *Online Learning Center*. Select a city or state to learn how the ozone levels vary over the ozone season. Summarize your findings.

Daily ozone maps also are available for Canada. As you can see from Figure 1.17, some of Canada's polluted air can originate in the United States. Predictably, many cities across the globe have high ozone levels. Couple vehicles with a sunny location anywhere on the planet, and you are likely to find unacceptable levels of ozone. Some places, though, are worse than others. Seattle, for example, with its cool rainy days has low ozone levels. In contrast, ozone is a serious problem in Mexico City.

Ironically, ozone attacks rubber, thus damaging the tires of the vehicles that led to its production in the first place. Should you park your car indoors in the garage to minimize possible rubber damage? In fact, should you park *yourself* indoors if the levels of ozone outside are unhealthy? The next section speaks to the quality of indoor air.

1.13 The Inside Story of Air Quality

Most of us sleep, work, study, and play indoors, often spending the majority of our time in our dorm rooms, classrooms, offices, shopping malls, restaurants, or health clubs. In spite of where we spend our time, standards have been established for outdoor air, but not for the air inside. Ironically, the levels of air pollutants indoors may far exceed those outdoors. Furthermore, those least able to handle poor air, the very old, newborns, and

those who are ill, may seldom get out-of-doors. Consequently, it makes sense to study the chemistry of indoor air pollution with an eye (or nose) to minimizing it.

Indoor air is a complex mixture; nearly a thousand substances typically are detectable at the parts per billion level or higher. If you are in a room where somebody is smoking, add another thousand or so. Although the list of chemicals in the air is lengthy, you will recognize some familiar culprits: VOCs, NO, NO_2, SO_2, CO, ozone, radon, and PM. Less familiar pollutants include chemicals such as formaldehyde, benzene, and acrolein. Some of these pollutants are present because they are brought in with the air from outside the building; others are generated right inside.

Let's begin by considering the question posed at the end of the previous section. Should you move indoors to escape ozone? To decide about ozone or any other pollutant, you need to answer two questions: (1) Is the pollutant generated indoors as well as outdoors? If so, by moving inside you will not escape it. (2) Assuming that the pollutant is generated *only* outdoors, how reactive is it? If very reactive, the pollutant may not persist long enough to be transported indoors.

For now, let us put aside item (1) when the pollutant in question is also generated indoors (we will discuss this case shortly). In assessing item (2), the chemical reactivity, assume that the more reactive the chemical species, the less likely it is to persist indoors. Thus, for reactive molecules such as O_3, NO_2, SO_2, we expect lower levels indoors. The numbers bear this out: The actual ratio of ozone levels outdoors to those indoors is typically between 0.1 to 0.3, meaning that the ozone levels inside are less by nearly a factor of ten. Thus, all things being equal, one way to escape an ozone alert is to move inside. Similarly, sulfur dioxide and nitrogen dioxide levels are expected to be lower indoors, although the ratio is not quite as favorable as that for ozone.

As a relatively unreactive pollutant, CO is a different story. This gas has a long enough atmospheric lifetime to move freely in and out of buildings, either through doors, windows, or through the ventilation system. The same is true for the less reactive VOCs, but not the more reactive ones such as those that give pine forests their scent. If your want to smell the volatile compounds from the Ponderosa pines, best to remain out-of-doors.

Finally, the air-handling system of a building can lower the level of indoor pollutants. This is especially true for the larger sized particulate matter. For example, heating and cooling units usually contain filters. People who suffer from pollen allergies often escape into air-conditioned buildings, since the pollen levels tend to be lower inside. Similarly, those living in areas where wild fires burn sometimes can escape the irritating smoke particles by staying indoors. Only to a certain extent, however, do buildings filter out the fine particles of smoke and ash. Those who lived in New York City in the aftermath of the 2001 attacks on the World Trade Center can attest to the fact that the smell of smoke pervaded their living and sleeping spaces for months.

Some copy machines and air cleaners generate ozone, in which case the indoor-to-outdoor ratio may not be as favorable.

Your Turn 1.31   Indoor Activities

Name five indoor activities that generate pollutants. To get you started, two are pictured in Figure 1.18. Remember that you cannot smell some pollutants, such as carbon monoxide.

Following up on Your Turn 1.31, we now will examine indoor sources of air pollution. As noted earlier, the decision to move inside to escape a particular pollutant is based in part on whether the pollutant is generated indoors. If so, both the rate of generation and the amount of ventilation will determine how quickly it builds up. As it turns out, several pollutants can be produced indoors at relatively high rates and will accumulate if the ventilation is insufficient.

For example, consider indoor volatile organic compounds (VOCs). We already have discussed these pollutants in Sections 1.11 and 1.12 using the compounds and molecular

(a) **(b)**

Figure 1.18
Activities that pollute indoor air.

fragments present in vehicle engine exhaust as an example. Because tailpipes are unlikely to be found indoors, we need to look for another source of VOCs. One possibility is environmental tobacco smoke (ETS), often referred to as "second-hand smoke." If you know any of the 40 million smokers in the United States (or smoke yourself), you know that the pollutants in cigarette smoke are easily noticeable indoors. As pointed out earlier, cigarette smoke contains over a thousand chemical substances that, taken as a whole, are **carcinogenic,** meaning capable of causing cancer. One VOC you probably recognize is nicotine; others include benzene and formaldehyde.

But cigarettes are not the only thing we burn indoors and hence not the only source of VOCs. Burning incense and candles also produce VOCs, often with a good deal of soot (particulate matter) as well. Again, the indoor pollutants can be carcinogenic; for example, epidemiological studies have linked the regular burning of incense with some childhood cancers. Wood stoves, fireplaces, and some appliances also emit a variety of VOCs. The risks are high if wood-burning appliances are improperly installed, improperly vented, or simply malfunctioning.

Cigarette smoking in particular and combustion in general are good examples of how indoor sources produce more than one type of pollutant. For example, by burning any carbon-containing fuel you would expect to produce carbon monoxide. Indeed, CO levels from cigarette smoking in bars can reach 50 ppm, well within the range considered unhealthy. Similarly, the high temperatures from burning tobacco or wood would be expected to produce nitrogen oxides, although in much smaller amounts. For cigarette smoke, NO_2 levels can exceed 50 ppb.

Other indoor activities add pollutants to the air as well. Often your nose alerts you to the source; sometimes your head as well, if you get a headache from the fumes. For example, when you paint, use a brush cleaner, or paint your fingernails, you can smell the VOCs. Paint thinner in particular carries the warning on the label to use with sufficient ventilation. If you use hairspray or something else in a spray can, the odor may linger and be especially noticeable to somebody who enters the room. New carpets and new furniture also emit their own characteristic odors. These and other sources of indoor air pollutants are listed in Table 1.9.

Radon, a noble gas, is a special case of indoor air pollution. It is a naturally occurring carcinogen, but only tends to build up in dwellings (particularly basements) and in some mines and caves. Like all noble gases, radon is colorless, odorless, tasteless, and chemically

Table **1.9**	**Selected Indoor Air Pollutants and Their Sources**	
Form	**Pollutant**	**Source**
Solid/particulate	Asbestos	Floor tile, insulation
	Pet dander, dust	Pets
	Molds, mildew, bacteria, viruses	Plants, bedding, furniture
Liquid/gas	Styrene	Carpet
	CO, benzene, nicotine, PM	Cigarette smoke
	Dry-cleaning fluid, moth balls	Clothes
	O_3	Electric arcing
	CO, NO, NO_2	Unvented space heaters
	Formaldehyde	Furniture
	Acetone, toluene	Glues and solvents
	Methanol, methylene chloride	Paint and paint thinners
	Radon	Soils/rocks under dwelling

Radioactive substances are explained in Chapter 7.

unreactive, but radon is unique in being a noble gas that is radioactive. It is generated as part of the nuclear decay series of another radioactive element, uranium. Because uranium occurs naturally and is ubiquitous at the surface of our planet, chances are that the rocks and soils beneath your home contain small amounts of uranium. Depending on how your dwelling is constructed, radon may seep into your basement and become trapped indoors. Although the inhalation of radon can produce lung cancer, the threshold for danger is controversial. Radon test kits can be used to determine the radon concentration in a basement or other living space (Figure 1.19).

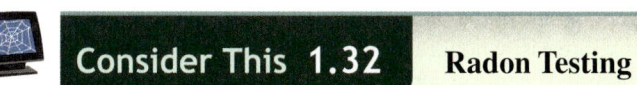

Consider This **1.32**   **Radon Testing**

As a public service, local and national agencies provide information about radon on the Web.

a. Find two Web sites about radon provided by government agencies. Cite the source and the URL for each. You might find it helpful to use the keywords *radon detection, air quality,* and *EPA* in your search.
b. Find a company on the Web that sells radon test kits. Describe the kit, including its price.
c. Compare the dangers of radon described on your Web sites from parts **a.** and **b.** Is commercial information about radon different from that provided as a public service? If so, report the differences and suggest reasons why.

Figure 1.19
A home radon test kit.

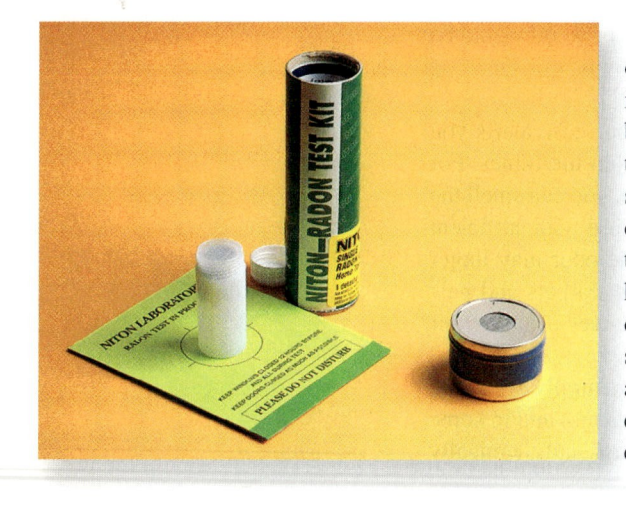

A further word on pollutants and building construction is in order. As you would expect, the rates at which outdoor air moves inside and indoor air moves out affects how quickly air pollutants build up indoors. An insufficient exchange of outside air can cause the concentration of indoor air pollutants to build up to troublesome levels. Consider the risk–benefit trade-off in which buildings constructed within the past two decades have been more airtight to increase energy efficiency. Although greater energy efficiency has decreased heating and cooling costs, it has been at the cost of decreasing the circulation of air between the building and the outside. When air exchange is reduced, the levels of indoor air pollutants increase. Therefore, initially what was a benefit (better energy efficiency) can turn into an increased risk (increased pollutant levels). Construction of some large office buildings has been so highly

energy-efficient that little exchange of outside air occurs within them. In some of these cases, the reduced air exchange has allowed indoor pollutants to reach levels hazardous to the health of some individuals, creating a condition known as "sick building syndrome."

Whether we breathe indoor or outdoor air, we inhale (and exhale) a truly prodigious number of molecules and atoms during a lifetime. On a molecular and atomic level, these particles have some fascinating properties, ones that we consider next.

1.14 Back to the Breath—at the Molecular Level

The maximum concentrations of pollutants specified in Table 1.5 seem very small, and they are. Nine CO molecules out of every 1 million molecules in air is a tiny fraction. But, as we will soon calculate, even this low concentration of CO contains a staggering number of carbon monoxide molecules. This seeming contradiction is a consequence of the minuscule size of molecules and their immense numbers. Recall Consider This 1.1: Take a Breath. If you are an average-sized adult in good physical condition, the total capacity of your lungs is between 5 and 6 L. You do not exchange this volume of air each time you take a breath. Rather, perhaps now as you are reading this, you are inhaling (and exhaling) only about 500 milliliters (mL, 0.500 L) of air, or approximately half a quart.

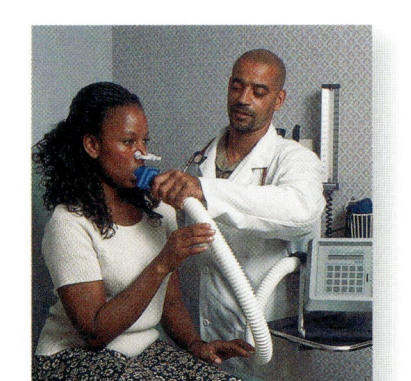

Figure 1.20
A spirometer is used for measuring an individual's breathing capacity.

Accurately measuring the volume of air you inhale and exhale can be done with the help of a spirometer (Figure 1.20). Determining the number of molecules and atoms in this volume of air is a harder task, but it can be done. From experiments (as well as from theory), we know that a typical breath of 500 mL contains about 2×10^{22} molecules and atoms. Remember that air is mostly N_2 and O_2 molecules together with atoms such as Ar and He.

Using this number of molecules and atoms in the air (2×10^{22}), we now can calculate the number of CO molecules in the breath you just inhaled. We will assume the breath contained 2×10^{22} molecules and atoms, and that the CO concentration in the air was the NAAQS of 9 ppm. Thus, out of every million (1×10^6) molecules and atoms in the air, nine will be CO molecules. To compute the number of CO molecules in a breath, multiply the total number of molecules and atoms in the air by the fraction that are CO molecules.

$$\frac{\text{\# of CO molecules}}{1 \text{ breath of air}} = \frac{2 \times 10^{22} \text{ molecules and atoms in air}}{1 \text{ breath of air}} \times \frac{9 \text{ CO molecules}}{1 \times 10^6 \text{ molecules and atoms in air}}$$

$$= \frac{2 \times 9 \times 10^{22}}{1 \times 10^6} \frac{\text{CO molecules}}{\text{breath of air}}$$

$$= \frac{18}{1} \times \frac{10^{22}}{10^6} \frac{\text{CO molecules}}{\text{breath of air}}$$

In writing this out, we carefully retain the units on the numbers. Not only does this remind us of the physical entities involved, but also it guides us in setting up the problem correctly. The labels "molecules and atoms in the air" cancel each other, and we are left with the label we want: CO molecules.

However, we need to divide 10^{22} by 10^6 to determine a final answer. To *divide* powers of 10, simply *subtract* the exponents. In this case,

$$\frac{10^{22}}{10^6} = 10^{(22-6)} = 10^{16}$$

Thus, a breath contains 18×10^{16} CO molecules.

The preceding answer is mathematically correct, but in scientific notation it is customary to have only one digit to the left of the decimal point. Here we have two: 1 and 8. Therefore,

our last step will be to rewrite 18×10^{16} as 1.8×10^{17}. We can make this conversion because $18 = 1.8 \times 10$, which is the same as 1.8×10^1. We *add* exponents to *multiply* powers of 10. Thus, 18×10^{16} CO molecules equals $(1.8 \times 10^1) \times 10^{16}$ CO molecules, which equals 1.8×10^{17} CO molecules in that last breath you inhaled. (If all of this use of exponents is coming at you a little too fast, consult Appendix 2.)

It may sound surprising, but it would be more accurate to round off the answer and report it as 2×10^{17} CO molecules. Certainly 1.8×10^{17} looks more accurate, but the data that went into our calculation were not very exact. The breath contains *about* 2×10^{22} molecules, but it might be 1.6×10^{22}, 2.3×10^{22}, or some other number. The jargon is that 2×10^{22} expresses a physically based property to "one **significant figure,**" that is, a number that correctly represents the accuracy with which an experimental quantity is known. Only one digit, the 2 from the value 2.3 is used, and so 2×10^{22} has only one significant figure. Accordingly, the number of molecules in the breath is closer to 2×10^{22} than to 1×10^{22} or to 3×10^{22}, but anything beyond this level of exactness we cannot say with certainty.

Similarly, the concentration of carbon monoxide is known to only one significant figure, 9 ppm. That 2×9 equals 18 is certainly correct mathematically, but our question about CO is based on physical data. The answer, 1.8×10^{17} CO molecules, includes two significant figures: the 1 and the 8. Two significant figures imply a level of knowledge that is not justified. The accuracy of a calculation is limited by the *least accurate* piece of data that goes into it. In this case, both the concentration of CO and the number of molecules and atoms in the breath were each known only to one significant figure (9 and 2, respectively); two significant figures in the answer are unjustified. The common-sense rule is that you cannot improve the accuracy of experimental measurements by ordinary mathematical manipulations like multiplying and dividing. Therefore, the answer must also contain only one significant figure; hence 2×10^{17}.

Your Turn 1.33   Ozone Molecules

The local news reports that today's ground-level ozone readings are right at the acceptable level, 0.12 ppm. How many ozone molecules do you inhale in each breath? Assume that one breath contains 2×10^{22} molecules and atoms.

Answer

Start with the number of molecules and atoms in a breath. If ozone is 0.12 ppm, this gives the ratio 0.12 O_3 molecules per 10^6 molecules and atoms in air. Multiply times this ratio.

$$\frac{2 \times 10^{22} \text{ molecules and atoms in air}}{1 \text{ breath of air}} \times \frac{0.12 \text{ } O_3 \text{ molecules}}{1 \times 10^6 \text{ molecules and atoms in air}}$$

$$= 2.4 \times 10^{15} \text{ } O_3 \text{ molecules / breath}$$

$$= 2 \times 10^{15} \text{ } O_3 \text{ molecules / breath (to one significant figure)}$$

You may well question the significance of all of this talk about significant figures, but these are important in interpreting numbers associated with physical quantities. It has been observed that "figures don't lie, but liars can figure." Numbers often lend an air of authenticity to newspaper or television stories, so popular press accounts are full of numbers. Some are meaningful and some are not, and the informed citizen must be able to discriminate between the two types. For example, the assertion that the concentration of carbon dioxide in the atmosphere is 385.2537 ppm should be taken with a rather large grain of sodium chloride (salt). Values such as 385 ppm or 385.6 ppm (three or four significant figures) better represent what we actually can measure; any assertion with seven significant figures simply is not valid.

Your Turn 1.34   CO Monitors

Carbon monoxide monitors are available for homes and businesses. Figure 1.21 shows a convenient handheld CO detector that reads 35 ppm.

a. Would it be more helpful to have a meter that read 35.0388217 ppm? Explain.
b. Would 35.0388217 ppm be more valid? Explain.

Answer
a. No, it wouldn't be more helpful. The issue with CO is whether it exceeds a certain value, such as 9 ppm over an 8-hour period or 35 ppm over a 1-hour period. The extra decimal places are of no use.

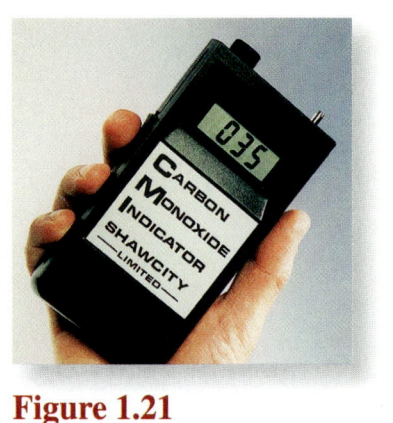

Figure 1.21
Carbon monoxide meter.

Numbers can introduce ambiguity in other ways as well. You have just encountered some conflicting information. The concentration of CO in air is very small, 9 ppm. Nevertheless, the number of CO molecules in a breath is still large, about 2×10^{17}. Both statements are true. The consequence of these numbers is that it is *impossible* to completely remove pollutant molecules from the air. "Zero pollutants" is an unattainable goal; using the most sophisticated detection methods you still could not even determine whether it had been achieved. At present, our most sensitive methods of chemical analysis are capable of detecting one target molecule out of a trillion. One part per trillion corresponds to: moving 6 inches in the 93 million-mile trip to the Sun; a single second in 320 centuries; or a pinch of salt in 10,000 tons of potato chips. A chemical could be undetectable at this level, and yet a breath might still include 2×10^{10} molecules of the substance.

> Absence of evidence is not the same as evidence of absence. The substance may be present, but in undetectable amounts.

Your Turn 1.35   CO Molecules in Perspective

To help you comprehend the magnitude of the 2×10^{17} CO molecules in just one of your breaths, assume that they were equally distributed among the 6 billion (6×10^9) human inhabitants of the Earth. Calculate each person's share of the 2×10^{17} CO molecules you just inhaled.

Answer
You are trying to distribute the huge number of molecules in a breath among all the human inhabitants of the Earth. This can be found by dividing the total number of CO molecules by the total number of humans:

$$\text{Each person's share is } \frac{2 \times 10^{17} \text{ CO molecules}}{6 \times 10^9 \text{ people}}$$

Thus, each person's share is 3×10^7, or 30,000,000, molecules of CO per person (to one significant figure).

A breath of air typically contains molecules of hundreds, perhaps thousands, of different compounds, most in minuscule concentrations. For almost all these substances, it is impossible to say whether the origin is natural or artificial. Indeed, many trace components, including the oxides of sulfur and nitrogen, come from both natural sources and those related to human activity. And, as with all chemicals, "natural" is not necessarily good and "human-made" is not necessarily bad. As you read in Section 1.4, what matters is exposure, toxicity, and the assessment of risk.

In addition to being extremely small, the particles in your breath possess other remarkable characteristics. In the first place, they are in constant motion. At room

temperature and pressure, a nitrogen molecule travels at about 1000 feet per second and experiences approximately 400 billion collisions with other molecules in that time interval. Nevertheless, relatively speaking, the molecules are quite far apart. The actual volume of the extremely tiny molecules making up the air is only about 1/1000th of the total volume of the gas. If the particles in your half-liter breath were squeezed together, their volume would be about 0.5 mL, less than a quarter of a teaspoon. Sometimes people mistakenly think that air is empty space. It's 99.9% empty space, but the matter that is in it is literally a matter of life and death!

Moreover, it is matter that we continuously exchange with other living things. The carbon dioxide we exhale is used by plants to make the food we eat, and the oxygen that plants release is essential for our existence. Our lives are linked together by the elusive medium of air. With every breath, we exchange millions of molecules with one another. As you read this, your lungs contain 4×10^{19} molecules that have been previously breathed by other human beings, and 6×10^8 molecules that have been breathed by some *particular* person, say Julius Caesar, Mahatma Gandhi, or Joan of Arc. Pick any person, your body almost certainly contains atoms that were once in his or her body. In fact, the odds are very good that right now your lungs contain one molecule that was in Caesar's *last* breath. The consequences are breathtaking!

Sceptical Chymist 1.36   Caesar's Last Breath

We just claimed that your lungs currently contain one molecule that was in Caesar's last breath. That assertion is based on some assumptions and a calculation. Are these assumptions reasonable? We are not asking you to reproduce the calculation, but rather to identify some of the assumptions and arguments we might have used. *Hint:* The calculation assumes that all of the molecules in Caesar's last breath have been uniformly distributed throughout the atmosphere.

Consider This 1.37   Air Quality Today

Pollution of our atmosphere has not occurred overnight. Rather, it has been a growing concern since at least the time of the Industrial Revolution. Why have we as a nation and as a world community become more concerned about air quality? Identify at least four factors that have brought air quality to the attention of voters.

Conclusion

The air we breathe has a personal and immediate effect on our health. Our very existence depends on having a large supply of relatively clean, unpolluted air with its essentials for life: two elements, oxygen and nitrogen, and two compounds, water and carbon dioxide. But the air you breathe may be polluted with carbon monoxide, ozone, sulfur oxides, and nitrogen oxides. This is true especially in the urban environments of our large cities, the very places where the majority of Americans live. The major pollutants are, for the most part, relatively simple chemical substances. They are produced often as unavoidable consequences of our dependence on coal for energy production in power plants and gasoline in internal combustion engines. Over the past 30 years, governmental regulations, industrial participation, modern technology, and green chemistry have resulted in large reductions in

many pollutants. But it is impossible to reduce pollutant concentrations to zero. Rather we must determine the risk from a given level of pollutant and then decide what level of risk is acceptable for various groups of people.

The oxygen-laden air we breathe, whether indoors or out, is, of course, very close to the surface of the Earth. But the Earth's atmosphere extends upward for considerable distance and contains other substances that are also essential for life on this planet. In the next two chapters, we consider two of these substances, ozone and carbon dioxide, and how they are changing as a result of human activities.

Chapter Summary

The numbers in parentheses indicate the sections in which the topics are introduced. Having studied this chapter, you should be able to:

- Describe air in terms of its major components, their relative amounts, and the local and regional variations in the composition of air (1.1–1.3)
- List major air pollutants and describe the effects of each on humans (1.3, 1.11–1.13)
- Compare and contrast indoor and outdoor air, in terms of which pollutants are likely to be present and their relative amounts (1.3, 1.13)
- Interpret values of the color-coded AQI and know how to assess local air quality data from the EPA (1.3)
- Understand the terms NAAQS, exposure, and toxicity, and why the NAAQS are set at different levels for different periods of time (1.3)
- Evaluate conditions significant in risk–benefit analysis (1.4)
- Identify the general regions of the atmosphere with respect to altitude (1.5)
- Interpret air quality data in terms of concentration units (ppm, ppb) and pollution levels, including unreasonableness of "pollution-free" levels (1.2–1.3, 1.12, 1.14)
- Relate these terms and differentiate among them: matter, pure substances, mixtures, elements, and compounds (1.6)

- Discuss the features of the periodic table, including the groups it contains and the locations of metals and nonmetals (1.6)
- Understand the difference between atoms and molecules, and between the symbols for elements and the formulas for chemical compounds (1.7)
- Name selected chemical elements and compounds (1.7)
- Write and interpret chemical formulas (1.8)
- Balance chemical equations (1.9–1.10)
- Understand oxygen's role in combustion, including how hydrocarbons burn to form carbon dioxide, carbon monoxide, and soot (1.9–1.10)
- Discuss the green chemistry initiative (1.11)
- Explain the different pollutants produced by burning coal and gasoline and how reductions in emissions have occurred (1.11)
- Describe how ozone forms, including how sunlight, NO, NO_2, and VOCs are involved (1.12)
- Identify the sources and nature of indoor air pollution (1.13)
- Interpret the nature of air at the molecular level (1.14)
- Use scientific notation and significant figures in performing basic calculations (1.4 and 1.14, respectively)

Questions

In each chapter, the questions are grouped into three categories:

- **Emphasizing Essentials** These questions give you the opportunity to practice fundamental skills. They most closely relate to the *Your Turn* exercises in the chapter.
- **Concentrating on Concepts** These questions ask you to integrate and apply the chemical concepts developed in the chapter and to relate them to societal issues. These questions most closely resemble the *Consider This* activities in the chapter.
- **Exploring Extensions** These questions challenge you to go beyond the information presented in the text. They provide an opportunity for you to extend and integrate the

facts, concepts, and communication skills from the chapter. Some questions closely relate to the type of analysis practiced in the *Sceptical Chymist* activities in the chapter.

See Appendix 5 for the answers to questions with numbers in blue. Questions marked with this icon require the resources of the Internet.

Emphasizing Essentials

1. Calculate the volume of air that an adult person exhales in an 8-hour day. Assume that each breath has a volume of about 0.5 L and that the person exhales 15 times a minute.

2. Given that dry air is 78% nitrogen by volume, how many liters of nitrogen are in 500 L?

3. A 5.0-L mixture of gases is prepared for a study of photosynthesis by combining 0.75 L of oxygen, 4.0 L of nitrogen, and 0.25 L of carbon dioxide. Compare the percentage of carbon dioxide gas in this mixture with that normally found in the atmosphere.

4. Give three examples of particulate matter found in air. What is the difference between $PM_{2.5}$ and PM_{10} in terms of their size? In terms of their health effects?

5. These gases are found in the troposphere: Rn, CO_2, CO, O_2, Ar, N_2.

   a. Rank them in order of their abundance in the troposphere.

   b. The concentration of one or more of these gases is more conveniently expressed in parts per million. Which one(s)?

   c. One or more of these gases is a criteria air pollutant. Which one(s)?

   d. One or more of these is a noble gas (Group 8A). Which one(s)?

6. a. The concentration of argon in air is approximately 9000 ppm. Express this value as a percent.

   b. The smoke inhaled from a cigarette contains 0.04–0.05% CO. Express these concentrations in parts per million.

   c. The concentration of water vapor in the atmosphere of a tropical rain forest may reach 50,000 ppm. Express this value as a percentage.

7. According to Table 1.2, the percentage of carbon dioxide in inhaled air is *lower* than it is in exhaled air, but the percentage of oxygen in inhaled air is *higher* than in exhaled air. How can you account for these relationships?

8. Cars don't inhale and exhale like humans do. Nonetheless, the air that goes into a car is different from what comes out. In Your Turn 1.26 you listed what comes out of a tailpipe. Now comment on the *differences* between the air that goes into the car engine and which comes out the tailpipe. For which chemicals have the concentrations noticeably increased or decreased?

9. Express each of these numbers in scientific notation.

   a. 1500 m, the distance of a foot race

   b. 0.0000000000958 m, the distance between O and H atoms in a water molecule

   c. 0.0000075 m, the diameter of a red blood cell

   d. 150,000 mg of CO, the approximate amount breathed daily

10. Write each of these values in standard notation.

   a. 8.5×10^4 g, the mass of air in an average room

   b. 1.0×10^7 gallons, the volume of crude oil spilled by the Exxon Valdez

c. 5.0×10^{-3} %, the concentration of CO in the air of a city street

d. 1×10^{-5} g, the recommended daily allowance of vitamin D

11. The value 0.00022 g/m^3 of air is the threshold for detecting NO_2 by smell.

   a. Express this value in scientific notation.

   b. Would you expect a similar value for the threshold of CO?

   c. Name another pollutant that has a sharp, easily detected odor.

12. Wildfires occur all over our planet. The one shown in the photograph here was taken on a commercial flight route north of Phoenix, AZ.

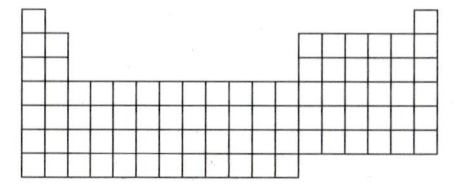

   a. Name two gases that you would expect as combustion products of wood.

   b. This fire is emitting at least three criteria pollutants. Which one(s) can you see? Which one(s) are not visible?

13. Consider this portion of the periodic table and the two groups shaded on it.

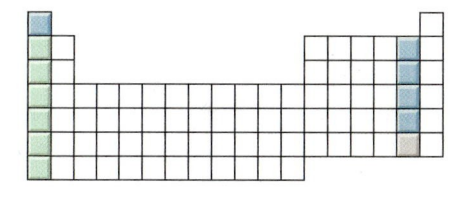

   a. What is the group number for each shaded region?

   b. Name the elements that make up each group.

   c. Give a general characteristic of the elements in each of these groups.

14. Consider the following blank periodic table.

   a. Shade the region of the periodic table where metals are found.

b. Six common metals are iron, magnesium, aluminum, sodium, potassium, and silver. Give the chemical symbol for each.

c. Give the name and chemical symbol for five nonmetals (elements that are not in your shaded region).

15. Classify each of these substances as an element, compound, or mixture.

a. a sample of "laughing gas" (dinitrogen monoxide, also called nitrous oxide)

b. steam coming from a pan of boiling water

c. a bar of deodorant soap

d. a sample of copper

e. a cup of mayonnaise

f. the helium filling a balloon

16. Each of the following is found in the atmosphere in small amounts: CH_4, SO_2 and O_3.

a. What information does each chemical formula convey in terms of the number and types of atoms present?

b. Give their chemical names.

17. Hydrocarbons are important fuels that we burn for many different reasons.

a. Explain the term *hydrocarbon*.

b. Put these hydrocarbons in order from smallest to largest, in terms of the number of carbons they contain: propane, methane, butane, octane, ethane.

c. We suggested "mother eats peanut butter" as a memory aid for the first four hydrocarbons. Come up with a new one that includes *pent-* (as in pentane), a prefix indicating five carbon atoms.

18. Write balanced chemical equations to represent these reactions. *Hint:* Nitrogen and oxygen are both diatomic molecules.

a. Nitrogen reacts with oxygen to form nitrogen monoxide.

b. Ozone decomposes into oxygen and atomic oxygen (O).

c. Sulfur reacts with oxygen to form sulfur trioxide.

19. Analogous to equation 1.8, draw models to represent the balanced chemical equations in question 18.

20. These questions relate to combustion of hydrocarbons.

a. LPG (liquid petroleum gas) is mostly propane, C_3H_8. Balance this equation.

$$C_3H_8(g) + O_2(g) \longrightarrow CO_2(g) + H_2O(g)$$

b. Cigarette lighters burn butane, C_4H_{10}. Write a balanced equation, assuming complete combustion, that is, plenty of oxygen.

c. With a limited supply of oxygen, both propane and butane can burn incompletely to form carbon monoxide. Write balanced equations for both reactions.

21. Balance these equations in which ethene, C_2H_4, burns in oxygen.

a. $C_2H_4(g) + O_2(g) \longrightarrow C(s) + H_2O(g)$

b. $C_2H_4(g) + O_2(g) \longrightarrow 2CO(g) + 2H_2O(g)$

c. $C_2H_4(g) + O_2(g) \longrightarrow CO_2(g) + H_2O(g)$

22. Compare the coefficient for oxygen in the equations from question 21. How does it vary, depending on the products formed?

23. Count the atoms on both sides of the arrow to demonstrate that these equations are balanced.

a. $2 C_3H_8(g) + 7 O_2(g) \longrightarrow 6 CO(g) + 8 H_2O(l)$

b. $2 C_8H_{18}(g) + 25 O_2(g) \longrightarrow 16 CO_2(g) + 18 H_2O(l)$

24. Platinum, palladium, and rhodium are used in the catalytic converters of cars.

a. What is the chemical symbol for each of these metals?

b. Where are these metals located on the periodic table?

c. What can you infer about the properties of these metals, given that they are useful in this application?

25. Nail polish remover containing acetone was spilled in a room 6 m × 5 m × 3 m, and 3600 mg of the acetone evaporated. Calculate the concentration of acetone in units of micrograms per cubic meter.

Concentrating on Concepts

26. In Section 1.1, air was referred to as ". . . that invisible stuff. . . ." Is this always true? What factors influence if air appears "invisible" or if you can "see" it?

27. In Consider This 1.1, you calculated the volume of air exhaled in a day. How does this volume compare with the volume of air in your chemistry classroom? Show your calculations. *Hint:* Think ahead about the most convenient unit to use for measuring or estimating the dimensions of your classroom.

28. a. Arrange these measurements in order of increasing distance: 1 m, 3.0×10^2 m, and 5.0×10^{-3} m.

b. You could also express 5.0×10^{-3} m as 0.5 cm or 5 mm. All of these are valid. Make an argument that in certain circumstances, you might want to select one unit over the others.

29. If risk is related to public perception, what is your feeling about the relative risks associated with each of these: inline-skating, eating raw cookie dough, driving on an interstate highway, breathing second-hand smoke, not wearing a bike helmet, taking aspirin, and drinking tap water. Rank them in order of your perception of the *most* risky to the *least* risky. Explain your rationale.

30. In these diagrams, two different types of atoms are represented using a different color and size. Characterize each

sample as an element, compound, or mixture. Explain your classification.

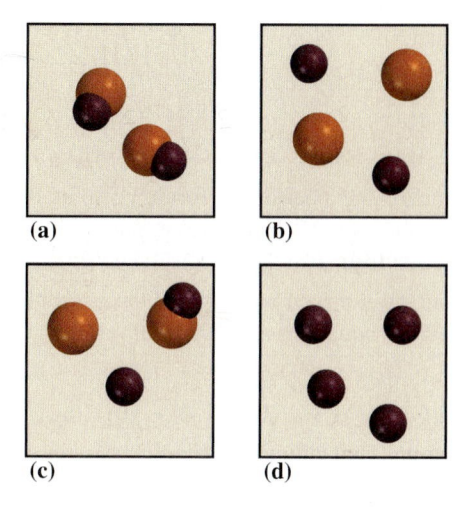

(a)   (b)

(c)   (d)

31. Consider this representation of the reaction between nitrogen and hydrogen to form ammonia (NH_3).

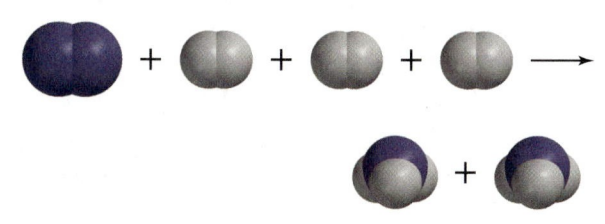

   a. Are the masses of reactants and products the same?

   b. Are the numbers of molecules of reactants and of products the same?

   c. Are the total number of atoms in the reactants and the total number of atoms in the products the same?

32. In Consider This 1.3, you considered how life on Earth would change if the concentration of oxygen were doubled. Now consider the opposite case; discuss how life on Earth would change if the concentration of O_2 were only 10%. Give some specific examples of how burning, rusting, and most metabolic processes in humans and plants would be affected.

33. Explain why CO is called the "silent killer." Select two other pollutants for which this name would not apply and explain why not.

34. Cigarette smoke is 2–3% carbon monoxide.

   a. How many parts per million is this?

   b. How does this value compare with the National Ambient Air Quality Standards for CO in both a 1-hour and an 8-hour period?

   c. Propose a reason why smokers do not succumb to carbon monoxide poisoning.

35. For many states, the ozone season runs from May 1 to October 1. Why are ozone levels typically not reported in the winter months?

36. The EPA characterizes ozone as "good up high, bad nearby." Explain.

37. Here are ozone air quality data for London, Ontario from June 8–20, 2005.

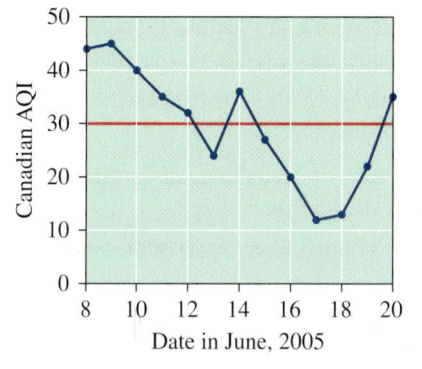

   a. In general, which groups of people are the most sensitive to ozone?

   b. Air rated above 30 is hazardous for some or all groups by the Canadian Environmental Assessment Agency. For the data shown, how many days was the air hazardous?

   c. Ozone levels drop off sharply at night. Explain why.

   d. During the daytime, the ozone dropped off sharply after June 10. Propose two different reasons that could account for this observation.

   e. Data for the month of December is not shown. Would you expect the ozone levels to be higher or lower than those in June? Explain.

38. Here are air quality data for April 1–10, 2005 in Beijing. The primary pollutant was PM_{10}.

   a. In general, which groups of people are the most sensitive to particulate matter?

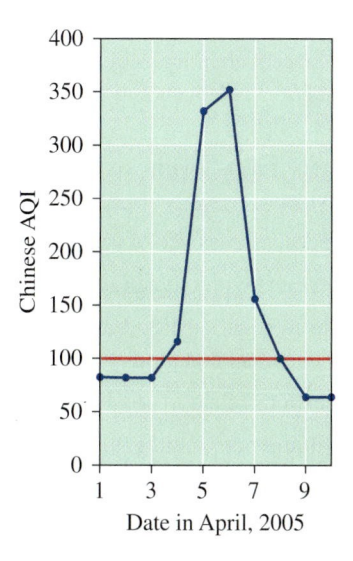

   b. Air rated above 100 is hazardous for some or all groups by the National Environmental Monitoring Centre in China. For the data shown, how many days was the air hazardous?

   c. The levels of PM do not necessarily drop off at night the way they do for ozone. Explain.

d. The levels of particulate matter increased sharply on April 5. Propose two different reasons that could account for this observation.

39. Young adults in Beijing, China, have gone to bars after work, not for glasses of beer or wine, but for fresh air. These "oxygen bars" provide a half-hour of deep breathing for the equivalent of $6.

    a. What does this tell you about air pollution in Beijing?

    b. If you wanted to establish "oxygen bars" in other cities of the world, which ones would you choose?

40. A certain city has an ozone reading of 0.13 ppm for 1 hour, and the permissible limit is 0.12 for that time. You have the choice of reporting that the city has exceeded the ozone limit by 0.01 ppm or saying that it has exceeded the limit by 8%. What are the advantages of each method?

41. Here is the air quality outlook for the United States for early August, 2005.

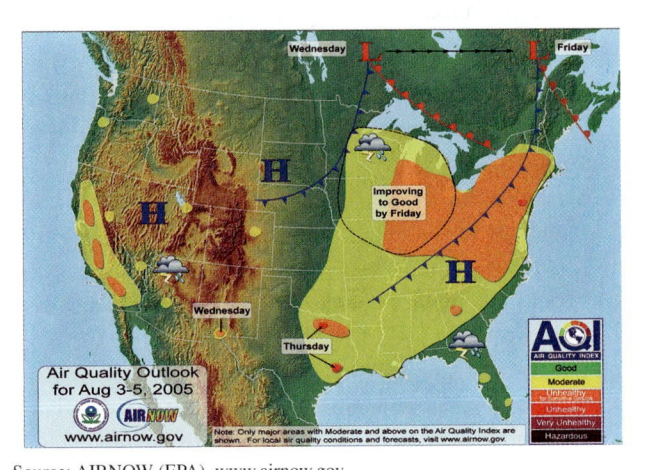

Source: AIRNOW (EPA), www.airnow.gov

    a. This forecast is typical in that most of the ozone pollution is expected in California, Denver, Texas, the Midwest and the East Coast. Why these regions of the country?

    b. Phoenix typically has high ozone levels in the summer, but not on this particular day. Offer a possible explanation.

    c. The air quality is forecast to improve in Illinois and Wisconsin. Offer a possible explanation.

    d. Why are inland areas in California, such as the Sacramento Valley which is shown as unhealthy for sensitive groups, likely to have worse air quality than the California coast?

42. Question 35 states that the ozone season runs from May through September. Look up the ozone air quality data for a state in the sunbelt. Does this state have a longer ozone season? From what you found, should it? *Hint:* Use the air quality data archives at the AIRNOW site. Consider This 1.30 has a link at the *Online Learning Center*.

43. Throughout most of the year, inhabitants of Santiago, Chile, breathe some of the worst air on the planet.

    a. Driving private cars has been severely restricted in Santiago. What pollutants will be lower if cars are kept off the roads?

    b. Although the population of Santiago is comparable to that in other cities, its air quality is much worse. Suggest geographical features that might be responsible.

    c. Which groups are most susceptible to the air pollution produced by automobiles?

44. Explain why jogging outdoors (as opposed to sitting outdoors) increases your exposure to pollutants.

45. Jogging indoors at home can decrease your exposure to some pollutants, but may increase your exposure to others. Explain.

46. In this chapter, we discussed what may come out of the tailpipe of a car. Fifty years ago when tetraethyl lead (TEL) was added to gasoline, tailpipe emissions also included lead. TEL has the chemical formula $Pb(C_2H_5)_4$.

    a. Name the elements present in the compound TEL.

    b. The lead in TEL burns to form lead oxide, a toxic compound with lead and oxygen in a 1:1 ratio. Write the chemical formula for lead oxide.

    c. Lead also can form another toxic compound with oxygen in which the lead-to-oxygen ratio is 1:2. Write the chemical formula.

    d. What compounds that do not contain lead also can be produced from the combustion of TEL?

47. In urban areas, the concentration of formaldehyde in *outdoor* air is typically about 0.01 ppm, assuming no smog formation. In contrast, the level of formaldehyde *indoors* can average 0.1 ppm, the level at which most people will smell its pungent odor. What factors can lead to formaldehyde accumulation indoors?

Exploring Extensions

48. The percentage of oxygen gas in the atmosphere (21%) is usually expressed as the volume of oxygen gas relative to the total volume of the atmosphere being considered. The percentage can also be reported as the mass of oxygen gas relative to the total mass of the atmosphere being considered; in this case, it is 23%. Offer a possible explanation why these two values are not the same.

49. 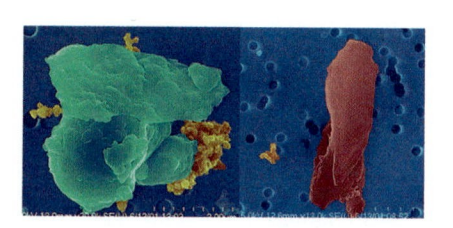 The EPA oversees the Presidential Green Chemistry Challenge Awards. Use the EPA Web site to find when the program started and to find the list of the most recent winners of the award. Pick one winner and summarize in your own words the green chemistry advance that merited the award.

50. Recreational scuba divers usually use compressed air that has the same composition as normal air. A mixture being used is called Nitrox. What is its composition, and why is it being used?

51. Here are two scanning electron micrograph images of particulate matter, courtesy of the National Science Foundation and researchers at Arizona State University. The first is of a soil particle and the second of a rubber particle, and each is about 10 μm in diameter.

a. Where do you think the rubber particle came from? Name some other common substances that might contribute to PM in the air.

b. The soil dust is composed mainly of silicon and oxygen. What other elements are commonly present in the rocks and minerals in Earth's crust?

c. What do you notice about the shapes of both that suggests that particles such as these would inflame your blood vessels?

52. Ultrafine particles have diameters less than 0.1 μm. In terms of their sources and health effects, how do these particles compare with $PM_{2.5}$ and PM_{10}? Use the resources of the Web to locate the most up-to-date information.

53. Most lawn mowers do not have catalytic converters (at least as this book went to press.) What comes out of the tailpipe of a gasoline-powered lawn mower? Why has adding a catalytic converter been so controversial?

54. An article in *USA Today* on January 12, 1999, is titled "Taking Technology from Here to the Infinitesimal." By the year 2020, the article predicts, "the age of atomic engineering, . . . a type of nanotechnology, will dawn." What does this term imply? What kinds of applications will be possible that are not now part of our technology?

55. Michael Crichton, in his best-seller *Prey* (2002), entertained his readers with a fearful tale of self-replicating nanorobots. Although his novel fell into the realm of science fiction, nonetheless his point is well taken that a new discovery can have unintended consequences. Along these lines, in 2003 Congress asked for

studies to determine the social, economic, and environmental impact(s) of nanotechnology. Find out what has happened since then. Are the reports uniformly positive or have some unfavorable effects of nanotechnology been reported?

56. A concept web or concept map is a convenient way to represent knowledge and connection among ideas. Concept webs are constructed by joining a word or expression to another one by means of linking words. For example, the atmosphere has three layers, the mesosphere, stratosphere, and troposphere.

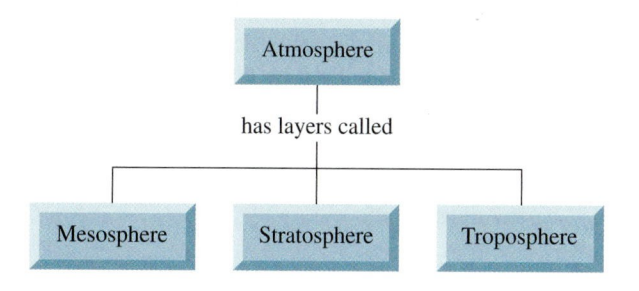

What advantages or disadvantages does this representation have compared with Figure 1.7? Explain your reasoning.

57. Consider this graph that shows the effects of carbon monoxide inhalation on humans.

a. Both the amount of exposure and the duration of exposure have an effect on CO toxicity in humans. Use the graph to explain why.

b. Use the information in this graph to prepare a statement to include with a home carbon monoxide detection kit about the health hazards of carbon monoxide gas.

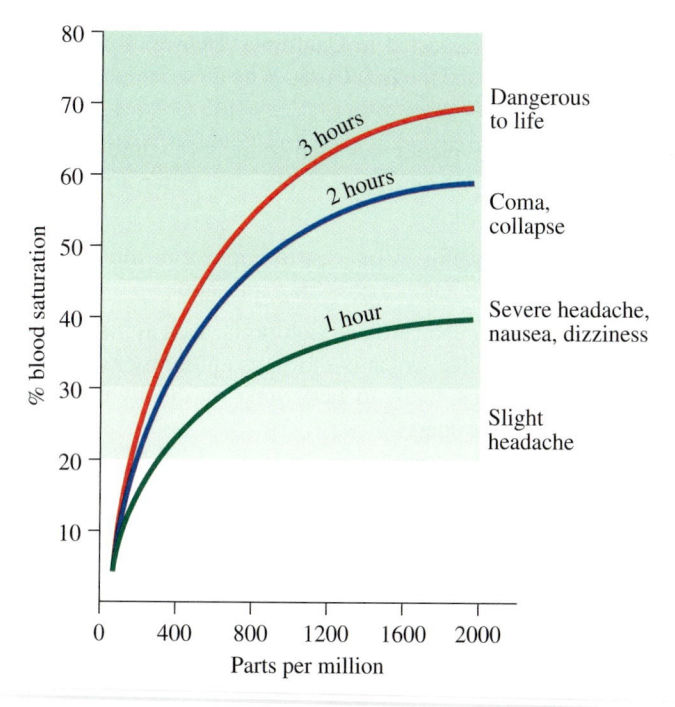

58. Consider This 1.3 asks you to consider how our world would be different if the oxygen content of the atmosphere were doubled. Develop your answer into an essay. Title your essay "An Hour in the Life of . . ." and describe how things would be different for a person of your choice. If an hour is too short to make your point, substitute "A Morning . . ." or "A Day . . .".

59. Mercury, another serious air pollutant, is not described in this chapter. Nevertheless, if you were a textbook author, what would you include about mercury emissions? Write several paragraphs in a style that would match that of this textbook. Perhaps even design a Consider This exercise to accompany it. Feel free to send these to one of the authors.

60. The dark color associated with heavy smog can be caused by the presence of particulate matter or a high concentration of brown NO_2 gas (or both). Once in the atmosphere, some NO_2 can form N_2O_4, a colorless gas.

a. Write a balanced equation for the reaction of two molecules of NO_2 to form N_2O_4, a reaction that releases heat energy. Use a double-headed reaction arrow in your equation to indicate that equilibrium is established between the formation of N_2O_4 and its decomposition back to NO_2.

b. Offer a reason why smog may be darker on a warm summer day than on a cold winter one, even if the levels of nitrogen oxides are the same in both cases.

Chapter 2

Protecting the Ozone Layer

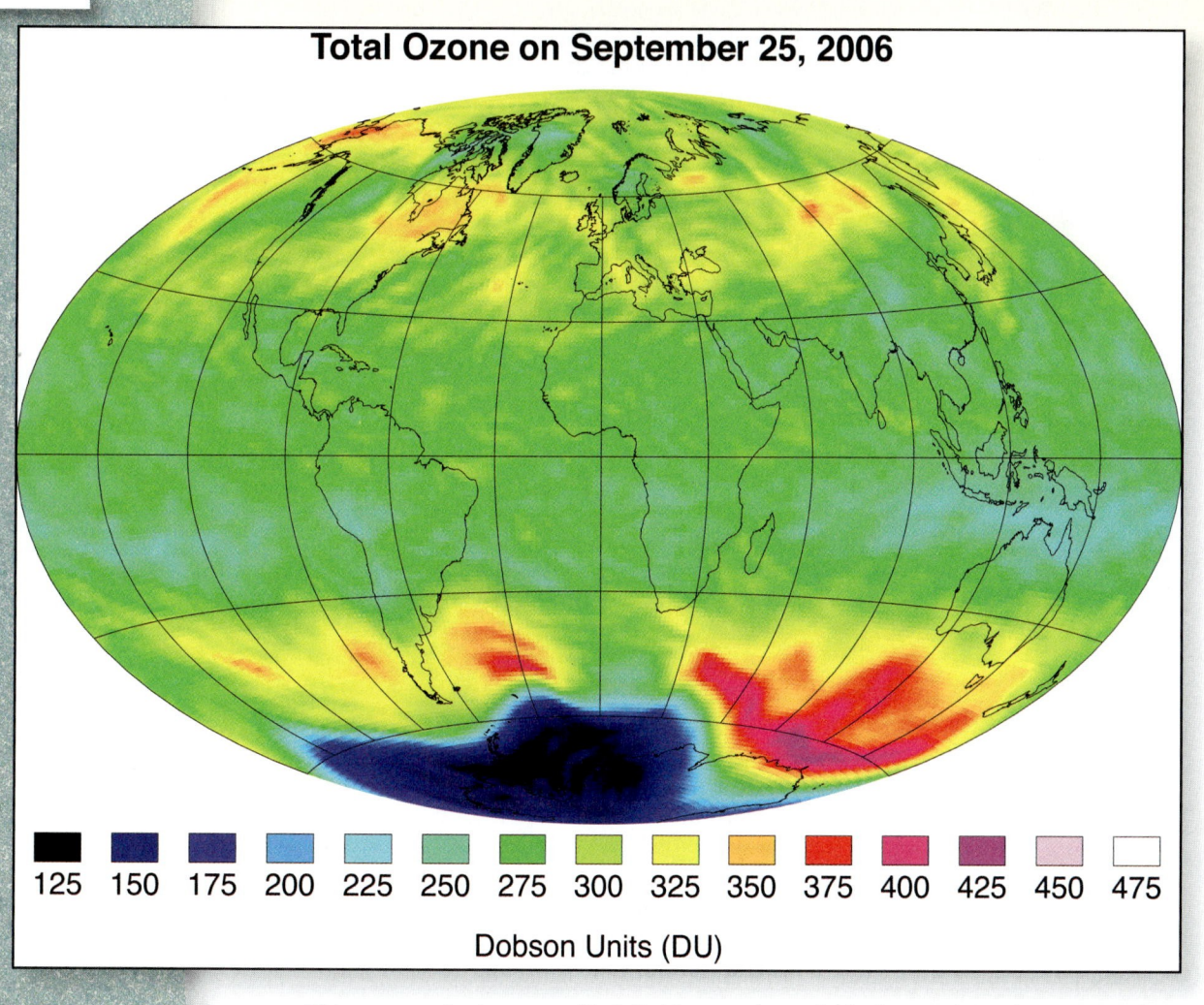

Total Ozone on September 25, 2006

Dobson Units (DU)

This stratospheric ozone "hole" (the purple and black areas) in the region of Antarctica was at its 2006 minimum of 105 DU on September 25th and covered a maximum area of 28.0 million km². The 2006 hole was almost as large as the record 2000 hole, which peaked at 28.4 million km².

Note: One Dobson unit (DU) corresponds to about one ozone molecule for every billion molecules and atoms present in air.

Stratospheric ozone plays a vital role in protecting Earth's surface and those who live here from damaging solar radiation. In the 1970s, it was discovered that certain chemicals could make their way into the upper atmosphere and destroy the protective ozone found there. Ever since, scientists, policy makers, and indeed concerned citizens worldwide have participated in efforts to control and reverse ozone destruction. Somewhat surprisingly, the most severe depletion has been over Antarctica, and the yearly images of the "ozone hole" have become some of the most widely recognized scientific graphics. Later in this chapter, you will have the opportunity to examine past trends and to bring the Antarctic ozone hole story up-to-date.

You may be wondering what this story has to do with you, because last time we checked, not many college students were living in Antarctica. Even though the phenomenon of ozone depletion was first documented in that faraway region, it also has been monitored and observed in many other locations on Earth, including over North America. Where you live and the season of the year both influence the amount of stratospheric ozone overhead and how well it provides its protective effects. Take a look at some of the important data for yourself.

Consider This 2.1 — Ozone Levels Above Your Spot on Earth

How much protective ozone is above you? How does it compare with the amount of ozone above Antarctica?

a. Use the NASA link at the *Online Learning Center* to access satellite data. Click on your location on the world map or enter your specific latitude and longitude to find the most recent data for total column ozone. Also request data for the same date and location for the last three years. Do you see any trend?

b. How do values at your location compare with those given for Antarctica on the same date each year?

We are now ready to consider many questions involving chemistry and its role in helping understand mechanisms affecting our protective ozone layer. Just what has caused the stratospheric ozone depletion that has already occurred? Why is this serious? What has been done to slow down or correct the problem? Are these measures working, and what are the economic and societal costs? And finally, are any new threats to the stratospheric ozone layer emerging that merit careful evaluation?

2.1 Ozone: What and Where Is It?

Ozone is an atmospheric gas found in both the troposphere and the stratosphere. If you have ever been near a sparking electric motor or been in a severe lightning storm, you may have smelled ozone. Ozone's odor is unmistakable, but difficult to describe. Some compare the odor to that of chlorine gas; others think the odor reminds them of newly mown grass. It is possible for humans to detect concentrations as low as 10 parts per billion (ppb), that is, 10 molecules out of 1 billion. Appropriately enough, the name *ozone* comes from a Greek word meaning "to smell."

Ozone is oxygen that has changed from the normal diatomic molecule, O_2, to a triatomic form, O_3. A simple chemical equation summarizes the reaction:

$$\text{energy} + 3\,O_2 \longrightarrow 2\,O_3 \qquad [2.1]$$

Energy must be absorbed for this reaction to occur. This helps explain why ozone forms when oxygen is subjected to electrical discharge, whether from an electric spark or lightning.

Ozone is an allotropic form of oxygen. **Allotropes** are two or more forms of the same element that differ in their chemical structure and therefore in their properties. The allotropes O_2 and O_3 obviously differ in molecular structure. This variance is

Diamond, graphite, and fullerenes (buckyballs) are all allotropic forms of carbon. They have different structures and different properties.

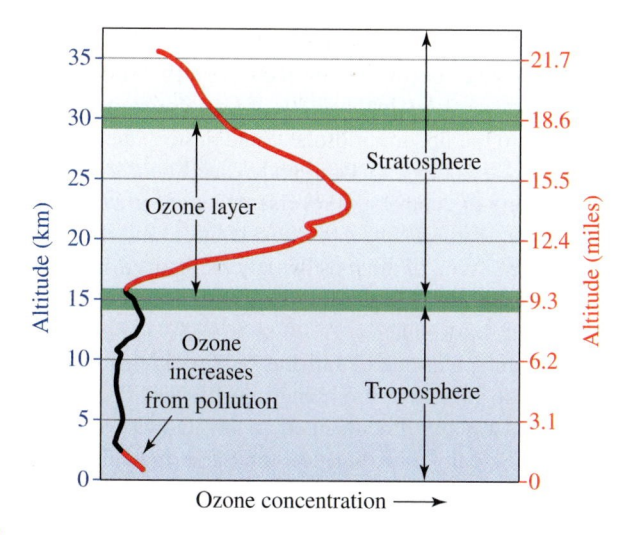

Figure 2.1

Ozone concentrations at different altitudes.

Source: *Scientific Assessment of Ozone Depletion: 2002;* World Meteorological Organization (WMO), United Nations Environmental Programme (UNEP), p. 3 of "Twenty Questions and Answers About the Ozone Layer" Note: Linked from http://www.epa.gov/ozone/science/index.html

responsible for differences in the physical and chemical properties of the two allotropes. For example, O_2 is odorless. At a pressure of 1 atmosphere (atm), it condenses from a colorless gas to a light blue liquid at -183 °C. Ozone changes its physical state from a gas to a dark blue liquid at -112 °C. Because O_3 is chemically more reactive than O_2, it is often used in the purification of water and to bleach paper pulp and fabrics. At one time it was even advocated as a deodorant for air in crowded interiors and continues to be used by some hotels to remove residual smoke from rooms.

In the troposphere, the region of the atmosphere in which we live, somewhere between 20 and 100 ozone molecules typically occur for each billion molecules and atoms that make up the air. The highest concentrations are found near Earth's surface, the result of photochemical smog mechanisms. In Chapter 1, we learned that the limits in ambient air are set at very low concentrations, only 0.08 ppm for an 8-hr average. But what is detrimental in one region of the atmosphere, even at very low concentrations, can be essential in another. The stratosphere, at an altitude of 15 to 30 km, is where ozone does most of its filtering of ultraviolet light from the Sun. The concentration of ozone in this region is somewhat greater than in the troposphere, but still very low. At most, there are about 12,000 ozone molecules for every billion molecules and atoms of gases that make up the atmosphere at this level. Most ozone, about 90% of the total, is found in the stratosphere. The term **ozone layer** refers to the stratospheric region of maximum ozone concentration. Figure 2.1 shows the relative location and concentration of ozone in the atmosphere.

> 0.080 parts per *million* is equivalent to 80 ozone molecules for every *billion* molecules and atoms found in air.

Your Turn 2.2   Finding the Ozone Layer

Use Figure 2.1 and values given in the text to answer these questions.

a. What is the altitude of maximum ozone concentration?

b. What is the range of altitudes in which ozone molecules are more concentrated than in the troposphere?

c. What is the maximum number of ozone molecules per billion molecules and atoms of all types found in the stratosphere?

d. What is the maximum number of ozone molecules per billion molecules and atoms of all types found in ambient air just meeting the EPA limit for an 8-hr average?

Answers

a. About 23 km (or 14 miles)

b. Between 17 and 35 km (or 10–22 miles)

Because the range of altitudes within the stratosphere in which one finds significant ozone is so broad, the concept of the "ozone layer" can be a little misleading. No thick, fluffy blanket of ozone exists in the stratosphere. At the altitudes of the maximum ozone concentration, the atmosphere is very thin, so the total amount of ozone is surprisingly small. If all the O_3 in the atmosphere could be isolated and brought to the average pressure and temperature at Earth's surface (1.0 atm and 15 °C), the resulting layer of gas would have a thickness of less than 0.5 cm, or about 0.25 inch. On a global scale, this is a minute amount of matter. Yet, this fragile shield protects the surface of the Earth and its inhabitants from the harmful effects of ultraviolet radiation.

Reliable information about atmospheric ozone concentrations can help us understand changes that may occur. The total amount of ozone in a vertical column of air of known volume can be determined with relative ease. The determination can be done from Earth's surface by measuring the amount of UV radiation reaching a detector; the lower the intensity of the radiation, the greater the amount of ozone in the column. G. M. B. Dobson, a scientist at Oxford University, pioneered this measurement method. In 1920, he invented the first instrument to quantitatively measure the concentration of ozone in a column of the Earth's atmosphere. Therefore, it is fitting that the unit of such measurements is named for him.

One Dobson unit (DU) is equivalent to about 3×10^{16} O_3 molecules in a column of the atmosphere with a cross section of 1 cm².

Consider This 2.3   **Interpreting Ozone Values**

A classmate used the NASA Web site to check the atmospheric ozone above her hometown in Ohio. She found the readings to be 417 DU on April 10 and 286 DU on May 10. The student was reassured by these findings, concluding there had been an improvement in protection from damaging UV radiation. Do you agree? Why or why not?

Scientists will continue to measure and evaluate ozone levels using ground observations, weather balloons, and high-flying aircraft. However, since the 1970s, measurements of total column ozone have also been made from the top of the atmosphere. Satellite-mounted detectors record the intensity of the ultraviolet radiation scattered by the upper atmosphere. The results are then related to the amount of O_3 present. What has proven more difficult is quantifying ozone concentrations at intermediate altitudes.

The Space Shuttle, *Columbia,* tested a new approach for monitoring ozone. Rather than looking directly downward toward Earth from a satellite, the equipment aboard the Shuttle looked sideways through the thin blue haze that rises above the denser regions of the troposphere and follows the curve of the Earth. This region is known as the Earth's "limb" and is responsible for the name of this new technique, called "limb viewing." Reliable information can be gathered at each level of the atmosphere, particularly allowing scientists to better understand chemistry taking place in the lower regions of the stratosphere. In January 2004, the National Aeronautics and Space Administration (NASA) launched a new mission called Earth Observing System (EOS) *Aura* that also uses limb viewing to gather additional data about changes in Earth's stratospheric ozone layer.

NASA's EOS *Aura* mission also is collecting data about tropospheric air quality (Chapter 1) and global warming (Chapter 3).

The process by which ozone protects us from damaging solar radiation involves the interaction of matter and energy from the Sun. Understanding this process requires knowledge about both of these fundamental topics. We turn first to a submicroscopic view of matter, and then examine the interaction of matter and energy from the Sun.

2.2 Atomic Structure and Periodicity

Both O_2 and O_3 are composed of oxygen atoms, but what do we know about these atoms? During the 20th century, chemists and other scientists made great progress in discovering details about the structure of atoms and the particles that make them up. The physicists have been almost too successful; they have found more than 200 subatomic particles. Fortunately, most chemical behavior can be explained with only three well-known particles.

Table 2.1	Properties of Subatomic Particles		
Particle	**Relative Charge**	**Relative Mass**	**Actual Mass, kg**
proton	+1	1	1.67×10^{-27}
neutron	0	1	1.67×10^{-27}
electron	−1	0*	9.11×10^{-31}

*The relative mass of the electron is not actually zero, but is so small that it appears as zero when expressed to the nearest whole number.

Every atom has at its center a **nucleus,** a minuscule and highly dense region composed of protons and neutrons. **Protons** are positively charged particles, and **neutrons** are electrically neutral particles, but both have almost exactly the same mass. Indeed, the protons and neutrons in the nucleus account for almost all of an atom's mass. Well beyond the nucleus are the electrons that define the outer boundary of the atom. An **electron** has a much smaller mass than a proton or neutron and a negative electric charge equal in magnitude to that of a proton, but opposite in sign. Therefore, in any electrically neutral atom, the number of electrons equals the number of protons. The charge and mass properties of these particles are summarized in Table 2.1.

Each element has a characteristic number of protons. We use the **atomic number** to refer to the number of protons in an atom of that element. This unique number characterizes the elemental identity of an atom. For example, the simplest atom is hydrogen (H), and every H atom contains one proton, and thus has an atomic number of 1. Every helium (He) atom contains two protons, and thus has an atomic number of 2. With each successive element in the periodic table, the atomic number increases, right up through element 111, whose atoms contain 111 protons.

Your Turn 2.4   Protons and Electrons

Using the periodic table as a guide, specify the number of protons and electrons in a neutral atom of each of these elements.

a. carbon (C)   **b.** calcium (Ca)
c. chlorine (Cl)   **d.** chromium (Cr)

Answers
a. 6 protons, 6 electrons   **b.** 20 protons, 20 electrons

We wish we could show you a picture of a typical atom. However, atoms defy easy representation, and depictions in textbooks are at best oversimplifications. Electrons are sometimes pictured as moving in orbits about the nucleus, but the modern view of electrons is a good deal more complicated and abstract. For one thing, the relative size of the nucleus and the atom creates serious problems for the illustrator. If the nucleus of a hydrogen atom were the size of a period on this page, the atom's single electron would most likely be found at a distance of about 10 feet that period. It is true that an atom is mostly empty space. Moreover, electrons do not follow specific circular orbits. In spite of what you may have learned early in your education, an atom is really not very much like a miniature solar system. Rather, the distribution of electrons in an atom is described best using concepts of probability and statistics.

If this sounds rather vague to you, you are not alone. Common sense and our experience of ordinary things are not particularly helpful in our efforts to visualize the interior of an atom. Instead, we are forced to resort to mathematics and metaphors. The mathematics required (a field called quantum mechanics) can be formidable. Chemistry majors do not normally encounter this field until rather late in their undergraduate study. We cannot fully share with you the strange beauties of the peculiar quantum world of the atom, although we can provide some useful generalizations.

In the periodic table, the elements are sequenced in order of increasing atomic number. The table also has elements arranged so that those with similar chemical properties fall in the same columns (groups). This array of elements reveals **periodic properties** that vary in a regular way with increasing atomic number and that repeat at regular intervals. Thus, lithium (Li, atomic number 3), sodium (Na, 11), potassium (K, 19), rubidium (Rb, 37), and cesium (Cs, 55) must share something besides their behavior as highly reactive metals. What fundamental feature accounts for these similar chemical properties?

Today we know that periodic properties are the consequence of the distribution of electrons in the atoms of the elements. Because the atomic number represents the number of protons in each atom and electrons in a neutral atom for each particular element, properties vary with atomic number. And when properties repeat themselves, it signals a repeat in electronic arrangement. The total number of electrons may vary, but it is the electrons farthest from the nucleus that are the main determinant of chemical properties.

Both experiment and calculation demonstrate that the electrons are arranged in certain energy levels about the nucleus. What we are calling "levels" used to be referred to as "shells," using the earlier solar system model of atomic structure. The electrons in the innermost level are the most strongly attracted by the positively charged protons in the nucleus. The greater the distance between an electron and the nucleus, the weaker the attraction between them. We say that the more distant electron is in a higher energy level, which means that the electron itself possesses more potential energy.

Each energy level has a maximum number of electrons that can be accommodated and is particularly stable when fully occupied. The innermost level, corresponding to the lowest energy, can hold only two electrons. The second level has a maximum capacity of eight, and the higher levels are also particularly stable when they contain eight electrons.

Table 2.2 shows some important information about electrons in neutral atoms of the first 18 elements. The total number of electrons in each atom is printed in blue and the number of outer electrons is printed in maroon. **Outer (valence) electrons** are found in the highest energy level and help to account for many of the observed trends in chemical properties. Observe that the group designation (1A, 2A, etc.) corresponds to the number of *outer* electrons for the A group elements, one of the great organizing benefits of the periodic table.

Take another look at the first column in Table 2.2. Lithium and sodium atoms both have one *outer* electron per atom, despite having different *total* numbers of electrons. This fact explains much of the chemistry that these two alkali metals have in common. It places them in Group 1A of the periodic table (the 1 indicates one outer electron). Moreover, we would be correct in assuming that potassium, rubidium, and the other

Table 2.2	**Total and Outer Electrons for Atoms of the First 18 Elements**							
								Noble Gases
Group 1A	**2A**	**3A**	**4A**	**5A**	**6A**	**7A**		**8A**
1								2
H								He
1								2
3	4	5	6	7	8	9		10
Li	Be	B	C	N	O	F		Ne
1	2	3	4	5	6	7		8
11	12	13	14	15	16	17		18
Na	Mg	Al	Si	P	S	Cl		Ar
1	2	3	4	5	6	7		8

• Number *above* the atomic symbol is the atomic number, the total number of protons. It also gives the total number of electrons in a neutral atom.

• Number *below* the atomic symbol is the number of **outer** electrons in a neutral atom.

elements in column 1A of the periodic table also have a single outer electron in each of their atoms. They are all metals that react readily with oxygen, water, and a wide range of other chemicals. In fact, chemical reactivity and the bonds that hold atoms together to form molecules and crystals are largely a consequence of the number of outer electrons in any element. Figure 2.2 shows photographs of some Group 1A elements.

The periodic table is a useful guide to electron arrangement in the various elements. In the families, or groups, of elements marked "A," the number that heads the column indicates the number of outer electrons in each atom. You have already seen that Group 1A elements are characterized by one outer electron. Similarly, the atoms of the Group 2A elements (the "alkaline earth metals") all have two outer electrons. The same pattern holds true for all the A groups in the periodic table. Group 3A elements have three outer electrons, Group 4A elements have four, and so on across the table. Seven outer electrons characterize the atoms that make up Group 7A, the "halogens": fluorine (F), chlorine (Cl), bromine (Br), iodine (I), and astatine (At). The next two exercises provide some practice with elements in the A groups.

> The group number does not *necessarily* indicate the number of outer electrons for elements in B groups, where the situation is a bit more complicated.

Lithium (stored in oil)

Sodium (removed from oil, being cut)

Potassium (in sealed glass tube)

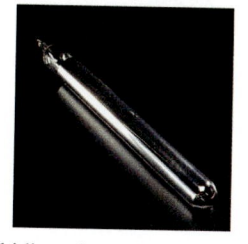

Rubidium (in sealed glass tube)

Figure 2.2

Selected Group 1A elements.

Your Turn 2.5   Outer Electrons

Using the periodic table as a guide, specify the group number and number of outer electrons in a neutral atom of each element.

a. sulfur (S)   b. silicon (Si)
c. nitrogen (N)   d. krypton (Kr)

Answers

a. Group 6A; 6 outer electrons   b. Group 4A; 4 outer electrons

---

Your Turn 2.6   Family Features

a. What feature of atomic structure is shared by fluorine (F), chlorine (Cl), bromine (Br), and iodine (I)? To which group do they belong?
b. Give the name and symbol for each element in the A group that has two outer electrons. To which A group do they belong?

Answer

a. These elements all have seven outer electrons. They belong to Group 7A, the halogen family.

In addition to electrons and protons, atoms also contain neutrons. The one (and only) exception is an atom of the most common form of hydrogen, which consists of one proton and one electron (if electrically neutral). But even in pure hydrogen, one atom out of 6700 also has a neutron in its nucleus. This naturally occurring form of hydrogen is called deuterium. Tritium, a radioactive form of hydrogen that is quite rare in nature, has two neutrons in its nucleus. Hydrogen, deuterium, and tritium are examples of **isotopes,** two or more forms of the same element (same number of protons) whose atoms differ in number of neutrons, and hence in mass.

An isotope is identified by its **mass number**—the sum of the number of protons and neutrons in the nucleus of an atom. The mass number, indicated by a superscript to the left of an atomic symbol, can vary for the same element. The atomic number, often included as a subscript to the left of the symbol, cannot vary for the same element. For example, the full atomic symbol ^1_1H represents the most common isotope of hydrogen. Because hydrogen *always* has an atomic number of 1, the subscript is often omitted, making the simplified symbol for the common isotope of hydrogen just ^1H. Written in text, you may read hydrogen-1, or H-1. Clearly there is redundancy in actually giving the atomic

number as well as the atomic symbol, even though it may be convenient for the reader. Table 2.3 summarizes information about the isotopes of hydrogen.

Table 2.3	Isotopes of Hydrogen			
Isotope	Atomic Symbol	Number of Protons	Number of Neutrons	Mass Number
hydrogen, H-1	$^{1}_{1}H$	1	0	1
deuterium, H-2	$^{2}_{1}H$	1	1	2
tritium, H-3	$^{3}_{1}H$	1	2	3

Your Turn 2.7   Protons, Electrons, and Neutrons

Specify the number of protons, electrons, and neutrons in a neutral atom of each isotope.

   a. carbon-14 ($^{14}_{6}C$)   **b.** uranium-235 ($^{235}_{92}U$)   **c.** iodine-131 ($^{131}_{53}I$)

Answers
   a. 6 protons, 6 electrons, 8 neutrons
   b. 92 protons, 92 electrons, 143 neutrons

All elements have isotopes, but the number of stable and unstable ones varies considerably. Each element's atomic mass, the number you see on every periodic table, takes the relative natural abundance of isotopes, as well as their masses, into account. Following our general rule of introducing information as needed, we will return to a discussion of atomic masses in Chapter 3.

Mass number is the total number of protons and neutrons in a specific isotope. **Atomic mass** refers to a **weighted average** of all naturally occurring isotopes of that element.

2.3   Molecules and Models

Having completed our excursion into the atomic realm, we come to our primary motivation for studying atoms—understanding molecular structure so that we can understand ozone depletion. The stability of filled electron shells helps to explain why atoms bond to one another to form molecules. The simplest case is H_2, a diatomic molecule. A hydrogen atom has only one electron. If two hydrogen atoms come together, the two electrons become common property. Each atom effectively has a share in both electrons. The resulting H_2 molecule has a lower energy than the sum of the energy in the two individual H atoms, and consequently the molecule with its bonded atoms is more stable than the separate atoms. The two electrons that are shared constitute a **covalent bond.** Appropriately, the name *covalent* implies "shared strength."

If we represent each atom by its symbol and each electron by a dot, the two individual hydrogen atoms might look something like this:

$$H\cdot \quad \text{and} \quad \cdot H$$

Bringing the two atoms together yields a molecule that can be represented this way.

$$H\!:\!H$$

A **Lewis structure** is a representation of an atom or molecule that shows its outer electrons. The name honors Gilbert Newton Lewis (1875–1946), an American chemist who pioneered its use. Lewis structures, also called dot structures, can be predicted

for many simple molecules by following a set of straightforward steps. We first illustrate the procedure with hydrogen fluoride, HF, a very reactive compound used to etch glass.

1. Starting with the chemical formula of the compound, note the number of outer electrons contributed by each of the atoms. (Remember that the periodic table is a useful guide for A group elements.)

    H·   1 H atom × 1 outer electron per atom = 1 outer electron

    :F·   1 F atom × 7 outer electrons per atom = 7 outer electrons

2. Add the outer electrons contributed by the individual atoms to obtain the total number of outer electrons available.

    $$1 + 7 = 8 \text{ outer electrons}$$

3. Arrange the outer electrons in pairs. Then distribute them in such a way as to maximize stability by giving each atom a share in enough electrons to fully fill its outer shell: two electrons in the case of hydrogen, eight electrons for most other atoms.

    H:F:

We surrounded the F atom with eight dots, organized into four pairs. The pair of dots between the H and the F represents the electron pair that forms the bond uniting the hydrogen and fluorine atoms. The other three pairs of dots are the three pairs of electrons that are not shared with other atoms and hence not involved in bonding. As such, they are called "nonbonding" electrons, or "lone pairs."

A **single covalent bond** is formed when only one pair of shared electrons forms the linkage between atoms. A line often replaces the electron pair forming a single covalent bond. This line connects the symbols for the two atoms.

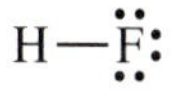

    H—F:

Sometimes the nonbonding electrons are removed from a Lewis structure, simplifying it still more. The result is called a **structural formula**, a representation that replaces each bonded electron pair in a Lewis structure with a line.

    H—F

Remember that the single line represents one pair of shared electrons. These two electrons plus the six electrons in the three nonbonding pairs mean that the fluorine atom is associated with a total of eight outer electrons, whether or not all the electrons are specifically shown. Remember that the hydrogen atom has no additional electrons other than the single pair shared with fluorine. It is at maximum capacity with two electrons, thanks to its small size.

The fact that electrons in many molecules are arranged so that every atom (except hydrogen) shares in eight electrons is called the **octet rule**. This generalization is useful for predicting Lewis structures and the formulas of compounds. Consider the Cl_2 molecule, the diatomic form of elemental chlorine. From the periodic table, we can see that chlorine, like fluorine, is in Group 7A, which means that its atoms each have seven outer electrons. Using the scheme given for HF earlier, we first count and add up the outer electrons for Cl_2.

    2 :Cl·   2 Cl atom × 7 outer electrons per atom = 14 outer electrons

For Cl_2 to exist, there must be a bond between the two atoms, which we show by a single line designating a shared electron pair: a single covalent bond. The remaining 12 electrons constitute six nonbonding pairs, distributed in such a way as to give each chlorine atom 8 electrons (2 bonding and 6 nonbonding). This meets the octet rule. Accordingly, this is the Lewis structure for Cl_2.

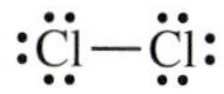

    :Cl—Cl:

Your Turn 2.8   Lewis Structures for Diatomic Molecules

Draw the Lewis structure for each molecule.

a. HBr   **b.** Br_2

Answer

a. H·   1 H atom × 1 outer electron per atom = 1 outer electron

·B̤r̈:   1 Br atom × 7 outer electrons per atom = 7 outer electrons

Total = 8 outer electrons

These are the Lewis structures for HBr.

H:B̈r̤:   or   H—B̈r̤:

So far we have dealt only with molecules having just two atoms. **Polyatomic molecules** consist of three or more atoms. The octet rule applies to many of these molecules as well. Here is another generalization just as useful as the octet rule in helping to predict Lewis structures. In most molecules where there is only one atom of one element bonded to two or more atoms of another element (or elements), *the single atom goes in the center of the Lewis structure.* You'll encounter exceptions to these generalizations, but this is a good place to begin to apply them. We start with a water molecule, H_2O, as an example.

Following the same procedures used for two-atom molecules, we first count and add up the outer electrons.

2 H·   2 H atoms × 1 outer electron per atom = 2 outer electrons

·Ö·   1 O atom × 6 outer electrons per atom = 6 outer electrons

Total = 8 outer electrons

We place the symbol O in the center. Each of the H atoms is bonded to the O atom with a pair of electrons, using four electrons. The remaining four electrons are also placed on the O atom, but as two nonbonding pairs. This is the result.

H:Ö:H

A quick count confirms that the O is surrounded by eight dots, representing the eight electrons predicted by the octet rule. Alternatively, we could use lines for the single bonds.

H—Ö—H

These Lewis structures provide more information than does the chemical formula, H_2O. The formula shows the types and ratio of atoms present, and so does the Lewis structure. The Lewis structure also indicates how the atoms are connected to one another and the nonbonding pairs of electrons, if present. On the other hand, Lewis structures do *not* directly reveal the shape of a molecule. From the structures for water given so far, it might appear that the atoms of the water molecule all fall in a straight line. In fact, the molecule is bent. It looks something like this.

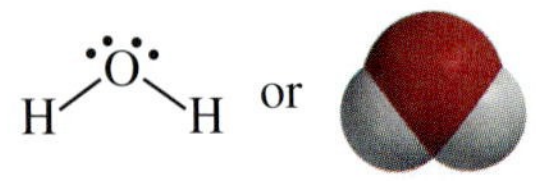

We will return to this discussion of shape in Chapter 3 and see how the Lewis structure can lead to the prediction of this bent structure. We will examine the experimental evidence for the shape of the water molecule in Chapter 5.

Another example of a molecule with more than two atoms is methane, CH_4. Using the rules and generalizations given earlier, we can write the Lewis structure of methane.

Each hydrogen atom forms only one bond (two shared electrons). Oxygen can form two bonds and is the central atom in H_2O.

The space-filling model of water was shown in Section 1.7.

The combustion of methane was discussed in Section 1.10. The geometry of the methane molecule is described in Section 3.3.

4 H·   4 H atoms × 1 outer electron per atom = 4 outer electrons

·C̈·   1 C atom × 4 outer electrons per atom = 4 outer electrons

Total = 8 outer electrons

The C representing a carbon atom goes in the center and is surrounded by the eight electrons, giving carbon an octet of electrons. Each of the four hydrogen atoms uses two of the electrons to form a shared pair with carbon, for a total of four single covalent bonds. This gives us the Lewis structure of methane.

$$\begin{matrix} & & H & & & & H & \\ & & \ddot{} & & & & | & \\ H & : & \ddot{C} & : H & \text{or} & H— & C & —H \\ & & \ddot{} & & & & | & \\ & & H & & & & H & \end{matrix}$$

Check the methane structure to be sure that the carbon atom has a share in eight electrons, as would be expected by the octet rule. Remember that H can only accommodate a pair of electrons.

Your Turn 2.9   Lewis Structures for Polyatomic Molecules

Draw the Lewis structures for each of these molecules. Both obey the octet rule.

a. hydrogen sulfide (H_2S)

b. dichlorodifluoromethane (CCl_2F_2)

Answer

a. 2 H·   2 H atoms × 1 outer electron per atom = 2 outer electrons

·S̈·   1 S atom × 6 outer electrons per atom = 6 outer electrons

Total = 8 outer electrons

These are the Lewis structures for H_2S.

$$H : \ddot{S} : H \quad \text{or} \quad H— \ddot{S} —H$$

Both S and O are in Group 6A. Therefore, the Lewis structures for H_2S and H_2O only differ in the identity of the central element.

In some structures, single covalent bonds do not allow the atoms to follow the octet rule. Consider, for example, the very important molecule O_2. Here we have 12 outer electrons to distribute, 6 from each of the Group 6A oxygen atoms. There are not enough electrons to give each of the atoms a share in eight electrons if only one pair is held in common. However, the octet rule can be satisfied if the two atoms share four electrons (two pairs). **A covalent bond consisting of two pairs of shared electrons is called a double bond.** This bond is represented by four dots or by two lines.

$$\ddot{O} :: \ddot{O} \quad \text{or} \quad \ddot{O} = \ddot{O}$$

Double bonds are shorter, stronger, and require more energy to break than single bonds involving the same atoms. The experimentally measured length and strength of the bond in the O_2 molecule correspond to a double bond. However, oxygen has a property that is not fully consistent with the Lewis structure just drawn. When liquid oxygen is poured between the poles of a strong magnet, it sticks there like iron filings. Such magnetic behavior implies the presence of unpaired electrons rather than the paired arrangement shown in the preceding Lewis structures. But this is hardly a reason to discard the useful generalizations of the octet rule. After all, simple

scientific models seldom if ever explain all phenomena, but they can be helpful approximations. There are other common examples in which the straightforward application of the octet rule leads to discrepancies in interpreting experimental evidence. Coming across data that do not seem to fit has led to the development of more sophisticated models.

A **triple bond** is a covalent linkage made up of three pairs of shared electrons. Triple bonds are even shorter, stronger, and harder to break than double bonds involving the same atoms. For example, the nitrogen molecule, N_2, contains a triple bond. Each Group 5A nitrogen atom contributes 5 outer electrons for a total of 10. These 10 electrons can be distributed in accordance with the octet rule if 6 of them (three pairs) are shared between the two atoms, leaving 4 of them to form two nonbonding pairs, one on each nitrogen atom.

The stability of the triple bond linking N atoms in N_2 gas helps explain nitrogen's relative inertness in the troposphere.

$$:N:::N: \quad \text{or} \quad :N\equiv N:$$

The ozone molecule introduces another structural feature. We again start with the octet rule. Each of the three oxygen atoms contributes 6 outer electrons for a total of 18. These 18 electrons can be arranged in two ways; each way gives a share in 8 outer electrons to each atom.

$$\ddot{O}::\ddot{O}:\ddot{O}: \qquad :\ddot{O}:\ddot{O}::\ddot{O}$$

a   **b**

Structures **a** and **b** predict that the molecule should contain one single bond and one double bond. In structure **a,** the double bond is shown to the left of the central atom; in **b** it is shown to the right. But experiments reveal that the two bonds in the O_3 molecule are identical, being intermediate between the length and strength of a single and double bond. Structures **a** and **b** are called **resonance forms,** Lewis structures that represent hypothetical extremes of electron arrangements in a molecule. For example, no single resonance form represents the electron arrangement in the ozone molecule. Rather, the actual structure is something like a hybrid of the two resonance forms. A double-headed arrow linking the different forms is used to represent the resonance phenomenon.

$$\ddot{O}=\ddot{O}-\ddot{O}: \longleftrightarrow :\ddot{O}-\ddot{O}=\ddot{O}$$

Resonance is just another modeling concept invented by chemists to represent the complex microworld of molecules. It is not intended to be the "truth," but rather just a way to describe the structures of molecules that do not exactly fit the octet rule model. Figure 2.3 compares the Lewis structures of several different oxygen-containing species relevant to the chemistry in this and other chapters.

A closer experimental inspection of that microworld reveals that the O_3 molecule is not linear as the simple Lewis structures just drawn would seem to indicate. Remember that Lewis structures tell us only what is connected to what and do not necessarily show the molecular geometry. The structure of the O_3 molecule is actually bent, as in this representation.

Observe that both H_2O and O_3 are bent molecules, with O as the central atom.

A more complete explanation of why the O_3 molecule is bent will have to wait until Chapter 3. At this point we are more concerned with how bonding in O_2 and O_3 influences their interaction with sunlight.

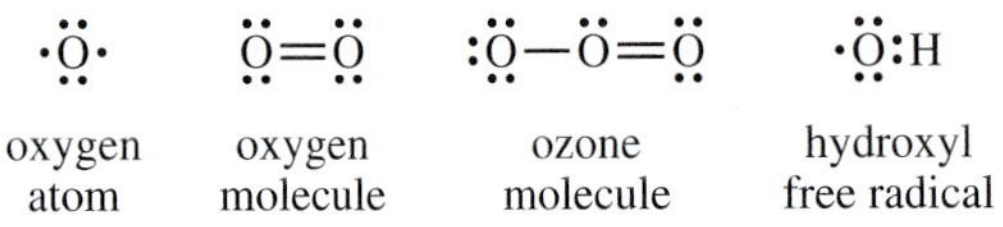

oxygen   oxygen   ozone   hydroxyl
atom   molecule   molecule   free radical

Figure 2.3

Lewis structures for several oxygen species. Only one resonance form of ozone is shown.

2.4   Waves of Light

The next step in building a better understanding of how stratospheric ozone screens out much of the Sun's harmful radiation is to learn something about the fundamental properties of light. The interaction of sunlight with matter is important in several processes, such as in photosynthesis or in the damage high-energy solar radiation can cause in living organisms. Every second, 5 million tons of the Sun's matter is converted into energy, which is radiated into space. The fact that our eyes are capable of detecting different colors is one indication that the radiation that reaches us is not all identical. Prisms and raindrops break sunlight into a spectrum of colors. Each of these colors can be identified by the numerical value of its wavelength. The word *wavelength* correctly suggests that light behaves something like a wave in the ocean. The **wavelength** is the distance between successive peaks. It is expressed in units of length and symbolized by the Greek letter lambda (λ). Waves are also characterized by a certain **frequency,** the number of waves passing a fixed point in 1 second. Frequency is symbolized by the Greek letter nu (ν). Figure 2.4 shows two waves of different wavelength and frequency.

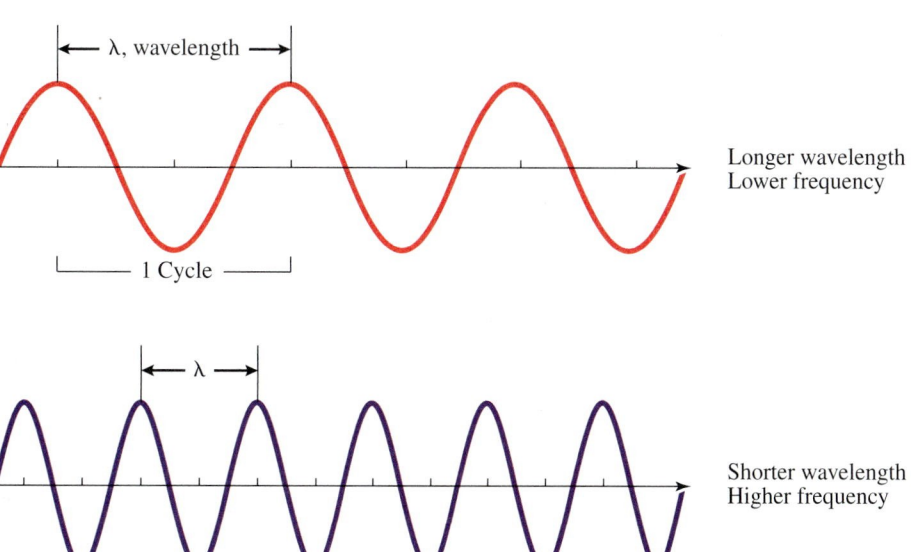

Figure 2.4

Comparison of two different waves.

The relationship between frequency and wavelength can be summarized in a simple equation where v is the frequency and c represents the constant speed at which visible light and other forms of electromagnetic radiation travel, 3.00×10^8 m·s^{-1}.

$$\text{frequency } (v) = \frac{\text{speed of light } (c)}{\text{wavelength } (\lambda)} \qquad [2.2]$$

The form of equation 2.2 indicates that wavelength and frequency are *inversely* related. As the value for λ decreases, the value for v increases, and vice versa.

As wavelength ↑, frequency ↓.

It is both interesting and humbling to realize that out of the vast array of radiant energies, our eyes are sensitive to only a very tiny portion of the total range: wavelengths between about 700×10^{-9} m (corresponding to red) and 400×10^{-9} m (corresponding to violet). These lengths are very short, so we typically express them in nanometers. One **nanometer (nm)** is defined as one-billionth of a meter (m).

$$1 \text{ nm} = \frac{1}{1{,}000{,}000{,}000} \text{ m} = \frac{1}{1 \times 10^9} \text{ m} = 1 \times 10^{-9} \text{ m}$$

We can use this equivalence to convert meters to nanometers. For example, how many nanometers are there in 700×10^{-9} m?

$$\text{wavelength } (\lambda) = 700 \times 10^{-9} \text{ m} \times \frac{1 \text{ nm}}{1 \times 10^{-9} \text{ m}} = 700 \text{ nm}$$

The units of meters cancel and we are left with nanometers.

Figure 2.5
A rainbow of color.

Consider This 2.11   **Analyzing a Rainbow**

Water droplets in a rainbow act as prisms to separate visible light into its component colors.

  a. Which color in the rainbow (Figure 2.5) has the longest wavelength?
  b. Which color in the rainbow has the highest frequency?
  c. Green light has a wavelength of 500 nm. Express this wavelength in meters.

Answer
  c. 500×10^{-9} m or, expressed in scientific notation, 5.00×10^{-7} m.

Scientists have devised a variety of detectors that are sensitive to many other wavelengths, even though these cannot be detected by our eyes. The continuum of waves, known as the **electromagnetic spectrum,** ranges from short and high-energy X-rays and gamma rays to long and low-energy radio waves. Visible light is only a narrow band in this entire range. The term **radiant energy** is used to refer to the entire collection of different wavelengths, each with its own energy. Figure 2.6 shows the continuum that makes up the electromagnetic spectrum, the relative wavelengths, and some examples to help you develop perspective on the range of wavelengths represented.

In this chapter we will consider the **ultraviolet (UV) region** that lies at wavelengths shorter than those of the visible color of violet. At still shorter wavelengths are the X-rays used in medical diagnosis and the determination of crystal structures, and gamma rays that are given off in processes of nuclear decay. At wavelengths longer than those of red visible light, one encounters **infrared (IR)**. We cannot see these wavelengths, but can feel their heating effect. The microwaves used in radar and to cook food quickly have wavelengths on the order of centimeters. At still longer wavelengths are the regions of the spectrum used to transmit your favorite AM and FM radio and television programs.

We will consider the IR region of the spectrum in Chapter 3.

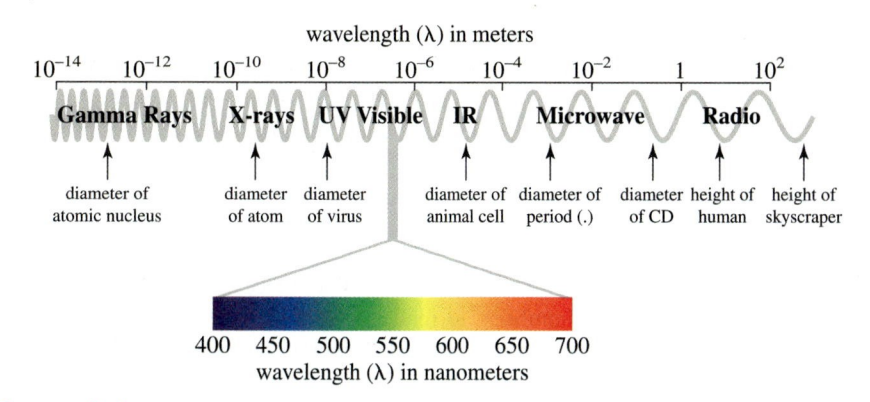

wavelength (λ) in meters

Figure 2.6

The electromagnetic spectrum. The wavelength variation from gamma rays to radio waves is not drawn to scale.

Figures Alive! Visit the *Online Learning Center* to learn more about relationships in the electromagnetic spectrum. Practice, using the interactive exercises. Look for the **Figures Alive!** icon elsewhere in this chapter.

Your Turn 2.12   Relative Wavelengths

Consider these four types of radiant energy from the electromagnetic spectrum: infrared, microwave, ultraviolet, visible.

  a. Arrange them in order of *increasing* wavelength.
  b. Approximately how many times longer is a wavelength associated with a radio wave than one associated with an X-ray? *Hint:* See Figure 2.6.

Answer
  a. ultraviolet < visible < infrared < microwave

Our local star, the Sun, emits many types of radiant energy but not with equal intensity. This is evident from Figure 2.7, a plot of the relative intensity of solar radiation as a function of wavelength. The curve represents the spectrum as measured *above* the atmosphere, before there has been opportunity for interaction of radiation with the molecules found in air. The peak indicating the greatest intensity is in the visible region. However, 53% of the total energy emitted by the Sun is radiated to Earth as infrared radiation. This is the major source of heat for the planet. Approximately 39% of the energy comes to us

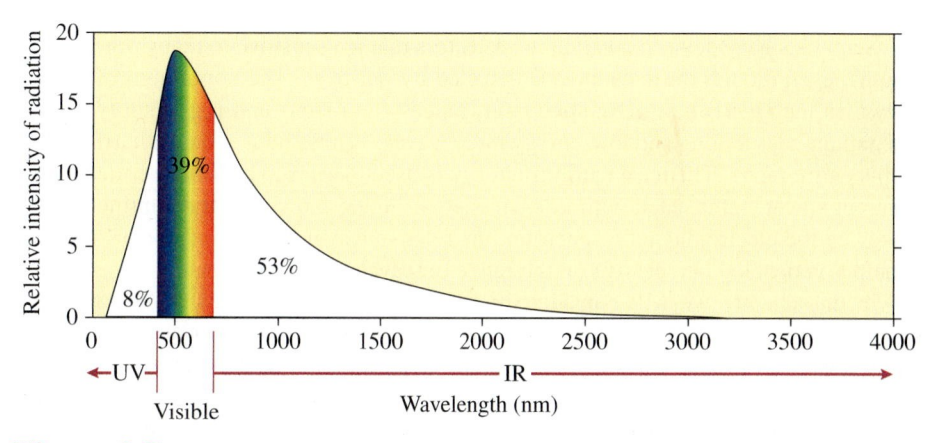

Figure 2.7

Wavelength distribution of solar radiation above Earth's atmosphere.

Source: From *An Introduction to Solar Radiation*, by Muhammad Iqbal, Academic Press, © 1983. Reprinted with permission from Elsevier.

as visible light and only about 8% as ultraviolet. (The areas under the curve give an indication of these percentages.) But in spite of its small percentage, the Sun's UV radiation is potentially the most damaging to living things. To understand why, we need to look at electromagnetic radiation in a different light, this time in terms of its energy.

2.5   Radiation and Matter

The idea that radiation can be described in terms of wave-like character is well established and very useful. However, around the beginning of the 20th century, scientists found a number of phenomena that seemed to contradict this model. In 1909, a German physicist named Max Planck (1858–1947) argued that the shape of the energy distribution curve pictured in Figure 2.7 could only be explained if the energy of the radiating body were the sum of many energy levels of minute but discrete size. In other words, the energy distribution is not really continuous, but consists of many individual steps. Such an energy distribution is called **quantized.** An often-used analogy is that the quantized energy of a radiating body is like steps on a staircase, which are also quantized (no partial steps allowed), not like a ramp, which allows any size stride. Albert Einstein (1879–1955), in the work that won him his 1921 Nobel Prize in physics, suggested that radiation itself should be viewed as constituted of individual bundles of energy called **photons.** One can regard these photons as "particles of light," but they are definitely not particles in the usual sense. For example, they have no mass. These ideas form the basis of modern quantum theory.

> Planck and Einstein were both amateur violinists who played duets together.

The wave model is still useful, even with the new development of the quantum theory to explain the particle-like property of energy. Both are valid descriptions of radiation. This dual nature of radiant energy seems to defy common sense. How can light be described in two different ways at the same time, both waves and particles? There is no obvious answer to that very reasonable question—that's just the way nature is. The two views are linked in a simple relationship that is one of the most important equations in modern science. It is also an equation relevant to the role of ozone in the atmosphere.

$$\text{energy } (E) = \frac{hc}{\lambda} \qquad [2.3]$$

Here E represents the energy of a single photon. Both symbols h and c represent constants. The symbol h is called Planck's constant and c is the speed of light. This equation therefore shows that energy, E, is *inversely* proportional to the wavelength, λ. Consequently, as the wavelength of radiation gets shorter, its energy increases. This qualitative relationship is important in the story of ozone depletion.

> As wavelength $\uparrow$, energy $\downarrow$.

Your Turn 2.13   Color and Energy Relationships

Arrange these colors of the visible spectrum in order of *increasing* energy per photon:

green, red, yellow, violet

Answer

red < yellow < green < violet

Using equation 2.3, one can calculate that the energy associated with a photon of UV radiation is approximately 10 million times larger than the energy of a photon emitted by your favorite radio station. A consequence of this large difference in energy is that you can damage your skin with exposure to UV radiation, but not by listening to the radio—unless you happen to be listening to it outside in the sunlight. Whether or not your radio is turned on, you are continuously bombarded by radio waves. Your body cannot detect them, but your radio can. The energy associated with each of the radio photons is very low and not sufficient to produce a local increase in the concentration of the skin pigment, melanin, as happens with exposure to UV. Producing melanin involves a quantum jump, an electronic transition that requires far more energy than radio wave photons can supply.

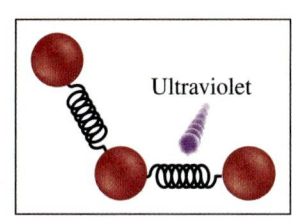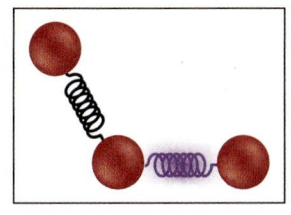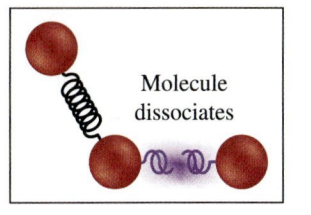

Figure 2.8

Ultraviolet radiation can break some chemical bonds. Bonds are represented as springs that hold the atoms together but allow the atoms to move relative to each other.

The Sun bombards Earth with countless photons—indivisible packages of energy. The atmosphere, the planet's surface, and Earth's living things all absorb these photons. Radiation in the infrared region of the spectrum warms Earth and its oceans, causing molecules to move, rotate, and vibrate. The cells of our retinas are tuned to the wavelengths of visible light. Photons associated with different wavelengths are absorbed, and the energy is used to "excite" electrons in biological molecules. Some electrons jump to higher energy levels, triggering a series of complex chemical reactions that ultimately lead to sight. Compared with animals, green plants capture photons in an even narrower region of the visible spectrum (corresponding to red light) and use the energy to convert carbon dioxide and water into food, fuel, and oxygen in the process of photosynthesis.

Remember that as the wavelength of light *decreases,* the energy carried by each photon *increases.* Photons in the UV region of the spectrum are sufficiently energetic to displace electrons within neutral molecules, converting them into positively charged species. Even shorter UV wavelength photons break bonds, causing molecules to come apart. In living things, such changes disrupt cells and create the potential for genetic defects and cancer. The interaction of UV radiation with chemical bonds is shown schematically in Figure 2.8.

It is part of the fascinating symmetry of nature that this interaction of radiation with matter explains both the damage ultraviolet radiation can cause and the atmospheric mechanism that protects us from it. We turn next to understanding the ultraviolet shield provided by oxygen and ozone in our stratosphere.

2.6   The Oxygen-Ozone Screen

Solar UV radiation is greatly diminished by passing through oxygen and particularly through ozone in the stratosphere. Different UV wavelengths and energies influence how much UV solar radiation reaches Earth and how much damage it can cause. Table 2.4 shows the characteristics of UV radiation coming from the Sun.

As we noted in Chapter 1, about 21% of the atmosphere consists of diatomic oxygen, O_2. The forms of life that inhabit our planet are absolutely dependent on the chemical properties of this gas and its interaction with ultraviolet radiation. The strong covalent bond holding the two O atoms together in the O_2 molecule can be broken by the absorption of a photon of the proper radiant energy. The photon excites a bonding electron to a higher energy level, causing the atoms to come apart. Only photons with energy corresponding to a wavelength of 242 nm or less have sufficient energy to break the bonds in an O_2 molecule. These wavelengths are found in the UV-C region.

$$O_2 \xrightarrow[\lambda \leq 242 \text{ nm}]{\text{UV photon}} 2\,O \qquad\qquad [2.4]$$

Consider This 2.14   The ABCs of Solar UV Radiation

a. Arrange the three regions of UV in order of increasing wavelength.
b. Will the order for increasing energy be the same as that for wavelength? Explain.
c. Would you buy a sunscreen that claims to protect against UV-C? Explain.

Table 2.4	Categories and Characteristics of UV Radiation		
Radiation	**Wavelength Range (nm)**	**Relative Energy**	**Comments**
UV-A	320–400	Least energetic of these three UV categories	Least damaging, reaches Earth's surface in greatest amount
UV-B	280–320	More energetic than UV-A, less energetic than UV-C	More damaging than UV-A, less damaging than UV-C, most absorbed by O_3 in the stratosphere
UV-C	200–280	Most energetic of these three categories	Most damaging of these three, but not a problem because totally absorbed by O_2 and O_3 in stratosphere

If O_2 were the only UV absorber in the atmosphere, Earth's surface and the creatures that live on it would still be subjected to damaging radiation in the 242- to 320-nm range. It is here that O_3 plays its important protective role. The O_3 molecule is more easily broken apart than O_2. Recall that the atoms in the O_2 molecule are connected with a double bond, but each of the bonds in O_3 is somewhere between a single and double bond in length and in strength. This makes the bonds in O_3 energetically weaker than the double bonds in O_2. Therefore, photons of a lower energy (longer wavelength) should be sufficient to separate the atoms in O_3. This is in fact the case, as radiation of wavelength 320 nm or less induces this reaction.

$$O_3 \xrightarrow[\lambda \leq 320 \text{ nm}]{\text{UV photon}} O_2 + O \qquad [2.5]$$

Sceptical Chymist 2.15   Energy and Wavelength

It has been stated that it takes higher energy UV photons, those with wavelengths ≤ 242 nm, to break the double bond in O_2. Given that the bonds in O_3 are somewhat weaker than those in O_2, photons of wavelengths ≤ 320 nm can break those bonds. The Sceptical Chymist does realize that wavelength is inversely proportional to energy, but how does the energy associated with a photon of each of these wavelength limits compare? Just how much greater is the energy of the 242-nm photon than that of a 320-nm photon?

Hint: One approach could be to calculate the ratio of the energies for a 242-nm photon and that of a 320-nm photon and then to compare that with the ratio of their wavelengths. Values for Planck's constant and for the speed of light can be found in Appendix 1.

Equations 2.4 and 2.5 are just parts of a natural cycle of reactions in the stratosphere. Every day, 300,000,000 (3×10^8) tons of stratospheric O_3 form and an equal mass decomposes. As with any chemical or physical change, new matter is neither created nor destroyed but merely changes its chemical or physical form. The overall concentration of ozone remains constant in the natural cycle. The process is an example of a

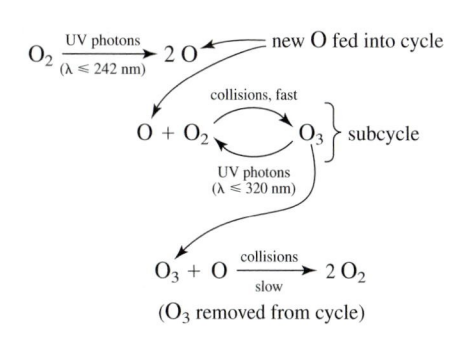

Figure 2.9
The Chapman cycle.

steady state, a condition in which a dynamic system is in balance so that there is no net change in concentration of the major species involved. A steady state arises when a number of chemical reactions, typically competing reactions, balance each other. The **Chapman cycle** (Figure 2.9) refers to the set of natural steady-state reactions for stratospheric ozone. This natural process shows both ozone formation and ozone decomposition. The "lifetime" of a given ozone molecule depends strongly on altitude, ranging from days to years. In the center of the ozone layer, an O_3 molecule can persist for several months before it dissociates into O_2 and O, producing a maximum concentration at altitudes within the stratosphere.

This set of reactions is named after Sydney Chapman, a physicist who first proposed it in 1929.

Your Turn 2.16   The Chapman Cycle

a. Use words or an equation to describe each step in which O_3 forms.
b. Use words or an equation to describe each step in which O_3 is removed.
c. What is the result if you apply the rules of algebra and add all three forward reaction equations together, canceling terms that are the same on each side?

In a later section, we will consider what happens when something disturbs the steady state of the Chapman cycle, leading to destruction of the protective ozone layer. Because of the O_2 and O_3 in the stratosphere, only a relatively small fraction of the Sun's UV radiation reaches Earth's surface. However, what does arrive can do significant damage, the topic of the next section.

2.7   Biological Effects of Ultraviolet Radiation

The consequences of UV radiation for plants and animals depend primarily on two factors: the energy associated with the radiation and the sensitivity of the organism to that radiation. We have seen that highly energetic photons can excite electrons and break bonds in biological molecules, rearranging them and altering their properties. Solar radiation at wavelengths below 320 nm is well screened out by O_3 in the stratosphere. This is most fortunate, because radiation in this region of the spectrum is particularly damaging to living things. This relationship is evident from Figure 2.10, where biological sensitivity is plotted versus wavelength. As defined here, biological sensitivity is based on experiments in which the damage to deoxyribonucleic acid (DNA), the chemical basis of heredity, is measured at various wavelengths. In the figure, the biological sensitivity is expressed in relative units, expressed on a logarithmic scale. On this scale, each mark on the y-axis represents a biological sensitivity value that is 10 times the value corresponding to the mark immediately below it on the vertical axis. Biological sensitivity at 320 nm is about 1×10^{-5}, or 0.00001 units. But at 280 nm, the sensitivity is 1×10^0, or 1 unit. This means that radiation at 280 nm is 100,000 times more damaging than radiation at 320 nm. As we have seen, this is because the energy per photon and the potential for biological damage increase as the wavelength decreases.

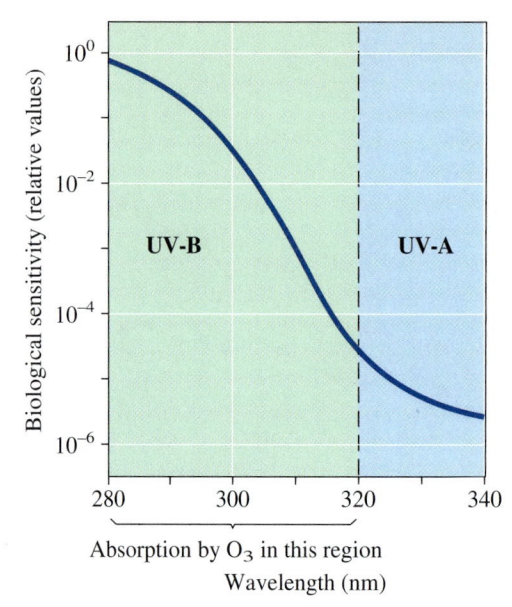

Figure 2.10

Variation of biological sensitivity of DNA with UV wavelength. The UV-C region is below 280 nm.

Source: Reprinted by permission of John E. Frederick, University of Chicago.

Consider This 2.17   Relative Biological Sensitivity

Figure 2.10 illustrates that DNA sensitivity falls with increasing wavelength of UV radiation.

a. What explanation can you propose for this phenomenon?
b. What does this graph tell you about DNA sensitivity for wavelengths longer than 340 nm? Explain.

All evidence shows that the average stratospheric ozone concentration has dropped significantly in the last 20–30 years. Although this has happened to varying extents in different regions, living things are now exposed to greater intensities of damaging radiation than in the past. Scientists have made calculations predicting that a given percent decrease in stratospheric ozone will increase the effects of biologically damaging UV radiation by twice that percentage. For example, a 6% decrease in stratospheric ozone could mean a 12% rise in skin cancer, especially the more easily treated forms such as basal cell and squamous cell carcinomas. These conditions are considerably more common among whites than among people with more heavily pigmented skin (Figure 2.11).

Good evidence links the incidence of the most deadly form of skin cancer, melanomas, with the intensity of UV radiation and the latitude at which you live. For example, the disease generally becomes more prevalent as one moves farther south in the Northern Hemisphere. Those who endure the long nights and short days of northern winters are compensated in general by a level of skin cancer that is only about half that of those who enjoy year-round sunshine. The geographical effect on radiation intensity and skin cancer is, at least to date, much greater than that caused by ozone depletion.

Consider This 2.18   Geography of Skin Cancer

a. Which five states in the United States have the highest incidence rates of melanoma?
b. Which five states have the highest mortality rates from melanoma?
c. Offer some possible reasons why these two sets of states are not the same.
d. Alaska and Hawaii both have very low mortality rates from melanoma. What factors help account for this?

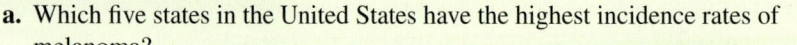

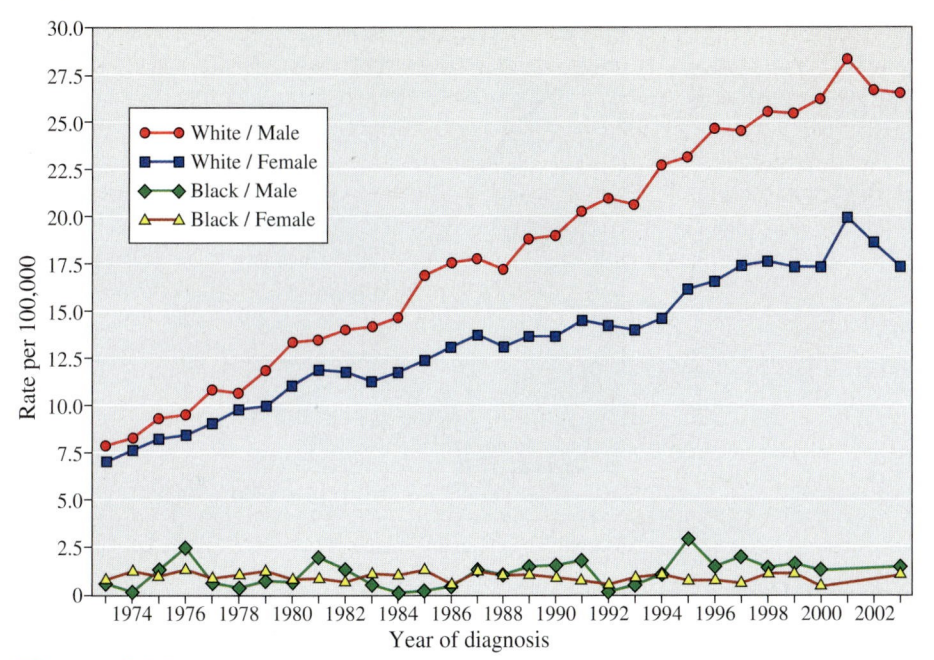

Figure 2.11

Increase in incidence of melanoma skin cancer in the United States, 1973–2003.

Source: Surveillance, Epidemiology, and End Results (SEER) Program of the National Cancer Institute.

Clearly, geographic location cannot be the only factor influencing development of skin cancer. Skin cancer rates generally continue to rise in all countries, despite increased awareness of the dangers of exposure to UV radiation. Changes in the natural protection afforded by the stratospheric ozone layer are only partially responsible for higher rates of skin cancer. About a million new cases of skin cancers occur each year in the United States, almost as many as the total number of cases of all other cancers. Although whites are about 40 times more likely to develop melanoma, blacks have a significantly lower long-term survival rate than whites, according to a 2005 report issued by the American Academy of Dermatology. Skin cancers can develop many years after repeated, excessive exposure has stopped. Skin cancers even have been linked to a single episode of extreme sunburn in adolescence with the effects showing up many years later. Public service campaigns center on early detection and promote regular inspection of suspicious moles for people of all ethnic backgrounds. Other possible causes need to be considered. One of these is tanning, either naturally or in a tanning bed. This practice presents a risk–benefit activity for everyone because skin cancer can strike people of all skin colors.

Consider This 2.19   Bronze by Choice—Tanning Salons

The indoor tanning industry maintains a constant public relations campaign that highlights positive news about indoor tanning, promoting it as part of a healthy lifestyle no matter the pigmentation of one's skin. Countering these claims are the studies published in scientific journals that support the view of dermatologists that there is no such thing as a "safe tan" for any skin type. Investigate at least two Web sites that present different points of view and list the specific claims made. Based on your findings, which criteria would you personally use to decide whether or not to go to an indoor tanning salon for health benefits?

Fair-haired, fair-skinned residents in Australia have the highest skin cancer rates in the world. Two of every three Australians will develop skin cancer sometime during his or her lifetime, and nearly 1300 per year die from this form of cancer. The Australian

government has acted in several ways to reverse this trend, including banning tanned models from all advertising media. National Skin Cancer Action Week is held each year at the start of their summer season in November, with the goals of raising awareness and urging the use of sun protection from clothing and lotions.

Wearing protective sunscreen is one way to reduce the risk of skin cancer. Such products contain compounds that absorb UV-B to some extent together with others for absorbing UV-A. The American Academy of Dermatology recommends a sunscreen with a skin protection factor (SPF) of 15 to 30. But wearing a sunscreen does not mean that you are without risk from the Sun's UV rays. Because sunscreens allow you to be exposed for a longer time without burning, they may ultimately cause greater skin damage. The Australian product Blue Lizard Suncream (Figure 2.12) uses "smart bottle" technology for containing and marketing their product. The bottle itself changes color from white to blue in UV light, sending an extra reminder that the dangers of UV light are still present, even if a sunscreen is being used.

Nanotechnology is influencing the development of sunscreen products too. The Western Australian-based company Advanced Powder Technologies (APT) produces nanosized zinc oxide particles 30 nm across that can be used in "see-through" sunscreen formulations. Because the nanoparticles of zinc oxide are so small, they do not scatter light, leaving the end product clear rather than white. Such sunscreens can be spread more evenly, cover better, be more cost-effective, and yet extremely effective at absorbing UV radiation. While capturing a large percent of market share in Australia, they are not yet in wide use in the United States or Europe until further studies on the safety of these ultrafine particles are conducted. A key question is the extent to which the particles are able to penetrate the skin. In the United Kingdom, the Royal Society and the Royal Academy of Engineering, in a comprehensive 2004 report on Nanoscience and Nanotechnologies, called for further studies before nanoparticles were even more widely used for sunscreens and cosmetics.

Because of the possible damage caused by exposure to UV radiation, the U.S. National Weather Service issues an Ultraviolet Index forecast that appears nationally in newscasts, in newspapers, and on the Web. UV Index values range from 0 to 15 and are based on how long it takes for skin damage to occur (Table 2.5). As is the case for many

Figure 2.12

Blue Lizard Suncream

Nanotechnology was defined in Section 1.7. Other examples will be discussed in Sections 8.8 and 10.8.

Table 2.5	The UV Index	
Exposure Category	**Index**	**Sun Protection Messages**
LOW	<2	Wear sunglasses on bright days. In winter, reflection off snow can nearly double UV strength. If you burn easily, cover up and use sunscreen SPF 15+.
MODERATE	3–5	Take precautions, such as covering up and using sunscreen SPF 15+, if you will be outside. Stay in shade near midday when the Sun is strongest.
HIGH	6–7	Protection against sunburn is needed. Reduce time in the Sun between 10 AM and 4 PM. Cover up, wear a hat and sunglasses, and use sunscreen SPF 15+.
VERY HIGH	8–10	Take extra precautions. Unprotected skin will be damaged and can burn quickly. Try to avoid the Sun between 10 AM and 4 PM. Otherwise, seek shade, cover up, wear a hat and sunglasses, and use sunscreen SPF 15+.
EXTREME	11+	Take all precautions. Unprotected skin can burn in minutes. Beachgoers should know that white sand and other bright surfaces reflect UV and will increase UV exposure. Avoid the Sun between 10 AM and 4 PM. Seek shade, cover up, wear a hat and sunglasses, and use sunscreen SPF 15+.

Source: The Environmental Protection Agency, 2006.

indices designed to communicate with the public, the UV Index may be color-coded to help provide visual information with the numerical value. Specific steps are suggested to protect eyes and skin from sun damage.

Consider This 2.20   UV Index Forecasts

The UV Index indicates the amount of UV radiation reaching Earth's surface at solar noon, the time when the Sun is highest in the sky.

    a. The UV Index depends on the latitude, the day of the year, time of day, amount of ozone above the city, elevation, and the predicted cloud cover. How is the UV Index affected by each of these?

    b. The UV Index forecast is available on the Web, compliments of the Environmental Protection Agency. Go to the *Online Learning Center* for a direct link. Account for the range of values that you see on today's map of the United States.

    c. Surfaces such as snow, sand, and water intensify your exposure to UV radiation, because they reflect it back at you. What outdoor activities might increase your risk from exposure?

Although the UV Index focuses on skin damage, that is not the only biological effect of UV radiation in humans. Everyone, no matter the pigmentation of their skin, can suffer eye problems caused by UV exposure. Retinal damage can take place, as can photokeratitis, which is sunburn to the eye. Cataracts, a clouding of the lens of the eye, can also be caused by excessive exposure to UV-B radiation. It has been estimated that a 10% decrease in the ozone layer could create up to 2 million new cataract cases globally. However, just as putting on sunscreen often and liberally can cut down on skin damage, wearing optical-quality sunglasses capable of blocking at least 99% of UV-A and UV-B is a sensible action for protecting the eyes. This is particularly important while taking part in water or snow sports, when the danger to unprotected eyes is greatest.

Consider This 2.21   Protecting Your Eyes from UV Rays

Sunglasses make far more than a fashion statement. They can protect your eyes from harmful UV rays. What characteristics would you look for in sunglasses to be used when water skiing or sailing? Check out two or three manufacturers to find out what virtues of their products are stressed in their advertising for this use. What materials are used to make sunglasses for water sports? Did you find that the price of sunglasses was related to the amount of UV protection? Is the same protection available for those wearing prescription glasses? Would you purchase and wear the sunglasses you found through your research? Explain your choices.

Human beings are not the only creatures on the globe affected by UV radiation. Increases in UV radiation will bring harm to young marine life, such as floating fish eggs, fish larvae, juvenile fish, and shrimp larvae. There is also experimental evidence of DNA damage in the eggs of Antarctic ice fish. Plant growth is suppressed by UV radiation, and experiments have measured the negative effect that increased UV-B radiation has on phytoplankton. These photosynthetic microorganisms live in the oceans where they occupy a fundamental niche in the food chain. Phytoplankton ultimately supply the food for all the animal life in the oceans, and any significant decrease in their number could have a

major effect globally. An international panel of scientists confirmed in 1999 that exposure to elevated levels of UV-B radiation affected phytoplankton movement (up and down in water) and their motility (moving through water). Without such movement and motility, phytoplankton cannot achieve proper position in the water and are unable to carry out photosynthesis as effectively. Moreover, these tiny plant-like organisms play an important role in the carbon dioxide balance of the planet by absorbing approximately 50% of the atmospheric CO_2 created by human activities. Thus, it is possible that ozone depletion may influence another atmospheric problem, global warming, the topic of the next chapter.

Decreasing stratospheric ozone and consequences of this reduction are cause for concern and action. But action requires knowledge of the chemistry that occurs in the stratosphere, 20–25 km above Earth's surface. We now turn to this topic.

2.8 Stratospheric Ozone Destruction: Global Observations and Causes

Stratospheric O_3 concentrations have been measured over the past 80 years at ground experimental stations spread over the planet and for more than 20 years by satellite-mounted detectors. The natural concentration of stratospheric O_3 is not uniform over all parts of the globe. On average, the total O_3 concentration increases the closer one gets to either pole (with the exception of the seasonal "hole" over the Antarctic). The formation of ozone in the Chapman cycle is triggered when an O_2 molecule absorbs a photon of UV light. Therefore, ozone production increases with the intensity of the radiation striking the stratosphere, an intensity that is not constant. Intensity varies with the seasons, reaching its maximum (in the Northern Hemisphere) in March and its minimum in October (just the reverse of the Southern Hemisphere). Consequently, stratospheric ozone concentrations also follow this seasonal pattern. In addition, the amount of radiation emitted by the Sun changes over an 11- to 12-year cycle related to sunspot activity. This variation also influences O_3 concentrations, but only by 1–2%. The winds blowing through the stratosphere cause other variations in ozone concentrations, some on a seasonal basis and others over a 28-month cycle. To further complicate matters, seemingly random fluctuations often occur. Finally, it is well established that certain gases from both natural and human-made sources are also responsible for the destruction of stratospheric ozone.

Extraordinary images of the Earth, such as the one that opens this chapter, are color-coded to show stratospheric ozone concentrations in Dobson units. The violet and purple regions indicate where the lowest concentrations of O_3 are observed. Total ozone levels above Earth's surface are expressed in Dobson units (DU). A value of 250–270 DU is typical at the equator. As we move north away from the equator, values are typically between 300 and 350 DU with seasonal variation. At the highest northern latitudes, values can be as high as 400 DU in some seasons. Of special interest is the region around Antarctica. From the mid-1990s on, the size of the depleted ozone region around Antarctica annually equaled nearly the total area of the North American continent, in some cases exceeding it. Yet, relatively close by, there are some regions that appear to be enriched in ozone. Figure 2.13 shows measurements of minimum ozone made in Antarctica from 1979 to 2006. The unique circumstances that produce the Antarctic ozone hole are discussed in Section 2.10.

Figure 2.13 shows the dramatic decline in ozone levels observed near the South Pole. The data displayed were collected in September and October of every year from 1979 through 2006 and illustrate the decline in the minimum amount of ozone detected. Indeed, these changes were so pronounced that, when the British monitoring team at Halley Bay in Antarctica first observed it in 1985, they thought their instruments were malfunctioning. The area covered by ozone levels less than 220 DU, defined as the "ozone hole," was larger in early September 2000, than in any year before or since, although the area approached that record in late September 2006. Total ozone destruction in the hole occurred from an altitude of 15 to 20 km, consistent with measurements taken in recent years. The minimum total ozone value of 88 DU, recorded in late September 1994, is the lowest recorded anywhere in the world in nearly 40 years of measurement. Keep in mind that seasonal variation has always occurred in ozone concentration

Recall that a Dobson unit corresponds to about one ozone molecule for every billion molecules and atoms of air.

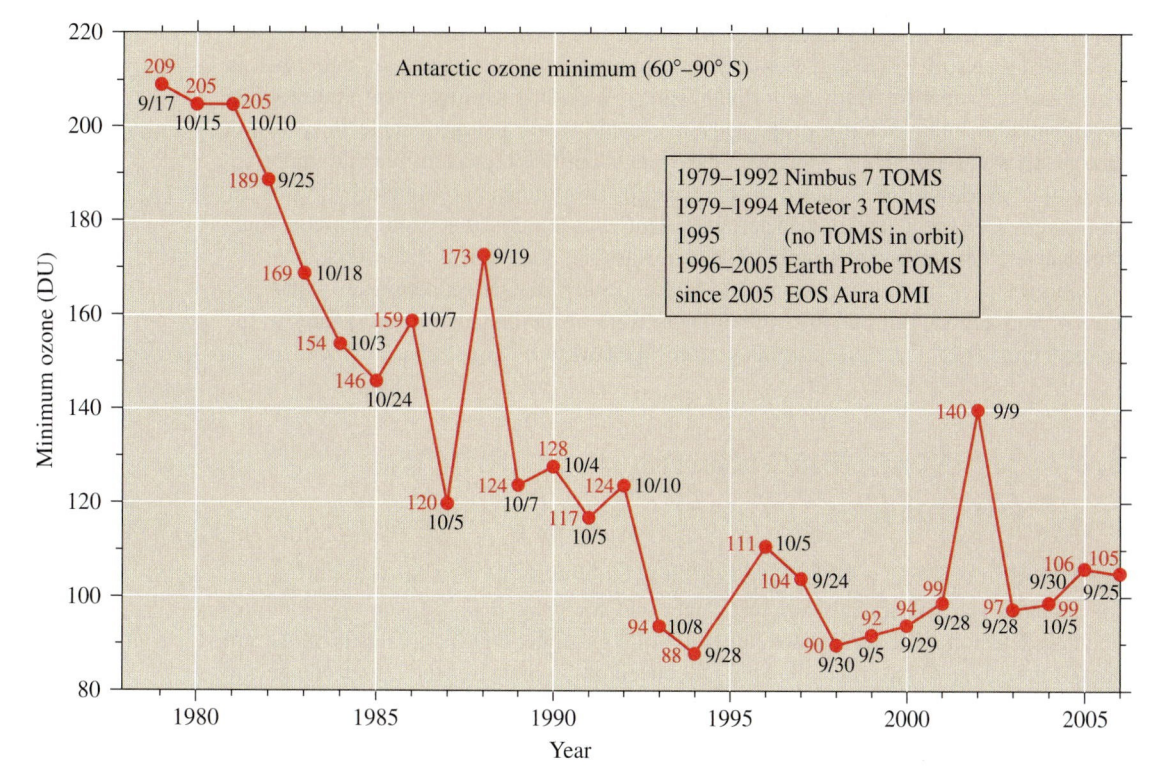

Figure 2.13

Minima in spring stratospheric ozone in Antarctica, 1979–2006. The whole number (in red) is the minimum reading in Dobson units. The date (in black) is when the minimum occurred that year. TOMS (Total Ozone-Measuring Spectrometer) and OMI (Ozone Monitoring Instrument) are analytical instruments.

Source: http://toms.gsfc.nasa.gov/ozone

over the South Pole, with a minimum in late September or early October—the Antarctic spring. Unprecedented is the striking decrease in this minimum that has been observed over the 40 years.

The major natural cause of ozone destruction, wherever it takes place around the globe, is a series of reactions involving water vapor and its breakdown products. The great majority of the H_2O molecules that evaporate from the oceans and lakes fall back to Earth's surface as rain or snow. But a few molecules reach the stratosphere, where the H_2O concentration is about 5 ppm. There, photons of UV radiation trigger the dissociation of water molecules into hydrogen atoms (H·) and hydroxyl free radicals (·OH). A **free radical** is a highly reactive chemical species with one or more unpaired electrons. An unpaired electron is often indicated with a dot, as it is here:

> Free radicals are also discussed in several other places in this text.
> Chapter 1: smog mechanisms
> Chapter 6: acid rain formation
> Chapter 9: addition polymerization
> Chapter 11: food spoilage

$$H_2O \xrightarrow{\text{photon}} H\cdot + \cdot OH \qquad [2.6]$$

Because of its unpaired electron, a free radical reacts readily. Thus, the H· and ·OH radicals participate in many reactions, including some that ultimately convert O_3 to O_2. This is the most efficient mechanism for destroying ozone at altitudes above 50 km.

Your Turn 2.22   Free Radicals

a. How many outer electrons are in the Lewis structure for the ·OH free radical? After deciding, draw the Lewis structure.

b. Can the reaction of H· and ·OH radicals completely account for depletion of ozone in the stratosphere? Explain.

Water molecules and their breakdown products are not the only agents responsible for natural ozone destruction. Another is NO (nitrogen monoxide, also called nitric oxide). Most of the NO in the stratosphere is of natural origin. It is formed when nitrous oxide, N_2O, reacts with oxygen atoms. The N_2O is produced in the soil and oceans by microorganisms and gradually drifts up to the stratosphere. There is really little that can or should be done to control this process. It is part of a cycle involving compounds of nitrogen and living things.

However, not all the nitric oxide in the atmosphere is of natural origin; human activities can alter steady-state concentrations. That is why, in the 1970s, chemists became concerned about the increase in NO that would result from developing and deploying a fleet of supersonic transport (SST) airplanes. These planes were designed to fly at altitudes of 15–20 km, the region of the ozone layer. The scientists calculated that much additional NO would be generated by the direct combination of nitrogen and oxygen.

$$N_2 + O_2 \xrightarrow{\text{high temperature}} 2\,NO \qquad [2.7]$$

This reaction requires large amounts of energy, which can be supplied by lightning or the high-temperature jet engines of the SSTs. To evaluate this risk–benefit situation, many experiments and calculations were carried out, leading to predictions about the net effect of a fleet of SSTs. The conclusion was that the risks outweighed the benefits, and the decision was made, partly on scientific grounds, not to build an American fleet. The Anglo–French Concorde was the only commercial plane that operated at this altitude. The Concorde took its last flight on October 24, 2003. Safety concerns and economic factors both played a role in ending the flights of these remarkable jets.

Even when the effects of water, nitrogen oxides, and other naturally occurring compounds are included in stratospheric models, the measured ozone concentration is still lower than predicted. Measurements worldwide indicate that the ozone concentration has been decreasing over the past 20 years. There is a good deal of fluctuation in the data, but the trend is clear. Stratospheric ozone concentration at midlatitudes (60° south to 60° north) has decreased by more than 8% in some cases. These changes cannot be correlated with changes in the intensity of solar radiation, so we must look elsewhere for a more complete explanation.

> NO has an odd number of outer electrons, 11 (5 from N and 6 from O), so it has an unpaired electron. As a free radical, NO reacts with additional oxygen to form NO_2 and N_2O_4, other oxides of nitrogen.

> Nitrogen oxides were discussed in Section 1.11 as air pollutants. They also will be featured in Section 6.12 for their role in forming acid rain.

Consider This 2.23   Up and Down the Latitudes

In an earlier exercise, you used the Web to get stratospheric ozone data at a location of your choice. Now, using the direct link on the *Online Learning Center*, go to NASA's archive of satellite data on total column ozone levels to find data from 2004 to 2007 over the lower Northern Hemisphere latitudes. You may wish to coordinate your efforts with other students by each picking one year to research.

  a. Obtain values of total column ozone levels at latitudes from +45° north to 0° (the equator) for September 15 of each year. Enter −90° west (roughly the middle of the United States) as the longitude. Obtain readings 5° apart. Make a table of the stratospheric ozone values and compute the average.
  b. Compare data over these 4 years with others in your class. Note that you may not always be able to use the average as a meaningful comparison, because satellite data may be missing at some latitudes.

We have seen much evidence for the abnormally large decrease in stratospheric ozone. Although natural processes can cause such changes, they are insufficient to cause or explain the magnitude of the depletion. It is time to turn our attention to understanding chlorofluorocarbons, compounds that play an important role in stratospheric ozone depletion.

Chlorine

Bromine

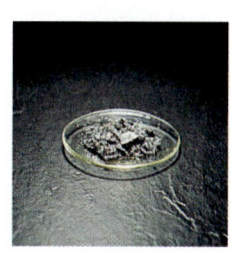

Iodine

Figure 2.14

Selected elements from Group 7A, the halogen family.

2.9 Chlorofluorocarbons: Properties, Uses, and Interactions with Ozone

A major cause of stratospheric ozone depletion was uncovered through the masterful scientific sleuthing for which F. Sherwood Rowland, Mario Molina, and Paul Crutzen won the 1995 Nobel Prize in chemistry. Vast quantities of atmospheric data have been collected and analyzed, hundreds of chemical reactions have been studied, and complicated computer programs have been written to identify the chemical culprit. As with most scientific results, some uncertainties remain, but there is now compelling evidence implicating an unlikely group of compounds: the chlorofluorocarbons.

As the name implies, **chlorofluorocarbons** (CFCs) are compounds composed of the elements chlorine, fluorine, and carbon. Fluorine and chlorine are members of the same chemical group, the halogens (Group 7A), as are bromine and iodine (Figure 2.14). At ordinary temperature and pressures, fluorine and chlorine exist as gases made up of diatomic molecules, F_2 and Cl_2. Bromine and iodine also form diatomic molecules, but the former is a liquid and the latter a solid at room temperature. Fluorine is one of the most reactive elements known. It combines with many other elements to form a wide variety of compounds, including those in polytetrafluoroethylene, otherwise known as Teflon, and other synthetic materials. Chlorine may be best known to you as a water purifier, but it is also a very important starting material in the chemical industry.

CFCs do not occur in nature; they are artificially produced. This is an important verification point in the debate over their role in stratospheric ozone depletion because there are no known natural sources of CFCs. Other contributors to the destruction of ozone, such as the •OH and •NO free radicals, are formed in the atmosphere from both natural sources and human activities.

Two of the most widely used CFCs were known by the trade names Freon 11 and Freon 12. They are commonly known as CFC-11 and CFC-12, respectively, following a naming scheme developed in the 1930s by chemists at DuPont. The chemical formula, chemical name, and Lewis structure is given for each of these CFCs in Table 2.6. Note that the scientific names for these two compounds are based on methane, CH_4. The prefixes *di-* and *tri-* specify the number of halogen atoms that substitute for hydrogen atoms usually found on methane. CFCs do not contain any hydrogen atoms.

The introduction of CFC-12 as a refrigerant in the 1930s was rightly hailed as a great triumph of chemistry and an important advance in consumer safety and environmental protection. This synthetic substance replaced ammonia or sulfur dioxide, two toxic and corrosive refrigerants that made leaks in refrigeration systems extremely hazardous. In many respects, CFC-12 was (and is) an ideal substitute. It has a boiling point in the right range, it is not poisonous, it does not burn, and the CCl_2F_2 molecule is so stable that it does not chemically react with much of anything.

The many desirable properties of CFCs soon led to other uses: propellants in aerosol spray cans, gases blown into polymer mixtures to make expanded plastic foams, solvents for oil and grease, and sterilizers for surgical instruments. **Halons** are compounds similar to CFCs, in which bromine or fluorine atoms replace some or all of the chlorine atoms. Like

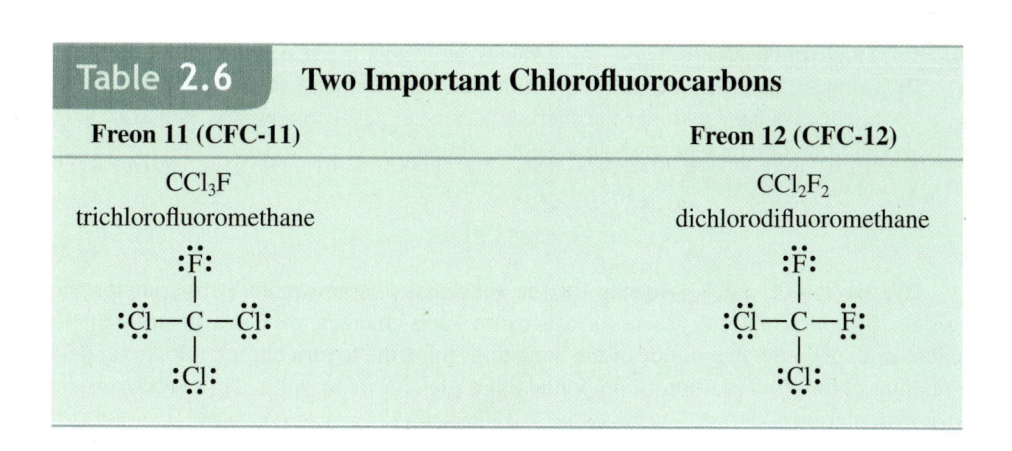

Table 2.6	Two Important Chlorofluorocarbons
Freon 11 (CFC-11)	**Freon 12 (CFC-12)**
CCl_3F	CCl_2F_2
trichlorofluoromethane	dichlorodifluoromethane

CFCs, halons do not contain any hydrogen atoms. Halons proved to be very effective fire extinguishers and are used to protect property that would be especially vulnerable to water and other conventional fire-fighting chemicals. Thus, they have found applications in electronics and computer installations, chemical storerooms, aircraft, and rare book rooms.

For better or worse, the synthesis of CFCs has had a major effect on our lives. Because CFCs are nontoxic, nonflammable, cheap, and widely available, they revolutionized air-conditioning, making it readily accessible in the United States for homes, office buildings, shops, schools, and automobiles. Oppressive summer heat and humidity became more manageable with the use of low-cost CFCs as coolants. Throughout the American South, beginning in the 1960s and 1970s, air-conditioning using CFCs helped to spur the booming growth of cities such as Atlanta, Dallas, and Houston, to be followed by others such as San Antonio, Austin, Charlotte, Phoenix, Memphis, Orlando, and Tampa. Some of these are now among our nation's most populated metropolitan areas. In effect, a major demographic shift occurred because of CFC-based technology that transformed the economy and business potential of an entire region of the country.

Ironically, the very property that makes CFCs ideal for so many applications—chemical inertness—ends up posing a threat to the environment. The C-to-Cl and C-to-F bonds in the CFCs are so strong that the molecules can remain intact for long periods. For example, it has been estimated that an average CCl_2F_2 molecule will persist in the atmosphere for 120 years before it is destroyed. In a much shorter time, typically about five years, many CFC molecules penetrate to the stratosphere with their structures intact.

In 1973, Rowland and Molina, motivated largely by intellectual curiosity, set out to study the fate of these stratospheric CFC molecules. They knew that as altitude increases and the concentrations of oxygen and ozone decrease, the intensity of UV radiation increases. Therefore, they reasoned that in the stratosphere high-energy photons, such as UV-C, corresponding to wavelengths of 220 nm or lower, can break carbon-chlorine bonds. Equation 2.8 shows that this reaction releases chlorine atoms from CFC-12.

> Under the provisions of the Clean Air Act, the term *halon* refers to these compounds used as fire extinguishing agents.

$$F-\underset{\underset{F}{|}}{\overset{\overset{Cl}{|}}{C}}-Cl \xrightarrow[\lambda \leq 220 \text{ nm}]{\text{UV photon}} F-\underset{\underset{F}{|}}{\overset{\overset{Cl}{|}}{C}}\cdot \; + \; \cdot Cl \qquad [2.8]$$

A chlorine atom has seven outer electrons, six of them paired and one unpaired, making atomic chlorine very reactive. In equation 2.8, we wrote the atom as $\cdot Cl$ to emphasize the unpaired electron that results from breaking the C-to-Cl bond. Commonly, free radicals are shown with the dot after the symbol, $Cl\cdot$. The free radical chlorine atom exhibits a strong tendency to achieve a stable octet by combining and sharing electrons with another atom. Rowland and Molina and subsequent researchers hypothesized that this reactivity would result in a chain of reactions. Although CFCs are known to destroy stratospheric ozone via several pathways, we will illustrate with a typical process known to take place in polar regions.

First, the $Cl\cdot$ free radical pulls an oxygen atom away from the O_3 molecule, forms chlorine monoxide, $ClO\cdot$, and leaves an O_2 molecule. Equation 2.9 shows this key first step in the cycle of ozone destruction. The repeated coefficient 2 is not canceled because we are anticipating the next step, shown in equation 2.10.

> The $Br\cdot$ free radical undergoes a comparable reaction, starting another cycle of ozone destruction. $Br\cdot$ is up to 10 times more effective than $Cl\cdot$ in destroying O_3.

$$2\,Cl\cdot + 2\,O_3 \longrightarrow 2\,ClO\cdot + 2\,O_2 \qquad [2.9]$$

The $ClO\cdot$ species is another free radical; it has 13 outer electrons ($7 + 6$). Recent experimental evidence indicates that 75–80% of stratospheric ozone depletion involves $ClO\cdot$ joining to form $ClOOCl$, as shown in equation 2.10.

> The term *cancellation* is often used for removing duplicate terms from two sides of a mathematical equation.

$$2\,ClO\cdot \longrightarrow ClOOCl \qquad [2.10]$$

In turn, $ClOOCl$ decomposes in the two-step sequence shown in equations 2.11 and 2.12.

$$ClOOCl \xrightarrow{\text{UV photon}} ClOO\cdot + \cdot Cl \qquad [2.11]$$

$$ClOO\cdot \longrightarrow Cl\cdot + O_2 \qquad [2.12]$$

We can treat this series of chemical equations (2.9–2.12) as if they were mathematical equations and add them together. Equation 2.13 is the result.

$$2\,Cl\cdot + 2\,O_3 + 2\,ClO\cdot + ClOOCl + ClOO\cdot \longrightarrow$$
$$2\,ClO\cdot + 2\,O_2 + ClOOCl + ClOO\cdot + Cl\cdot + Cl\cdot + O_2 \qquad [2.13]$$

Many of the same species are found on both sides of the combined equation 2.13. Just as is done with mathematical equations, we can eliminate the duplicate $Cl\cdot$, $ClO\cdot$, and $ClOOCl$ species from both sides of the chemical equation. The terms for O_2 on the right side of the equation, $2\,O_2$ and O_2, can be combined into $3\,O_2$. What remains, is the net equation showing the conversion of ozone into oxygen gas, equation 2.14.

$$2\,O_3 \longrightarrow 3\,O_2 \qquad [2.14]$$

Thus, the complex interaction of ozone with atomic chlorine provides a pathway for the destruction of ozone.

The fact that $Cl\cdot$ appears as a reactant (equation 2.9) and then as a product (equations 2.11 and 2.12) is important. This indicates that $Cl\cdot$ is both consumed and regenerated in the cycle, so there is no net change in its concentration. Such behavior is characteristic of a **catalyst**, a chemical substance that participates in a chemical reaction and influences its speed without undergoing permanent change. Atomic chlorine acts catalytically by being regenerated and recycled to remove more ozone molecules. On average, a single $Cl\cdot$ atom can catalyze the destruction of as many as 1×10^5 O_3 molecules before it is carried back to the lower atmosphere by winds.

Interestingly, the mechanism just described for ozone destruction by CFCs in the stratosphere was not the one first proposed by Rowland and Molina. Their initial hypothesis was that $Cl\cdot$ reacted with O_3 to form $ClO\cdot$ and O_2. The second step proposed was that $ClO\cdot$ reacted with oxygen atoms to form O_2 and regenerate $Cl\cdot$ radicals.

$$Cl\cdot + O_3 \longrightarrow ClO\cdot + O_2 \qquad [2.15]$$

$$ClO\cdot + O \longrightarrow Cl\cdot + O_2 \qquad [2.16]$$

Although this mechanism did not prove to be the correct one in the stratosphere, it did provide a reasonable explanation for why recycling a limited number of chlorine atoms could be responsible for the destruction of a large number of ozone molecules. As is often true in science, hypotheses need to be recast in light of experimental evidence.

Atomic chlorine can also become incorporated into stable compounds that do not react to destroy ozone. Hydrogen chloride, HCl, and chlorine nitrate, $ClONO_2$, are two of these "safe" compounds that are quite readily formed at altitudes below 30 km. Thus, chlorine atoms are fairly effectively removed from the region of highest ozone concentration (about 20–25 km). Maximum ozone destruction by chlorine atoms appears to occur at about 40 km, where the normal ozone concentration is quite low.

Rowland, a professor at the University of California at Irvine, and Molina, then a postdoctoral fellow in Rowland's laboratory, published their first paper on CFCs and ozone depletion in 1974 in the scientific journal *Nature.* At about the same time, other scientists were obtaining the first experimental evidence of stratospheric ozone depletion and CFCs in the stratosphere. Since then, the correctness of the Rowland–Molina hypothesis has been well established. Perhaps the most compelling evidence for the involvement of chlorine and chlorine monoxide in the destruction of stratospheric ozone is presented in Figure 2.15. This figure contains two plots of Antarctic data: one of O_3 concentration and the other of $ClO\cdot$ concentration. Both are plotted versus the latitude at which samples were measured. As stratospheric O_3 concentration decreases, the $ClO\cdot$ concentration increases; the two curves mirror each other almost perfectly. The major effect is a decrease in ozone and an increase in chlorine monoxide as the South Pole is approached. Because $ClO\cdot$, $Cl\cdot$, and O_3 are linked by equation 2.9, the conclusion is compelling. Figure 2.15 is sometimes described as the "smoking gun," the clinching evidence.

Not all of the chlorine implicated in stratospheric ozone destruction comes from CFCs. Other chlorinated carbon compounds come from natural sources, such as seawater and volcanoes. However, the majority of atmospheric scientists agree that most chlorine from natural sources is in water-soluble forms. Therefore, any natural chlorine-containing substances are

The term *catalyst* was first defined in Section 1.11.

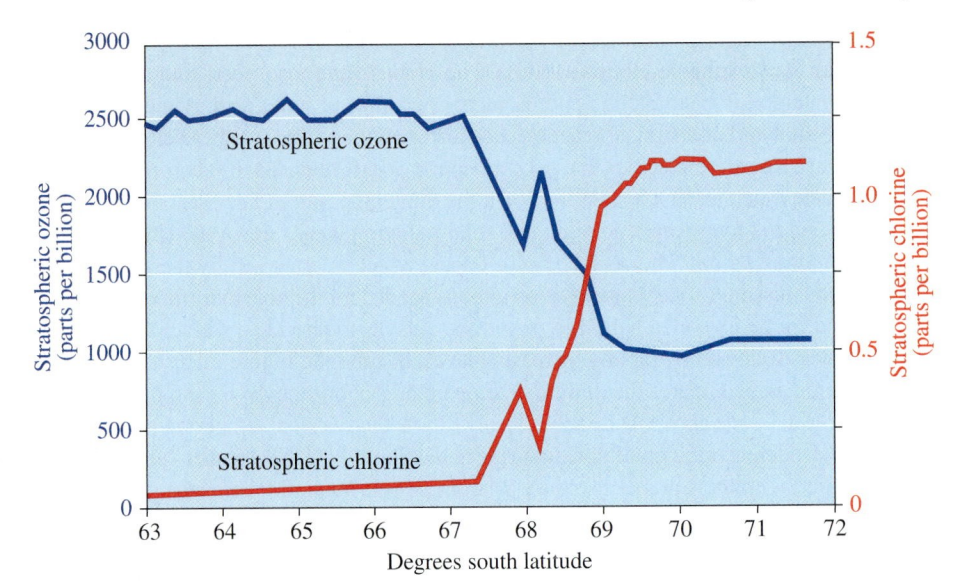

Figure 2.15

Antarctic stratospheric concentrations of ozone and reactive chlorine (from a flight into the Antarctic ozone hole, 1987).

Source: United Nations Environmental Programme. Data from http://ozone.unep.org/Public_Information/ press_backgrounder.pdf

Dr. Susan Solomon, a chemist, headed the team that first gathered stratospheric ClO• and ozone data over Antarctica. The data solidified the causal connection between CFCs and the ozone hole. She was just 30 years old at the time.

washed out of the atmosphere by rainfall, long before they would reach the stratosphere. Of particular significance are the data gathered by NASA and by international researchers that establish that high concentrations of HCl (hydrogen chloride) and HF (hydrogen fluoride) always occur together. Although some of the HCl might conceivably arise from a variety of natural sources, the only reasonable origin of significant stratospheric HF is CFCs.

Consider This 2.24	Talk Radio Opinion

"And if prehistoric man merely got a sunburn, how is it that we are going to destroy the ozone layer with our air conditioners and underarm deodorants and cause everybody to get cancer? Obviously we're not . . . and we can't . . . and it's a hoax. Evidence is mounting all the time that ozone depletion, if occurring at all, is not doing so at an alarming rate."*

Consider the first thing you would ask this talk-show host about these statements. Remember that you need to formulate a short and focused question to get any airtime!

*Limbaugh, R. 1993. *See, I Told You So.* New York: Pocket Books.

2.10 The Antarctic Ozone Hole: A Closer Look

A particularly intriguing question is why the greatest losses of stratospheric ozone have occurred over Antarctica when ozone-depleting gases are present throughout the stratosphere. Why is the effect greatest in polar regions, although less pronounced in the Arctic than in the Antarctic? Given that ozone-depleting gases are emitted mainly in the more developed Northern Hemisphere, why are their effects felt most strongly in the Southern Hemisphere?

Evidence suggests that CFCs are present in comparable abundance in lower parts of the atmosphere over both hemispheres, driven by global wind circulation patterns. A special mechanism is operative in Antarctica. This mechanism is related to the fact that the lower stratosphere over the South Pole is the coldest spot on Earth. From June to September, during the Antarctic winter, winds circulating around the South Pole prevent warmer air from entering the region. Temperatures get as low as −90 °C. Under these conditions, the small

amount of water vapor present freezes into ice crystals, forming thin stratospheric clouds, called **polar stratospheric clouds (PSCs).** The clouds have also been found to contain particles containing the sulfate ion and droplets or crystals of nitric acid. Atmospheric scientists have shown that chemical reactions occurring on the surface of these cloud particles convert otherwise safe molecules that do not deplete ozone, like $ClONO_2$ and HCl, to more reactive species such as HOCl and Cl_2. When the Sun comes out in late September or early October to end the long Antarctic night, the solar radiation breaks down the HOCl and Cl_2, releasing chlorine atoms. The destruction of ozone, which is catalyzed by these atoms, accounts for the missing ozone. Notice the conditions needed for the hole to form: extreme cold and no wind for an extended period to permit ice crystals to provide a surface for the reactions; darkness followed by rapidly increasing levels of sunlight. Figure 2.16 shows the seasonal variation and compares the minimum temperatures above the Arctic and the Antarctic.

Changes in ozone above Antarctica closely follow the seasonal temperatures. Typically a rapid ozone decline takes place during spring at the South Pole (September– early November) compared with the summer (January–March). As the sunlight warms the stratosphere, the ice clouds evaporate, halting the chemistry that occurs on the PSCs. Then air from lower latitudes flows into the polar region, replenishing the depleted ozone levels. By the end of November, the hole is largely refilled. Although the deepest decrease in the ozone layer over Antarctica occurs during the spring, recent discoveries by British Antarctic Survey researchers indicate that the ozone depletion may begin earlier, as early as midwinter at the edges of the Antarctic, including over populated southern areas of South America.

There is already evidence that the ozone reduction over the Southern Hemisphere is greater than would be predicted solely on the basis of the midlatitude chlorine cycle. Australian scientists believe that wheat, sorghum, and pea production have already been lowered as a result of increased UV radiation. We have already noted that Australian health officials

> Decreased stratospheric ozone over the South Pole leads to increased UV-B levels reaching the Earth, causing increased skin cancer rates in Australia and southern Chile.

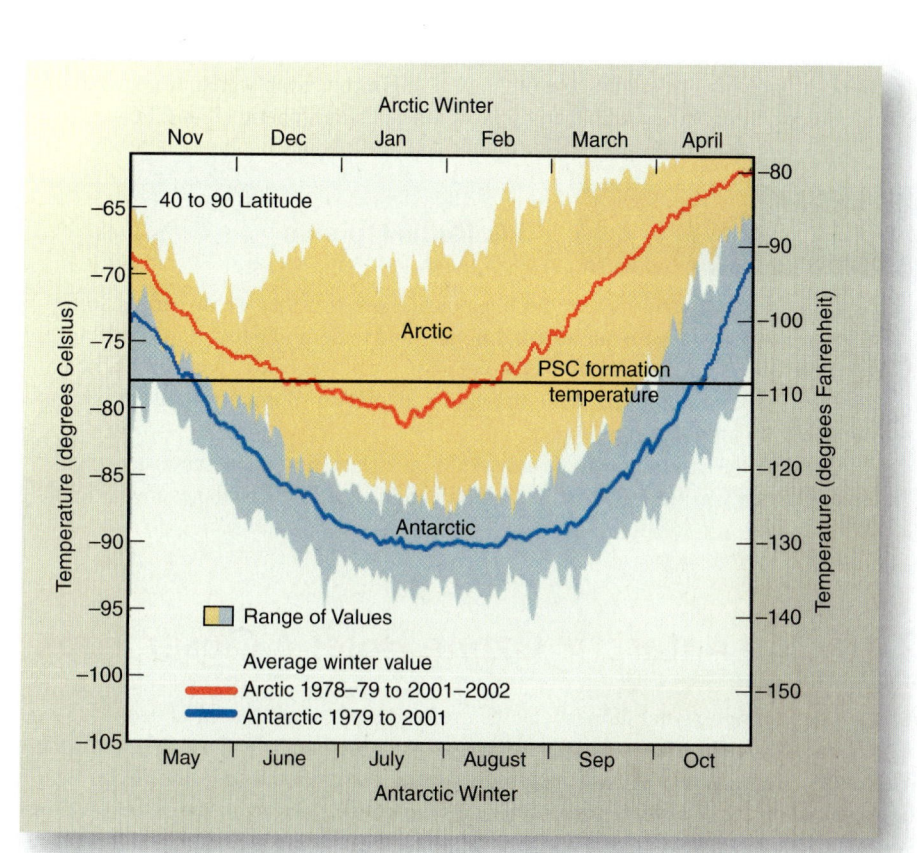

Figure 2.16

Minimum air temperatures in the polar lower stratosphere. Polar stratospheric clouds (PSCs) are thin clouds of ice crystals, formed at very low temperatures.

Source: *Scientific Assessment of Ozone Depletion: 2002,* World Meteorological Organization (WMO), United Nations Environmental Programme (UNEP).

observed significant increases in skin cancers despite a very active public health campaign to alert the population to the danger of exposure to UV radiation. UV alerts have even been issued in Australia. Similar effects are also being felt in southern Chile in the area around Punta Arenas, and on the island of Tierra del Fuego at the southernmost tip of South America. Chile's health minister has warned the 120,000 residents of Punta Arenas not to be out in the Sun between 11 AM and 3 PM during the spring, when ozone depletion is greatest.

The general extent of the depletion in the Northern Hemisphere is not as severe as in the Southern Hemisphere. Scientists have not classified the ozone depletion over the North Pole as a "hole," but are carefully monitoring the location and intensity of UV-B radiation being received. We have already observed that the main reason for the observed difference between the total ozone changes in the two hemispheres is that the atmosphere above the North Pole usually is not as cold as that over its Southern Hemisphere counterpart. Although polar stratospheric clouds have been repeatedly observed in the Arctic, the air trapped over the Arctic generally begins to diffuse out of the region before the Sun gets bright enough to trigger as much ozone destruction as has been observed in Antarctica. During the winter of 2004–2005, temperatures in the ozone layer above the Arctic were the lowest in 50 years and stayed low for more than three months. The extreme conditions led to uncommonly high ozone depletion during that winter. NOAA scientists report that in portions of the Arctic region the average values for total ozone were as much as 50% lower than comparable values during the 1980s. Although the area of low ozone readings was larger than that for the previous two winters, it was not as large as the record area of depletion in the Arctic observed during the winter of 1999–2000. The situation in the middle latitudes of the Arctic was quite different. There, higher than average values for ozone were recorded, reversing the previously observed downward trend. PSCs are regularly observed in Arctic regions, such as these "mother-of-pearl" clouds above Porjus, a village in Swedish Lapland, in the northern part of Sweden (Figure 2.17). The colors are caused by diffraction around the ice particles in the clouds. Other types of PSCs contain a mixture of frozen water and nitric acid, HNO_3.

Figure 2.17

Arctic polar stratospheric clouds.

| Consider This 2.25 | Northern Hemisphere Ozone Maps |

Use the Web resources of NOAA's Climate Prediction Center or Environment Canada to find seasonal images of total ozone concentrations over the Northern Hemisphere. How do the ozone concentrations vary with the seasons? Do these match the seasonal variations for the Southern Hemisphere?

2.11　Responses to a Global Concern

Once the role of synthetic CFCs in ozone destruction was understood, the response was surprisingly rapid. Some of the first steps toward reversing ozone depletion were taken by individual countries. For example, the use of CFCs in spray cans was banned in North America in 1978, and their use as foaming agents for plastics was discontinued in 1990. The problem of CFC production and subsequent release, however, was a global one, and it required international cooperation for actions to be effective. The sequence of events provides a model for solving problems cooperatively before a full-scale global crisis results.

In 1977, in response to growing experimental evidence, the United Nations Environmental Program (UNEP) convened a conference that adopted a World Plan of Action on the Ozone Layer and established a Coordinating Committee to guide future international actions. In 1985, a number of world governments participated in the Vienna Convention on the Protection of the Ozone Layer. Through action taken at the convention, these nations committed themselves to protecting the ozone layer and to conducting scientific research to better understand atmospheric processes. A major breakthrough came in 1987 with the signing of the Montreal Protocol on Substances That Deplete the Ozone Layer.

An important provision of the agreement was to hold future meetings to revise goals as scientific knowledge evolved. Therefore, with growing knowledge of the cause of the ozone hole and the potential for global ozone depletion, atmospheric scientists, environmentalists, chemical manufacturers, and government officials soon agreed that the original Montreal Protocol was not sufficiently stringent. In 1990, representatives of approximately 100 nations met in London and decided to ban the production of CFCs by the year 2000. Delegates meeting in 1992 in Copenhagen and in Montreal in 1997 enacted more stringent controls. The Beijing Amendments in 1999 added bromine-containing halons to the schedule for phaseout, and revised controls on short-term CFC substitutes. Subsequent meetings have been held in Nairobi, Kenya in 2003; in Prague, the Czech Republic, in 2004; and in Dakar, Senegal, in 2005. Participants at each conference considered various questions of implementation of the Montreal Protocol, including essential and critical use exemptions and methods to control illegal trade in ozone-depleting substances. Meetings were held in New Delhi, India, in 2006 and in Montreal, Canada, in 2007.

CFCs used as propellants in medical inhalers, such as those used by asthmatics, were exempt from the ban on CFCs until 2008.

| Consider This 2.26 | Graffiti with a Message |

a. Explain the source of humor in this cartoon when it originated in the mid-1970s.
b. Is this cartoon still relevant to the problem of ozone depletion today? Explain.
c. Create your own cartoon dealing with the issue of ozone layer depletion. Be sure that the chemistry is correct!

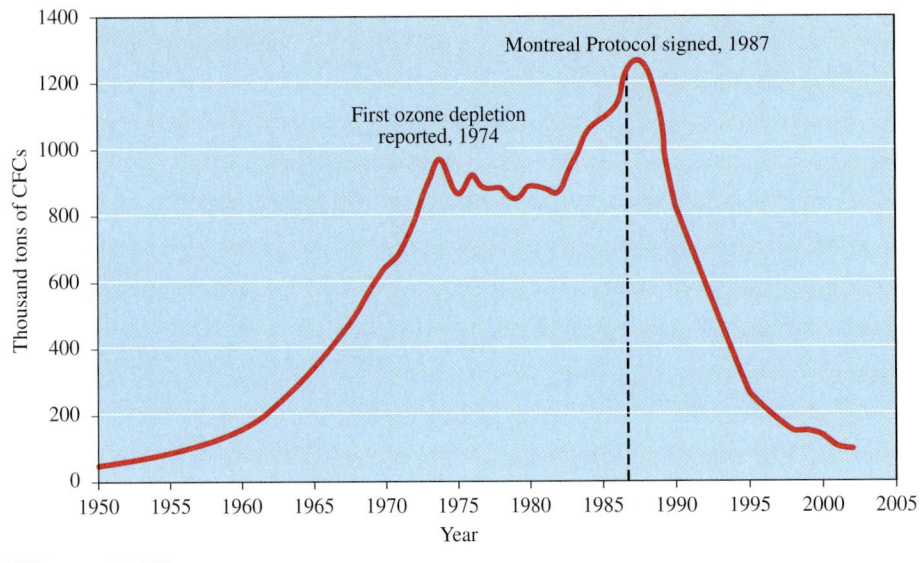

Figure 2.18

Global production of CFCs, 1950–2002.

Source: United Nations Environmental Programme and http://www.defra.gov.uk/environment/statistics/globalatmos/kf/gakfll.htm

The key initial strategy for reducing chlorine in the stratosphere was to stop production of CFCs. The United States and 140 other countries agreed to a complete halt in CFC manufacture after December 31, 1995. Figure 2.18 indicates that the decline in global CFC production has been dramatic. By 1996, production of CFCs had dropped to 1960 levels. Production and consumption of CFCs fell by 86% overall between 1986 and 1996, and by 95% in industrialized countries by the end of 1998. World production of CFCs fell by 91% between 1986 and 2002 and is now nearly at 1950 levels. Without the international action required by the Montreal Protocol, stratospheric abundances of chlorine found in the stratosphere could have tripled by the middle of the 21st century. A total of 189 countries have now ratified the Montreal Protocol. They have reaffirmed that the production of CFCs and other fully halogenated CFCs is to be eliminated by 2010 by all parties, no matter the basic domestic economic needs.

Just because new production of CFCs has been stopped and uses of CFCs have been restricted does not mean that the stratospheric concentration of chlorine instantly will drop. Most Earth systems, including those in the atmosphere, are very complex and respond slowly to change. Slow change, rather than rapid response, is usually beneficial. These observations hold true in the case of CFCs. In fact, atmospheric concentration of ozone-depleting gases continued to rise steadily through the 1990s, despite the restrictions of the Montreal Protocol and its subsequent amendments. Many of the CFCs have very long lifetimes in the atmosphere, estimated to be 100 years or more in some cases. However, there are encouraging signs that the Montreal Protocol has already had a significant effect. Decreases are now being observed in the amount of **effective stratospheric chlorine,** a term reflecting both chlorine and bromine-containing gases in the stratosphere. The values take into account the greater effectiveness but lower concentration of bromine relative to chlorine in depleting stratospheric ozone. Figure 2.19 shows one prediction of the future abundance of effective chlorine.

Analysis of trends in chlorine levels indicates that stratospheric chlorine peaked in the late 1990s and then diminished slowly. This is taken as evidence that the Montreal Protocol and its amendments have slowed the release of CFCs and related ozone-depleting materials. But we are not completely in the clear. Scientists estimate that even under the most stringent international controls on the use of ozone-depleting chemicals, the stratospheric chlorine concentration would not drop to 2 ppb (2000 ppt, parts per trillion) for some years to come. That concentration is significant because the Antarctic ozone hole first appeared when effective stratospheric chlorine increased to that level.

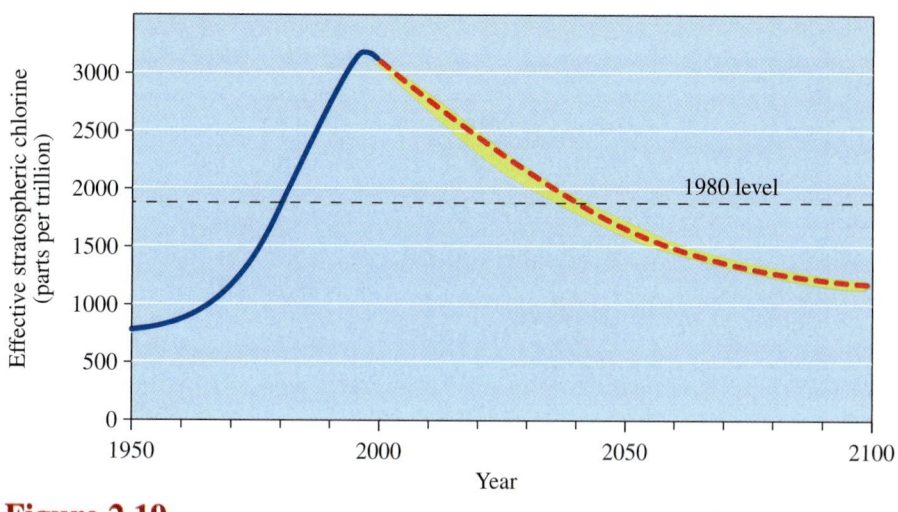

Figure 2.19

Concentrations of effective chlorine, 1950–2100. The height of the yellow band for any year is an estimate of the uncertainty in the prediction.

Source: *Scientific Assessment of Ozone Depletion: 2002,* World Meteorological Organization (WMO), United Nations Environmental Programme (UNEP).

Consider This 2.27   **Past and Future Effective Chlorine Levels**

Use Figures 2.18 and 2.19 to help answer these questions.

a. In approximately what year did effective chlorine concentration peak? What was the reading in that year?

b. Is the peak year for effective chlorine concentration the same as the peak year for CFC production? Why or why not?

c. In approximately what year will the effective chlorine level return to 1980 levels? What will the reading be in that year?

Although the Montreal Protocol and its subsequent adjustments set dates for the halt of all CFC production, the sale of existing stockpiles and recycled materials will remain legal until phaseout dates in the future. In the United States alone, 140 million car air conditioners and the majority of home air conditioners are designed to use these compounds. As a result, both the paperwork and the price of legally obtained CFCs have risen sharply. Nevertheless, many U.S. trade groups promote converting to less harmful substitute refrigerants when repairs must be made on older systems. Until 2010, small amounts of CFCs may be produced in developing countries, keeping some CFCs in legal circulation. Unfortunately, this transition period also encourages an increase in the black market. Bootleggers have been tempted to smuggle CFCs into the United States, largely from the Russian Federation, China, India, Eastern Europe, and Mexico. China now holds the dubious honor of being the leading supplier of black market CFCs. According to U.S. law enforcement officers, CFCs are second only to illicit drugs as the most lucrative illegal import.

Sceptical Chymist 2.28   **Black Market CFCs**

A study reported in the May 2000 issue of *Atmospheric Environment* concluded that illegal trade in CFCs is only a "small threat" to ozone layer recovery. Although the Sceptical Chymist would *like* for this to be true, *is* it? Please provide some current information to either support or refute this statement.

2.12 Replacements for CFCs

No one seriously advocates returning to ammonia and sulfur dioxide in home refrigeration units or giving up air-conditioning as solutions to necessary restrictions on CFCs. Instead, chemists responded by synthesizing new compounds, concentrating on those similar to the CFCs but without their long-term effects on stratospheric ozone. Substitute molecules might include one or two carbon atoms, at least one hydrogen atom, and often fluorine or more tightly bound chlorine atoms. The rules of molecular structure limit the options. For example, each carbon atom forms single bonds to four other atoms in the molecules under consideration.

In synthesizing substitutes for CFCs, chemists weighed three undesirable properties—toxicity, flammability, and extreme stability—and attempted to achieve the most suitable compromise. Compounds containing only carbon and fluorine (fluorocarbons) are neither toxic nor flammable, and they are not decomposed by UV radiation, even in the stratosphere. Consequently, they would not catalyze the destruction of ozone. This would be ideal, were it not for the fact that the fluorocarbons would eventually build up in the atmosphere and contribute to the global warming effect by absorbing infrared radiation. Therefore, fluorocarbons were not suitable replacement compounds.

Introducing hydrogen atoms in place of one or more of the halogen atoms reduces molecular stability and promotes destruction of the compounds at low altitudes, long before they enter the ozone-rich regions of the atmosphere. However, too many hydrogen atoms increase flammability. Moreover, if a hydrogen atom replaces a halogen atom, the total mass of the molecule is decreased. This results in a decrease in boiling point, making the compounds less ideal for use as refrigerants. A boiling point in the -10 to $-30\,°C$ range is an important property for a refrigerant. Too many chlorine atoms seem to increase toxicity, and therefore chloroform, $CHCl_3$, would not be a good CFC substitute. The relationships among composition, molecular structure, boiling point, and proposed use must all be considered along with toxicity, flammability, and stability for any substitute.

Fortunately, chemists already know a good deal about how these variables are related, and they have used this knowledge to synthesize some promising replacements for CFCs. Table 2.7 shows the formulas, names, and structures for two **hydrochlorofluorocarbons, (HCFCs),** compounds of hydrogen, chlorine, fluorine, and carbon. HCFC-22 ($CHClF_2$) is the most widely used HCFC and is suitable both for air conditioners and as a blowing agent to make fast-food containers. Its ozone-depleting potential is about 5% that of CFC-12 and its estimated atmospheric lifetime is only 20 years, compared with 111 years for CFC-12. HCFC-141b ($C_2H_3Cl_2F$) is also used as a blowing agent to make foam insulation. HCFCs decompose in the troposphere more readily than CFCs, and hence do not accumulate to the same extent in the stratosphere.

Because HCFCs themselves have some adverse effects on the ozone layer, they are regarded only as an interim solution in the industrialized countries. The U.S. phaseout of HCFCs will hit a major milestone at the end of this decade, when all production and importation of HCFC-22 and HCFC-141b will stop. By 2015, production or importation of *all* HCFCs will end in the United States. Any continued demand to service refrigerating equipment manufactured prior to those deadlines must be met with recovered HCFCs. By 2030, there will be a 100% reduction and HCFCs will no longer be in use in the United States.

> Global warming is the topic of Chapter 3.

> Foamed fast-food containers and blowing agents will be discussed in Section 9.4.

> The use of HCFC-22 is still growing in China, typical of developing countries.

Table 2.7	**Two Important Hydrochlorofluorocarbons**
HCFC-22	**HCFC-141b**

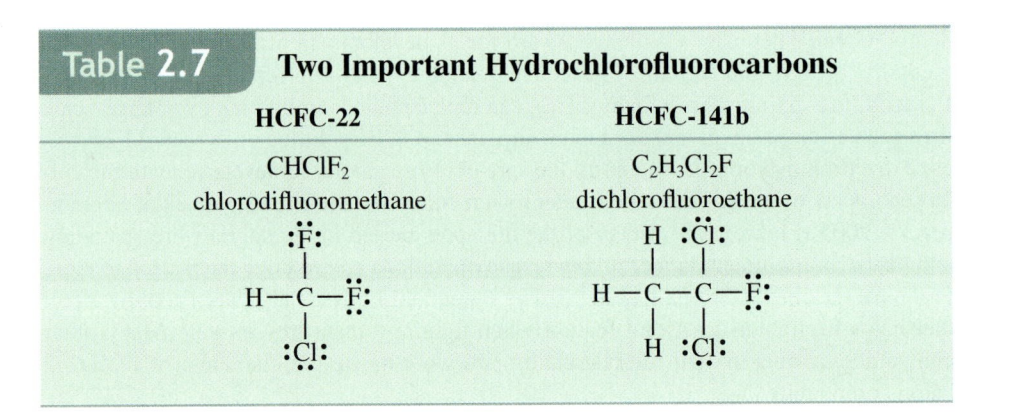

$CHClF_2$
chlorodifluoromethane

$C_2H_3Cl_2F$
dichlorofluoroethane

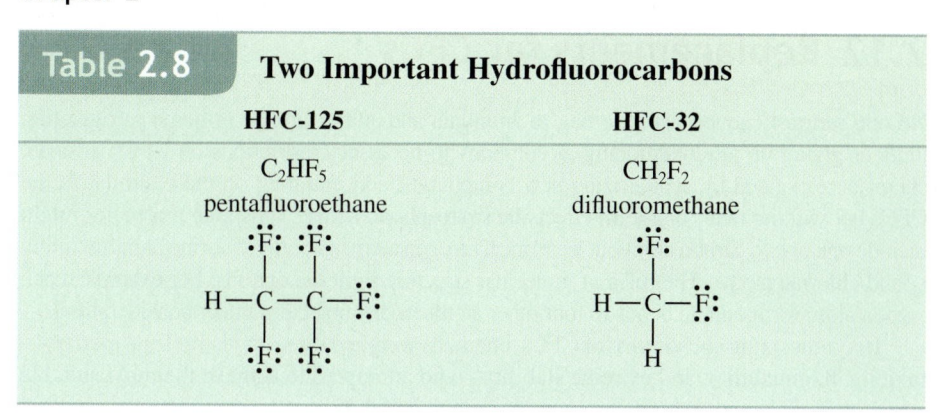

Table 2.8	Two Important Hydrofluorocarbons	
	HFC-125	**HFC-32**
	C_2HF_5 pentafluoroethane	CH_2F_2 difluoromethane

During the phaseout period, the price of available HCFCs will undoubtedly rise dramatically. One factor will be the limited supply of HCFCs for existing equipment but another is change in air-conditioning energy-efficiency regulations mandated by the Department of Energy's new Seasonal Energy Efficiency Rating (SEER) standards for residential air-conditioner manufacturers. Starting in January 2006, SEER ratings of new air-conditioning units were required to show a 30% increase in energy efficiency, a performance that HCFCs helped manufacturers achieve.

If HCFCs are being phased out, what will take their place? In the long run, refrigerant machinery will likely depend on **hydrofluorocarbons, (HFCs)**, compounds of hydrogen, fluorine, and carbon. HFCs have no chlorine atoms to interact with ozone, and their hydrogen atoms facilitate decomposition in the lower atmosphere without being flammable under normal conditions. Table 2.8 shows the formulas, names, and structures for two HFCs that are often blended into a refrigerant known as R-410A. Newer designs for air conditioners will be engineered to use this blend as a replacement for HCFC-22.

Consider This 2.29   Blended HFCs

Another HFC mixture under development to replace HCFCs is R-407c. It blends HFC-125, HFC-32, and HFC-134a. Table 2.8 gives the formula, name, and Lewis structure for the first two components of R-407c. The formula for HFC-134a is $C_2H_2F_4$ and its name is tetrafluoroethane.

a. Why are HFC-125, HFC-32, and HFC-134a *not* classed as CFCs? As HCFCs?
b. Draw the Lewis structure for HFC-134a. *Hint:* There are two F atoms on each C atom.

The role of halons in global warming will be discussed in Section 3.8.

One more class of compounds that must be replaced under the Montreal Protocol is the bromine-containing halons, discussed earlier in Section 2.9. Although very effective in fighting fires, halons are even more successful than CFCs in causing destruction to the ozone layer. Pyrocool Technologies of Monroe, Virginia, won a 1998 Presidential Green Chemistry Challenge Award for its development of foam that is environmentally benign and yet more effective than the halons it replaces. The product, Pyrocool fire-extinguishing foam (FEF), can replace halons in fighting even large-scale fires such as those on oil tankers and jet airplanes. A 0.4% solution of Pyrocool FEF was used to extinguish or at least control the spread of fires in the sublevels beneath the collapsed towers of the World Trade Center towers following the terrorist attack of September 11, 2001 (Figure 2.20). Many of the hot spots buried in the debris were spreading and posed an imminent danger to any rescue operations and to huge tanks used to store Freon for the air-conditioning systems. The Pyrocool FEF foam also has a cooling effect that helps firefighters, a useful feature when fighting brush fires as well. Many other companies, nationally and internationally, are working on the challenge of replacing halon compounds.

Figure 2.20

Pyrocool FEF being applied to subterranean fires at Ground Zero, of the north tower of the World Trade Center, September 30, 2001.

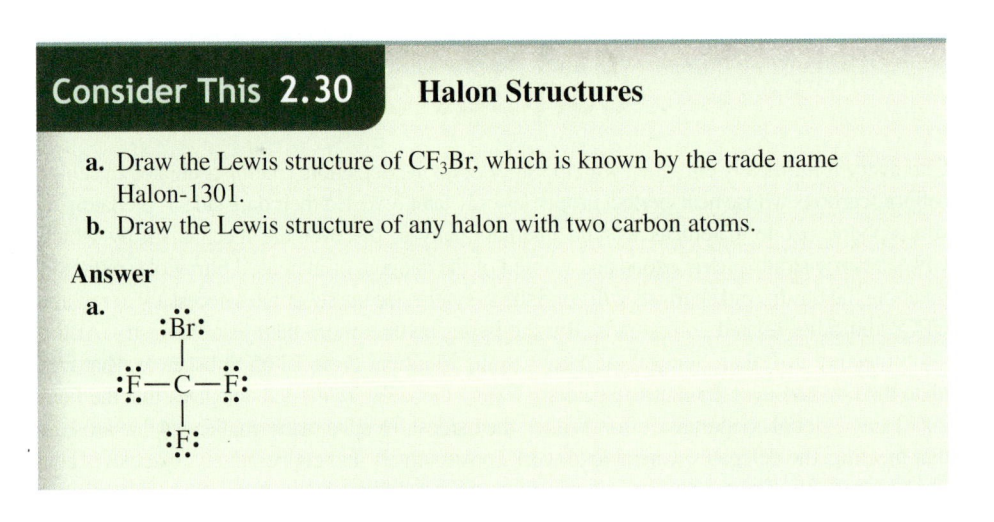

Consider This 2.30 **Halon Structures**

a. Draw the Lewis structure of CF_3Br, which is known by the trade name Halon-1301.
b. Draw the Lewis structure of any halon with two carbon atoms.

Answer

a.

The phaseout of CFCs and the continuing development of alternative materials are not without major economic considerations. At its peak, the annual worldwide market for CFCs reached $2 billion, but that was only the tip of a very large financial iceberg. In the United States alone, CFCs were used in or used to produce goods valued at about $28 billion per year. Although the conversion to CFC replacements has had some additional costs associated with it, the overall effect on the U.S. economy actually has been minimal. Companies that produce refrigerators, air conditioners, insulating plastics, and other goods have adapted to using the new compounds. Some substitutes for CFC refrigerants are less energy-efficient, hence increasing energy consumption somewhat. But the conversions provide a market opportunity for innovative syntheses using green chemistry to produce environmentally benign substances.

Developing countries face another set of economic problems and priorities. CFCs have played an important role in improving the quality of life in the industrialized nations. Few would be willing to give up the convenience and health benefits of refrigeration or the comfort of air-conditioning. It is understandable that millions of people over the globe aspire to the lifestyle of the industrialized nations. As an example, over the past decade, the annual production of refrigerators in China has increased from 500,000 to over 8 million tons. But, if the developing nations are banned from using the relatively inexpensive CFC-based technology, they may not be able to afford alternatives. "Our development strategies cannot be sacrificed for the destruction of the environment caused by the West,"

asserts Ashish Kothari, a member of an Indian environmental group. Both India and China originally refused to sign the Montreal Protocol because they felt that it discriminated against developing countries. To gain the participation of these highly populated nations, the industrially developed nations created a special fund that is administered through the World Bank. The goal of the fund is to help countries phase out their use of ozone-depleting materials without stifling their economic development.

Consider This 2.31   Montreal Protocol Ratification

a. How many nations have currently ratified the Montreal Protocol?
b. Does that number include India and China? If so, in what year did they do so?
c. Have the signatories to the Montreal Protocol also ratified the subsequent amendments?
d. How many countries have *not* signed the Montreal Protocol?

Clearly, an understanding of chemistry is necessary to protect the ozone layer, but it is not sufficient. Chemists can help unravel the causes of ozone depletion and develop alternative materials to replace CFCs, but the debate among governments about how best to protect the stratospheric ozone layer continues in the global political arena.

Conclusion

Chemistry is intimately entwined with the story of ozone depletion. Chemists created the chlorofluorocarbons whose near-perfect properties only later revealed their dark side as predators of stratospheric ozone. Chemists worked internationally to discover the mechanism by which CFCs destroy stratospheric ozone and warned of the dangers of increased ultraviolet radiation reaching the Earth. And chemists will continue to synthesize the substitutes necessary to replace CFCs and other related compounds. But the issues involve more than just chemistry. At the 2005 meeting in Dakar, Senegal, of parties to the Montreal Protocol on Substances That Deplete the Ozone Layer, Executive Secretary Marco González reminded delegates that the final 20% of any global cooperative effort is often the hardest. Despite many challenges that arose at that meeting, the delegates were able to work constructively and cooperatively to achieve their short-term goals. There are concerns that fundamental differences in domestic regulatory approaches will deplete the stockpiles of goodwill more swiftly than those of controlled ozone-depleting substances, placing long-term goals in jeopardy.

The "Action on Ozone" report for 2000 from the Ozone Secretariat of the United Nations Environmental Program sums up the global experience with ozone depletion story in this manner:

> Perhaps the most important feature of the ozone regime is the way in which it has brought together an array of different participants in pursuit of a common end. Scientists have provided the information, with steadily increasing degrees of precision, on the causes and effects of ozone depletion. Industry, responding to the stimulus provided by the control measures, has developed alternatives far more rapidly and more cheaply than initially thought possible, and has participated fully in the debates over further phaseout. NGOs (nongovernmental organizations) and the media are the essential channels of communication, and education, with the peoples of the world in whose name the measures have been taken. . . . Governments have worked well together in patiently negotiating agreements acceptable to a range of countries with widely varying circumstances, aims, and resources—and showed courage and foresight in putting the precautionary principle into effect before the scientific evidence was entirely clear.

These are lessons to remember as we turn to our next topic, the chemistry of global warming.

Chapter Summary

Having studied this chapter, you should be able to:

- Differentiate between harmful ground-level ozone and beneficial stratospheric ozone layer (2.1)

- Describe the chemical nature of ozone, location of the ozone layer, and factors affecting its existence (2.1, 2.6, 2.8–2.10)

- Apply the basics of atomic structure to atoms of certain elements (2.2)

- Understand the organization of the periodic table (2.2)

- Relate an element's atomic number to its position in the periodic table (2.2)

- Differentiate atomic number from mass number and apply the latter to isotopes (2.2)

- Write Lewis structures using the octet rule for molecules with single, double, and triple covalent bonds (2.3)

- Given a Lewis structure, be able to identify the covalent bonds present in a molecule (2.3)

- Describe the electromagnetic spectrum in terms of frequency, wavelength, and energy (2.4, 2.5)

- Interpret graphs related to wavelength and energy, radiation and biological damage, and ozone depletion (2.4–2.8)

- Understand the natural Chapman cycle of stratospheric ozone depletion (2.6)

- Understand how the stratospheric ozone layer protects against harmful ultraviolet radiation (2.6, 2.7)

- Compare energies and biological effects of UV-A, UV-B, and UV-C radiation (2.6, 2.7)

- Discuss the interaction of radiation with matter and changes caused by such interactions, including biological sensitivity (2.6, 2.7)

- Relate the meaning and the use of the UV Index (2.7)

- Recognize the complexities of collecting accurate data for stratospheric ozone depletion and interpreting them correctly (2.8, 2.9)

- Understand the chemical nature and role of CFCs in stratospheric ozone depletion (2.9, 2.10)

- Explain the unique circumstances responsible for seasonal ozone depletion in the Antarctic (2.10)

- Summarize the scientific and political dimensions of the Montreal Protocol and its amendments (2.11, 2.12)

- Evaluate articles on green chemistry alternatives to stratospheric ozone-depleting compounds and recognize the effect that market forces have on the success of these innovations (2.12)

- Discuss the factors that will help lead to the recovery of the ozone layer (2.11, 2.12)

Questions

Emphasizing Essentials

1. The text states that the odor of ozone can be detected in concentrations as low as 10 ppb. Will you be able to detect the odor of ozone in either of these air samples?

    a. 0.118-ppm ozone, a concentration reached in the troposphere

    b. 25-ppm ozone, a concentration reached in the stratosphere

2. a. What is a Dobson unit?

    b. Does a reading of 320 DU or 275 DU indicate more total column ozone overhead?

3. How does ozone differ from oxygen in its chemical formula? In its properties?

4. Which of these pairs are allotropes?

    a. diamond and graphite

    b. water, H_2O, and hydrogen peroxide, H_2O_2

    c. white phosphorus, P_4, and red phosphorus, P_8

5. Where is the ozone layer found? Answer by giving a range of altitudes.

6. Assume there are 2×10^{20} CO molecules per cubic meter in a sample of tropospheric air. Furthermore, assume there are 1×10^{19} O_3 molecules per cubic meter at the point of maximum concentration of the ozone layer in the stratosphere.

    a. Which cubic meter of air contains the larger number of molecules?

    b. What is the ratio of CO to O_3 molecules in a cubic meter?

7. Using the periodic table as a guide, specify the number of protons and electrons in a neutral atom of each of these elements.

    a. oxygen (O)   b. nitrogen (N)

    c. magnesium (Mg)   d. sulfur (S)

8. Consider this periodic table.

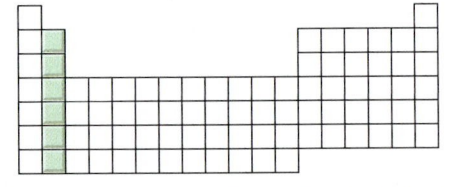

    a. What is the group number of the shaded column?

    b. Which elements make up this group?

    c. What is the number of electrons for a neutral atom of each element in this group?

    d. What is the number of outer electrons for a neutral atom of each element of this group?

9. Give the name and symbol for the element with this number of protons.

   a. 2   **b.** 19   **c.** 29

10. Give the number of protons, neutrons, and electrons in each of these.

    a. oxygen-18 ($^{18}_{8}O$)   **b.** sulfur-35 ($^{35}_{16}S$)

    c. uranium-238 ($^{238}_{92}U$)   **d.** bromine-82 ($^{82}_{35}Br$)

    e. neon-19 ($^{19}_{10}Ne$)   **f.** radium-226 ($^{226}_{88}Ra$)

11. Give the symbol showing the atomic number and the mass number for the element that has:

    a. 9 protons and 10 neutrons (used in nuclear medicine).

    b. 26 protons and 30 neutrons (the most stable isotope of this element).

    c. 86 protons and 136 neutrons (the radioactive gas found in some homes).

12. Write the Lewis structure for each of these atoms.

    a. calcium   **b.** nitrogen

    c. chlorine   **d.** helium

13. Assuming that the octet rule applies, write the Lewis structure for each of these molecules. Start by counting the number of available outer electrons. Write both the complete electron dot structure and the structure representing shared pairs with a dash, showing nonbonding electrons as dots.

    a. CCl_4 (carbon tetrachloride, a substance formerly used as a cleaning agent)

    b. H_2O_2 (hydrogen peroxide, a mild disinfectant; the atoms are bonded in this order: H-to-O-to-O-to-H)

    c. H_2S (hydrogen sulfide, a gas with the unpleasant odor of rotten eggs)

    d. N_2 (nitrogen gas, the major component of the atmosphere)

    e. HCN (hydrogen cyanide, a molecule found in space and a poisonous gas)

    f. N_2O (nitrous oxide, "laughing gas"; the atoms are bonded N-to-N-to-O)

    g. CS_2 (carbon disulfide, used to kill rodents; the atoms are bonded S-to-C-to-S)

14. Several different oxygen species are related to the story of ozone in the stratosphere. These include oxygen atoms, oxygen gas, ozone, and hydroxyl radicals. Compare and contrast the Lewis structure for each of these species.

15. Consider these two waves representing different parts of the electromagnetic spectrum.

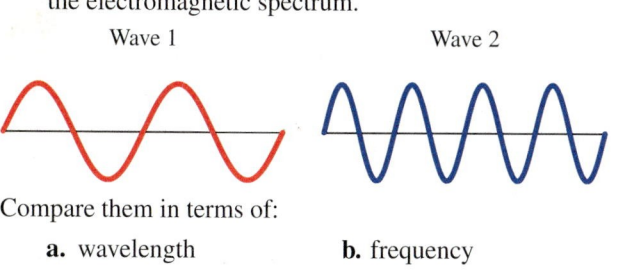

   Wave 1   Wave 2

   Compare them in terms of:

   a. wavelength   **b.** frequency

   c. forward speed

16. Use Figure 2.6 to specify the region of the electromagnetic spectrum where radiation of each wavelength is found. *Hint:* Change each wavelength to meters before making the comparison.

    a. 2.0 cm   **b.** 400 nm

    c. 50 μm   **d.** 150 mm

17. Arrange the wavelengths in question 16 in order of *increasing* energy. Which wavelength possesses the most energetic photons?

18. Arrange these types of radiation in order of *increasing* energy per photon: gamma rays, infrared radiation, radio waves, visible light.

19. The microwaves in home microwave ovens have a frequency of 2.45×10^9 s^{-1}. Is this radiation more or less energetic than radio waves? Than X-rays?

20. Ultraviolet radiation coming from the Sun is categorized as UV-A, UV-B, and UV-C. Arrange these three regions in order of their increasing:

    a. wavelength   **b.** energy

    c. potential for biological damage

21. Consider the Chapman cycle in Figure 2.9. Will this cycle take place in the troposphere as well as the stratosphere? Explain.

22. These free radicals all play a role in catalyzing ozone depletion reactions: Cl, NO_2, ClO, and HO.

    a. Count the number of outer electrons available and then draw a Lewis structure for each free radical.

    b. What characteristic is shared by these free radicals that makes them so reactive?

23. In Chapter 1, the role of nitrogen monoxide, NO, in forming photochemical smog was discussed. What role, if any, does NO play in stratospheric ozone depletion? Are NO sources the same in the troposphere and in the stratosphere?

24. **a.** How were the original measurements of increases in chlorine monoxide and the stratospheric ozone depletion over the Antarctic obtained?

    b. How are these measurements made today?

25. Which graph shows how measured increases in UV-B radiation correlate with percent reduction in the concentration of ozone in the stratosphere over the South Pole?

26. a. Can there be any H atoms in a CFC molecule?

b. What is the difference between an HCFC and an HFC?

27. a. Most CFCs are based either on methane, CH_4, or ethane, C_2H_6. Use structural formulas to represent these two compounds.

b. Substituting chlorine or fluorine (or both) for hydrogen atoms, how many different CFCs can be formed from methane?

c. Which of the substituted CFC compounds in part **b** has been the most successful?

d. Why weren't all of these compounds equally successful?

Concentrating on Concepts

28. The allotropes oxygen and ozone differ in molecular structure. What differences does this produce in their properties, uses, and significance?

29. Explain why is it possible to detect the pungent odor of ozone after a lightning storm or around electrical transformers.

30. The EPA has used the slogan "Ozone: Good Up High, Bad Nearby" in some of its publications aimed at the public. What message is this slogan trying to communicate?

31. How do *allotropes* of oxygen and *isotopes* of oxygen differ? Explain your reasoning.

32. Consider the Lewis structures for SO_2. How are they similar to or different from the Lewis structures for ozone?

33. It is possible to write three resonance structures for ozone, not just the two shown in the text. Verify that all three structures satisfy the octet rule and offer an explanation as to why the triangular structure is not reasonable.

34. The average length of an oxygen-to-oxygen single bond is 132 pm. The average length of an oxygen-to-oxygen double bond is 121 pm. What do you predict the oxygen-to-oxygen bond lengths will be in ozone? Will they all be the same? Explain your predictions.

35. Consider the graph in Figure 2.1 showing ozone concentrations at various altitudes.

a. What does this graph tell you about the concentration of ozone as you travel upward from the surface of the Earth? Write a brief description of the trends shown in the graph, describing the location of the ozone layer.

b. The *y*-axis in this graph starts at zero. Why doesn't the *x*-axis appear to start at zero?

36. Which of these forms of electromagnetic radiation from the Sun has the lowest energy and therefore the least potential for damage to biological systems? infrared radiation, ultraviolet radiation, visible radiation, radio waves

37. Even if you have skin with little pigment, why can't you get a suntan from standing in front of your radio in your living room or dorm room?

38. The morning newspaper reports a UV Index of 6.5. What should that mean to you as you plan your daily activities?

39. UV-C has the shortest wavelengths of all UV radiation and therefore the highest energies. All the reports of the damage caused by UV radiation focus on UV-A and UV-B radiation. Why is the focus of attention not on the damaging effects that UV-C radiation can have on our skin?

40. If all 3×10^8 tons of stratospheric ozone that are formed every day are also destroyed every day, how is it possible for stratospheric ozone to offer any protection from UV radiation?

41. Find the Material Safety Data Sheet (MSDS) for Freon-12. What does the MSDS say about the stability and toxicity of Freon-12? How are these properties related to both the usefulness and the problems associated with this compound?

42. Explain the significance of the information in this graph to a classmate who is not taking your course.

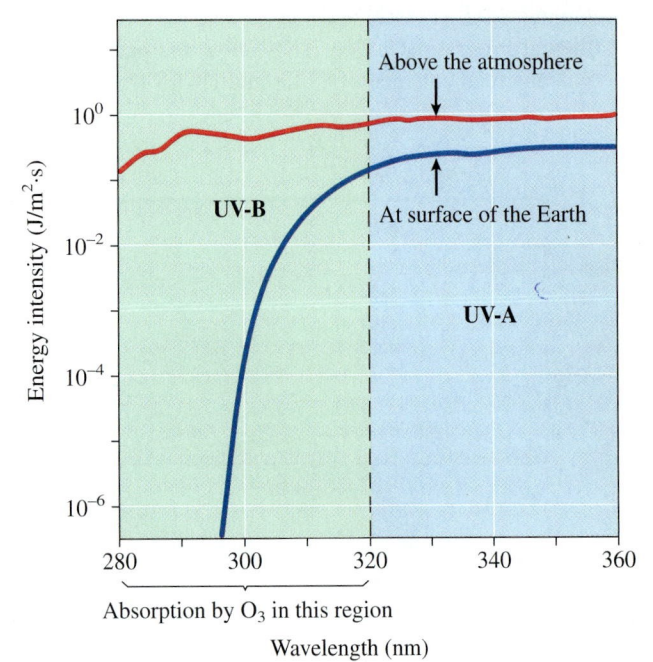

Source: Reprinted by permission of John E. Frederick, University of Chicago.

43. Explain how the small changes in ClO• concentrations (measured in parts per billion) can cause the much larger changes in O_3 concentrations (measured in parts per million).

44. Development of the stratospheric ozone hole has been most dramatic over Antarctica. What set of conditions exist over Antarctica that help to explain why this area is well-suited to studying changes in stratospheric ozone concentration? Are these same conditions not operating in the Arctic? Why or why not?

45. Prepare a graph to communicate this information about HCFC phaseout requirements. EPA has been directed

by the U.S. Congress to make reductions based on data for 1999, when 15,240 metric tons of HCFCs were produced or imported for use in the United States.

Year	2004	2010	2015	2020	2030
Percent Reduction	35	65	90	99.5	100

46. The free radical $CF_3O\cdot$ is produced during the decomposition of HFC-134a.

    a. Propose a Lewis structure for this free radical.

    b. Offer a possible reason why this free radical does not cause ozone depletion.

47. One of the mechanisms that helps to break down ozone in the Antarctic region involves the $BrO\cdot$ free radical. Once formed, it reacts with $ClO\cdot$ to form BrCl and O_2. BrCl in turn reacts with sunlight to break into $Cl\cdot$ and $Br\cdot$, both of which react with O_3 and form O_2.

    a. Represent this information with a set of equations similar to those shown for the Chapman cycle.

    b. What is the net equation for this cycle?

48. Consider these graphs from the World Meteorological Organization (WMO) and the United Nations Environmental Program (UNEP). They show the pattern of atmospheric abundance of CFCs and HCFS from 1950 to 2100.

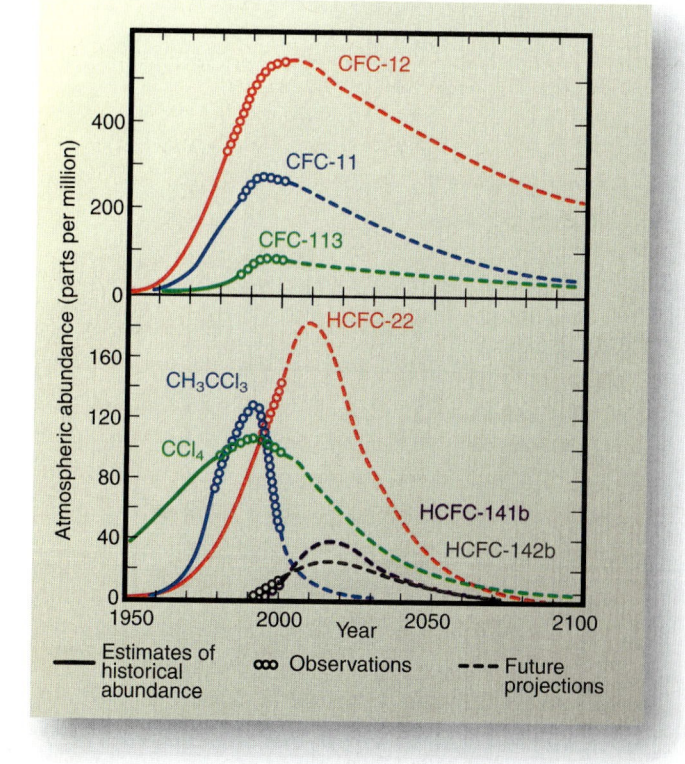

Source: *Scientific Assessment of Ozone Depletion: 2002,* World Meteorological Organization, United Nations Environmental Program.

    a. Compare the peaking patterns for CFCs and for HCFCs. Offer possible reasons for any similarities or differences.

    b. Compare the peaking patterns for CFCs and for CCl_4 and CH_3CCl_3. Offer possible reasons for any similarities or differences.

    c. Which halogen-containing compound has shown the largest reduction? Explain.

49.  One of the most striking series of images of Antarctic ozone depletion covered the period from 1979 to 1996 and was known as "Purple Octobers." Find such a series of images on the Web and explain what information they convey.

Exploring Extensions

50. What are some of the reasons that the solution to ozone depletion proposed in this Sydney Harris cartoon will not work?

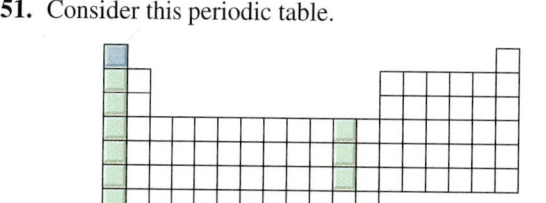

"OH, FOR PETE'S SAKE, LET'S JUST GET SOME OZONE AND SEND IT BACK UP THERE!"

Source: Reprinted with permission, www.ScienceCartoonsPlus.com.

51. Consider this periodic table.

Two groups are highlighted; the first is Group 1A and the second is Group 1B. The text states that although A groups have very regular patterns, the "situation gets a bit more complicated with the B groups." Use other resources to find out which of these predictions becomes more complicated.

    a. number of electrons for the elements in each group

    b. number of outer electrons for the elements of each group

    c. formula when each element combines chemically with chlorine

52. Resonance structures can be used to explain the bonding in charged groups of atoms as well as in neutral molecules, such as ozone. The nitrate ion, NO_3^-, has one additional electron plus the outer electrons contributed by nitrogen and oxygen atoms. That extra electron gives the ion its charge. Draw the resonance structures, verifying that each obeys the octet rule.

53. Although oxygen exists as O_2 and O_3, nitrogen exists only as N_2. Propose an explanation for these facts. *Hint:* Try drawing a Lewis structure for N_3.

54. It has been suggested that the term *ozone screen* would be a better descriptor than *ozone layer* to describe ozone in the stratosphere. What are the advantages and disadvantages to each term?

55. Many different types of ozonators are on the market for sanitizing air, water, and even food. They are often sold with a slogan such as this one from a pool store. "Ozone, world's most powerful sanitizer!"

    a. Find out how these devices work.

    b. What claims are made for ozonators intended to purify air?

    c. What claims are made for ozonators designed to purify water?

    d. How do medical ozonators differ from other models?

56. The effect a chemical substance has on the ozone layer is measured by a value called its *ozone-depleting potential,* ODP. This is a numerical scale that estimates the lifetime potential stratospheric ozone that could be destroyed by a given mass of the substance. All values are relative to CFC-11, which has an ODP defined as equal to 1.0. Use those facts to answer these questions.

    a. What factors do you think will influence the ODP value for a chemical? Why?

    b. Most CFCs have ODP values ranging from 0.6 to 1.0. What range do you expect for HCFCs? Explain your reasoning.

    c. What ODP values do you expect for HFCs? Explain your reasoning.

57. Recent experimental evidence indicates that $ClO\cdot$ initially reacts to form Cl_2O_2.

    a. Predict a reasonable Lewis structure for this molecule. Assume the order of atom linkage is Cl-to-O-to-O-to-Cl.

    b. What effect does this evidence have on understanding the mechanism for the catalytic destruction of ozone by $ClO\cdot$?

58. Chemical formulas for individual CFCs, such as CFC-11 (CCl_3F), can be figured out from their code numbers. A quick way to interpret the code number for CFCs is to add 90 to the number. In this case, $90 + 11 = 101$. The first number in this sum is the number of carbon atoms, the second is the number of hydrogen atoms, and the third is the number of fluorine atoms. CCl_3F has one carbon, no hydrogen, and one fluorine atom. All remaining bonds are assumed to be chlorine until carbon has the required four single covalent bonds to satisfy the octet rule.

    a. What is the chemical formula for CFC-12?

    b. What is the code number of CCl_4?

    c. Will this "90" method work for HCFCs? Use HCFC-22, which is $CHClF_2$, to explain your answer.

    d. Will this method work for halons? Use Halon-1301, which is CF_3Br, to explain your answer.

59. The graph shows the atmospheric abundance of bromine-containing gases from 1950 to 2100.

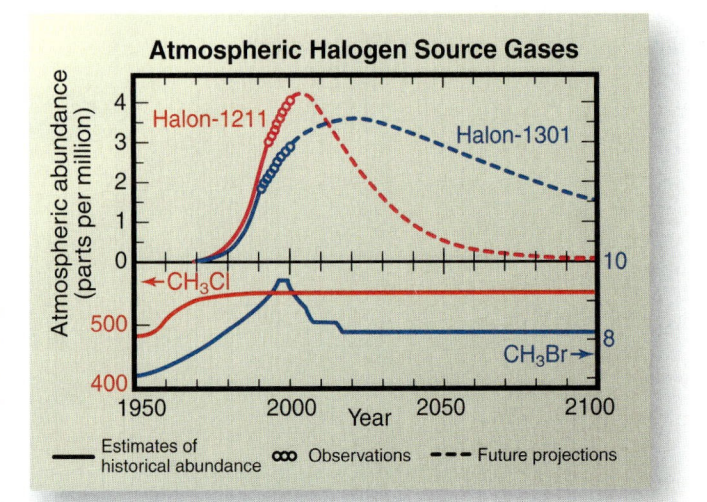

Source: *Scientific Assessment of Ozone Depletion: 2002,* World Meteorological Organization (WMO), United Nations Environmental Programme (UNEP), p. 29 in "Twenty Questions and Answers About the Ozone Layer."

    a. Compare the patterns for Halon-1211 and Halon-1301. Explain why their concentrations are not peaking at the same time.

    b. Are both Halon-1211 and Halon-1301 still legally manufactured and used in the United States? In Mexico?

    c. Explain why the level of CH_3Br is predicted to remain high and constant through 2100.

60. a. What factors account for the fact that Australia has the highest incidence of skin cancer in the world?

    b. Why is their government actively involved in changing this statistic?

    c. What facts are being stressed in the public education campaign?

    d. Is the rate of skin cancer the same for migrants coming to Australia as it is for white Australians?

    e. Is the rate the same for Australian Aborigines as it is for white Australians? Explain.

Chapter 3

The Chemistry of Global Warming

Dawn strikes the mountains rising above St. Mary's Lake in Montana's Glacier National Park. When the park was created in 1910, it had 150 glaciers. Now there are 27. If warming trends continue, all those glaciers are likely to disappear in the next 25 years.

"There are some things that are absolutely incontrovertible...: that greenhouse gases are increasing, that they are increasing because of human activity, that the planet is actually getting warmer, and that some part of that warming is due to greenhouse gases."

Gavin Schmidt, Climate Scientist
Goddard Institute for Space Studies

Global warming is a popular term used to describe the increase in average global temperatures. What is the scientific evidence supporting this phenomenon? What role does chemistry play in understanding global warming? What effects are linked to the observed increase in Earth's average temperature? What is the role of human activity in producing global warming? Why are the temperature changes more pronounced in the Antarctic and Arctic? To answer these and other questions, we need to understand a bit about how Earth's climate system is regulated and responds to change. There are many factors to consider, including incoming and outgoing solar radiation, wind and water currents, the role of atmospheric gases, clouds, snow and ice, and atmospheric haze. We also need to consider the rate at which temperature change is taking place and if the climate system can respond at a similar rate. This chapter will help you to understand and connect all of these considerations.

Carbon dioxide is a major player in the debate about global warming, yet the reason is far from obvious. After all, CO_2 is an essential component of the atmosphere, a gas that all animals exhale and green plants absorb. Central to understanding global warming is studying the molecular mechanism by which CO_2 and other compounds absorb the infrared radiation emitted by the planet, helping to keep it warm. Some knowledge of molecular structure and shape is necessary to understand this mechanism. Global warming has a significant quantitative component; we need numbers to help assess the seriousness of the situation. Recognizing that global warming has international implications, we will look at parallels between responses to protecting the ozone layer (the Montreal Protocols) and to slowing global warming (the Kyoto Conference Protocols). Current policies for restricting emissions of CO_2 and other gases implicated in global warming will be examined. Developing understanding of these issues will lead us on a journey into the realm of chemical knowledge and its connections with public policy around the world.

Consider This 3.1   Shrinking Glaciers Worldwide

Many of the world's freshwater glaciers are shrinking. Search or use the direct links provided at the *Online Learning Center* to learn about the activity of glaciers in two parts of the world outside of the continental United States.

a. Where is each glacier located?
b. What changes are taking place with each glacier?
c. Will the effects of these changes be the same at each location?

3.1   In the Greenhouse: Earth's Energy Balance

The brightest and most beautiful body in the night sky, after our own moon, is considered by many to be Venus (Figure 3.1). It is ironic that the planet named for the goddess of love is a most unlovely place by earthly standards. Spacecraft have revealed a desolate, eroded surface with an average temperature of about 450 °C (840 °F). The beautiful blue-green ball we inhabit has an average annual temperature of 15 °C (59 °F). The atmosphere surrounding Venus has a pressure 90 times greater than that of Earth, and it is 96% carbon dioxide, with clouds of sulfuric acid. It makes the worst smog-bound day anywhere on Earth seem like a breath of fresh country air.

The point of this little astronomical digression is that both Venus and Earth are warmer than one would expect based solely on their distances from the Sun and the amount of

Figure 3.1

Computer-generated image of Maat Mons, a volcano on Venus. The image is based on radar data collected by the U.S. space probe *Magellan*. The vertical scale has been exaggerated 10-fold.

solar radiation they receive. If distance were the *only* determining factor, the temperature of Venus would average approximately 100 °C, the boiling point of water. Earth, on the other hand, would have an average temperature of −18 °C (0 °F), and the oceans would be frozen year-round.

The idea that Earth's atmospheric gases might somehow be involved in trapping some of the Sun's heat was first proposed around 1800 by the French mathematician and physicist, Jean-Baptiste Joseph Fourier (1768–1830). Fourier compared the function of the atmosphere to that of the glass in a "hothouse" (his term), what we would call today a greenhouse. Although he did not understand the mechanism or know the identity of the gases responsible for the effect, his metaphor has persisted. Some 60 years later, the Irish physicist John Tyndall (1820–1893) experimentally demonstrated that carbon dioxide and water vapor absorb heat radiation. In addition, he calculated the warming effect that would result from the presence of these two compounds in the atmosphere. In the 1890s, Swedish scientist Svante Arrhenius (1859–1927) considered the potential problems that could be caused by CO_2 building up in the atmosphere. Observed warming of surface air temperatures between the 1890s and 1940 led some scientists to suggest that the American Dust Bowl was an early sign of the greenhouse effect.

U.S. oceanographer Roger Revelle (1909–1991) suggested in 1957 that ever-increasing amounts of **greenhouse gases,** those gases capable of absorbing and reemitting infrared radiation to the atmosphere, could cause rising temperatures. Since that time, there has been a steady increase in the amount and reliability of data gathered about the role that CO_2 and other gases play in global warming. We know that molecules of CO_2 absorb heat. We know that the concentration of CO_2 in the atmosphere has increased over the past 150 years, and we know that Earth's average temperature has not remained constant. As we move through the chapter, we will investigate how these observations are interrelated and what other factors come into play.

You may not have personally experienced the warmth of a greenhouse nurturing your seedlings or prize tropical plants on a cold winter's day. Almost everyone, however, has had the experience of returning to a car after it has been sitting closed in direct sunlight. The windows of the car allow visible and a relatively small amount of ultraviolet light from the Sun to pass through into the car. Energy is absorbed by the interior of the car, particularly by dark fabrics and surfaces. Some of that energy is reemitted as longer wavelength infrared radiation (IR), but these wavelengths cannot escape back through the windows. The heat builds up in the car until, when you return,

"Dust Bowl" describes a period of severe drought in the Midwest and southern plains during the 1930s.

Water vapor is the most abundant greenhouse gas in our atmosphere. However, contributions of H_2O from human activity are negligible compared with natural sources.

the meaning of the term *hothouse* is very clear. In sunny climates, temperatures in a closed car can quickly exceed 49 °C (120 °F). No person or pet should be left in cars under these conditions.

Your Turn 3.2   **Wavelength and Energy Relationships**

Consider these three types of radiant energy from the electromagnetic spectrum: infrared, ultraviolet, and visible.

   a. Arrange them in order of *increasing* wavelength.
   b. Arrange them in order of *increasing* energy.
   c. How are these arrangements related to the Sun's ability to heat a closed car? Explain your reasoning.

Answers
   a. ultraviolet, visible, infrared
   b. infrared, visible, ultraviolet

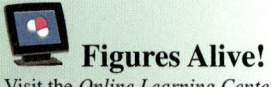

Figures Alive!
Visit the *Online Learning Center* to learn more about the electromagnetic spectrum, Earth's energy balance, and the greenhouse effect. Look for the **Figures Alive!** icon elsewhere in this chapter.

Is the process of heat building up in your car different from heat building up in Earth's greenhouse, its atmosphere? There are many similarities and overall, the energy exchange between our Earth and its atmosphere is both natural and beneficial, helping to maintain the existence of life on our planet. Without the protective layer of our atmosphere, Earth could become very hot if it received all the incoming radiation from the Sun. However, without the atmosphere's ability to reflect Earth's radiated heat back toward the surface, our lovely orb could become an ice planet because of the direct loss of heat into space. The current average temperature of our planet, about 15 °C (59 °F), is about 33 °C warmer than what would be expected from its distance from the Sun. It is also much higher than the −270 °C of outer space. Consequently, the Earth acts overall like a global radiator, radiating heat to its frigid surroundings. Figure 3.2 is a schematic representation of our Earth's energy balance.

Several important relationships are shown in Figure 3.2. Energy from the Sun to Earth is absorbed by the atmosphere (23%) and by Earth's continents and oceans (46%), warming them. Some of the incoming energy (25%) is reflected from the molecules, dust, and aerosol particles that make up our envelope of air or from the Earth's surface (6%). These processes account for 100% of the incoming radiation from the Sun. Earth, in turn, radiates some of its absorbed energy back into the atmosphere (37%), where greenhouse gases such as H_2O and CO_2 are very efficient absorbers of this longer wavelength IR radiation. Much of this heat is redirected and comes back through the lower regions of our atmosphere toward the Earth, rather than being directly lost to space. Heat is transferred by collisions between neighboring molecules, and these molecules are found in greater abundance in the denser regions of the lower atmosphere. A small percentage of the absorbed terrestrial radiation goes directly into space from the surface (9%).

Figure 3.2 also illustrates that of the 46% of incoming solar energy that is absorbed by the Earth, 37% is absorbed in the atmosphere when Earth radiates longer wavelength heat energy. Dividing 37 by 46 and changing to percent, it is easy to calculate that about 80% of incoming solar radiation striking the Earth remains in the atmosphere and does not directly escape into space. This is known as the **greenhouse effect,** the process by which atmospheric gases trap and return a major portion of the heat (infrared radiation) radiated by the Earth. Because of the constant, dynamic exchange between Earth, its atmosphere, and space, a steady state is established, with a more or less constant average terrestrial temperature being the result.

Another steady-state process, the Chapman cycle, was discussed in Section 2.6.

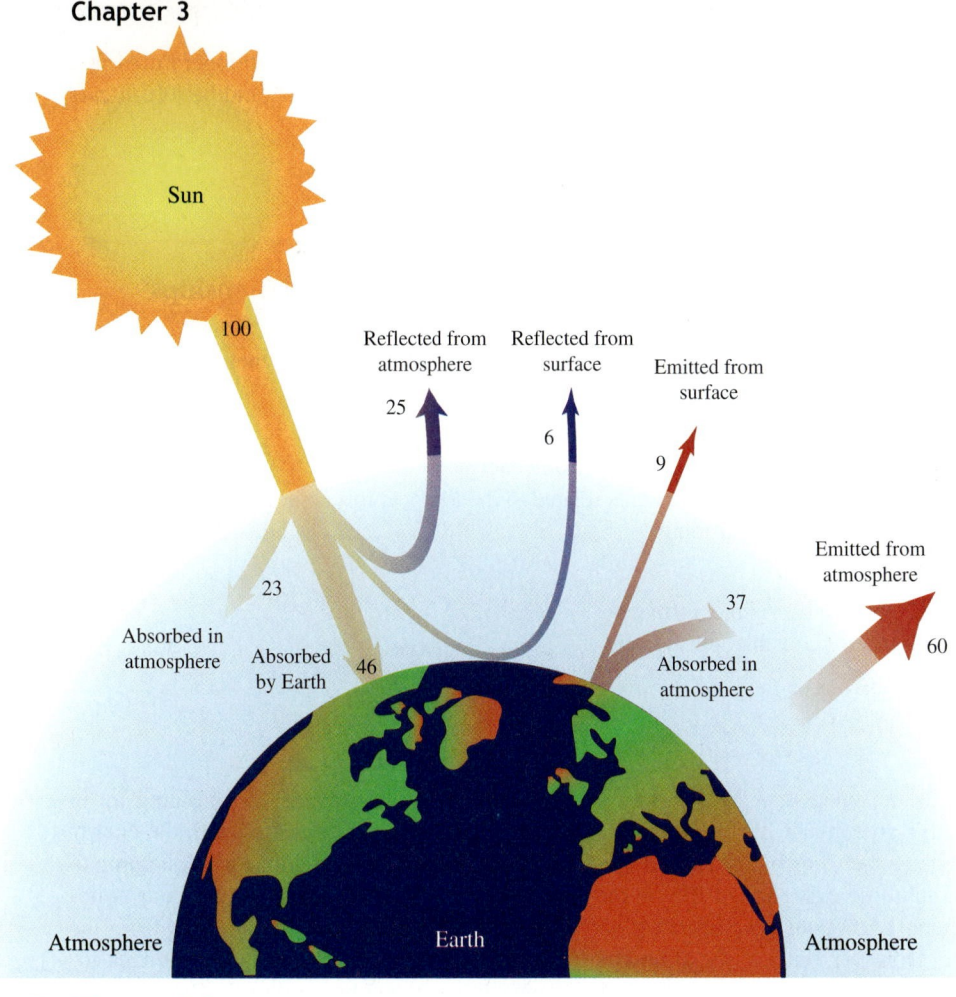

Figure 3.2

Earth's energy balance by percent. Yellow represents a mixture of wavelengths. Shorter wavelengths of radiation are shown in blue, longer in red.

Your Turn 3.3   Earth's Energy Balance

a. As seen in Figure 3.2, processes of absorption and reflection account for 100% of incoming solar radiation. Show that energy from the Earth into space also sums to 100%, required for energy balance.

b. What total percentage of solar radiation is either directly absorbed in the atmosphere or absorbed after being radiated from Earth's surface? How does that balance with the percentage of energy emitted from the atmosphere?

c. Explain the meaning of the different colors used for radiation.

Consider This 3.4   Science Fiction Story

Successful writers of science fiction sometimes begin their careers as science majors. Their best work reveals a sound understanding of scientific phenomena and principles. Often a good science fiction story assumes a slightly different scientific reality than the one we know. For example, *Dune,* by Frank Herbert, takes place on a desert planet.

    Here is an opportunity to exercise your imagination in a different climate. Make the assumption that the planet has an average temperature of −18 °C (0 °F). What would human life be like? Write a brief description of a day on a frozen planet. Residents of northern climates should have a great advantage here.

Obviously, the greenhouse effect is essential in keeping our planet habitable. However, if having some greenhouse gases in the atmosphere is a good thing, having more is not necessarily better. The term **enhanced greenhouse effect** refers to the process in which atmospheric gases trap and return *more than* 80% of the heat energy radiated by the Earth. An increase in the concentration of infrared absorbers will very likely mean that even more than 80% of the radiated energy will be returned to Earth's surface, with an attendant increase in average temperature. Back in 1898, Arrhenius estimated the extent of this effect. He calculated that doubling the concentration of CO_2 would result in an increase of 5–6 °C in the average temperature of the planet's surface. Writing in the *London, Edinburgh, and Dublin Philosophical Magazine* to announce his findings, Arrhenius dramatically described the phenomenon: "We are evaporating our coal mines into the air." At the end of the 19th century, the Industrial Revolution was already well under way in Europe and America, and it was "picking up steam" as well as generating it (and CO_2 also).

Consider This 3.5   Evaporating Coal Mines

Although the Arrhenius statement about "evaporating our coal mines into the air" certainly was effective in grabbing attention in 1898, what process do you think he really was referring to in discussing the amount of CO_2 being added to the air? Explain your reasoning.

3.2 Gathering Evidence: The Testimony of Time

In the 4.5 billion years that our planet has existed, its atmosphere and climate have varied widely. Evidence from the composition of volcanic gases suggests the concentration of carbon dioxide in Earth's early atmosphere was perhaps 1000 times greater than it is today. Much of the CO_2 that dissolved in the oceans became incorporated in rocks such as limestone, which is calcium carbonate, $CaCO_3$. High concentrations of carbon dioxide all those years ago also made possible a significant event in the history of our planet—the development of life on Earth. Even though the Sun's energy output was 25–30% less than it is today, the ability of CO_2 to trap heat kept Earth sufficiently warm to permit life to develop. As early as 3 billion years ago, the oceans were filled with primitive plants such as cyanobacteria (blue-green bacteria). Like their more sophisticated descendants, these simple plants were capable of photosynthesis. They were able to use chlorophyll to capture sunlight and use this energy to combine carbon dioxide gas and water, forming more complex molecules such as glucose and releasing oxygen (equation 3.1).

$$6\,CO_2 + 6\,H_2O \xrightarrow{\text{chlorophyll}} \underset{\text{glucose}}{C_6H_{12}O_6} + 6\,O_2 \qquad [3.1]$$

Photosynthesis dramatically reduced the concentration of atmospheric CO_2 and increased the amount of O_2 present. The microbiologist Lynn Margulis has called this "the greatest pollution crisis the Earth has ever endured." We, and all past and future generations, are the unknowing beneficiaries of this long-ago pollution crisis. The increase in oxygen concentration helped make possible the evolution of animals. But even 100 million years ago, in the age of dinosaurs and well before humans walked the Earth, the average temperature is estimated to have been 10–15 °C warmer than it is today, and the CO_2 concentration is assumed to have been considerably higher.

How do we know such numbers? Deeply drilled cores from the ocean floor give us a slice through time. The number and nature of the microorganisms present at any particular level provide one indication of the temperature when they lived. Supplementing this, the alignment of the magnetic field in particles in the sediment provides an independent measure of time. Other relevant information comes from the analysis of ice cores. Starting in

Figure 3.3

Scientists use data from ice cores to determine changes in temperature and levels of carbon dioxide over time.

Isotopes of hydrogen were discussed in Section 2.2.

1957, the Russian Federation's drilling project at the Vostok Station in Antarctica yielded over a mile of ice cores taken from the snows of 160 millennia. The Antarctic Treaty, signed in 1959, reserves the region beyond 60° south latitude for peaceful scientific purposes with international cooperation. Japanese scientists announced in 2006 that they have drilled more than 3 km into Antarctica's ice sheet, coming up with million-year-old ice core samples. Research carried out in the Antarctic led to the conclusion that the concentrations of carbon dioxide and methane are far higher now than at any time in the last 800,000 years. Figure 3.3 shows drilling for these icy record keepers of the past.

Ice cores provide data for estimating past temperatures because of the isotopes of hydrogen found in the frozen water. Water molecules containing the most abundant form of hydrogen atoms, ^{1}H, are lighter than those that contain deuterium, ^{2}H. The lighter H_2O molecules evaporate just a bit more readily than the heavier ones. As a result, there is more ^{1}H and less ^{2}H in the water vapor of the atmosphere, compared with the amounts in the oceans. However, the heavier H_2O molecules in the atmosphere condense just a bit more readily than the lighter ones. Therefore, snow that condenses from atmospheric water vapor will be enriched in ^{2}H. The degree of enrichment is dependent on temperature. The ratio of ^{2}H to ^{1}H in the ice core can be measured and used to estimate the temperature at the time the snow fell.

Both carbon dioxide and temperature data are incorporated in Figure 3.4. The upper curve, with its concentration scale on the left, is a plot of parts per million of carbon dioxide in the atmosphere versus time over a span of 160,000 years. The lower plot and the left-hand scale indicate how the average global temperature has varied over the same period. For example, the figure shows that 20,000 years ago, during the last ice age, the average temperature of Earth was about 9 °C below the 1950–1980 average. At the other extreme, a maximum temperature (just over 16 °C) occurred approximately 130,000 years ago.

Particularly striking in Figure 3.4 is that temperature values and CO_2 concentrations follow the same pattern. When the CO_2 concentration was high, the temperature was high. Other measurements show that periods of high temperature also have been characterized by high atmospheric concentrations of methane (CH_4). Such correlations do not necessarily *prove* that elevated atmospheric concentrations of CO_2 and CH_4 *caused* the temperature increases. Presumably, the converse could have taken place. But both these compounds trap heat, and without doubt, they can and do contribute to global warming.

To be sure, other mechanisms also are involved in the periodic fluctuations of global temperature. Some propose that Earth's climate can sometimes behave "more like a switch

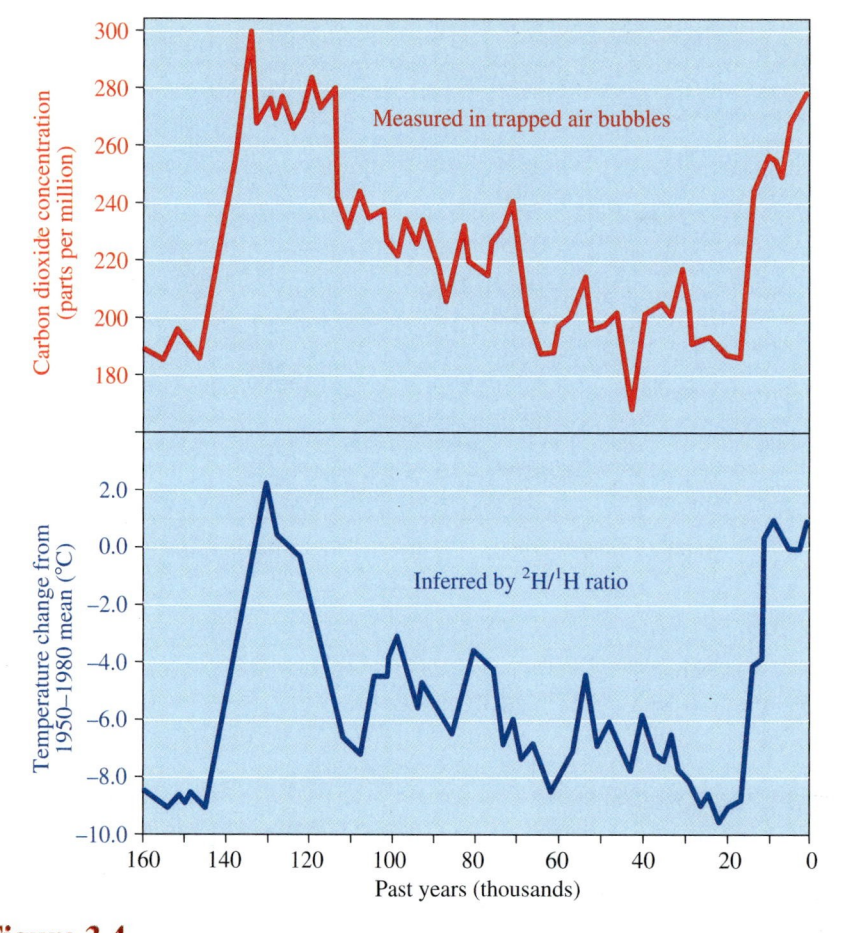

Figure 3.4

Atmospheric CO_2 concentration (red) and temperature change from the 1950–1980 mean (blue) over 160,000 years (ice core data).

than a dial," with abrupt changes taking place over a relatively short period. Temperature maxima seem to come at roughly 100,000-year intervals, with interspersed major and minor ice ages. Over the past million years, Earth has experienced 10 major periods of glacier activity and 40 minor ones. Some of this temperature variation probably is caused by minor changes in Earth's orbit that affect the distance from Earth to the Sun and the angle at which sunlight strikes the planet. However, this hypothesis cannot fully explain the observed temperature fluctuations. Orbital effects most likely are coupled with terrestrial events such as changes in reflectivity, cloud cover, airborne dust, and CO_2 and CH_4 concentration. These factors can diminish or enhance the orbit-induced climatic changes. The feedback mechanism is complicated and not well understood. One thing is clear: Earth is a far different place now than it was at the time of our last temperature maximum 130,000 years ago. Our ancestors had discovered fire by then, but they had not learned to exploit it as we have.

The past has provided its testimony, but more recent trends in atmospheric CO_2 and average global temperatures are important for assessing the current status of the greenhouse effect. There is compelling evidence that CO_2 concentrations have increased significantly in the past century. The best data are those acquired at Mauna Loa in Hawaii. Figure 3.5 presents values from Antarctic ice cores taken from 1860 to the 1950s, and then adds the Mauna Loa data to show the continuation of the trend. The series of vertical lines from 1960 until 2005 indicates seasonal variation, but the general increase in average annual values from 315 ppm in 1960 to about 385 ppm in 2005 is clear. Later in this chapter we will examine why scientists believe that much of the added carbon dioxide has come from the burning of fossil fuels. We also will discuss future projections of changes in CO_2 based on computer modeling.

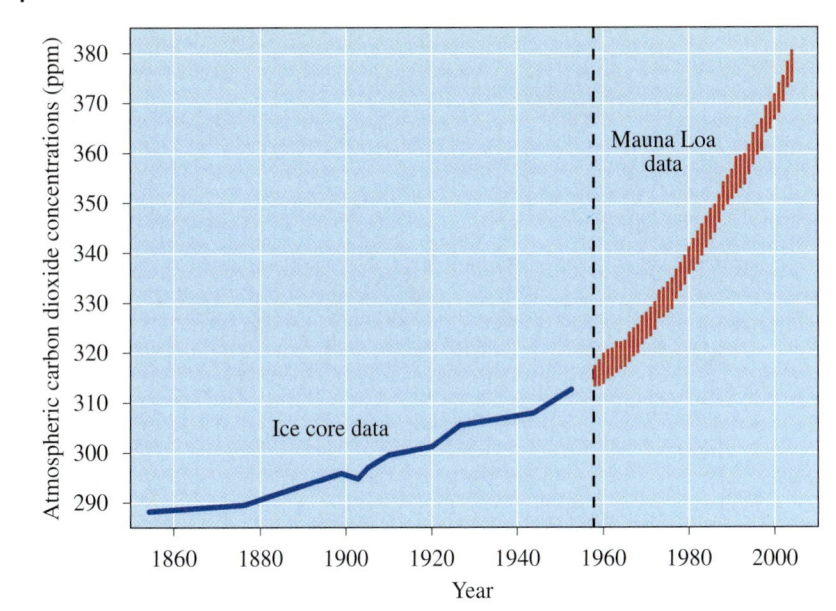

Figure 3.5

The rise in atmospheric levels of carbon dioxide between 1855 and 2005.

Source: From "The Greenhouse Effect and Historical Emissions," figure 4. Taken from http://clinton2.nara.gov/Initiatives/climate/greenhouse.html.

Consider This 3.6   The Cycles of Mauna Loa

The pattern observable in Figure 3.5 after 1960 is due to "seasonal variation."

a. Estimate the variation in parts per million (ppm) CO_2 within each year.
b. Explain these seasonal variations in CO_2 concentrations.

Sceptical Chymist 3.7   Checking the Facts on CO_2 Increases

a. A recent government report states that the atmospheric level of CO_2 has increased 30% since 1860. Use the data in Figure 3.5 to either prove or disprove this statement.
b. A global warming skeptic states that the percent increase in the atmospheric level of CO_2 since 1957 has been only about half as great as the percent increase from 1860 to the present. Comment on the accuracy of that statement and how it could affect policy on global warming.

Other measurements indicate that during the past 120 years or so, the average temperature of the planet has increased somewhere between 0.4 and 0.8 °C. Figure 3.6 shows the changes in the air temperature at Earth's surface from 1855 to 2005. Some scientists correctly point out that a century or two is an instant in the 4.5-billion-year history of our planet. They caution restraint in reading too much into short-term temperature fluctuations. In fact, although some areas such as in Alaska and northern Eurasia have warmed by up to 6 °C, cooling has occurred in the North Atlantic and in the central North Pacific. Short-term changes in atmospheric circulation patterns are thought to cause some observed temperature anomalies.

Figure 3.6 also shows the variability in temperatures from year to year, as well as the longer term trends. Many scientists conclude that the global temperature climbed upward only since about 1970. From Figure 3.6 we see that the average temperature of

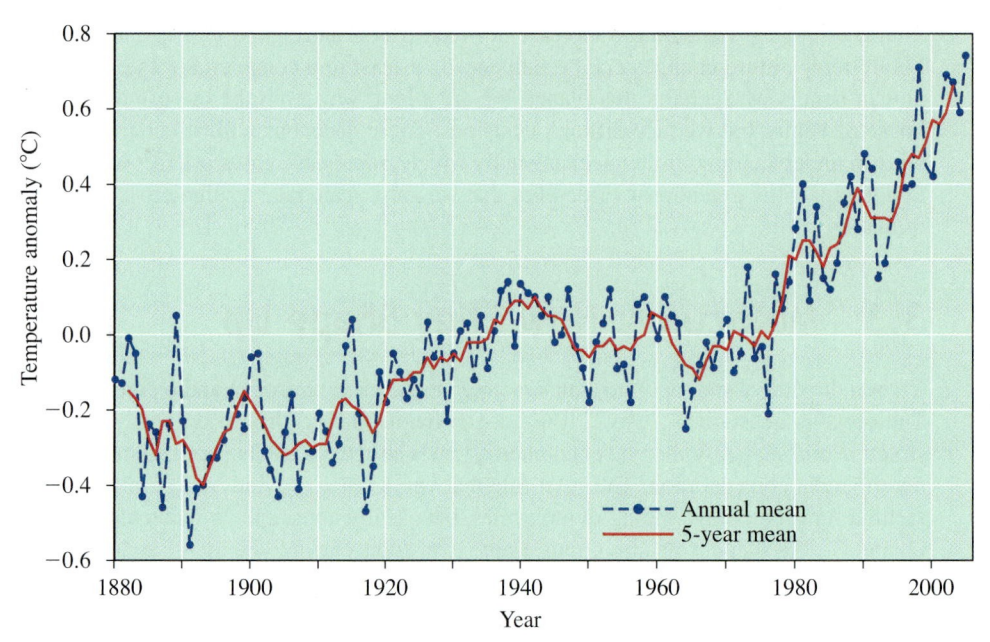

Figure 3.6

Global surface temperature change (1880–2005).

Source: http://www.data.giss.nasa.gov

the Earth is about 0.6 °C higher now than it was in 1880. Whether this temperature increase is a consequence of the increased CO_2 concentration cannot be concluded with absolute certainty. Nevertheless, experimental evidence implicates carbon dioxide from human-related sources as a cause of recent global warming.

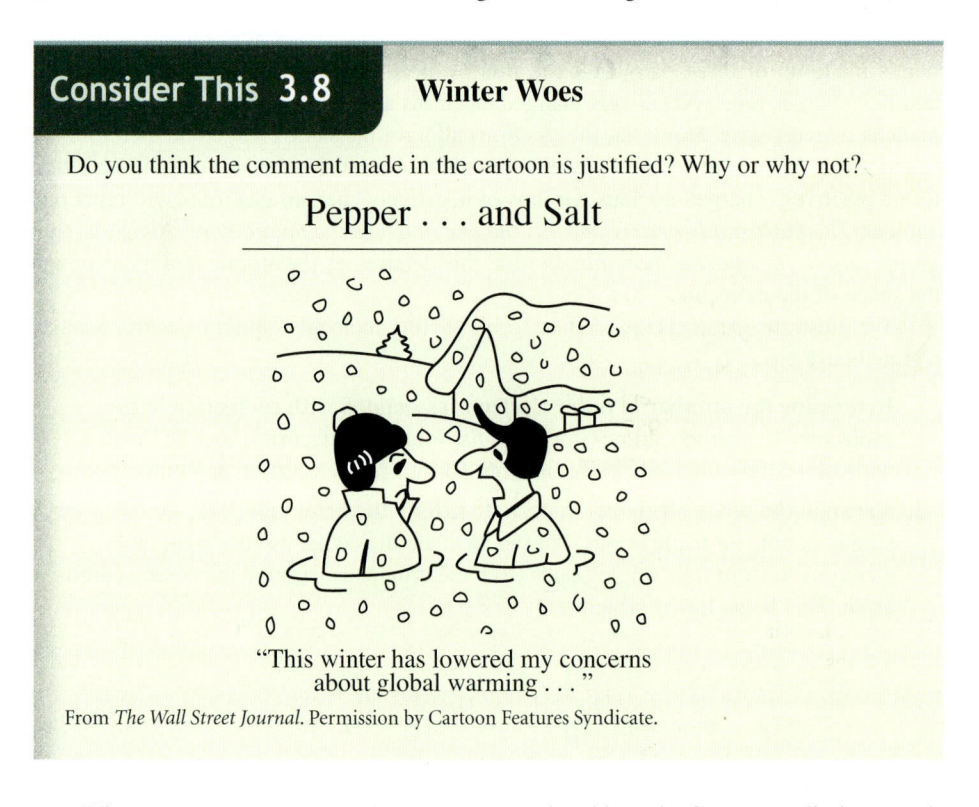

When temperature measurements are extrapolated into the future, predictions made by Arrhenius of a 5–6 °C rise in the average temperature of the planet's surface may need to be revised. Current estimates from the United Nations predict that average temperatures will increase somewhere between 1.4 °C and 5.8 °C (2.5 °F and 10.4 °F) by the year 2100. Other scientists, looking at a possible doubling of CO_2 emissions in the

future, estimate a temperature increase between 1.0 °C and 3.5 °C (1.8 °F and 6.3 °F). Future temperature changes can be influenced, at least to a considerable extent, by the human beings who inhabit this planet. We are a long way from the out-of-control hot-house of Venus, but we face difficult decisions. These decisions will be better informed with an understanding of the mechanism by which greenhouse gases interact with radiation to create the greenhouse effect. For that we must again take a submicroscopic view of matter.

3.3   Molecules: How They Shape Up

Carbon dioxide, water, and methane are greenhouse gases; nitrogen and oxygen are not. The obvious question is "why?" The not-so-obvious answer has to do with molecular structure and shape. When you encountered Lewis structures in Chapter 2, geometry was not the main consideration. The octet rule that you learned provides a generally reliable method for predicting bonding in molecules, but usually not shape. In molecules such as O_2 and N_2, the shape is unambiguous, as the two atoms can only be in a straight line.

$$:N:::N: \quad \text{or} \quad :N\equiv N: \quad \text{or} \quad N\equiv N$$

$$\ddot{O}::\ddot{O} \quad \text{or} \quad \ddot{O}=\ddot{O} \quad \text{or} \quad O=O$$

Different shapes become possible with molecules of more than two atoms. Fortunately, knowing where the outer electrons are located within a molecule provides insight into molecular shape. Therefore, the first step in predicting molecular shape is to write the Lewis structure for the molecule. If the octet rule is obeyed throughout the molecule, each atom (except hydrogen) will be associated with four pairs of electrons. Some molecules include nonbonding lone-pair electrons, but all molecules contain some bonding electrons or they would not be molecules!

Bonding electrons can be paired to form single bonds. In other molecules, the bonding electrons are involved in double bonds consisting of two pairs of electrons, or in triple bonds made up of three pairs. A basic rule of electricity is that unlike charges attract and like charges repel. Negatively charged electrons are attracted to a positively charged nucleus in every case. However, the electrons all have the same charge and therefore are found as far from each other in space as possible while still maintaining their attraction to the positively charged nucleus. Groups of negatively charged electrons will repel one another. *The most stable arrangement is the one in which the mutually repelling electron groups are as far apart as possible.* In turn, this determines the atomic arrangement and the shape of the molecule.

We illustrate a stepwise procedure for predicting molecular structure with methane, a greenhouse gas.

1. **Determine the number of outer electrons associated with each atom in the molecule.** The carbon atom (Group 4A) has four outer electrons; each of the four hydrogen atoms contributes one electron. This gives $4 + (4 \times 1)$, or 8 outer electrons.

2. **Arrange the outer electrons in pairs to satisfy the octet rule.** This may require single, double, or triple bonds. For the methane molecule, use the eight outer electrons to form four single bonds (four electron pairs) around the central carbon atom. This is the Lewis structure.

Although this structure seems to imply that the CH_4 molecule is flat, it is not. In fact, the methane molecule is tetrahedral, as we will see in the next step.

3. **Assume that the most stable molecular shape has the bonding electron pairs as far apart as possible.** (*Note:* In other molecules we will need to consider nonbonding electrons as well, but CH_4 has none.) The four bonding electron pairs around

the carbon atom in CH_4 repel one another, and in their most stable arrangement they are as far from one another as possible. As a result, the four hydrogen atoms also are as far from one another as possible. This shape is tetrahedral, because the hydrogen atoms correspond to the corners of a **tetrahedron,** a four-cornered figure with four equal triangular sides.

One way to describe the shape of a CH_4 molecule is by analogy to the base of a folding music stand. The four C-to-H bonds correspond to the three evenly spaced legs and the vertical shaft of the stand (Figure 3.7). The angle between each pair of bonds is 109.5°. The tetrahedral shape of a CH_4 molecule has been experimentally confirmed. Indeed, it is one of the most common atomic arrangements in nature, particularly in carbon-containing molecules.

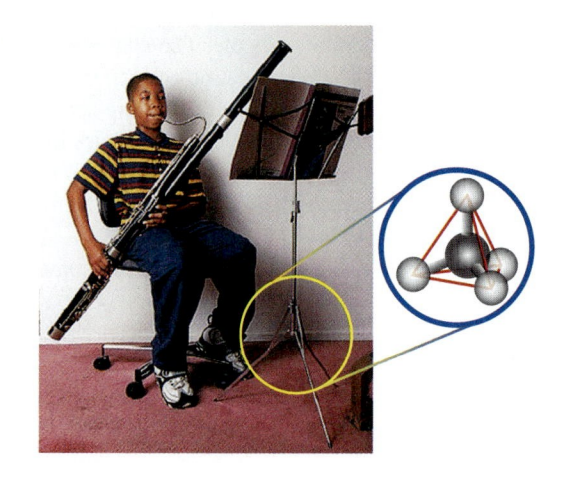

Figure 3.7

The legs and the shaft of this music stand approximate the arrangement of the bonds in a tetrahedral molecule like methane.

Consider This 3.9   Flat or Tetrahedral Methane?

  a. If the methane molecule were really two-dimensional, as the Lewis structure representation seems to indicate, what would the H-to-C-to-H bond angle be?
  b. Offer a reason why the tetrahedral shape, not the two-dimensional flat shape, is more advantageous for this molecule.
  c. Consider the music stand shown in Figure 3.7. In the analogy of shape using a music stand, where would the carbon atom be located? Where would each of the hydrogen atoms be?

Answer
  a. 90° (and 180° for H atoms across from one another)

Chemists represent molecules in several different ways. The simplest, of course, is the formula itself. In the case of methane, that is simply CH_4. We know that Lewis structures, without further interpretation, provide information on bonding but only two-dimensional information for most molecules. Other representations in Figure 3.8 show some generally accepted methods used by chemists to convey the three-dimensional structure of methane. For example, the third structure in part a of Figure 3.8 contains a wedge-shaped line that represents a bond coming out of the paper at an angle generally toward the reader. The dashed wedge in the same structural formula represents a bond pointing away from the reader. The two solid lines lie in the plane of the paper. This is an improvement over the two-dimensional structure, but a better way to visualize molecules is with a molecular modeling program, as in Figure 3.8, parts b and c. You will have a chance to see the results of a

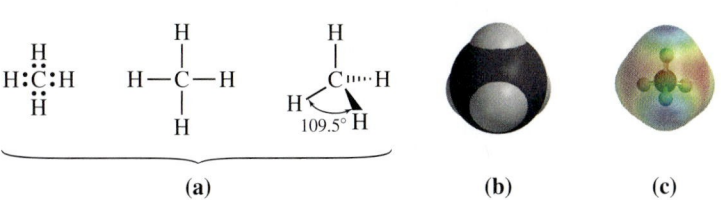

(a)   **(b)**   **(c)**

Figure 3.8

Representations of CH_4.
(a) Lewis structures and structural formula; **(b)** Space-filling model; **(c)** Charge-density model.

The charge-density model for H_2O is discussed in Section 5.5.

modeling program in Consider This 3.12. Seeing and manipulating physical models, either in the classroom or laboratory, can also help you visualize the structure of molecules.

Space-filling models and computer-generated charge-density models both enclose the volume occupied by electrons in a molecule. The charge-density model displays an internal ball-and-stick model to show the location of nuclei. The colors in the charge-density model will help you visualize how the electrons are arrayed within the molecule. Overall, the molecule is neutral. Within the molecule, red hues indicate regions of higher electron density. At the other end of the spectrum, blue hues represent lower electron density. The intensity of the colors reflects how greatly the electrons are pulled from one region of the molecule to another.

Not all outer electrons must reside in bonding pairs. In some molecules, the central atom has nonbonding electron pairs, also called lone pairs. For example, Figure 3.9 shows the ammonia molecule in which nitrogen completes its octet with three bonding pairs and one nonbonding pair.

A nonbonding pair effectively occupies greater space than a bonding pair of electrons. Consequently, the nonbonding pair repels the bonding pairs somewhat more strongly than the bonding pairs repel one another. This stronger repulsion forces the bonding pairs closer to one another, creating an H-to-N-to-H angle slightly less than the predicted 109.5° associated with a regular tetrahedron. The experimental value of 107.3° is close to the tetrahedral angle, again indicating that our model is reasonably reliable.

The shape of a molecule is described in terms of its arrangement of atoms, not electrons. The hydrogen atoms of NH_3 form a triangle with the nitrogen atom above them at the top of the pyramid. Thus, ammonia is said to be a triangular pyramid; it has a *trigonal pyramidal* shape. Going back to the analogy of the folding music stand (see Figure 3.7), you could expect to find hydrogen atoms at the tip of each leg of the music stand. This places the nitrogen atom at the intersection of the legs with the shaft, with the nonbonded electron pair forming around the shaft of the stand.

Section 2.9 discussed replacement of NH_3 as a refrigerant gas by CFCs. Section 11.8 will describe the importance of NH_3 for agriculture.

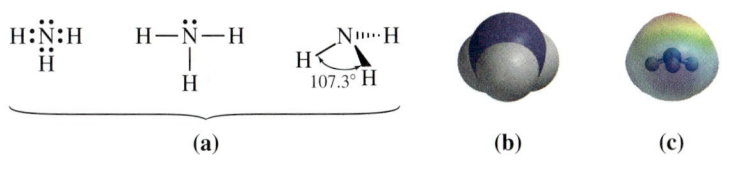

(a)   **(b)**   **(c)**

Figure 3.9

Representations of NH_3.
(a) Lewis structures and structural formula; **(b)** Space-filling model; **(c)** Charge-density model.

Water vapor is a greenhouse gas.

The water molecule illustrates yet another shape. There are eight outer electrons: one from each of the two hydrogen atoms plus six from the oxygen (Group 6A). Its Lewis structure discloses how the eight electrons on the central oxygen atom are distributed: two bonding pairs and two lone pairs (Figure 3.10, part a).

If these four pairs of electrons were arranged as far apart as possible, we might predict the H-to-O-to-H bond angle to be the same as the H-to-C-to-H bond angle in methane, namely, 109.5°. However, unlike methane, water has two bonding and two nonbonding pairs. The repulsion between the two nonbonding pairs and, in turn, their repulsion of the bonding pairs cause the bond angle to be less than 109.5°. Experiments indicate a value of approximately 104.5°. A water molecule has a *bent* shape.

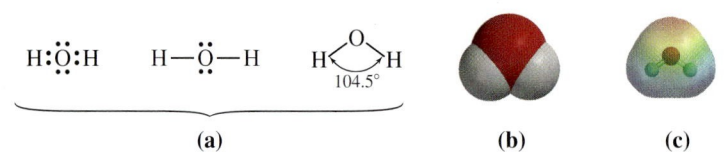

Figure 3.10

Representations of H_2O.
(a) Lewis structures and structural formula; **(b)** Space-filling model; **(c)** Charge-density model.

Your Turn 3.10 **Predicting Molecular Shapes, Part 1**

Using the strategies just described, predict and sketch the shape of each of these molecules.

a. CCl_4 (carbon tetrachloride)
b. CCl_2F_2 (Freon-12; dichlorodifluoromethane)
c. H_2S (hydrogen sulfide)

Answer

a. Total the outer electrons: $4 + 4(7) = 32$. Eight of these go around the central C atom to form four single bonds, one to each Cl. The other 24 outer electrons are nonbonding pairs on the Cl atoms. The bonding electron pairs on C arrange themselves so that their separation is maximized. The CCl_4 molecule is tetrahedral, the same as CH_4.

We have already looked at the structures of several molecules important for understanding the chemistry of global warming. What about the structure of the carbon dioxide molecule? It has 16 outer electrons: The carbon atom contributes four electrons and six come from each of the two oxygen atoms. If only single bonds were involved, there would not be enough electrons to provide eight electrons for each atom. That would require 20 electrons. However, with 16 electrons the octet rule still can be obeyed if the central carbon atom forms a double bond with each of the two oxygen atoms, thus sharing four electrons.

What is the shape of the CO_2 molecule? Again, groups of electrons repel one another, and the most stable configuration provides the furthest separation of the negative charges. In this case, the groups of electrons are the double bonds, and these are furthest apart with an O-to-C-to-O bond angle of 180°. The model predicts that all three atoms in a CO_2 molecule will be in a straight line and the molecule will be linear. This is, in fact, the case as shown in Figure 3.11.

We applied the idea of electron pair repulsion to molecules in which there are four groups of electrons (CH_4, NH_3, and H_2O) and two groups of electrons (CO_2). Electron

> Lewis structures with double bonds were introduced in Section 2.3.

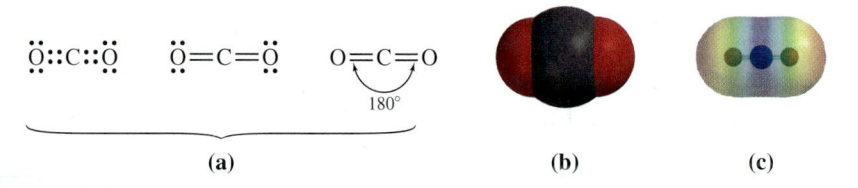

Figure 3.11

Representations of CO_2
(a) Lewis structures and structural formula; **(b)** Space-filling model; **(c)** Charge-density model.

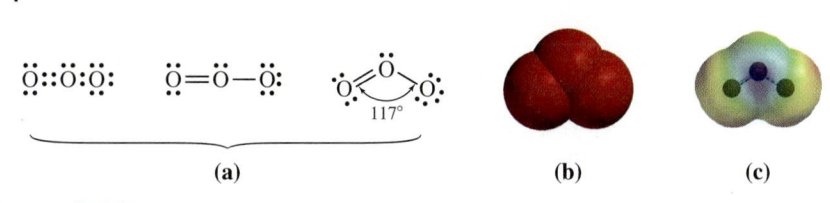

Figure 3.12
Representations of O_3.
(a) Lewis structures and structural formula for one resonance form; (b) Space-filling model;
(c) Charge-density model.

The resonance forms and bent shape of the O_3 molecule were discussed in Section 2.3.

pair repulsion also applies reasonably well to molecules that include three, five, or six groups of electrons. In most molecules, the electrons and atoms are still arranged to keep the separation of the electrons at a maximum. This logic accounts for the bent shape we associated with the ozone molecule.

The O_3 molecule (18 total outer electrons) contains a single bond and a double bond, and the central oxygen atom carries a nonbonding lone pair of electrons. Thus, there are three groups of electrons on this central atom: the pair that makes up the single bond, the two pairs that constitute the double bond, and the lone pair. These three groups of negatively charged electrons repel one another, and the minimum energy of the molecule corresponds to the furthest separation of these electron groups. This will occur when the electron groups are all in the same plane and at an angle of about 120° from one another. We predict, therefore, that the O_3 molecule should be bent, and the angle made by the three atoms should be approximately 120°. Experiments show the O-to-O-to-O bond angle to be 117°, just slightly smaller than the prediction (Figure 3.12). The nonbonding electron pair on the central oxygen atom occupies an effectively greater volume than bonding pairs of electrons, causing a greater repulsion force responsible for the slightly smaller bond angle.

Your Turn 3.11   Predicting Molecular Shapes, Part 2

Using the strategies just described, predict and sketch the shapes of these molecules.

   a. SO_2 (sulfur dioxide)
   b. SO_3 (sulfur trioxide)
   Hint (part a): Since S and O are in the same group on the periodic table, the structures for SO_2 and O_3 will be closely related.

Consider This 3.12   Molecules in Motion

Three-dimensional representations of molecules can be viewed on the Web using several different molecular-modeling programs. Use a program available to you to view some of the molecules discussed in this chapter. Has your mental picture of these molecules changed after working with the 3-D representations? Explain.

3.4   Vibrating Molecules and the Greenhouse Effect

You learned in Chapter 2 that if a photon is part of the UV region of the spectrum, it has sufficient energy to disrupt the arrangement of electrons within some molecules. This can cause covalent bonds to break, as in the dissociation of O_2 and O_3 by UV-B and UV-C radiation. Fortunately, this is not the case with photons in the IR range.

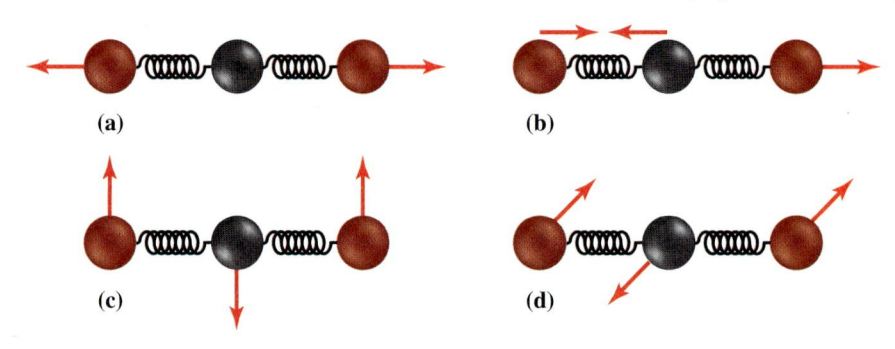

Figure 3.13

Molecular vibrations in CO_2.
Each spring represents a C-to-O double bond. Vibrations **a** and **b** are stretching vibrations; **c** and **d** are bending vibrations.

Photons in the IR range of the spectrum are not sufficiently energetic to break bonds. However, a photon of IR radiation can add energy to the vibrations in a molecule. Depending on the molecular structure, only certain vibrations are possible. The energy of the photon must correspond exactly to the vibration energy of the molecule for the photon to be absorbed. This means that different molecules absorb IR radiation at different wavelengths and thus vibrate at different energies.

We illustrate these ideas with the CO_2 molecule, representing the atoms as balls and the covalent bonds as springs. A molecule of CO_2 can vibrate in the four ways pictured in Figure 3.13. The arrows indicate the direction of motion of each atom when the molecule is vibrating. The atoms can move forward and backward along the arrows. Vibrations **a** and **b** are called stretching vibrations. In vibration **a,** the central carbon atom is stationary and the oxygen atoms move back and forth (stretch) in opposite directions away from the central atom. Alternatively, the oxygen atoms can move in the same direction and the carbon atom in the opposite direction (vibration **b**). Vibrations **c** and **d** look very much alike. In both cases, the molecule bends from its normal linear shape. The bending counts as two vibrations because it can occur in either of two possible planes. Vibration **c** is shown bending in an *xy*-plane, up and down on the plane of the paper on which the diagram is printed. Vibration **d** is moving in an *xz*-plane, in front and in back of the plane of the paper.

If you have ever examined a spring, you have probably observed that more energy is required to stretch than bend it. Similarly, more energy is required to stretch a CO_2 molecule than to bend it. This means that more energetic photons, those with shorter wavelengths, are needed to add energy to stretching vibrations **a** or **b** than to add energy to bending vibrations **c** or **d.** When the molecule absorbs IR radiation with a wavelength of 15.00 micrometers (μm), bending motions (**c** and **d**) take place. Stretching vibration **b** will occur only if radiation of wavelength of 4.26 μm is absorbed. Together, vibrations **b, c,** and **d** account for the greenhouse properties of carbon dioxide.

Stretching vibration **a** cannot be triggered by the direct absorption of IR radiation. In a CO_2 molecule, the average concentration of electrons is greater on the oxygen atoms than on the carbon atom. This means that the oxygen atoms carry a partial negative charge relative to the carbon atom. As the bonds stretch, the positions of the electrons change, and therefore the charge distribution in the molecule changes as well. Because of the linear shape and symmetry of CO_2, the changes in charge distribution during vibration **a** cancel each other and no infrared absorption occurs.

The infrared (heat) energies absorbed or transmitted by molecules can be measured with an instrument called an infrared spectrometer. Heat radiation from a glowing filament is passed through a sample of the compound to be studied, in this case gaseous CO_2. A detector measures the amount of radiation, at various frequencies, transmitted by the sample. High transmission means low absorbance, and vice versa. This information is displayed graphically, where the relative intensity of the transmitted radiation is plotted versus wavelength. The result is called the *infrared spectrum* of the compound. Figure 3.14 shows the infrared spectrum of CO_2.

A micrometer is equal to one-millionth of a meter: $1 \text{ μm} = 1 \times 10^{-6} \text{ m}$

The property of electronegativity, a measure of an atom's ability to attract bonded electrons, will be discussed in Section 5.5.

Spectroscopy is the field of study that examines matter by passing electromagnetic energy through a sample.

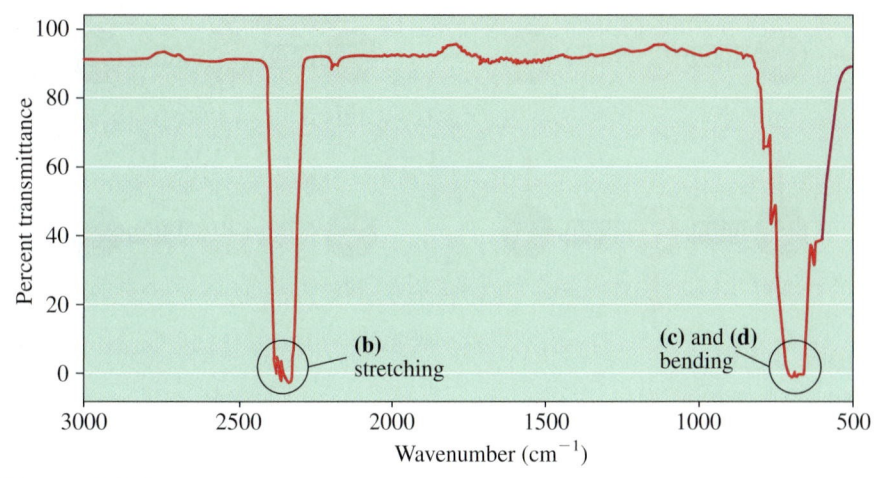

Figure 3.14

Infrared spectrum of carbon dioxide.
The letters (**b**), (**c**), and (**d**) refer to the molecular vibrations shown in Figure 3.13.

Understanding this spectrum and others like it will take just a bit more explanation. The units of the *y*-axis are percent transmittance, as described earlier. The *x*-axis values are expressed in the unit called **wavenumber,** a number that is inversely proportional to wavelength. Although it may seem logical to use familiar units of wavelength, most IR spectra are reported with units of wavenumber. Fortunately, a simple relationship relates wavelength, expressed in micrometers, to the wavenumber, expressed in cm^{-1}.

$$\text{wavenumber (cm}^{-1}) = \frac{10{,}000}{\text{wavelength (}\mu\text{m)}} \qquad [3.2]$$

The infrared spectrum shown in Figure 3.14 was determined in the laboratory, but the same absorption phenomenon takes place in the atmosphere. CO_2 molecules absorb specific wavelengths of infrared energy, vibrate for a while, and then reemit the energy as heat and return to their normal unexcited, or "ground," state. This is how CO_2 captures and returns the infrared radiation coming from Earth's surface, preventing our planet from becoming too cold. This is what makes CO_2 a greenhouse gas.

Your Turn 3.13   Relating Wavelength to Wavenumber

Use the relationship given in equation 3.2 to convert each of these wavelengths to wavenumbers.

  a. 4.26 μm
  b. 15.00 μm
  c. How do the two values just calculated compare with the locations of lowest transmittance (greatest absorbance) in Figure 3.14?

Answer

  a. $\dfrac{10{,}000}{4.26 \ \mu\text{m}} = 2350 \text{ cm}^{-1}$

Nitrous oxide, N_2O, is also called dinitrogen monoxide. You will encounter this gas again in Chapter 6.

Any molecule that can vibrate in response to the absorption of specific photons of IR radiation can behave as a greenhouse gas. There are many such molecules. CO_2 and H_2O are the most important in maintaining Earth's temperature. Figure 3.15 shows the IR spectrum of H_2O molecules absorbing IR radiation. However, methane, nitrous oxide, ozone, and chlorofluorocarbons (such as CCl_3F) are among the other substances that help retain planetary heat.

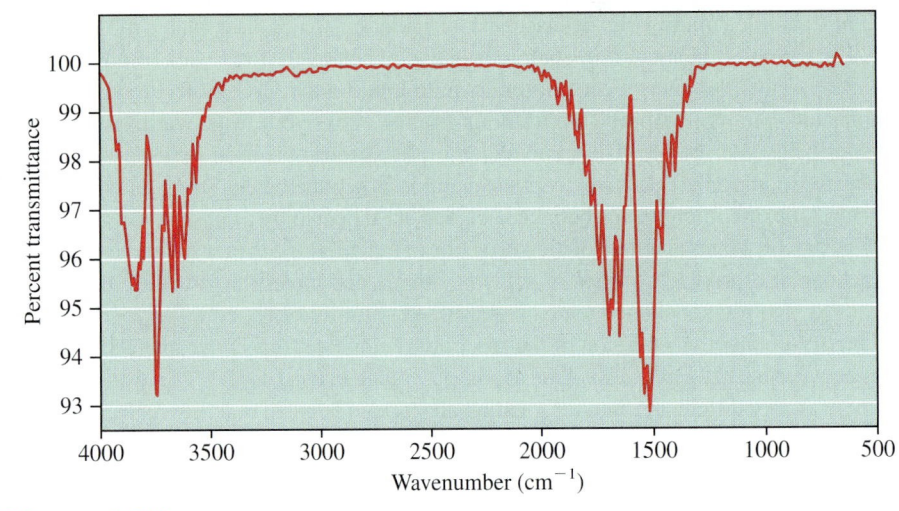

Figure 3.15

Infrared spectrum of water vapor.

Diatomic gases, such as N_2 and O_2, are not greenhouse gases. Although molecules consisting of two identical atoms do vibrate, the overall electric charge distribution does not change during these vibrations. Hence, these molecules cannot be greenhouse gases. Earlier we discussed this lack of overall electric charge distribution as the reason why stretching vibration **a** in Figure 3.13 was not responsible for the greenhouse gas behavior of CO_2.

Consider This 3.14   Bending and Stretching Water Molecules

a. Use Figure 3.15 to estimate the wavenumber (cm^{-1}) for the two maximum absorbancies of IR energy of the water molecule.
b. Change each value to wavelength (μm).
c. Which wavelength do you predict represents bending vibrations and which represents stretching? Explain the basis of your predictions.
Hint: Compare the IR spectrum of CO_2 with that for H_2O.

So far, you have encountered two ways that molecules respond to radiation. Highly energetic photons with high frequencies and short wavelengths (such as UV radiation) can break bonds within molecules. The less energetic photons (such as IR radiation) cause many molecules to vibrate. Both processes are depicted in Figure 3.16, but the figure also includes another response of molecules to radiant energy that is probably a good deal more familiar to you. Longer wavelengths than those in the IR range have only enough energy to cause molecules to rotate or spin, not vibrate or dissociate.

Section 2.5 discussed how UV radiation can break chemical bonds.

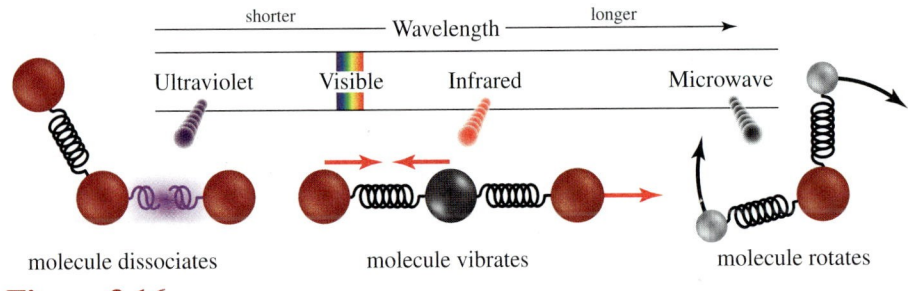

Figure 3.16

Molecular response to types of radiation.

For example, microwave ovens generate radiation that causes water molecules to spin. The radiation generated in such a device is of relatively long wavelength, about a centimeter. Thus the energy per photon is quite low. As the H_2O molecules absorb the photons and spin more rapidly, the resulting friction cooks your food, warms up the left-overs, or heats your coffee. The same region of the spectrum is used for radar. Beams of microwave radiation are sent out from a generator. When the beams strike an object such as an airplane, the microwaves bounce back and are detected by a sensor.

The practical consequences of the interaction of radiation and matter are immense. These interactions also provide a means of studying atomic and molecular structure. Electronic, vibrational, and rotational energy are all **quantized,** meaning only certain energy levels are permitted. No matter what region of the spectrum is employed, spectroscopy reveals differences between energy levels. Using the appropriate mathematical model, scientists can translate these energy differences into information about bond lengths, bond strengths, and bond angles. A consequence of looking through a spectroscopic window into atoms and molecules is that chemists can describe the microscopic world with great confidence.

3.5   The Carbon Cycle: Contributions from Nature and Humans

Levi's book, *The Periodic Table,* was written in 1975. About 6 billion people now inhabit Earth.

In his book *The Periodic Table,* the late chemist, author, and World War II concentration camp survivor Primo Levi, wrote eloquently about CO_2.

> "This gas which constitutes the raw material of life, the permanent store upon which all that grows draws, and the ultimate destiny of all flesh, is not one of the principal components of air but rather a ridiculous remnant, an 'impurity' thirty times less abundant than argon, which nobody even notices. . . . [F]rom this ever renewed impurity of the air we come, we animals and we plants, and we the human species, with our four billion discordant opinions, our millenniums of history, our wars and shames, nobility and pride."

In the essay from which this quotation is taken, Levi traces a brief portion of the life history of a carbon atom from a piece of limestone (calcium carbonate, $CaCO_3$), where it lies "congealed in an eternal present," to a CO_2 molecule, to a molecule of glucose in a leaf, and ultimately to the brain of the author. And yet that is not the final destination. "The death of atoms, unlike our own," writes Levi, "is never irrevocable." That carbon atom, already billions of years old, will continue to persist into the unimagined future.

This marvelous continuity of matter, a consequence of its conservation, is beautifully illustrated by the carbon cycle. Even without being described with Primo Levi's poetic gifts, the story is fascinating and one that is important to understand to comprehend the danger of the cycle's being changed by human activities. It is certain that without the proper functioning of the carbon cycle, every aspect of life on Earth could undergo dramatic change. Figure 3.17 is one representation of this important cycle.

Consider This 3.15   Understanding the Carbon Cycle

a. What processes add carbon (in the form of CO_2) to the atmosphere?
b. Compare the processes of deforestation and combustion of fossil fuels. Which adds more carbon to the atmosphere?
c. How is carbon removed from the atmosphere?
d. What are the two largest reservoirs of carbon?
e. Which parts of the carbon cycle can be changed by human activities?

The carbon cycle is a dynamic system. All processes illustrated are happening simultaneously, but at far different rates. Plants die and decay, releasing CO_2. Other plants

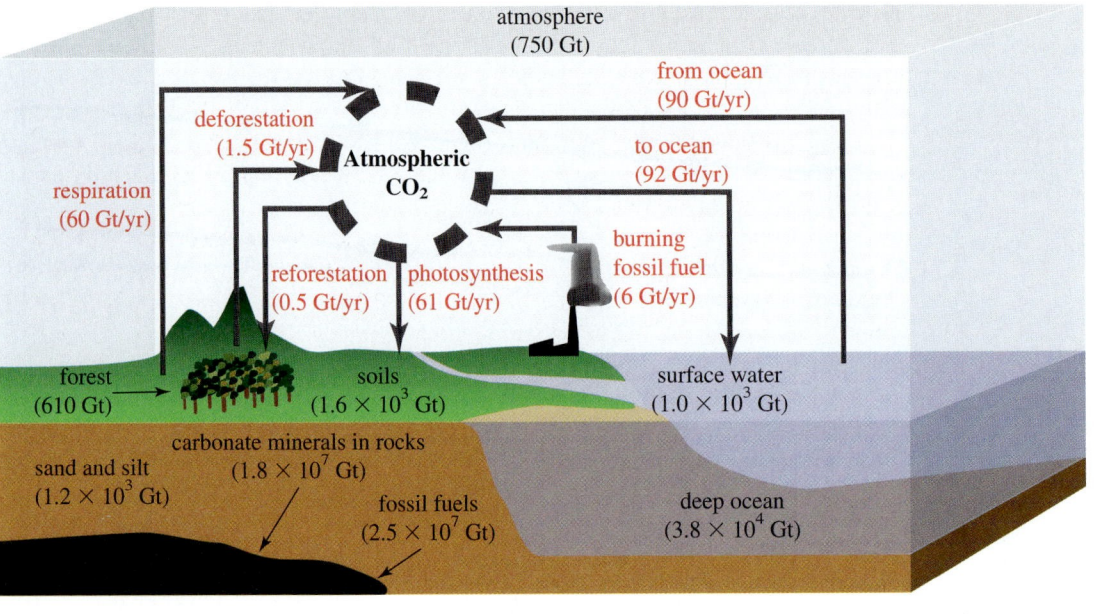

Figure 3.17

The global carbon cycle. The numbers show the quantity of carbon, expressed in gigatonnes (Gt), that is stored in various carbon reservoirs (black numbers) or moving through the system per year (red numbers).

Source: From Purves, Orians, Heller and Sadava, *Life: The Science of Biology,* 5th edition, 1998 page 1186. Reprinted with permission of Sinauer Associates, Inc.

A gigatonne (Gt) is a billion metric tons (a billion tonnes) or about 2200 billion pounds. For comparison, a fully loaded 747 jet weighs about 800,000 lb. It would take nearly 3 million 747s to have a total mass of 1 Gt.

enter the food chain where their complex molecules are broken down into CO_2, H_2O, and other simple substances. Animals exhale CO_2, carbonate rocks decompose, and carbon dioxide escapes through the vents of volcanoes. And the cycle goes on and on. Michael B. McElroy of Harvard University estimated, "The average carbon atom has made the cycle from sediments through the more mobile compartments of the Earth back to sediments, some 20 times over the course of Earth's history." CO_2 in the air today may have come from campfires more than a thousand years ago.

As members of the animal kingdom, we *Homo sapiens* participate in the carbon cycle along with our fellow creatures. But we do more than our share. As is true for any animal, we inhale and exhale, ingest and excrete, live and die. But we also have developed processes that permit us to significantly perturb the system. The Industrial Revolution, which began in Europe in the late 18th century, was fueled largely by coal. Coal powered steam engines in mines, factories, locomotives, ships, and later, electrical generators. The subsequent discovery and exploitation of vast deposits of petroleum made possible the development of automobiles and other types of transportation. To a very considerable extent, the Industrial Revolution was a revolution in energy sources and energy transfer. But another transfer takes place. Burning fossil fuels transfers carbon from one of the largest underground carbon reservoirs into the atmosphere. Natural removal processes may not respond quickly enough to the increased amount of CO_2, leading to an accumulation in the atmosphere.

Consider This 3.16	Earth's Small Atmospheric Carbon Reservoir

Carbon dioxide is dominant in the atmospheres of Mars and Venus but a minor component of Earth's atmosphere. The atmosphere on Mars has about 30 times more CO_2 than Earth's atmosphere, and on Venus it is about 300,000 times more.

a. How small is the Earth's atmospheric carbon reservoir compared with other carbon reservoirs? *Hint:* Consider Figure 3.17.
b. Offer a possible reason why the atmospheric carbon reservoirs of Mars and Venus are so much greater.

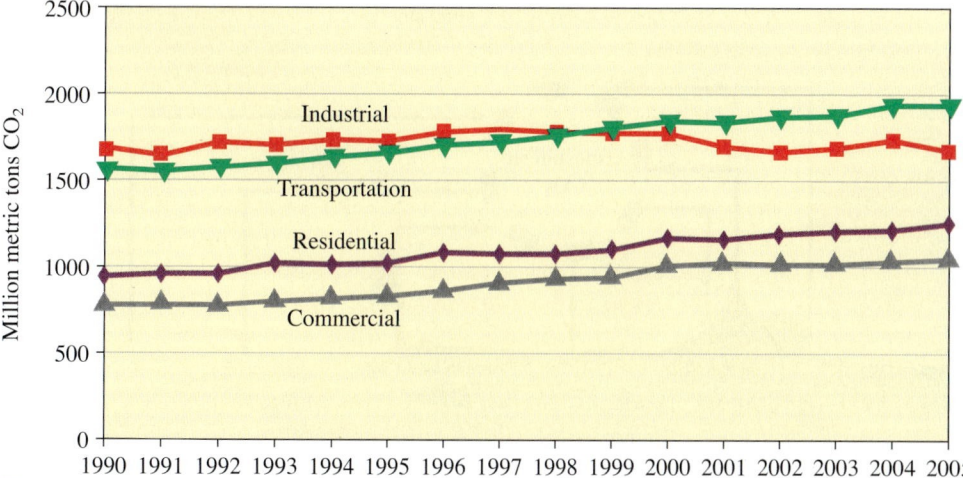

Figure 3.18

U.S. carbon dioxide emissions by sector, 1990–2005.

Source: EIA, *U.S. Carbon Dioxide Emissions from Energy Sources, 2006.*

1 metric ton (1 tonne)
= 10^3 kg
= 2200 lb

The U.S. Energy Information Administration (EIA) is required by the Energy Policy Act of 1992 to prepare a yearly updated report on greenhouse gas emissions. Prior to 2001, CO_2 emission sources from fossil-fuel consumption were broken into five categories: power utilities, transportation, industrial, commercial, and residential emitters. As part of its continuing review of information and methodology, the EIA in 2001 removed power utilities as a separate category. EIA explained the change this way. "Energy-related CO_2 emissions now have been revised as part of an agency-wide adjustment to energy consumption data and sectoral allocation." In practical terms, this means that all emissions from power utilities were reassigned based on the end use of the energy. All data since 1990, set as the index year, have recalculated using this new approach and are shown in Figure 3.18.

Your Turn 3.17   Comparing Sector Emissions

a. In 2005, which sector is responsible for the largest source of CO_2 emissions? Was that always the case from 1990 to 2005?

b. Emissions from the commercial sector, although the smallest in tons, have been increasing at a rate of about 2% a year since 1990. Verify that from Figure 3.18.

c. Suggest some factors that can affect energy-related CO_2 emissions in both the long and short term.

One other human influence on CO_2 emissions is deforestation by burning, a practice that releases 0.6–2.6 Gt of carbon to the atmosphere each year. It is estimated that forested land the size of two football fields is lost every second of the day from the rain forests of the world. Although firm numbers are rather elusive, Brazil continues as the country with the greatest loss of rain forest acreage. In Brazil alone, over 5.4 million acres of Amazon rain forest is vanishing each year. Trees, those very efficient absorbers of carbon dioxide, are removed from the cycle through deforestation. If the wood is burned, vast quantities of CO_2 are generated; if it is left to decay, that process also releases carbon dioxide, but more slowly. Even if the lumber is harvested for construction purposes and the land is replanted in cultivated crops, the loss in CO_2-absorbing capacity may approach 80%.

Systematic deforestation is not a new phenomenon, nor is it limited to tropical forests. Logging practices worldwide have altered the natural landscape. There are actually

more trees in the United States now than there were in colonial times and in the later 1800s, although the same cannot be said for European countries. The focus in the 20th century shifted from the heavily deforested regions of Europe and North America to the tropical rain forests of Central and South America, Africa, and Asia.

The total quantity of carbon released by the human activities of deforestation and burning fossil fuels is 6.0–8.2 Gt per year. About half of this is recycled into the oceans and the biosphere, which serve as **carbon sinks,** natural processes that remove CO_2 from the atmosphere. These processes do not always remove CO_2 with the speed required by the ever-increasing concentrations of CO_2. Much of the CO_2 emitted stays in the atmosphere, adding between 3.1 and 3.5 Gt of carbon per year to the existing base of 750 Gt noted in Figure 3.17. We are concerned primarily with this *increase* in atmospheric carbon dioxide, because the *excess* CO_2 is implicated in global warming. Therefore, it would be useful to know the mass (Gt) of CO_2 added to the atmosphere each year. In other words, what mass of CO_2 contains 3.3 Gt of carbon, the midpoint between 3.1 Gt and 3.5 Gt? To answer this question will require another scenic tour into the land of chemistry, one you also will find useful later in this text.

3.6 Quantitative Concepts: Mass

To solve the problem just posed, we need to know how the mass of C is related to the mass of CO_2. Regardless of the source of CO_2, its chemical formula is stubbornly the same. The mass percent of C in CO_2 is also unwavering and therefore we must calculate the mass percent of C in CO_2, based on the formula of the compound. As you work through this and the next section, keep in mind that we are seeking a value for that percentage.

The approach requires the use of the atomic masses of the elements involved. But this raises an important question: How much does an individual atom weigh? Most of the mass of an atom is attributable to the neutrons and protons in the nucleus. Thus, elements differ in atomic mass because their atoms differ in composition. Rather than use absolute masses of individual atoms, chemists have found it convenient to employ relative atomic masses—in other words, to relate all atomic masses to some convenient standard. The internationally accepted atomic mass standard is carbon-12, the isotope that makes up 98.90% of all carbon atoms. C-12 has a mass number of 12 because each atom has a nucleus consisting of 6 protons and 6 neutrons plus 6 electrons outside the nucleus. The mass of one of these atoms is arbitrarily assigned a value of exactly 12 atomic mass units (amu). We can thus define the **atomic mass** of an element as the average mass of an atom of that element as compared with an atomic mass of exactly 12 amu for C-12. Atoms are so small that an atomic mass unit represents a very small mass: 1 amu = 1.66×10^{-24} g.

The periodic table in the text shows that the atomic mass of carbon is 12.01, not 12.00. This is not an error; it reflects the fact that carbon exists naturally as three isotopes. Although C-12 predominates, 1.10% of carbon is C-13, the isotope with six protons and *seven* neutrons. In addition, natural carbon contains a trace of C-14, the isotope with six protons and *eight* neutrons. The tabulated atomic mass value of 12.01 is often called by the name atomic weight, an average that takes into consideration the masses and percent natural abundance of all naturally occurring isotopes of carbon. This isotopic distribution and average atomic mass of 12.01 characterize carbon obtained from any chemical source—a graphite ("lead") pencil, a tank of gasoline, a loaf of bread, a lump of limestone, or your body.

The radioactive isotope carbon-14, although present only in trace amounts, plays a key role in determining the origin of the increasing atmospheric carbon dioxide. In all living things, only one out of 10^{12} carbon atoms is a C-14 atom. A plant or animal constantly exchanges CO_2 with the environment, and this maintains the C-14 concentration in the organism at a constant level. However, when the organism dies, the biochemical processes that exchange C stop functioning and the C-14 is no longer replenished. This means that after the death of the organism, the concentration of C-14 decreases with time because it undergoes radioactive decay to form N-14. Coal and oil are the fossilized remains of plant life that died millions of years ago. Hence, the level of C-14 is extremely low in fossil fuels, and in the carbon dioxide released when fossil fuels burn. Careful experiments

Remember that the natural "greenhouse effect" makes life on Earth possible. Problems occur when the amount of greenhouse gases *increases* faster than the sinks can accommodate the increases. The result is the *enhanced* greenhouse effect.

Isotopes and relative mass of subatomic particles were discussed in Section 2.2.

The term *atomic weight* is a familiar (but not technically correct) term used for the relative scale of atomic masses.

You will learn to write equations for nuclear reactions in Section 7.2.

show that the concentration of C-14 in atmospheric CO_2 has recently decreased. This strongly suggests that the origin of the added CO_2 is indeed the burning of fossil fuels, a decidedly human activity.

Your Turn 3.18   Isotopes of Nitrogen

Nitrogen (N) is an important element in the atmosphere and in biological systems. It has two naturally occurring isotopes: N-14 and N-15.

a. Use the periodic table to find the atomic number and atomic mass of nitrogen.
b. What is the number of protons, neutrons, and electrons in a neutral atom of N-14?
c. Compare your answers for part **b** with those for a neutral atom of N-15.
d. Given the atomic mass of nitrogen, which isotope has the greatest natural abundance?

Having reviewed the meaning of isotopes and atomic mass, we return to the matter at hand—the masses of atoms and particularly the atoms in CO_2. Not surprisingly, it is impossible to weigh a single atom because of its extremely small mass. A typical laboratory balance can detect a minimum mass of 0.1 mg; that corresponds to 5×10^{18} carbon atoms, or 5,000,000,000,000,000,000 carbon atoms. An atomic mass unit is far too small to measure in a conventional chemistry laboratory. Rather, the gram is the chemist's mass unit of choice. Therefore, scientists use exactly 12 g of carbon-12 as the reference for the atomic masses of all the elements. **Atomic mass** can therefore be alternatively defined as the mass (in grams) of the same number of atoms that are found in exactly 12 g of carbon-12. This number of atoms is, of course, *very* large. This important chemical number is named after an Italian scientist with the impressive name of Count Lorenzo Romano Amadeo Carlo Avogadro di Quaregna e di Ceretto (1776–1856). (His friends called him Amadeo.) **Avogadro's number** is the number of atoms in exactly 12 g of C-12. Avogadro's number, if written out, is 602,000,000,000,000,000,000,000. It is more compactly written in scientific notation as 6.02×10^{23}. This is the incredible number of atoms in 12 g of carbon, no more than a tablespoonful of soot!

Avogadro's number counts a large collection of atoms, much like the term *dozen* counts a collection of eggs. It does not matter if the eggs are large or small, brown or white, "organic" or not. No matter, for if there are 12 eggs, they are still counted as a dozen. A dozen ostrich eggs will have a greater mass than a dozen quail eggs. Figure 3.19 illustrates this point with a half-dozen tennis and a half-dozen golf balls. Like atoms of different elements, the masses of a tennis ball and a golf ball differ. The number of balls is the same—six in each bag, a half dozen.

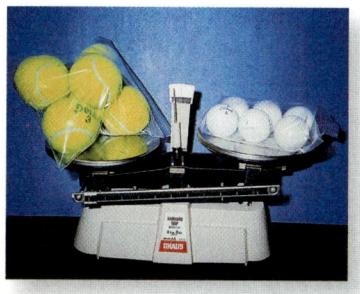

Figure 3.19
Six tennis balls has a greater mass than six golf balls.

Sceptical Chymist 3.19   Marshmallows and Pennies

Avogadro's number is so large that about the only way to hope to comprehend it is through analogies. For example, one Avogadro's number of regular-sized marshmallows, 6.02×10^{23} of them, would cover the surface of the United States to a depth of 650 miles. Or, if you are more impressed by money than marshmallows, assume 6.02×10^{23} pennies were distributed evenly among the more than 6 billion inhabitants of the Earth. Every man, woman, and child could spend $1 million every hour, day and night, and half of the pennies would still be left unspent at death.

Can these fantastic claims be correct? Check one or both, showing your reasoning. Come up with an analogy of your own.

Knowledge of Avogadro's number and the atomic mass of any element permit us to calculate the average mass of an individual atom of that element. Thus, the mass of 6.02×10^{23} oxygen atoms is 16.00 g, the atomic mass from the periodic table. To find the average mass of just one oxygen atom, we must divide the mass of the large collection of atoms by the size of the collection. In chemist's terms, this means dividing the atomic mass by Avogadro's number. Fortunately, calculators help make this job quick and easy.

$$\frac{16.00 \text{ g oxygen}}{6.02 \times 10^{23} \text{ oxygen atoms}} = 2.66 \times 10^{-23} \text{ g oxygen/oxygen atom}$$

This very small mass confirms once again why chemists do not generally work with small numbers of atoms. We manipulate trillions at a time. Therefore, practitioners of this art need to measure matter with a sort of chemist's dozen—a very large one, indeed. To learn about it, read on . . . but only after stopping to practice your new skill.

Your Turn 3.20   Calculating Mass of Atoms

a. Calculate the average mass (in grams) of an individual atom of nitrogen.
b. Calculate the mass (in grams) of 5 trillion nitrogen atoms.
c. Calculate the mass (in grams) of 6×10^{15} nitrogen atoms.

Answer

a. $\dfrac{14.01 \text{ g nitrogen}}{6.02 \times 10^{23} \text{ nitrogen atoms}} = 2.34 \times 10^{-23}$ g nitrogen/nitrogen atom

> **Calculation tip**
> *Predict:*
> Will the answer be a large number? A small number?
>
> *Check:*
> Does the answer match your prediction? Is it reasonable?

3.7 Quantitative Concepts: Molecules and Moles

Chemists have another way of communicating the number of atoms, molecules, or other small particles present. This is to use the term **mole (mol),** defined as containing an Avogadro's number of objects. The term is derived from the Latin word to "heap" or "pile up." Thus, 1 mol of carbon atoms consists of 6.02×10^{23} C atoms, 1 mol of oxygen gas is made up of 6.02×10^{23} oxygen molecules, and 1 mol of carbon dioxide molecules corresponds to 6.02×10^{23} carbon dioxide molecules.

> When used together with a number, *mol* is an abbreviation for *mole.*

As you already know from previous chapters, chemical formulas and equations are written in terms of atoms and molecules. For example, reconsider the equation for the complete combustion of carbon in oxygen.

$$C + O_2 \longrightarrow CO_2 \qquad\qquad [3.3]$$

This equation tells us that one atom of carbon combines with one molecule of oxygen to yield one molecule of carbon dioxide and reflects the *ratio* in which the particles interact. Thus, it would be equally correct to say that 10 C atoms react with 10 O_2 molecules (20 O atoms) to form 10 CO_2 molecules. Or, putting the reaction on a grander scale for that matter, we could say 6.02×10^{23} C atoms combine with 6.02×10^{23} O_2 molecules (12.0×10^{23} O atoms) to yield 6.02×10^{23} CO_2 molecules. The last statement is equivalent to saying: "one *mole* of carbon plus one *mole* of oxygen yields one *mole* of carbon dioxide." The point is that *the numbers of atoms and molecules taking part in a reaction are proportional to the numbers of moles of the same substances.* The ratio of two oxygen atoms to one carbon atom remains the same regardless of the number of carbon dioxide molecules, as summarized in Table 3.1.

> There are 2 mol of oxygen atoms, O, in every mole of oxygen molecules, O_2.

In the laboratory and the factory, the quantity of matter required for a reaction is often measured by mass. The mole is a practical way to relate number of particles to the

Table 3.1	Different Interpretations of a Chemical Equation	
C +	**O$_2$**	**CO$_2$**
1 atom	1 molecule	1 molecule
6.02×10^{23} atoms	6.02×10^{23} molecules	6.02×10^{23} molecules
1 mol	1 mol	1 mol

more easily measured mass. The **molar mass** is the mass of one Avogadro's number, or mole, of whatever particles are specified. For example, the mass of a mole of carbon atoms, rounded to the nearest tenth of a gram, is 12.0 g. A mole of oxygen atoms has a mass of 16.0 g. But we can also speak of a mole of O$_2$ molecules. Because there are two oxygen atoms in each oxygen molecule, there are two moles of oxygen atoms in each mole of molecular oxygen, O$_2$. Consequently, the molar mass of O$_2$ is 32.0 g, twice the molar mass of O. Some refer to this as the molecular mass or molecular weight of O$_2$, emphasizing its similarity to atomic mass or atomic weight.

The same logic for the molar mass of the element O$_2$ applies to compounds, which brings us, at last, to the composition of carbon dioxide. The formula, CO$_2$, reveals that each molecule contains one carbon atom and two oxygen atoms. Scaling up by 6.02×10^{23}, we can say that each mole of CO$_2$ consists of 1 mol of C and 2 mol of O atoms (see Table 3.1). But remember that we are interested in the mass composition of carbon dioxide—the number of grams of carbon per gram of CO$_2$. This requires the molar mass of carbon dioxide, which we obtain by adding the molar mass of carbon to twice the molar mass of oxygen:

$$1 \text{ mol CO}_2 = 1 \text{ mol C} + 2 \text{ mol O}$$

$$= \left(1 \text{ mol C} \times \frac{12.0 \text{ g C}}{1 \text{ mol C}} \right) + \left(2 \text{ mol O} \times \frac{16.0 \text{ g O}}{1 \text{ mol O}} \right)$$

$$= 12.0 \text{ g C} + 32.0 \text{ g O}$$

$$1 \text{ mol CO}_2 = 44.0 \text{ g CO}_2$$

This procedure is routinely used in chemical calculations, where molar mass is an important property. Some examples are included in the next activity. In every case, you multiply the number of moles of each element by the corresponding atomic mass in grams and add the result.

Your Turn 3.21   Molecular Molar Mass

Calculate the molar mass of each of these greenhouse gases.

a. O$_3$ (ozone)
b. N$_2$O (dinitrogen monoxide or nitrous oxide)
c. CCl$_3$F (Freon-11; trichlorofluoromethane)

Answer
a. $1 \text{ mol O}_3 = 3 \text{ mol O}$

$$= 3 \text{ mol O} \times \frac{16.0 \text{ g O}}{1 \text{ mol O}}$$

$$= 48.0 \text{ g O}_3$$

We started out on this mathematical excursion so that we could calculate the mass of CO_2 produced from burning 3.3 Gt of carbon. We now have all the pieces necessary to solve the problem. Out of every 44.0 g of CO_2, 12.0 g is C. This mass ratio holds for all samples of CO_2, and we can use it to calculate the mass of C in any known mass of CO_2. More to the question at hand, we can use it to calculate the mass of CO_2 released by any known mass of carbon. It only depends on how we arrange the ratio. The C-to-CO_2 ratio is $\dfrac{12.0 \text{ g C}}{44.0 \text{ g } CO_2}$, but it is equally true that the CO_2-to-C ratio is $\dfrac{44.0 \text{ g } CO_2}{12.0 \text{ g C}}$.

For example, we could compute the number of grams of C in 100.0 g CO_2 by setting up the relationship in this manner.

$$100.0 \text{ g } \cancel{CO_2} \times \frac{12.0 \text{ g C}}{44.0 \text{ g } \cancel{CO_2}} = 27.3 \text{ g C}$$

The fact that there is 27.3 g of carbon in 100.0 g of carbon dioxide is equivalent to saying that the mass percent of C in CO_2 is 27.3%. Note that carrying along the labels "g CO_2" and "g C" helps you do the calculation correctly. The label "g CO_2" can be canceled, and you are left with the desired label, "g C." Keeping track of the labels and canceling where appropriate are useful strategies in solving many problems. This method is sometimes called "unit analysis."

Your Turn 3.22　　Mass Ratios and Percents

a. Calculate the mass ratio of S in SO_2.
b. Find the mass percent of S in SO_2.
c. Calculate the mass ratio and the mass percent of N in N_2O.

Answers

a. The mass ratio is found by comparing the molar mass of S to the molar mass of SO_2.

$$\frac{32.1 \text{ g S}}{64.1 \text{ g } SO_2} = \frac{0.501 \text{ g S}}{1.00 \text{ g } SO_2}$$

b. To find the mass percent of S in SO_2, multiply the mass ratio by 100.

$$\frac{0.501 \text{ g S}}{1.00 \text{ g } SO_2} \times 100 = 50.1\% \text{ S in } SO_2$$

Calculation tip

Predict:
Will the answer be larger or smaller than the given value? How will you label the answer?

Check:
Does the answer match your prediction? Have labels canceled, leaving the label needed for the answer?

To find the mass of CO_2 that contains 3.3 gigatons (Gt) of C, we use a similar approach. We could convert 3.3 Gt to grams, but it is not necessary. As long as we use the same mass unit for C and CO_2, the same numerical ratio holds. Compared with our last calculation, this problem has one important difference in how we use the ratio. We are solving for the mass of CO_2, not the mass of C. Look carefully at the cancellation of labels this time.

$$3.3 \text{ } \cancel{\text{Gt C}} \times \frac{44.0 \text{ Gt } CO_2}{12.0 \text{ } \cancel{\text{Gt C}}} = 12 \text{ Gt } CO_2$$

The question to be answered is: What mass of CO_2 contains 3.3 Gt of carbon?

Once again the labels cancel and the answer comes out with the needed label, Gt CO_2.

Our burning question, "What is the mass of CO_2 added to the atmosphere each year from the combustion of fossil fuels?" has finally been answered: 12 gigatons. Of course, our not-so-hidden agenda was to demonstrate the problem-solving power of chemistry and to introduce five of its most important ideas: atomic mass, molecular mass, Avogadro's number, mole, and molar mass. The next few activities provide opportunities to practice your skill with these concepts.

Your Turn 3.23   SO$_2$ from Volcanoes

a. It is estimated that volcanoes globally release about 19×10^6 t (19 million metric tons, tonnes) of SO$_2$ per year. Calculate the mass of sulfur in this amount of SO$_2$.

b. If 142×10^6 t of SO$_2$ is released per year by fossil-fuel combustion, calculate the mass of sulfur in this amount of SO$_2$.

Answer

a. The mass ratio of S to SO$_2$ is known from Your Turn 3.22.

$$19 \times 10^6 \text{ t SO}_2 \times \frac{32.1 \times 10^6 \text{ t S}}{64.1 \times 10^6 \text{ t SO}_2} = 9.5 \times 10^6 \text{ t S}$$

If you know how to apply these ideas, you have gained the ability to critically evaluate media reports about releases of C or CO$_2$ (and other substances as well) and judge their accuracy. One can either take such statements on faith or check their accuracy by applying mathematics to the relevant chemical concepts. Obviously, there is insufficient time to check every assertion, but we hope that readers develop questioning and critical attitudes toward all statements about chemistry and society, including those found in this book.

Sceptical Chymist 3.24   Checking Carbon from Cars

A clean-burning automobile engine will emit about 5 lb of C in the form of CO$_2$ for every gallon of gasoline it consumes. The average American car is driven about 12,000 miles per year. Using this information, check the statement that the average American car releases its own weight in carbon into the atmosphere each year. List the assumptions you make in solving this problem. Compare your list and your answer with those of your classmates.

3.8   Methane and Other Greenhouse Gases

Concerns about an enhanced greenhouse effect are based primarily, but not solely on increases in atmospheric CO$_2$. Total greenhouse gas emissions have risen 16% from 1990–2005, and the dominant gas emitted was CO$_2$, mainly from fossil-fuel combustion. However, other gases play a role. Methane, CH$_4$, is present in the atmosphere in a much lower concentration than CO$_2$, but is at least 20 times more effective than CO$_2$ in its ability to trap infrared energy. It has a relatively short average atmospheric lifetime of 12 years. The **global atmospheric lifetime** characterizes the time required for a gas added to the atmosphere to be removed. It is also referred to as the "turnover time." In the case of CH$_4$ added to the air in a given year, it will be gone from the atmosphere on average 12 years later. Fortunately, CH$_4$ is quite readily converted to less harmful chemical species by interaction with tropospheric •OH free radicals. Compare that situation with CO$_2$, a gas with far slower removal mechanisms. Atmospheric concentration of CH$_4$ at this time is relatively low, but its current level is estimated to be more than twice that before the Industrial Revolution. Table 3.2 gives a comparison of changes in methane concentration with those of carbon dioxide and another greenhouse gas, nitrous oxide.

Methane has a variety of sources. Although the CH$_4$ cycle is not as well understood as the carbon cycle discussed earlier, about 40% of CH$_4$ emissions are thought to be from natural sources. Some of the natural sources have been magnified by human

Table 3.2	Greenhouse Gases–Concentration Changes and Lifetimes		
	CO_2	CH_4	N_2O
Preindustrial concentration (1750)	278 ppm	0.700 ppm	0.270 ppm
2005 concentration	385 ppm	1.75 ppm	0.314 ppm
Average rate of concentration change, 1990–2005	1.5 ppm/year	0.007 ppm/year	0.0008 ppm/year
Global atmospheric lifetime	50–200 years*	12 years	114 years

*A single value for the atmospheric lifetime of CO_2 is not possible. Different removal mechanisms take place at different rates, leading to variation in atmospheric lifetime.

Global atmospheric lifetime values, although useful for comparison, are best thought of as approximations. There are many variables in their determination.

activities. For example, because CH_4 is a major component of natural gas, some has always leaked into the atmosphere from rock fissures. But the exploitation of these deposits and the refining of petroleum have led to increased emissions. Similarly, CH_4 has always been released by decaying vegetable matter in wetlands. Its early name, "marsh gas," reflects this origin. Thus, the decaying organic matter in landfills and from the residue of cleared forests generates CH_4. Methane formed in the major New York City landfill at Fresh Kills, Staten Island is used for residential heating, and manufacturing sites such as the BMW factory in South Carolina regularly use landfill gases to reduce their energy expenditures. However, at most landfills CH_4 simply escapes into the atmosphere.

Another major source of CH_4 is agriculture, particularly cultivated rice paddies. Rice is grown with its roots under water where **anaerobic bacteria,** those that can function without the use of molecular oxygen, produce methane. Most of this is released to the atmosphere. Additional agricultural CH_4 comes from an increasing number of cattle and sheep. The digestive systems of these ruminants (animals that chew their cud) contain bacteria that break down cellulose. In the process, methane is formed and released through belching and flatulence—about 500 L of CH_4 per cow per day. The ruminants of the Earth release a staggering 73 million tonnes of CH_4 each year. A similar chemistry is carried on in the guts of termites, making them a major source of CH_4. The sheer number of termites is staggering, estimated to be more than half a tonne for every man, woman, and child on the planet!

There is a possibility that global warming exacerbates the release of CH_4 from ocean mud, bogs, peatlands, and even the permafrost of northern latitudes. In these areas, a substantial amount of methane appears to be trapped in "cages" made of water molecules. Such deposits are referred to as methane hydrates. As the temperature increases, the escape of CH_4 becomes more likely. Australia's Commonwealth Scientific and Industrial Research Organization (CSIRO) has been taking a series of ocean core drillings to gather evidence about methane hydrate and its role in global warming. Its findings link periods of historic global warming with the release of methane (Figure 3.20).

The complex details of the generation and fate of atmospheric methane make it difficult to speak with certainty about its future effect on the average temperature of the planet. There was a 10% decrease in CH_4 emissions in the United States between 1990 and 2005, although atmospheric concentrations, which typically lag behind changes in emissions, have not yet shown the same decline. Globally, methane's effect will be less pronounced than temperature changes caused by CO_2, adding only perhaps a few tenths of a degree to the average temperature of Earth in the next 100 years. This is in sharp contrast to the major effect predicted for CO_2, a temperature rise of at least 1.0–3.5 °C by the end of this century.

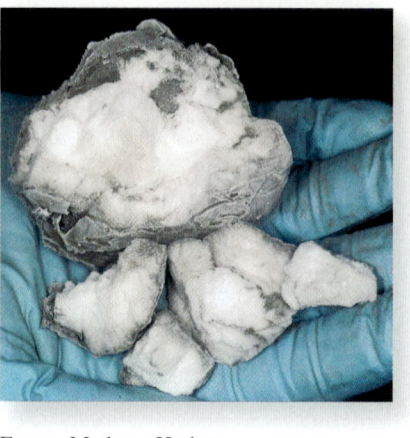

Ocean Drilling Program   Frozen Methane Hydrate

Figure 3.20

The Ocean Drilling Program of CSIRO obtained this sample of frozen methane hydrate from the continental shelf off the coast of Florida.

Another gas under study for its contributions to global warming is nitrous oxide, also known as "laughing gas." It has been used as an inhaled anesthetic for dental and medical purposes. Its sources and sinks are not as well established as are those for other greenhouse gases. The majority of N_2O molecules in the atmosphere come from the bacterial removal of nitrate ion (NO_3^-) from soils, followed by removal of oxygen. Agricultural practices, again linked to population pressures, can speed up the removal of reactive compounds of nitrogen from soils. Other sources include gases from ocean upwelling, and stratospheric interactions of nitrogen compounds with high-energy oxygen atoms. Major anthropogenic, or human-caused, sources of N_2O are automobile catalytic converters, ammonia fertilizers, burning of biomass, and certain industrial processes (nylon and nitric acid production). In the atmosphere, a typical N_2O molecule persists for about 114 years, absorbing and emitting infrared radiation. Over the past decade, global atmospheric concentrations of N_2O have shown a slow but steady rise. There has been a slight decrease in U.S. emissions from 1990 to 2005.

Ozone itself also can act like a greenhouse gas, but its efficiency depends very much on altitude. It appears to have its maximum warming effect in the upper troposphere, around 10 km above the Earth. Depletion of ozone has a slight cooling effect in the stratosphere and it may also promote slight cooling at Earth's surface. Although ozone is a part of the global warming story, depletion of the ozone layer in the stratosphere is clearly not a principal cause of climate change. However, stratospheric ozone depletion and climate change are linked in another important way, through ozone-destroying substances. CFCs, HCFCs, and halons, all implicated in the destruction of stratospheric ozone, also absorb infrared radiation and are greenhouse gases. Emissions of these synthetic gases have risen by 58% from 1990–2005, although their concentrations are still very low.

Section 2.8 discussed the role of N_2O in destroying stratospheric ozone.

In addition to the variation in atmospheric lifetime among greenhouse gases, they also vary in their effectiveness in absorbing infrared radiation. This is quantified by the **global warming potential (GWP),** a number that represents the relative contribution of a molecule of the atmospheric gas to global warming. This number is assigned only for greenhouse gases with relatively long lifetimes. Carbon dioxide is assigned the reference value of 1; all other greenhouse gases are indexed with respect to it. Gases with relatively short lifetimes, such as water vapor, tropospheric ozone, tropospheric aerosols, and other ambient air pollutants, are distributed unevenly around the world. It is difficult to quantify their effect, and therefore GWP values are not usually assigned. Values of the global warming potential for the three most common long-lived greenhouse gases and their average concentrations in the troposphere are given in Table 3.3.

Table 3.3	Global Warming Potential for Three Greenhouse Gases	
Substance	**Global Warming Potential (GWP)***	**Tropospheric Abundance (ppm)**
CO_2	1	385
CH_4	23	1.8
N_2O	296	0.31

*GWP values are given for the estimated direct and indirect effects over a 100-year period and are relative to the assigned value of 1 for CO_2.

Your Turn 3.25   Comparing Greenhouse Gas Effectiveness

Multiplying GWP by tropospheric abundance provides a number that can be used to compare the effectiveness of a greenhouse gas.

a. HFC-134a (CF_3CH_2F) has a GWP value of 1300 and a tropospheric abundance of 7.5 parts per trillion (1998 data). Compare its effectiveness as a greenhouse gas with that of CO_2.
Hint: Use the same unit of tropospheric abundance for both gases.

b. Freon-12 (CCl_2F_2) has a GWP of 10,600 and a tropospheric abundance of 553 parts per trillion (1998 data). Comment on its effectiveness as a greenhouse gas relative to that of both CO_2 and HFC-134a.

c. HFC-134a (lifetime = 13.8 years) is a replacement for Freon-12 (lifetime = 100 years). Why are their global atmospheric lifetimes important to their overall effectiveness as greenhouse gases?

HFCs were discussed in Section 2.12.

Other anthropogenic greenhouse gases that are assigned GWP values include several hydrofluorocarbons (HFCs), two perfluorocarbons (CF_4 and C_2F_6), and sulfur hexafluoride (SF_6). Perfluorocarbons (PFCs) are emitted as a by-product of aluminum smelting and used in the manufacture of semiconductors. Both CF_4 and C_2F_6 have long lifetimes and high GWP values. However, their concentration in the atmosphere is very low at this time, but rising. Sulfur hexafluoride (SF_6), used for electrical insulation in transformers and a cover gas for smelting operations, has a tropospheric lifetime of 3200 years. It is over 22,000 times more potent as a greenhouse gas than CO_2, but its atmospheric concentration is extremely low, measured in parts per trillion.

3.9   Gathering Evidence: Projecting into the Future

Understanding evidence from the past is important. So is making sense of recent trends. The real challenge, however, lies in understanding the complexities well enough to *predict* climate change. In all computer models, the assumption is that rising concentrations of greenhouse gases will increase the average global temperatures (Figure 3.21). Rising temperatures, in turn, may produce changes in weather patterns, land use, human health, and alterations in Earth's ecosystems. To accurately model global climate, one must include a number of often incompletely understood astronomical, meteorological, geological, and biological factors. Even the most sophisticated computer program can succeed only if the important factors are identified and weighted in importance. The situation is greatly complicated because many of these variables are interrelated and cannot be studied independently. Dr. Michael Schlesinger, who directs climate research

Figure 3.21
Studying a computer simulation of future climate change.

at the University of Illinois, remarked: "If you were going to pick a planet to model, this is the *last* planet you would choose." Despite these difficulties, policy decisions must be made on the best possible models. The usefulness and limitations of all models needs to be clearly understood and the level of uncertainty in them quantified. These are tall tasks, to be sure, but essential ones for making informed assessments of climate change.

Important in any model for climate change is the role of the oceans, where over 97% of water on Earth is found. Much of the heat radiated by the greenhouse gases may be going into the oceans, which act as a thermal buffer. Although the oceans are very important in moderating the temperature of the planet, their capacity to do so is limited. We know that increasing the temperature of the oceans will decrease the solubility of CO_2, thus releasing more of it into the atmosphere. You may have witnessed the same effect when a glass of cold sparkling water or soda warms up to room temperature, becoming "flat" as its dissolved gases escape. An increase in the temperature of the oceans may promote the growth of tiny photosynthetic plants called phytoplankton, and hence increase CO_2 absorption. But the result could be just the opposite. Water in a warmer ocean will not circulate as well as it does now, which may inhibit plankton growth and CO_2 removal. Processes in the ocean vary considerably by depth, with the deep ocean being one of the largest global carbon reservoirs. Turnover time for carbon in the deep ocean is in the range of 2000–5000 years, but 0.1–1 year for marine biomass. These and other rate factors must make their way into the computer model.

> The unique properties of water, including its ability to absorb heat, will be discussed in Chapter 5.

Consider This 3.26   Climate Questions

Climate-modeling sites on the Web may deluge you with technical terms and numerical analyses. A good place to begin your understanding of climate modeling is to visit the National Climatic Data Center (NCDC), billed as "the world's largest active archive of weather data." A direct link is provided at the *Online Learning Center.* What types of data are provided by NCDC? Propose two or three questions that you might like to investigate using these data.

Smoke and haze from all sources may be clouding our view of global warming. One group of researchers, led by Benjamin Santer of Lawrence Livermore National Laboratory, has found that predictions agree more closely with observations if the model includes the cooling effect of atmospheric aerosols. Aerosols are a complex group of materials that include dust, sea salt, smoke, carbon, and compounds containing nitrogen and sulfur. One of the most common aerosols consists of tiny particles of ammonium sulfate, $(NH_4)_2SO_4$. This compound can form when sulfur dioxide (SO_2) reacts with ammonia (NH_3). Both compounds can be released by natural or human-influenced sources. Burning crop wastes, rain forest trees, low-grade fuels such as charcoal, and fossil fuels all produce aerosols capable of blocking sunlight.

> Aerosols were defined in Section 1.11 and will be discussed further in Section 6.5.

Many particles in aerosols are smaller than about 4 μm in diameter and are efficient at scattering incoming solar radiation with wavelengths close to their size. Thermal radiation coming from the Earth has wavelengths in the infrared part of the spectrum, ranging from 4 to 20 μm. The smaller aerosol particles are not effective in scattering these wavelengths, allowing the greenhouse gases to continue to absorb terrestrial radiation but at a reduced level because less solar radiation is reaching the surface to be radiated back into space. In addition, aerosol particles serve as nuclei for the condensation of water droplets and hence cloud formation. Thus, aerosols counter the warming effects of greenhouse gases.

In December 2005, the journal *Nature* reported results from an international scientific team's study about the consequences of aerosol concentration for global warming. They confirmed earlier research that aerosols have helped dramatically to counter the effects of global warming. The observed temperature increase of 0.7 °C over the last century might well have been over 2 °C without the steadily increasing concentration of aerosols. Have

we minimized global warming by emitting increasing amounts of smoke and soot in the atmosphere? We have much to learn, such as understanding the role of water droplets themselves in haze that contains aerosols. In any case, the protection provided by anthropogenic aerosols may only be temporary, given that the health effects of many aerosols in the troposphere will mandate declining emissions.

Although the influence of aerosols on climate change has been widely studied, aerosols are far from the only factor. Change in land use is another major driver of climate change. The influences of deforestation or crop change on atmospheric concentrations of CO_2 and CH_4 have been recognized, but other effects are being studied. One of the most important is the effect on **albedo,** the ratio of electromagnetic radiation *reflected* relative to the amount of radiation *incident* on the surface. Thus albedo is a measure of the reflectivity of a surface. Changes in albedo can alter regional temperatures, precipitation, vegetation, and other climate variables. If a snow-covered area warms and the snow melts, the albedo decreases, more sunlight is absorbed, and the temperature tends to increase further. This effect helps to explain the greater increases in average temperature observed in the Arctic, for example, where the amount of sea ice is decreasing. Similarly, if a glacier retreats leaving exposed darker rock, the albedo decreases and the surface temperature increases.

> Earth has an average albedo of 39%. The albedo of the Moon is about 12%.

The albedo–temperature effect is actually much stronger in tropical regions of the Earth, despite the lack of snow. The sunlight is more consistent in the tropics, and changes in land use produce great change in albedo. If expanses of deep green tropical rain forest trees are removed to expose darker soil, the albedo decreases. Studies have shown an average temperature increase in Brazilian rain forests converted to agriculture of about 3 °C (5 °F) year-round, a significant change.

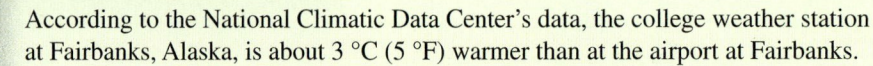

Consider This 3.27 Frozen Fairbanks

According to the National Climatic Data Center's data, the college weather station at Fairbanks, Alaska, is about 3 °C (5 °F) warmer than at the airport at Fairbanks.

a. Give possible reasons for this observed difference.
b. Why is the difference greater during the winter months?

A significant uncertainty in any projection about global warming is whether the rate of population growth will stabilize during the next 100 years. During the 20th century, worldwide population increased from 2 billion people to approximately 6 billion. By January 2006, approximately 6.5 billion people inhabited our world. Because increased numbers of people translate into increased energy use and greater greenhouse gas emissions through burning fossil fuels, global warming scenarios incorporate different assumptions about population growth rates and economic growth. A low-end projection assumed a world population in the year 2100 of 6.4 billion and an annual economic growth rate of 1.2%. A midrange scenario is based on 11.3 billion people with a 2.3% annual economic growth rate, nearly twice that of the low-end projection. In a high-end projection, the year 2100 population is 11.3 billion as in the midrange calculation, but annual economic growth would occur at 3.0%.

Scientists have developed increasingly detailed computer programs to model Earth's climate. With a new generation of parallel computers, many researchers are working to improve climate models and resolve some of the observed conflicting data. Studying the climate is difficult not only because the system itself is incredibly complex, but also because no possibility exists for a controlled experiment. There will remain uncertainty and room for scientific dissent, as well as differences in political interpretation. Simulating Earth's temperature variations and comparing the results with measured changes can provide insight into the underlying causes of major changes. Climate scientists call both natural and anthropogenic causes by the term **forcings,** factors that affect the annual global mean surface temperature (Figure 3.22).

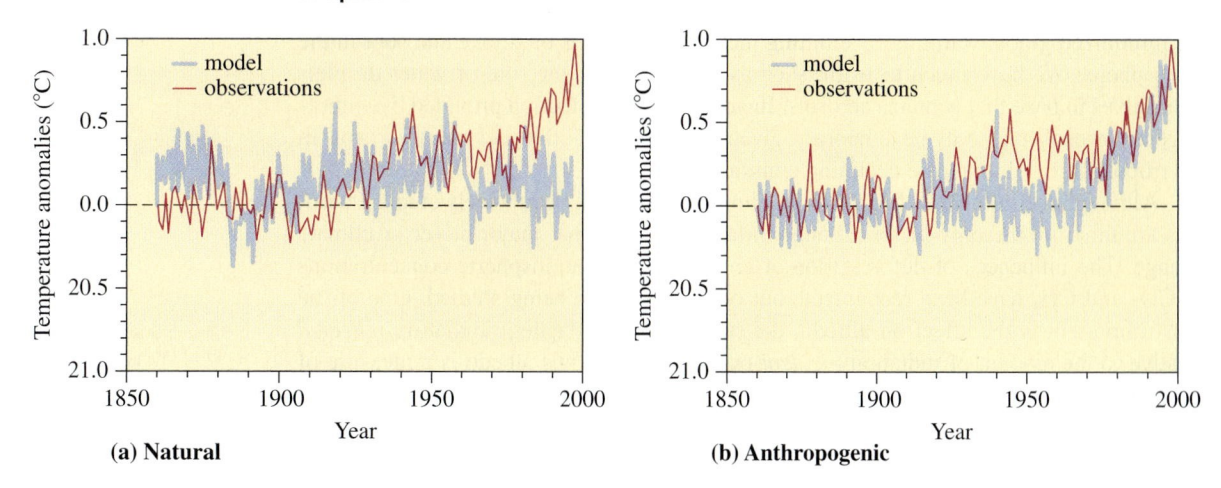

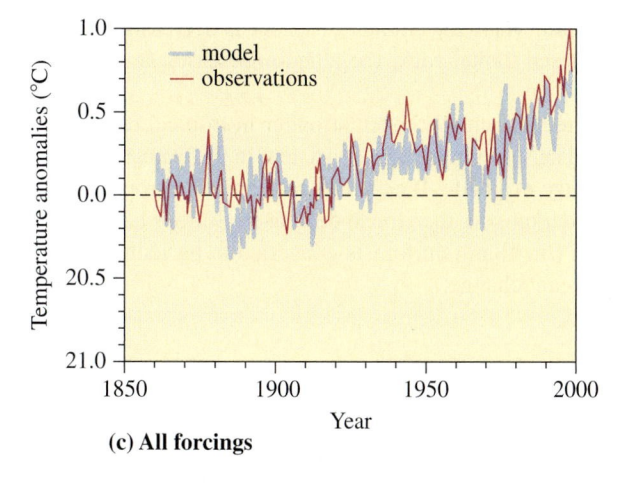

Figure 3.22

Simulated annual global mean surface temperatures of the Earth.

Source: *Intergovernmental Panel on Climate Change (IPCC) Report, 2001.*

Consider This 3.28	**Comprehending Computer Models**

a. How successful was the IPCC model shown in Figure 3.22 in correlating natural forcings with observed temperature change?

b. How successful was the IPCC model shown in Figure 3.22 in correlating anthropogenic forcings with observed temperature change?

c. What are some of the forcings included in the part (c) graph?

Many hundreds of scientists from all over the world participated in preparing the 2001 IPCC report. They used words to help policy makers and the public at large better understand the inherent uncertainty and reliability of the data. Table 3.4 gives these terms and the scientists' definitions that continue to be used in all subsequent updates to this report.

The 2001 IPCC report came to conclusions that utilized these terms. For example, it was judged very unlikely that all of the observed global warming was due to natural climate variability. Rather, the scientific evidence strongly supports the position that human activity is a significant factor causing the increase in average global temperature observed over the last century. The 2007 IPCC report stated that the scientific evidence for global warming was *unequivocal* and that human activity is the main driver. Some of IPCC's conclusions, together with their assigned probabilities, are shown in Table 3.5.

The World Meteorological Organization reported that 2004, 2005, and 2006 were among the hottest years on record; 1998 remains the hottest. Nine of the 10 hottest years on record occurred in the decade from 1995 to 2005. The last six years were among the

Table 3.4	Judgmental Estimates of Confidence	
Term	**Probability That a Result Is True**
Virtually certain	> 99%
Very likely	90–99%
Likely	66–90%
Medium likelihood	33–66%
Unlikely	10–33%
Very unlikely	1–10%

Source: *Summary for Policymakers, A Report of Working Group 1 of the Intergovernmental Panel on Climate Change,* Shanghai: IPCC, January 1, 2001.

seven hottest years on record since reliable data began being kept in 1861. "Temperatures are warming, and they've been warming over the past century," said Jay Lawrimore, head of the climate-monitoring branch at National Climatic Data Center. "There's pretty much a consensus that there will be continued warming over the next century."

What are the anticipated effects of projected warming? Recent models predict that the top 11 ft of surface permafrost in the Arctic could disappear by 2100. Changes in sea ice, snow, and glaciers all can contribute to changes in sea level. Global mean sea level is projected to rise by 9–88 cm (3.5–34.6 in.) between 1990 and 2100. Predictions of rising sea levels are caused by thermal expansion of warmer water as well as by melting of frozen precipitation. Sea level rises of this magnitude would endanger New York, New Orleans, Miami, London, Venice, Bangkok, Taipei, and other coastal cities. Millions of people might have to relocate; millions more could drown. Dr. Richard Williams of the U.S. Geological Survey states: "If the ice sheets on Greenland melted, that alone could raise sea level by 20 feet." It is far from certain that major increases in sea level will occur. Even if they did, they would take place over many years, providing considerable time for preparation and protection.

Table 3.5	IPCC Conclusions

Very Likely
- Human-caused emissions are the main factor in causing warming since 1950.
- Higher maximum temperatures are observed over nearly all land areas.
- Snow cover decreased about 10% since the 1960s (satellite data); in the 20th century there was a reduction of about two weeks in lake and river ice cover in the middle and high latitudes of the Northern Hemisphere (independent ground-based observations).
- Increased precipitation has been observed in most of the Northern Hemisphere continents.

Likely
- Temperatures in the Northern Hemisphere during the 20th century have been the highest of any century during the past 1000 years.
- Arctic sea ice thickness declined about 40% during late summer to early autumn in recent decades.
- An increase in rainfall, similar to that in the Northern Hemisphere, has been observed in tropical land areas falling between 108° N and 108° S.
- Increased summer droughts.

Very Unlikely
- The observed warming over the past 100 years is due to climate variability alone, providing new and even stronger evidence that changes must be made to stem the influence of human activities.

Consider This 3.29   **A Paradise Drowning**

The Maldives, a chain of tropical islands southwest of India, may be the first modern nation to be drowned by rising sea levels.

a. Why are these islands particularly susceptible to rising sea levels?

b. What are the implications for the United States if the Maldives were lost to the sea?

c. Are any other islands currently in danger of the same fate?

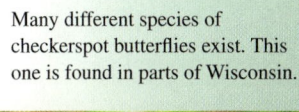

Many different species of checkerspot butterflies exist. This one is found in parts of Wisconsin.

G8 countries are Canada, France, Germany, Italy, Japan, the U.K., the U.S., and Russia.

Some climatologists also feel that an increase in the average ocean temperature could cause more weather extremes, including storms, floods, and droughts. In the Northern Hemisphere, summers are predicted to be drier and winters wetter. The regions of greatest agricultural productivity could change. Drought and high temperatures could reduce crop yields in the American Midwest, but the growing range might extend farther into Canada. It is also possible that some of what is now desert could get sufficient rain to become arable. One region's loss may well become another locale's gain, but it is too early to tell.

Global warming already is having observable effects on plants, insects, and animal species around the world. Species as diverse as the California starfish, Alpine herbs, ants, and checkerspot butterflies have all exhibited changes in either their ranges or their habits. Dr. Richard P. Alley, a Pennsylvania State University expert on past climate shifts, sees particular significance in the fact that animals and plants that rely on each other will not necessarily change ranges or habits at the same rate. Referring to affected species, he said, "You'll have to change what you eat, or rely on fewer things to eat, or travel farther to eat, all of which have costs." The result in decades to come could be substantial ecological disruption, local losses of wildlife, and possible extinctions.

In other respects, we may all be losers in a warmer world. Recently, physicians and epidemiologists have attempted to assess the costs of global warming in terms of public health. An increase in average temperatures is expected to increase the geographical range of mosquitoes, tsetse flies, and other insects. The result could be a significant increase in diseases such as malaria, yellow and dengue fevers, and sleeping sickness in new areas, including Asia, Europe, and the United States. Indeed, it has been suggested that the deadly 1991 outbreak of cholera in South America is attributable to a warmer Pacific Ocean. The bacteria that cause cholera thrive in plankton. The growth of plankton and bacteria are both stimulated by higher temperature. We even may feel the effects of global warming in our own backyards, if predictions for more ticks, more ants, more mosquitoes, more pollen, and faster growing and more potent poison ivy are correct.

The leaders of the Group of Eight (G8) countries, meeting in Scotland in the summer of 2005, issued a communiqué on climate change, clean energy, and sustainable development. They declared "climate change is a serious and long-term challenge that has the potential to affect every part of the globe." They concluded that scientific evidence now warrants a new sense of urgency. The next section will consider how public policy around the world has been and will continue to be influenced by our scientific knowledge.

3.10   Strategies for Change

The debate over climate change has shifted in the last 15 years. The discussion is no longer about "is there a problem" but rather about "what should we do?" There is a growing scientific consensus that global warming is occurring and that humans are the cause. The focus has switched to understanding the causes of the warming and the means of slowing or preventing projected climate changes. There are two related

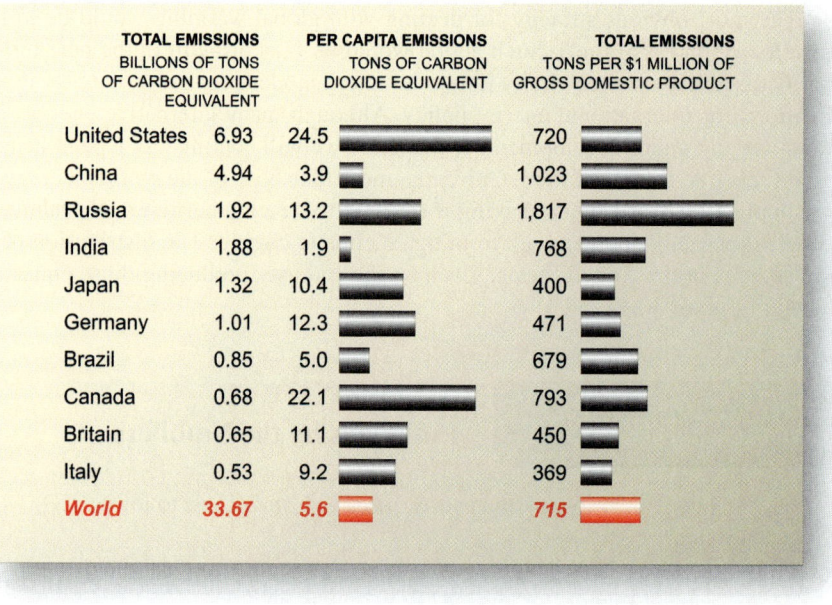

	TOTAL EMISSIONS BILLIONS OF TONS OF CARBON DIOXIDE EQUIVALENT	PER CAPITA EMISSIONS TONS OF CARBON DIOXIDE EQUIVALENT		TOTAL EMISSIONS TONS PER $1 MILLION OF GROSS DOMESTIC PRODUCT	
United States	6.93	24.5		720	
China	4.94	3.9		1,023	
Russia	1.92	13.2		1,817	
India	1.88	1.9		768	
Japan	1.32	10.4		400	
Germany	1.01	12.3		471	
Brazil	0.85	5.0		679	
Canada	0.68	22.1		793	
Britain	0.65	11.1		450	
Italy	0.53	9.2		369	
World	*33.67*	*5.6*		*715*	

Figure 3.23

Countries that are the largest producers of greenhouse gases from fossil-fuel combustion.
Note: Data exclude CO_2 emissions resulting from land use change and deforestation. If included, the amounts would probably rise significantly for Brazil.

Source: From "In Russia, Pollution Is Good for Business," by Andrew E. Kramer, *New York Times,* December 28, 2005. Copyright ©2005 The New York Times. Reprinted with permission.

questions. What *can* we do and what *should* we do about the possibility of significant climate change caused by global warming? One thing is clear: Given the recent results from improved climate modeling, we will start seeing even more definitive climatic changes within a decade or so. But can we prudently wait that long or is prompt action essential? Whether to act and how to act are not just scientific issues. What determines our response is a complicated mix of science, perception of risk, societal values, politics, and economics.

In total emissions and on a per capita basis, the United States leads the world in carbon emissions. Russia leads the world in emissions based on economic output. Therefore both countries have a mandate to help lead the way in reducing emissions, but to do so without sacrificing the economy. Comparisons of total emissions, per capita emissions, and emissions based on economic output are found in Figure 3.23.

Consider This 3.30   The Top Emitters

Use Figure 3.23 and the direct link to the Carbon Dioxide Information Analysis Center (CDIAC) Web site provided on our *Online Learning Center* to answer these questions.

a. The United States was the leader in total CO_2 emissions and per capita emissions in 2005. Which countries rank second and third in each category?

b. Russia is the leader in total CO_2 emissions based on economic output. Which countries rank second and third in that category?

c. Compare the lists from parts **a** and **b.** Offer reasons for the observed differences.

d. From the CDIAC Web site, pick any country *not* shown in Figure 3.23. Where is that country located? How do the total emissions compare with those of the United States? How have this country's CO_2 emissions changed over time?

Alternatives to fossil fuels will be discussed in Chapters 7 and 8.

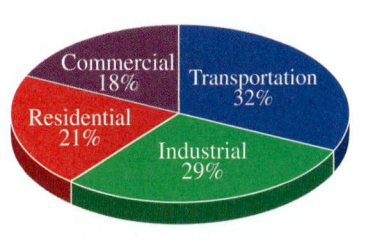

Figure 3.24

U.S. CO_2 emissions from fossil fuels, by end-use sectors, 2004.

Source: EIA, *Emission of Greenhouse Gases in the U.S., 2004.*

Clean coal technology is discussed in Sections 4.6 and 6.14.

The most obvious strategy for dealing with global warming would be to reduce our reliance on fossil fuels. Such action would be very difficult in the near term. Not only is this energy source vitally important to our modern economy, but also it has implications for international energy policy. Although many alternative energy sources are in use and under development, realistically we will continue to depend heavily on fossil fuels into the near future. Currently, more than 85% of the world's energy needs are supplied by fossil fuels, making it difficult to develop sufficient alternative technological capacity soon enough to mitigate climate change. The distribution of emission sources in the United States, given by the end uses producing those emissions, is shown in Figure 3.24.

Consider This 3.31   Emissions by the Numbers

a. Use Figure 3.24 to rank the end-use sectors from highest to lowest CO_2 emissions.
b. Is the sector with the highest emissions the one receiving the most government support for change? Why or why not?

To just keep using fossil fuels at the same rate until we run out is not a reasonable approach either. In fact, it is already too late for this to be a viable option. We saw in Chapter 2 that although CFCs are no longer being added to the atmosphere, it will be many years before their concentration in the stratosphere is reduced to earlier desirable levels. Similarly, significant time lags occur from the production of CO_2, until its accumulation in the atmosphere, and to the eventual return of the CO_2 into an environmental sink. Thus, waiting until we run out of fossil fuels to take action will not solve the problem.

Some still advocate delaying action regarding global warming, believing that further study is needed. They argue that uncertainties in our predictive powers and climate models are so great that money and effort would be wasted now to undertake preventive or ameliorative action. The U.S. National Research Council report generally agreed with the assessment of human-caused climate change presented in the 2001 IPCC scientific report, but issued this caution. "Because there is considerable uncertainty in current understanding of how the climate system varies naturally and reacts to emissions of greenhouse gases and aerosols, current estimates of the magnitude of future warming should be regarded as tentative and subject to future adjustments (either upward or downward)."

The developed world has often adopted policies that depend on "technological fixes" for a problem. Increasing the efficiency of fossil-fuel plants through clean coal technology is one successful approach. Rather than stopping or slowing the emission of greenhouse gases, some advocate capturing and isolating the gases after their emission. One such method is **sequestration**, which literally means keeping something apart. If CO_2 is properly sequestered, it cannot reach the troposphere and contribute to global warming. However, the scale for sequestering all CO_2 produced from fossil-fuel power plants is quite daunting, and likely not attainable.

Successful approaches involve linking technological advances, economic development, and environmental needs. In 2005, a U.S. Department of Energy (DOE) project in Kansas successfully demonstrated the feasibility of "flooding" an oil field with waste CO_2 produced from the fermentation of corn to produce ethanol. The CO_2 was injected into the Hall-Guerney oil field in Kansas, allowing recovery of oil that otherwise might never have been produced. The benefits from integrating ethanol production, enhancing oil recovery, and CO_2 sequestration could become a model for other types of projects.

Another proposed form of carbon sequestration is based on capturing CO_2 from a power plant or removed from natural gas. The CO_2 can then be liquefied and pumped deep into the ocean. This type of sequestration has been implemented off the coast of

Figure 3.25

Injecting CO_2 beneath the floor of the North Sea. The green cable carries electricity, the blue pipeline transports CO_2 to the injection sites, and the red "haltenpipe" transports natural gas from the offshore Heidrun platform to a methanol plant onshore at Tjeldbergodden.

Source: Image courtesy of Statoil.

Norway since 1996 in the Sleipner natural gas field. Here CO_2 is being pumped to a depth of over 100 m below the ocean's surface (Figure 3.25). This project is driven by concern for global warming and by financial incentives. Norway's state oil company Statoil, is projected to save millions of dollars by using carbon sequestration. The company built an $80 million dollar at-sea facility to separate CO_2 from natural gas for two reasons: first because they could not sell their natural gas to European customers without first removing carbon dioxide from it, and second because sequestering the CO_2 avoids a stiff "carbon tax" imposed by Norway. This tax would have cost Statoil about $50 for every ton of CO_2 emitted, thus saving about $50 million a year in taxes. The injection site is being monitored for possible seismic effects. A similar project is under consideration by a consortium involving Exxon and the Indonesian State Oil Company in an offshore gas field in the South China Sea.

Increasing rates of carbon sequestration may mean planting trees as environmental sinks that absorb CO_2. For example, British and Malaysian scientists and volunteers planted 120,000 trees in logged-over rain forest in Malaysian Borneo during 2003. This is the largest-ever experiment to explore how tree diversity can influence both timber production and the storage of carbon. Although reforestation, saving old-growth forests, and improved land management are good for humankind, they may turn out to have little long-term effect on the CO_2 in the atmosphere. This is because much of the carbon tied up in forests and soils moves back into circulation in 30–60 years, according to Dale Simbeck, Vice President of Technology for the consulting firm SFA Pacific.

Consider This 3.32   **Trees as Carbon Sinks**

Some researchers have concluded that new forest plantations are not very efficient at sequestering carbon. What evidence exists for this conclusion? Does it make a difference if the new plantings replace other trees or cropland? Present your findings in a report.

3.11   Beyond the Kyoto Protocol on Climate Change

More than a century ago Arrhenius first proposed that carbon dioxide emissions could accumulate in the atmosphere and lead to global warming. The world may have been slow in responding, but the last few decades have seen considerable progress. Rising concentrations of CO_2 were detected in the early 1960s, and data were gathered for other greenhouse gas concentrations in the 1970s. Primarily only atmospheric scientists knew of these trends before the mid-1980s, when international workshops and conferences brought the situation to the attention of United Nations agencies, particularly the UN Environmental Program and the World Meteorological Organization (WMO). The stage was set in 1988 when the IPCC was established to gather available scientific research on climate change and provide advice to policy makers. A series of international conferences led to the "Earth Summit" in Rio de Janeiro, Brazil, in June of 1992. More than 160 countries, including the United States, adopted the Framework Convention on Climate Change (FCCC) by the close of that meeting. This document presented scientific evidence that increasing temperatures were a global concern and suggested effective ways of responding.

In 1997, nearly 10,000 participants from 161 countries gathered in Kyoto, Japan. They established goals to stabilize and reduce atmospheric greenhouse gas concentrations to more environmentally responsible levels. The result is what has come to be known as the Kyoto Protocol to the Framework Convention, or simply the Kyoto Protocol. The Framework Convention divided all the signing countries into three groups. Annex I countries are industrialized. Annex II countries are developed countries that pay for costs of developing countries. Developing countries have no immediate restrictions on emissions and are eligible to receive money and technology from Annex II countries.

The U.S. is an **Annex I** and **Annex II** country, as are the U.K., Japan, Australia, the Russian Federation, and the countries of the European Union. China and India are examples of **developing countries** under the Framework Convention.

Binding emission targets based on five-year averages were set for 38 Annex I nations to reduce their emissions of six greenhouse gases from 1990 levels. Accomplishing these goals between the years 2008 to 2012 could decrease emissions from industrialized nations overall by about 5%. Under the Kyoto Protocol, the United States was expected to reduce emissions to 7% below its 1990 levels, the European Union (EU) nations 8%, and Canada and Japan 6%.

No binding emission targets were established for the developing countries, a contentious issue then and now. Developed nations are permitted to trade emission credits to meet their targets. That is, countries that have emissions lower than their targets can sell the residual amounts to countries exceeding their targets. Developed nations also can receive further credits for investments and projects to help developing countries reduce their emission of greenhouse gases through better technologies. The gases regulated include carbon dioxide, methane, nitrous oxide, hydrofluorocarbons (HFCs), perfluorocarbons (PFCs), and sulfur hexafluoride.

To take effect, the Kyoto Protocol had to be ratified by at least 55 countries and by enough Annex I countries to account for 55% of their total 1990 CO_2 emissions. By June 2003, 110 countries had ratified or otherwise accepted the protocol. The sticking point was that ratifying Annex I countries only accounted for 44.2% of total Annex I emissions, not the required 55%. With the United States firmly reiterating that it would not ratify the protocol, it was left to Russia to bring the protocol into effect. The European Union announced in mid-2004 that it would back the Russian Federation's entry into the World Trade Organization (WTO), an important step for Russia's economic growth. The negotiation hinged on Russia's promise to support environmental policies of the EU, particularly agreeing to work toward ratification of the Kyoto Protocol. "Russia clearly traded its support for Kyoto in exchange for some concessions on WTO entry terms," said Christopher Weafer, chief equity strategist with Alfa Bank in Moscow. Both houses of the Russian Parliament ratified the Kyoto Protocol in late 2004, and the protocol finally came into effect in February 2005.

The overarching goal of the Kyoto Protocol is to substantially reduce the amount of greenhouse gases released into the atmosphere. However, despite the European Union

targeted reduction of 8% below 1990 levels, the emissions as of December 2005 are 6% *above* their 1990 levels. This creates a situation in which the "cap-and-trade" mechanism is allowed. Russia is in an excellent position to take part in such trades, because by a historical twist of fate, 1990 was a pivotal year for Russia. That was the last year that the factories were operating at full capacity before the collapse of the former Soviet Union. Since then, their greenhouse gas emissions have dropped by about 43%, leaving Russia with emissions credits to sell. Many countries in the EU and also Japan are eager customers for these "carbon credits." Canada, Switzerland, and even California are interested in being part of a global carbon credit system.

Trading credits is a strategy used to reduce acid rain and will be discussed in Section 6.15.

Consider This 3.33   **The British Experience**

The British Labour Party in 1997, under the leadership of Tony Blair, boldly committed to cutting 20% off British greenhouse gas emissions by 2010. This is significantly more than the 12.5% required by the Kyoto treaty. Has progress been made toward that goal? Research this question and write a short report on the British experience in reducing greenhouse gases.

As of 2007, the United States has continued to opt out of the Kyoto Protocol. President George W. Bush does not support the agreements reached in Kyoto, calling them "fatally flawed." One reason was the belief that meeting the reductions required by the protocol would cause serious harm to the U.S. economy. Another reason for not ratifying the protocol was concern about the lack of restrictions for developing nations. Although more than 150 nations agreed in late 2005 to launch formal talks on mandatory post-2012 reductions in greenhouse gases, the United States has only agreed to a nonbinding dialogue to respond to climate change. The agreement among Kyoto parties now commits most of the world's most influential nations to moving ahead, with or without the active involvement of the United States. With the long delay in ratification and implementation of the Kyoto Protocol, the targets for 2012 most likely cannot be met without further restrictions.

The Bush administration is pursuing several alternatives to participation in the Kyoto Protocol. President Bush announced a Global Climate Change Initiative in 2002. The proposal sets greenhouse gas emission goals based on units of gross domestic product. From 2002 to 2012, a reduction of 18% based on this measure is predicted, attained by totally voluntary efforts. This initiative and its subsequent amendments also provide increased funding for technological changes, such as development of the hydrogen economy.

The hydrogen economy will be discussed in Section 8.8.

Individual states have taken matters into their own hands, believing that reducing their own emissions could have a significant effect on global emissions. One example is the Regional Greenhouse Gas Initiative (RGGI). Ten northeastern and mid-Atlantic states are now working together to develop a cap-and-trade program for greenhouse gas emissions (Figure 3.26). In addition, Pennsylvania, Maryland, the District of Columbia, and the Eastern Canadian Provinces are observers in the process. The initial focus is reducing CO_2 emissions from power plants, while maintaining energy affordability for consumers. Developing successful programs could provide models for implementation elsewhere.

Following the Northeastern lead, three Western states—California, Oregon, and Washington—are uniting to combat the effects of greenhouse gases. Proposals in California are the boldest and therefore the most controversial. They seek not only to reduce greenhouse gas emissions from power plants, but also plan reductions from cars and light trucks. New York is adopting California's ambitious new regulations to reduce automotive emissions as well. Development of renewable sources, tax incentives for solar and wind power, and laws promoting the production of ethanol are

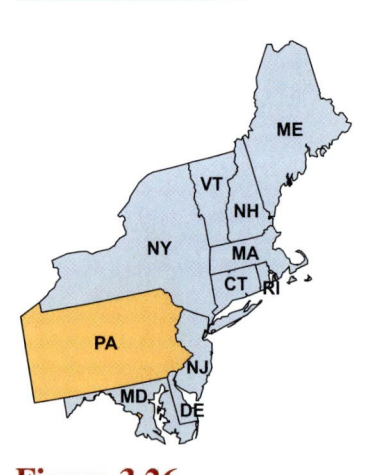

Figure 3.26

States agreeing to participate in the Regional Greenhouse Gas Initiative. Participating states shown in blue; an observer state is shown in orange.

all part of the needed strategies. In late 2006, the U.S. Mayors Climate Protection Agreement included 227 cities committed to cutting emissions of greenhouse gases to 7% below 1990 levels by 2012. The cities include some of the largest in the Northeast, the Great Lakes region, and West Coast and their mayors represent some 44 million Americans.

Sceptical Chymist 3.34   Drop in the Bucket?

Critics suggest that state actions, even if successful, cannot possibly have a significant effect on global emissions of greenhouse gases. Proponents, on the other hand, point out that Texas has higher greenhouse gas emissions than Canada or the U.K. In fact, if Texas were a country (and some think it is), it would be the seventh largest emitter of greenhouse gases in the world. Use the resources of the Web to prove or disprove these statements.

Developed countries are the largest emitters of greenhouse gases and have a historic responsibility to reduce their emissions. However, developing countries are predicted to become the major producers of carbon dioxide and other greenhouse gases in the not-too-distant future. The developed countries have a prodigious lead, but the rates of greenhouse gas emissions of developing nations are increasing faster than those of industrialized countries and are predicted to grow even faster in the future (Figure 3.27).

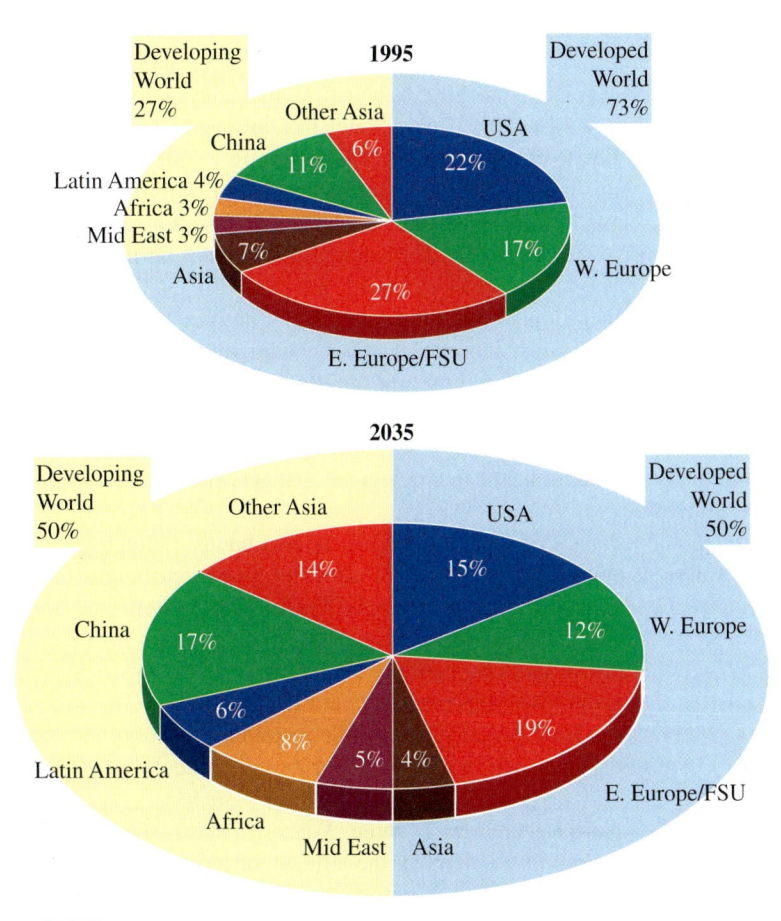

Figure 3.27

Total world CO_2 emissions for 1995 (6.46 billion tonnes) and 2035 (projected to be 11.71 billion tonnes). Contributions from the developed world are shown in light blue, the developing world in light yellow. Notice that parts of Asia fall into each category. The abbreviation FSU stands for the former Soviet Union, now called the Russian Federation.

The rate of emissions growth from the rapidly developing economies of China and India are estimated to be 4.5% per year, far higher than the growth rate for established economies. Furthermore, the Kyoto Protocol does not cap emissions from developing countries.

Your Turn 3.35   Changing Contributions

Use Figure 3.27 to answer these questions.

a. How are the tonnes of total world CO_2 emissions projected to change from 1995 to 2035?
b. How is the percentage of total world CO_2 emissions from the developed world projected to change from 1995 to 2035?
c. How are the tonnes of total world CO_2 emissions from the developed world projected to change from 1995 to 2035?
d. How do you think these changes in quantity and relative percentage of emissions will influence future policy? Explain.

3.12   Global Warming and Ozone Depletion

Global warming and ozone depletion both involve the atmosphere, and both are much in the news. The casual reader of newspaper accounts may easily mix them up. Sometimes, the authors of the articles themselves get confused! We want to avoid such mix-ups and for that reason, will conclude this chapter by returning to one of the common misconceptions. The question is often posed this way: Is the depletion of the ozone layer the principal cause of climate change? Perhaps this seems logical because if stratospheric ozone is destroyed, more solar radiation might be able to reach Earth and warm it. This is not the case, however. In fact, loss of stratospheric ozone causes a small negative effect, cooling the Earth. This contrasts with the effect of increasing tropospheric ozone, showing a larger positive effect, raising the temperature of the Earth. This negative effect arises from the absorption of UV radiation by O_3 during its destruction in the stratosphere. Both effects are small compared with the observed effects of the major greenhouse gases. Figure 3.28 shows the relative contributions, or forcings, from changes in atmospheric gases.

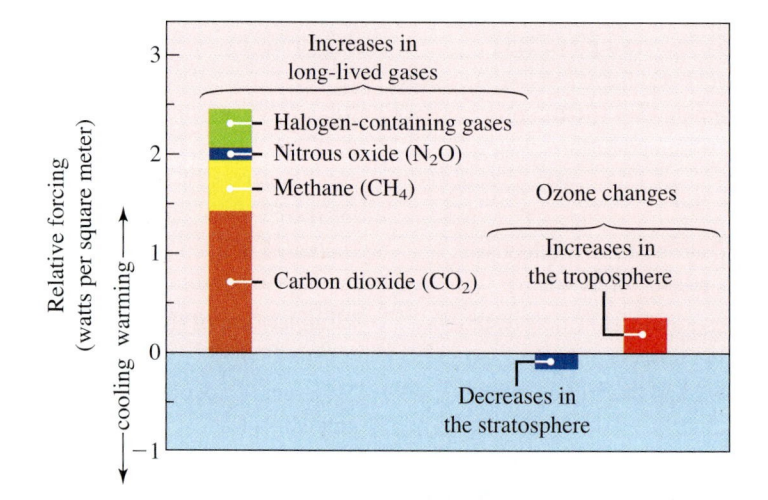

Figure 3.28

Radiative forcing of climate change from atmospheric gas changes (1750–2000).

Source: *Scientific Assessment of Ozone Depletion: 2002;* World Meteorological Organization (WMO), United Nations Environmental Programme (UNEP).

Even though the ozone in the stratosphere contributes little to global warming, nonetheless the issues of ozone depletion and global warming are linked. Human activities have led to the accumulation in the atmosphere of several long-lived greenhouse gases. This list includes ozone and many of the CFCs that have been held responsible for stratospheric ozone depletion. It even includes some of the compounds used to replace CFCs, such as HFCs. We end this chapter by summarizing in Table 3.6 some of the important differences between global warming and stratospheric ozone destruction. Such a tabulation invites oversimplification, but it can be a useful summary of some of the important aspects of these two environmental problems and the responses to them by the United States and the international community. The next Consider This activity provides an opportunity to express your informed opinion about their relative significance.

Consider This 3.36   Air Quality, Ozone Depletion, or Global Warming

Now that you have studied air quality (Chapter 1), stratospheric ozone depletion (Chapter 2), and global warming (Chapter 3), which do you believe poses the most serious problem for you in the short run? In the long run? Discuss your reasons with others and draft a short report on this question.

Table 3.6   Global Warming and Ozone Depletion: Some Characteristics

	Global Warming	Stratospheric Ozone Depletion
Region of atmosphere involved	Mostly troposphere	Stratosphere
Major substances involved	H_2O, CO_2, CH_4, N_2O	O_3, O_2, CFCs
Interaction with radiation	When molecules absorb IR radiation, they vibrate and return heat energy to Earth.	When molecules absorb UV radiation, they break apart into smaller molecules or atoms.
Nature of problem	Increasing concentrations of greenhouse gases are apparently increasing average global temperature.	Decreasing concentration of O_3 is increasing exposure to UV radiation.
Major sources	Release of CO_2 from burning fossil fuels, deforestation; CH_4 from agriculture. Natural sources of H_2O	Release of long-lived CFCs from past uses as solvents, foaming agents, air conditioners. CFCs release Cl• that destroys O_3.
Credible consequences	Altered climate and agricultural productivity, increased sea level, effects on health	Increased incidence of skin cancer, damage to phytoplankton
Possible remedies	Decrease use of fossil fuels, slow deforestation; change agricultural practices	Eliminate use of CFCs, find suitable replacements
International response	Kyoto Protocol, 1997 and later amendments	Montreal Protocol, 1987 and later amendments
U.S. response	Signed Kyoto Protocol in 1998; not submitted to Senate for ratification, therefore not bound by its provisions; alternative proposals; actions by states	Signed Montreal Protocol in1987; full participation in protocol and its amendments

Conclusion

Is our world warming to unintended climatic effects? To assess and reverse such effects, much will depend on the quality of information gathered and how it is used to make sound economic and environmental decisions. The development of technology to exploit fossil-fuel energy resources has been a double-edged sword, leading both to dramatic improvements in our lives and to a whole new generation of problems. During the 20th century, the atmosphere on which our very existence depends was subjected to repeated assaults. The fact that most of these environmental insults were unintentional and, in some cases, the unexpected consequences of social progress does not alter the problems we face. We have only relatively recently recognized the harm that air pollution, stratospheric ozone depletion, and global warming can bring to our personal, regional, national, and global communities. To reverse the damage already done and to prevent more, all these communities must respond with intelligence, compassion, commitment, and wisdom. Many suggestions for change would contribute to sound, prudent, and responsible stewardship of our planet.

A central theme to many of the issues explored in *Chemistry in Context* is the production of electricity. In an editorial comment in *Science* (4 April 2003), Editor emeritus Philip Abelson reminds us of the importance of looking at the larger picture and longer timeline of our dependence on fossil fuels.

> "The United States has large resources of coal that now supply the energy for more than half of its electricity. Nuclear energy furnishes another 20%. In principle, future U.S. needs for liquid fuels and electricity could be met. However, . . . the greenhouse effect will cause added future problems for the use of coal. Construction and testing of new electrical generating plants cannot be achieved quickly."

We will return to the theme of energy in the next chapter, titled "Energy, Chemistry, and Society." Nuclear energy is considered in Chapter 7, and alternative energy sources such as fuel cells and solar power are explored in Chapter 8. The many energy-related challenges and uncertainties faced by our modern society are daunting. They also are the wages of our success.

Chapter Summary

Having completed this chapter you should be able to:

- Understand the different processes that take part in Earth's energy balance (3.1)

- Realize the difference between Earth's natural greenhouse effect and the enhanced greenhouse effect (3.1)

- Understand the major role that certain atmospheric gases play in the greenhouse effect (3.1–3.2)

- Explain the methods used to gather past evidence for global warming (3.2)

- Relate Lewis structures to molecular geometry, including bond angles (3.3)

- Understand how molecular geometry is related to absorption of infrared radiation (3.4)

- List the major greenhouse gases and explain why each has the appropriate molecular geometry to be a greenhouse gas (3.4)

- Explain the roles that natural processes play in the carbon cycle and through it, in global warming (3.5)

- Summarize how human activities contribute to the carbon cycle and through it, to global warming (3.5)

- Understand how molar mass is defined and used (3.6)

- Use Avogadro's number to calculate the average mass of an atom (3.6)

- Understand the chemical mole and explain how it is useful (3.7)

- Assess the sources, relative emission quantities, and effectiveness of greenhouse gases other than CO_2 (3.8)

- Recognize the successes and limitations of computer-based models in predicting climatic change (3.9)

- Summarize the different levels of confidence in drawing conclusions about climate change (3.9)

- Consider the global and national implications of a rise in Earth's average temperature (3.10)

- Identify the single most effective strategy for reducing CO_2 emissions (3.10)

- Describe various technological approaches toward reducing CO_2 emissions (3.10)

- Explain world and U.S. policy concerning the now ratified Kyoto Protocol (3.11)

- Give reasons for projected changes in the relative amounts of CO_2 emissions for developed and developing countries (3.11)

- Compare how the issue of global warming is both similar to and different from the issue of ozone depletion (3.12)

- Read and hear news stories on global warming with some measure of confidence in your ability to interpret the accuracy and conclusions of such reports (3.1–3.12)

- Take an informed position with respect to issues surrounding global warming (3.1–3.12)

Questions

Emphasizing Essentials

1. **a.** Is the greenhouse effect taking place now? Explain your reasoning.

   b. Is the enhanced greenhouse effect taking place now? Explain your reasoning.

2. Concentrations of CO_2 in Earth's early atmosphere were much higher than those of today. What happened to this CO_2?

3. Using the analogy of a greenhouse to understand the energy radiated by Earth, of what are the "windows" of Earth's greenhouse made?

4. Consider equation 3.1 for the photosynthetic conversion of CO_2 and H_2O to form glucose, $C_6H_{12}O_6$, and O_2.

   a. Demonstrate that the equation is balanced by counting atoms of each element on either side of the arrow.

   b. Is the number of molecules on either side of the equation the same? Why or why not?

5. What is the difference between climate and weather?

6. **a.** It is estimated that 29 MJ of energy per square meter comes to the top of our atmosphere from the Sun each day, but only 17 MJ/m^2 reaches the surface. What happens to the rest?

   b. Under steady-state conditions, how much energy would leave the top of the atmosphere?

7. Consider Figure 3.4.

   a. How does the present concentration of CO_2 in the atmosphere compare with its concentration 20,000 years ago? With its concentration 120,000 years ago?

   b. How does the present temperature of the atmosphere compare with the 1950–1980 mean temperature? With the temperature 20,000 years ago? How does each of these values compare with the average temperature 120,000 years ago?

   c. Do your answers to parts **a** and **b** indicate causation, correlation, or no relation? Explain.

8. Understanding Earth's energy balance is essential to understanding the issue of global warming. For example, the solar energy striking the Earth's surface averages 168 watts per square meter (W/m^2), but the energy leaving Earth's surface averages 390 W/m^2. Why isn't the Earth cooling rapidly?

9. Explain each of the observations.

   a. A car parked in a sunny location may become hot enough to endanger the lives of pets or small children left in it.

   b. Clear winter nights tend to be colder than cloudy ones.

   c. A desert shows much wider daily temperature variation than a moist environment.

   d. People who wear dark clothing in the summertime put themselves at a greater risk of heatstroke than those who wear white clothing.

10. Using the Lewis structures for H_2 and H_2O as examples, show how the Lewis structure for a molecule can allow an unambiguous prediction of molecular geometry for some molecules, but not for others.

11. Use a molecular model kit to build a methane molecule, CH_4. (If a kit is not available, this model can be made using Styrofoam balls or gumdrops to represent the atoms and toothpicks to represent the bonds.) Demonstrate that the hydrogen atoms are farther from one another in a tetrahedron than they would be if all the atoms of methane were in the same plane (square planar).

12. Draw the Lewis structure and structural formulas for CCl_3F. Sketch (or use a computer program) the space-filing model and the charge-density model for this molecule.

13. **a.** Draw the Lewis structure for H_3COH, which is methanol (wood alcohol).

    b. Based on this structure, predict the H-to-C-to-H bond angle. Explain the reason for your prediction.

    c. Based on this structure, predict the H-to-O-to-C bond angle. Explain the reason for your prediction.

14. a. Draw the Lewis structure for H_2CCH_2, ethene, a small hydrocarbon with a C-to-C double bond.

  b. Based on this structure, predict the H-to-C-to-H bond angle. Explain the reason for your prediction.

  c. Sketch the molecule showing the predicted bond angles.

15. The text states that a UV photon can break chemical bonds, but that an IR photon can cause only vibration in the bonds. Assuming a wavelength of 320 nm for the UV photon and a wavelength of 5000 nm for the IR photon, how do the energies of these compare?

16. Three different modes of vibration of a water molecule are shown. Which of these modes of vibration contributes to the greenhouse effect? Explain.

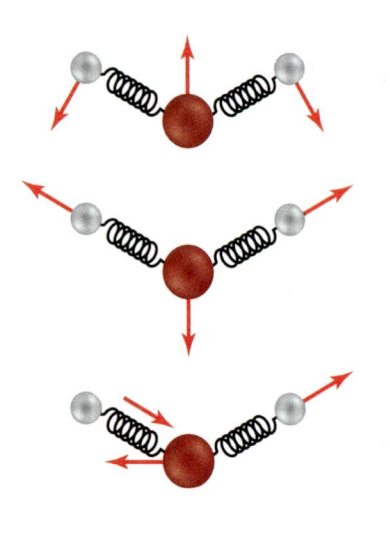

17. If a carbon dioxide molecule interacts with certain photons in the IR region, the molecule vibrates. For CO_2, the major wavelengths and their corresponding wavenumbers of absorption occur at 4.26 μm (2350 cm^{-1}) and 15.00 μm (667 cm^{-1}).

  a. What is the energy corresponding to each of these IR photons?

  b. What happens to the energy in the vibrating CO_2 species?

18. Water vapor makes up about 1% of our atmosphere, but it is not regulated as a greenhouse gas. Explain.

19. Explain how each of these relates to global climate change.

  a. volcanic eruptions

  b. CFCs in the troposphere

  c. CFCs in the stratosphere

20. A biochemical process that releases carbon dioxide to the atmosphere is the fermentation of sugar to produce alcohol. For example, glucose ($C_6H_{12}O_6$) ferments in the presence of yeast to produce ethanol (C_2H_5OH) and CO_2. Write a balanced chemical equation for this reaction.

21. Consider Figure 3.18.

  a. Which sector has the highest CO_2 emission from fossil-fuel combustion?

  b. What alternatives exist for each of the major sectors of CO_2 emissions?

22. Silver has an atomic mass of 107.9 and an atomic number of 47.

  a. What is the number of protons, neutrons, and electrons in a neutral atom of the most common isotope, Ag-107?

  b. How do the numbers of protons, neutrons, and electrons in a neutral atom of Ag-109 compare with those of Ag-107?

23. Silver has only two naturally occurring isotopes: Ag-107 and Ag-109. Why isn't the atomic mass of silver given on the periodic table just the average, 108?

24. a. Calculate the average mass (in grams) of an individual atom of silver.

  b. Calculate the mass (in grams) of 10 trillion silver atoms.

  c. Calculate the mass (in grams) of 5.00×10^{45} silver atoms.

25. Calculate the molar mass of these molecules. Each plays a role in atmospheric chemistry.

  a. H_2O

  b. CCl_2F_2 (Freon-12)

  c. N_2O

26. a. Calculate the mass percent of chlorine in CCl_3F (Freon-11).

  b. Calculate the mass percent of chlorine in CCl_2F_2 (Freon-12).

  c. What is the maximum mass of chlorine that could be released in the stratosphere by 100 g of each compound?

  d. How many atoms of chlorine correspond to the masses calculated in part **c**?

27. The total mass of carbon in living systems is estimated to be 7.5×10^{17} g. Given that the total mass of carbon on Earth is estimated to be 7.5×10^{22} g, what is the ratio of carbon atoms in living systems to the total carbon atoms on Earth? Report your answer in percent and in ppm.

28. Consider the information presented in Table 3.2.

  a. Calculate the percent increase in CO_2 when comparing 1998 concentrations with preindustrial concentrations.

  b. Considering CO_2, CH_4, and N_2O, which has shown the greatest percentage increase when comparing 1998 concentrations with preindustrial concentrations?

29. Consider the information presented in this graph.

Percent contribution to global warming

CO_2 CFCs CH_4 N_2O

a. Which gas makes the largest percent contribution to global warming?

b. Use these percentages together with the global warming potentials for CO_2, CH_4, and N_2O given in Table 3.3 to decide which of these gases has the largest effect on global warming. Explain your reasoning. *Hint:* One approach is to calculate "net effectiveness" by finding the product of percent contribution and GWP for each gas.

30. Total greenhouse gas emissions in the United States rose 16% from 1990 to 2005, growing at a rate of 1.3% a year since 2000. How is this possible when CO_2 emissions grew by 20% in the same time period? *Hint*: See Table 3.2.

Concentrating on Concepts

31. The text makes a distinction between the *correlation* of two events and the *causation* of one by the other. Identify each of these pairs as an example of correlation, causation, or no relationship. Explain.

a. tons of coal burned; tons of CO_2 emitted

b. national per capita income; per capita emission of CO_2

c. number of cigarettes smoked per day; increase in the incidence of lung cancer

d. number of bonds between two O atoms; length of O-to-O bond

e. building a greenhouse; successfully raising seedlings

f. buying a pair of roller blades; breaking your leg

32. Friends sometimes bring living plants, rather than cut flowers, to someone recovering from an illness. Which gift is thought to provide more oxygen to benefit the patient? Explain.

33. Given that direct measurements of Earth's atmospheric temperature over the last several thousands of years are not available, how can scientists estimate past fluctuations in the temperature?

34. Consider Figure 3.5, showing atmospheric concentrations of CO_2 at Mauna Loa, and Figure 3.6, showing temperature changes on Earth's surface. Why is the pattern

with the trend so regular in Figure 3.5, but not as regular in Figure 3.6?

35. Consider this figure showing sources of CO_2 emissions in the United States for 1998.

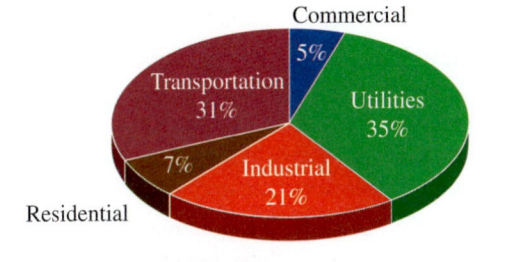

a. Compare it with Figure 3.24 in the text, which gives CO_2 emissions for 2004. How do these two graphs differ?

b. Can you directly compare the information in the two graphs? Why or why not?

c. If you wanted to reduce your personal contribution to CO_2 emissions, what changes could you make that would be the most effective? Explain.

36. The text states that Lewis structures show linkages (what is connected to what) but they do not show shape. Explain this statement, using the molecule H_2O as an example.

37. Carbon dioxide gas and water vapor both absorb IR radiation. Do they also absorb visible radiation? Offer some evidence based on your everyday experiences to help explain your answer.

38. How would the energy required to cause IR-absorbing vibrations in CO_2 change if the carbon and oxygen atoms were connected with single rather than with double bonds?

39. Explain why water in a glass cup is quickly warmed in a microwave oven, but the glass cup warms much more slowly, if at all.

40. Ethanol, C_2H_5OH, can be produced from sugars and starches in crops such as corn or sugar cane. The ethanol is used as a gasoline additive and when burned, it combines with O_2 to form H_2O and CO_2.

a. Write a balanced equation for the complete combustion of C_2H_5OH.

b. How many moles of CO_2 are produced from each mole of C_2H_5OH completely burned?

c. How many moles of O_2 are required to burn 10 mol of C_2H_5OH?

41. Hexane, C_6H_{14}, and octane, C_8H_{18}, will burn with oxygen gas, O_2. The products of complete combustion are H_2O and CO_2.

a. Write a balanced equation for the complete combustion of hexane.

b. Repeat for octane.

c. Compare the number of moles of CO_2 produced when 1 mol of each hydrocarbon burns.

42. Why is the atmospheric lifetime of a greenhouse gas important?

43. CO_2 has a greater density than N_2. Why don't these gases settle out into layers in the atmosphere?

44. One of the first radar devices developed during World War II used microwave radiation of a specific wavelength that triggers the rotation of water molecules. Why was this design not successful?

45. It is estimated that Earth's ruminants, such as cattle and sheep, produce 73 million tonnes of CH_4 each year. How many tonnes of carbon are present in this mass of CH_4?

46. Nine out of 10 of the warmest years in the United States in the last century have been in the decade from 1995 to 2005. Does this *prove* that the enhanced greenhouse effect (global warming) is taking place? Explain.

47. A possible replacement for CFCs is HFC-152a, with a lifetime of 1.4 years and a GWP of 120. Another is HFC-23, with a lifetime of 260 years and a GWP of 12,000. Both of these possible replacements have a significant effect as greenhouse gases and are regulated under the Kyoto Protocol.

    a. Based on the given information, which appears to be the better replacement? Consider only the potential for global warming.

    b. What other considerations are there in choosing a replacement?

48. The quino checkerspot butterfly is an endangered species with a small range in northern Mexico and southern California. Evidence reported in 2003 indicates that the range of this species is even smaller than previously thought.

    a. Propose an explanation why this species is being pushed north, out of Mexico.

    b. Propose an explanation why this species is being pushed south, out of southern California.

    c. Propose a plan to prevent further harm to this endangered species.

49. This figure shows global emissions of CO_2 in metric tons (tonnes) per person per year.

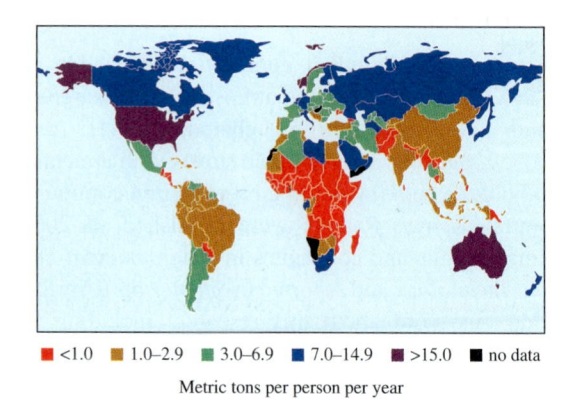

■ <1.0 ■ 1.0–2.9 ■ 3.0–6.9 ■ 7.0–14.9 ■ >15.0 ■ no data

Metric tons per person per year

Source: From *New Scientist Supplement,* April 28, 2001. Reprinted with permission.

a. Which countries lead the world in CO_2 emissions?

b. The values in this figure are reported as tonnes of CO_2, not as tonnes of C in CO_2. How are these values related? *Hint:* Think about the mass relationships developed in Section 3.6.

c. Find the value for U.S. CO_2 emissions from a source other than shown in this figure. Do the values agree? Explain.

50. This figure shows who uses the most energy and who gains the greatest value for money spent on energy.

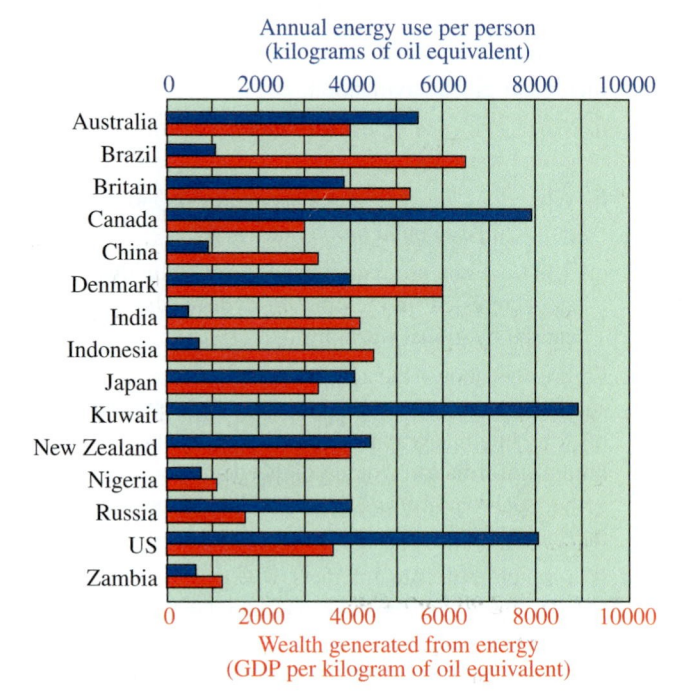

Note: GDP = gross domestic product

Source: From *New Scientist Supplement,* April 28, 2001. Reprinted with permission.

a. What units are used to report annual energy use per person?

b. Which countries lead the world in energy use?

c. What units are used to report wealth generated from energy?

d. Which countries lead the world in wealth generated from energy?

e. What implications for U.S. policy are suggested by this figure? Explain your reasoning.

Exploring Extensions

51. China's growing economy is fueled largely by its dependence on coal, described as China's "double-edged sword." Coal is both the new economy's "black gold" and the "fragile environment's dark cloud."

    a. What are some of the consequences of dependence on high-sulfur coal?

b. Sulfur pollution from China may slow global warming, but only temporarily. Explain.

c. Why is China not bound by the restrictions of the Kyoto Protocol?

d. What other country is rapidly stepping up its construction of coal-fired power plants and is expected to have a larger population than China by the year 2030?

52. Former Vice President Al Gore says this about global warming in his 2006 book and film, *An Inconvenient Truth*:

"We can no longer afford to view global warming as a political issue—rather, it is the biggest moral challenge facing our global civilization."

a. What evidence does Mr. Gore cite that supports global warming?

b. Why does he feel global warming is no longer a political issue but a moral one?

c. What actions are recommended for individuals who are interested in helping to reduce the problems caused by global warming?

53. Figure 3.6 shows the temperature increase on Earth's surface from 1880 to 2005. Imagine you are in charge of extending this graph to include the present year. What kind of information would you need and how might you gather such data? *Hint:* Consider the source of the data in Figure 3.6.

54. If a water molecule interacts with certain photons in the IR region, the molecule vibrates (see Figure 3.16). In Consider This 3.16, you estimated the wavenumber, expressed per centimeter, for water's two maximum absorbancies. Then, you converted those values to wavelengths, expressed in millimeters. (Complete this if you have not already done so.)

a. Calculate the energy associated with each maximum absorbance. *Hint*: See equation 2.3 for the appropriate relationship.

b. Which absorption peak represents higher energy?

c. Does the higher energy absorption correspond to bending or stretching in the water molecule?

55. Data taken over time reveal an increase in CO_2 in the atmosphere. The large increase in the combustion of hydrocarbons since the Industrial Revolution is often cited as a reason for the increasing levels of CO_2. However, an increase in water vapor has *not* been observed during the same period. Remembering the general equation for the combustion of a hydrocarbon, does the difference in these two trends *disprove* any connection between human activities and global warming? Explain your reasoning.

56. In the energy industry, 1 standard cubic foot (SCF) of natural gas contains 1196 mol of methane (CH_4) at 15.6 °C (60 °F).

a. How many moles of CO_2 could be produced by the complete combustion of 1 SCF of natural gas?

b. How many kilograms (kg) of CO_2 could be produced?

c. How many tonnes of CO_2 could be produced? *Hint:* See Appendix 1 for conversion factors.

d. How many pounds (lb) of CO_2 could be produced? *Hint:* See Appendix 1 for conversion factors.

57. Consider how your position on controlling emissions of carbon dioxide might change if you were a student in a developing country rather than in the United States. To help your consideration, pick a country in the developing world. Then search or use the direct access provided on the *Online Learning Center* to find out if that country has signed and ratified the Kyoto Protocol. Also find the total tonnes of carbon dioxide emitted and the per capita emission for that country. Compare these data with those of the United States and comment on the differences and their possible effect on policy.

58. The 2004 CO_2 emissions for several countries are shown in Figure 3.23. This list gives the estimated 2004 gross domestic product (GDP) per capita for some of those countries.

U.S.	$40,100
Canada	$31,500
Japan	$29,400
Germany	$28,700
Italy	$27,700
Russia	$ 9,800
Brazil	$ 8,100
China	$ 5,600
India	$ 3,100

a. How is the GDP related to the total emissions? To the per capita emissions? To the emissions based on economic output? Offer some reasonable explanations for the observed relationships.

b. Considering the relationship between GDP per capita and CO_2 emissions per capita, what are the policy implications for implementing the Kyoto agreement?

59. Incomplete fuel combustion also adds greenhouse gases to the atmosphere. Researchers are studying emissions from charcoal stoves, ceramic stoves, and simple three-stone stoves with a metal grate to hold the cooking pot in an agricultural community in central Kenya. Results were published in 2003 by Robert Bailis and colleagues in *Environmental Science and Technology* and reported in *Science* on 1 April 2005. Find out more about this research, including which greenhouse gases and air pollutants are emitted and which stoves are best. Also discuss the policy implications of this research.

60. The world community responded differently to the atmospheric problems described in Chapters 2 and 3. The evidence of ozone depletion was met with the Montreal Protocol, a schedule for decreasing the production of ozone-depleting chemicals. The evidence of global warming was met with the Kyoto Protocol, a plan calling for targeted reduction of greenhouse gases.

  a. Suggest reasons why the world community dealt with the issue of ozone depletion *before* that of global warming.

  b. Compare the current status of the two responses. When was the latest amendment to the Montreal Protocol? How many nations have ratified it? Has the level of chlorine in the stratosphere dropped as a result of the Montreal Protocol? How many nations have ratified the Kyoto Protocol? What has happened since it went into effect? Have any other initiatives been proposed? Have levels of greenhouse gases dropped as a result of the Kyoto Protocol?

Chapter 4

Energy, Chemistry, and Society

UNESCO ISSUES WARNING RE CHEVIOT and WORLD HERITAGE SITE

"The United States cannot afford to wait for the next energy crisis to marshal its intellectual and industrial resources. . . . Our growing dependence on increasingly scarce Middle Eastern oil is a fool's game—there is no way for the rest of the world to win. Our losses may come suddenly through war, steadily through price increases, agonizingly through developing-nation poverty, relentlessly through climate change—or through all of the above."

James Woolsey, U.S. Director of Central Intelligence (1993–1995)

In terms of energy, the United States is number one: the largest producer, the leading consumer, and the biggest importer in the world. The citizens of this country generate a tremendous need for energy and do so with an almost casual attitude about its availability and cost. That need is met predominantly by fossil fuels—coal, oil, and natural gas. Collectively, they account for about 70% of U.S. electricity and almost 85% of all of the nation's energy requirements.

The opening images remind us of various facets of energy production and use, but may not bring to mind some important challenges related to our dependence on fossil fuels. One of these challenges is securing an ever-increasing supply. Precipitated by political turmoil in the Middle East during the mid-1970s, the United States experienced shortages of petroleum products and an accompanying spike in prices. Our national policy response focused on decreasing our reliance on oil from that part of the world. The changes that resulted from new policies did produce a lower demand overall and a redistribution of our suppliers. However, in the last two decades our imports have risen from about 35% of our needs to over 50%. As the opening quotation suggests, ensuring the continuing availability of more and more energy in the coming years is a matter of national security.

Simply using fossil fuels creates a second challenge. Combustion, of course, is what releases the huge energy content of the molecules. However, burning carbon-based fuels also releases carbon dioxide, water, and often soot, carbon monoxide, and oxides of sulfur and nitrogen as well. The link between these combustion products and serious environmental concerns such as global warming, acid rain, and the deterioration of air quality is undeniable.

> We examined air quality in Chapter 1, the contribution of carbon dioxide to global warming in Chapter 3, and will discuss acid rain in Chapter 6.

A third challenge arises from the fact that fossil fuels are in limited supply. Recent estimates suggest that at current rates of consumption, oil reserves will be depleted in less than 50 years and coal in over a century and a half. Renewable energy sources continue to attract the attention of policy makers and the energy industry alike. For example, traditional suppliers of fossil fuels are the largest investors in wind power, electricity generated by "farms" of windmills. Furthermore, increasing amounts of agricultural products are finding their way into alternative fuels. As a gasoline additive, ethanol produced from the fermentation of corn reduces harmful emissions. Biodiesel produced from soybean oil can be used in normal diesel engines.

Consider This 4.1   Energy in the News

  a. Examine the Web site of your local newspaper for recent articles about local, regional, and national aspects of energy production, use, and policies.
  b. Search national or international news sites for similar articles.
  c. Do the items in both sources overlap? Do the "big picture" items find their way into the local news?

Given these challenges, what courses of action do we need to consider? Conservation measures seem rational, especially if the negative economic effects are minimal. Increasing our reliance on renewable energy sources would certainly have positive outcomes. However, major transitions in the way we use fuels and the ways in which they are supplied raise questions about who is willing and able to bear the costs.

To better answer these questions, we need to understand the fundamental chemistry that underlies fuels and the energy they release. We begin by defining some important terms (energy, heat, and work) and a law that tells us that energy is never created or destroyed, but just changes forms. Inefficiency becomes an issue in our ability to harness the energy or produce it completely in a form we want. We will explore the properties of fuels, their chemical composition, and the structures of the molecules they contain. We will learn how these molecules store energy and how to write the chemical reactions that describe its release. Once we understand the chemical principles, we will turn our attention to energy consumption and to two major fossil fuels—coal and petroleum—and

to methods for manipulating them into useful forms. The chapter ends with a discussion about energy policies and the possibilities for conservation.

4.1   Energy, Work, and Heat

Energy is a word we hear every day, but whose precise chemical meaning is not well understood by the public. The scientific definition of **energy**, the capacity to do work or supply heat contains two words whose colloquial use does not quite match the chemical one. To a scientist, **work** is done when movement occurs against a restraining force. Mathematically, work is equal to the force multiplied by the distance over which the motion occurs. Thus, when you lift a book against the force of gravity, you are doing work.

Much of the work done on our planet comes from the most common form of chemical work, the expansion of gases like those produced in an internal combustion engine, or from another familiar form of energy—heat. The formal definition of the latter sounds a little strange: **heat** is energy that flows from a hotter to a colder object. When we grab a hot pan on the stove, we immediately experience heat.

Temperature is a property of heat; it defines the degree of hotness (or coldness) on a specified scale. Another definition of temperature sounds awkward and circular: **temperature** is a property of matter that determines the direction of heat flow. However, we know that temperature and heat are not the same thing. When two bodies are in contact, heat always flows from the object at the higher temperature to that at the lower temperature. Your bottle of water and the Pacific Ocean may be at the same temperature, but the ocean contains and can transfer far more heat than the bottle of water. Indeed, bodies of water can affect the climate of an entire region as a consequence of their ability to absorb and transfer heat.

In order to continue our discussion of energy, we need a unit in which to express it. Originally, the **calorie** was defined as the amount of heat necessary to raise the temperature of exactly one gram of water by one degree Celsius. It has been redefined as exactly 4.184 J. To put this unit into context, one joule (1 J) is approximately equal to the energy required to raise a 2-lb book 4 in. against the force of gravity. On a more personal basis, each beat of the human heart requires about 1 J of energy. In the next few sections, we will use energy units that range from kilojoules (1 kJ = 1000 J) in discussions about chemical bonds to exajoules (1 EJ = 10^{18} J) when we consider annual energy production worldwide.

The calorie was introduced as a measure of heat with the metric system in the late 18th century. Calories are perhaps most familiar when used to express the energy released when foods are metabolized. The values tabulated on package labels and in cookbooks are, in fact, kilocalories (1 kcal = 1000; cal = 1 Calorie); when Calorie is capitalized it generally means kilocalorie (Figure 4.1). Thus, the energetic equivalent of a donut is 425 Cal (425 kcal, 425,000 calories). For most purposes we will use joules and kilojoules in this chapter, but when it seems more appropriate or more easily understandable, we will express energy in calories or kilocalories.

Metabolism is discussed in detail in Section 11.6

Figure 4.1
The energy content of foods is usually listed in Calories.

Your Turn 4.2   Energy Calculations

a. When a donut is metabolized, 425 kcal (425 Cal) are released. Express this value in kilojoules.

b. Calculate the number of 2-lb books you could lift to a shelf 6 ft off the floor with the amount of energy from metabolizing one donut.

c. A 12-oz serving of a soft drink has an energy equivalent of 92 kcal. In kilojoules, what is the energy released when metabolizing this beverage?

d. Assume that you use this energy to lift concrete blocks that weigh 22 lb (10 kg) each. How many blocks could you lift to a height of 4 ft with the energy calculated in part **c**?

Answers

a. Recall that 1 kcal is equivalent to 4.184 kJ.

$$425 \text{ kcal} \times \frac{4.184 \text{ kJ}}{1 \text{ kcal}} = 1.78 \times 10^3 \text{ kJ}$$

b. Earlier the text stated that 1 J is approximately equal to the energy required to raise a 2-lb book a distance of 4 in. against Earth's gravity. We can use this information to calculate the number of 2-lb books that could be lifted 6 ft. First note that 6 ft is 72 in. Next, calculate the energy (joules) required to lift one 2-lb book the whole 6 ft.

$$72 \text{ in.} \times \frac{1 \text{ J}}{4 \text{ in.}} = 18 \text{ J}$$

Then, express this value in kilojoules.

$$18 \text{ J} \times \frac{1 \text{ kJ}}{10^3 \text{ J}} = 0.018 \text{ kJ}$$

Use this value to make the final calculation.

$$1.78 \times 10^3 \text{ kJ} \times \frac{1 \text{ book}}{0.018 \text{ kJ}} = 9.9 \times 10^4 \text{ books}$$

To work off one donut requires lifting almost 100,000 books!

Sceptical Chymist 4.3 | Checking Assumptions

A simplifying (and erroneous) assumption was made in doing the calculations in parts **b** and **d** of Your Turn 4.2. What was the assumption, and is it reasonable? Is the answer based on this assumption too large or too small? Explain your answer.

4.2 Energy Transformation

We introduced the law of conservation of mass when balancing chemical reactions in Chapter 1. Here, we incorporate energy into another of the "conservation laws." The **first law of thermodynamics,** also called the law of conservation of energy, states that energy is neither created nor destroyed. But how can we ever experience an energy crisis if the energy of the universe is constant? The answer lies in understanding that energy can be converted into different types. When we use energy, whether burning wood in the fireplace, driving our car, or turning on a light, we neither create nor destroy it. However, we do transform it from one type to another, losing a bit along the way with each transformation.

There are two main types of energy. **Potential energy,** as the name suggests is stored energy or the energy of position. For example, energy can be stored in the position of a book lifted against the force of gravity. The heavier the book and the higher you lift it, the more potential energy it has. Another type of energy is called **kinetic energy,** defined as the energy of motion. The heavier an object is and the faster it is moving, the more kinetic energy it possesses. Which would you rather be hit by, a baseball traveling at 90 mph or a ping-pong ball traveling at 90 mph? The baseball has considerably more kinetic energy because of its larger relative mass.

To better understand how energy is transformed from one type to another, consider the desktop toy known as a Newton's cradle shown in Figure 4.2. To start, you must do work on the system by lifting one of the balls against the force of gravity. Because of its position relative to the other spheres, you have given potential energy to that ball.

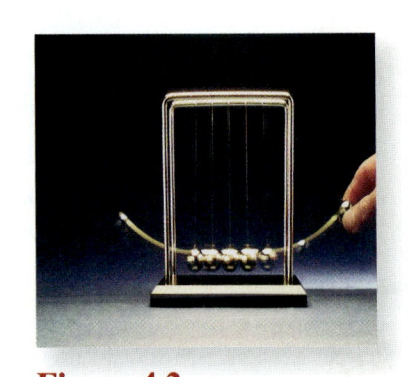

Figure 4.2
A Newton's cradle.

When you release the ball, it falls back toward its starting place. The potential energy of position is converted into kinetic energy of motion. When the moving ball hits the stationary ones, the kinetic energy is transferred from one ball to the next, and the one on the other end begins to move *up*. Its kinetic energy is gradually converted into potential energy as it rises and slows. When that conversion is complete, the second ball begins to fall, and the process repeats. However, with each successive cycle the balls don't rise quite as high. Eventually, the balls all come to rest at their original position.

Why do they stop? Doesn't this violate the law of conservation of energy? Where did the energy go? In each collision, some of the energy is used to make sound, and some is used to generate heat. If we could measure precisely enough, we would observe the balls heating up slightly; this heat is then transferred to the surrounding atoms and molecules in the air. Therefore, when all the balls finally come to rest, the energy of the universe has been conserved, but all the work you initially put into the system has been dissipated as random motion of the atoms and molecules in the surrounding air. In essence, the device is a fun way of dissipating a little bit of *work into heat*.

Conversely, the industrialization of the world's economy began with the invention of devices to convert *heat into work*. Essentially all the energy from all of the fuels we will consider in this chapter—coal, oil, alcohol, and garbage—is extracted through combustion. Each of the fuels burns to generate heat and at the same time generates products like carbon dioxide and water. Chief among these devices was the steam engine, developed in the latter half of the 18th century. The heat from burning wood or coal was used to vaporize water, which in turn was used to drive pistons and turbines. The resulting mechanical energy was used to power pumps, mills, looms, boats, and trains. The smoke-belching mechanical monsters of the English midlands soon replaced humans and horses as the primary source of motive power in the Western world.

A second energy revolution occurred early in the 1900s with the commercialization of electrical power. Today, most of the electrical energy produced in the United States is generated by the descendants of those early steam engines. Figure 4.3 illustrates a modern power plant. Heat from the burning fuel is used to boil water, usually under high pressure. The elevated pressure serves two purposes: it raises the boiling point of the water and it compresses the water vapor. The hot, high-pressure vapor is directed at the fins of a

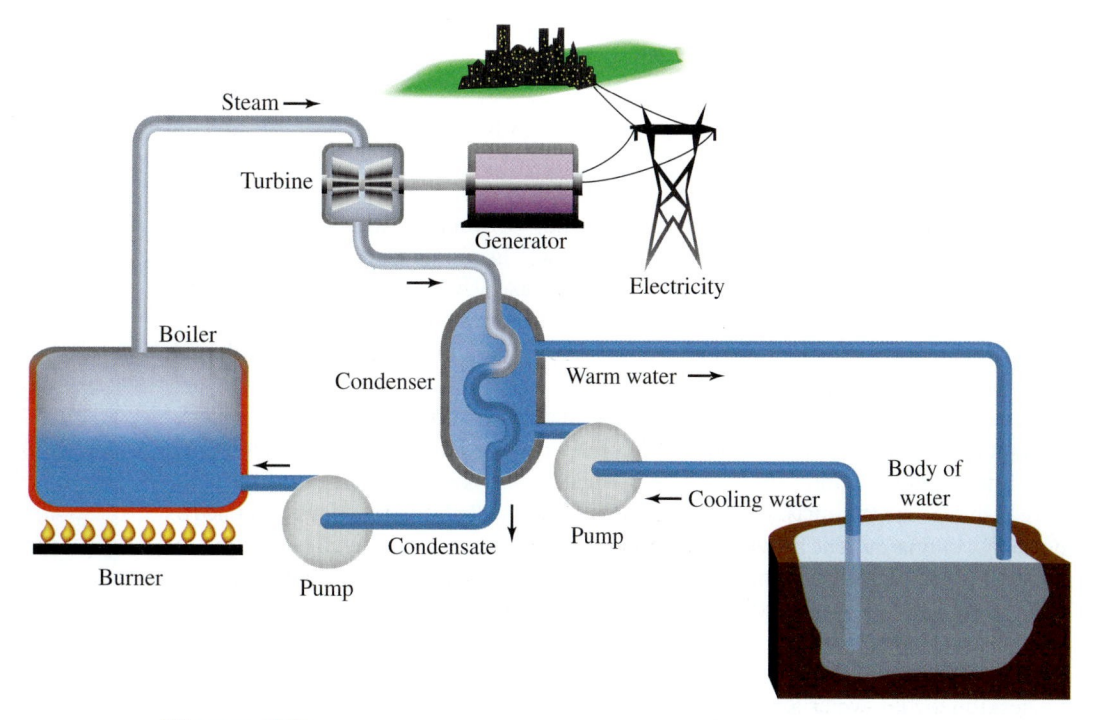

Figure 4.3

Diagram of a power plant for the conversion of chemical fuels to electricity.

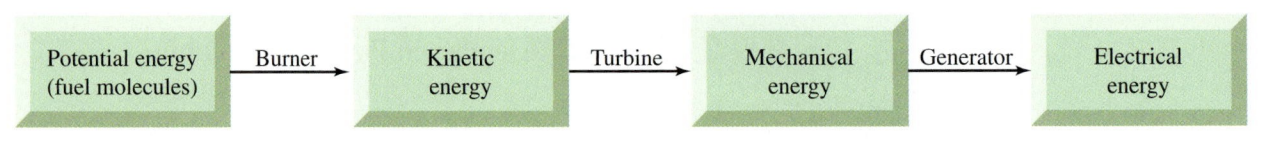

Figure 4.4

Energy transformation in a fossil-fuel electrical power plant.

turbine. As the gas expands and cools, it gives up some of its energy to the turbine, causing it to spin like a pinwheel in the wind. The shaft of the turbine is connected to a large coil of wire that rotates within a magnetic field. The turning of this dynamo generates an electric current, a particularly convenient form of energy. Meanwhile, the water vapor leaves the turbine and continues in its closed cycle. It passes through a heat exchanger where a stream of cooling water carries away the remainder of the heat energy originally acquired from the fuel. The water condenses into its liquid state and reenters the boiler, ready to resume the energy transfer cycle.

This amazing process of energy transformation is summarized in Figure 4.4. Molecules that make good fuels have high potential energy; those that make poor fuels have lower ones. The process of combustion converts some of the potential energy of the fuel molecules into heat, which in turn is absorbed by the water in the boiler. As the water molecules absorb the heat, they move faster and faster in all directions; their kinetic energy has increased. This chaotic, molecular level motion is in fact the origin of what we call *heat,* whereas *temperature* is a statistical measure of the average speed of that motion. Hence, the temperature increases as the amount of kinetic energy of the molecules increases. When the water is vaporized to steam, the water molecules acquire a tremendous amount of kinetic energy. That energy is transformed into *mechanical energy* in the spinning turbine that turns the generator that converts the mechanical energy into *electrical energy.*

Consider This 4.4   **Energy Conversion**

List at least three other technologies or devices that convert energy from one form to another. Include in your list the types of energy involved and speculate on the efficiency of each type of transformation.

In compliance with the first law of thermodynamics, energy is conserved throughout these transformations. To be sure, no new energy is created, but none is lost, either. We may not be able to win, but we can at least break even . . . or can we? The question is not as facetious as it might sound. In fact, we *cannot* break even. No power plant, no matter how well designed, can completely convert heat into work. Inefficiency is inevitable, in spite of the best engineers and the most sincere environmentalists. There are, of course, energy losses due to friction and heat leakage that can be corrected, but these are not the major problems. The chief difficulty is nature; more specifically, the nature of heat and work.

Table 4.1 lists the efficiencies of a number of steps in production of electricity. The overall efficiency is the product of the efficiencies of the individual steps. The net result is that today's most advanced power plants operate at an overall efficiency of only about 42%. Consider, for example, the case of electrical home heating, sometimes advertised as being clean and efficient. We will assume that the electricity is produced by a methane-burning power plant with a maximum theoretical efficiency of 60%. The efficiencies of the boiler, turbine, electrical generator, and power transmission lines are given in Table 4.1; converting the electrical energy back into heat in the home is 98% efficient.

Table 4.1	Typical Efficiencies in Power Production
Maximum theoretical efficiency	55–60%
Efficiency of boiler	90%
Mechanical efficiency of turbine	75%
Efficiency of electrical generator	95%
Efficiency of power transmission	90%

To find the overall efficiency of the electricity generation to home-heating sequence, we multiply the efficiencies of the individual steps, expressed as their decimal equivalents. Note that for the methane-burning power plant, we use the maximum theoretical efficiency, not its actual operating efficiency.

$$\text{overall efficiency} = \text{efficiency of (power plant)} \times \text{(boiler)} \times \text{(turbine)}$$
$$\times \text{(electrical generator)} \times \text{(power transmission)} \times \text{(home electric heater)}$$
$$= 0.60 \times 0.90 \times 0.75 \times 0.95 \times 0.90 \times 0.98$$
$$= 0.34$$

The overall efficiency of 0.34 indicates that only 34% of the total heat energy derived from the burning of methane at the power plant is available to heat the house. If the electrically heated house requires 3.5×10^7 kJ (a typical value for a northern city in January), how much methane (in grams) has to be burned at the power plant? The combustion of 1 gram of methane releases 50.1 kJ. Remember that only 34% of the energy from the burned methane is available to heat the house. So, because of inefficiencies, far more methane has to be burned than the amount to release 3.5×10^7 kJ. We now calculate the total quantity of heat that must be used.

$$\text{heat used} \times \text{efficiency} = \text{heat needed}$$
$$\text{heat used} \times 0.34 = 3.5 \times 10^7 \text{ kJ}$$
$$\text{heat used} = \frac{3.5 \times 10^7 \text{ kJ}}{0.34} = 1.0 \times 10^8 \text{ kJ}$$

Because each gram of methane burned yields 50.1 kJ, 2.0×10^6 g of methane must be burned to furnish 1.0×10^8 kJ.

$$1.0 \times 10^8 \text{ kJ} \times \frac{1 \text{ g CH}_4}{50.1 \text{ kJ}} = 2.0 \times 10^6 \text{ g CH}_4$$

You can compare the efficiency of heating a home with electricity versus natural gas by completing the following activity.

Sceptical Chymist 4.5   Clean Electric Heat

Is electric heat clean and efficient? The electricity must first be generated, usually by a fossil-fuel power plant.

a. The house also could have been heated directly with a gas furnace burning methane at 85% efficiency. Calculate the number of grams of methane required using this method of heating.

b. On the basis of your answer to part a and the discussion preceding this exercise, comment on the claim that electric heat is "clean" and efficient.

Answer

a. Because the only inefficiency is that of the furnace, we can do a similar calculation using 0.85 as the efficiency. The energy required is 4.1×10^7 kJ, which corresponds to 8.2×10^5 g of methane.

Your Turn 4.6   Energy and Efficiency

A coal-burning power plant generates electrical power at a rate of 500 megawatts (MW) (5.00×10^8 J/s). The plant has an overall efficiency of 0.375 for the conversion of heat to electricity.

a. Calculate the total quantity of electrical energy (in joules) generated in 1 year of operation and the total quantity of heat energy used for that purpose.

b. Assuming the power plant burns coal that releases 30 kJ/g, calculate the mass of coal (in grams and metric tons, tonnes) that will be burned in 1 year of operation. *Hint:* 1 tonne = 1×10^3 kg = 1×10^6 g.

Answers

a. 1.58×10^{16} J generated; 4.20×10^{16} J used

b. 1.40×10^{12} g; 1.40×10^6 tonnes

Consider again the Newton's cradle. You would never expect the balls at rest to start knocking into one another on their own, right? For that to occur, all the heat energy that was dissipated when the balls were knocking into one another would have to be gathered together again. Both the inability of a power plant to convert heat into work with 100% efficiency and the inability of a Newton's cradle to start up on its own result from the influence of entropy. We define **entropy** as randomness in position or energy level. The **second law of thermodynamics** has many versions, the most general of which is that the entropy of the universe is constantly increasing. Naturally occurring changes always result in an increase in the disorder, or randomness, of the universe. This means that the most useful, organized kind of energy for doing work is always being transformed into the chaotic (and hence more randomized) motion of heat energy and not the other way around. An important consequence is that it is impossible to completely convert heat into work without making some other changes in the universe.

A helpful way to look at the increase in universal entropy that characterizes all changes is in terms of probability. Disordered states are more probable than ordered ones, and natural change always proceeds from the less probable to the more probable. Let's suppose you define perfect order as a beautifully organized sock drawer, all the socks matched, folded, and placed in rows. This would represent a condition of low entropy (little randomness). If you are like most people, this is probably a rather unlikely arrangement. It certainly did not occur by itself; it took work (an input of energy) to organize the socks. Without the continuing work of organization, it is quite possible that, over the course of a week, a month, or a semester, the entropy and disorder of that sock drawer would increase. The point is that there are a lot more ways that socks can be *mixed up* than there are ways for the socks to be paired; disorder is more probable than order. Conversely, it is not very likely that you would open your drawer some morning and find that the previously jumbled socks were in perfect order and the entropy in that particular part of the universe had suddenly and spontaneously decreased without any external intervention.

That sort of improbable change from disorder to order is essentially what is involved in the conversion of heat to work. Henry Bent, a renowned chemist, has estimated that the probability of the complete conversion of one calorie of heat to work is about the same as the likelihood of a bunch of monkeys typing Shakespeare's complete works 15 quadrillion times (15×10^{15}) in succession without a mistake. Clearly, we need not wait around to observe the Newton's cradle running by itself; similarly, we cannot expect the conversion of energy between forms to be completely efficient.

Consider This 4.7   **Entropy Decrease–Entropy Increase**

Processes that result in a decrease in "local" entropy (like arranging the sock drawer) require an input of energy. Identify another process in which entropy appears to decrease but is actually coupled with an increase in entropy elsewhere in the universe. *Hint:* Think about the entropy changes associated with energy production.

4.3   Measuring Energy Changes

What makes substances such as coal, gas, oil, or wood usable as fuels, whereas others are not? To answer this question let us consider combustion, the most common energy-generating process. **Combustion** is a chemical process in which a fuel combines rapidly with oxygen to release energy and form products, often carbon dioxide and water. In such a chemical transformation, the potential energy of the reactants is greater than that of the products. Because energy is conserved, the difference in energy between products and reactants is given off, primarily as heat.

We illustrate the process with the combustion of methane, CH_4, the principal component of natural gas, a major home-heating fuel. The products are carbon dioxide and water vapor. In Chapter 1, you encountered this combustion equation.

$$CH_4(g) + 2\,O_2(g) \longrightarrow CO_2(g) + 2\,H_2O(g) + \text{energy} \qquad [4.1]$$

Although chemical reactions are often written without including "energy" on either side, here we do so to emphasize the energy change associated with it. This reaction is **exothermic,** a term applied to any chemical or physical change accompanied by the release of heat. Not surprisingly, the amount of heat generated depends on the amount of fuel burned.

The quantity of heat energy released in a combustion reaction can be determined experimentally with a device called a **calorimeter** (Figure 4.5). A known mass of fuel and an excess of oxygen are introduced into a heavy-walled stainless steel "bomb." The bomb is then sealed and submerged in a bucket of water. The reaction is initiated with an electric current that burns through a fuse wire. The heat evolved by the exothermic reaction flows from the bomb to the water and the rest of the apparatus. As a consequence, the temperature of the entire calorimeter system increases. The quantity of heat given off

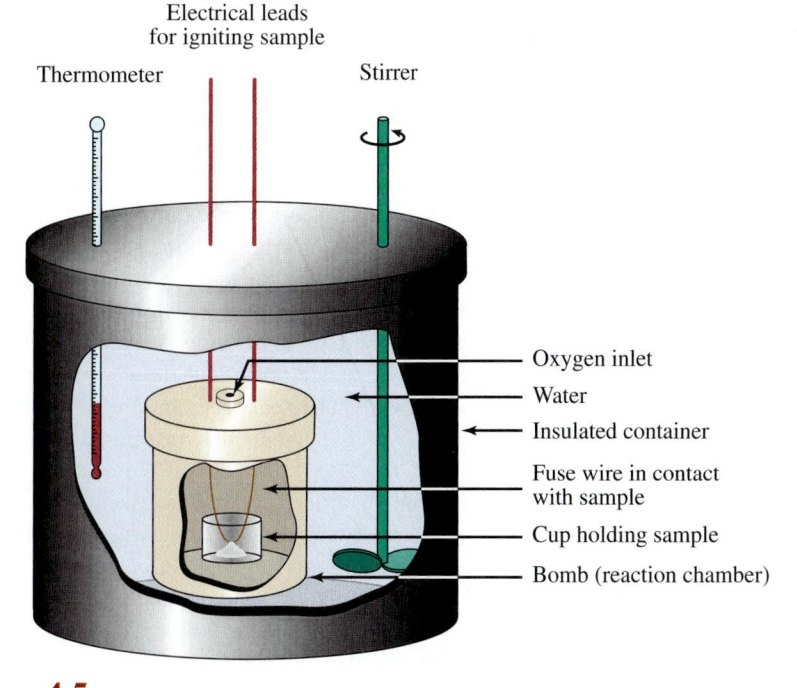

Figure 4.5
Schematic drawing of a bomb calorimeter.

by the reaction can be calculated from this temperature rise and the known heat-absorbing properties of the calorimeter and the water it contains. The greater the temperature increase, the greater the quantity of energy evolved.

Experimental measurements of this sort are the source of most tabulated values of heats of combustion. As the name suggests, the **heat of combustion** is the quantity of heat energy given off when a specified amount of a substance burns in oxygen. Heats of combustion are typically reported as positive values in kilojoules per mole (kJ/mol), kilojoules per gram (kJ/g), kilocalories per mole (kcal/mol), or kilocalories per gram (kcal/g). The energy equivalents of various foods also usually are determined using calorimetry. The experimentally determined heat of combustion of methane is 802.3 kJ. This means that 802.3 kJ of heat is given off when 1 mol of $CH_4(g)$ reacts with 2 mol of $O_2(g)$ to form 1 mol of $CO_2(g)$ and 2 mol of $H_2O(g)$ (see equation 4.1). We also can calculate the number of kilojoules released when one gram of methane is burned. The molar mass of CH_4, calculated from the atomic masses of carbon and hydrogen, is 16.0 g/mol. The heat of combustion per gram of methane (kJ/g) is obtained as follows:

$$\frac{802.3 \text{ kJ}}{1 \text{ mol } CH_4} \times \frac{1 \text{ mol } CH_4}{16.0 \text{ g } CH_4} = 50.1 \text{ kJ/g } CH_4$$

The heat evolved signals a decrease in the energy of the chemical system during the reaction. In other words, the reactants (methane and oxygen) are at higher potential energy than the products (carbon dioxide and water vapor). Therefore, the burning of methane is analogous to water going over a falls or to any falling object. In all these changes, potential energy decreases and is converted into other forms of energy (heat, sound, light, etc.). The negative sign traditionally attached to the energy change for all exothermic reactions signifies this decrease. For example, the energy change for the combustion of methane is −802.3 kJ/mol. Figure 4.6 is a schematic representation of this process. The downward arrow indicates that the energy associated with 1 mol of $CO_2(g)$ and 2 mol of $H_2O(g)$ is less than the energy associated with 1 mol of $CH_4(g)$ and 2 mol of $O_2(g)$. The energy difference between the products and the reactants is thus a negative quantity, as is the case for all exothermic reactions. In the combustion of methane, the energy difference is −802.3 kJ.

> Heats of combustion, by convention, are tabulated as positive values even though all combustion reactions *release* heat.

> You will learn about food Calories in Section 11.1.

> $E_{products} - E_{reactants} < 0$ for an exothermic reaction

---

Your Turn 4.8   **Methane by the Cubic Foot**

According to information in this section, the heat of combustion of methane is 802.3 kJ/mol. Methane is usually sold by the standard cubic foot (SCF). One SCF contains 1.250 mol of methane. Calculate the energy (in kJ) released by burning 1.000 SCF of methane.

Answer
1003 kJ released

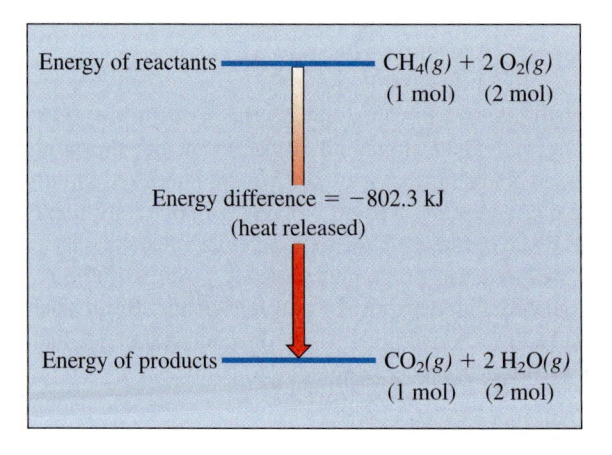

Figure 4.6
Energy difference in the combustion of methane, an exothermic reaction.

These reactions were discussed in Sections 2.6 and 2.8.

Although all chemical reactions used to generate energy are exothermic, many natural reactions, such as photosynthesis, absorb energy as they occur. You already encountered two important examples in atmospheric chemistry. One is the decomposition of O_3 to yield O_2 and O, and the other is the combination of N_2 and O_2 to yield two molecules of NO. Both reactions require energy that can be in the form of an electrical discharge, a high-energy photon, or a high temperature. These reactions are **endothermic,** the term applied to any chemical or physical change that absorbs energy. This situation arises when the potential energy of the products is higher than the potential energy of the reactants. The energy change for an endothermic reaction is always positive.

Endothermic Reaction	Exothermic Reaction
$Energy_{products} > Energy_{reactants}$	$Energy_{products} < Energy_{reactants}$
Energy change is positive.	Energy change is negative.
Energy is absorbed.	Energy is released.

The potential energy of any specific chemical species is related to the chemical bonds involved. In the following section, we illustrate how knowledge of molecular structure can be used to calculate heats of combustion and allows us to understand some of the differences between fuels.

4.4   Energy Changes at the Molecular Level

In the previous section, we learned we can quantify experimentally the energy changes associated with many reactions, either exothermic or endothermic. Now we turn our attention to explaining the origin of the energy changes. Chemical reactions involve a rearrangement of atoms; chemical bonds are broken and formed. Energy is required to *break bonds,* just as energy is required to break chains or tear paper. In contrast, the formation of chemical bonds is an exothermic process in which energy is released. The overall energy change associated with a chemical reaction depends on the net effect of the bond breaking and bond making. If the energy required to break the bonds in the reactants (endothermic) is greater than the energy released (exothermic) when the products form, the overall reaction is *endothermic;* energy is absorbed. If, on the other hand, the exothermic bond-making energy of the products is greater than the endothermic bond breaking in the reactants, then the net energy change is *exothermic;* energy is released by the reaction.

Look for more on hydrogen as a fuel in Sections 8.5–8.8.

As an example, consider the combustion of hydrogen. There is much interest in hydrogen as a fuel because of the large amount of energy per gram released when it burns. We can calculate the total energy change associated with the combustion of hydrogen to form water vapor, as represented by equation 4.2.

$$2\,H_2(g) + O_2(g) \longrightarrow 2\,H_2O(g) + energy \qquad [4.2]$$

The approach we will take is to assume that all the bonds in the reactant molecules are broken and then the individual atoms are reassembled into the product molecules. In fact, the reaction does not occur that way. But we are interested in only the overall (net) change, not the details. Therefore, we will proceed with our convenient plan and see how well our calculated result agrees with the experimental value.

The numbers we need for the computation are given in Table 4.2, a listing of the bond energies associated with a variety of covalent bonds. **Bond energy** is the amount of energy that must be absorbed to break a specific chemical bond. Thus, because energy must be absorbed, breaking bonds is an endothermic process, and all the bond energies in Table 4.2 are positive. Obviously, the amount of energy required depends on the number of bonds broken: more bonds take more energy. Typically, bond energies are expressed in kilojoules per mole of bonds. Note that the atoms appear both across the top and down the left side of Table 4.2. The number at the intersection of any row and column is the

Table 4.2	Bond Energies (in kJ/mol)								
	H	**C**	**N**	**O**	**S**	**F**	**Cl**	**Br**	**I**
Single Bonds									
H	436								
C	416	356							
N	391	285	160						
O	467	336	201	146					
S	347	272	—	—	226				
F	566	485	272	190	326	158			
Cl	431	327	193	205	255	255	242		
Br	366	285	—	234	213	—	217	193	
I	299	213	—	201	—	—	209	180	151

Multiple Bonds					
C=C	598	C=N	616	C=O	803 in CO_2
C≡C	813	C≡N	866	C≡O	1073
N=N	418	O=O	498		
N≡N	946				

Source: Data from Darrell D. Ebbing, *General Chemistry,* Fourth Edition, 1993 Houghton Mifflin Co. Data originally from *Inorganic Chemistry: Principles of Structure and Reactivity,* Third Edition, by James E. Huheey, 1983, Addison Wesley Longman.

energy (in kilojoules) needed to *break a mole of bonds* linking the two atoms. For example, the energy of a H-to-H bond, as in the H_2 molecule, is 436 kJ/mol. Similarly, the energy required to break 1 mol of O-to-O double bonds is 498 kJ, as noted from the bottom part of the table. Bond energies for other double bonds, as well as for some triple bonds, also are given in the table.

We need to keep track of the energy change involved in each bond-breaking or bond-making process and whether the energy is taken up or given off. To do this, we assume that the energy absorbed carries a positive sign, like a deposit to your checkbook. On the other hand, energy given off is like money spent; it bears a negative sign. Bond energies are positive because they represent energy absorbed when bonds are broken. But the formation of bonds releases energy, and hence the associated energy change is negative. For example, the bond energy for the O-to-O double bond is 498 kJ/mol. This means that when 1 mol of O-to-O double bonds is broken, the energy change is +498 kJ; correspondingly, when 1 mol of O-to-O double bonds is formed, the energy change is −498 kJ.

Now we are finally ready to apply these concepts and conventions to the burning of hydrogen gas, H_2. First, we need to determine how many moles of bonds are broken and how many moles of bonds are formed. We can do so by drawing the Lewis structures of the species as shown in equation 4.3.

$$2\,H{-}H \;+\; \overset{..}{\underset{..}{O}}{=}\overset{..}{\underset{..}{O}} \longrightarrow 2\;{\underset{H}{}}\!\!\overset{\overset{..}{O}{\,\cdot\cdot}}{\diagup}\!\!{\underset{H}{\diagdown}} \qquad\qquad [4.3]$$

Remember that chemical equations can be read in terms of moles. Both equation 4.2 and 4.3 indicate "2 mol of $H_2(g)$ plus 1 mol of $O_2(g)$ yields 2 mol of gaseous water (water vapor)." But to use bond energies, we need to count the number of moles of *bonds* involved. Because each H_2 molecule contains one H-to-H bond, 1 mol of H_2 must contain 1 mol of H-to-H bonds. Similarly, equation 4.3 indicates that 1 mol of O_2 contains 1 mol of O-to-O double bonds. Each mole of water contains 2 mol of H-to-O bonds; thus, 2 mol of water contain 4 mol of H-to-O bonds. Therefore, we now have the total number of moles of bonds to be broken (2 mol of H-to-H and 1 mol of O-to-O double) and those to be formed (4 mol of H-to-O). This number of bonds is then *multiplied* by the representative bond energy, using the appropriate sign convention (+ for bonds broken; − for bonds formed).

Molecule	Bonds per Molecule	Moles	Total Number of Moles of Bonds	Bond Process	Energy per Moles of Bonds	Total Energy
H—H	1	2	$1 \times 2 = 2$	Breaking	$+436$ kJ	$2 \times (+436) = +872$ kJ
O=O	1	1	$1 \times 1 = 1$	Breaking	$+498$ kJ	$1 \times (+498) = +498$ kJ
H—O—H	2	2	$2 \times 2 = 4$	Making	-467 kJ	$4 \times (-467) = -1868$ kJ

Consequently, the *overall* energy change in breaking bonds (872 kJ + 498 kJ = 1370 kJ) and forming new ones (−1868 kJ) is −498 kJ.

A schematic representation of this calculation is presented in Figure 4.7. The energy of the reactants, 2 H$_2$ and O$_2$, is set at zero, an arbitrary but convenient value. The green arrows pointing upward signify energy absorbed to break bonds and convert the reactant molecules into individual atoms: 4 H and 2 O. The red arrow on the right pointing downward represents energy released as these atoms are reconnected with new bonds to form the product molecules: 2 H$_2$O. The shorter red arrow corresponds to the net energy change of −498 kJ signifying that the overall combustion reaction is strongly exothermic. The *release* of heat corresponds to a *decrease* in the energy of the chemical system, which explains why the energy change is *negative*. The net result is the evolution of energy, mostly in the form of heat. Another way to look at such exothermic reactions is as a conversion of reactants involving weaker bonds to products involving stronger ones. In general, the products are more stable and less reactive than the starting substances.

We also can use bond energies from Table 4.2 to calculate the energy change for the combustion of methane. Again, it is useful to write the Lewis structures for each species in the reaction as in equation 4.4.

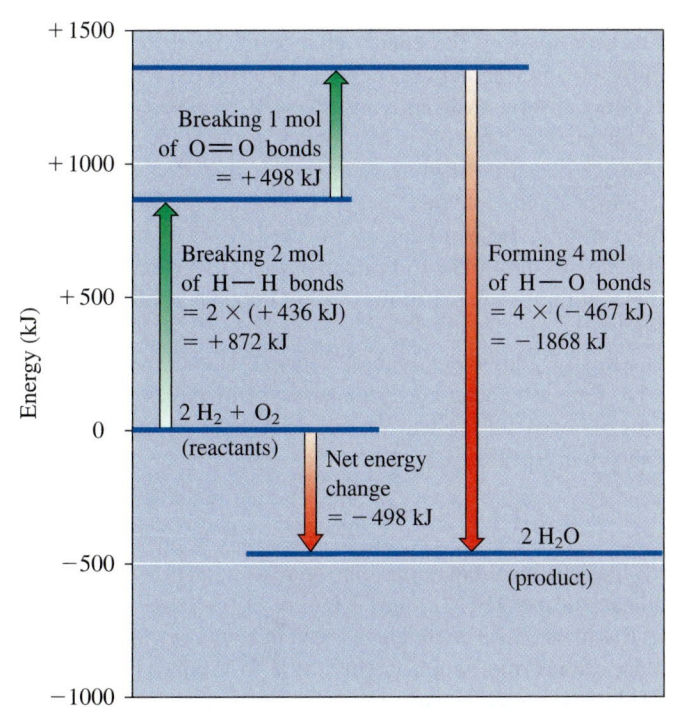

Figure 4.7

The energy changes during the combustion of hydrogen to form water.

 Figures Alive! Visit the *Online Learning Center* to learn more about the energy changes of this reaction.

$$CH_4(g) + 2\,O_2(g) \longrightarrow CO_2(g) + 2\,H_2O(g) + energy$$

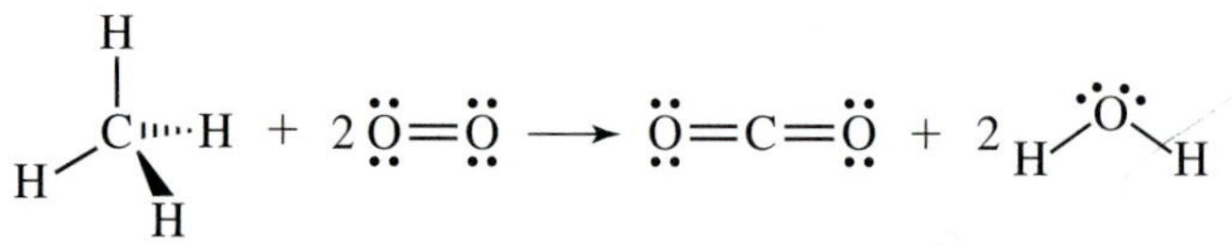

[4.4]

One mole of methane contains 4 mol of C-to-H bonds, each with a bond energy of 416 kJ. Breaking 2 mol of O-to-O double bonds requires 996 kJ (2 × 498 kJ). To form the products, 2 mol of C-to-O double bonds in 1 mol of CO_2 (2 × −803 kJ), and 4 mol of H-to-O bonds in 2 mol of water (4 × −467 kJ) are required. Note again that bond formation is exothermic, and the associated bond energies have minus signs.

Molecule	Bonds per Molecule	Moles	Total Number of Moles of Bonds	Bond Process	Energy per Moles of Bonds	Total Energy
H—C—H (with H above and below C)	4	1	4 × 1 = 4	Breaking	+416 kJ	4 × (+416) = +1664 kJ
O=O	1	2	1 × 2 = 2	Breaking	+498 kJ	2 × (+498) = +996 kJ
O=C=O	2	1	2 × 1 = 2	Making	−803 kJ	2 × (−803) = −1606 kJ
H—O—H	2	2	2 × 2 = 4	Making	−467 kJ	4 × (−467) = −1868 kJ

Total energy change in breaking bonds = (+1664 kJ) + (+996 kJ) = +2660 kJ

Total energy change in making bonds = (−1606 kJ) + (−1868 kJ) = −3474 kJ

Net energy change = (+2660 kJ) + (−3474 kJ) = −814 kJ

Heats of combustion, by convention, are listed as positive values. Thus, the heat of combustion of methane calculated using bond energies is +814 kJ.

The energy changes we just calculated from bond energies, −498 kJ for burning 1 mol of hydrogen and −814 kJ for burning 1 mol of methane, compare favorably with the experimentally determined values. This agreement justifies our rather unrealistic assumption that all the bonds in the reactant molecules are first broken, then all the bonds in the product molecules are formed. This is not what actually happens. But the energy change that accompanies a chemical reaction depends on the energy *difference* between the products and the reactants, not on the particular process, mechanism, or individual steps that connect the two. This is an extremely powerful idea when doing calculations related to energy changes in reactions.

Not all calculations come out as well as this one. For one thing, the bond energies of Table 4.2 apply only to gases, so calculations using these values agree with experiment only if all the reactants and products are in the gaseous state. Moreover, tabulated bond energies are average values. The strength of a bond depends on the overall structure of the molecule in which it is found; in other words, on what else the atoms are bonded to. Thus, the strength of an O-H bond is slightly different in HOH, HOOH, and CH_3OH. Nevertheless, the procedure illustrated here is a useful way of estimating energy changes in a wide range of reactions. The approach also helps illustrate the relationship between bond strength and chemical energy.

This analysis also helps clarify why the H_2O or CO_2 formed in combustion reactions cannot be used as fuels. There are no substances into which these compounds can be converted that have stronger bonds and are lower in energy; we cannot run a car on its exhaust.

Generally, experimental values differ somewhat from those calculated using bond energies. The experimental heat of combustion of methane is 802.3 kJ; the calculated value, 814 kJ, differs by 1.5%.

Fossil fuels as a source of greenhouse gases were discussed in Sections 1.10 and 3.5.

Your Turn 4.9 Heat of Combustion for Acetylene

Use the bond energies in Table 4.2 to calculate the heat of combustion for acetylene, C_2H_2. Report your answer both in kilojoules per mole (kJ/mol) C_2H_2 and kilojoules per gram (kJ/g) C_2H_2. The balanced equation for the reaction is:

$$2\ H{-}C{\equiv}C{-}H\ +\ 5\ \ddot{O}{=}\ddot{O}\ \longrightarrow\ 4\ \ddot{O}{=}C{=}\ddot{O}\ +\ 2\ H{-}\ddot{O}{-}H$$

Answer

Energy change $= -397$ kJ/mol C_2H_2 or -15.3 kJ/g C_2H_2

Heat of combustion $= 397$ kJ/mol C_2H_2 or 15.3 kJ/g C_2H_2

Your Turn 4.10 O_2 versus O_3

O_2 can absorb shorter wavelength radiation than O_3 can. Why? Use the bond energies in Table 4.2 plus information from Chapter 2 to explain.

Hint: You may want to consider the resonance structures for ozone.

4.5 Our Need for Fuel

It is nearly impossible to comprehend the vastness of the quantities of fossil fuels we burn worldwide to generate energy. What probably is apparent, however, is that energy is not consumed equally across the globe. For example, the 5% of the world's population living in North America consumes roughly 30% of the world's energy supply. Figure 4.8 shows trends in energy consumption for the past several decades, as well as projected rates for the next 20 years. The data are grouped into established market economies (the United States, Canada, and Western Europe), emerging economies (India, China, but also including Africa and Central and South America), and the transitional economies of

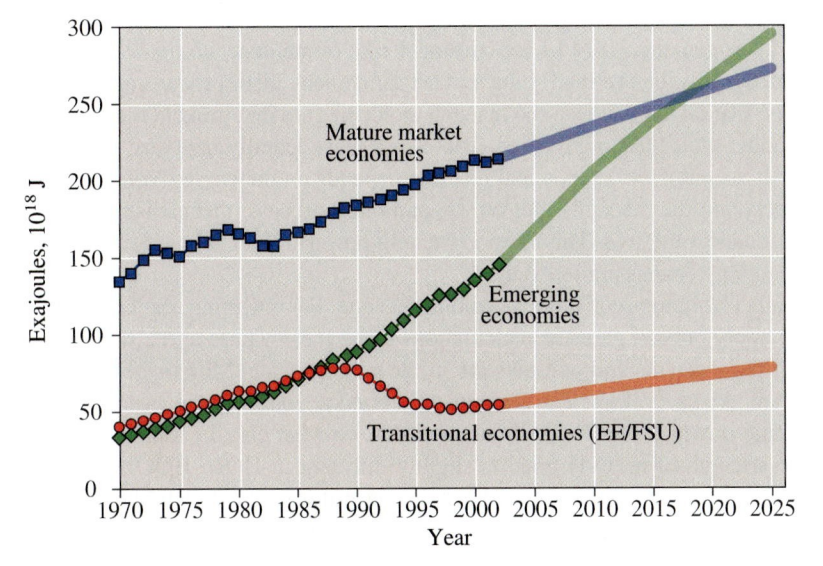

Figure 4.8

The history (data points) and projected future (solid lines) of energy consumption worldwide. EE = Eastern Europe, FSU = former Soviet Union.

Source: *Annual Energy Review 2005,* Department of Energy/EIA.

Eastern Europe and the former Soviet Union. Energy is a primary driver of industrial and economic progress, and therefore it is not surprising that gross national product correlates well with energy production and use. In addition, so do life expectancy, infant mortality, and literacy. Notice also that although energy use in each of the economic sectors is projected to increase in the future, the 4.5% rate of increase in the emerging economies (dominated by the tremendous growth in China and India) far exceeds that of either the established (1.2%) or transitional economies (1.7%). Our global thirst for energy won't be quenched in the near future.

The great burst in energy consumption is of relatively recent origin. Two million years ago, before our ancestors learned to use fire, the sources of energy available to an individual were that of his or her own body or that from the Sun. Earliest hominids probably consumed the equivalent of 2000 kcal per day and expended most of it finding food. This daily energy use corresponds to that used by a 100-W (watt) light bulb burning for 24 hr. The discovery of fire and the domestication of beasts of burden increased the energy available to an individual by about six times. Hence, we estimate that about 2000 years ago, a farmer with an ox or donkey had roughly 12,000 kcal available each day.

The Industrial Revolution brought another five- or sixfold increase in the energy supply, most of it from coal via steam engines. Yet another energy jump occurred during the 20th century. By the year 2000, the total energy used in the United States (from all sources and for all purposes) corresponded to about 650,000 kcal per person per day. This translates to an annual equivalence of 65 barrels of oil or 16 tons of coal for each American. In human terms, the energy available to each resident of the United States would require the physical labor of 130 workers. Yet, there are still people on the planet whose energy use and lifestyle closely approximate those of 2000 years ago.

The history of increasing energy consumption is closely related to changing energy sources and the development of devices for extracting and transforming that energy. Figure 4.9 displays the average American energy consumption from a variety of sources since 1800. The graph indicates that wood was originally the major energy source in the United States, and it continued to be so until the late 1880s when wood was surpassed by coal. Coal provided more than 50% of the nation's energy from then until about 1940. By 1950, oil and gas were the source of more than half of the energy used in this country. Nuclear fission, once hailed as an almost limitless source of energy, has not achieved its full potential for a variety of reasons. Falling water has long been used to power mills and, more recently, to generate electricity, but it provides only a small percentage of our total energy. Waste, alcohol, geothermal, wind, solar sources and other "renewables" combined would barely be visible on this graph, and that only in the last 10 years.

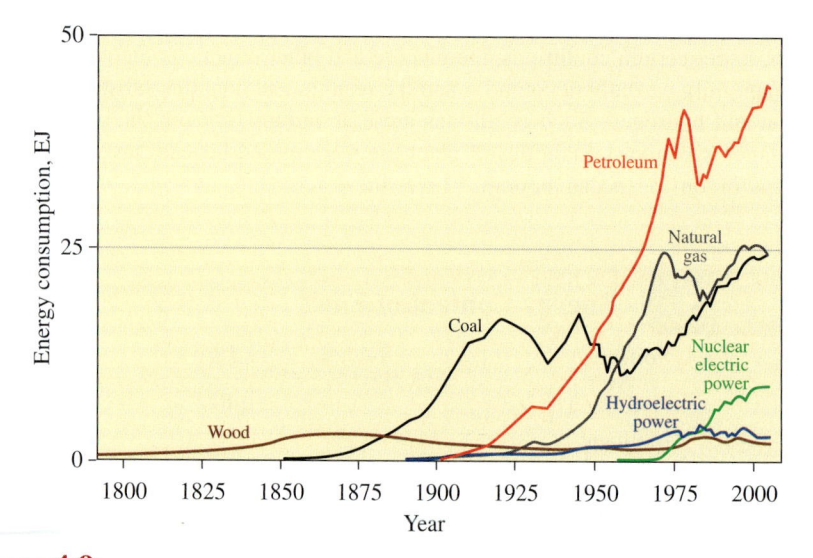

Figure 4.9

History of U.S. energy consumption by source, 1800–2005 (1 EJ = 10^{18}J).

Source: *Annual Energy Review 2005*, Department of Energy/EIA.

Consider This 4.11   **U.S. Sources of Energy over Time**

Figure 4.9 shows the history of U.S. energy consumption. In the period from 1950 to 2005, which sources of energy have shown steady growth and which have not? Propose reasons for the observed trends.

How do we select fuels? Materials such as coal, oil, and natural gas possess many of the properties needed in a fuel. They contain substantial energy content; values for several fuels appear in Table 4.3. Vast quantities appear to be available as "natural" resources, to be harvested almost at will. From where do these reserves come? In a very real sense, these fossil fuels are sunshine in the solid, liquid, and gaseous state. Most of the energy that drives the engines of our economy comes from these remnants of the past. The sunlight was captured millions of years ago by green plants that flourished on the prehistoric planet. The same reaction is carried out by plants today.

$$2800 \text{ kJ} + 6 \text{ CO}_2(g) + 6 \text{ H}_2\text{O}(l) \xrightarrow{\text{chlorophyll}} \underset{\text{glucose}}{\text{C}_6\text{H}_{12}\text{O}_6(s)} + 6 \text{ O}_2(g) \qquad [4.5]$$

This conversion of carbon dioxide and water to glucose and oxygen is endothermic. It requires the absorption of 2800 kJ of sunlight per mole of $C_6H_{12}O_6$ or 15.5 kJ/g of glucose formed. The reaction could not occur without the absorption of energy and the participation of a green pigment molecule called chlorophyll. The chlorophyll interacts with photons of visible sunlight and uses their energy to drive the photosynthetic process, a very energetically uphill reaction.

You already are aware of the essential role of photosynthesis in the initial generation of the oxygen in Earth's atmosphere, in maintaining the planetary carbon dioxide balance, and in providing food and fuel for creatures like us. During **respiration, the process by which humans and animals exchange the oxygen necessary for metabolism with the carbon dioxide produced by it,** we in essence run photosynthesis backward.

$$C_6H_{12}O_6(s) + 6 \text{ O}_2(g) \longrightarrow 6 \text{ CO}_2(g) + 6 \text{ H}_2\text{O}(l) + 2800 \text{ kJ} \qquad [4.6]$$

We extract the 2800 kJ released per mole of glucose "burned" and use that energy to power our muscles and nerves, though we do not do it with perfect efficiency (see Sceptical Chymist 4.3). The same overall reaction occurs when we burn wood, which is primarily cellulose, a polymer composed of repeating glucose units.

When plants die and decay, they also are largely transformed into CO_2 and H_2O. However, under certain conditions, the glucose and other organic compounds that make up the plant only partially decompose and the residue still contains substantial amounts of carbon and hydrogen. Such conditions arose at various times in the prehistoric past of our planet, when vast quantities of plant life were buried beneath layers of sediment in swamps or on the ocean bottom. There, these remnants of vegetable matter were

Polymers will be discussed in Chapter 9.

Table 4.3	**Energy Content of Fuels**
Source	**kJ/g**
Hydrogen	140
Methane	56
Propane	51
Gasoline	48
Coal (hard)	31
Ethanol	30
Wood (oak)	14

protected from atmospheric oxygen, and the decomposition process was halted. However, other chemical transformations occurred in Earth's high-temperature and high-pressure reactor. Over millions of years, the plants that captured the rays of a young Sun were transmuted into the fossils we call coal and petroleum. Jacob Bronowski, in his book *Biography of an Atom—And the Universe,* aptly describes the cycle by saying, "You will die but the carbon will not; its career does not end with you . . . it will return to the soil, and there a plant may take it up again in time, sending it once more on a cycle of plant and animal life." So, in a sense, fossil fuels are renewable, but not anywhere in the time frame that is helpful to human beings.

4.6 Coal

The great exploitation of fossil fuels began with the Industrial Revolution, about two centuries ago. The newly built steam engines consumed large quantities of fuel, but in England, where the revolution began, most of the forests already had been cut down. Coal turned out to be an even better energy source than wood because it yielded more heat per gram (see Table 4.3). By the 1960s, most coal was used for generating electricity and by 2004, the electric power sector accounted for 92% of all coal consumption.

Coal is a complex mixture of substances. Although not a single compound, coal can be approximated by the chemical formula $C_{135}H_{96}O_9NS$. This formula corresponds to a carbon content of 85% by mass. The carbon, hydrogen, oxygen, nitrogen, and sulfur atoms come from the original prehistoric plant material. In addition, some samples of coal typically contain small amounts of silicon, sodium, calcium, aluminum, nickel, copper, zinc, arsenic, lead, and mercury.

Coal occurs in varying grades, but in whatever grade, coal is a better fuel than wood because it contains a higher percentage of carbon and a lower percentage of oxygen and water. Soft lignite, or brown coal, is the lowest grade (Figure 4.10). The vegetable matter that makes it up has undergone the least amount of change, and its chemical composition is similar to that of wood or peat. Consequently, the heat of combustion of lignite is only slightly greater than that of wood (Table 4.4). The higher grades of coal, bituminous and anthracite, have been exposed to higher pressures in the Earth. In the process, they lost more oxygen and moisture and became a good deal harder—more mineral than vegetable (see Figure 4.10). The percentage of carbon is higher as is the heat of combustion. Anthracite has a particularly high carbon content and a low concentration of sulfur, both of which make it the most desirable grade of coal. Unfortunately, the deposits of anthracite are relatively small, and in the United States the supply is almost exhausted. We now rely more heavily on bituminous and subbituminous coal.

Figure 4.10
Samples of anthracite (*left*) and lignite coal (*right*).

Table 4.4	Energy Content of Various U.S. Coals	
Type of Coal	**State of Origin**	**Energy Content (kJ/g)**
Anthracite	Pennsylvania	30.5
Bituminous	Maryland	30.7
Subbituminous	Washington	24.0
Lignite (brown coal)	North Dakota	16.2
Peat	Mississippi	13.0
Wood	Various	10.4–14.1

Generally speaking, the less oxygen a compound contains, the more energy per gram it will release on combustion because such compounds lie higher on the potential energy scale. As a specific example, burning 1 mol of carbon to form carbon dioxide yields about 40% more energy than is obtained from burning 1 mol of carbon monoxide. To be sure, coal is a mixture, not a compound, but the same principles apply. Anthracite and bituminous coals consist primarily of carbon. Their heat of combustion is, gram for gram, about twice that of lignite, which contains a much lower percentage of carbon.

Your Turn 4.12   Calculations Concerning Coal

a. Assuming the composition of coal can be approximated by the formula $C_{135}H_{96}O_9NS$, calculate the mass of carbon (in tons) in 1.5 million tons of coal. This quantity of coal might be burned by a typical power plant in 1 year.
b. Compute the amount of energy (in kilojoules) released by burning this mass of coal. Assume the process releases 30 kJ/g of coal. Recall that 1 ton = 2000 lb and that 1 lb = 454 g.
c. What mass of CO_2 is formed by the complete combustion of 1.5 million tons of this coal?

Answers

a. Calculate the approximate molar mass of coal. The subscripts for each element give the number of moles:

$$135 \text{ mol C} \times \frac{12.0 \text{ g C}}{1 \text{ mol C}} = 1620 \text{ g C}$$

$$96 \text{ mol H} \times \frac{1.0 \text{ g H}}{1 \text{ mol H}} = 96 \text{ g H}$$

$$9 \text{ mol O} \times \frac{16.0 \text{ g O}}{1 \text{ mol O}} = 144 \text{ g O}$$

$$1 \text{ mol N} \times \frac{14.0 \text{ g N}}{1 \text{ mol N}} = 14.0 \text{ g N}$$

$$1 \text{ mol S} \times \frac{32.1 \text{ g S}}{1 \text{ mol S}} = 32.1 \text{ g S}$$

The sum of these elemental contributions for $C_{135}H_{96}O_9NS$ is 1906 g/mol. Therefore, every 1906 g of coal contains 1620 g C. The mass-to-mass relationship stays the same as long as the same mass unit is used for both; the ratio is just as useful expressed in tons.

$$\text{Mass of carbon} = 1.5 \times 10^6 \text{ tons } C_{135}H_{96}O_9NS \times \frac{1620 \text{ tons } C}{1906 \text{ tons } C_{135}H_{96}O_9NS}$$

$$= 1.3 \times 10^6 \text{ tons } C = 1.3 \text{ million tons } C$$

b. 4.1×10^{13} kJ   **c.** 4.8 million tons

Although the global supply of coal is large and it remains a widely used fuel, coal has some serious drawbacks. It is difficult to obtain, and underground mining is both dangerous and expensive. Since 1900, more than 100,000 workers have been killed in American coal mines by accidents, cave-ins, fires, explosions, and poisonous gases. Many more have been injured or incapacitated by respiratory diseases. Mine safety has dramatically improved in the United States in recent years, but the industry remains a dangerous one in countries like China, where over 4,700 mining deaths were reported in 2006 alone.

Safer mining techniques can be employed if the coal deposits lie sufficiently close to the surface. The overlying vegetation, soil and rock are cleared to reveal the coal seam, which is then removed by heavy machinery. This "strip mining" must be done carefully to prevent serious environmental deterioration. Current regulations in the United States require the replacement of earth and topsoil and the planting of trees and vegetation at former mine sites. In the past, however, these regulations were not in place to prevent the great holes in the earth and heaps of eroding soil that still dot regions of abandoned strip mines. Once the coal is out of the ground, its transportation is both difficult and expensive because it is a solid. Unlike gas and oil, coal cannot be pumped unless it is finely divided and suspended in a water slurry.

Another disadvantage is that coal is a dirty fuel. It is, of course, physically dirty, but its dirty combustion products are more serious. The unburned soot from countless coal fires in the 19th and early 20th centuries blackened both buildings and lungs in many industrial cities. Less visible but equally damaging are the oxides of nitrogen (a consequence of the high temperatures involved) and the oxides of sulfur (arising from any sulfur present in coal). In the United States, coal-burning power plants are responsible for two thirds of the sulfur dioxide emissions and one fifth of the nitrogen oxide emissions. These gases are the principal culprits of acid rain. Although coal contains only minor amounts of mercury (50–200 ppb), mercury is concentrated in the fly ash that escapes as particulate matter into the atmosphere or the "bottom" ash that remains. In the United States, coal-fired power plants emit over 48 tonnes of mercury to the environment each year. Coal also suffers from the same drawback of all fossil fuels; the greenhouse gas carbon dioxide is an inescapable product of its combustion. Because of the lower energy content per gram of coal compared with other fossil fuels (see Table 4.3), *more* carbon dioxide must be released to generate the same amount of energy with coal.

Because of these less-than-desirable properties and the fact that coal reserves are relatively plentiful in the United States, significant research efforts are underway aimed at developing new coal technologies. Though it may sound like an oxymoron, "clean coal" is promoted by its supporters as one important step toward decreasing our reliance on energy imports and minimizing environmental effects. The two main thrusts of clean coal technologies are to increase the efficiency of coal-fired power plants while diminishing the environmentally damaging emissions of sulfur oxides and nitrogen oxides. For example, in fluidized-bed power plants, pulverized coal is burned in a blast of air. The large surface area of the very fine coal dust means that it reacts rapidly and completely with oxygen. The combustion actually occurs at a lower temperature than that required to ignite larger pieces of coal, so that the generation of nitrogen oxides is minimized. If finely divided limestone (calcium carbonate) is mixed with the powdered coal, sulfur dioxide is also removed from the effluent gas. Thus, the amount of pollution is reduced while the efficiency of coal combustion is enhanced.

With coal reserves in the United States far outweighing all other fossil fuels, coal usage will only increase in the future. It is, however, possible that coal will not be burned in its familiar form, but rather converted to cleaner and more convenient liquid and gaseous fuels. After we discuss petroleum, we will turn our attention to some of these alternative uses of coal.

Look for more on clean coal technology in Section 6.14.

4.7   Petroleum

Most people in an average American city or town would be hard-pressed to find lumps of coal. Indeed, many of you may never have seen coal, but you undoubtedly have seen gasoline. Around 1950, petroleum surpassed coal as the major energy source in the United States. The reasons are relatively easy to understand. Petroleum, like coal, is partially decomposed organic matter. However, it has the distinct advantage of being a liquid, making it easily pumped to the surface from its natural, underground reservoirs, transported via pipelines, and fed automatically to its point of use. Moreover, petroleum is a more concentrated energy source than coal, yielding approximately 40–60% more energy per gram. Typical figures are 48 kJ/g for petroleum and 30 kJ/g for coal.

The major component extracted from petroleum (crude oil) is gasoline. Although it has been extracted since the mid-1800s, gasoline became valuable and important only with the advent of the automobile and the internal combustion engine early in the 20th century. That fuel and engine partnership has led to our seemingly insatiable appetite for gasoline. In 2005, 117 billion gallons of gasoline was burned in more than 220 million American automobiles, SUVs, and light trucks traveling an astounding 2.6 trillion miles. In this country, our capacity to consume gasoline has far outstripped our ability to produce it from crude oil. With 5% of the world's population, we consume 25% of the oil produced worldwide. Prior to the 1950s, we imported almost no oil. By the mid-1970s, the United States was producing only about two thirds of the crude oil it required to power its automobiles and factories, heat its homes, and lubricate its machines (Figure 4.11).

Our strong dependency on oil from abroad continues today, increasing from 4.3 million barrels imported per day in 1985 to 13 million barrels per day in 2004, and now accounts for 60% of our total oil usage. In 2004, the United States imported oil from 89 different countries, the top 8 of which are shown in Figure 4.12. A significant fraction of these imports comes from politically volatile regions, including the Persian Gulf and Venezuela.

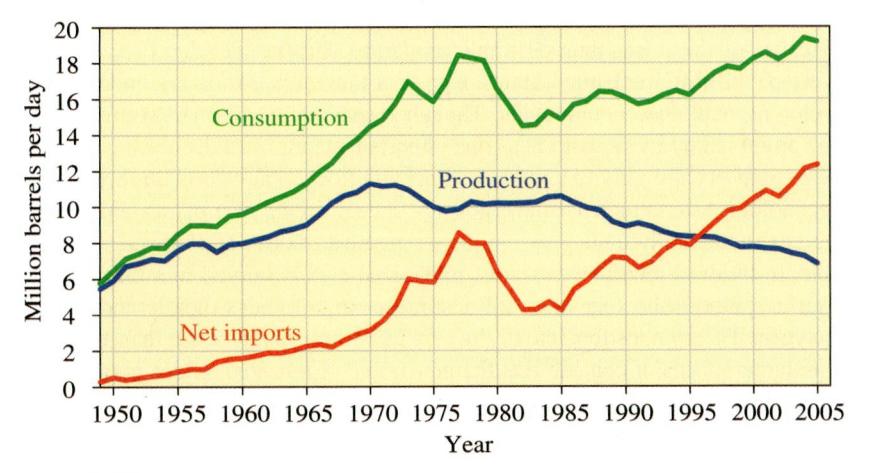

Figure 4.11

U.S. petroleum product use, domestic production, and imports. At present, more than 60% of the total oil used in the United States is imported, and projections show oil imports will continue to increase.

Source: Department of Energy, Energy Information Administration, *Annual Energy Review 2005.*

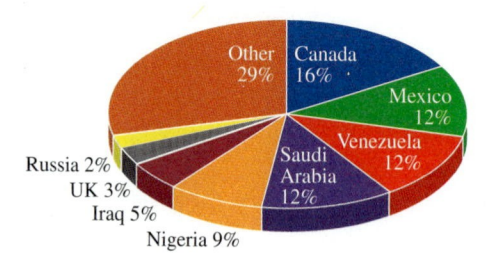

Figure 4.12

Sources of crude oil and petroleum products imported by the United States in 2004.

Source: EIA/DOE.

Consider This 4.14   Shipping Oil

Figure 4.12 identifies the major countries from which the United States imports oil.

a. Which countries, if any, surprised you as sources of our imported oil? Explain.
b. How is oil transported to the United States from these countries?
c. What are the risks associated with oil transport? Comment on how these risks compare with the benefits.

As a nation, our voracious appetite for oil was met by consuming an average of nearly 22 million barrels of oil *daily* in 2005, enough oil to cover a football field with a column of oil over 2500 ft tall. Two thirds of this was for transportation. Unlike coal, however, crude oil is not ready for immediate use when it is extracted from the ground. Crude oil must first be refined, a process that has given gainful employment to many chemists and chemical engineers (and quite a few others). It has also provided an amazing array of products.

Petroleum is a complex mixture of thousands of different compounds. The great majority are **hydrocarbons,** molecules consisting of only hydrogen and carbon atoms. Hydrocarbons in petroleum can contain from 1 to as many as 60 carbon atoms per molecule. A set of **alkanes,** hydrocarbons with only single bonds between carbon atoms, is shown in Table 4.5. Concentrations of sulfur and other contaminating elements are generally quite low, minimizing emissions of environmentally damaging gases.

The oil refinery has become an icon of the petroleum industry (Figure 4.13). During one step in the refining process, the crude oil is separated into fractions that consist of compounds with similar properties. A physical process called distillation accomplishes this fractionation. **Distillation** is a separation process in which a solution is heated to its boiling point and the vapors are condensed and collected. To distill (fractionate) the crude oil, it is pumped into an industrial-sized container (still) and heated. As the temperature increases, the components with the lowest boiling points are the first to vaporize. The molecules of these low-boiling components escape from the liquid and travel high up the tall distillation tower. With further temperature increases, the higher boiling fractions of the mixture vaporize, but their molecules do not travel as high up the tower. At different levels in the tower, each fraction is condensed back into the liquid state.

Figure 4.14 illustrates a distillation tower and lists some of the fractions obtained. These include gases such as methane, liquids such as gasoline and kerosene, and waxy solids such as paraffin. Note that the boiling point increases with increasing number of carbon atoms in the molecule and hence with increasing molecular mass and size.

Table 4.5   Selected Alkanes

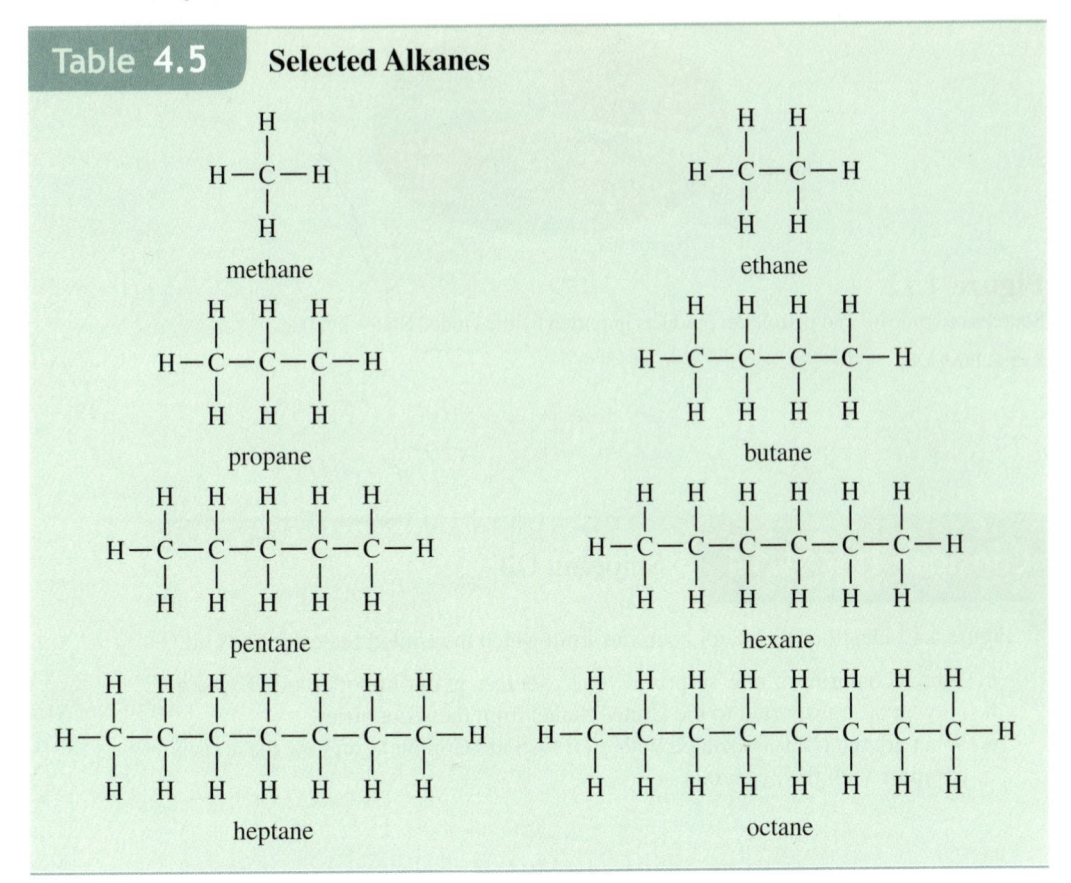

methane

ethane

propane

butane

pentane

hexane

heptane

octane

As of July 2007, there were 143 oil refineries operating in the United States. The last one was completed in 1976.

Figure 4.13

An oil refinery, the symbol of the petroleum industry, at night.

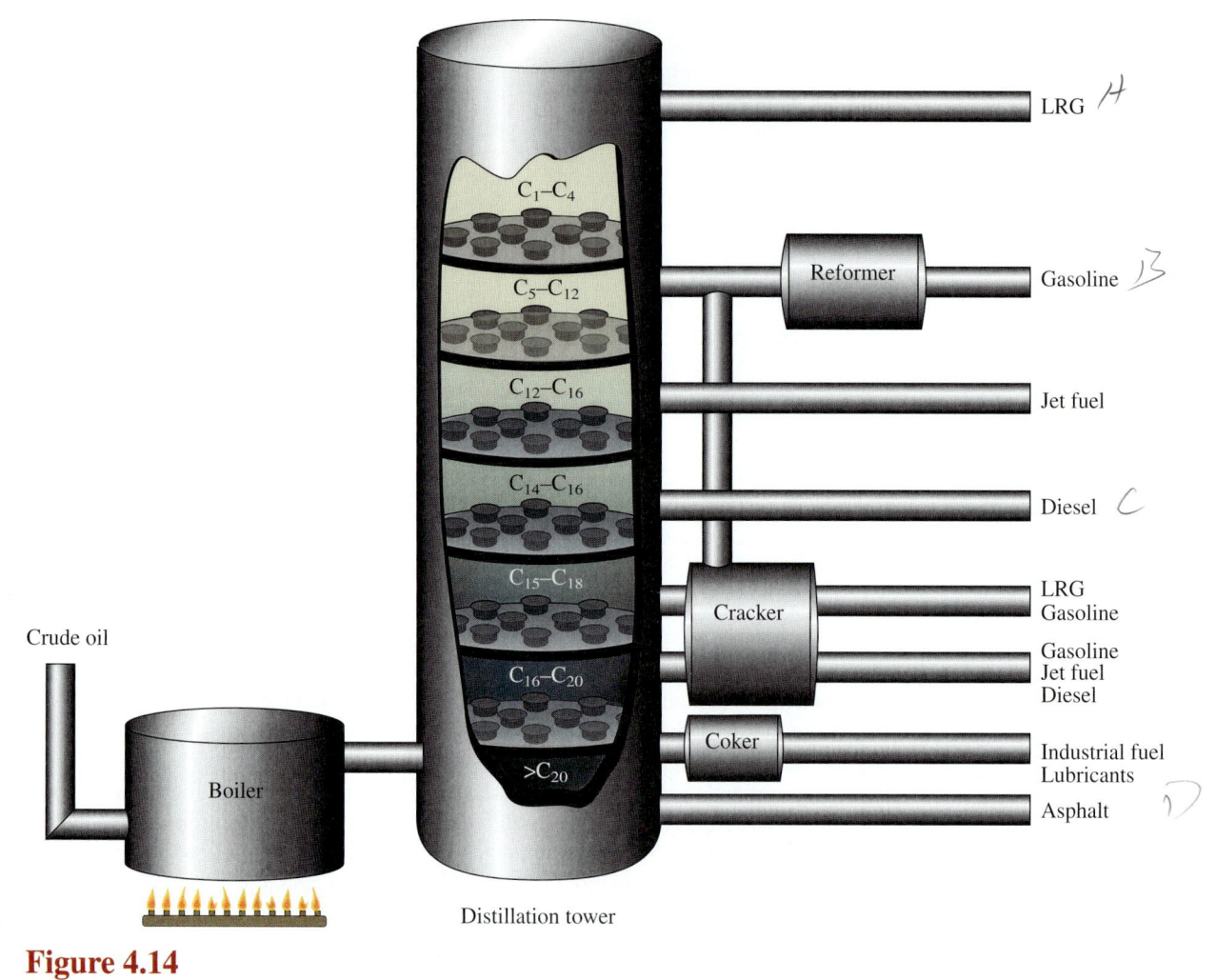

Figure 4.14

Diagram of a crude oil distillation tower showing various fractions and some typical uses. LRG = liquefied refinery gas.

Heavier, larger molecules are attracted to one another more than are the lighter, smaller molecules. In turn, higher temperatures are required to vaporize the compounds of higher molecular mass and size.

The various fractions distilled from crude oil have different properties and hence different uses. Indeed, the great diversity of products obtained has made petroleum a particularly valuable source of matter and energy. The most volatile components of petroleum, refinery gases, boil far below room temperature. Refinery gases are often used as fuels to operate the distillation towers. They also can be liquefied refinery gas (LRG) and sold for home use or used to synthesize other molecules by chemical manufacturers. The gasoline fraction, containing hydrocarbons with 5–12 carbon atoms per molecule, is particularly important to our automotive civilization. Efforts at designing and mass producing automobiles were largely unsuccessful until petroleum provided a convenient and relatively safe liquid fuel. Higher boiling fractions are used to fuel diesel engines and jet planes. Still higher boiling fractions are used as industrial heating oil and lubricating oils.

Refining a barrel of crude oil provides an impressive array of products, with gasoline constituting the largest fraction (Figure 4.15). A staggering 37 of the almost 45 gal in a typical barrel of refined crude oil is simply burned for heating and transportation. The remaining 7.6 gal is used for nonfuel purposes, including only 1.25 gal set aside to serve as nonrenewable starting materials (reactants, commercially called feedstocks) to make the myriad of plastics, pharmaceuticals, fabrics, and other carbon-based industrial products so common in our society.

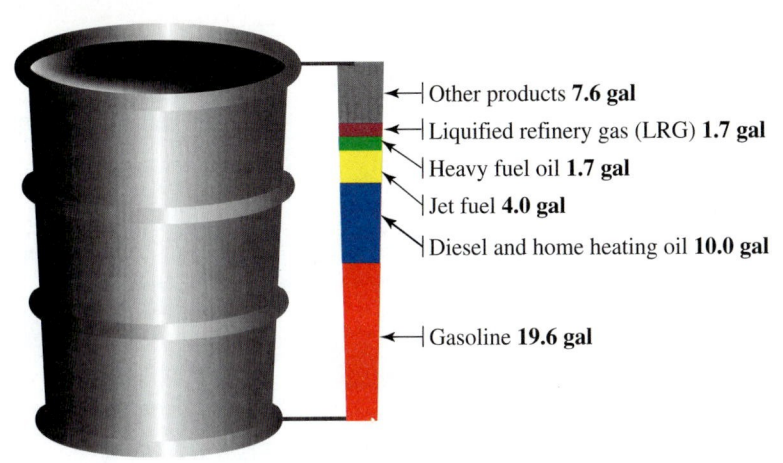

Figure 4.15

Products (in gallons) from the refining of 1 barrel of crude oil.

Note: A barrel of crude oil contains 42 gal. However, the process of refining adds slightly to the volume so that the products total more than 42 gal.

A discussion of petroleum also should include natural gas. This fuel (typically containing 87–96% methane, 2–6% ethane, and smaller quantities of larger hydrocarbons, nitrogen, carbon dioxide, and oxygen) currently provides heat for two thirds of the single-family homes and apartment buildings in the United States. Recently, interest has increased in using natural gas as an energy source for generating electricity and for powering cars and trucks. A distinct advantage of natural gas is that it burns much more completely and cleanly than other fossil fuels. Because of its purity, it releases essentially no sulfur dioxide when burned. Natural gas emits only very low levels of unburned volatile hydrocarbons, carbon monoxide, and nitrogen oxides, and it leaves no residue of ash or toxic metals like mercury. Moreover, per joule of energy produced, burning natural gas produces 30% less carbon dioxide than oil and 43% less carbon dioxide than coal.

> ### Consider This 4.15   Products from a Barrel of Crude
>
> Chemical research has helped increase the amount of gasoline derived from a barrel of crude oil. For example, in 1904 a barrel of crude oil produced 4.3 gal of gasoline, 20 gal of kerosene, 5.5 gal of fuel oil, 4.9 gal of lubricants, and 7.1 gal of miscellaneous products. By 1954, the products were 18.4 gal of gasoline, 2.0 gal of kerosene, 16.6 gal of fuel oil, 0.9 gal of lubricants, and 4.1 gal of miscellaneous products. Compare these values with those shown in Figure 4.15 and offer some reasons why the distribution of products has changed over time.

4.8   Manipulating Molecules to Make Gasoline

The distribution of compounds obtained by distilling crude oil does not correspond to the prevailing commercial use pattern. For example, the demand for gasoline is considerably greater than that for higher boiling fractions. Several chemical processes can be employed after fractionation to change the natural distribution and to obtain more gasoline of higher quality. These include cracking, combining, and re-forming (see Figure 4.14).

Cracking is a chemical process by which large molecules are broken into smaller ones suitable for use in gasoline. For example, a hydrocarbon with 16 carbons can be cracked into two almost equal fragments,

$$C_{16}H_{34} \longrightarrow C_8H_{18} + C_8H_{16} \tag{4.7}$$

or into different sized ones.

$$C_{16}H_{34} \longrightarrow C_{11}H_{22} + C_5H_{12} \qquad [4.8]$$

Note that the total number of carbon and hydrogen atoms is unchanged from reactants to products. The larger reactant molecules simply have been fragmented into smaller, more economically important molecules. Representing equation 4.8 with space-filling models shows the size difference more clearly. The model of $C_{11}H_{22}$ also shows a "bend" where a C-to-C double bond is located.

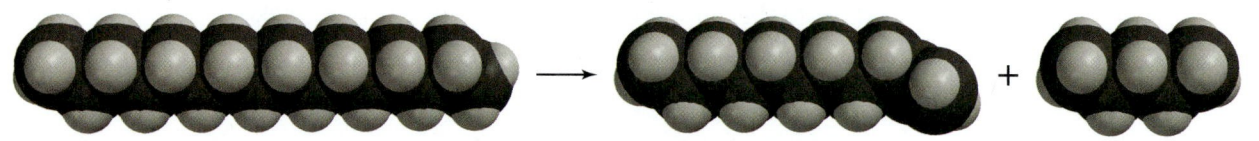

Historically, **thermal cracking** was achieved by heating the starting materials to a high temperature. The extreme example of thermal cracking is called coking. In this process, the very heaviest fractions are heated to between 400 and 450 °C. Coking converts the heaviest tarry crude oil "bottoms" into useful gasoline and diesel fuel and leaves behind a residue of almost pure carbon.

Valuable energy is saved when catalysts are used to promote molecular breakdown at lower temperatures in an operation called **catalytic cracking.** Important cracking catalysts have been developed by chemists at all major oil companies, and researchers continue to find more selective and inexpensive processes. We will discuss how catalysts affect the rates of chemical reactions in Section 4.10.

> The catalysts employed in cracking are chemically similar to the ion-exchange resins used in water softening.

If simple distillation produces more small molecules than needed but not enough intermediate-sized ones essential for gasoline, catalytic combination can be used. In this process, smaller molecules are joined to form useful intermediate-sized molecules.

$$4\,C_2H_4 \xrightarrow{\text{catalyst}} C_8H_{16} \qquad [4.10]$$

> Extending this process to make very large molecules will be discussed in Section 9.2.

In another part of the refining process called re-forming, the atoms within a molecule can be rearranged. It turns out that not all the molecules with the same chemical formula are necessarily identical. For example, *n*-octane, an important component of gasoline, has the formula C_8H_{18}. Careful analysis reveals 18 different compounds with this formula. Different compounds with the same chemical formula are called **isomers.** Isomers differ in molecular structure—the way in which the constituent atoms are arranged. In *n*-octane (normal octane) all the carbon atoms are in an unbranched ("straight") line as shown in Figure 4.16a. In isooctane, the carbon chain has several branch points as shown in Figure 4.16b. Although the chemical and physical properties of these two isomers are similar, they are not identical. For example, the boiling point of *n*-octane is 125 °C, compared with 99 °C for isooctane.

The heats of combustion for *n*-octane and isooctane also are nearly identical, but the more compact shape of the latter compound imparts more controllable combustion. In a well-tuned car engine, gasoline vapor and air are drawn into a cylinder, compressed by a piston, and ignited by a spark. Normal combustion occurs when the spark plug ignites the fuel–air mixture and the flame front travels across the combustion chamber rapidly and

> The heat of combustion for a branched hydrocarbon is 2–4% more exothermic than their straight-chain isomers.

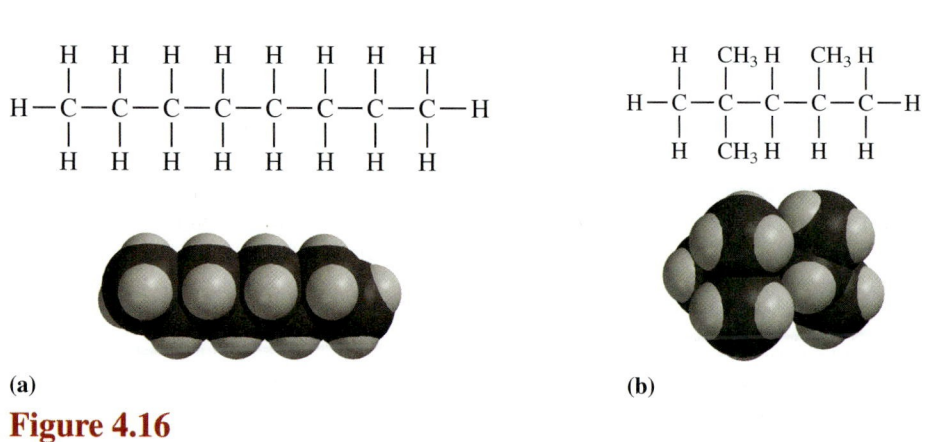

(a)   **(b)**

Figure 4.16

Line drawings and space-filling representations of **(a)** *n*-octane and **(b)** isooctane.

Table 4.6	Octane Ratings of Several Compounds	
	Compound	**Octane Rating**
	n-Octane	−20
	n-Heptane	0
	Isooctane	100
	Methanol	107
	Ethanol	108
	MTBE	116

smoothly until the fuel is consumed. However, compression alone may be often enough to ignite the fuel before the spark occurs. The term for this premature firing is *preignition*. It results in lower engine efficiency and higher fuel consumption because the piston was not in its optimal location when the burned gases expanded. Knocking, a violent and uncontrolled pressure that may be several times the usual value for the engine, occurs after the spark ignites the fuel, causing the unburned mixture to burn at supersonic speed with an abnormal rise in pressure. Knocking produces an objectionable metallic sound, loss of power, overheating, and engine damage when severe.

In the 1920s, knocking was shown to depend on the chemical composition of the gasoline. The "octane rating" was developed to designate a particular gasoline's resistance to knocking. Isooctane performs exceptionally well in automobile engines and arbitrarily has been assigned an octane rating of 100. Like *n*-octane, *n*-heptane is a straight-chain hydrocarbon, but with one fewer —CH_2 groups. It also has a high tendency to cause knocking and has been assigned an octane rating of 0 (Table 4.6). When you go to the gasoline pump and fill up with 87 octane, you are buying gasoline that has the same knocking characteristics as a mixture of 87% isooctane (octane number 100) and 13% heptane (octane number 0). Higher grade gasolines are also available: 89 octane (regular plus) and 92 octane (premium); these contain a greater percentage of compounds with higher octane ratings (Figure 4.17).

Although *n*-octane has a poor rating, it is possible to rearrange or "re-form" *n*-octane to isooctane, thus greatly improving its performance. This rearrangement is accomplished by passing *n*-octane over a catalyst consisting of rare and expensive elements such as platinum (Pt), palladium (Pd), rhodium (Rh), or iridium (Ir). Reforming isomers to improve the octane rating became important starting in the late 1970s because of the nationwide efforts to ban the use of tetraethyllead (TEL) as an antiknock additive. Fuels such as methanol, ethanol, and MTBE contain oxygen and have octane ratings even higher than isooctane. These additives are the topic of the next section.

Figure 4.17

Gasoline is available in a variety of octane ratings.

TEL was discussed in Section 1.11.

Consider This 4.16	Getting the Lead Out

The United States completed the ban on leaded gasoline in 1996 as a result of the increased health risks associated with lead exposure. But sources of exposure other than leaded gasoline still exist. Be a detective on the Web to identify:

   a. an occupational source of lead exposure.
   b. a hobby that is a source of lead exposure.
   c. a source of lead exposure that particularly affects children.

The *Online Learning Center* provides helpful links to aid your search.

4.9   Oxygenated Gasoline

The ubiquitous role of the automobile in U.S. culture and the consequential need for gasoline has given rise to some additional issues. Elimination of TEL as an octane enhancer necessitated finding substitutes that were inexpensive, easy to produce, and environmen-

tally friendly. Several are used, including ethanol and MTBE (*methyl tertiary-butyl ether*), each with an octane rating greater than 100 (see Table 4.6).

ethanol   MTBE

Fuels with these additives are referred to as **oxygenated gasolines,** blends of petroleum-derived hydrocarbons with added oxygen-containing compounds such as MTBE, ethanol, or methanol (CH_3OH). Because they already contain oxygen, oxygenated gasolines burn more cleanly by producing less carbon monoxide than their nonoxygenated counterparts, thereby reducing CO emissions. The Winter Oxyfuel Program, originally implemented in 1992 as part of the Clean Air Act Amendments, targeted cities with excessive wintertime CO emissions to use oxygenated gasolines that contain 2.7% oxygen by weight. Ethanol is the primary oxygenate used in this program. About 40 cities were mandated to participate, but by 2005 over two thirds were no longer implementing the Winter Oxyfuel Program.

Your Turn 4.17   Percent Oxygen in Fuels

The chemical formula of MTBE is $C_5H_{12}O$; that of ethanol is C_2H_6O. Calculate the percent (by mass) of oxygen in each of these two oxygenated fuels.
Hint: Section 3.7 contains examples of such calculations.

Since 1995, about 90 cities and metropolitan areas with the worst ground-level ozone levels have adopted the Year-Round Reformulated Gasoline Program mandated by the Clean Air Act Amendments of 1990. This program requires the use of **reformulated gasolines (RFGs),** which are oxygenated gasolines that also contain a lower percentage of certain more volatile hydrocarbons such as benzene found in nonoxygenated conventional gasoline. RFGs cannot have greater than 1% benzene (C_6H_6) and must be at least 2% oxygenates. Because of their composition, reformulated gasolines evaporate less easily than conventional gasolines, and produce less carbon monoxide emissions. The more volatile hydrocarbons in conventional gasoline have been implicated in tropospheric ozone formation, especially in high-traffic metropolitan areas. Currently, about 35% of U.S. gasoline is reformulated (i.e., RFG) of which nearly 45% contains MTBE.

The use of RFGs and oxygenated gasolines exemplifies a risk–benefit situation. The potential benefits are considerable. Replacing conventional gasolines with RFGs has resulted in substantial benefits. Starting with the second phase of the RFG program in January 2000, the EPA estimates an annual reduction of at least 100 thousand tons of smog-forming pollutants and over 20 thousand tons of toxics through the use of RFGs. Yet, these environmentally friendly fuels may not be risk-free; they have not been used long enough for possible long-term adverse effects, if any, to arise. The health concerns regarding MTBE center on its considerable solubility in water. Unfortunately, MTBE has leaked from underground storage tanks at gas stations, dissolved into ground water and has found its way into water supplies all over the country. The EPA drinking water advisory states that there is little likelihood that MTBE will cause adverse health effects at concentration of about 40 ppb or below; above these levels most people can detect its presence by taste or odor. In January 2004, the National Institute of Environmental Health Sciences reported the human health effects of short-term exposure to large or small amounts of MTBE are unknown. Animal studies at high doses (much higher than human exposure) have shown adverse effects on the nervous system ranging from hyperactivity and loss of coordination to convulsions and unconsciousness and also have shown some instances of cancer.

The molecular structure of benzene is discussed in Section 10.3.

Volatile organic compounds and other air pollutants were discussed in Section 1.11.

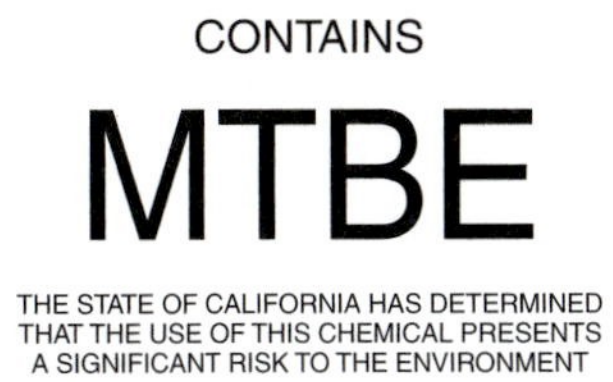

CONTAINS

MTBE

THE STATE OF CALIFORNIA HAS DETERMINED
THAT THE USE OF THIS CHEMICAL PRESENTS
A SIGNIFICANT RISK TO THE ENVIRONMENT

Figure 4.18
Sign posted at a California gasoline
station.

Considering the financial stakes involved and the possible risk to human health, it will come as no surprise that the use of gasoline oxygenates comes with significant political and legal consequences. As of 2005, 140 lawsuits were pending against the oil industry for environmental damage and adverse health effects caused by spills and seepage of MTBE. The Energy Policy Act, signed by President George W. Bush in 2005, bans MTBE in fuels beginning in 2014. The bill also establishes a "transition assistance" program giving MTBE manufacturers $1.75 billion dollars of federal aid to move into other businesses. As of 2005, however, 25 states have stepped in and passed their own MBTE bans or strict limitations on its use (Figure 4.18). Because of the ongoing litigation and public pressure, many oil companies discontinued the use of MTBE in 2006, eight years before the federal ban was to take effect.

Consider This 4.18   RFGs in the USA

Contamination of municipal water supplies is forcing RFG producers to switch from MTBE to ethanol.

a. Which regions of the country produce the most MTBE and ethanol?
b. Which regions require the most usage of RFGs?
c. Comment on the possible implications caused by a transition from MTBE to ethanol in RFGs nationwide.

4.10   New Fuels, New Sources

World supplies of coal are predicted to last for at least the next 150 years, much longer than current estimates of remaining available oil reserves. Unfortunately, the fact that coal is a solid is inconvenient for many applications, especially transportation; liquid or gaseous fuels are required for internal combustion engines. Therefore, research and development projects are underway aimed at converting solid coal into fuels with characteristics similar to petroleum products.

Before large supplies of natural gas were discovered and exploited, cities were lighted with water gas. This is a mixture of carbon monoxide and hydrogen, formed by blowing steam over hot coke (the impure carbon that remains after volatile components have been distilled from coal):

$$\underset{\text{coke}}{C(s)} + H_2O(g) \longrightarrow \underset{\text{water gas}}{CO(g) + H_2(g)} \qquad [4.11]$$

This same reaction is the starting point for the Fischer–Tropsch process for producing synthetic gasoline. German chemists Emil Fischer and Hans Tropsch developed this technology during the 1920s. It is economically feasible only where coal is plentiful and cheap, and oil is scarce and expensive. This is the case in South Africa today, where 40% of gasoline is obtained from coal. Such economic factors may become a reality in the United States in the near future.

The Fischer–Tropsch process can be described by this general reaction.

$$n\,CO(g) + (2n + 1)\,H_2(g) \longrightarrow C_nH_{2n+2}(g,l) + n\,H_2O(g) \qquad [4.12]$$

The hydrocarbon products can range from small gas molecules like methane, CH_4 ($n = 1$) to the medium-sized molecules ($n = 5$–8) typically found in gasoline. Reaction [4.12] proceeds when the carbon monoxide and hydrogen are passed over a catalyst containing iron or cobalt. What role does the metal catalyst have in this reaction? It appears neither on the reactant side, nor on the product side.

To understand this process, consider a typical exothermic reaction as shown in Figure 4.19. Notice that the potential energy of the reactants (left side) is higher than the potential energy of the products (right side) because it is an exothermic reaction. Now examine the pathways that connect the reactants and products. The green line indicates

Catalytic converters in
automobiles were introduced in
Section 1.11, and will be further
described in Section 6.14.

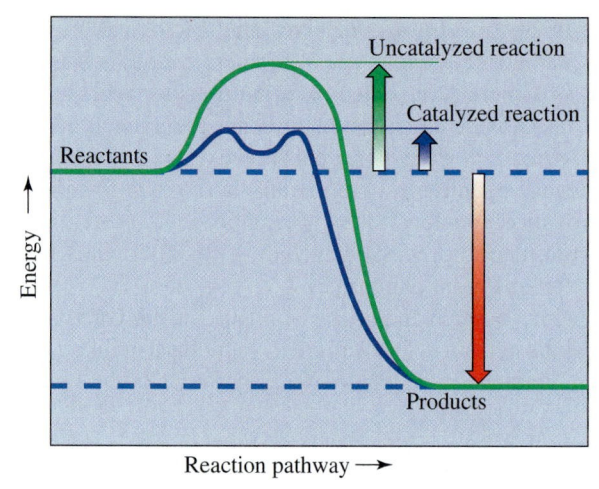

Figure 4.19

Energy–reaction pathway diagram for an uncatalyzed reaction (*green line*) and a catalyzed one (*blue line*). The green and blue arrows represent the activation energies for the uncatalyzed and catalyzed reactions, respectively. The red arrow represents the overall energy change for either pathway.

the energy changes during a reaction in the absence of a catalyst. Overall, this reaction gives off energy, but the energy initially goes *up* because some bonds break (or start to break) first. The energy necessary to initiate a chemical reaction is called its **activation energy** and is indicated by the green arrow. Although energy must be expended to get the reaction started, energy is given off as the process proceeds to a lower potential energy state. Generally, reactions that occur rapidly have low activation energies; slower reactions have higher activation energies. However, there is no direct relationship between the height of the activation barrier and the net energy change in the reaction. In other words, a highly exothermic reaction can have a large or a small activation energy.

Increasing the temperature often results in increased reaction rates; when molecules have extra energy, more collisions can overcome the required activation energy. Sometimes, however, increasing the temperature isn't a practical solution. The blue line shows how a catalyst can provide an alternative reaction pathway and thus a lower activation energy (represented by the blue arrow) without raising the temperature.

To understand how a catalyst can function, consider the reaction represented by equation 4.12 on the molecular level. The carbon monoxide molecule contains a very strong C-to-O triple bond that must be broken for the products to form. Breaking this bond corresponds to an activation energy so large the reaction simply does not proceed. Here's where the metal catalyst enters the reaction. CO molecules can form bonds with the metal surface, and when this happens, the C-to-O bonds weaken. As an analogy, imagine someone hanging by both hands to the edge of a cliff. In order to be rescued, she or he has to let go with one hand to grab onto a person at the top and be pulled to safety. The metal catalyst plays the role of the hero in the chemical reaction. The hydrogen molecules also attach to the metal surface, completely breaking the H-to-H single bonds. The rest of the reaction proceeds quickly as the two highly reactive hydrogen atoms are able to tear the oxygen atom away from the carbon atom, forming water. Other hydrogen atoms then form bonds with the leftover carbon atom. If there are additional carbon atoms on the surface, C-to-C single bonds can form, producing the higher molecular weight hydrocarbons.

The recent spike in gas prices may spark increased use of the Fischer–Tropsch process in the United States. In 2005, the governors of two coal-rich states, Pennsylvania and Montana, independently announced ventures to build coal-to-liquid plants that will convert so-called waste coal (leftovers from the mining process) into low-sulfur diesel fuel. Recent work by the National Renewable Energy Laboratory, however, indicates that greenhouse gas emissions over the entire fuel cycle for producing coal-based fuels are nearly twice as high as their petroleum-based equivalent.

Coal, like petroleum, is a nonrenewable resource, and therefore coal-based fuels can be only a temporary solution; sustainable energy sources must be found. A possible solution lies in the conversion of **biomass,** the general term for plant matter

such as trees, grasses, agricultural crops, or other biological material, into a variety of usable fuels. The most common type of biomass, wood, is insufficient to meet the energy demands of our modern society. Cutting trees for fuel also destroys effective absorbers of carbon dioxide and adds that greenhouse gas and other pollutants to the atmosphere. In Africa and elsewhere, entire ecosystems are being lost to massive deforestation because people must continually burn wood for cooking and heat. Instead of relying on direct combustion of natural products, current research focuses on creating fuels from natural processes, including ethanol made via fermentation and biodiesel made from different plant oils.

The fermentation of starch and sugars in grains such as corn has been known since ancient times. Making ethanol in this way creates a fuel that, unlike gasoline, is renewable because crops can continue to be planted. Enzymes released by yeast cells catalyze the reaction typified by this equation.

Enzymes (biological catalysts) will be discussed in Sections 11.4 and 12.4.

$$\underset{\text{glucose}}{C_6H_{12}O_6} \longrightarrow \underset{\text{ethanol}}{2\ C_2H_5OH} + 2\ CO_2 \qquad [4.13]$$

Ethanol also can be prepared commercially in large quantities by the reaction of water (steam) with ethylene, C_2H_4 ($H_2C{=}CH_2$).

$$CH_2CH_2(g) + H_2O(g) \longrightarrow CH_3CH_2OH(l) \qquad [4.14]$$

When the second method is used to produce ethanol for oxygenated fuels, any residual water must scrupulously be removed so that it will not create problems in an automobile engine.

The burning of ethanol releases 1367 kJ/mol of C_2H_5OH, or 29.7 kJ/g.

$$C_2H_5OH(l) + 3\ O_2(g) \longrightarrow 2\ CO_2(g) + 3\ H_2O(l) + 1367\ kJ \qquad [4.15]$$

This value is less than the 47.8 kJ/g produced by burning C_8H_{18}, because the ethanol already contains some oxygen. Nevertheless, ethanol is being mixed with gasoline to form "gasohol." At the usual concentration of 10% ethanol, gasohol can be used without modifying standard automobile engines. Both in the United States and elsewhere, there is significant interest and investment in producing cars and trucks that use a much higher percentage of ethanol. Of the 13 million vehicles in Brazil, more than 4 million use pure ethanol (made from fermented sugar cane), and the remainder of the cars operate on a mixture of ethanol and gasoline. More than 3 million flexible fuel vehicles (FFVs) already sold in the United States can use E-85 (85% ethanol and 15% gasoline), gasoline or any mixture of the two. It is likely that the buyers of many of those 3 million FFVs, which include sedans, mini-vans, SUVs, and pickup trucks, remain unaware that they can fuel with E-85. A record 3.9 billion gallons of ethanol was produced in the United States in 2005. The industry clearly anticipates a shift to a greater use of ethanol as a fuel; additional plant constructions will bring the annual production capacity to over 6 million gallons by 2008.

However, this renewable fuel source is not without its critics. The main issue is that the Sun is not the only energy source involved in ethanol manufacturing. Energy is required to plant, cultivate, and harvest the corn; to produce and apply the fertilizers; to distill the alcohol from the fermented mash; and to manufacture the tractors and other necessary farm equipment. Because of the numerous assumptions involved, accurate energy balances are difficult to obtain. Some studies estimate that for every joule put into ethanol production 1.2 J are recovered, but others conclude that the combined energy inputs outweigh the energy content of the ethanol produced.

In any event, all those energy inputs make a gallon of ethanol more expensive to produce than a gallon of gasoline. Each year, the federal government provides more than $2B in subsidies to ethanol producers, the majority of which go to large agricultural corporations not small farmers. Furthermore, a gram of ethanol does not produce as much energy as a gram of gasoline, so that while the octane rating of gasohol is higher than regular unleaded, gas mileage is slightly lower with the alcohol blends.

In addition, those opposed to ethanol as a fuel question whether valuable farmland, normally used to grow crops such as corn that feed people and animals, should be used to produce grain for ethanol. Currently, the United States produces a significant surplus of

Figure 4.20

An advertisement for gasohol that contains ethanol.

corn and other grains that could be converted to ethanol, but using that excess for ethanol decreases the amount of grains for export. A final point of controversy lies in the very nature of growing corn. Producing corn requires heavy use of fertilizers, herbicides, and pesticides, all of which result in deterioration of both soil and water quality.

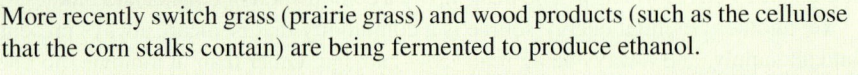

Consider This 4.19   The Switch to Switch Grass

More recently switch grass (prairie grass) and wood products (such as the cellulose that the corn stalks contain) are being fermented to produce ethanol.

a. These newer sources of ethanol are less controversial than corn. Explain why.
b. Search the Web for information about switch grass and ethanol. How does the energy required to produce one liter of ethanol from corn compare with that from switch grass? Why is it more difficult to produce ethanol from switch grass and other cellulose-based feedstocks than from corn?

The battle lines regarding the use of ethanol as a fuel are clearly drawn, largely based on self-interest. Supporting the greater use of ethanol are the EPA, Archer Daniels Midland (ADM, an agribusiness), and over 20 farm groups, including the National Corn Growers Association (Figure 4.20). In opposition are the American Petroleum Institute, the petroleum refiners and gasoline companies, and the Sierra Club. Depending on whether they are from agricultural states or ones tied closely to oil, U.S. senators and representatives carve out positions on the issue. There is a good deal at stake—among other things, 100–200 million bushels of corn per year.

Consider This 4.20   Ethanol and Politics

Senators Barack Obama (D-Illinois) and Jim Talent (R-Missouri) cosponsored amendments to the Energy Bill passed in 2005 to increase the availability of fuels blended with 85% ethanol. The legislation provided a tax credit incentive of 30% through 2010 for filling stations switching one or more traditional petroleum pumps to E-85 fueling systems. Why might these two senators have spearheaded this particular legislation?

Sceptical Chymist 4.21   Keeping Track of Ethanol

Senator Tom Harkin (D–Iowa), chair of the Senate Agriculture Committee, predicted a need for 400 million additional gallons of ethanol per year to replace MTBE for auto use in California. It is estimated that the resulting gasohol mixture would be sufficient to power California vehicles for about 8 billion miles of driving per year, or over 4600 miles per vehicle. California has approximately 17 million passenger vehicles. Assume that the needed ethanol is used exclusively for gasohol and that 10 gal of gasohol contains 1 gal of ethanol and 9 gal of gasoline. State any additional assumptions you make and then show calculations to support or refute this estimate.

The production of biodiesel as an alternative fuel has grown dramatically during the last few years (Figure 4.21). Biodiesel is made from natural, renewable resources such as new and used vegetable oils and animal fats. It can be burned as a pure fuel or blended with petroleum products and used in diesel engines that have no major modifications. It significantly reduces most regulated emissions and is nontoxic and biodegradable.

Notably, biodiesel releases much more energy when combusted than it costs to produce. In 1998, the U.S. Department of Energy and the U.S. Department of Agriculture performed the prevailing life cycle study of the energy balance (energy in versus energy out) of biodiesel. The study concluded that for every one unit of fossil energy used in the entire biodiesel production cycle, 3.2 units of energy are gained when the fuel is burned; a positive energy balance of 320%. The Energy Policy Act of 2005 (EPACT) contains a provision that requires refiners to produce 4 billion gallons of renewable fuel annually in 2006 and double that amount by 2012, which will be made up by a combination of ethanol and biodiesel.

Yet another potential energy source is a commodity that is cheap, always present in abundant supply, and always being renewed—garbage. Other than in a movie, no one is likely to design a car that will run on orange peels and coffee grounds, but approximately 140 power plants in the United States do just that. One of these, pictured in Figure 4.22, is the Hennepin Energy Resource Company (HERC) in Minneapolis, Minnesota. Hennepin County produces about 1 million tons of solid waste each year. One truckload of garbage (about 27,000 lb) generates the same quantity of energy as 21 barrels of oil. HERC converts 365,000 tons of garbage per year into enough to provide power to the equivalent of 25,000 homes. In addition, over 11,000 tons of iron-containing metals are recovered from the garbage and recycled. Elk River Resource Recovery Facility, the second in Hennepin County, converts another 235,000 tons of garbage to electricity. The emissions at both sites are significantly below state and federal standards.

Figure 4.21
Biodiesel at the pump (*left*) and being formulated by a regional vendor from recycled restaurant vegetable oil (*right*).

Figure 4.22

Hennepin County Resource Recovery Facility, a garbage-burning power plant.

This resource recovery approach, as it is sometimes called, simultaneously addresses two major problems: the growing need for energy and the growing mountain of waste. The great majority of the trash is converted to carbon dioxide and water, and no supplementary fuel is needed. The unburned residue is disposed of in landfills, but it represents only about 10% of the volume of the original refuse. Although some citizens have expressed concern about gaseous emissions from garbage incinerators, the incinerator's stack effluent is carefully monitored and must be maintained within established limits. Both Japan and Germany are making considerably greater use of waste-to-energy technology than the United States.

Methane generators provide another good example of using waste as an energy source. Rural China and India have over one million reactors in which animal and vegetable wastes are fermented to form biogas. This gas, which is about 60% CH_4, can be used for cooking, heating, lighting, refrigeration, and generating electricity. The technology lends itself very well to small-scale applications. The daily manure from one or two cows can generate enough methane to meet most of the cooking and lighting needs of a farm family. Two thirds of China's rural families use biogas as their primary fuel.

Consider This 4.22   **Building a Waste-Burning Plant**

Imagine you were the administrator of a city of a million residents charged with drafting a proposal to your city council outlining the pros and cons of a waste-burning plant. Use the Hennepin County facilities in Minnesota as a model and their Web site to collect information and examples. The pros and cons might center on issues such as resource recovery, pollution, or other potential concerns of local residents.

4.11   The Case for Conservation

A fundamental feature of the universe is that energy and matter are conserved. As we have seen, however, the process of combustion converts both matter and energy into less useful forms. For example, when we burn hydrocarbon molecules, we eventually

dissipate their energy as heat. The products of combustion—carbon dioxide and water—are not usable as fuels, and increasing concentrations of atmospheric CO_2 are linked to potentially severe global climate changes. Furthermore, the supplies of fossil fuels themselves are limited and nonrenewable. Given these underlying constraints, it would appear that changes in our energy policy are inevitable. When and how these changes occur is up to us.

In his 2006 State of the Union address, President George W. Bush observed that the United States is "addicted to oil." In fact, it is a global dependence. The U.S. Energy Information Administration projects a 60% increase in global demand for oil to a whopping 4.0×10^{10} barrels per year by 2020. Not surprisingly, the greatest increases will occur in the developing countries where growing populations, migration to the cities, and industrialization will drive up the energy demand to unprecedented highs. By 2010, emerging economies will account for over half of the energy consumed.

The demands of conventional power plants for coal, oil, and natural gas are tremendous, but fossil fuels are also vital feedstocks for chemical synthesis. Late in the 19th century, Dmitri Mendeleev, the great Russian chemist who proposed the periodic table of the elements, visited the oil fields of Pennsylvania and Azerbaijan. He is said to have remarked that burning petroleum as a fuel "would be akin to firing up a kitchen stove with bank notes." Mendeleev recognized that oil could be a valuable starting material for a wide variety of chemicals and the products made from them. He would, no doubt, be amazed at the fibers, plastics, rubber, dyes, medicines, and pharmaceuticals currently produced from petroleum. Yet, we continue to ignore Mendeleev's warning and burn nearly 85% of the oil pumped from the ground.

Consider This 4.23   Now and in the Future

Dr. Ronald Breslow, a Columbia University chemistry professor and a former president of the American Chemical Society, speaking at a February 2001 symposium on "Sustainability Through Science," remarked: "Succeeding generations are going to curse us for burning their future raw materials, and they are right. Not only are we using up valuable resources—petroleum and coal—but we are adding pollution and carbon dioxide which may be contributing to global warming." Comment on Dr. Breslow's remarks.

Although we do not know with great certainty the amount of recoverable oil remaining in the Earth, we do know it is limited. In the mid-1950s, yearly global oil consumption was 4 billion barrels, and over 30 billion barrels of new deposits were being found annually. Today, those numbers are nearly reversed. Experts therefore predict that sometime in the near future, oil production will peak and then decline as we deplete the most easily recoverable deposits. In fact, this has already happened in the United States where oil production has slowly but steadily declined since 1970. Assuming consumption increases 2% per year, the Energy Information Agency (EIA) developed three possible "peak oil" scenarios. Figure 4.23 displays predicted future levels of oil production assuming a low, an average, and a high value for recoverable reserves worldwide. Even the most optimistic estimates predict maximum oil production before 2050. Other organizations adopting more pessimistic values of available reserves claim that "peak oil" will occur within the next several years. Notice that we won't abruptly "run out" of oil, but dramatically higher prices and increasing scarcity will characterize the era preceding the peak.

The way forward is clear. As a global community, we must find ways to use less oil. Countries around the world are facing this inevitable challenge in many different ways. The construction of the immense and controversial Three Gorges Dam and hydroelectric power station in China is one attempt by that country to meet its exploding electrical

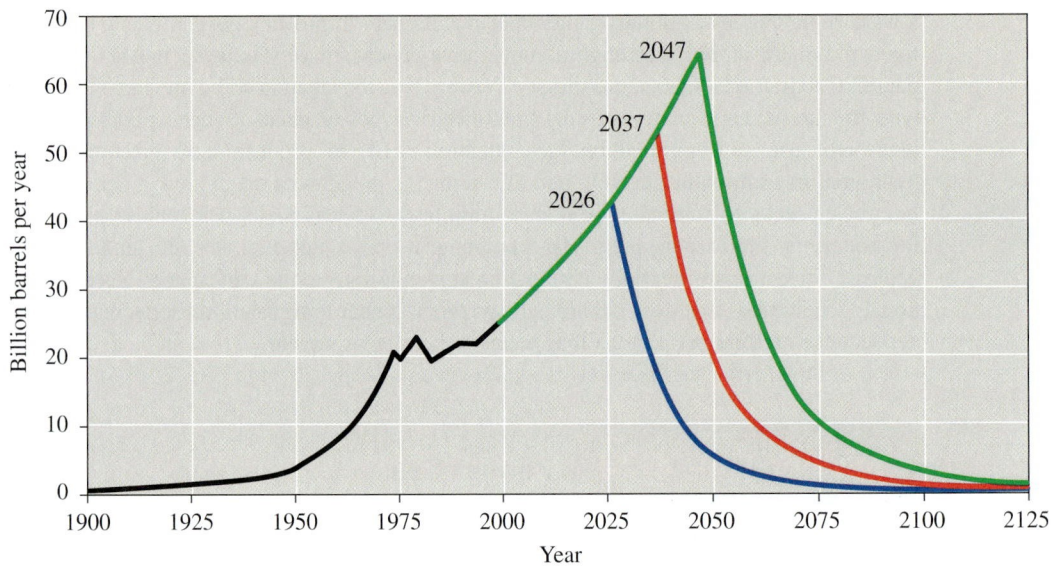

Figure 4.23

Worldwide oil production since 1900 (black) and three possible peak oil scenarios. The blue, red, and green lines correspond to low, average, and high estimates of recoverable oil, respectively.

Source: EIA/DOE.

energy demands. Denmark leads the way in wind power (Figure 4.24). Many countries are exploring other clean, renewable energy sources including geothermal, tidal, and solar energy. Of course making these technologies competitive will require significant investments. A great concern in this country is the decline of resources allocated to energy research. From 1980 to 2005, research spending on energy-related topics in the United States dropped fivefold, from 10% to just 2% of the total.

> Sections 8.9 and 8.10 describe more details of solar energy technology.

In addition to exploring new energy sources, it is important to use efficiently the ones we have. Considerable savings have already been realized both on the production side and on the consumption side by improving the efficiency of energy transformation. The production of electricity by power plants is the major use of energy in the United States, making up 38% of the total. The second law of thermodynamics limits the conversion of heat to work, but power plants currently operate well below the thermodynamic maximum efficiency.

In 1999, the U.S. Department of Energy announced an aggressive new research and development program, Vision 21, whose lofty goals include dramatically increased power plant efficiencies with essentially zero emissions. Vision 21 activities are oriented

Figure 4.24

A satellite view of the Three Gorges Dam (*left*) that began to supply China with significant hydroelectric power in 2003, and a wind farm off the coast of Denmark (*right*).

toward achieving revolutionary rather than evolutionary improvements that would be ready for deployment in 2015, through a cooperative effort between industry, universities and national laboratories. One major thrust of Vision 21 is the creation of FutureGen, the prototype for the next generation of coal-fired electrical power plants. When operational, the plant will use coal gasification (see equation 4.10) to produce both electricity and hydrogen. In addition to the NO_x- and SO_2-reducing procedures mentioned in Section 4.6, FutureGen seeks to eliminate (or at least minimize) CO_2 emissions. The current plan calls for liquefying the carbon dioxide and pumping it under high pressure into underground geologic formations, essentially eliminating any emissions to the atmosphere. Such transitional technologies that use coal are perhaps the best near-term solution for decreasing our dependence on foreign oil, as well as protecting our environment.

The use of hydrogen as a fuel will be discussed in Section 8.8.

Consider This 4.24   FutureGen

As of the writing of this book, only potential sites for the first FutureGen plant had been established. Check the current status of FutureGen, courtesy of a Web site posted by the Department of Energy.

Improving end-use efficiency is perhaps the best way to save energy and energy resources. Estimates of the technically feasible savings in electricity range from 10 to 75%. In a *Scientific American* article published in September 2005, Amory B. Lovins asserted, "With the help of efficiency improvements and competitive renewable energy sources, the U.S. can phase out oil use by 2050." Driving this push for efficiency is simple economics, as individuals and corporations realize that Mendeleev was right; it is much cheaper to save fossil fuels than it is to burn them. Recent advances in building design, information technology, and data processing make sizeable energy savings possible. "Smart" office buildings or homes feature a complicated system of sensors, computers, and controls that maintain temperature, airflow, and illumination at optimum levels for comfort and energy conservation. Similarly, the computerized optimization of energy flow and the automation of manufacturing processes have brought about major transformations in industry. Over the past 20 years, industrial production in the United States has increased substantially, but the associated energy consumption has actually gone down. In one remarkable example, carpet manufacturers have developed enzyme-based dyeing procedures allowing them to color fibers at room temperature instead of the normal 100–140 °C, saving about 90% in electricity costs.

Conservation also results from recycling materials, especially aluminum. Because extracting aluminum metal from its ore is extremely energy intensive, recycling the metal yields an energy saving of about 70%. To put things in perspective, you could watch television for 3 hr on the energy saved by recycling just one aluminum can.

Nowhere could conservation and increased efficiency have greater effect than in transportation. Over 70% of U.S. oil production goes to power motor vehicles, and accounts for about one third of the our carbon dioxide emissions. In response to the energy crisis in the early 1970s, automakers, under pressure from the federal government and the consuming public, significantly improved the fuel economies of all types of vehicles. From the mid-1970s to the early 1990s, gasoline consumption in the United States dropped by one half as average fuel economies dramatically improved. Yet this environmentally friendly trend is under assault as the 21st century begins. Since 1988, there has been an overall *decline* in vehicle fuel economy (Figure 4.25). It currently stands at 24.6 miles per gallon (mpg), less than the 25.9 mpg highest value obtained in 1988.

A major factor for the decline in fuel economy in the United States has been the increase in the proportion of vehicles classified as light-duty trucks—sport utility vehicles (SUVs), vans, and pickup trucks—that now command almost half of the traditional car market. Although car manufacturers must meet the federally mandated 27.5 mpg average fuel consumption for cars, the requirement is only 20.7 mpg for light-duty trucks. SUVs are classified

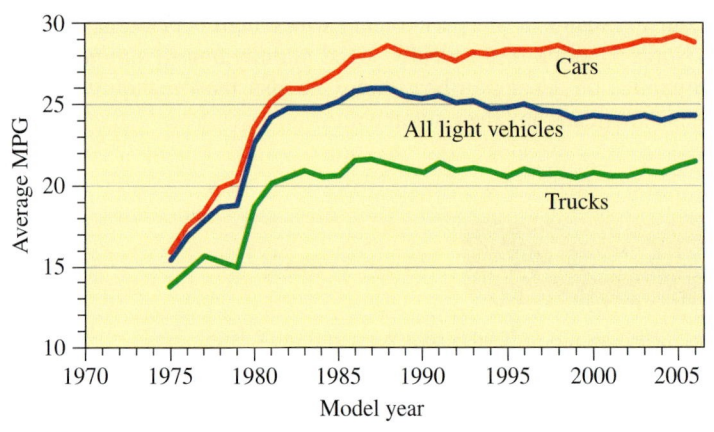

Figure 4.25

Fuel economy trends for U.S. cars and light trucks since 1975.

Source: U.S. EPA. Taken from www.epa.gov/otag/cert/mpg/retrends.

as light trucks, not as cars, and therefore need not meet the higher mpg requirements for cars. Though the United States is the leader in many aspects of energy, fuel economy is not one of them. The average European car gets about 40 mpg, and those in Japan get about 45 mpg.

The political and legal battles over gas mileage standards are just beginning. Neither the Senate nor the House version of the bill that would become the Energy Policy Act passed in July 2005 contained any significant increases in gasoline efficiency standards for American automobiles. However, individual states are stepping in. In 2003, Californians passed a law placing significant restrictions on greenhouse gas emissions from automobiles. Its practical effect is to require significant improvements in gas mileage over the next decade. It is worth noting that if the new stricter regulations in California are fully implemented (in 2016), they will still be below the average mileage of cars and trucks in China. Even those relatively moderate steps sparked a lawsuit brought by the automobile industry against the State of California. The plaintiffs argued that only the federal government, not individual states, has the authority to regulate fuel economy.

In 2006, the National Highway Traffic Safety Administration (NHTSA) for the first time announced fuel economy standards applicable to all trucks. The standards were expected to go into effect in 2008 (22.5 mpg) and increase to about 24 mpg in 2011. Though this marks the first year that heavy SUVs like the Hummer H2 will be included in mileage standards, several states believe the new regulations are not tough enough. Massachusetts, joined by the attorneys general of 11 other states and several environmental groups, filed a lawsuit against the federal government. The states claim that NHTSA failed to address the effects that car emissions have on global warming and the environment. The controversy centers on the Bush administration's stance that CO_2 is not a pollutant, and therefore it cannot be regulated by the Environmental Protection Agency under the Clean Air Act. That position was dealt a blow in April 2007 when the Supreme Court ruled 5-4 that the EPA must reconsider its position and that in fact the EPA *does* have the authority to regulate CO_2 emissions. The ramifications of the ruling will be widespread; fuel economy standards and power plant emissions are likely first targets.

Consider This 4.25   **Gasoline Efficiency**

Cars in Europe get about 50% better gas mileage than those in the United States, yet no governmental regulations set particular standards in European countries like those in the United States. List other factors that may explain this difference.

The fact remains that the automobile will remain an energy-intensive means of transportation. A mass transit system is far more economical, provided it is heavily used. In Japan, 47% of travel is by public transportation, compared with only 6% in the United

States. Of course, Japan is a compact country with a high population density. The great expanse of North America is not ideally suited to mass transit, although some regions, such as the population-dense Northeast are. One also must reckon with the long love affair between Americans and their automobiles.

Consider This 4.26   Mass Transit

Advocates propose mass transit as a way to reduce fossil-fuel consumption in the United States. Consider, for example, the Midwest. One suggestion is to use a light rail to interconnect a set of cities like Milwaukee, Kenosha, and Madison. Another is to use a light rail to connect a metropolitan center like Chicago with a new airport built far beyond the suburbs.

a. List reasons for and against the two suggestions.
b. What arguments would you expect from citizens concerning more extensive use of mass transit?
c. What other creative measures could be taken to reduce the use of personal vehicles?

Conclusion

Both our bodies and our society require a continual supply of energy for survival. Most of the food we eat is a renewable resource, but our industrialized way of life, even our very existence, currently relies on the consumption of nonrenewable fossil fuels. During the 1970s, a series of energy crises occurred because of a dramatic rise in the cost of imported crude oil, principally from the Middle East. The next energy crisis, when it comes, will be fundamentally different from those of the past. There is little doubt that overcoming the challenges it will bring will require more drastic measures.

At the most basic level, we have no option. Alternative sources like ethanol, biodiesel, wind, and hydroelectric must become more prevalent. We will discuss nuclear energy in Chapter 7 and solar energy in Chapter 8 as other possible ways of satisfying our ever-increasing appetite for energy. As individuals and as a society, we must decide what sacrifices we are willing to make in speed, comfort, and convenience for the sake of our dwindling fuel supplies and the good of the planet. One thing is clear: the sooner we honestly examine our options, our priorities, and our will, the better. Energy, chemistry, and society are closely intertwined, and this chapter is an attempt to untangle them.

Chapter Summary

Having studied this chapter, you should be able to:

- Distinguish between energy and heat, and be able to convert among energy units: joules, calories, Calories (4.1)

- Relate the energy theoretically available from a process with the efficiency of that process (4.2)

- Understand the difference between kinetic and potential energy, both on the macroscopic and molecular level (4.2)

- Use the concept of entropy to explain the second law of thermodynamics (4.2)

- Apply the terms *exothermic* and *endothermic* to chemical systems (4.3–4.5)

- Interpret chemical equations and basic thermodynamic relationships to calculate heats of reaction, particularly heats of combustion (4.4–4.5)

- Use bond energies to describe the energy content of materials, and to calculate energy changes in reactions (4.4)

- Describe the factors related to the United States' dependency on fossil fuels for energy (4.5)

- Evaluate the risks and benefits associated with petroleum, coal, and natural gas as fossil-fuel energy sources (4.6–4.7)
- Relate energy use to atmospheric pollution and global warming (4.7–4.8)
- Understand the physical and chemical principles associated with petroleum refining (4.7–4.8)
- Describe how octane ratings are assigned and explain how the refining process, and use of gasoline additives such as lead, ethanol, and MTBE affect octane ratings (4.8)
- Describe why reformulated and oxygenated gasolines are used (4.9–4.10)

- Understand activation energy, and how it relates to the rates of reaction (4.10)
- Compare and contrast ethanol and biodiesel as fuels (4.10)
- Take an informed stand on what energy conservation measures are likely to produce the greatest energy savings (4.11)
- With confidence, examine news articles on energy crises and energy conservation measures and interpret the accuracy of such reports (4.11)

Questions

Emphasizing Essentials

1. **a.** List three fossil fuels.

    b. What is the origin of fossil fuels?

    c. Are fossil fuels a renewable resource?

2. The Calorie, used to express food heat values, is the same as a kilocalorie of heat energy. If you eat a chocolate bar from the United States with 600 Calories of food energy, how does the energy compare with eating a Swiss chocolate bar that has 3000 kJ of food energy? (*Note:* 1 kcal = 4.184 kJ)

3. A single serving bag of Granny Goose Hawaiian Style Potato Chips has 70 Cal. Assuming that all of the energy from eating these chips goes toward keeping your heart beating, how long can these chips sustain a heartbeat of 80 beats per minute? *Note:* 1 kcal = 4.184 kJ, and each human heart beat requires approximately 1 J of energy.

4. Three power plants have been proposed that operate at these efficiencies.

Plant	Power Plant Efficiency (%)
I	81
II	66
III	41

    a. Calculate the overall efficiency of each plant (not the maximum theoretical efficiency) using the other efficiencies given in Table 4.1.

    b. Identify the factors that affect the efficiency.

    c. Discuss the practical limits that govern such efficiencies. Which plant would be most likely to be built? If plant III only costs half of plant I or II to operate, which would be most likely to be built?

5. Equation 4.1 shows the complete combustion of methane.

    a. By analogy, write a similar chemical equation using ethane, C_2H_6.

    b. Represent this equation with Lewis structures.

6. The heat of combustion for ethane, C_2H_6, is 52.0 kJ/g. How much heat would be released if 1 mol of ethane undergoes complete combustion?

7. **a.** Write the chemical equation for the complete combustion of heptane, C_7H_{16}.

    b. The heat of combustion for heptane is 4817 kJ/mol. How much heat would be released if 250 kg of heptane undergoes complete combustion?

8. Figure 4.7 shows energy differences for the combustion of hydrogen, an exothermic chemical reaction. The combination of nitrogen gas and oxygen gas to form nitrogen monoxide is an example of an endothermic reaction:

    $$180 \text{ kJ} + N_2(g) + O_2(g) \longrightarrow 2 \text{ NO}(g)$$

    Sketch an energy diagram for this reaction.

9. One way to produce ethanol for use as a gasoline additive is the reaction of water vapor with ethylene:

    $$CH_2CH_2(g) + H_2O(g) \longrightarrow CH_3CH_2OH(l)$$

    a. Rewrite this equation using Lewis structures.

    b. Is this reaction endothermic or exothermic?

    c. In your calculation, was it necessary to break *all* the chemical bonds in the reactants to form the product ethanol? Explain your answer.

10. From personal experience, state whether these processes are endothermic or exothermic. Give a reason for each.

    a. A charcoal briquette burns.

    b. Water evaporates from your skin.

    c. Ice melts.

    d. Wood burns.

11. Use the bond energies in Table 4.2 to explain why:

    a. chlorofluorocarbons, CFCs, are so stable.

    b. it takes less energy to release Cl atoms than F atoms from CFCs.

12. Use the bond energies in Table 4.2 to calculate the energy changes associated with each of these reactions. Lewis structures of the reactants and products may be

useful for determining the number and kinds of bonds. Label each reaction as endothermic or exothermic.

a. $N_2(g) + 3 H_2(g) \longrightarrow 2 NH_3(g)$

b. $2 C_5H_{12}(g) + 11 O_2(g) \longrightarrow 10 CO(g) + 12 H_2O(l)$

c. $H_2(g) + Cl_2(g) \longrightarrow 2 HCl(g)$

13. Use the bond energies in Table 4.2 to calculate the energy changes associated with each of these reactions. Label each reaction as endothermic or exothermic.

a. $2 H_2(g) + CO(g) \longrightarrow CH_3OH(g)$

b. $H_2(g) + O_2(g) \longrightarrow H_2O_2(g)$

c. $2 BrCl(g) \longrightarrow Br_2(g) + Cl_2(g)$

14. Use Figure 4.9 to compare the sources of U.S. energy consumption. Arrange the sources in order of decreasing percentage and comment on the relative rankings.

15. Table 4.3 lists the energy content of some fuels in kilojoules per gram (kJ/g). Calculate the fuel energy in kilojoules per mole (kJ/mol) for methane CH_4, propane C_3H_8, hydrogen H_2, coal mostly C, and ethanol C_2H_6O. Make a generalization regarding the chemical composition of fuels and their respective energy contents. Visit *Figures Alive!* at the *Online Learning Center* for related activities.

16. Mercury is present in minor amounts (50–200 ppb) in coal. Use the amount of coal burned by a power plant in Your Turn 4.12 to determine how much Hg is released by that plant. Calculate the amount based on the lower (50 ppb) and higher (200 ppb) limits.

17. Fossil-fuel consumption of 650,000 kcal per person per day in the United States is equivalent to an annual personal consumption of 65 barrels of oil or 16 tons of coal. Use this information to calculate the amount of energy available in each of these quantities.

a. one barrel of oil

b. 1 gal of oil (42 gal per barrel)

c. one ton of coal

d. 1 lb of coal (2000 lb per ton)

18. Use the information in question 17 to find the ratio of the quantity of energy available in 1 lb of coal to that in 1 lb of oil. *Hint:* One pound of oil has a volume of 0.56 qt.

19. Consider the data for three hydrocarbons shown in the table.

Compound, Formula	Melting Point (°C)	Boiling Point (°C)
Pentane, C_5H_{12}	−130	36
Triacontane, $C_{30}H_{62}$	66	450
Octane, C_8H_{18}	−57	125

Predict the physical state (solid, liquid, or gas) of each hydrocarbon at a temperature of 25°C.

20. Table 4.5 shows the structural formulas of alkanes containing one to eight carbons.

a. Draw the structural formula for decane, $C_{10}H_{22}$.

b. Use Table 4.5 to predict the structural formulas for nonane, the alkane with 9 carbons, and dodecane, the alkane with 12 carbons.

c. The structural formulas in Table 4.5 are two-dimensional. Use the bond angle information in Chapter 3 to predict the C-to-C-to-C and H-to-C-to-H bond angles in decane.

21. Consider this equation representing the process of cracking.

$$C_{16}H_{34} \longrightarrow C_5H_{12} + C_{11}H_{22}$$

a. Which bonds are broken and which bonds are formed in this reaction? Use Lewis structures to help answer this question.

b. Use the information from part **a** and Table 4.2 to calculate the energy change during this cracking reaction.

22. This is the ball-and-stick representation of one isomer of butane (C_4H_{10}).

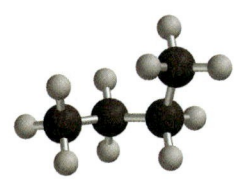

a. Draw the Lewis structure for this isomer. *Hint:* Show how atoms are linked, but not their spatial arrangement.

b. Draw the Lewis structure for each additional isomer, being careful not to repeat isomers.

c. What is the total number of isomers of C_4H_{10}?

23. A premium gasoline available at most stations has an octane rating of 92. What does that tell you about:

a. the knocking characteristics of this gasoline?

b. whether the fuel contains oxygenates?

Concentrating on Concepts

24. How might you explain the difference between temperature and heat to a friend? Use some practical, everyday examples.

25. Write a response to this statement: "Because of the first law of thermodynamics, there can never be an energy crisis."

26. Candle wax is composed of straight-chain hydrocarbons of about 50 carbon atoms.

a. Make a general statement describing how the number of carbon atoms affects the physical state of normal hydrocarbons.

b. Write a chemical formula for the alkane having 50 carbon atoms.

27. A friend tells you that hydrocarbon fuels containing larger molecules liberate more heat than those containing smaller molecules.

a. Use the data on the next page, together with appropriate calculations, to discuss the merits of this statement.

Hydrocarbon	Heat of Combustion
Octane, C_8H_{18}	5450 kJ/mol
Butane, C_4H_{10}	2859 kJ/mol

b. Considering your answer to part **a,** do you expect the heat of combustion per gram of candle wax, $C_{25}H_{52}$, to be more or less than the heat of combustion per gram of octane? Do you expect the molar heat of combustion of candle wax to be more or less than the molar heat of combustion of octane? Justify your predictions.

28. Halons are synthetic chemicals similar to CFCs, but they also include bromine. Although halons are excellent materials for fire fighting, they more effectively deplete ozone than CFCs. Here is the Lewis structure for halon-1211.

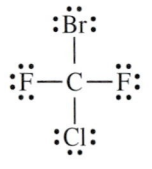

a. Which bond in this compound is broken most easily? How is that related to the ability of this compound to interact with ozone?

b. C_2HClF_4 is a compound being considered as a replacement for halons as a fire extinguisher. Draw the Lewis structure for this compound and identify the bond broken most easily. How is the structure related to the ability of this compound to interact with ozone?

29. The Fischer–Tropsch conversion of hydrogen and carbon monoxide into hydrocarbons and water was given in equation 4.12:

$$n\, CO + (2n + 1)\, H_2 \longrightarrow C_nH_{2n+2} + n\, H_2O$$

a. Determine the heat evolved by this reaction when $n = 1$.

b. Without doing a calculation, do you think that more or less energy will be given off in the formation of larger hydrocarbons ($n > 1$)? Explain your reasoning.

30. During the distillation of petroleum, kerosene and hydrocarbons with 12–18 carbons used for diesel fuel will condense at position C marked on this diagram.

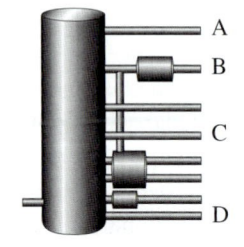

a. Separating hydrocarbons by distillation depends on differences in a specific physical property. Which one?

b. How will the number of carbon atoms in the hydrocarbon molecules separated at A, B, and D compare with those separated at position C? Explain your prediction.

c. How will the uses of the hydrocarbons separated at A, B, and D differ from those separated at position C? Explain your reasoning.

31. Imagine you are at the molecular level, looking at what happens when liquid ethylene, C_2H_4, boils. Consider a collection of four ethylene molecules.

$$C_2H_4 =$$

a. Draw a representation of ethylene in the liquid state and then in the gaseous state. How will the two differ?

b. Estimate the temperature at which the transition from liquid to gas is taking place. What is the basis for your estimation?

32. Explain why cracking is necessary in the refinement of crude oil.

33. Catalysts speed up cracking reactions in oil refining and allow them to be carried out at lower temperatures. What other examples of catalysts were given in the first three chapters of this text?

34. Octane ratings of several substances are listed in Table 4.6.

a. What evidence can you give that the octane rating is or is not a measure of the energy content of a gasoline?

b. Octane ratings are measures of a fuel's ability to minimize or prevent engine knocking. Why is the prevention of knocking important?

c. Why are higher octane rating gasolines more expensive than lower ones?

35. The octane rating describes a fuel's resistance to preignition. Considering Figure 4.19, how do the activation energies for combustion of *n*-octane and isooctane compare? Explain.

36. The combustion of ethanol produces about 40% less energy per gram than normal hydrocarbon fuels. In 2006, the Indy car racing circuit will begin to use engines that have been modified to run on pure ethanol, replacing the current engines that run on methanol, CH_3OH. Why do you think these high-performance vehicles are switching to ethanol?

37. One risk of depending on foreign oil is periodic gasoline shortages due to unfavorable international events. Does a gasoline shortage affect only individual motorists?

Name some ways that a gasoline shortage could affect your life.

38. These three structure have the chemical formula C_8H_{18}. The hydrogen atoms and C-to-H bonds have been omitted for simplicity.

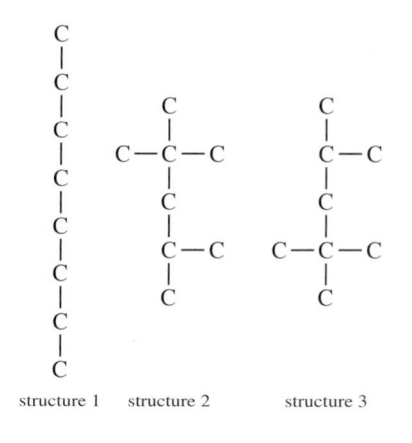

structure 1   structure 2   structure 3

a. Fill in the missing hydrogen atoms and C-to-H bonds and confirm that these structures all represent C_8H_{18}.

b. Are any of these representations identical isomers? If so, which ones?

c. Obtain a model kit and construct one of the structures. What are the C-to-C-to-C bond angles?

d. If you were to build a different one, would the C-to-C-to-C bond angles change? Explain.

e. Draw the structural formula of two more isomers of C_8H_{18}.

39. A ball-and-stick model of ethanol, C_2H_6O, is shown here. Dimethyl ether also has the formula C_2H_6O. Rearrange the atoms in ethanol to draw the Lewis structure of dimethyl ether. *Hint*: Remember to complete the octet for the carbon and oxygen atoms.

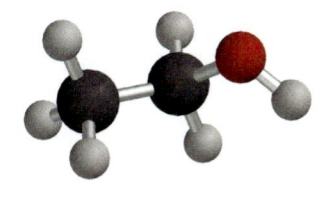

40. How is the growth in oxygenated gasolines related to:

a. restrictions on the use of lead in gasoline?

b. federal and state air quality regulations?

41. Do oxygenated fuels have a higher energy content than nonoxygenated fuels? Use the bond energies in Table 4.2 to calculate the heat of combustion of MTBE.

42. Your neighbor is shopping for a new family vehicle. The salesperson identified a van of interest as a flexible fuel vehicle (FFV).

a. Explain what is meant by FFV to your neighbor.

b. What does it mean for the van to be able to use E-85 fuel?

c. Would your neighbor and his family be particularly interested in using E-85 fuel depending on what region of the country they live?

43. Find information about the availability of biodiesel fuel distributors in the United States.

a. Why are a majority of the distributors located where they are?

b. According the National Biodiesel Board, their distributors will ship the fuel anywhere in the country, particularly to operators of fleets of trucks or cars. Would trucking companies in Florida and Oklahoma both be equally interested? List factors that would be important in such a decision.

44. China's large population has increased energy consumption as the standard of living increases.

a. Report on China's increasing number of automobiles over the last 10 years.

b. What evidence suggests that the increase in the number of vehicles has affected air quality? What interventions, if any, does the Chinese government have underway?

Exploring the Extensions

45. The concept of entropy and disorder is used in games like poker. Describe how the rank of hands (from a simple high card to a royal flush) is related to entropy.

46. Another claim in the *Scientific American* article by Lovins referenced in Section 4.11 was that replacing an incandescent bulb (75 W) with a compact fluorescent bulb (18 W) would save about 75% in the cost of electricity. Electricity is generally priced per kilowatt-hour (kwh). Using the price of electricity where you live, calculate how much money you would save over the life of one compact fluorescent bulb (about 10,000 hr).

47. Section 4.9 states that RFGs burn more cleanly by producing less carbon monoxide than nonoxygenated fuels. What evidence supports this statement?

48. Another type of catalyst used in the combustion of fossil fuels is the catalytic converter that was discussed in Chapter 1. One of the reactions that these catalysts speed up is the conversion of $NO(g)$ to $N_2(g)$ and $O_2(g)$.

a. Draw a diagram of the energy of this reaction similar to the one shown in Figure 4.19.

b. Why is this such an important reaction? *Hint*: See Sections 1.9 and 1.11.

49. Chemical explosions are *very* exothermic reactions. Describe the relative bond strengths in the reactants and products that would make for a good explosion.

50. Bond energies such as those in Table 4.2 are sometimes found by "working backward" from heats of reaction. A reaction is carried out, and the heat absorbed or evolved

is measured. From this value and known bond energies, other bond energies can be calculated. For example, the energy change associated with the combustion of formaldehyde (H_2CO) is -465 kJ.

$$H_2CO(g) + O_2(g) \longrightarrow CO_2(g) + H_2O(g)$$

Use this information and the values found in Table 4.2 to calculate the energy of the C-to-O double bond in formaldehyde. Compare your answer with the C-to-O bond energy in CO_2 and speculate on why there is a difference.

51. You may have seen some General Motors advertisements using the slogan "Live Green by Going Yellow" for their FlexFuel vehicles that can use E-85 gasoline. To what do the colors in this slogan refer?

52. Explain why a distillation tower can separate a mixture of hydrocarbons into different fractions, but it is not possible to separate seawater, also a complex mixture, into all of its different fractions in the same manner.

53. Section 4.8 states that both *n*-octane and isooctane have essentially the same heat of combustion. How is that possible if they have different structures? Explain.

54. Why do you think that countries are willing to go to war over energy issues, but not over other environmental issues? Write a brief op-ed piece for your school newspaper discussing this issue.

55. Since the inception of reformulated gasoline requirements, MTBE has been preferred over ethanol by oil companies. One reason for this is the infrastructure requirements. MTBE can be blended with gasoline at the refinery, whereas ethanol must be shipped separately and mixed with the fuel at the filling station. Why the difference? *Hint*: A detailed look at solubility will be given in the next chapter.

56. What relative advantages and disadvantages are associated with using coal and with using oil as energy sources? Which do you see as the better fuel for the 21st century? Give reasons for your choice.

57. What are the advantages and disadvantages of replacing gasoline with renewable fuels such as ethanol? Indicate your personal position on the issue and state your reasoning.

58. According to the EPA, driving a car is "a typical citizen's most 'polluting' daily activity."

a. Do you agree? Why or why not?

b. What pollutants do cars emit? *Hint:* Information on automobile emissions provided by the EPA (together with the information in this text) can help you fully answer this question.

c. RFGs play a role in reducing emissions. Where in the country are RFGs required? Check the current list published on the Web by the EPA.

d. Explain which emissions RFGs are supposed to lower.

59. It was stated in Section 4.11 that the Three Gorges Dam in China is a controversial project. Use the resources of the Web to investigate some of the major issues concerning this dam.

60. C. P. Snow, a noted scientist and author, wrote an influential book called *The Two Cultures,* in which he stated: "The question, 'Do you know the second law of thermodynamics?' is the cultural equivalent of 'Have you read a work of Shakespeare's?'" How do you react to this comparison? Discuss these questions in light of your own educational experiences.

Chapter

5

The Water We Drink

"Water has never lost its mystery. After at least two and a half millennia of philosophical and scientific inquiry, the most vital of the world's substances remains surrounded by deep uncertainties. Without too much poetic license, we can reduce these questions to a single bare essential: What exactly is water?"

Philip Ball, in *Life's Matrix: A Biography of Water*, University of California Press, Berkeley, CA, 2001, p. 115

Arguably, water is the most important chemical compound on the face of the Earth. In fact, it covers about 70% of that face, giving the planet the lovely blue color in the famous "blue marble" photos taken from outer space. Water is essential to all living species. A child's body is about 75% water, and adult bodies are approximately 50–65% water. A human brain is 75% water, blood is 83% water, and lungs are approximately 90% water. Even our bones, seemingly so solid, are 22% water. Water is so important to life that our speculation about life on other planets depends primarily on whether water is present. Water refreshes and sustains us, governs the weather, and gives us relaxing pleasures as well as recreational opportunities.

In this chapter, we will consider water from the perspective of those who drink it. There is more to know about water, however, than can be seen or tasted. Unseen impurities in water, depending on their identities and amounts, can impart a crisp, fresh taste or produce an unpleasant illness. Water is incredibly versatile, dissolving many substances and suspending others. To better appreciate how water works its magic, we will use chemical concepts such as electronegativity, polarity, and hydrogen bonding to understand the properties of water molecules.

Some of the most critical questions about drinking water deal with its safety and how that safety is ensured. The Safe Drinking Water Act, as amended in 1996, mandates that each water supplier deliver a right-to-know report once a year to consumers from any community water system, public or private. These reports, also called Consumer Confidence Reports, may be distributed with water bills or made available online. As a person studying chemistry, you are in a good position to understand the meaning of the measurements being reported and the standards of quality that must be met. Knowing the quality of tap water can then help consumers to make wise choices about purchasing bottled or filtered water as alternatives. You will have the opportunity to explore the quality of water at your college or in your hometown, once we have considered many of the questions of chemistry and public policy involved in safe drinking water. But first, we invite you to raise your glass of tap water, bottled water, or filtered tap water and prepare to "take a drink."

Consider This 5.1 Take a Drink of Water

Obtain samples of tap water, bottled water, and filtered tap water if available.

a. Carefully describe each water sample, including taste, odor, appearance, and any other characteristics. A table will help to organize your observations.
b. What do you like or dislike about each water sample?
c. List five qualities that you expect in your drinking water.
d. What concerns, if any, do you have about the safety of each water sample?

5.1 Water from the Tap or Bottle

Chapter 1 began with an invitation to "Take a breath of air." Without thinking, we do it automatically about 15 times a minute. We would die quickly without air and its life-supporting oxygen. We generally don't have any choice about *what* air we breathe; we must rely on that which surrounds us. On the other hand, we generally do have choices with water. We can make decisions about *how frequently* we drink, *how much* we drink, and about the *source* of the water we drink. Municipal tap water may be consumed with or without further home filtering. Some may have access to well water. We may choose bottled water or beverages containing water. If out on the trail, we may purify water from nearby streams or collect rainwater. If **potable water,** water that is fit for human consumption, is not available, we are in danger. Our bodies can go weeks without food but only days without water. If the water in our bodies is reduced by just 1%, thirst will develop. When the loss reaches 5%, muscle strength declines. At a 10% loss, delirium and blurred vision occur, and a 20% reduction results in death.

Restoration of safe municipal drinking water was a major concern after numerous hurricanes of the 2003, '04, and '05 seasons, including hurricane Katrina that devastated the Gulf Coast and New Orleans in 2005.

(a)

(b)

Figure 5.1

(a) We usually take the safety of drinking water for granted in the United States. (b) Some prefer the taste and convenience of bottled water.

Nothing could be more familiar than this clear, colorless, and (usually) tasteless liquid. In fact, we generally take water and water quality for granted in most regions of this country. Unless a water emergency occurs, brought on by drought or contamination of our municipal water supply, we seldom think about where the water comes from, what it contains, how pure it is, or how long the supply will last. We turn on a faucet for a drink or a shower and simply expect a sufficient quantity of water to come flowing out of the tap. Most Americans obtain their drinking water from a water faucet or a drinking fountain (Figure 5.1a). This marvelous liquid is remarkably inexpensive, costing only about 1/10 of a penny per quart.

But not everyone drinks tap water. An increasing number of Americans are drinking bottled water instead (Figure 5.1b). Bottled water is big business and now the fastest growing segment of the beverage industry in the world. According to the Beverage Marketing Corporation, Americans consumed 7.4 billion gallons of bottled water at a cost of $9.8 billion dollars in 2005. It is a common sight on campuses worldwide to see students carrying bottled water. Consumers 18–24 years old are the major users of bottled water.

You have only to walk down a grocery aisle in stores anywhere in the world to take a tour of bottled water brands. Advertisements for bottled water tout its purity, often using words or images that conjure up nature, purity, and pristine beauty. For example, the label for Dasani bottled water, the brand owned by Coca-Cola, makes these statements.

> "Purified Water. Enhanced with Minerals for a Pure, Fresh Taste."

Web sites for bottled water also try to convey images of purity and natural processes. This is Evian's statement.

> "Every drop of Evian Natural Spring Water begins as rain and snow falling high in the pristine and majestic French Alps."

Some Web sites, such as that for LeBleu UltraPure Drinking Water, even try to teach a little chemistry!

> "Water, the fluid of life and the shaper of the earth, is made from the simplest and most abundant element in the universe, hydrogen, joined to the vital gas oxygen. Two atoms pair with a single oxygen atom to establish the triple structure water, H_2O. Water, the universal solvent, given sufficient time, will dissolve or suspend almost any material on earth."

Although highly popular, bottled water is also very expensive relative to tap water. Typically, bottled water in large-volume containers costs up to $3 or more per gallon. The price can be much higher for individually sized bottles. These costs are approximately 1000 times higher than for the same volume of tap water. Bottled water is far more expensive, drop for drop, than milk, lemonade (Figure 5.2), or West Texas crude oil. "Ounce for ounce it costs more than gasoline, even at today's high gasoline prices; depending on the brand, it costs 250 to 10,000 times more than tap water" according to the *New York Times* 2005 article "Bad to the Last Drop."

Why are consumers willing to ante up so much more for bottled water? Should you save your money and drink tap water? How can you tell if your tap water is safe to drink, and what are some common contaminants found in municipal water supplies? One of the goals of this chapter is to help you learn enough about drinking water quality to make intelligent choices.

Growth of the bottle industry has had a direct influence on CO_2 emissions (Chapter 3), energy costs for production and transportation (Chapter 4), and plastics production (Chapter 9).

Consider This 5.2   Bottled Water and You

Many people buy bottled water despite its high cost relative to tap water.

a. List what you perceive to be the advantages and disadvantages of drinking bottled water.

b. Rank the items on your list in order of importance for your personal decision about whether to drink bottled water.

Consider This 5.3   **Finding out About Bottled Water**

If you search for "bottled water" on the Web, you will get over a million hits. Select two sites to explore, one provided by a supplier and the other provided as a source of consumer information. The former may flood you with statistics about the benefits of bottled water; the latter may raise questions, such as "Is bottled water safer?" or "Is it worth the cost?" For each site, list the title, source, URL, and two things that you learned about water.

Figure 5.2

Bottled water and Dennis the Menace.

Source: DENNIS THE MENACE used by permission of Hank Ketcham Enterprises and © by North American Syndicate.

5.2   Where Does Drinking Water Come From?

What journey does water take to get from its natural source to your tap or bottle? Water is widely distributed on planet Earth. On the surface, it is found in oceans, lakes, rivers, snow, and glaciers (Figure 5.3). In the atmosphere it exists as water vapor and as tiny droplets in clouds that replenish surface water by means of rain and snow.

Across the world, water is found underground in **aquifers,** great pools of water trapped in sand and gravel 50–500 ft below the surface. Some aquifers are enormous, such as the Ogallala Aquifer that underlies parts of eight states from South Dakota to Texas (Figure 5.4). Keeping these underground resources free from contamination is an important consideration for future generations. If an aquifer becomes contaminated, it may take decades to become clean again.

Water that can be made suitable for drinking comes from either surface water or groundwater. **Surface water,** water from lakes, rivers, and reservoirs, frequently contains substances that must be removed before it can be used as drinking water. By contrast, **groundwater,** water pumped from wells that have been drilled into underground aquifers, is usually free of harmful contaminants. Large-scale water supply systems for cities tend to rely on surface water resources. For example, Chicago residents drink water from Lake Michigan that has been purified. Smaller cities, towns, and private wells tend to rely on groundwater, the source of drinking water for a little over half of the U.S. population.

Figure 5.3

Lakes and reservoirs provide much of our drinking water.

Figure 5.4

The Ogallala Aquifer is shown in dark blue on this map.

Consider This 5.4   Your Home's Water Source

When you turn on the tap at home, where does your drinking water come from? You can find out by accessing local drinking water information from the EPA. A link is provided at the *Online Learning Center*. Search for the source of your home's water by clicking on the state in which you live. Consumer Confidence Reports and other information for your state will help you answer these questions.

    a. What is the name of the water system that services your home or closest reporting region?

    b. What is the primary water source for that system?

    c. What information from the past five years is given about the quality of that system's drinking water?

The Earth was not always as wet as it is today. Scientists believe that much of the water now on this planet originally was spewed as vapor from thousands of volcanoes that pocked Earth's surface. The vapor condensed as rain and the process repeated over the ages. Water molecules cycled from sea to sky and back again. About 3 billion years ago, primitive plants, and then animals, extracted water from and contributed water to the cycle—one that continues today. During an average year, enough precipitation falls on the continents to cover all the land area to a depth of more than 2.5 ft. As we are well aware, this precipitation does not fall all at once, nor is it distributed uniformly. The wettest place on record is Mount Waialeale on Kauai, Hawaii, which receives an annual average rainfall of 460 inches (12 m). In contrast is the yearly average of 0.03 inch (0.08 cm) of rain in Arica, Chile. Data have been gathered in this Chilean desert for the past 64 years. For 14 consecutive years, it did not rain at all! Closer to home, an average of 1.5×10^{13} L of water falls daily on the continental United States—enough to fill about 400 million swimming pools.

This sounds like a great deal of water, but on a global scale, the amount of fresh water is relatively small. The great majority of Earth's water, 97.4% of the total, is in the salty oceans, water that is undrinkable without expensive purification. The remaining 2.6% is all the fresh water we have. The majority of even this relatively small supply of fresh water is frozen in glaciers and polar ice caps. Only about 0.01% of Earth's total water is conveniently located in lakes, rivers, and streams as fresh water (Figure 5.5). Consequently, the world's drinking

> Desalination of seawater is discussed in Section 5.15.

> Lake Superior holds over 10% of the world's fresh water supply.

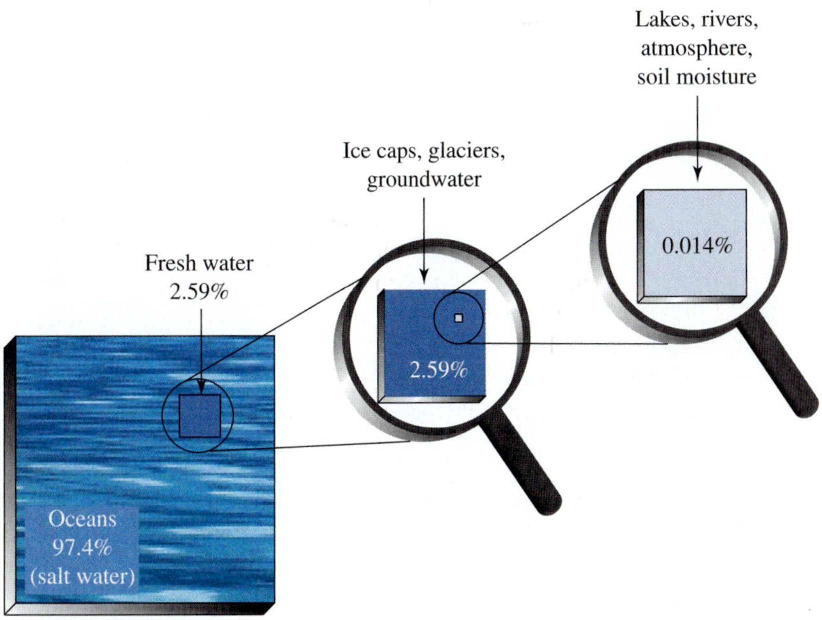

Lakes, rivers, atmosphere, soil moisture

Ice caps, glaciers, groundwater

0.014%

Fresh water 2.59%

2.59%

Oceans 97.4% (salt water)

Figure 5.5
Distribution of water on Earth.

water supplies are quite limited, varying widely depending on locale. In the United States, 80% of the fresh water is used to irrigate crops and to cool electrical power plants.

Sceptical Chymist 5.5	Swimming in the Rain

We just stated that the daily amount of rainfall in the continental United States is sufficient to fill about 400 million swimming pools. If the average swimming pool contains 20,000 gallons of water, does this assertion check out? Explain.

5.3 Water as a Solvent

A major reason we must consume water is that it is an excellent solvent for many of the chemicals that make up our bodies. In this capacity, water acts as a **solvent,** a substance capable of dissolving other substances. **Solutes** are those substances that dissolve in a solvent. The resulting mixture is called a **solution,** a homogeneous mixture of uniform composition. Furthermore, **aqueous solutions** are solutions in which water is the solvent. Later in Sections 5.9 and 5.10 we will examine why certain kinds of substances dissolve in water and others do not. For now, we simply note that a remarkable variety of substances can dissolve in water and that this has important consequences for living organisms as well as for the environment.

Your Turn 5.6	Common-Sense Solubility

Based on your experience, classify these substances as to how well they dissolve in water. Use the relative terms *very soluble, partially soluble,* or *insoluble.*

a. salt   **b.** sugar   **c.** sidewalk chalk
d. grape Kool-Aid   **e.** cooking oil   **f.** aspirin

Because water is such a good solvent, drinking water is never "pure" water. It always contains other substances. Municipal water companies provide information about the dissolved mineral content, the solutes, for tap water. A **mineral** is a naturally occurring element or compound that usually has a definite chemical composition, a crystalline structure, and is formed as a result of geological processes. An analysis of tap water in a Midwest home yielded the information in Table 5.1. Similar information for commercial bottled water can be found on the label or at the bottler's Web site. For example, a label on Evian bottled water provides the information shown in Table 5.2.

All of the mineral solutes except silica in Tables 5.1 and 5.2 are in ionic form and will be discussed in Section 5.7. Calcium, magnesium, and sodium are present as Ca^{2+}, Mg^{2+}, and Na^+, respectively. Some ions are listed in the plural because they may form compounds with various metal ions. The number given with each dissolved ion indicates how much of that substance (in milligrams) is present in 1 L of water. This raises a reasonable question: Should we be worried about these tiny amounts? Calcium ions, for example, have a definite health benefit in producing stronger bones. Milk and milk products, not

1.00 liter (L) = 1.06 quart

Table 5.1	Mineral Composition of Tap Water, mg/L		
calcium	66	sulfates	42
magnesium	24	chlorides	48
sodium	18	nitrates	6
		fluorides	1

Table 5.2	Mineral Composition of Evian, mg/L		
calcium	78	bicarbonates	357
magnesium	24	sulfates	10
silica	14	chlorides	4
		nitrates	1

Evian water, are the preferred sources for calcium ion; you would have to drink 4 L of Evian water to get the same amount of calcium ion as that in one 8-oz glass of milk. In contrast, the nitrate ion, depending on its concentration, can be dangerous, especially for infants. The other substances listed for Evian bottled water are not likely to cause a health problem. Elsewhere on the label it is noted that sodium (Na^+), a health concern for some people, is present at less than 5 mg per 500-mL bottle.

Your Turn 5.7   Bottled Water for Dietary Calcium?

One 500-mL bottle of Evian water provides 4% of the recommended daily requirement of calcium (in the form of calcium ion, Ca^{2+}).

a. Use this information to calculate the approximate number of milligrams of calcium recommended per day.
b. Will this recommended value apply to everyone? Explain.
c. How many 500-mL bottles of Evian water would you have to drink to obtain your recommended daily supply of calcium?

Sceptical Chymist 5.8   Bottled Water and Claims of Purity

The Web site for Penta Ultra Premium Purified Drinking Water states:

"Penta ultra-premium purified drinking water is the cleanest-known bottled water."

Consider the claims made by the company about the product of their 13-step purification process:

- Studies on human cells (in vitro) show that Penta water increases cell survivability by 266%.
- Penta water differs from water in having a higher boiling point, a higher surface tension, and a lower viscosity.
- Studies on human cells (in vitro) show that DNA chromosomal mutation rates were 271% greater in lab distilled water than in Penta water.
- Penta water is a new composition of matter.

As a Sceptical Chymist, evaluate each of these claims. Be sure to explain your reasoning.

Perhaps you have never considered drinking a glass of water as a risk–benefit activity, yet it is. We usually assume water that has been chemically analyzed and treated to have important benefits with very low risk. Overwhelmingly, this is a valid assumption. Out of necessity, however, each label is incomplete. As already noted, no information appears about health risks for the given concentrations of solutes. No information appears about whether other substances, if any, are present in the water or if these substances might be harmful. For example, even though tiny amounts of lead are found in most water samples, the amount is usually too low to be a health problem. If the water has been chlorinated for purification, it almost certainly has trace amounts of some chlorination by-products. In

Section 5.12 some of these by-products will be examined. Indeed, we rarely stop to think about what trace amounts of substances may be in the water, because we tend to assume that the water is safe to drink. In part, this is because extensive federal and state regulations and standards govern municipal water quality to protect the public. Most bottled water is regulated as well, often by self-imposed industry standards.

In assessing the risks of drinking water, it is not sufficient to know what substances are present in the water and how toxic they are. We also need to know how much of each substance is present in a particular amount of the water. In other words, we need to understand what is meant by the concentration of a solute and the typical ways of expressing it. We now turn to these topics.

5.4   Solute Concentration in Aqueous Solutions

The concept of concentration was first introduced in Chapter 1 in relation to the composition of air. We revisited concentration again in Chapters 2 and 3, looking at the concentrations of chlorine compounds in the stratosphere and at greenhouse gases in the troposphere. Now we will examine this concept in terms of substances dissolved in water.

Although concentrations of components found in air might be a bit hard to visualize, solute concentrations in aqueous solutions are more familiar and therefore more easily imagined. For example, if a recipe called for you to dissolve 1 teaspoon of an ingredient in 1 cup of water, a solution of a specific concentration would result: 1 tsp per cup (1 tsp/cup). Note that you would have the same 1 tsp/cup concentration if you also dissolved 2 tsp of the ingredient in 2 cups of water, 4 tsp in 4 cups, or ½ tsp in ½ cup. Even though you used larger or smaller quantities of the ingredient, the number of cups of water increased or decreased proportionally. Therefore, the **concentration,** the ratio of amount of solute to amount of solution—or for this recipe, the ratio of amount of ingredient to amount of water solution—would be the same in each case: 1 tsp per 1 cup (1 tsp/cup).

Solute concentrations in aqueous solutions follow the same pattern, but usually are expressed in different units. We will use four ways of expressing concentration: percent; parts per million; parts per billion; and molarity. Three of these are familiar to you from earlier chapters; the fourth, molarity, uses the mole concept introduced in Chapter 3. We now will discuss each in turn.

The most familiar way of expressing concentration is **percent,** defined in Chapter 1 as parts per hundred. For example, a solution containing 5 g of sodium chloride (NaCl) in 100 g of solution would be a five percent (5%) solution by weight. Hydrogen peroxide (H_2O_2) solutions, often found as an antiseptic in medicine cabinets, are usually 3% H_2O_2, indicating that they contain 3 g of H_2O_2 in 100 g of solution (or 6 g in 200 g of solution, etc.). Percent concentrations are used most often for solutions of high solute concentration. When solute concentrations are much lower, two commonly used units are ppm and ppb.

Concentrations of dissolved substances in drinking water are normally far lower than 1%, so they are often described in terms of **parts per million (ppm,** 1 part per million). A 1-ppm solution of calcium ion in drinking water contains 1 g of calcium ion in 1 million (1,000,000, or 10^6) g of that sample. The same concentration, 1 ppm, could be applied to a solution with 2 g of calcium ion in 2×10^6 g of water, 5 g in 5×10^6 g of water, or 5 mg (5×10^{-3} g) in 5000 (5×10^3) g of water. Although parts per million is a very useful concentration unit, measuring 1 million grams of water is not very convenient. So we employ an easier but equivalent way to establish ppm. We use the unit **liter (L),** the volume occupied by 1000 g of water at 4 °C. Now we can say that 1 ppm of any substance in water equals 1 mg of that substance per 1 L of water. Notice that the units cancel.

> The mass of the solution is determined by the mass of the solvent for low-solute concentrations.

$$1 \text{ ppm} = \frac{1 \text{ g solute}}{1 \times 10^6 \text{ g water}} \times \frac{1000 \text{ mg solute}}{1 \text{ g solute}} \times \frac{1000 \text{ g water}}{1 \text{ L water}} = \frac{1 \text{ mg solute}}{1 \text{ L water}}$$

> Although strictly true only at 4 °C, 1 L is close to being the volume of 1000 g (1×10^3 g) of H_2O throughout its liquid range, including at room temperature.

Drinking water contains substances naturally present at concentrations in the parts per million range, as illustrated on the Evian bottled water label. Pollutants also may be present in the parts per million range. For example, the acceptable limit for nitrate ion, often found in well water in some agricultural areas, is 10 ppm; the limit for the fluoride ion is 4 ppm.

Some pollutants are of concern at concentrations much lower than parts per million and are reported as **parts per billion** (ppb, 1 part per billion). One part per billion of mercury (Hg) in water means 1 g Hg in 1 billion (1×10^9) g of water. In more convenient terms, this means 1 microgram (1×10^{-6} g, or 1 μg) Hg in 1 L of water. For example, the acceptable limit for mercury in drinking water is 2 ppb.

$$2 \text{ ppb Hg} = \frac{2 \text{ g Hg}}{1 \times 10^9 \text{ g H}_2\text{O}} \times \frac{1 \times 10^6 \text{ μg Hg}}{1 \text{ g Hg}} \times \frac{1000 \text{ g H}_2\text{O}}{1 \text{ L H}_2\text{O}} = \frac{2 \text{ μg Hg}}{1 \text{ L H}_2\text{O}}$$

Convince yourself that the units cancel as in the previous example.

One part per million is a tiny concentration. Several analogies to a concentration of 1 ppm were given in Section 1.2, including that 1 ppm corresponds to 1 second in nearly 12 days. A similar analogy can be offered for parts per billion: 1 ppb corresponds to 1 second in 33 years, or approximately 1 inch on the circumference of the Earth.

Your Turn 5.9   Lead Ion Concentrations

a. 80 μg of lead were detected in 5 L of water. What is the concentration of lead? Express your answer in both ppm and ppb.
b. If the maximum lead concentration in drinking water allowed by the federal government were 15 ppb, would the sample in part **a** be in compliance with federal limits? Explain.

Molarity (M), another useful concentration unit, is defined as the number of moles of solute present in one liter of solution.

$$\text{Molarity (M)} = \frac{\text{moles of solute}}{\text{liter of solution}}$$

The great advantage of molarity is that solutions of the same molarity all contain exactly the same number of chemical units (atoms, ions, or molecules). For example, the mass of solute may vary depending on the molar mass, but for all 1 M solutions the number of chemical units will be the same. Methods of chemical analysis of water frequently use molarity to express concentration. For now, we simply want to develop some familiarity with the concept of molarity itself.

As an example, consider a solution of NaCl in water. The molar mass of NaCl is 58.5 g; therefore, 1 mol of NaCl weighs 58.5 g. If we were to dissolve 58.5 g of NaCl in some water and then add enough water to make exactly 1.00 L of solution, we would have a 1.00 M NaCl solution (Figure 5.6), and we could say that we have prepared a one molar solution of sodium chloride. Note the use of a **volumetric flask,** a type of glassware that contains a precise amount of solution when filled to the mark on its neck. But because concentrations are simply ratios of solute to solvent, there are many ways to make a 1 M NaCl solution. Another possibility would be to use 0.500 mol NaCl (29.2 g) in 0.500 L of solution. This would require the use of a 500.0-mL volumetric flask, rather than the 1.00-L flask shown in Figure 5.6.

$$1 \text{ M NaCl} = \frac{1 \text{ mol NaCl}}{1 \text{ L solution}} \quad \text{or} \quad \frac{0.5 \text{ mol NaCl}}{0.5 \text{ L solution}}, \text{etc.}$$

Let's say you have a water sample that has 150 ppm of mercury in it. What would this concentration be if expressed in molarity?

$$150 \text{ ppm Hg} = \frac{150 \text{ mg Hg}}{1 \text{ L H}_2\text{O}} \times \frac{1 \text{ g Hg}}{1000 \text{ mg Hg}} \times \frac{1 \text{ mol Hg}}{200.6 \text{ g Hg}}$$

And when the units cancel:

$$150 \text{ ppm Hg} = \frac{150 \text{ mg Hg}}{1 \text{ L H}_2\text{O}} \times \frac{1 \text{ g Hg}}{1000 \text{ mg Hg}} \times \frac{1 \text{ mol Hg}}{200.6 \text{ g Hg}} = \frac{7.5 \times 10^{-4} \text{ mol Hg}}{1 \text{ L H}_2\text{O}} = 7.5 \times 10^{-4} \text{ M Hg}$$

We have just shown that a sample of water containing 150 ppm of mercury can also be expressed as 7.5×10^{-4} M Hg.

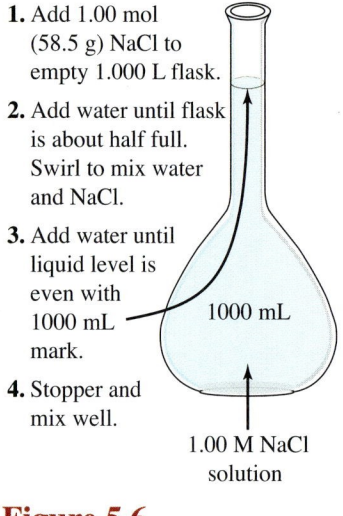

1. Add 1.00 mol (58.5 g) NaCl to empty 1.000 L flask.

2. Add water until flask is about half full. Swirl to mix water and NaCl.

3. Add water until liquid level is even with 1000 mL mark.

4. Stopper and mix well.

1000 mL

1.00 M NaCl solution

Figure 5.6

Preparing a 1.00 M NaCl solution.

Your Turn 5.10   Moles and Molarity

a. For 1.5 M and 0.15 M NaCl, how many moles of solute are present in 500 mL of each?

b. A solution is prepared by adding enough water to 0.50 mol NaCl to form 250 mL of solution. A second solution is prepared by adding enough water to 0.60 mol NaCl to form 200 mL of solution. Which solution is more concentrated? Explain.

c. A student was asked to prepare 1.0 L of a 2.0 M NaOH solution. The student measured and placed 80.0 g of NaOH pellets in a beaker and then added 1000 mL of water. Was the resulting solution 2.0 M? Explain.

d. A technician was cleaning up a chemical stockroom when she found two bottles of hydrochloric acid, HCl(*aq*); one bottle contained 450 mL of 3.0 M HCl and the other contained 250 mL of 1.5 M HCl. She decided to combine both solutions in a 1 L bottle. What was the resulting concentration of the new solution?

So far in this chapter we have developed some ideas about drinking water, some of the substances that may be present in it, and how to express the concentrations of those substances. We shift now to a more detailed examination of water at the molecular level. Our aim is to understand water's unique properties, including its excellence as a solvent.

5.5 The Molecular Structure and Physical Properties of Water

What *is* water? It is clear that water is essential to our lives and that water is an excellent solvent. What may not be as clear is that our dependence on water is possible only because water has a number of unusual properties. In fact, the physical properties of water are quite peculiar, and we are very fortunate that they are. If water were a more conventional compound, we would not exist.

This most common of liquids is full of surprises, so let's start with its physical state. Water is a liquid at room temperature (about 25 °C) and normal atmospheric pressure. This is surprising because almost all other compounds with similar molar masses (18.0 g/mol) are gases under similar conditions of temperature and pressure. Consider three common atmospheric gases (N_2, O_2, and CO_2) whose molar masses are 28, 32, and 44 g/mol, respectively. All have molar masses greater than that of water, yet they are gases rather than liquids.

Not only is water a liquid under these conditions, but also it has an anomalously high boiling point of 100 °C. This temperature is one of the reference points for the Celsius temperature scale. The other is the freezing point of water, 0 °C. And when water freezes, it exhibits another somewhat bizarre property—it expands. Most liquids contract when they solidify. These and other unusual properties derive from water's chemical composition and molecular structure.

To better understand the chemical and physical properties of water, we need information about its chemical composition and molecular structure. The chemical composition is known to practically everyone. Indeed, the formula for water, H_2O, is very likely the world's most widely known bit of chemical information. Recall from Chapter 2 that water is a covalently bonded molecule (Section 2.3). In Chapter 3 (Section 3.3), the molecular shape of water was illustrated by means of a ball-and-stick model and a space-filling model. These representations are shown again in Figure 5.7.

In general for covalent substances, as molar mass increases the boiling point also increases. See Section 4.7 for an example using distillation.

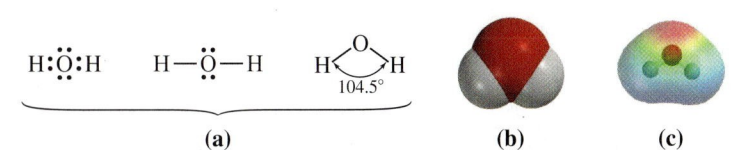

Figure 5.7

Representations of H_2O. (a) Lewis structures and structural formula; (b) Space-filling model; (c) Charge-density model.

The electrons between the oxygen and hydrogen atoms that form the covalent bond are not shared equally. Experimental evidence indicates that the oxygen atom attracts the shared electron pair more strongly than does the hydrogen atom. To use the appropriate technical term, oxygen is said to have a higher electronegativity than hydrogen. **Electronegativity (EN) is a measure of an atom's attraction for the electrons it shares in a covalent bond.** The greater the electronegativity, the more an atom attracts bonding electrons to itself. Table 5.3 shows a periodic table of electronegativity values for the first 18 elements, all in "A" subgroups. The concepts of electronegativity and electronegativity values were introduced by the great American quantum chemist and biochemist Linus Pauling (1901–1994).

An examination of Table 5.3 reveals some useful generalizations about electronegativity. The highest electronegativity values are associated with nonmetallic elements such as fluorine and oxygen. The halogens, members of Group 7A, have atoms with seven outer electrons. Recall from Section 2.3 that each of these atoms has a strong tendency to bond with another atom in such a way as to acquire a share in an additional electron, thus completing a stable octet of electrons. A similar argument explains the high electronegativity values of other nonmetals. For example, oxygen, with six outer electrons per atom, also exhibits a relatively strong attraction for shared electrons and will typically interact with other atoms in a manner that will allow it to share in two additional electrons. Conversely, the lowest electronegativity values are associated with the metals found in Groups 1A and 2A. Atoms of these metallic elements have much weaker attractions for electrons than do nonmetals. In general, electronegativity values increase as you move across a row of the periodic table from left to right (from metals to nonmetals) and decrease as you move down a group of the table.

According to Table 5.3, the electronegativity of oxygen is 3.5; that of hydrogen is 2.1. Because of this electronegativity difference, the shared electrons are pulled closer to the more electronegative oxygen and away from the less electronegative hydrogen. This unequal sharing gives the oxygen end of the O-to-H bond a partial negative charge and the hydrogen end a partial positive charge. The result is a **polar covalent bond, a covalent bond in which the electrons are not equally shared, but rather displaced toward the more electronegative atom.** The greater the electronegativity difference of the elements involved, the more polar the bond. A polar covalent bond is an example of an **intramolecular force, a force that exists within a molecule.** In Figure 5.8, an arrow is used to indicate the direction in which the electron pair is displaced. The δ^+ and δ^- symbols indicate partial positive and partial negative charges, respectively.

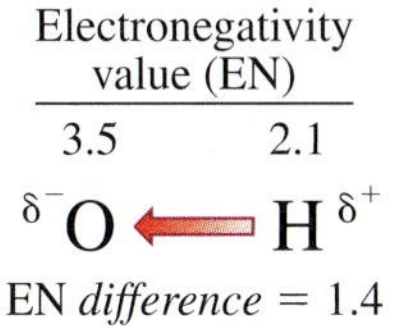

Electronegativity
value (EN)

3.5	2.1

$$\delta^- O \Longleftarrow H^{\delta^+}$$

EN *difference* = 1.4

Figure 5.8

Polar covalent bond between hydrogen and oxygen atoms. The electrons are displaced toward the more electronegative oxygen atom.

Table 5.3		**Electronegativity Values, Arranged by Group Number**					
1A	**2A**	**3A**	**4A**	**5A**	**6A**	**7A**	**8A**
H							He
2.1							—
Li	Be	B	C	N	O	F	Ne
1.0	1.5	2.0	2.5	3.0	3.5	4.0	—
Na	Mg	Al	Si	P	S	Cl	Ar
0.9	1.2	1.5	1.8	2.1	2.5	3.0	—

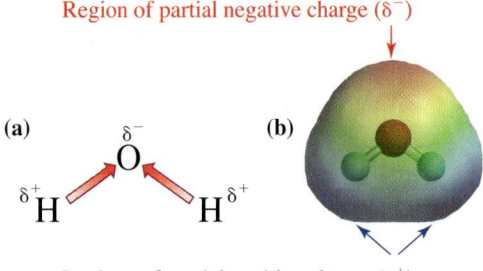

Region of partial negative charge (δ^-)

Regions of partial positive charge (δ^+)

Figure 5.9

(a) H_2O, a polar covalent molecule with polar covalent bonds. **(b)** The charge-density drawing shows the partial positive and negative charges in a water molecule.

Your Turn 5.11   Polar Bonds

For each pair, which is the more polar bond? To which of the atoms in the bond will the electron pair be more strongly attracted?
Hint: Use the electronegativity values given in Table 5.3.

  a. H-to-F or H-to-Cl
  b. N-to-H or O-to-H
  c. N-to-O or S-to-O

Answer
  a. The H-to-F bond is more polar than the H-to-Cl bond; the electron pair
     forming the H-to-F bond will be more strongly attracted by the fluorine.

A molecule that contains only nonpolar covalent bonds must be nonpolar. This is why diatomic molecules such as Cl_2 or H_2 are nonpolar. However, a molecule that contains polar covalent bonds may or may not be polar. The polarity depends on the geometry of the molecule. The specific case of the water molecule is shown in Figure 5.9. Note that the hydrogen atoms have a partial positive charge and the oxygen atom a partial negative charge. Many of the unique properties of water are a consequence of both the polarity of the molecule and its overall shape.

Consider This 5.12   Polar or Nonpolar Molecules

The H_2O molecule contains polar bonds and is a polar molecule. What about CO_2?

  a. Are the covalent bonds in CO_2 polar or nonpolar? *Hint:* Use the
     electronegativity values given in Table 5.3.
  b. Draw a figure similar to Figure 5.9(a) for CO_2. Be sure to use the correct
     bond angle.
  c. Offer a possible explanation why the H_2O molecule is polar, but the CO_2
     molecule is not.

5.6   The Role of Hydrogen Bonding

Polar covalent bonds can help us understand some of the unusual properties of water. Consider what happens at the molecular level when two water molecules approach each other. Because opposite charges attract, one of the partially positive-charged hydrogen atoms of one water molecule is attracted to the region of partial negative

Compare:
• *Intermolecular* forces are *between* molecules.
• *Intercollegiate* sports are played *between* colleges.

charge associated with the nonbonding electron pairs on the oxygen of the other water molecule. This is an **intermolecular force,** a force that occurs between molecules. Each H_2O molecule has two hydrogen atoms and two nonbonding pairs of electrons that allow for multiple intermolecular attractions (Figure 5.10). The bonds that form are known as **hydrogen bonds,** electrostatic attractions between a hydrogen atom bearing a partial positive charge in one molecule and an O, N, or F atom bearing a partial negative charge in a neighboring molecule. Hydrogen bonds typically are only about one tenth as strong as the covalent bonds that connect atoms together *within* molecules; they are also longer than the covalent bonds. The effect of hydrogen bond formation is central to understanding water's unusual properties.

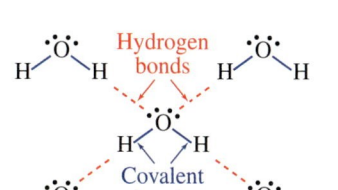

Figure 5.10
Hydrogen bonding in water (distances not to scale).

 Figures Alive! Visit the *Online Learning Center* to learn more about hydrogen bonds.

Your Turn 5.13   Water's Hydrogen Bonds

a. How many hydrogen bonds are shown in Figure 5.10 around the central water molecule?
b. Are hydrogen bonds intermolecular or intramolecular forces? Explain.

Although hydrogen bonds are not as strong as covalent bonds, hydrogen bonds are quite strong compared with other types of intermolecular forces. For example, to boil water, the H_2O molecules must be separated from their relatively close contact in the liquid state and moved into the gaseous state, where they are much farther apart. In other words, their intermolecular hydrogen bonds must be broken. If the hydrogen bonds in water were weaker, water would have a much lower boiling point and require less energy to boil. If water had no hydrogen bonding at all, it would boil at about $-75\ °C$, a prediction based on its molar mass. Because of hydrogen bonding, almost all of our body's water, whether in cells, blood, or other bodily fluids, is in the liquid state, well below the boiling point. Our very existence depends on hydrogen bonding.

Consider This 5.14   Bonds Within and Between Water Molecules

What bonds are broken when water boils? Explain with drawings.
Hint: Start with molecules of water in the liquid state as shown in Figure 5.10. Make a second drawing to show what happens when water boils.

DNA molecules (Section 12.2) form hydrogen bonds between *different* strands of DNA, and proteins can form hydrogen bonds within different regions of the *same* molecule.

The molecular structures of proteins and nucleic acids such as DNA are discussed in Sections 11.4 and 12.1–12.4.

The phenomenon of hydrogen bonding is not restricted to water. There is evidence for similar intermolecular attraction in many molecules that contain hydrogen atoms covalently bonded to oxygen, nitrogen, or fluorine atoms. In each of these cases, polar covalent bonds form within each molecule, the necessary requirement for the formation of intermolecular hydrogen bonds. Hydrogen bonding is also important in stabilizing the shape of large biological molecules, such as proteins and nucleic acids. In proteins, the major components of skin, hair, and muscle, hydrogen bonding occurs between hydrogen atoms and oxygen or nitrogen atoms. The coiled, double-helical structure of DNA (deoxyribonucleic acid) is stabilized by thousands of hydrogen bonds formed between particular segments of the linked DNA strands. So in this respect, too, hydrogen bonding plays an essential role in life processes.

Hydrogen bonding also explains why ice cubes and icebergs float in water. Ice is a regular array of water molecules in which every H_2O molecule is hydrogen-bonded to four others. The pattern is shown in Figure 5.11. Note that the pattern includes a good

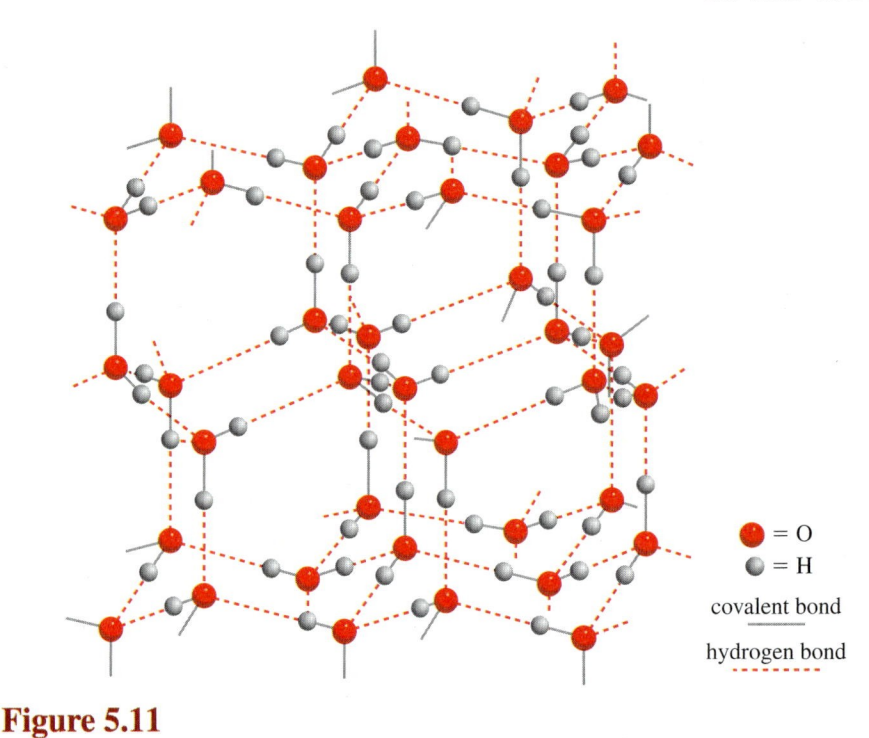

= O
= H

covalent bond

hydrogen bond

Figure 5.11

The hydrogen-bonded lattice structure of the common form of ice. Note the open channels between "layers" of water molecules that cause ice to be less dense than water.

deal of empty space in the form of hexagonal channels. When ice melts, this regular array begins to break down, and individual H_2O molecules can enter the open channels. As a result, the molecules in the liquid state are, on the average, more closely packed than in the solid state. Thus, a volume of one cubic centimeter (1 cm^3) of liquid H_2O contains more molecules than 1 cm^3 of ice. Consequently, liquid water has a greater mass per cubic centimeter than ice. This is simply another way of saying that the **density, the ratio of mass per unit volume,** of liquid water is greater than that of ice.

For water, mass is often expressed in grams and volume in cubic centimeters (cm^3), [identical to milliliters (mL)]. Furthermore, 1.00 cm^3 of water weighs 1.00 g. In other words, its density is 1.00 g/cm^3, or 1.00 g/mL. On the other hand, 1 cm^3 of ice weighs 0.92 g, so its density is 0.92 g/cm^3, or 0.92 g/mL.

People often confuse density with mass. For example, you may hear someone say that iron is "heavy" or that lead is "very heavy." Large pieces of iron and lead are indeed often quite heavy, but it is more accurate to say that iron has a high density (7.9 g/cm^3) and lead an even higher density (11.3 g/cm^3). Then again, popcorn has a low density; we are likely to say that even a large bag of popcorn feels "light."

For the great majority of substances, the solid state is denser than the liquid. The fact that water shows the reverse behavior means that lakes freeze from the top down, not the bottom up. This topsy-turvy behavior is convenient for aquatic plants, fish, and ice skaters. Of course, it is not so convenient for people whose water pipes and car radiators burst when the water inside them expands as it freezes.

> For any liquid at any temperature,
>
> $$1 \text{ cm}^3 = 1 \text{ mL}$$
>
> The statement that 1.00 cm^3 of liquid water has a mass of 1.00 g is only true for water. Strictly speaking, it is valid only at 4 °C, but is a useful approximation for water at room temperature.

Consider This 5.15   Oil and Water

Relative densities have practical consequences for water when it mixes (but does not dissolve) with other substances in the environment. Crude oil has a density of approximately 0.8 g/mL; salt water has a density more than 1.0 g/mL. What are the implications for cleaning up oil spills in the ocean?

Joules and calories, units of heat energy, were defined in Section 4.1.

Global warming was discussed in Chapter 3.

Finally, we want to examine another of water's unusual properties, namely, its uncommonly high capacity to absorb and release heat. This property is expressed by **specific heat,** the quantity of heat energy that must be absorbed to increase the temperature of 1 g of a substance by 1 °C. The specific heat of liquid water is 1.00 cal/g · °C, which means that 1 cal of energy will raise the temperature of 1 g of liquid water by 1 °C. In fact, the calorie was originally defined in this manner. Conversely, when the temperature of 1 g of liquid water falls 1 °C, 1 cal of heat is given off. The specific heat of water also can be expressed as 4.18 J/g · °C. Water has one of the highest specific heats of any known liquid. Because of this, it is an exceptional coolant used to carry away excess heat in chemical industry, power plants, and the human body. Most other compounds have significantly lower specific heats.

On a global scale, water's high specific heat helps determine climate. By absorbing vast quantities of heat, the oceans and the droplets of water in clouds help mediate global warming. These processes are complicated and thus it is difficult to create accurate global warming models. We do know that heat is absorbed when water evaporates from seas, rivers, and lakes. Heat also is released when water condenses as rain or snow. Since water has a higher capacity to store heat than does earth, when the weather turns cold, the ground cools more quickly. Water retains more heat and is able to provide more warmth for a longer time to the areas bordering it. Such properties should be familiar to anyone who has ever lived in a northern coastal region.

The unusually high specific heat of water is a consequence of strong hydrogen bonding and the resultant degree of order that exists in the liquid. When molecules are strongly attracted to one another, a good deal of energy is required to overcome these intermolecular forces and enable the molecules to move more freely. Such is the case with water. On the other hand, intermolecular forces are much weaker in nonhydrogen-bonded liquids such as the hydrocarbon benzene (C_6H_6) and the forces are much easier to overcome. Consequently, the specific heat for benzene is only 0.406 cal/g · °C, less than half that of water.

5.7   A Close Look at Solutes

As discussed in Section 5.3, water is an excellent solvent for a wide variety of substances. A lot of chemistry occurs in aqueous solution so it's important to have an understanding of the substances that dissolve in water and how that process happens.

The behavior of aqueous solutions of sugar and salt illustrate two main classes of aqueous solutions. A significant difference between the two can be demonstrated experimentally with a **conductivity meter,** an apparatus that produces a signal to indicate that electricity is being conducted (Figure 5.12). Two wires attach a battery to a lightbulb. As

(a)   (b)   (c)

Figure 5.12
Conductivity experiments.
(a) Distilled water (nonconducting). **(b)** Sugar dissolved in water (nonconducting). **(c)** Salt dissolved in water (conducting).

long as the two separate wires do not touch, the electrical circuit is not completed. If the separated wires are placed into distilled water or a solution of sugar in distilled water, the bulb will not be illuminated. However, if the separate wires are placed into an aqueous solution of salt, the bulb illuminates. Perhaps the light has also gone on in the mind of the experimenter! Pure water or a solution of sugar in water do not conduct electricity and therefore do not complete the electrical circuit; the light does not glow. Sugar is a **nonelectrolyte,** a nonconducting solute when in aqueous solutions. But an aqueous solution of common table salt, NaCl, conducts electricity, and the lightbulb lights. Sodium chloride and other conducting solutes are classified as **electrolytes,** conducting solutes in aqueous solution.

What makes salt in solution behave any differently from sugar in solution or pure water? The observed flow of electric current through a solution involves the transport of electric charge. Therefore, the fact that aqueous NaCl solutions conduct electricity suggests they contain some charged species capable of moving electrons through the solution. When solid NaCl dissolves in water, it separates into $Na^+(aq)$ and $Cl^-(aq)$. An **ion** is an atom or group of atoms that has acquired a net electric charge as a result of gaining or losing one or more electrons. The term is derived from the Greek for "wanderer." Na^+ is an example of a **cation,** a positively charged ion. Cl^- is an example of an **anion,** a negatively charged ion. No such separation occurs with covalently bonded sugar or water molecules, making these liquids unable to carry electric charge. Although many hydrogen bonds are present in both the water and sugar solutions, even polar covalent bonds do not have enough charge separation to allow the transport of electric charge.

It may be surprising to learn that Na^+ and Cl^- ions exist both in crystals of salt (such as those in a saltshaker) and in an aqueous solution of salt. Solid sodium chloride is a three-dimensional cubic arrangement of sodium and chloride ions occupying alternating positions. An **ionic bond** is a chemical bond formed by the attraction between oppositely charged ions. In the case of NaCl, ionic bonds hold the crystal together; there are no covalently bonded atoms, only positively charged cations and negatively charged anions held together by electrical attractions. **Ionic compounds** are made up of electrically charged ions that are present in fixed proportions and are arranged in a regular, geometric pattern. In the case of NaCl, each Na^+ ion is surrounded by six oppositely charged Cl^- ions. Likewise, each Cl^- ion is surrounded by six positively charged Na^+ ions. A single, tiny crystal of sodium chloride consists of many billions of Na^+ and Cl^- ions in the arrangement shown in Figure 5.13.

We have described the structure and some of the properties of ionic compounds, but have not explained *why* certain atoms lose or gain electrons to form ions. Not surprisingly, the answer involves the distribution of electrons within atoms. Recall that a sodium atom (atomic number 11) has 11 electrons and 11 protons. Sodium, like all metals in Group 1A, has one valence electron. This electron is rather loosely attracted to the nucleus and can be easily lost. When this happens, the Na atom becomes a Na^+ ion.

$$Na \longrightarrow Na^+ + e^- \qquad [5.1]$$

The Na^+ ion contains 11 protons but only 10 electrons, hence it has a 1+ charge. These 10 electrons configured similarly to those in the inert element neon (Ne), thus Na^+ has a complete octet. Table 5.4 shows the comparison.

Ions in aqueous solution are indicated with *(aq)*.

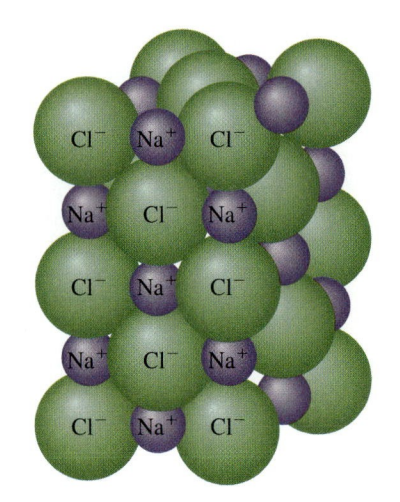

Figure 5.13

The arrangement of Na^+ and Cl^- ions in a crystal of sodium chloride.

Valence electrons were discussed in Section 2.2.

Table **5.4**	**Electronic Bookkeeping for Cation Formation**	
Sodium Atom	**Sodium Ion**	**Neon Atom**
Na	Na^+	Ne
11 protons	11 protons	10 protons
11 electrons	10 electrons	10 electrons
Net charge: zero	*Net* charge: 1+	*Net* charge: zero

Table 5.5	Electronic Bookkeeping for Anion Formation	
Chlorine Atom	**Chloride Ion**	**Argon Atom**
Cl	Cl$^-$	Ar
17 protons	17 protons	18 protons
17 electrons	18 electrons	18 electrons
Net charge: zero	*Net* charge: 1$-$	*Net* charge: zero

A Na$^+$ ion, like a Ne atom, has two inner electrons and eight outer electrons. We may generalize by saying that **metals** tend to form cations by losing their valence electrons. Metals are the largest category of elements and are found in the left and middle blocks of the periodic table.

By contrast, a chlorine atom has a tendency to gain an electron. Recall that a chlorine atom (atomic number 17) has 17 electrons and 17 protons. Chlorine, like all nonmetals in Group 7A, has seven valence electrons. Because of the stability associated with eight outer electrons, it is energetically favorable for a Cl atom to acquire an extra electron, equation 5.2 shows this change.

$$Cl + e^- \longrightarrow Cl^- \tag{5.2}$$

The chloride ion (Cl$^-$) has 18 electrons and 17 protons; thus the net charge is 1$-$ (Table 5.5). Because elemental chlorine consists of diatomic Cl$_2$ molecules, we also can write this gain of electrons in the following fashion.

$$Cl_2 + 2\,e^- \longrightarrow 2\,Cl^- \tag{5.3}$$

In general, **nonmetals** are found on the right-hand side of the periodic table and gain electrons to form anions. The elements in Group 8A, the noble gases, are exceptions. Some Group 8A elements, such as helium and neon, do not combine chemically with any elements.

When sodium metal and chlorine gas react, electrons are transferred from sodium atoms to chlorine atoms with the release of a considerable amount of energy. The result is the aggregate of Na$^+$ ions and Cl$^-$ ions known as sodium chloride. In the formation of an ionic compound such as sodium chloride, the electrons are actually transferred from one atom to another, not simply shared as they would be in a covalent compound.

Is there evidence for electrically charged ions in pure sodium chloride? Experimental tests show that crystals of sodium chloride do not conduct electricity, but when these crystals are melted, the resulting liquid conducts electricity. This provides evidence that Na$^+$ and Cl$^-$ ions from the solid NaCl also exist in the liquid state, without the presence of water. Crystals of NaCl and other ionic compounds are hard yet brittle. When hit sharply, they shatter rather than being flattened. This suggests the existence of strong forces that extend throughout the ionic crystal. Strictly speaking, there is no such thing as a specific, localized "ionic bond" analogous to covalent bonds in molecules. Rather, generalized ionic bonding holds together a large assembly of ions.

Other elements form ions and ionic compounds, not just sodium and chlorine. Electron transfer to form cations and anions is likely to occur between metallic elements and nonmetallic elements, respectively. Sodium, lithium, magnesium, and other metallic elements have a strong tendency to give up electrons and form positive ions. On the other hand (or the other side of the periodic table), chlorine, fluorine, oxygen, and other nonmetals have a strong attraction for electrons and readily gain them to form negative ions. Potassium chloride (KCl) and sodium iodide (NaI) are two of many such compounds. Because ordinary table salt (NaCl) is such an important example of an ionic compound, chemists frequently refer to other ionic compounds simply as "salts," meaning ionic crystalline solids.

Refer to Section 1.8 for information about naming covalent compounds.

Your Turn 5.16 **Predicting Ionic Charge**

Predict the ion that will form from each of these atoms. Draw the Lewis structure for both the atom and the ion, clearly labeling the charge on the ion.
Hint: Use the periodic table to find the number of outer electrons. Then determine how many electrons must be lost or gained to achieve stability with an octet of electrons.

a. Br   **b.** Mg   **c.** O   **d.** Al

Answer

a. Bromine is in Group 7A and gains one electron to form a stable ion with a charge of $1-$, just as was the case for chlorine. These are the Lewis structures.

$$:\!\overset{..}{\underset{..}{Br}}\!\cdot \quad \text{and} \quad \left[:\!\overset{..}{\underset{..}{Br}}\!:\right]^{-}$$

5.8 Names and Formulas of Ionic Compounds

In this section, we will work on the "vocabulary" you need in order to work with ionic compounds. As we pointed out in Chapter 1, chemical symbols are the alphabet of chemistry, and chemical formulas are the words. Earlier, we helped you to "speak chemistry"; that is, to correctly use chemical formulas and names for the substances in the air you breathe. Now we'll do the same for the substances in the water you drink. Again we follow the "need-to-know" philosophy, helping you learn what you need to understand the topic at hand.

Let's begin with the ionic compound formed from the elements calcium and chlorine: $CaCl_2$. The explanation for the 1:2 ratio of Ca to Cl lies in the charges of the two ions. Calcium, a member of Group 2A, readily loses its two outer electrons to form Ca^{2+}.

$$Ca \longrightarrow Ca^{2+} + 2\,e^{-} \qquad\qquad [5.4]$$

Chlorine, as we saw in equation 5.2, gains an outer electron to form Cl^{-}. Two Cl^{-} ions are required to electrically balance each Ca^{2+} ion. Hence, the formula for this compound is $CaCl_2$. In an ionic compound, the sum of the positive charges equals the sum of the negative charges.

The logic is the same with MgO and Al_2O_3, two other ionic compounds. These both contain oxygen, but in different ratios. Recall that oxygen, Group 6A, has six outer electrons. Thus a neutral oxygen atom can gain two electrons to form the O^{2-} ion. The magnesium atom loses two electrons to form Mg^{2+}. These two ions must then combine in a 1:1 ratio so the overall charge will be zero; the chemical formula is MgO. Note that although the charge *always* must be written on an individual ion, we omit the charges in the chemical formulas of ionic compounds. Thus, it is *not* correct to write the chemical formula as $Mg^{2+}O^{2-}$. The charges are implied by the chemical formula.

Here is another example. Armed with the knowledge that aluminum tends to lose three electrons to form the Al^{3+} ion, you can write the chemical formula of the ionic compound formed from Al^{3+} and O^{2-} ions as Al_2O_3. Here, a 2:3 ratio of ions is needed so that the overall electric charge on the compound will be zero. Again, it is *not* correct to write the chemical formula as $Al_2^{3+}O_3^{2-}$.

Earlier in the chapter, we referred to several ionic compounds by their names, including sodium chloride, sodium iodide, and potassium chloride. Observe the pattern: name the cation first, then the anion, modified to end in the suffix *-ide*. Thus, the name of $CaCl_2$ is calcium chloride. Similarly, NaI is sodium iodide and KCl is potassium chloride. Always leave a space between the two names.

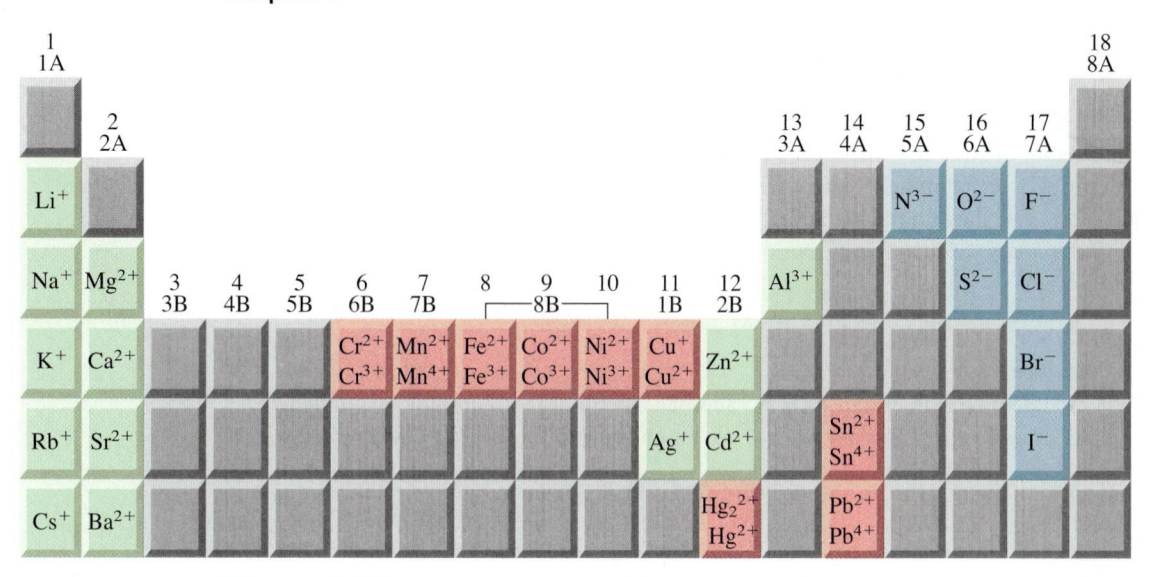

Figure 5.14

Common ions formed from their elements. Ions in green (cations) or blue (anions) have only one charge. Ions in red (cations) have more than one possible ionic charge.

The elements presented thus far formed only one type of ion. Group 1A and 2A elements only form 1+ and 2+ ions, respectively. The halogens form only 1− ions. Lithium bromide is LiBr. The ratio of 1:1 is understood because lithium only forms Li^+ and bromine only forms Br^-. There is no need to call it monolithium monobromide. $AlCl_3$ is aluminum chloride, not aluminum trichloride. Aluminum *only* forms the Al^{3+} ion, and the ratio of 1:3 is understood and so does not need to be stated. Note that the prefixes *mono-, di-, tri-,* and *tetra-* are *not* used when naming ionic compounds such as these.

But some elements do form more than one ion, as you can see in Figure 5.14. Prefixes still are not used, but rather the charge on the ion must be specified using a Roman numeral. Take copper for example. If your instructor asks you to head down to the stockroom and grab some copper oxide, what will you do? You will ask if what is wanted is copper(I) oxide or copper(II) oxide, right? Similarly, iron can form different oxides. The two possible combinations are FeO (formed from Fe^{2+}) and Fe_2O_3 (formed from Fe^{3+}, commonly called rust). The names for FeO and Fe_2O_3 are iron(II) oxide and iron(III) oxide, respectively. Note the space after, but not before the parenthesis enclosing the Roman numeral.

Again compare. The name $CuCl_2$ is copper(II) chloride, but the name of $CaCl_2$ is calcium chloride. Calcium only forms one ion (Ca^{2+}), whereas copper can form two. The activity that follows offers an opportunity to practice.

Your Turn 5.17   Ionic Compounds

Write formulas for the ionic compound (in some cases more than one) that will form from each pair of elements. Also name each compound.

a. Ca and S   **b.** F and K   **c.** Mn and O   **d.** Cl and Al   **e.** Co and Br

Answer

e. Co forms both the Co^{2+} and the Co^{3+} ion. Br forms the Br^- ion. The chemical formula could be $CoBr_2$, cobalt(II) bromide, or $CoBr_3$, cobalt(III) bromide.

Table 5.6	Common Polyatomic Ions		
Name	**Formula**	**Name**	**Formula**
acetate	$C_2H_3O_2^-$	nitrite	NO_2^-
bicarbonate*	HCO_3^-	phosphate	PO_4^{3-}
carbonate	CO_3^{2-}	sulfate	SO_4^{2-}
hydroxide	OH^-	sulfite	SO_3^{2-}
hypochlorite	ClO^-	ammonium	NH_4^+
nitrate	NO_3^-		

*Also called the hydrogen carbonate ion.

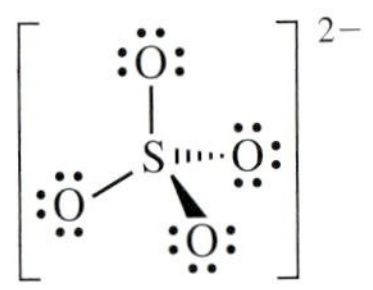

Ionic compounds may contain **polyatomic ions,** ions that are made up of two or more atoms covalently bound together. An example is the sulfate ion, SO_4^{2-}, with four oxygen atoms covalently bonded to a central sulfur atom. The Lewis structure shown in Figure 5.15 reveals that there are 32 electrons, 2 more than the 30 valence electrons provided by one S atom (6) and four O atoms ($4 \times 6 = 24$). The "extra" two electrons give the sulfate ion a charge of $2-$.

Table 5.6 lists common polyatomic ions. Most are anions, but polyatomic cations also are possible, as in the case of the ammonium ion, NH_4^+. Note that some elements (carbon, sulfur, and nitrogen) form more than one polyatomic anion with oxygen.

The rules for naming ionic compounds containing polyatomic ions are similar to those for ionic compounds of two elements. Consider, for example, aluminum sulfate, an ionic compound used in water purification. The compound is formed from Al^{3+} and SO_4^{2-} ions. As is true for all ionic compounds, the name of the cation is given first. Note that the name of the anion is sulfate and does not end in *-ide*. Rather, use the names in Table 5.6. Also note that prefixes such as *di-* and *tri-* are not used in the name and there are no Roman numerals unless the cation has more than one possible charge.

The chemical formulas for ionic compounds containing polyatomic ions are also based on balancing electric charges. Charges on the positive ions must equal those on the negative ions. So for aluminum sulfate, the Al^{3+} and SO_4^{2-} ions must be in the ratio of 2:3, and Table 5.7 shows this ratio. When you see $Al_2(SO_4)_3$, "read" this chemical formula as a compound containing the two types of ions: aluminum and sulfate.

The parentheses in $Al_2(SO_4)_3$ can help you. The subscript 3 applies to the *entire* SO_4^{2-} ion that is enclosed in parentheses. Accordingly, you are to "read" this as three sulfate ions, not as one larger unit composed of three sulfate ions. Similarly, in the ionic compound ammonium sulfide (see Table 5.7), the NH_4^+ ion is enclosed in parentheses. The subscript of 2 indicates that there are two ammonium ions for each sulfide ion.

In some cases, though, the polyatomic ions will *not* be enclosed in parentheses. Table 5.7 shows two examples. The PO_4^{3-} ion in aluminum phosphate has no parentheses; similarly, the NH_4^+ ion in ammonium chloride has no parentheses. Nonetheless, you still have to "read" the chemical formula of $AlPO_4$ as containing the phosphate ion, and you have to "read" NH_4Cl as containing the ammonium ion. Parentheses are omitted when the subscript of the polyatomic ion is 1.

Figure 5.15

Structure of the sulfate ion, SO_4^{2-}.

Table 5.7	Formulas of Ionic Compounds with Polyatomic Ions			
Chemical Formula	$Al_2(SO_4)_3$	$(NH_4)_2S$	$AlPO_4$	NH_4Cl
Cation(s)	Al^{3+} Al^{3+}	NH_4^+ NH_4^+	Al^{3+}	NH_4^+
Anion(s)	SO_4^{2-} SO_4^{2-} SO_4^{2-}	S^{2-}	PO_4^{3-}	Cl^-

The activities that follow will give you practice with the names and chemical formulas for ionic compounds that contain polyatomic ions.

Your Turn 5.18   Polyatomic Ions I

Write the chemical formula for the ionic compound formed from each pair of ions.

a. Na^+ and SO_4^{2-} b. Mg^{2+} and OH^-
c. Al^{3+} and $C_2H_3O_2^-$ d. OH^- and NH_4^+

Answers
a. Na_2SO_4 b. $Mg(OH)_2$

Your Turn 5.19   Polyatomic Ions II

Name each of these compounds:

a. KNO_3 b. $(NH_4)_2SO_4$ c. $NaHCO_3$
d. $CaCO_3$ e. $Mg_3(PO_4)_2$

Answers
a. potassium nitrate   b. ammonium sulfate

Your Turn 5.20   Polyatomic Ions III

Write the formula for each of these compounds.

a. calcium hypochlorite (used in bleaches)
b. lithium carbonate (treatment of bipolar disorders)
c. potassium nitrate (matches and fireworks)
d. barium sulfate (medical X-rays)

Answer
a. $Ca(ClO)_2$. Two hypochlorite ions (ClO^-) are needed to equal the charge on one calcium ion (Ca^{2+}).

5.9   Water Solutions of Ionic Compounds

We are now in a position to understand one of the most important properties of ionic compounds, namely, why many are quite soluble in water. Recall from Section 5.6 that water molecules are polar. When a solid sample of an ionic compound is placed in water, the polar H_2O molecules are attracted to the individual ions. The partial negative charge on the oxygen atom of a water molecule is attracted to the positively charged cations of the solute. At the same time, hydrogen atoms in H_2O, with their partial positive charges, are attracted to the negatively charged anions of the solute. Thus, the ions are separated and then surrounded by water molecules, as the anion–cation attraction in the solid is diminished. Equation 5.5 and Figure 5.16 represent this process for sodium chloride and water.

$$NaCl(s) \xrightarrow{H_2O} Na^+(aq) + Cl^-(aq) \qquad [5.5]$$

When compounds containing polyatomic ions dissolve in water, the polyatomic ions remain intact. For example, when sodium sulfate dissolves in water, the sodium ions and sulfate ions simply separate.

$$Na_2SO_4(s) \xrightarrow{H_2O} 2\,Na^+(aq) + SO_4^{2-}(aq) \qquad [5.6]$$

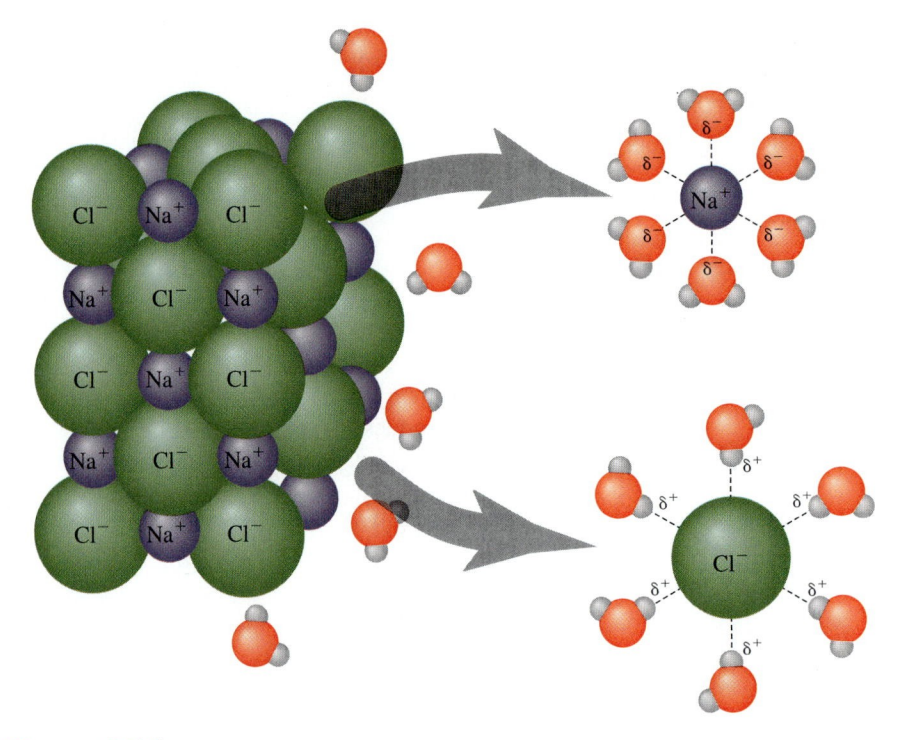

Figure 5.16

Dissolving sodium chloride in water.

What has just been described for sodium chloride and sodium sulfate dissolving in water is true for many other ionic compounds. Indeed, this behavior is so common that the chemistry of ionic compounds is largely that of their behavior in aqueous solutions. Conversely, almost all naturally occurring water samples contain various amounts of ions. Even our body fluids contain significant concentrations of ions.

Consider This 5.21   **Electricity and Water Don't Mix**

Small electric appliances, such as a hair dryer or curling iron, carry a warning label prominently advising the consumer of the hazard associated with using the appliance near water. Why is this a problem if water does not conduct electricity? What is the best course of action if a plugged-in hair dryer does accidentally fall into a sink full of water?

In principle, the dissolving process in water, as just described, ought to be true for any ionic compound. Indeed, many ionic compounds are highly soluble in water. But some are at best only slightly soluble; others have extremely low solubilities. The reasons for this range of behavior involve the sizes and charges of the ions, how strongly the ions attract one another, and how strongly the ions are attracted to water molecules. A few generalizations are quite useful for predicting the solubility of common ionic compounds (Table 5.8).

You can use Table 5.8 to determine the solubility (or insolubility) of many compounds. For example, calcium nitrate, $Ca(NO_3)_2$, is soluble in water as are all compounds containing the nitrate ion. Calcium carbonate, $CaCO_3$, is insoluble as most carbonates are, and calcium is not one of the exceptions for carbonates. By similar reasoning, copper(II) hydroxide, $Cu(OH)_2$, is insoluble, but copper(II) sulfate, $CuSO_4$, is soluble.

Table 5.8	Water Solubility of Ionic Compounds		
Ions	Solubility of Compounds	Solubility Exceptions	Examples
sodium, potassium, and ammonium	All soluble	None	$NaNO_3$ is soluble KBr is soluble
nitrates	All soluble	None	$LiNO_3$ is soluble $Mg(NO_3)_2$ is soluble
chlorides	Most soluble	Silver and some mercury chlorides	$MgCl_2$ is soluble AgCl is insoluble
sulfates	Most soluble	Strontium, barium, and lead sulfate	K_2SO_4 is soluble $BaSO_4$ is insoluble
carbonates	Mostly insoluble*	Group IA and NH_4^+ carbonates are soluble	Na_2CO_3 is soluble $CaCO_3$ is insoluble
hydroxides and sulfides	Mostly insoluble*	Group IA and NH_4^+ hydroxides and sulfides are soluble	KOH is soluble $Al(OH)_3$ is insoluble

*Insoluble means that the compounds have extremely low solubilities in water (less than 0.01 M). All ionic compounds have at least a very small solubility in water.

Your Turn 5.22   Solubility of Ionic Compounds

From the solubility generalizations in Table 5.8, which of these compounds are soluble?

a. ammonium nitrate, NH_4NO_3 (used in fertilizers)
b. sodium sulfate, Na_2SO_4 (an additive in detergents)
c. mercury(II) sulfide, HgS (the mineral cinnabar)
d. aluminum hydroxide, $Al(OH)_3$ (used in some antacid tablets)

Answer
a. Soluble. All ammonium compounds and all nitrates are soluble.

The landmasses on Earth are made up largely of minerals consisting of ionic compounds that have extremely low solubility in water. If that were not the case, most would have dissolved long ago. Table 5.9 summarizes some environmental consequences of the differing solubility of minerals and other substances in water.

5.10 Covalent Compounds and Their Solutions

From the previous discussion, you might get the impression that only ionic compounds dissolve in water. But, other kinds of compounds dissolve as well. Common experience tells us that ordinary table sugar dissolves readily in water. But table sugar, chemically known as sucrose, contains no ions; it is a covalent compound. Like water, carbon dioxide, chlorofluorocarbons, and many of the other compounds you have been reading about, table sugar molecules consist of covalently bonded atoms. The formula for sucrose is $C_{12}H_{22}O_{11}$, and it exists as individual covalently bonded molecules consisting of 45 atoms (Figure 5.17).

Table **5.9**	**Environmental Consequences of Solubility**	
Source	**Ions**	**Solubility and Consequences**
Salt deposits	sodium and potassium halides*	These salts are soluble. Over time, they dissolve from the land and wash into the sea. Thus, oceans are salty and sea water cannot be used for drinking without expensive purification.
Agricultural fertilizers	nitrates	All nitrates are soluble. The runoff from fertilized fields carries nitrates into surface and groundwater. Nitrates are toxic, especially for infants.
Metal ores	sulfides and oxides	Most sulfides and oxides are insoluble. Minerals containing iron, copper, and zinc are often sulfides and oxides. If these minerals had been soluble in water, they would have washed out to sea long ago.
Mining waste	mercury, lead	Most mercury and lead compounds are classed as insoluble. However, they are leached slowly from waste piles into rivers and lakes where they contaminate water supplies.

*Halides are the anions in Group 7A, such as Cl^- and I^-.

When sugar dissolves in water, its molecules become uniformly dispersed among the H_2O molecules. As in all true solutions, the mixing is at the most fundamental level of the solute and solvent—the molecular or ionic level. The $C_{12}H_{22}O_{11}$ molecules remain intact and do not separate into ions. Evidence for this is the fact that aqueous sucrose solutions do not conduct electricity, as was shown in Figure 5.12. However, the sugar molecules do interact with the water molecules. In fact, solubility is always promoted when a net attraction exists between the solvent molecules and the solute molecules or ions. This suggests a general solubility rule: *Like dissolves like.* Compounds with similar chemical composition and molecular structure tend to form solutions with each other. The intermolecular attractive forces between similar molecules are high, promoting solubility. Dissimilar compounds do not dissolve in each other.

"Like dissolves like" is a useful generalization.

Consider, for example, three familiar covalently bonded compounds, all of which are highly soluble in water: sucrose; ethylene glycol (the main ingredient in antifreeze); and ethanol (ethyl alcohol, the "grain alcohol" found in alcoholic beverages). Like all alcohols, they contain one or more —OH groups (Figures 5.18 and 5.19).

We start with the simplest, ethanol, C_2H_5OH. The —OH group of a C_2H_5OH molecule can form hydrogen bonds with H_2O molecules (see Figure 5.19). This hydrogen bonding is the reason that water and ethanol have a great affinity for each other, a conclusion consistent with the fact that they form solutions in all proportions. Ethylene glycol is also an alcohol but has two —OH groups available for hydrogen bonding with H_2O. Therefore, ethylene glycol is highly water-soluble, a necessary property for an antifreeze ingredient.

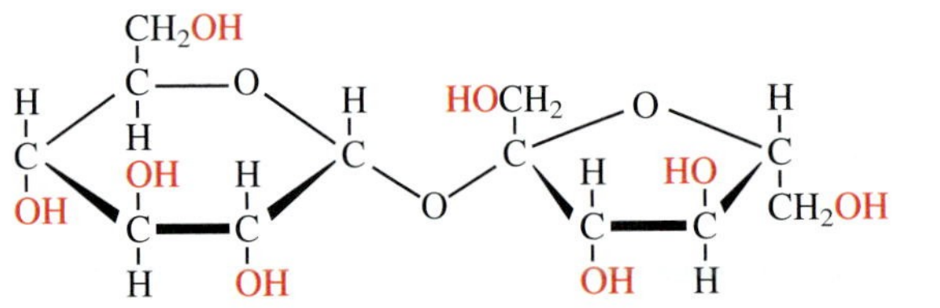

Figure 5.17

Molecular structure of sucrose. The —OH groups are shown in red.

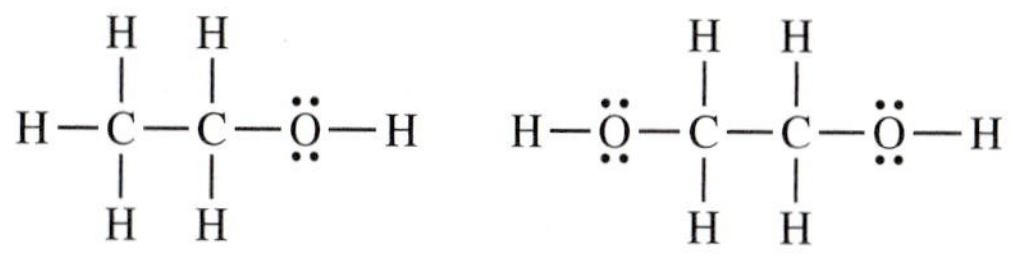

H—C—C—O—H   H—O—C—C—O—H

Ethanol   Ethylene glycol

Figure 5.18

Lewis structures of ethanol and ethylene glycol.

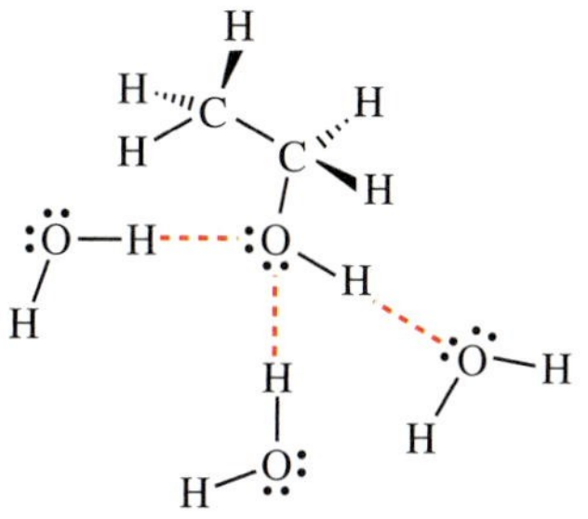

— covalent bond
---- hydrogen bond

Figure 5.19

Hydrogen bonding between an ethanol molecule and water molecules.

Your Turn 5.23   Hydrogen Bonding—Ethylene Glycol and Water

Make a sketch to show hydrogen bonding between ethylene glycol and water.

Finally, we consider sucrose, the compound that introduced this section. Examination of its structure (see Figure 5.17) shows that the sucrose molecule contains eight —OH groups and three additional oxygen atoms that can participate in hydrogen bonding. These help explain the high solubility of sugar in water.

Consider This 5.24   Three-Dimensional Representations of Molecules

Three-dimensional representations of molecules can be viewed on the Web using several different molecular-modeling programs. Use a program available to you to view ethanol, ethylene glycol, and sucrose. Use these molecular representations to identify the places in each compound where hydrogen bonding occurs. Has your mental picture of these molecules changed after seeing these 3-D representations? Explain.

"Like dissolves like" is a useful generalization. Implied is the fact that covalently bonded compounds that differ in composition and molecular structure do not attract each other strongly. It has often been observed that "oil and water don't mix." They don't mix because they are structurally very different. Water molecules are highly polar, whereas oil consists of nonpolar hydrocarbon molecules. When in contact, these molecules stick with their own, like rain water splattered across oil-covered pavement (Figure 5.20). The

Figure 5.20

Oil and water do not dissolve in each other.

water molecules bead up together in small puddles along the oily surface. You may have also seen the same effect when water hits a freshly waxed automobile hood, as wax also consists of nonpolar hydrocarbons. But greasy, nonpolar compounds generally dissolve readily in hydrocarbons or chlorinated hydrocarbons. For this reason, the latter have often been used in dry-cleaning solvents.

The tendency of nonpolar compounds to mix with other nonpolar substances affects how fish and animals store certain highly toxic substances such as PCBs (polychlorinated biphenyls) or the pesticide DDT. PCB and DDT molecules are nonpolar, and so when fish absorb them from water, the molecules are stored in body fat (also nonpolar) rather than in the blood (a highly polar aqueous solution).

Solvents used to dry-clean clothes are usually chlorinated compounds such as tetrachloroethylene, $Cl_2C{=}CCl_2$, also known as "perc" (perchloroethylene). Perc is a human **carcinogen**, a compound capable of causing cancer. These materials also have serious environmental consequences. Dr. Joe DeSimone of the University of North Carolina–Chapel Hill has discovered a substitute for chlorinated compounds by synthesizing cleaning detergents that work in liquid carbon dioxide. Key to the process are detergents, molecules designed so that one end of the molecule is soluble in nonpolar substances like grease and oil stains, while the other end dissolves in the liquid CO_2. The new method recycles carbon dioxide produced as a waste product from industrial processes. Replacing large volumes of perc by using recycled CO_2 reduces perc's negative effect on the workplace and the environment. The breakthrough process is paving the way for designing replacements for conventional halogenated solvents currently used in manufacturing and industries making coatings. For his work, Professor DeSimone received the 1997 Presidential Green Chemistry Challenge Award.

> PCBs are organochlorine chemicals that were widely used as cooling agents in electrical transformers. Careless disposal has caused serious environmental problems.

5.11 Protecting Our Drinking Water: Federal Legislation

We can now apply to drinking water what we know about the structure and properties of pure water and aqueous solutions. What dissolves in drinking water determines its quality and the potential for adverse health effects. Keeping public water supplies safe has long been recognized as an important public health issue. In 1974, the U.S. Congress passed the Safe Drinking Water Act (SDWA) in response to public concern about harmful substances in drinking water supplies. The aim of the SDWA, as amended in 1996, is to provide public health protection to all Americans who get their water from community water supplies (over 250 million people). Contaminants that may be health risks are regulated by EPA as required by the SDWA. The EPA sets legal limits for such contaminants according to their levels of adverse risk (Table 5.10). These limits also take into account the practical realities the water utilities face in trying to remove the contaminants by using available technology.

For each contaminant, the EPA has established a maximum contaminant level goal (MCLG). The **MCLG** is the maximum level of a contaminant in drinking water at which

Table 5.10	MCLGs and MCLs (in ppm) for Drinking Water	
Pollutant	**MCLG**	**MCL**
cadmium (Cd^{2+})	0.005	0.005
chromium (Cr^{3+}, CrO_4^{2-})	0.1	0.1
lead (Pb^{2+})	0	0.015
mercury (Hg^{2+})	0.002	0.002
nitrate (NO_3^-)	10	10
benzene (C_6H_6)	0	0.005
trihalomethanes ($CHCl_3$, etc.)	0	0.080

no known or anticipated adverse effect on the health of persons would occur. They are considered to be the level, expressed in parts per million or parts per billion, at which a person weighing 70 kg (154 lb) could drink 2 L (about 2 qt) of water containing the contaminant every day for 70 years without suffering any ill effects. Each MCLG includes built-in safety factors accounting for uncertainties in collection data and for how different people might react to each contaminant. An MCLG is not a legal limit with which water systems must comply; it is a *goal,* based on considerations of human health. For known carcinogens, the EPA has set the health goal at zero, under the assumption that *any* exposure to the substance could present a cancer risk.

Before any regulatory action is taken against a water utility, the concentration of an impurity must exceed the maximum contaminant level (MCL). The **MCL** sets the legal limit for the concentration of a contaminant. It is expressed in parts per million or parts per billion. The EPA sets legal limits for each impurity as close to the MCLG as possible, keeping in mind the practical realities of technical and financial barriers that may make it difficult to achieve the goals. Except for contaminants regulated as carcinogens, for which the MCLG is zero, most legal limits and health goals are the same. Even when they are less strict than the MCLGs, the MCLs provide substantial public health protection.

Consider This 5.25   Understanding MCLGs and MCLs

Most people are unfamiliar with these terms from the Safe Drinking Water Act. Assume you are making a presentation to explain what these acronyms mean and how the information helps to safeguard our drinking water. Prepare a short outline of what you will say. Be prepared to answer questions from the audience, particularly dealing with why MCLs are not set to zero for all carcinogens.

Because of improved detection and quantitative analytical methods, the number of regulated contaminants in drinking water increases each time Congress updates the legislation. Lower limits for MCL values have been established as more accurate risk information has become available. Currently, more than 80 contaminants are regulated; they fall into several major categories: metals (for instance, cadmium, chromium, copper, mercury, and lead), a few nonmetallic elements (such as, fluorine and arsenic), pesticides, industrial solvents, compounds associated with plastics manufacturing, and radioactive materials. Depending on the particular contaminant, MCLs vary from around 10 ppm to less than 1 ppb. Some contaminants interfere with liver or kidney function. Others can affect the nervous system if ingested over a long period at levels consistently above the legal limit (MCL). Pregnant women and infants are at particular risk for some contaminants because of their effects on a developing fetus or the digestive system of an infant.

In addition to contaminants that can pose chronic health problems, other substances in drinking water present acute health risks. For example, nitrate (NO_3^-) and nitrite (NO_2^-) ions limit the blood's ability to carry oxygen. Even when consumed in tiny doses, these ions cause immediate health effects for infants. Therefore, the EPA limit for nitrate and nitrite ions in drinking water specifically protects infants. Another acute health risk is biological, not chemical—from bacteria, viruses, and other microorganisms, including *Cryptosporidium* and *Giardia*. News media warnings announcing a "boil-water emergency" are typically the result of a "total coliform" violation. Coliforms are a broad class of bacteria, most of which are harmless, that live in the digestive tracts of humans and other animals. The presence of high coliform concentration in water usually indicates that the water-treatment or distribution system is not working properly. Diarrhea, cramps, nausea, and vomiting, the symptoms of coliform-related illness, are not serious for a healthy adult, but can be life-threatening for the very young, the elderly, or those with weakened immune systems.

Consider This 5.26   A Drink of Water—What's in It?

Table 5.10 is merely a starting point for the wealth of information available about possible pollutants in drinking water. The EPA Office of Ground Water and Drinking Water, has a consumer fact sheet on dozens of pollutants. A consumer version and a technical version are available, and the latter is recommended. Look up a pollutant listed in Table 5.10 to find out how it gets into the water supply and how you would know if it were in your drinking water. Is your state listed as one of the top states that release the contaminant?

Hint: Arsenic, cadmium, lead, chromium, mercury, and nitrate/nitrite ions are found under the section on Inorganic Chemicals. Benzene is listed under Volatile Organic Chemicals. No trihalomethane (THM) such as $CHCl_3$ is currently listed, but you can find other chlorinated compounds such as CCl_4 and CH_2Cl_2.

Consider This 5.27   Water Emergency Relief

In 2006, at the 231st national meeting of the American Chemical Society in Atlanta, Georgia, a great buzz of excitement surrounded a new, tiny water purification system. Called PUR Purifier of Water, the chemical systems are made by Proctor and Gamble. Producing clean water that rivals what you would get from a modern treatment facility, 40 million of the small packets were distributed worldwide for sustained water remediation and emergency relief. Able to remove toxic metals, pesticides, and deadly pathogens, the chemical filters the size of a ketchup packet are also inexpensive. Use the Web to see if the water purifiers have lived up to their hype. Write a short report detailing their contents and effectiveness, and describe situations where they have proven useful, if at all.

In addition to the Safe Drinking Water Act, other federal legislation also controls pollution of surface waters, including lakes, rivers, and coastal areas. The Clean Water Act (CWA), passed by Congress in 1972 and amended several times, provided the foundation for dramatic progress in reducing surface water pollution over the past three decades. The CWA establishes limits on the amounts of pollutants that industries can discharge into surface waters, resulting in actions that have removed over a billion pounds of toxic pollution from U.S. waters every year. Improvements in surface water quality have at least two major beneficial effects: They reduce the amount of clean-up needed for public drinking water supplies, and they result in a more healthful natural environment for aquatic organisms. In turn, a more healthful aquatic ecosystem has many indirect benefits for humans. In keeping with the new trend toward green chemistry, industries are finding ways to convert these waste materials into useful products, as well as to initially design processes so that they neither use nor produce substances that degrade water quality.

Consider This 5.28   *Cryptosporidium*

As of January 1, 2002, EPA's surface water-treatment rules require large systems using surface water, or groundwater under the direct influence of surface water, to remove or deactivate 99% of *Cryptosporidium*.

a. What is *Cryptosporidium?*
b. What are the sources of this contaminant in drinking water?
c. What are its potential health effects?
d. What is the LT2 Enhanced Surface Water Treatment Rule, and when did it take effect?
e. Why are some cities actively fighting the EPA on this ruling?

5.12 Treatment of Municipal Drinking Water

Just because a supply of water is large is no guarantee that it is fit to drink. Coleridge's shipwrecked ancient mariner knew this all too well, surrounded as he was by "Water, water everywhere, Nor any drop to drink." So how is water treated to make it potable, that is, fit for human consumption?

The first step in a typical municipal drinking water-treatment plant (Figure 5.21) is to pass the water through a screen that excludes larger objects both natural (fish and sticks) and artificial (tires and beverage cans). The usual next step is to add two chemicals, aluminum sulfate, $Al_2(SO_4)_3$, and calcium hydroxide, $Ca(OH)_2$. These compounds are called flocculating agents and react to form a sticky floc, or gel, of aluminum hydroxide, $Al(OH)_3$, that collects suspended clay and dirt particles on its surface (equation 5.7). The $Al(OH)_3$ gel settles, slowly carrying with it the suspended particles down into a settling tank. Any remaining particles are removed as the water is filtered through coal or gravel and then sand.

$$Al_2(SO_4)_3(aq) + 3\,Ca(OH)_2(aq) \longrightarrow 2\,Al(OH)_3(s) + 3\,CaSO_4(aq) \qquad [5.7]$$

The filtered water is then pumped to the next step—disinfection to kill disease-causing organisms. This is the most crucial one for making drinking water safe. In the United States, this is most commonly done by chlorination. Chlorine is usually added in one of three forms: chlorine gas, Cl_2; sodium hypochlorite, $NaClO$; or calcium hypochlorite, $Ca(ClO)_2$. The antibacterial agent generated in solution by all three substances is hypochlorous acid, $HClO$. The degree of chlorination is adjusted so that a very low concentration of $HClO$, between 0.075 and 0.600 ppm, remains in solution to protect the water against further bacterial contamination as it passes through the pipes to the user. In some parts of the country, sodium fluoride, NaF, is added to the treated water to help protect against tooth decay. This step is discussed in more detail at the end of this section.

Before chlorination was used, thousands died in epidemics spread via polluted water. In a classic study, John Snow was able to trace a mid-1800s cholera epidemic in London to water contaminated with the excretions of victims of the disease. A more contemporary example occurred in Peru in 1991. This cholera epidemic was traced to bacteria in shellfish growing in estuaries polluted with untreated fecal matter. The bacteria found their way into the water supply, where they continued to multiply because of the absence of chlorination.

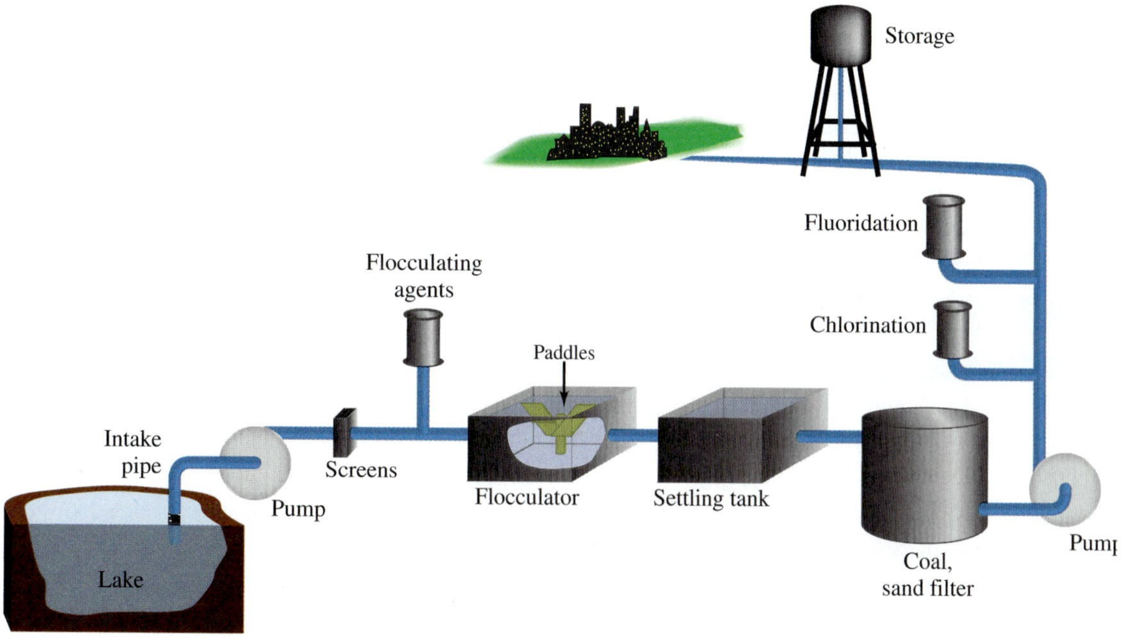

> $NaClO$ is used in Clorox and other brands of laundry bleach. $Ca(ClO)_2$ is commonly used to disinfect swimming pools.

Figure 5.21

Typical municipal water-treatment facility.

Chlorination, however, has drawbacks. The taste and odor of residual chlorine may be objectionable to some and is a reason commonly cited as why people drink bottled water or use home water filters to remove residual chlorine at the tap. A possibly more serious drawback is the reaction of residual chlorine with other substances in the water to form by-products at potentially toxic levels. The most widely publicized of these are trihalomethanes (THMs) such as chloroform, $CHCl_3$.

Many European and a few U.S. cities use ozone (O_3) to disinfect their water supplies. Chapter 1 discussed tropospheric ozone as a serious air pollutant. Chapter 2 described the beneficial effects of the stratospheric ozone layer. In water treatment, the toxic property of ozone is used for a beneficial purpose. The degree of antibacterial action necessary can be achieved with a smaller concentration of ozone than chlorine, and ozone is more effective than chlorine against water-borne viruses. But ozonation is more expensive than chlorination and becomes economical only for large water-treatment plants. An additional major drawback of ozone is that it decomposes quickly and hence does not protect the water from contamination after the water leaves the treatment plant. Consequently, a low dose of chlorine is added to ozonated water as it leaves the treatment plant.

Another disinfection method gaining in popularity is the use of ultraviolet (UV) radiation. In Chapter 2 it was pointed out that UV radiation is dangerous for living species, including bacteria. UV disinfection is very fast; leaves no residual by-products, and is economical for small installations (including rural homes with unsafe well water). Like ozonation, UV disinfection does not protect the water from contamination after it leaves the treatment site unless a low dose of chlorine is added.

Depending on local conditions, one or more additional purification steps may be carried out at the water-treatment facility after disinfection. Sometimes the water is sprayed into the air to remove volatile chemicals that create objectionable odors and taste. If the water is sufficiently acidic to cause problems such as corrosion of pipes or leaching of heavy metals from pipes, calcium oxide (lime) is added to partially neutralize the acid. If little natural fluoride is present in the water supply, many municipalities have added about 1 ppm of fluoride (as NaF) to protect against tooth decay. In water, sodium fluoride, NaF, dissociates into $Na^+(aq)$ and $F^-(aq)$ ions. In teeth, fluoride ions may be incorporated into a calcium compound called fluorapatite, which is more resistant to dental decay than apatite, the usual tooth material. However, there have been reports that the addition of fluoride is not a good idea.

Consider This 5.29	Oops . . . It Happened Again

Community water fluoridation was cited as one of 10 great public health achievements of the twentieth century by the Centers for Disease Control and Prevention (CDC). In 2005 the American Dental Association celebrated "20 years of community water fluoridation." But in 2006 the American Academies of Science published a report claiming that this practice can damage bones and teeth, and that federal standards may put children at risk. Who is to be believed? Use the Web to find sources that may help to settle this controversy. Perhaps the class could use this topic for a lively debate.

5.13 Is There Lead in Your Drinking Water?

It should be clear by now that water is an excellent solvent for many different substances, which may not always be a good thing. Some solutes are highly toxic and are cause for concern. Lead is one of the most serious pollutants that can make its way into drinking water. Concentrations may be low, but still cause serious harm. Lead and most of the metals close to it on the periodic table like cadmium and mercury are toxic. Their cations (Pb^{2+}, Hg^{2+}, and Cd^{2+}) form ionic compounds that are water-soluble and deadly. Because Pb^{2+} is the most common of these three and poses the most serious health risk

The symbol Pb comes from the Latin name for lead, *plumbum,* the origin of our word *plumbing.*

due to its widespread occurrence, let's examine its story in some detail. You may find the source of the problem often comes from your own home. Unless proper precautions are taken, lead from drinking water can have serious long-term health effects, especially tragic for young children.

Consider This 5.30	Lead, Mercury, or Cadmium

Find out whether lead, mercury, or cadmium ions are a significant problem in drinking water where you live or on your campus. You might begin with the map of local drinking water systems provided by EPA's Office of Ground Water and Drinking Water. Your local water utility company or state drinking water program should be able to provide information as well.

a. If these ions are present, what are some likely sources?
b. Are the concentrations of these ions in your water above the MCLG or MCL values? Compare the values reported for your water with those in Table 5.10.

In its metallic form, lead is 50% more dense than iron or steel. Because lead is an abundant, soft, and easily worked metal that does not rust, it has been used since ancient times for water pipes and roofs. Romans were likely the first to use lead for water pipes and as a lining for wine casks. Some historians attribute lead poisoning from such extensive use as a major factor contributing to the fall of the Roman Empire.

In more modern times, most U.S. homes built before 1900 had lead water pipes, now replaced over time by copper or plastic ones. Until 1930, lead pipes were commonly used to connect homes to public water mains. There is no accurate way of knowing how many people suffered permanent health damage from living in residences with lead pipes. But, there are a few recorded cases of fatalities caused by lead poisoning in which the victim over many years habitually prepared a morning beverage using the "first draw" of water that had been standing in lead pipes overnight.

Some Pb^{2+} can get into drinking water even where there are no lead pipes. Solder used to join copper pipes contains 50–75% lead. Some drinking fountains were designed with a holding tank to store chilled water, and the seams in the tank and connections from it to the fountain may have been made with lead-based solder. Water for drinking fountains may stand in the tank for many hours, thus providing more contact time for lead from the solder to dissolve into the water.

When ingested, lead causes severe and permanent neurological problems in humans. This is particularly tragic for children, who may suffer mental retardation and hyperactivity as a result of lead exposure, even at relatively low concentrations. Severe exposure in adults causes irritability, sleeplessness, and irrational behavior, including loss of appetite and eventual starvation. Unlike many other toxic substances, lead is a cumulative poison and is not transformed into a nontoxic substance. Once it enters the body, it accumulates in bones and the brain.

Lead toxicity is a particular problem for children because Pb^{2+} can be incorporated rapidly into bone along with Ca^{2+}. In children, who have less bone mass than adults, the Pb^{2+} remains in the blood longer, where it can damage cells, especially in the brain. Besides lead in drinking water, young children are exposed to large amounts of lead from chewing on paint that contains lead. This is especially the case in older houses where the paint is chipped and flaking. A national program monitoring blood lead levels in children is aimed at identifying children at risk. Health officials are required to investigate cases in which children are known to exceed the currently acceptable blood level of 15 µg/dL (micrograms per deciliter). U.S. children's blood lead levels significantly decreased during the 1970s and 1980s. However, according to the CDC, almost a million children under six still have elevated blood lead levels, with a

1 deciliter (dL) = 0.1 L

disproportionate number of them living in inner cities; thus, lead poisoning is still a major concern.

Your Turn 5.31   **Comparing Lead Content**

Two samples of drinking water were compared for their lead content. One had a concentration of 20 ppb and the other had a concentration of 0.003 mg/L.

a. Explain which one contains the higher concentration of lead.
b. Compare each sample with the current acceptable limit.

Since the 1970s, the federal government has had regulations for acceptable levels of lead in water and foods. These limits have gradually become more restrictive with the development of better analytical methods for measuring extremely low concentrations and as more has been learned about the health effects of lead. Lead is so widespread in the environment that older measurements suffered from unintentional contamination of both the equipment and the reference standards. Until recently, the MCL for Pb^{2+} in drinking water was 15 ppb. In 1992, the EPA converted this to an "action level," meaning that the EPA will take legal action if 10% of tap water samples exceed 15 ppb. The hazard from lead is so great that the EPA has established an MCLG of 0, even though lead is not a carcinogen.

The good news is that very little lead is present in most public water supplies. Amounts exceeding allowable limits are estimated to be present in less than 1% of public water supply systems and they serve less than 3% of the U.S. population. Most lead in drinking water comes from corrosion of plumbing systems, not from the source water itself. When lead is reported, consumers are advised to take simple steps to minimize exposure, such as letting water run before using and using only cold water for cooking. Both actions minimize the chances of ingesting dissolved Pb^{2+}.

> $PbCl_2$ is three times more soluble in hot water than in cold water.

In 2004, timely notification of consumers of high lead levels in drinking water was an issue in our nation's Capitol. Lawmakers faulted the Army Corps of Engineers, operator of the reservoir and water-treatment plants; EPA, monitor of water quality; and the District of Columbia Water and Sewage Authority, the water distributor, for being negligent in informing consumers that tests found over two thirds of more than 6000 homes had unacceptable lead levels, some as high as 20 times the 15-ppb limit. The major cause of contamination was aging lead pipes, although the ensuing scandal was caused by the failures of the three agencies to notify consumers and promptly work toward remediation of the problems.

Consider This 5.32   **Regulating Arsenic in Drinking Water**

Another toxic metal that can find its way into public water supplies is arsenic. In January 2001, the Clinton administration issued a 10-ppb standard for arsenic in drinking water, replacing the standard of 50 ppb set in 1962. The Bush administration soon after recalled the rule before it could take effect, thus reverting to the 50-ppb standard, a controversial decision.

a. What was the reasoning behind each administration's decision?
b. What has been the response to each administration's decision?
c. Determine whether 50 ppb is still the standard for arsenic.

> A spectrophotometer often is simply called a spectrometer. Its use was discussed in Section 2.8 for total ozone measurements and in Section 3.4 to measure infrared radiation.

The almost universal method for Pb^{2+} analysis in water utilizes a spectrophotometric technique. The general features of a spectrophotometer are shown in Figure 5.22. Light

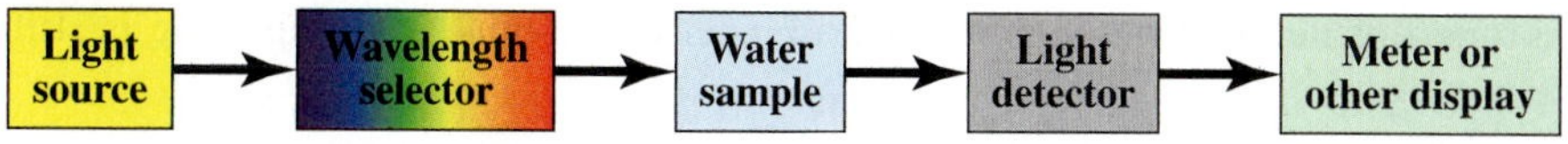

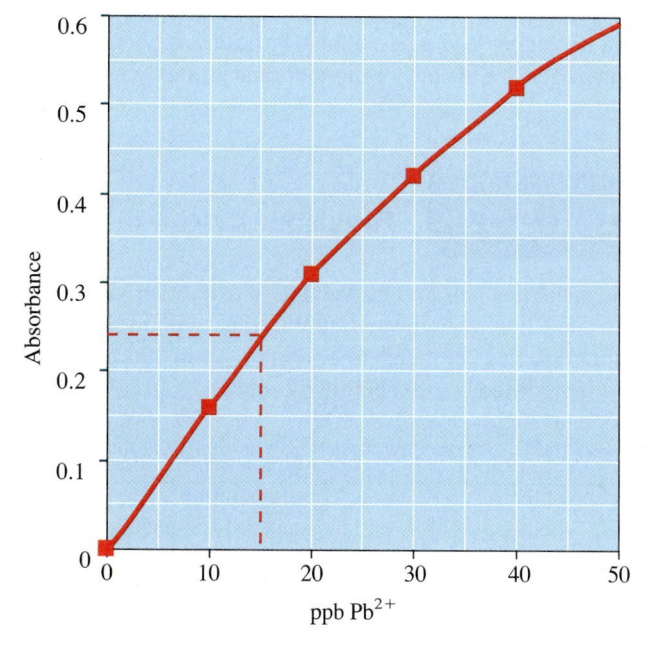

Figure 5.22

Features common to spectrophotometers used for water analysis.

of a specific wavelength passes through the sample and strikes a special detector where the light intensity is converted to a voltage. The voltage is displayed on a meter or sent to a computer or other recording device. The amount of light absorbed by the solution, and which therefore does not reach the detector, is proportional to the concentration of the species being tested. The higher the concentration of the species, the more light absorbed by the sample.

Low concentrations of Pb^{2+} can be analyzed using furnace atomic absorption (AA) spectrophotometry. A small water sample is vaporized at a very high temperature into a beam of UV light coming from a lead-containing lamp. Radiation unique to lead atoms is emitted from the hot lead atoms in the lamp and absorbed by lead atoms in the vaporized water sample. Conventional versions of AA spectrophotometers, in which the Pb^{2+} is heated in a flame, can measure Pb^{2+} concentrations in the parts per million range but cannot gather data in the range of 15 ppb, the current action level for Pb^{2+} in drinking water. More sophisticated AA spectrophotometers can measure lead at well below 1 ppb. However, many smaller communities cannot afford the appropriate equipment to make such sensitive measurements.

Spectrophotometric measurements from the sample being tested must be compared with absorbance data taken for known concentrations of the same species. This is done by use of a **calibration graph,** a graph that is made by carefully measuring the absorbancies of several solutions of known concentration for the species being analyzed. An example of a calibration graph for Pb^{2+} analysis at low ranges of concentration is shown in Figure 5.23. The Pb^{2+} concentration is shown on the horizontal axis and absorbance at a wavelength of 283.3 nm is shown on the vertical axis. For example, if a water sample gives an absorbance measurement of about 0.24, an analyst can use that value to read directly from the graph that the concentration of Pb^{2+} is just at the 15-ppb regulatory limit (see dashed lines, Figure 5.23).

Figure 5.23

Calibration graph for furnace AA spectrophotometric analysis of Pb^{2+} at a wavelength of 283.3 nm.

Your Turn 5.33   **Using the Pb^{2+} Calibration Graph**

Use Figure 5.23 to estimate the concentration of Pb^{2+} in each water sample being analyzed.

   a. Absorbance = 0.50
   b. Absorbance = 0.05
   c. Absorbance = 0.30

Answer

   a. Approximately 37–38 ppb Pb^{2+}

Figure 5.23 illustrates a caution about water analysis: The accuracy of the analysis is only as good as the accuracy of the calibration graph. Some uncertainty is present in each of the measurements, which leads to a small uncertainty in the analysis of any water sample compared with a calibration graph.

Consider This 5.34   **Shifting Limits for Lead**

Before 1962, the recommended limit for lead in drinking water was 100 ppb. In 1962, the limit was lowered to 50 ppb. In 1988, the MCL was lowered to 15 ppb, where it remains today.

In addition to the current action level of 15 ppb of lead in tap water in residences, the EPA recommends that source water from water utilities should contain no more than 5 ppb of lead, and water in school drinking fountains should contain no more than 20 ppb.

   a. Suggest possible reasons for these differences.
   b. If stricter limits are set, who do you think should pay the costs? Give the reasons behind your opinions.

5.14   Consumer Choices: Tap Water, Bottled Water, and Filtered Water

You now have enough information to allow you to make good choices about the water you drink. At your disposal you have the kinds of facts that will come in handy when you assess risk–benefit analyses about your drinking water. Let's consider some relevant questions pertaining to each choice.

Tap Water

Is safe tap water generally available in the United States? The answer, a resounding "yes," is due to high standards mandated by federal regulations for public water supply utilities. The treatment technology is available to achieve these high standards; without such technology, standards would be merely hollow gestures. Very few people in our country suffer acute illness from drinking contaminated water unless they are using water from a private well that has not been properly tested. The Safe Drinking Water Act Amendments of 1996 enhanced protection, including increased requirements for notifying consumers promptly of any problems with water safety.

Is tap water "pure" water? Certainly not; it almost surely contains small amounts of sodium, calcium, magnesium, chloride, sulfate, and bicarbonate ions, as well as trace amounts of other ions. Tap water also contains dissolved air, which is a mixture that includes N_2, O_2, CO_2, and airborne particles.

What problems are likely to exist? Some tap water may contain dangerous Pb^{2+} concentrations, although lead is normally an issue only in buildings with lead pipes. Other heavy-metal elements, such as mercury and cadmium, may be present at dangerous concentrations, although this is extremely unlikely. Chlorinated tap water from surface water sources will contain a small amount of residual chlorine. It may also contain small amounts of THMs, by-products of chlorination. Depending on its source, the water may contain low concentrations of mercury, nitrate, pesticide residues, PCBs, and industrial solvents. By now you should understand that the presence of such substances in drinking water is not likely a cause for alarm. Rather, the crucial question is "How much?" If pollutant concentrations are below the MCLs, the EPA regards the water as safe, with an adequate margin of safety.

Bottled Water

Is bottled water safe? The laws that are applied to public water supplies are not considered for bottled water. However, other regulations, both government- and industry-imposed, are enforced. Considered a food, bottled water is regulated by the Food and Drug Administration (FDA). Bottled water must meet standards of quality, comply with labeling regulations, and meet good manufacturing practices. A provision of the SDWA amendments of 1996 requires the FDA to develop bottled water standards that are equal to EPA drinking water standards. In years past, critics have questioned the safety of bottled water. However, member companies of the International Bottled Water Association (IBWA) produce more than 85% of the bottled water currently sold in the United States. The member companies must meet higher water-quality standards than those imposed by the FDA (Figure 5.24). Springs and underground aquifers that do not require disinfection are the principal sources of bottled water. If disinfection is

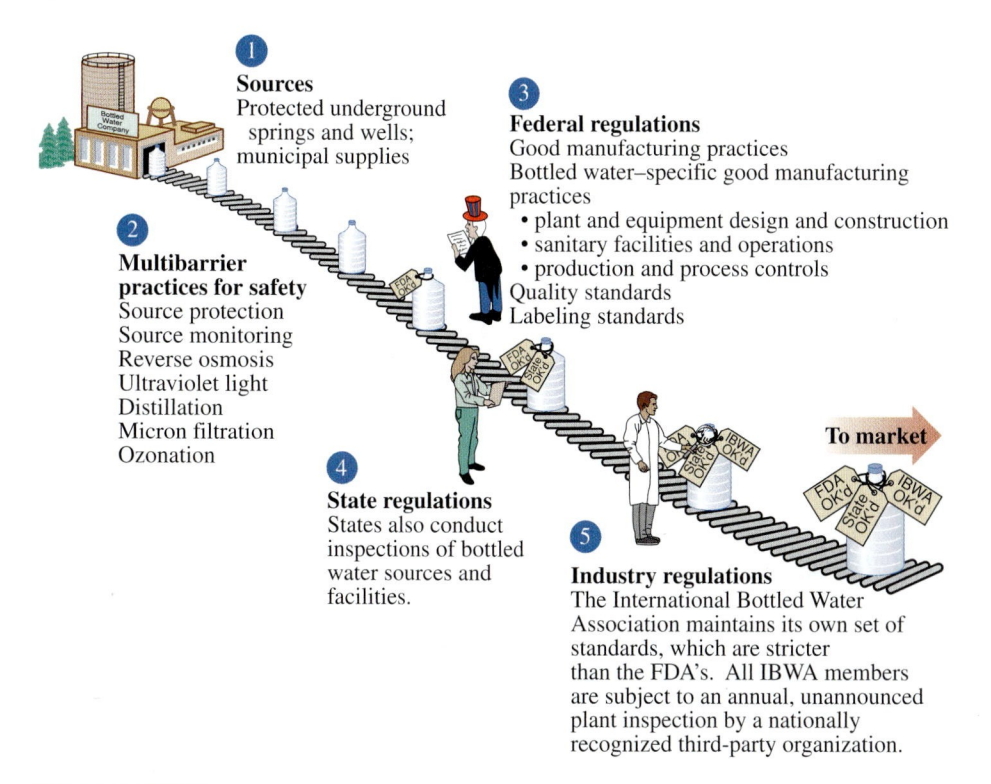

1 **Sources**
Protected underground springs and wells; municipal supplies

2 **Multibarrier practices for safety**
Source protection
Source monitoring
Reverse osmosis
Ultraviolet light
Distillation
Micron filtration
Ozonation

3 **Federal regulations**
Good manufacturing practices
Bottled water–specific good manufacturing practices
• plant and equipment design and construction
• sanitary facilities and operations
• production and process controls
Quality standards
Labeling standards

4 **State regulations**
States also conduct inspections of bottled water sources and facilities.

5 **Industry regulations**
The International Bottled Water Association maintains its own set of standards, which are stricter than the FDA's. All IBWA members are subject to an annual, unannounced plant inspection by a nationally recognized third-party organization.

To market

Figure 5.24

Bottled water's path to market.
The International Bottled Water Association illustrates the process its members' products follow from the source to the consumer's satisfaction. Federal, state, and industry regulations guarantee safety and quality.

Source: © International Bottled Water Association. Reprinted by permission.

required, it is done with ozone or UV radiation, rather than with chlorine, thus leaving no objectionable taste and unwanted by-products. In addition, most bottled water is subjected to filtration, reverse osmosis, or distillation (see Section 5.15). The absence of chlorine, and various trace pollutants found in surface water provide much of the argument for bottled water as a more healthful alternative to tap water. In the majority of all bottled water sold in the United States, the source is municipal tap water that has been subjected to further purification. Interestingly enough, if the municipal water meets processing standards allowing it to be labeled "distilled" or "purified," the water does not need to divulge its municipal tap water source.

Is bottled water pure? Because bottled water often comes from springs or wells, we can be sure that it contains ions, dissolved as the water percolates through the surrounding rocks. In fact, bottled water from some well-known spas, such as Bath in England, Baden-Baden in Germany, and White Sulfur Springs in West Virginia, contains relatively large amounts of calcium and other ions, as well as dissolved carbon dioxide. In a few cases, dissolved hydrogen sulfide gas provides a characteristic "sulfur" odor, thought to be a positive virtue by some connoisseurs of bottled water.

Filtered Water

Is filtered water safe? Yes, certainly it is as safe as the tap water supply being filtered. Water from such a unit is free of objectionable taste and odor and should be free of most hazardous substances. Most filters reduce the concentrations of toxic metal ions (Pb^{2+}, Cu^{2+}). Although these ions are not necessarily totally removed, their concentrations will be well below those of concern for human health.

How do the filters work? These units generally attach to a kitchen faucet, purifying water for drinking or cooking using two methods. The first is "activated carbon," a special form of charcoal with a very high surface area that absorbs most of the molecular solutes, including residual chlorine, pesticide residues, solvents, and other similar substances. The second component is an ion-exchange resin that removes Ca^{2+} and Mg^{2+} ions (the ones responsible for "hard" water) or those that can cause toxicity.

Are filters cost-effective? Tap water remains the least expensive choice for drinking water. Filtered water systems generally treat only the water for drinking and cooking, bringing costs to less than 20% of the costs for purchasing bottled water used for the same purposes.

Consider This 5.35　　　**Evaluating Your Drinking Water Choices**

In Consider This 5.1 and 5.2 activities, you were asked to think about which characteristics of drinking tap, bottled, and filtered water were important to you. Having now studied this chapter, check your lists. Would the order of importance be the same? Explain how your reasoning may have changed based on the information and understanding gained in the study of this chapter.

5.15　International Needs for Safe Drinking Water

Those who live in the United States are privileged to have choices in drinking water. We can select from tap, bottled, or filtered water, all generally of high quality. Such is not the case for people in most of the rest of the world. The reality is that more than a billion people (one in six), principally in developing nations, lack access to safe drinking water. About 1.8 billion people do not have adequate sanitary facilities. One estimate, made by *Scientific American,* is that it would cost $68 billion dollars over the next 10 years to provide safe water and decent sanitation facilities to everyone. Lack

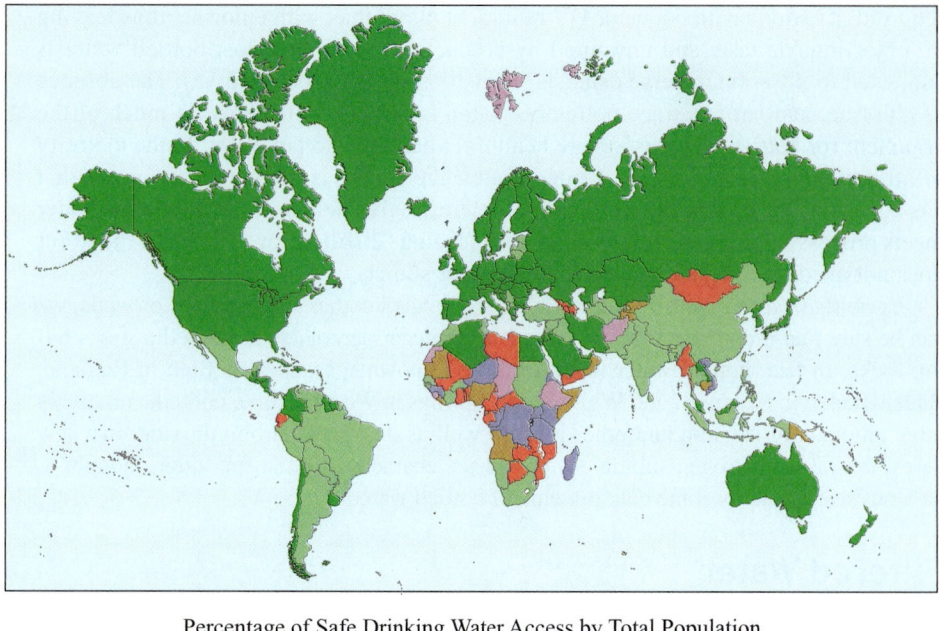

Percentage of Safe Drinking Water Access by Total Population

■	■	■	■	■	■	☐
Over 90%	75 - 90%	60 - 75%	45 - 60%	30 - 45%	Under 30%	No Data

Figure 5.25

Access to safe drinking water varies widely across the world.

Source: © 2006 Compare Infobase Limited.

of access to safe water poses a particular risk to infants and young children. Whereas bottled water is a discretionary option for many in the United States, the majority of the world's population does not have that option. Figure 5.25 shows how access to clean water varies worldwide.

For those living in arid regions, such as the Middle East, fresh water is scarce. Sea water is readily available in many such areas, but its high salt concentration makes it unfit for human consumption. Coleridge's ancient mariner's complaint is more than just a poetic fantasy; it is a physiological reality. Ocean water contains 3.5% salt compared with only about 0.9% salt in body cells. Consequently, sea water can be drunk only after most of the salt is removed. Fortunately, there are ways to do this, but they require large amounts of energy. Collectively, the methods are known as **desalination,** a broad general term describing any process that removes ions from salty water.

One desalination method is distillation, an old and remarkably simple way of purifying water for laboratory and other uses. **Distillation** is a separation process in which a solution is heated to the boiling point and the vapors are condensed and collected. Distilled water is used in steam irons, some car batteries, and other devices whose operation can be impaired by dissolved ions. An apparatus such as that shown in Figure 5.26 is used. Impure water is put into a flask, pot, or other container and heated to its boiling point, 100 °C. As the water vaporizes, it leaves behind most of its dissolved impurities. The water vapor passes through a condenser where it cools and reverts back into a liquid, now free of contaminants. If distillation is done very carefully, extremely pure water, with no detectable amounts of contaminants, is produced.

Energy is required for the distillation of any liquid, and recall from Section 5.5 that water has an unusually high specific heat and an unusually large amount of heat required for evaporation. Both result from the uniquely extensive hydrogen bonding in water. High energy costs for water purification by distillation suggests that it is economically practical only for countries or regions with abundant and cheap energy.

Another desalination technique gaining in popularity is reverse osmosis. To understand this method, we need to know that **osmosis** is the natural tendency for a solvent to

A process similar to distillation takes place as part of the natural hydrologic cycle. Water evaporates and then condenses and falls as rain or snow.

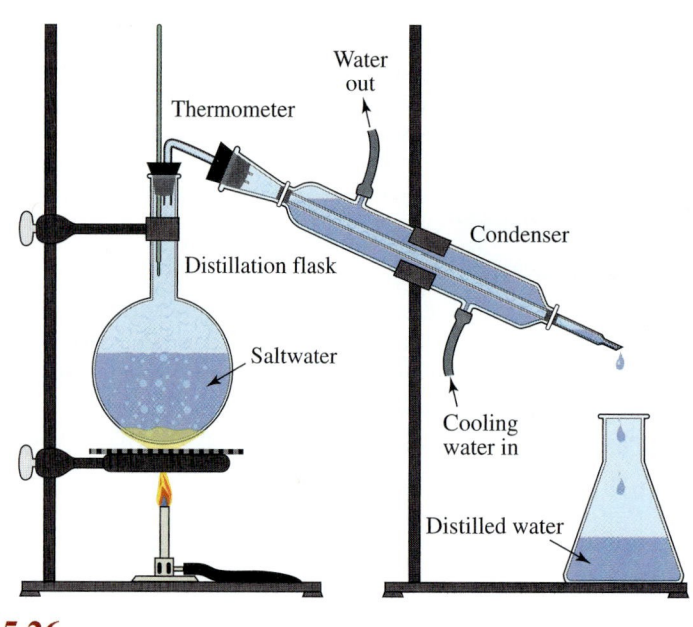

Figure 5.26
Water purification by distillation.

move through a membrane from a region of higher solvent concentration to a region of lower solvent concentration. This tendency to equalize concentrations is involved in many cellular processes, where the net effect is water loss from cells. However, osmosis can be reversed. **Reverse osmosis** is using pressure to force the movement of a solvent through a semipermeable membrane from a region of high solute concentration to a region of lower solute concentration. When using this process to purify water, pressure is applied to the saltwater side, forcing water through the membrane, leaving ions behind. Figure 5.27 is a schematic representation of this process.

The world's largest desalination plant, located at Ashkelon, Israel, was completed in 2005. It is expected to purify 100 million m^3 (68 billion gallons) of water annually, enough to meet about 15% of Israel's domestic consumer demand. Although most such installations are in the Middle East, the number of reverse osmosis plants is increasing in the United States. Florida has over 100 reverse osmosis desalination facilities, including the one that furnishes the city of Tampa Bay with 95,000 m^3 (25 million gallons) of fresh water every day. Small reverse osmosis installations are used in spot-free car washes and individual units are available for boaters. Figure 5.28 shows a small unit, suitable for use

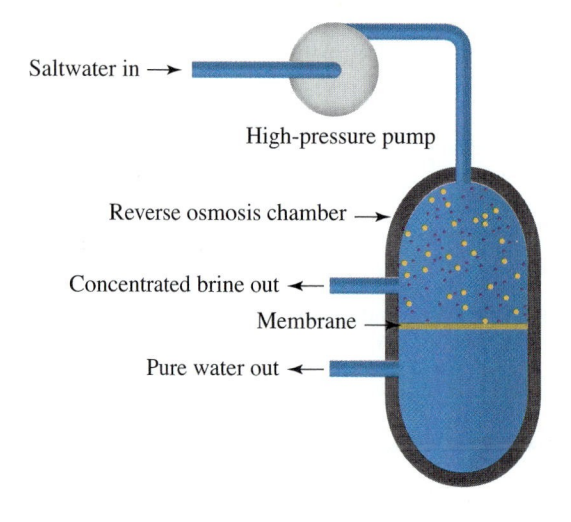

Figure 5.27
Water purification by reverse osmosis.

Figure 5.28
A small reverse osmosis apparatus for converting sea water to potable water.

on a sailboat. Generally reverse osmosis desalination is too expensive for use in most developing nations. It is an often-used method of purification for bottled water, particularly high-end "designer" waters. Using reverse osmosis is meant to impress customers and help them justify the cost because the quality of the water is so high.

Consider This 5.36   Water from the Sun

Solar-powered reverse osmosis desalination units providing 400 L of water per day were developed at Murdoch University in Perth, Australia. Twenty-five units are now in operation in Australia and Asia. Larger units providing 15,000 L/day are being tested. Use the Web to find other examples where potable water is being produced using the power of our nearest star. Be prepared to present your findings to your class.

Conclusion

Water is a very unusual substance, with many unique properties that contribute to its life-supporting role. Like the air we breathe, water is central to life, and we humans require large quantities of it. We sometimes take for granted that our drinking water, whether it is straight from the tap, filtered tap water, or bottled water, is free of harmful contaminants. This chapter has focused almost exclusively on the quality of drinking water—its sources, substances dissolved in it, and potential contaminants and determination of their concentrations. Federal and state regulations help make our drinking water safe. We considered how particular substances in water can be analyzed and treated. In the next chapter, we examine rainwater and the ways in which substances dissolved in rain can adversely affect the environment.

Chapter Summary

Having studied this chapter you should be able to:

- Describe the desirable properties of drinking water (5.1, 5.14)
- Explain some of the reasons why bottled water is so popular (5.2, 5.14)
- Recognize the sources and distribution of water (5.2)
- Discuss why water is such an excellent solvent for some ionic and some covalent compounds (5.3, 5.7–5.10)
- Describe the factors involved in providing pure drinking water (5.3, 5.11–5.15)
- Use concentration units: percent, ppm, ppb, and molarity (5.4)
- Discuss the relationship between the properties of water and its molecular structure (5.5–5.6)
- Describe the specific heat of water and compare it with that of other substances (5.5–5.6)
- Understand how electronegativity and bond polarity are related to the structure of water (5.5)

- Describe hydrogen bonding and its importance to the properties of water (5.6)
- Describe how the densities of ice and water are related to the structure of the water molecule (5.6)
- Determine the formulas for ionic compounds, including those with common polyatomic ions (5.7–5.8)
- Provide the names of simple ionic compounds, given their formulas (5.7–5.8)
- Explain how ionic substances dissolve in water (5.9)
- Explain how covalent substances dissolve in water (5.10)
- Understand the role of federal legislation in protecting safe drinking water (5.11)
- Discuss the maximum contaminant level goal (MCLG) and the maximum contaminant level (MCL) established by the EPA to ensure water quality (5.11)
- Discuss how drinking water is made safe to drink (5.11–5.15)
- Relate chlorination with water purification (5.11)

- Describe atomic absorption spectrophotometry as a method for analyzing contaminants in water (5.13)
- Explain how lead can be ingested and how it affects humans. Be able to use a calibration graph to determine the lead concentration of a water sample (5.13)
- Compare and contrast tap water, bottled water, and filtered water in terms of water quality (5.14)
- Appreciate the relative availability of pure drinking water in the United States and compare with international needs (5.15)
- Understand the processes of distillation and reverse osmosis for producing potable water (5.15)

Questions

Emphasizing Essentials

1. **a.** The text states that 50–65% of adult body weight is water. How many pounds is this for a 150-lb adult? Report your answer as a range of values.

   b. Given that a gallon of water weighs about 8 lb, how many gallons of water will this be for a 150-lb adult? Report your answer as a range of values.

2. **a.** What is an aquifer?

   b. Why is it important to prevent unwanted substances from reaching a clean aquifer?

3. If the water in a 500-L drum were representative of the world's total supply, how many liters would be suitable for drinking? *Hint:* See Figure 5.5.

4. Based on your experience, what is the solubility of each of these substances in water? Use terms such as *very soluble, partially soluble,* or *not soluble.* Cite supporting evidence.

   a. orange juice concentrate

   b. liquid laundry detergent

   c. household ammonia

   d. chicken broth

   e. chicken fat

5. **a.** Bottled water consumption was reported to be 21 gal per person in the United States in 2002. The last census reported 2.9×10^8 people in the United States. Given this, estimate the total bottled water consumption.

   b. If the per capita consumption of bottled water increased 20% in the last 10 years, what was the per capita consumption 10 years ago?

6. **a.** A certain bottled water lists a calcium concentration of 55 mg/L. What is its calcium concentration expressed in parts per million?

   b. How does this concentration compare with that for Evian listed in Table 5.2?

7. One particular vitamin tablet contains 162 mg of calcium and supplies 16% of the recommended daily amount of calcium required by a person on a typical 2000-Calorie diet. How many 500-mL bottles of Evian bottled water would you have to drink each day to obtain the same mass of calcium? *Hint:* See Your Turn 5.7 and Table 5.2.

8. The acceptable limit for nitrate, often found in well water in agricultural areas, is 10 ppm. If a water sample is found to contain 350 mg/L, does it meet the acceptable limit? Show a calculation to support your answer.

9. One reagent bottle on the shelf in a laboratory is labeled 12 M H_2SO_4 and another is labeled 12 M HCl.

   a. How does the number of moles of H_2SO_4 in 100 mL of 12 M H_2SO_4 solution compare with the number of moles of HCl in 100 mL of 12 M HCl solution?

   b. How does the number of grams of H_2SO_4 in 100 mL of 12 M H_2SO_4 solution compare with the number of grams of HCl in 100 mL of 12 M HCl solution?

10. A student weighs out 5.85 g of NaCl to make a 0.10 M solution. What size volumetric flask does she need? *Hint:* See Figure 5.6.

11. Both methane, CH_4, and water are compounds in which hydrogen atoms are bonded with a nonmetallic element. Yet, methane is a gas at room temperature and pressure and water is a liquid. Offer a molecular explanation for the difference in properties.

12. Explain why the term *universal solvent* is applied to water.

13. Here are four sets of atoms. Consult Table 5.3 to answer these questions.

   N and C   N and H

   S and O   S and F

   a. What is the electronegativity difference between the atoms?

   b. Assume that a single covalent bond forms between each pair of atoms. Which atom attracts the electron pair in the bond more strongly?

   c. Arrange the bonds in order of increasing polarity.

14. NaCl is an ionic compound, but $SiCl_4$ is a covalent compound.

   a. Use Table 5.3 to determine the electronegativity difference between chlorine and sodium, and between chlorine and silicon.

   b. What correlations can be drawn about the difference in electronegativity between bonded atoms and their tendency to form ionic or covalent bonds?

   c. How can you explain on the molecular level the conclusion reached in part **b**?

15. Consider a molecule of ammonia, NH_3.

  a. Draw its Lewis structure.

  b. Does the NH_3 molecule contain polar bonds?

  c. Is the NH_3 molecule polar? *Hint:* Consider its geometry.

  d. Is NH_3 soluble in water? Explain.

16. This diagram represents two water molecules in a liquid state. What kind of bonding force does the arrow indicate? Is this an *inter*molecular or *intra*molecular force?

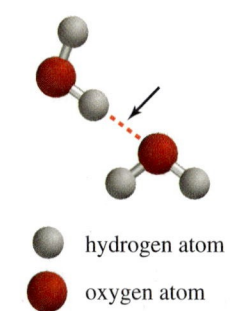

  ⬤ hydrogen atom

  🔴 oxygen atom

17. The density of liquid water at 0 °C is 0.9987 g/cm³; the density of ice at this same temperature is 0.917 g/cm³.

  a. Calculate the volume occupied at 0 °C by 100.0 g of liquid water and by 100.0 g of ice.

  b. Calculate the percentage increase in volume when 100.0 g of water freezes at 0 °C.

18. Consider these liquids.

Liquid	Density, g/mL
dishwashing detergent	1.03
maple syrup	1.37
vegetable oil	0.91

  a. If you pour equal volumes of these three liquids into a 250-mL graduated cylinder, in what order will you add the liquids to create three separate layers? Explain your reasoning.

  b. If a liquid were poured into the cylinder and it formed a layer that was on the bottom of the other three layers, what can you tell about one of the properties of this liquid?

  c. What would happen if a volume of water equal to the other liquids were poured into the cylinder in part **a** and then the contents are mixed vigorously? Explain.

19. Why is there the possibility of a water pipe breaking if the pipe is left full of water during extended frigid weather?

20. What ions typically form from these atoms? Draw Lewis structures for each atom and its corresponding ion. Use the octet rule to explain why each particular ion forms. *Hint:* Consider Tables 5.4 and 5.5.

  a. Cl

  b. Ba

  c. S

  d. Li

  e. Ne

21. Give the chemical formula and name of the ionic compound formed by the reaction of each pair of elements.

  a. Na and S

  b. Al and O

  c. Ga and F

  d. Rb and I

  e. Ba and Se

22. Write the chemical formula for each compound.

  a. calcium bicarbonate

  b. calcium carbonate

  c. magnesium chloride

  d. magnesium sulfate

23. Name each compound.

  a. $KC_2H_3O_2$

  b. $Ca(OCl)_2$

  c. $LiOH$

  d. Na_2SO_4

24. Name each compound.

  a. CoO

  b. $MnCl_3$

  c. ZnS

  d. $SnBr_4$

25. Solutions can be tested for conductivity using this type of apparatus.

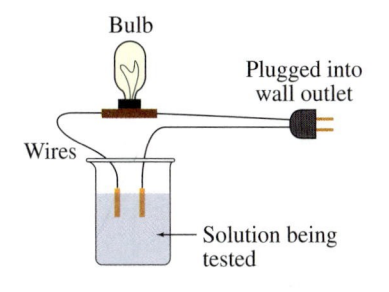

Predict what will happen when each of these dilute solutions is tested for conductivity. Explain your predictions briefly.

  a. $CaCl_2(aq)$

  b. $C_2H_5OH(aq)$

  c. $H_2SO_4(aq)$

26. What ions are present in each of these solutions?
    a. Ca(OCl)$_2$(aq)
    b. C$_2$H$_5$OH(aq)

27. Based on the generalizations in Table 5.8, which compounds are likely to be water-soluble?
    a. KC$_2$H$_3$O$_2$
    b. Ca(NO$_3$)$_2$
    c. LiOH
    d. Na$_2$SO$_4$

28. For a 2.5 M solution of Mg(NO$_3$)$_2$, what is the concentration of each ion present?

29. Explain how you would prepare these solutions from powdered reagents and whatever glassware you needed:
    a. 2.0 L of 1.5 M KOH
    b. 1.0 L of .05 M NaBr
    c. 0.10 L of 1.2 M Mg(OH)$_2$
    d. 300 mL of 3.0 M Ca(Cl)$_2$

30. Explain why desalination techniques, despite proven technological effectiveness, are not used more widely to produce potable drinking water.

Concentrating on Concepts

31. Consider the statement made by the company that makes LeBleu UltraPure Drinking Water: "Water, the universal solvent, given sufficient time, will dissolve or suspend almost any material on earth." Do you agree with this statement? Explain your answer.

32. Why is the concentration of calcium often given on the label for bottled water?

33. The label on Evian bottled water lists a magnesium concentration of 24 mg/L. The label of a popular brand of multivitamins lists the magnesium content as 100 mg per tablet. Which do you think is a better source of magnesium? Explain your reasoning.

34. A new sign is posted at the edge of a favorite fishing hole that says "Caution: Fish from this lake may contain over 1.5 ppm Hg." Explain to a fishing buddy what this unit of concentration means, and why the caution sign should be heeded.

35. This periodic table contains four elements identified by numbers.

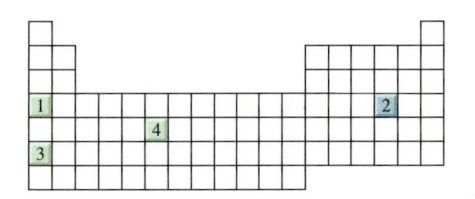

    a. Based on trends within the periodic table, which of the four elements would you expect to have the highest electronegativity value? Explain.

    b. Based on trends within the periodic table, rank the other three elements in order of decreasing electronegativity values. Explain your ranking.

36. A diatomic molecule XY that contains a polar bond *must* be a polar molecule. However, a triatomic molecule XY$_2$ that contains a polar bond *does not necessarily* form a polar molecule. Use some examples of real molecules to help explain this difference.

37. Imagine you are at the molecular level, watching water vapor condense.
    a. Sketch four water molecules using a space-filling representation similar to this one.

    Sketch them in the gaseous state and then in the liquid state. How does the collection of molecules change when water vapor condenses to a liquid?

    b. What happens at the molecular level when water changes from a liquid to a solid?

38. Propose an explanation for the fact that NH$_3$, like H$_2$O, has an unexpectedly high specific heat. *Hint*: See question 15 for the Lewis structure and H-to-N-to-H bond geometry in NH$_3$.

39. a. What type of bond holds together the two hydrogen atoms in the hydrogen molecule, H$_2$?
    b. Explain why the term *hydrogen bonding* does *not* apply to the bond within H$_2$.

40. Hydrogen bonding has been offered as a reason why ice cubes and icebergs float in water. Consider ethanol, C$_2$H$_5$OH.
    a. Draw its Lewis structure and use it to decide if pure ethanol will exhibit hydrogen bonding.
    b. A cube of solid ethanol sinks rather than floats in liquid ethanol. Explain this behavior in view of your answer in part **a**.

41. The unusually high heat capacity of water is very important in regulating our body temperature and keeping it within a normal range despite time, age, activity, and environmental factors. Consider some of the ways that the body produces heat, and some of the ways that it loses heat. How would these functions differ if water had a much lower heat capacity?

42. Suppose that you are in charge of regulating an industry in your area that manufactures agricultural pesticides. How will you decide if this plant is obeying necessary environmental controls? What criteria affect the success of this plant?

43. Health goals for contaminants in drinking water are expressed as MCLG, or maximum contaminant level goals. Legal limits are given as MCL, or maximum contaminant levels. How are MCLG and MCL related for a given contaminant?

44. Provide an explanation why $CoCl_2$ is called cobalt(II) chloride, whereas $CaCl_2$ is called calcium chloride (no Roman numeral in the latter).

45. Use the calibration graph in Figure 5.23 to determine the Pb^{2+} concentration (in M) of 5.0 mL of a $PbSO_4$ solution that has an absorbance reading of 0.4 at a wavelength of 283.8 nm.

46. Use the calibration graph in Figure 5.23 to find the absorbance of a solution resulting from the addition of 5.0 mL of a 20 ppb $PbCl_2$ solution to 10.0 mL of a 16 ppb $PbSO_4$ solution.

47. How can you purify your water when you are hiking? Use the Web to explore some of the possibilities. What are the relative costs and effectiveness of these alternatives? Are any of the methods similar to those used to purify municipal water supplies? Why or why not?

48. Water quality in the chemistry building on a campus was continuously monitored because testing indicated water from drinking fountains in the building had dissolved lead levels above those established by the Safe Drinking Water Act.

  a. What is the likely major source of the lead in the drinking water?

  b. Does the chemical research carried out in this chemistry building account for the elevated lead levels found in the drinking water? Why or why not?

Exploring Extensions

49. Most people turn on the tap with little thought about where the water comes from. In Consider This 5.4, you investigated the source of your drinking water. Now take a more global view. Where does drinking water come from in other areas of the world? Investigate the source of drinking water in a desert country, in a developed European country, and in an Asian country. How do these sources differ?

50. One of the large aquifers in the United States is under the pine barrens of New Jersey.

  a. Where are the pine barrens in New Jersey?

  b. Why are there increasing political pressures to use the water in this aquifer?

51. Aquifers are also important in providing clean drinking water in other parts of the world. Recently four countries in South America have reached a historic agreement to share the immense Guarani aquifer. This is particularly significant because although surface waters often provoke discord and result in agreements, underground sources are routinely not considered at least by international law.

  a. Where is the Guarani aquifer and what four countries are part of this international "underground concordat"?

  b. What concerns did each country have about this aquifer that led to the agreement?

52. Is there any such thing as "pure" drinking water? Discuss what is implied by this term, and how the term's meaning might change in different parts of the world.

53. In the mid-1990s, researchers in Canada and Australia reported that consumption of drinking water with more than 100 ppb aluminum can lead to neurological damage, such as memory loss and perhaps to a small increase in the incidence of Alzheimer's disease. Has further research substantiated these findings? Find out more about this topic, and write a brief summary of your findings. Be sure to cite the sources of your information.

54. The text states that hydrogen bonds are only about one tenth as strong as the covalent bonds that connect atoms within molecules. Check out that statement with this information. Hydrogen bonds vary in strength from about 4 to 40 kJ/mol. Given that the hydrogen bonds between water molecules are at the high end of this range, how does the strength of a hydrogen bond between water molecules compare with the strength of a hydrogen-to-oxygen covalent bond within a water molecule? *Hint:* Consult Table 4.2 for covalent bond energies.

55. The text states that mass and density are often confused. Here is an example of that potential misunderstanding of terms.

  a. What do you think the term *heavy metal* implies when talking about elements on the periodic table?

  b. Compare the scientific definitions of this term that you may find in different sources, and discuss whether each definition is related to relative density or to relative mass.

56. We all have the amino acid glycine in our bodies. This is its structural formula.

  a. Is glycine a polar or nonpolar molecule? Use electronegativity differences to help answer this question.

  b. Can glycine exhibit hydrogen bonding? Explain your answer.

  c. Is glycine soluble in water? Explain.

57. Hard water is defined as having high concentrations of Mg^{2+} and Ca^{2+} ions. The process of water softening involves removing these ions.

  a. How hard is the water in your local area? One way to answer this question is to determine the number of water-softening companies in your area. Use the resources of the Web, as well as ads in your local newspapers and yellow pages, to find out if your area is targeted for marketing water-softening devices.

b. Use the Web to research methods for treating hard water, and explain how an ion-exchange process is used for this purpose.

58. The calibration curve shown in Figure 5.23 is useful for Pb^{2+} concentrations between 0 and 40 ppm. What are your options if a water sample is expected to contain a much higher concentration of Pb^{2+}?

59. Some areas have a higher than normal amount of THMs in the drinking water. Suppose that you are considering moving to such an area. Write a letter to the local water district asking relevant questions to be answered before deciding to move.

60. PCBs are very useful chemicals that may end up in the wrong place, causing long-term damage to birds and mammals. What are the uses of PCBs that made them desirable, and what are some of the negative effects of these materials?

Chapter 6

Neutralizing the Threat of Acid Rain

"Stop five people on the street and chances are they will be able to tell you that carbon dioxide emissions cause global warming. Stop another five and ask them about nitrogen emissions, and they will probably stare at you blankly."

The New Scientist January 21, 2006

Why doesn't nitrogen get the same attention as carbon? As we saw in Chapter 3, carbon dioxide emissions are a hot topic. But when was the last time you heard somebody arguing that we needed to reduce our emissions of nitrogen oxides? As this book went to press, carbon emissions definitely held the spotlight on the global stage. In contrast, nitrogen emissions were waiting in the wings.

While you are puzzling over nitrogen's lack of notoriety, take a breath. No matter where you are on Earth, you will be inhaling trillions upon trillions of nitrogen molecules. Each time you breathe, N_2 molecules constitute roughly 80% of the air going in and out of your lungs. Recall that nitrogen is relatively unreactive as an element. Some might even go as far as to label N_2 a lackluster little molecule, as seemingly it does so little of interest. In contrast, the O_2 molecule is involved in high-profile reactions such as combustion, respiration, rusting, and photosynthesis. Wherein, then, lies the source of urgency that we need to turn our attention to nitrogen emissions?

In addition to the elemental form of nitrogen in our atmosphere, compounds of nitrogen are found widely dispersed on our planet. Furthermore, with human activity, their concentration in the biosphere is increasing, especially in some parts of the country. For example, nitrogen monoxide (which subsequently forms nitrogen dioxide) is an air pollutant formed wherever there is a source of high heat. As you saw in Chapter 1, NO and NO_2, whether from the engines of jet aircraft or wild brush fires, can lower the quality of the air you breathe. Fertilizers such as ammonia and ammonium nitrate end up not only on fields, but also in nearby streams. The nitrate ion can reach dangerously high concentrations in water supplies. Nitrous oxide in the air comes from the removal of nitrate from soils by bacteria. Nitrous oxide also is produced from catalytic converters, the burning of biomass, and the industrial processes that synthesize nylon and nitric acid.

For more about the nitrate ion, see Section 5.8.

For more about nitrous oxide, see Section 3.8.

| Your Turn 6.1 | Nitrogen Inventory |

As noted in the previous paragraph, the biosphere contains nitrogen in many different chemical forms. Several nitrogen compounds have been mentioned in previous chapters. Select any five and create a table with the names, chemical formulas, Lewis structures, and points of interest. Elemental nitrogen is done for you as an example.

Hint: Use the index at the back of the book and resources of the Web to complete the last column.

Name	Formula	Lewis Structure	Point(s) of Interest
Nitrogen ("nitrogen gas")	N_2	:N≡N:	The major component of our atmosphere and much of what you breathe (Section 1.1)

One way or another, these nitrogen compounds all are linked to acid rain and its cousins: acidic snows, fogs, and dry depositions. Acidic precipitation is not a new phenomenon. The acidity of rain apparently was first studied back in 1852 by a British chemist named Robert Angus Smith. Twenty years later, Smith wrote a book entitled *Air and Rain,* but his ideas soon fell into obscurity. Then, in the 1950s, the effects of acid rain were rediscovered by scientists working in several parts of the world, including the Northeastern United States, Scandinavia, and the English Lake District.

Reports of damage attributed to acidic precipitation grew dramatically over the next three decades. Dozens of books, scientific papers, and popular articles described the damage (Figure 6.1). Reports came from almost every part of the world. Lakes in Norway and Sweden were reported as being effectively "dead," without fish or any other living things. The sculptures adorning the exteriors of cathedrals in Europe and prehistoric sites in Mexico and Central America were eroding away. In all these instances, acid rain was blamed as a major cause of the damage.

Figure 6.1

Acid rain has been in the news (*The Cleveland Press,* June 23, 1980).

As the quote that opens this chapter states, nitrogen emissions in general (and acid rain in particular) are not in the public eye. Currently these topics seem to be out of vogue. However, many scientists feel strongly that nitrogen emissions *should* be discussed, warning that these emissions may pose a far greater global threat to human welfare than carbon emissions. Comparing nitrogen emissions to those of carbon, biologist Rowan Hooper comments "This one could be even worse."

Is he right? Do nitrogen emissions warrant the same attention as those of carbon? To assess the issues, we need to begin with a study of acids, bases, and their concentrations in the environment.

6.1   What Is an Acid?

Figure 6.2

Citrus fruit contains both citric acid and ascorbic acid.

The carbonate ion is CO_3^{2-}. See Table 5.6.

Acid rain nicely links the topics of atmospheric pollution and water chemistry from Chapters 1 and 5. To better understand this connection, we first need to discuss the term *acid.* Most definitions either cite the observable properties of acids or describe their behavior at the molecular level. Either way, the information is useful to our discussion.

Historically, chemists identified acids by their properties—sour taste, color changes with indicators, and reactions with certain minerals. Although tasting is not usually a smart way to identify chemicals, you undoubtedly know the sour taste of acetic acid in vinegar. The sour taste of lemons comes from acids as well (Figure 6.2). You may even be a fan of those incredibly sour candies. If you check the ingredient list, you will see that what makes your mouth pucker is citric acid, malic acid, or both (Figure 6.3).

Acids also display common chemical properties. For example, litmus, a plant dye, changes from blue to pink in the presence of an acid. Indeed, the term *litmus test* is so well known that you may hear it used as a figure of speech. Thus a "litmus test" for a political candidate would be something that quickly reveals this person's point of view.

Another property common to acids is that they can, under certain conditions, dissolve materials such as marble or eggshell. Both of these contain the carbonate ion, either as calcium or magnesium carbonate. The action of acids on carbonates releases carbon dioxide. This is the "fizz" (or burp) that accompanies some stomach antacids. We will return to this chemical reaction later.

At the molecular level, an **acid** is a compound that releases hydrogen ions, H^+, in aqueous solution. Remember that a hydrogen atom is electrically neutral and consists of one electron and one proton. If the electron is lost, it becomes positively charged (H^+). Since only a proton remains, sometimes the H^+ is simply called a proton.

Ingred.: cane sugar, water, malic acid, citric acid, natural and artificial flavors, sodium benzoate, FD&C Yellow 5, Yellow 6, Blue 1, Yellow 5 Lake, Blue 2 Lake, Red 40 Lake
Mfg. by Squire Boone Village
New Albany, IN 47150
For nutrition info, call 1-800-234-1804

Figure 6.3

A sour candy that contains malic acid and citric acid.

Consider the gas hydrogen chloride, which at room temperature consists of HCl molecules. This gas dissolves readily in water to release two ions that we represent as $H^+(aq)$ and $Cl^-(aq)$. The notation *(aq)* is short for *aqueous*.

$$HCl(g) \xrightarrow{H_2O} H^+(aq) + Cl^-(aq) \qquad [6.1]$$

We also could say that HCl *dissociates* into H^+ and Cl^-, or that HCl *ionizes* to form H^+ and Cl^-. Either way, essentially no undissociated HCl molecules remain in solution. Thus, HCl is an acid that ionizes (dissociates) completely.

There is a slight complication with the definition of acids as substances that release H^+ ions (protons) in aqueous solutions. By themselves, H^+ ions are much too reactive to exist as such. Rather, they attach to something else, such as water molecules. When dissolved in water, each HCl donates a proton (H^+) to an H_2O molecule, forming H_3O^+, a hydronium ion. The overall reaction can be represented like this.

$$HCl(g) + H_2O(l) \longrightarrow H_3O^+(aq) + Cl^-(aq) \qquad [6.2]$$

The solution represented on the product side in *both* equations 6.1 and 6.2 is called hydrochloric acid. It has the characteristic properties of an acid because of the presence of H_3O^+ ions. Chemists often simply write H^+ when referring to acids (for example, in equation 6.1), but understand this to mean H_3O^+ (hydronium ion) in aqueous solutions.

The Lewis structure for the hydronium ion is

$$\left[\begin{array}{c} H \\ H \!:\!\overset{\cdot\cdot}{\underset{\cdot\cdot}{O}}\!:\! H \end{array} \right]^+$$

It obeys the octet rule.

Your Turn 6.2　　　**Acidic Solutions**

For each of these aqueous acids, write a chemical equation that shows the release of a hydrogen ion. *Hint:* Remember to include the charges on the ions.

a. HI(*aq*), hydroiodic acid　　**b.** HNO_3(*aq*), nitric acid　　**c.** H_2SO_4(*aq*), sulfuric acid

Answer

c. $H_2SO_4(aq) \longrightarrow H^+(aq) + HSO_4^-(aq)$

Consider This 6.3　　　**Are All Acids Harmful?**

Although the word *acid* may conjure up all sorts of pictures in your mind, every day you eat or drink various acids. Check the labels of foods or beverages and make a list of the acids you find. Speculate on the purpose of each acid.

Notably, we have only encountered one acid in this section that contains nitrogen—nitric acid. Before we say more about this acid and its atmospheric sources, we first turn to a related topic of interest.

6.2   What Is a Base?

No discussion of acids would be complete without mentioning their chemical counterparts—bases. For our purposes, a **base** is a compound that produces hydroxide ions, OH⁻, in aqueous solution. For example, sodium hydroxide (NaOH), an ionic compound, dissolves in water to produce sodium ions and hydroxide ions.

$$\text{NaOH}(s) \xrightarrow{\text{H}_2\text{O}} \text{Na}^+(aq) + \text{OH}^-(aq) \qquad [6.3]$$

Bases have their own characteristic properties attributable to the presence of OH⁻(aq). Unlike acids, bases generally taste bitter and do not lend an appealing flavor to foods. When dissolved in water, bases have a slippery, soapy feel. Common examples of bases include household ammonia (an aqueous solution of NH₃) and NaOH, sometimes called lye. The cautions on oven cleaners (Figure 6.4) warn that lye can cause severe damage to eyes, skin, and clothing.

> The soapy feel of dilute basic solutions is exactly that. Bases can react with the oils of your skin to produce a tiny bit of soap.

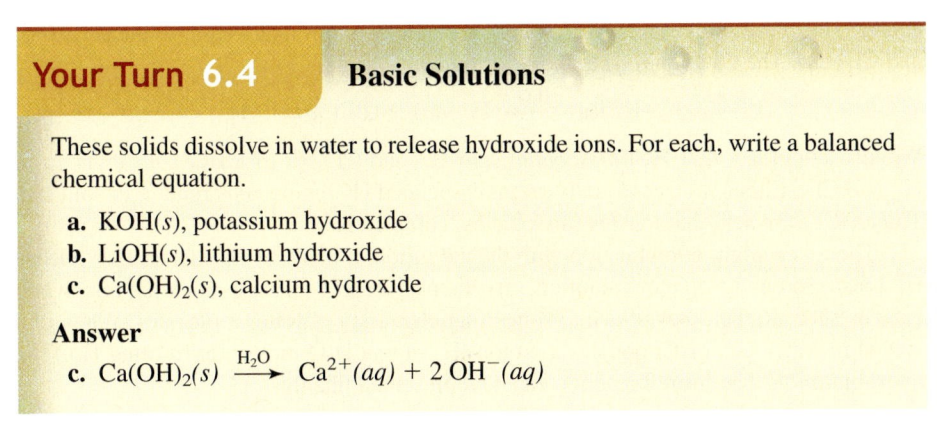

Your Turn 6.4   **Basic Solutions**

These solids dissolve in water to release hydroxide ions. For each, write a balanced chemical equation.

    a. KOH(s), potassium hydroxide
    b. LiOH(s), lithium hydroxide
    c. Ca(OH)₂(s), calcium hydroxide

Answer

    c. $\text{Ca(OH)}_2(s) \xrightarrow{\text{H}_2\text{O}} \text{Ca}^{2+}(aq) + 2\,\text{OH}^-(aq)$

Ammonia, a nitrogen-containing base, is of particular interest to the topics in this chapter. As you may remember, ammonia is a gas with a distinctive sharp odor. Aqueous ammonia is made by dissolving this gas in water. We represent this solution as NH₃(aq).

$$\text{NH}_3(g) \xrightarrow{\text{H}_2\text{O}} \text{NH}_3(aq) \qquad [6.4a]$$

The solution called "household ammonia" is about 5% NH₃ by mass. Although it can be unpleasant to work with, it is not highly concentrated.

Figure 6.4
Oven cleaner may contain NaOH, commonly called lye.

Given what we said about bases releasing hydroxide ions in solution, it may not be readily apparent why aqueous ammonia is a basic solution. We explain this by noting that a water molecule can transfer a hydrogen ion to $NH_3(aq)$ to form an ammonium ion, $NH_4^+(aq)$.

$$NH_3(aq) + H^+(aq) \longrightarrow NH_4^+(aq) \qquad [6.4b]$$

Assuming that the H^+ in this equation came from a water molecule, the reaction of aqueous ammonia with water can be represented as the formation of NH_4OH, ammonium hydroxide.

$$NH_3(aq) + H_2O(l) \longrightarrow NH_4OH(aq) \qquad [6.4c]$$

The source of the hydroxide ion in household ammonia now should be apparent. Ammonium hydroxide dissociates to form hydroxide and ammonium ions. This reaction occurs only to a limited extent; that is, only tiny amounts of the two ions are formed in an aqueous solution of ammonia. Nevertheless, this is enough to produce a basic solution.

> The ammonium ion, NH_4^+ is formed analogously to the hydronium ion, H_3O^+.

6.3   Neutralization: Bases Are Antacids

Acids and bases react with each other. Not only will this happen in test tubes in the laboratory, but also in your home and in almost every ecological niche of our planet. For example, if you put lemon juice on fish, you run an acid–base reaction. The acids found in lemons neutralize the ammonia-like compounds that produce the "fishy smell." Similarly, if ammonia fertilizer on the fields hits the acidic emissions of a nearby power plant, neutralization occurs. Most acid–base reactions occur readily and almost instantaneously.

Let us first examine the acid–base reaction of solutions of hydrochloric acid and sodium hydroxide. If equal volumes of solutions of equal concentration are mixed, the products are sodium chloride and water.

$$HCl(aq) + NaOH(aq) \longrightarrow NaCl(aq) + H_2O(l) \qquad [6.5]$$

This is an example of **neutralization,** a chemical reaction in which the hydrogen ions from an acid combine with the hydroxide ions from a base to form water molecules. The formation of water can be represented like this.

$$H^+(aq) + OH^-(aq) \longrightarrow H_2O(l) \qquad [6.6]$$

> Recall from Section 5.8 that NaCl is an ionic compound that dissolves in water to produce $Na^+(aq)$ and $Cl^-(aq)$.

What about the sodium and chloride ions? Recall from equations 6.1 and 6.3 that HCl and NaOH completely dissociate into ions when dissolved in water. We can rewrite equation 6.5 to show this.

$$H^+(aq) + Cl^-(aq) + Na^+(aq) + OH^-(aq) \longrightarrow Na^+(aq) + Cl^-(aq) + H_2O(l) \quad [6.7]$$

The $Na^+(aq)$ and $Cl^-(aq)$ don't take part in the neutralization reaction and remain unchanged. Canceling these ions from both sides produces equation 6.6.

Your Turn 6.5   **Neutralization Reactions**

For each acid–base pair, write a neutralization reaction. Then rewrite the equation in ionic form and eliminate ions common to both sides.

a. $HNO_3(aq)$ and $KOH(aq)$
b. $H_2SO_4(aq)$ and $NH_4OH(aq)$
c. $HBr(aq)$ and $Ba(OH)_2(aq)$

Answer
c. $2 HBr(aq) + Ba(OH)_2(aq) \longrightarrow BaBr_2(aq) + 2 H_2O(l)$

$2 H^+(aq) + 2 Br^-(aq) + Ba^{2+}(aq) + 2 OH^-(aq) \longrightarrow$
$$Ba^{2+}(aq) + 2 Br^-(aq) + 2 H_2O(l)$$

$2 H^+(aq) + 2 OH^-(aq) \longrightarrow 2 H_2O(l)$

Simplify this last equation by dividing both sides by 2.

$$H^+(aq) + OH^-(aq) \longrightarrow H_2O(l)$$

Neutral solutions are neither acidic nor basic, that is, they have equal concentrations of H^+ and OH^- ions. Pure water is a neutral solution. Some salt solutions also are neutral, such as the one formed by dissolving NaCl in pure water. In contrast, acidic solutions contain a higher concentration of H^+ than OH^- ions, and basic solutions a higher concentration of OH^- than H^+ ions.

It may seem strange that acidic solutions contain some OH^- and likewise that basic solutions contain H^+. But when water is involved, it is not possible to have H^+ without OH^- or vice versa. A simple, useful, and very important relationship exists between the concentration of hydrogen ions and hydroxide ions in any aqueous solution.

> The product [H⁺][OH⁻] is dependent on temperature. The value 1×10^{-14} is valid at 25 °C.

$$[H^+][OH^-] = 1 \times 10^{-14} \qquad [6.8]$$

The square brackets indicate that the ion concentrations are expressed in molarity, and $[H^+]$ is read as "the hydrogen ion concentration." When $[H^+]$ and $[OH^-]$ are multiplied together, the product is a constant with a value of 1×10^{-14} as shown in mathematical expression 6.8. This expression also tells us that the concentrations of H^+ and OH^- depend on each other. When $[H^+]$ increases, $[OH^-]$ decreases. And when $[H^+]$ decreases, $[OH^-]$ increases. Both ions are always present in aqueous solutions.

Knowing the concentration of H^+, we can use expression 6.8 to calculate the concentration of OH^- (or vice versa). For example, if a rain sample has a H^+ concentration of 1×10^{-5} M, we can calculate the OH^- concentration by substituting in 1×10^{-5} M for $[H^+]$.

$$1 \times 10^{-5} \times [OH^-] = 1 \times 10^{-14}$$

$$[OH^-] = \frac{1 \times 10^{-14}}{1 \times 10^{-5}}$$

$$[OH^-] = 1 \times 10^{-9}$$

Since the hydroxide ion concentration (1×10^{-9} M) is smaller than the hydrogen ion concentration (1×10^{-5} M), the solution is acidic.

In pure water or in a neutral solution, the molarities of the hydrogen and hydroxide ions both equal 1×10^{-7} M. Applying mathematical expression 6.8, we can see that $[H^+][OH^-] = (1 \times 10^{-7})(1 \times 10^{-7}) = 1 \times 10^{-14}$.

> Acidic solution
> $[H^+] > [OH^-]$
> Neutral solution
> $[H^+] = [OH^-]$
> Basic solution
> $[H^+] < [OH^-]$

Your Turn 6.6   Acidic and Basic Solutions

Classify these solutions as acidic, neutral, or basic at 25 °C. Then, for parts **a** and **c**, calculate $[OH^-]$. For **b**, calculate $[H^+]$.

   a. $[H^+] = 1 \times 10^{-4}$ M   **b.** $[OH^-] = 1 \times 10^{-6}$ M   **c.** $[H^+] = 1 \times 10^{-10}$ M

Answer
   a. The solution is acidic because $[H^+] > [OH^-]$.

$$[H^+][OH^-] = 1 \times 10^{-14}. \text{ Solving, } [OH^-] = 1 \times 10^{-10} \text{ M}.$$

Your Turn 6.7   Ions in Acidic and Basic Solutions

Classify each solution as acidic, basic, or neutral. Then list all of the ions present in order of decreasing concentration.

   a. KOH*(aq)*   **b.** HNO_2*(aq)*   **c.** H_2SO_3*(aq)*   **d.** $Ca(OH)_2$*(aq)*

Answer
   d. When calcium hydroxide dissociates, two hydroxide ions are released for every calcium ion. The basic solution contains much more OH^- than H^+.

$$OH^-(aq) > Ca^{2+}(aq) > H^+(aq)$$

To discuss acid rain, we will need a convenient way of reporting how acidic or basic a solution is. The pH scale is just such a tool, as it relates the acidity of a solution to its H^+ concentration. We now turn to the topic of pH.

6.4   Introducing pH

The term "pH" already may be familiar to you. Test kits for soils and for the water in aquariums and swimming pools report the acidity in terms of pH. Shampoos claim to be pH-balanced (Figure 6.5). And, of course, articles about acid rain make reference to pH. The notation pH is always written with a small p and a capital H and stands for "power of hydrogen." In the simplest terms, **pH is a number, usually between 0 and 14, that indicates the acidity of a solution.**

As the midpoint on the scale, pH 7 separates acidic from basic solutions. Solutions with a pH less than 7 are acidic, and those with a pH greater than 7 are alkaline, or basic. Figure 6.6 shows that "normal" rain is naturally slightly acidic, with a pH value between 5 and 6. Since pure water is neutral and has a pH of 7.0, the obvious inference is that rain is not pure H_2O. **Acid rain is more acidic than "normal" rain and has a lower pH value.** In the next section, you will see what "impurities" make *all* raindrops acidic, and some even more acidic than others.

Figure 6.6 also displays the pH values of common substances. You may be surprised that you eat and drink so many acids. Acids occur naturally in foods and contribute distinctive tastes. For example, the tangy taste of McIntosh apples comes from malic acid. Yogurt gets its sour taste from lactic acid, and cola soft drinks contain several acids, including phosphoric acid. Tomatoes are well known for their acidity, but with a pH of about 4.5, they are in fact less acidic than many other fruits.

Consider This 6.8	**Acidity of Foods**

   a. Rank tomato juice, lemon juice, milk, cola, and pure water in order of increasing acidity. Check your order against Figure 6.6.
   b. Pick any other five foods and make a similar ranking. Look up the actual pH values. A helpful link is provided at the *Online Learning Center*.
   c. Why do you suppose so few foods have a pH greater than 7?

As you might suspect, pH values are related to the hydrogen ion concentration, which in turn is related to the hydroxide ion concentration. For solutions in which $[H^+]$ is 10 raised to some power, the pH value is this power (the exponent) with its sign changed. For

Figure 6.5
This shampoo claims to be "pH-balanced," that is, adjusted to be closer to neutral. Soaps tend to be basic, which can be irritating to the skin.

For highly acidic or basic solutions, the pH may lie outside of the 0 to 14 range.

The mathematical relationship is
$$pH = -\log[H^+].$$
More information can be found in Appendix 3.

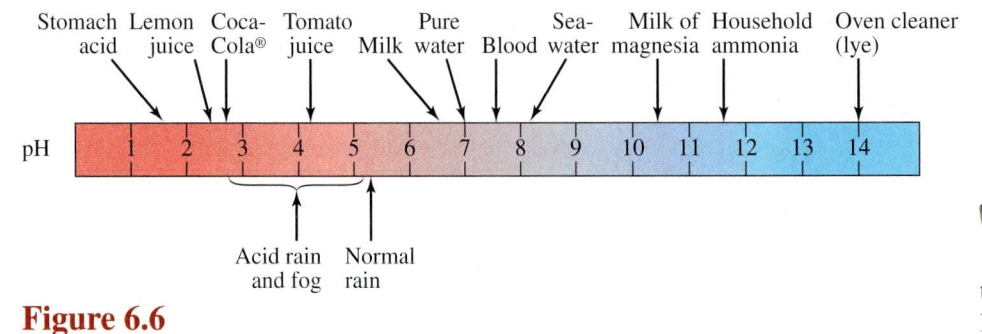

Figure 6.6
Common substances and their pH values.

   Figures Alive! Visit the *Online Learning Center* to learn more about acids, bases, and the pH scale.

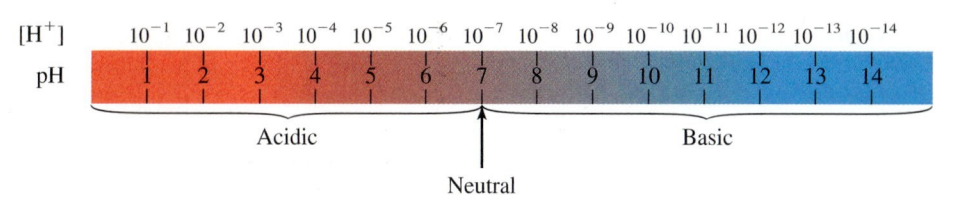

Figure 6.7

The relationship between pH and the concentration of H^+. As pH increases, $[H^+]$ decreases.

example, if $[H^+] = 1 \times 10^{-3}$ M, then the pH is 3. Similarly, for $[H^+] = 1 \times 10^{-9}$ M, the pH is 9. Appendix 3 describes the relation between pH and $[H^+]$ in more detail.

One aspect of the pH scale may be confusing to you. As the pH value *decreases,* the acidity *increases.* For example, a sample of water with a pH of 5.0 is *less* acidic than one with a pH of 4.0. This is because a pH of 4 means that the $[H^+]$ is 0.0001 M. By contrast, a solution with a pH of 5 is more dilute with a $[H^+] = 0.00001$ M. This second solution is *less* acidic with only 1/10 the concentration of hydrogen ion as a solution of pH 4. Figure 6.7 shows the relationship between pH and the hydrogen ion concentration.

Your Turn 6.9   Small Changes, Big Effects

Compare equal volumes of the samples below. Which one is more acidic? How much more acidic?

   a. Rain sample, pH = 5 and lake water sample, pH = 4.
   b. Tomato juice sample, pH = 4.5 and milk sample, pH = 6.5.

Answer
   b. Although the pH values differ only by 2, the sample of tomato juice is 100 times more acidic and has 100 times more H^+ than the sample of milk.

Consider This 6.10   On the Record

A legislator from the Midwest is on record with an impassioned speech in which he argued that the environmental policy of the state should be to bring the pH of rain all the way down to zero. Assume that you are an aide to this legislator. Draft a tactful memo to your boss to save him from additional public embarrassment.

Having established the pH scale as a measure of acidity, we now turn to acid rain and its causes.

6.5   The Challenges of Measuring the pH of Rain

Rain is only one of several ways that acids can be delivered to Earth's surface and waters. Snow and fog obviously are others. The term **acid deposition** includes wet forms such as rain, snow, fog, and cloud-like suspensions of microscopic water droplets often more acidic and damaging than acid rain. It also includes the "dry" forms of acids. For example, during dry weather, tiny solid particles (aerosols) of the acidic

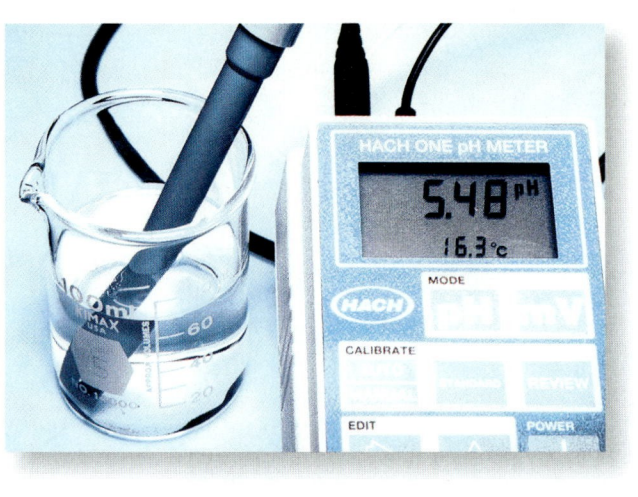

Figure 6.8

A pH meter with a digital display.

compounds ammonium nitrate (NH_4NO_3) and ammonium sulfate ((NH_4)$_2SO_4$) can settle on surfaces. Dry deposition can be just as significant as the wet deposition of the acids in rain, snow, and fog. These aerosols also contribute to haze, as we will see in Section 6.11.

What are the levels of acidity across the mainland United States, Alaska, Hawaii and Puerto Rico? To answer this question, we need an analytical tool, the pH meter. Many types of pH meters are available, depending both on the conditions under which you wish to use them and how much you are willing to pay. The pH meter that you are most likely to encounter has a special probe capped with a membrane that is sensitive to H^+. When the probe is immersed in a sample, H^+ ions create a voltage across the membrane. The meter measures this voltage, converts it to pH, and indicates the pH value on a dial or digital display, such as the one shown in Figure 6.8.

It is straightforward to measure the pH of a rain sample, although certain procedures, such as calibrating the electrode, are necessary to ensure accurate results. More challenging is to collect the rain samples without contaminating them. For example, the collection containers must be scrupulously clean and free of oils from your hands or minerals from the water in which they were washed. When a container is placed on site, it must be high enough to prevent splash contamination either from the ground or surrounding objects. Even if elevated, contamination may still occur from the pollen of nearby plants, insects, bird droppings, leaves, soil dust, or even the ash of a fire.

One way to minimize contamination is to fit a rain collection bucket with a lid and a moisture sensor that opens this lid when it begins to rain. This is the case for samples collected at the approximately 250 sites of the National Atmospheric Deposition Program/National Trends Network (NADP/NTN). Figure 6.9a shows the sensor and the two buckets at a NADP/NTN monitoring station in Illinois that has been in operation for over 25 years. One bucket is for dry deposition (open when it is not raining) and the other is covered. A sensor opens this bucket (closing the other) when it rains.

Deciding where to locate the collection sites also is a challenge. Due to budgetary constraints, the test sites cannot go in as many places as might be desired. Researchers may have to weigh the relative advantages of widely dispersing the sites versus putting several nearby in specialized ecosystems such as those in a national park. Currently there are more collection sites in the eastern United States, as historically the acidity levels have been higher there.

Rain samples have been collected routinely in the United States and Canada since about 1970. Since 1978, NADP/NTN has collected over 250,000 samples, analyzing them for pH and for these ions: SO_4^{2-}, NO_3^-, Cl^-, NH_4^+, Ca^{2+}, Mg^{2+}, K^+, and Na^+. Figure 6.9b shows the five active NADP/NTN sites in the state of Illinois. How many sites are in your state? Complete the following activity to find out.

Section 8.5 describes how hydrogen fuel cells also create a voltage across a membrane.

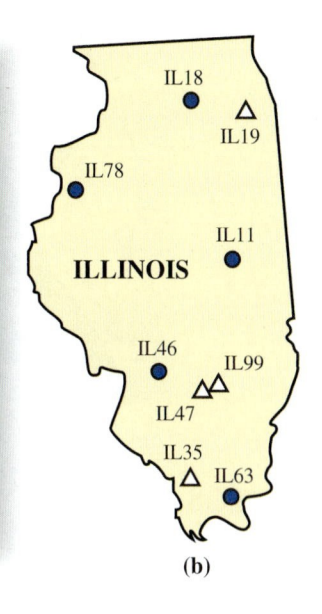

(a) (b)

Figure 6.9

(a) The Bondville Monitoring Station in central Illinois (IL11) has been in operation since 1979. The black moisture sensor connected to the left of the table controls which bucket is open. As it is not raining, the right bucket for wet deposition is closed. **(b)** The five active NTN precipitation monitoring sites in Illinois, including IL11 in Bondville. The sites marked with triangles are inactive.

Source: *National Atmospheric Deposition Program* 2006. NADP Program Office, Illinois State Water Survey, http://nadp.sws.uiuc.edu/sites/sitemap.asp?state=il

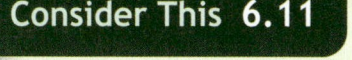

Consider This 6.11   The Rain in Maine . . . Oregon or Florida

Thanks to the NADP/NTN, almost every state plus Puerto Rico and the Virgin Islands has one or more precipitation monitoring sites.

a. In Figure 6.9a, name the precautions you see taken to preserve the integrity of the rain samples.

b. How many monitoring sites are in your state? A map with links is provided at the *Online Learning Center*.

c. Do you think the number and placement of collection sites in your state fairly represent the acidic deposition?

d. On the Web, select a collection site in your state (or in a neighboring one) that provides a photograph. Compare the picture with Figure 6.9a. What additional ways of minimizing contamination (for example, a fence or signage), if any, can you spot?

Answer

a. The collection buckets are located up off the ground, one is fitted with a lid that opens when it rains, and the area around the site is mowed. Also, the location is far from people and roads.

Each week, researchers at the Central Analytical Laboratory in Champaign, Illinois, receive hundreds of rain samples. The photographs assembled for Figure 6.10 give an indication of the magnitude of the operation. At the top left are sample collection buckets waiting to be cleaned prior to being shipped back to the collection sites. The top right photo shows a set of rain samples in the queue to be analyzed, each assigned an alphanumeric label. A small portion of each sample is saved after analysis and stored under refrigeration. The bottom left shows Karen Harlin, director of the laboratory, standing by the door to the cold room that contains the archived samples. Bottom right allows you to see some of the samples inside the cold room. Samples are available for researchers, including students.

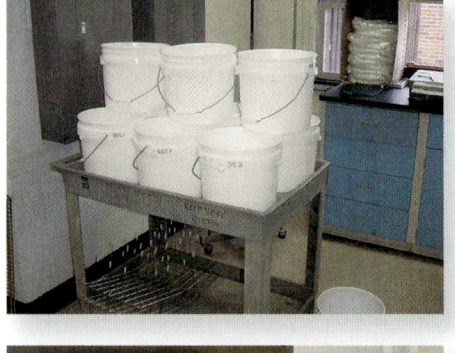
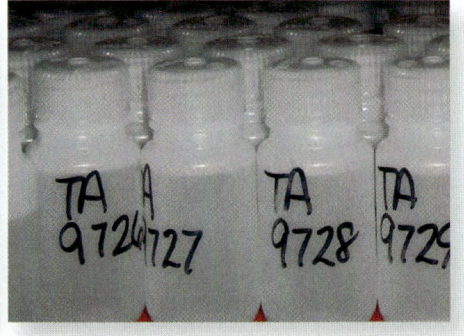

Figure 6.10

Photographs from the Central Analytical Laboratory (CAL), Champaign, Illinois.
Top left: Sample collection buckets waiting to be cleaned.
Top right: Rain samples in the queue to be analyzed.
Bottom left: Karen Harlin, former Director of the CAL, in front of the cold room.
Bottom right: Archived samples inside the cold room.

Rain samples used to be analyzed immediately in the field as well. This duplication served both as a check on the data and indicated the degree of deterioration of the samples during transport. The latter showed that small but nonetheless measurable changes can take place over time. For example, bacteria may consume the tiny amounts of natural acids present in rainwater (for example, formic acid and acetic acid) leading to a decrease in the acidity. Temperature changes also may lead to the loss of dissolved gases in the sample. Both because these effects were small, and because the measurements in the central laboratory were more easily standardized, the field measurements were discontinued in 2005.

Each year, researchers at the Central Analytical Laboratory use the analytical data to construct maps like the one shown in Figure 6.11. From these maps, we can confirm what we already knew, that all rain is slightly acidic. As mentioned previously, "pure rain" always contains a small amount of dissolved carbon dioxide. Recall that CO_2 is a natural component of Earth's atmosphere present in low concentration—about 385 ppm or 0.0385%. A tiny amount of carbon dioxide dissolves in water to produce a weakly acidic solution.

$$CO_2(g) + H_2O(l) \longrightarrow H^+(aq) + HCO_3^-(aq) \qquad [6.9]$$

This reaction occurs only to a limited extent; that is, only tiny amounts of H^+ and HCO_3^- (the hydrogen carbonate ion) are formed. But these small amounts are enough. At 25 °C, a sample of water exposed to atmospheric carbon dioxide has a pH of 5.6.

If you were to examine maps like the one in Figure 6.11 over the past decade, you would observe several trends. Generally speaking, the acidity has lessened slightly; that is, across the country, pH values are not quite as low as they once were. But as it turns out, pH is not the fundamental issue. Rather, it is the different chemicals *in the rain* that lower the pH. Since normal rain has a pH of about 5.3 (see Figure 6.6), CO_2 cannot be the sole source of H^+ in rainwater. Tiny amounts of other natural acids also contribute to its acidity. However, even these additional acids cannot account for the pH values below 5 that we observe in the Midwest and on the east coast (see Figure 6.11). Thus we must look elsewhere.

> Carbonated water is more acidic, because more CO_2 is forced to dissolve under pressure. The pH is about 4.7.

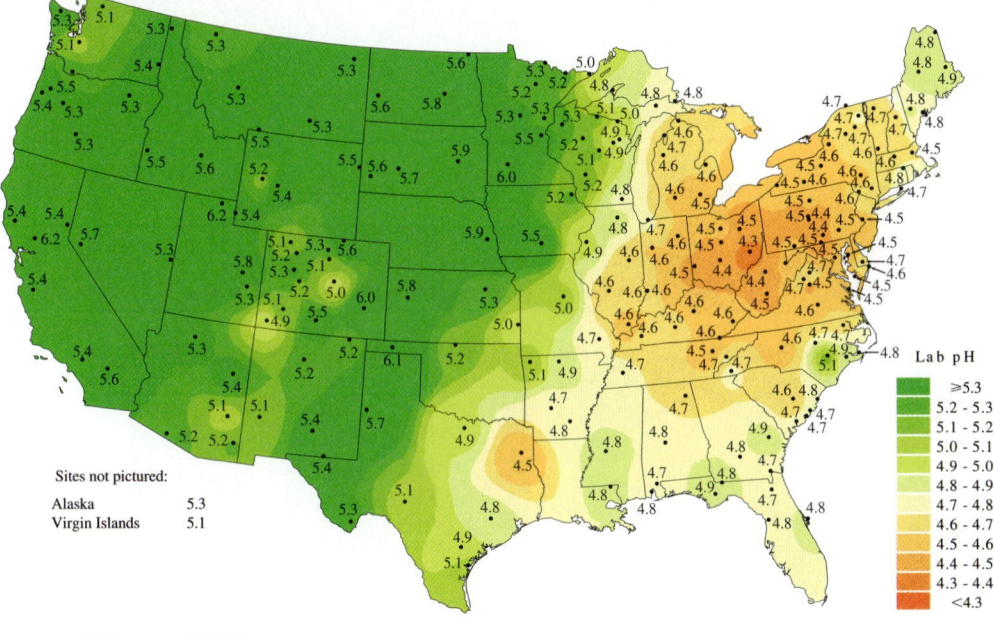

Figure 6.11

The pH of rain samples. Measurements made at the Central Analytical Laboratory, 2005. Values at stations in Alaska and the Virgin Islands are given at the lower left. Hawaii data not available.

Source: *National Atmospheric Deposition Program 2006*. Illinois State Water Survey, http://nadp.sws.uiuc.edu/isopleths/maps2005/phlab.gif

6.6   In Search of the Extra Acidity

According to Figure 6.11, acidic rain falls in the eastern third of the United States, especially in the Ohio River valley. What causes the extra acidity? Chemical analysis of the rain confirms that the chief culprits are sulfur dioxide (SO_2), sulfur trioxide (SO_3), nitrogen monoxide (NO), and nitrogen dioxide (NO_2). These compounds are collectively designated SO_x and NO_x, better known as "sox and nox."

At this stage the Sceptical Chymist should be raising an important question. Given the definition of an acid as a substance that contains and releases H^+ ions in water, how can SO_2, SO_3, NO, and NO_2 qualify? These compounds don't contain hydrogen! The explanation is that SO_x and NO_x dissolve in water to form acids that release H^+ ions. Although not acids themselves, the oxides of sulfur and nitrogen are **acid anhydrides,** literally "acids without water." When an acid anhydride is added to water, an acid is generated. For example, sulfur dioxide dissolves in water to form sulfurous acid.

Equations 6.10 and 6.11 are analogous to the reaction of CO_2 with water.	

$$SO_2(g) + H_2O(l) \longrightarrow \underset{\text{sulfurous acid}}{H_2SO_3(aq)} \qquad [6.10]$$

Similarly, sulfur trioxide dissolves in water to form sulfuric acid.

$$SO_3(g) + H_2O(l) \longrightarrow \underset{\text{sulfuric acid}}{H_2SO_4(aq)} \qquad [6.11]$$

In water, sulfuric acid is a source of H^+ ions.

$$H_2SO_4(aq) \longrightarrow H^+(aq) + \underset{\text{hydrogen sulfate ion}}{HSO_4{}^-(aq)} \qquad [6.12a]$$

The dissociations in equations 6.12b and 6.12c actually are more complex than shown here.

The hydrogen sulfate ion also can dissociate to yield another H^+ ion.

$$HSO_4{}^-(aq) \longrightarrow H^+(aq) + \underset{\text{sulfate ion}}{SO_4{}^{2-}(aq)} \qquad [6.12b]$$

Adding equations 6.12a and 6.12b shows that sulfuric acid dissociates to yield two hydrogen ions and a sulfate ion (SO_4^{2-}).

$$H_2SO_4(aq) \longrightarrow 2\,H^+(aq) + SO_4^{2-}(aq) \qquad [6.12c]$$

Your Turn 6.12   Sulfurous Acid

Write equations for the formation of two H^+ ions from sulfurous acid, analogous to chemical equations 6.12a, 6.12b, and 6.12c for sulfuric acid. Visit the *Online Learning Center* for interactive activities relating to acids and bases.

In a similar but more complicated way, NO_2 reacts in moist air to form nitric acid. This reaction is a simplification of the atmospheric chemistry that takes place.

$$4\,NO_2(g) + 2\,H_2O(l) + O_2(g) \longrightarrow 4\,HNO_3(aq) \qquad [6.13]$$
$$\text{nitric acid}$$

Like sulfuric acid, nitric acid also dissociates to release the H^+ ion.

$$HNO_3(aq) \longrightarrow H^+(aq) + NO_3^-(aq) \qquad [6.14]$$
$$\text{nitrate ion}$$

Geographically, then, regions with acidic rain should show elevated levels of the sulfate ion and the nitrate ion, from SO_x and NO_x, respectively. As we mentioned earlier, this acid deposition can be either wet or dry. Figure 6.12 shows wet deposition, usually called "acid rain," but also includes the other forms of precipitation that would land on your umbrella such as snow, sleet, or even hail.

Both Chapter 1 and Chapter 4 described the link between burning coal and SO_2 emissions. As you might then suspect, sulfur dioxide emissions are highest in states with many coal-fired electric power plants, steel mills, and other heavy industries that rely on coal. Ohio is one such state. In 2004 (as well as in years past), Ohio, followed by Pennsylvania and Indiana, lead the nation in SO_2 emissions. These same three states lead in NO_x emissions as well. But high NO_x emissions also are found in large urban areas with high population densities and heavy automobile traffic. Therefore, it is not surprising that in 1990 (and still today) the highest levels of atmospheric NO_2 were measured over Los Angeles County, the car capital of the country. Figure 6.12a does not show these high levels because the deposition in the arid west is often dry. Nonetheless, the emissions are significant. For example, the vegetation in Joshua Tree National Park, east of Los Angeles County, has been damaged by the dry deposition.

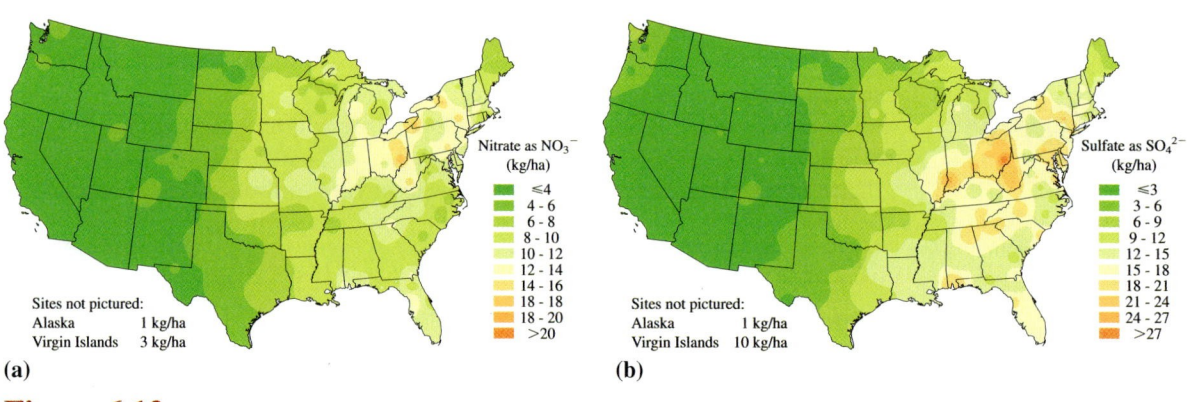

(a)   (b)

Figure 6.12

(a) 2005 wet deposition of nitrate ion in kilograms per hectare.
(b) 2005 wet deposition of sulfate ion in kilograms per hectare.

Source: *National Atmospheric Deposition Program* 2006. NADP Program Office, Illinois State Water Survey, Champaign, IL, http://nadp.sws.uiuc.edu/lib/data/2005as.pdf

Knowing now that oxides of sulfur and nitrogen contribute to acid rain formation, we need to get a closer look at how these oxides are formed and released into the atmosphere.

6.7 Sulfur Dioxide and the Combustion of Coal

Thus far, we have established a relationship between coal burning, atmospheric sulfur dioxide, and acid rain formation. Moreover, it is indisputable that both SO_2 and SO_3 react with water to yield an acidic solution. At this point, let's look more closely at coal and its combustion products. At first glance, coal may not appear much different from charcoal or black soot, both of which are essentially pure carbon. When carbon is burned with plenty of oxygen, it forms carbon dioxide and liberates large amounts of heat (which of course is the reason for burning it).

$$C\text{(in coal)} + O_2(g) \longrightarrow CO_2(g) \qquad [6.15]$$

As you learned in Chapter 4, coal is a complex substance. We can approximate its composition with the chemical formula $C_{135}H_{96}O_9NS$. Coal also contains small amounts of elements such as silicon, sodium, calcium, aluminum, nickel, copper, zinc, arsenic, lead, and mercury. Coal burns to release the elements it contains, primarily in the form of oxides. Because carbon and hydrogen are present in the largest quantity, large quantities of CO_2 and H_2O are produced. But burning coal also releases mercury, arsenic, and lead into the environment—definitely a cause for concern, but not one that we will pursue here. At the moment, sulfur is our primary element of interest.

How did the sulfur get in the coal? Several hundred million years ago, coal formed from decaying vegetation such as that found in swamps or peat bogs. Because sulfur is present in all living things, in part the sulfur originated in the ancient vegetation. However, most of the sulfur in coal came from the sulfate ion (SO_4^{2-}) naturally present in sea water. Millions of years ago, bacteria on the sea floors utilized sulfate as an oxygen source, removing the oxygen and releasing the sulfide ion (S^{2-}). In turn, the sulfide ion became incorporated into the ancient rocks (including coal) that were in contact with sea water. In contrast, the coal formed in fresh water peats has a lower sulfur content. Thus, the percent of sulfur in coal can vary from less than 1% to as much as 6%.

Burning sulfur in oxygen produces sulfur dioxide, a poisonous gas with an unmistakable choking odor (Figure 6.13).

$$S(s) + O_2(g) \longrightarrow SO_2(g) \qquad [6.16]$$

Because the sulfur content of coal varies, burning coal produces sulfur dioxide in varying amounts. This fact is central to the acid rain story. When coal is burned, the sulfur dioxide produced goes right up the smokestack along with the carbon dioxide, water vapor, and small amounts of metal oxide ash. Emission control measures can, of course, reduce the amount of SO_2, as we will see in Section 6.14. Thus depending on how coal-burning electrical utility plants are equipped, you will find varying levels of SO_2 emissions.

Once in the atmosphere, SO_2 can react with oxygen to form sulfur trioxide, SO_3. Sulfur trioxide plays a role in aerosol formation, as we will see in Section 6.11.

$$2\,SO_2(g) + O_2(g) \longrightarrow 2\,SO_3(g) \qquad [6.17]$$

This reaction is fairly slow, but is accelerated by the presence of finely divided solid particles, such as the ash that goes up the stack along with the SO_2. Once SO_3 is formed, it reacts rapidly with water vapor in the atmosphere to form sulfuric acid (see equation 6.11). Other pathways also are available for the conversion of sulfur dioxide into sulfuric acid. One of particular importance involves the hydroxyl radical ($\cdot OH$) that is formed from ozone and water in the presence of sunlight. The reaction of $\cdot OH$ with SO_2 accounts for 20–25% of the sulfuric acid in the atmosphere. The reaction goes faster in intense sunlight and thus is more important in summer and at midday.

The movement of sulfur through the biosphere should remind you of the carbon cycle described in Section 3.5.

Figure 6.13
Sulfur burns in air to produce SO_2. This gas produces an acidic solution when dissolved in water.

In ancient times sulfur was known as brimstone, thus the biblical admonition about "fire and brimstone."

A chemical calculation can help us better appreciate the vast quantities of SO_2 produced by coal-burning power plants. Such plants typically burn 1 million metric tons of coal a year, where a metric ton is equivalent to 1000 kg, or 1×10^3 g.

$$1 \times 10^6 \text{ metric tons coal/yr} = 1 \times 10^9 \text{ kg coal/yr} = 1 \times 10^{12} \text{ g coal/yr}$$

We will assume a low-sulfur coal that contains 2.0% sulfur; that is, 2.0 g sulfur per 100 g coal. First we can calculate the grams of sulfur released each year from 1 million metric tons (1×10^{12} g) of coal.

$$\frac{1 \times 10^{12} \text{ g coal}}{\text{yr}} \times \frac{2.0 \text{ g S}}{100 \text{ g coal}} = \frac{2.0 \times 10^{10} \text{ g S}}{\text{yr}}$$

Next, we use the fact that 1 mol of sulfur reacts with oxygen to form 1 mol of SO_2 (see equation 6.16). The molar mass of sulfur is 32.1 g, and the molar mass of SO_2 is 64.1 g, that is, 32.1 g + 2(16.0 g). Therefore, 32.1 g of sulfur burn to produce 64.1 g of SO_2.

$$\frac{2.0 \times 10^{10} \text{ g S}}{\text{yr}} \times \frac{64.1 \text{ g } SO_2}{32.1 \text{ g S}} = \frac{4.0 \times 10^{10} \text{ g } SO_2}{\text{yr}}$$

This mass of SO_2 is equivalent to 40,000 metric tons or 88 million pounds of SO_2 per year. Power plants burning higher sulfur coal may emit more than twice this!

The connection between burning coal and sulfur dioxide emissions in the United States is evident in Figure 6.14. Most of the emissions arise from power plants ("fuel combustion") in which coal or other fossil fuels are burned to generate electricity for public or industrial consumption. Transportation is responsible only for a small percent of the emissions because gasoline and diesel fuel contain relatively low amounts of sulfur. Industrial processes, such as the producing of metals from their ores, account for the remainder of the emissions. For example, the ores of both copper and nickel are sulfides. When nickel sulfide is heated to a high temperature in a smelter, the ore decomposes and sulfur dioxide is released. Similarly, smelting copper sulfide releases SO_2. Although the large-scale production of nickel and copper contributes only a few percent to the total emissions, huge quantities of SO_2 are generated in particular regions.

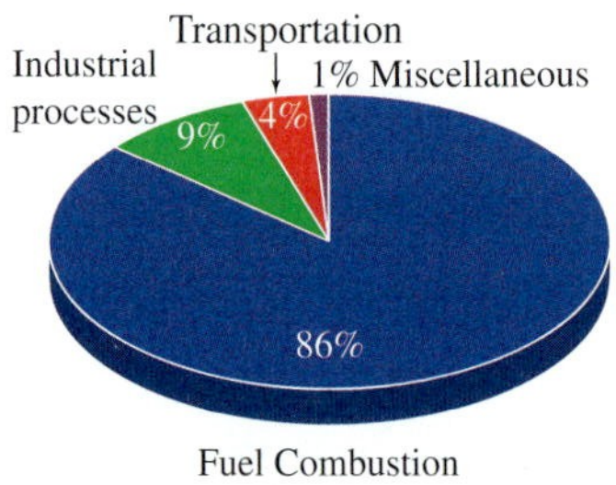

The world's largest smelter in Sudbury, Ontario, produces nickel from an ore that contains sulfur. The bleak, lifeless landscape in the immediate vicinity of the plant stands in mute testimony to earlier uncontrolled releases of SO_2. Today, after a major renovation in 1993, the two major smelters in the area have reduced their sulfur dioxide emissions substantially. Nonetheless, in 2003 over 200,000 metric tons of SO_2 was released, some of it up a tall smokestack. The fact that this is the world's tallest smokestack (equal in height to the Empire State Building) simply means that the emissions were carried farther away from Sudbury by the prevailing winds (Figure 6.15). Lest we point any fingers, Canadians report that more than half of acid deposition in the eastern portion of their country originates in the United States. The quantity of sulfur dioxide that drifts northward over the border into Canada is estimated to be 4 million tons per year.

Emissions data are given both in metric tons (1000 kg, 2200 lb) and in short tons (2000 lb). To add to the confusion, short tons usually are simply called tons; metric tons are also called tonnes.

See Section 3.7 to review mole calculations.

Transportation
Industrial processes
1% Miscellaneous
9%
4%
86%

Fuel Combustion

Figure 6.14

U.S. sulfur dioxide emission sources, 2003.

Source: EPA, Air Trends, http://www.epa.gov/air/airtrends/aqtrnd04/econemissions.html

---

Your Turn 6.13   Coal Calculations

a. Assume that coal can be represented by the chemical formula $C_{135}H_{96}O_9NS$. Calculate the fraction and percent (by mass) of sulfur in the coal.

b. A certain power plant burns 1.00×10^6 tons of coal per year. Assuming the sulfur content calculated in part **a**, calculate the tons of sulfur released per year.

c. Calculate the tons of SO_2 formed from this amount of sulfur.

d. Once released into the atmosphere, the SO_2 is likely to react with oxygen to form SO_3. What can happen next if SO_3 encounters water droplets?

Answers
a. 0.0168, or 1.68%   **b.** 1.68×10^4 tons S

Figure 6.15

The smokestack in Sudbury, Ontario, is the world's tallest at 1250 ft (381 m).

6.8   Nitrogen Oxides and the Acidification of Los Angeles

Coal has been indicted as a major environmental offender because, when burned, it produces sulfur dioxide. But SO_2 is not the only cause of acid precipitation; another guilty party has been identified. Consider, for example, the smoggy air that may settle into the Los Angeles basin. Although the concentration of SO_2 is relatively low, the rain is still very acidic. In January 1982, the fog near the Rose Bowl in Pasadena was found to have a pH of 2.5. Breathing it must have been like inhaling a fine mist of vinegar! The acidity of this fog exceeded that of normal precipitation by at least 500 times. In this same year, the fog at Corona del Mar on the coast south of Los Angeles was 10 times more acidic than near the Rose Bowl, registering a pH of 1.5. In both cases, something other than sulfur dioxide was involved.

To solve the mystery, we turn to the cars and trucks that jam the Los Angeles freeways day and night. At first glance, it may not be obvious how these thousands of vehicles contribute to acid precipitation. Gasoline burns to form CO_2 and H_2O, together with small amounts of CO, unburned hydrocarbons, and soot. But gasoline contains very little sulfur. Consequently, we must look for another source of acidity.

Nitrogen oxides have already been identified as contributors to acid rain, but gasoline does not contain nitrogen. Therefore, logic (and chemistry) asserts that nitrogen oxides cannot be formed from burning gasoline. Literally, this is correct. Remember, however, that about 80% of air consists of N_2 molecules. These molecules are remarkably stable and for the most part are unreactive. Nevertheless, if the temperature is high enough, nitrogen can and does react directly with a few elements. One of these is oxygen. Recall from Chapter 1 that with sufficient energy, nitrogen and oxygen combine to form nitrogen monoxide (nitric oxide).

$$N_2(g) + O_2(g) \xrightarrow{\text{high temperature}} 2\,NO(g) \qquad [6.18]$$

The energy necessary for this reaction can come from lightning or from the "lightning" inside an internal combustion engine. In an automobile, gasoline and air are drawn into the cylinders and compressed, bringing the N_2 and O_2 molecules closer together. The gasoline, once ignited, burns rapidly. The energy released powers the vehicle. But the unfortunate truth is that the energy also triggers chemical equation 6.18.

The reaction of N_2 with O_2 to form NO is not limited to automobile engines. The same reaction occurs when air is heated to a high temperature in the furnace of a coal-burning electrical power plant. Hence, such plants contribute vast amounts of both sulfur oxides and nitrogen oxides that acidify precipitation. On a national basis, the combustion of fuel (e.g., coal) in electrical utility plants and by industry releases just over a third of the nitrogen oxides (Figure 6.16). Transportation sources such as motor vehicles, aircraft, and trains account for over half. When in an urban environment, an even greater proportion of NO arises from motor vehicles.

In the early 1990s, a green chemistry solution to reducing NO emissions and energy consumption was introduced into U.S. glass manufacturing by Praxair Inc. of Tarrytown, NY. Their award-winning technology substitutes 100% oxygen for air in the large furnaces used to melt and reheat glass. Switching from air (78% nitrogen) to pure oxygen reduces NO production by 90% and cuts energy consumption by up to 50%. Glass manufacturers using the Praxair Oxy-Fuel technology save enough energy annually to meet the daily needs of 1 million Americans.

Once formed, nitrogen monoxide is highly reactive. As we noted in Chapter 1, through a series of steps it reacts with oxygen, the hydroxyl radical, and volatile organic compounds (VOCs) to form NO_2.

$$VOC + \cdot OH \longrightarrow A + O_2 \longrightarrow A' + NO \longrightarrow A'' + NO_2 \qquad [6.19]$$

Gasoline is a mixture of hydrocarbons. See Section 4.8.

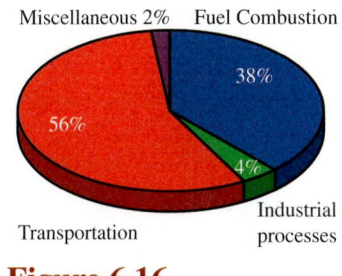

Figure 6.16

U.S. nitrogen oxide emission sources, 2003.

Source: EPA, Air Emission Trends, http://www.epa.gov/air

See Section 1.11 for more about equation 6.19.

The reactive intermediate species A, A′, and A″, present in trace amounts, are synthesized from the VOC molecules. The production of acid rain is connected to these same trace compounds in the atmosphere.

Nitrogen dioxide is a highly reactive, poisonous, red-brown gas with a nasty odor. For our purposes, the most significant reaction of NO_2 is the one that converts it to nitric acid, HNO_3. Earlier, equation 6.13 was a simplification of this conversion. Actually a series of reactions occurs in the presence of sunlight. These take place in the air surrounding Los Angeles, Phoenix, Dallas, and other sunny metropolitan areas. A key player is the hydroxyl radical. Once formed in the atmosphere, the hydroxyl radical can rapidly react with nitrogen dioxide to yield nitric acid.

$$NO_2(g) + \cdot OH(g) \longrightarrow HNO_3(l) \qquad [6.20]$$

As you have already seen in equation 6.14, HNO_3 dissociates completely in water to release H^+ and NO_3^-. The result is the alarmingly low pH values occasionally found in the rain and fog of Los Angeles.

6.9   SO$_2$ and NO$_x$—How Do They Stack Up?

Having identified SO_2 and NO_x as the two major contributors to acid precipitation, we will now examine their sources. In the United States, the annual anthropogenic (human) emissions are on the order of millions of tons—roughly 15 and 20 million for SO_2 and NO_x, respectively. Most of the sulfur dioxide emissions can be traced to coal-burning electrical utility plants. But these same utilities only account for a little over a third of the nitrogen oxides released (see Figure 6.16). The combustion engines that power cars, trucks, planes, and trains emit more than half of the NO_x.

The levels of these pollutants have changed dramatically over time. Before 1950, relatively small amounts of NO_x were present in rain, fog, and snow. Figure 6.17a shows that our current NO_x levels are the result of a relentless increase in emissions. These emissions, unlike SO_2 emissions, have leveled off in recent years. In contrast, SO_2 emissions have decreased substantially since their peak in 1974 (Figure 6.17b), a tribute to many things, including the 1990 Clean Air Act Amendments. In the final two sections of this chapter, we will examine how costs, control strategies, and politics have influenced SO_2 and NO_x emissions in the United States.

> 1 ton (short ton)
> = 2000 lb
> = 0.9072 metric tons (tonnes)

> **Your Turn 6.14**   **SO$_2$ and NO$_x$ Emissions**
>
> Figure 6.17a shows four sources of NO_x emissions in the United States. Which two did not change much between 1940 and 1995? Which did? For SO_2 in this same period, which sources led to the large increase in emissions in the 1970s?

Globally, the levels of SO_2 and NO_x also are changing over time. The latter are difficult to track. These originate from millions of small, unregulated, and mobile sources of NO_x across the globe. In contrast, emissions of SO_2 can be estimated with a reasonable degree of accuracy. National data on fossil-fuel consumption and the refining of metal ores that contain sulfur make this possible. To get an estimate, researchers start with the amount of fossil fuels (together with their sulfur content) produced in a country, then add in imports of fossil fuels, and finally subtract out exports. Metal refining is a bit trickier to estimate, as the amount of sulfur released depends on the technologies used (which are not always known). Nonetheless, it is possible to reach conclusions using these types of data.

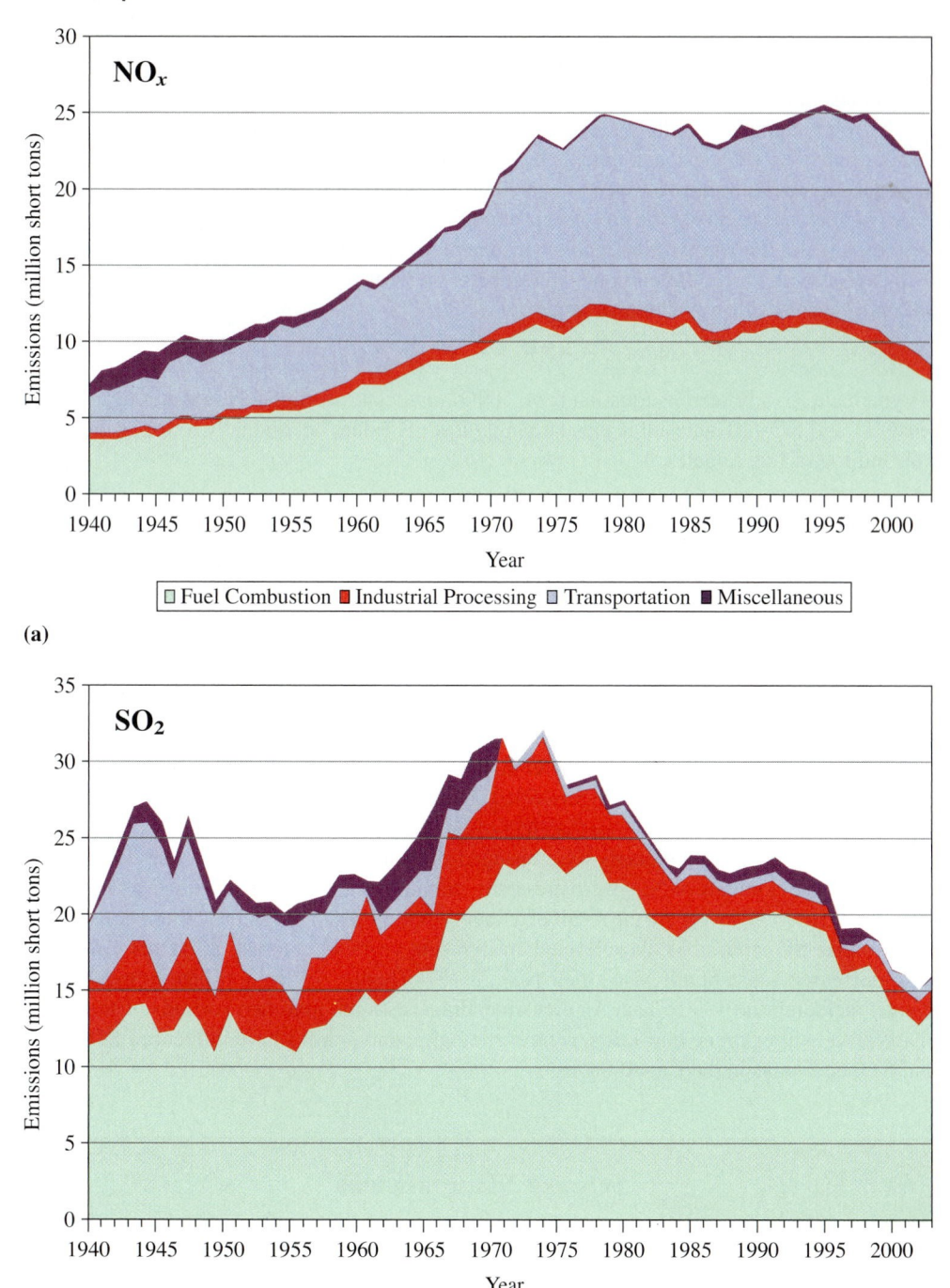

(a)

(b)

Figure 6.17

(a) U.S. nitrogen oxide emissions 1940–2003. **(b)** U.S. sulfur dioxide emissions 1940–2003.

Note: Fuel combustion refers to fossil-fuel combustion, such as coal.
Source: EPA/OAR, *National Air Pollutant Emission Trends, 1900–1998*, with recent data added.

One such estimate published in 2004 shows good news—a decline in world SO_2 emissions over the past decade. Back in the 1970s, Western Europe and North America shared the title of the world's largest emitters. As we saw earlier, the U.S. emission levels then declined rapidly and those of Western Europe followed suit. Eastern Europe took over the lead role, reaching its peak in 1989, and now likewise the emission levels are dropping. These decreases occurred for different reasons: environmental regulations in Europe in contrast to economic depression in Eastern Europe.

Table 6.1	Estimated Global Emissions of Sulfur Dioxide and Nitrogen Oxides	
	SO_2*	NO_x†
Natural Sources		
Oceans‡	25	
Soil		5.6
Volcanoes	10	
Lightning		5.0
Subtotal	35	10.6
Anthropogenic Sources		
All sources	69	
Fossil-fuel combustion		33.0
Biomass combustion		7.1
Aircraft		0.7
Subtotal	69	40.8
Total	104	51.4

*In units of 10^{12} g sulfur/year.

†In units of 10^{12} g nitrogen/year.

‡Sulfur is emitted from oceans in the form of dimethyl sulfide rather than SO_2. This compound is naturally converted to sulfur dioxide by the hydroxyl radical, $\cdot$OH.

Source: *Climate Change 2001: The Scientific Basis, Contribution of Working Group I to the Third Assessment Report of the Intergovernmental Panel on Climate Change,* Cambridge University Press, 2001, p. 315 and p. 260. Reprinted with permission.

Today, the continent of Asia leads in SO_2 emissions. In 1970, the United States emitted about 30 million tons of sulfur dioxide and China about 10 million tons. In 1990, both countries released about 22 million tons. With the start of the year 2000, China emerged as the clear leader in SO_2 emissions. However, with the closing of some older inefficient coal plants, the emissions from China have not risen as quickly as they could have. Time will tell.

Table 6.1 presents a global view of SO_2 and NO_x emissions from both natural and anthropogenic sources. Clearly, humans are not the only generators of sulfur and nitrogen oxides. Nonetheless, the amount of sulfur added to the atmosphere by humans is twice that of volcanoes, oceans, and other natural sources. The amount of nitrogen added as NO_x by humans is roughly four times that of natural sources such as lightning and the bacteria found in soils. The nitrogen cycle, which we will describe in Section 6.12, shows the complexities of the natural pathways of nitrogen in the biosphere.

Interpret the data of Table 6.1 with care. Natural emissions are inherently variable and difficult to estimate. For example, research studies have shown that the tons of NO_x formed by lightning vary widely by region (tending to be higher near the equator) and by month (higher during July in the Northern Hemisphere, January in the Southern). In addition, the local concentrations of NO_x are subject to the updrafts and downdrafts of storms.

Occasionally, major geological events alter the pattern. The June 1991 eruption of Mount Pinatubo in the Philippines is a case in point. This eruption, the largest in a century, injected between 15 and 30 million tons of sulfur dioxide into the stratosphere. At this elevation, SO_2 reacts to form tiny droplets and frozen crystals of sulfuric acid. For many months, this H_2SO_4 aerosol remained suspended in the atmosphere, reflecting and absorbing sunlight. The temporary drop in average global temperature observed in late 1991 that continued through 1992 has been attributed to the effects of the eruption. Indeed, when the cooling effects of Mount Pinatubo are included in the computer programs used to model global temperature changes, the predictions agree well with the observations. Evidence also indicates that the frozen crystals of H_2SO_4 provided many new catalytic sites for chemical reactions that lead to the destruction of stratospheric ozone. Quite obviously, the topics of this text are tightly interwoven.

6.10 Acid Deposition and Its Effects on Materials

As we have seen, much of the rain, mist, and snow in the United States is more acidic than unpolluted precipitation. On a regional basis, the acidity of precipitation has increased significantly since the Industrial Revolution. In the worst cases, fog and dew can have a pH of 3.0 or lower. But does this all really matter? To answer this question, we need to know something about the effects of acid deposition and how serious they really are.

National studies can help us. During the 1980s the U.S. Congress funded a national research effort called the National Acid Precipitation Assessment Program (NAPAP). Over 2000 scientists were involved, with a total expenditure of $500 million. The project was completed in 1990, and the participating scientists prepared a 28-volume set of technical reports (NAPAP, *State of the Science and Technology,* 1991). Some of the material in the remainder of this chapter is drawn from the NAPAP report and from a report from a conference in 2001. This conference was entitled "Acid Rain: Are the Problems Solved?" and was sponsored by the Center for Environmental Information. Its purpose was to "put the acid rain problem squarely back on the forefront of the public agenda.

And we agree—acid rain should remain on the public agenda. One reason is the damage done by acid rain (Table 6.2). In this section, we describe the effects of acid rain on metals, statues, and buildings. The effects of acid deposition on human health will be explored in the section that follows.

Table 6.2 Effects of Acid Rain and Recovery Benefits

Effects	Recovery Benefits
Materials Acid deposition contributes to the corrosion and deterioration of buildings, cultural objects, and cars. This decreases their value and increases the cost of correcting and repairing damage.	Less damage to buildings, cultural objects, and cars, thus lowering the future costs of correcting and repairing such damage. See Section 6.10.
Human Health Sulfur dioxide and nitrogen oxides in the air increase deaths from asthma and bronchitis and impair the cardiovascular system.	Fewer visits to the emergency room, fewer hospital admissions, and fewer deaths. See Section 6.11.
Visibility In the atmosphere, sulfur dioxide and nitrogen oxides form sulfate and nitrate aerosols that impair visibility and affect enjoyment of national parks and other scenic views.	Reduced haze, therefore the ability to view scenery at a greater distance and with greater clarity. See Section 6.11.
Surface Waters Acidic surface waters decrease the survivability of animal life in lakes and streams. In more severe instances, acidity eliminates some or all types of fish and organisms.	Lower levels of acidity in the surface waters and a restoration of animal life in the more severely damaged lakes and streams. See Section 6.13.
Forests Acid deposition contributes to forest degradation by impairing the growth of trees and increasing their susceptibility to winter injury, insect infestation, and drought. It also causes leaching and depletion of natural nutrients in forest soil.	Less stress on trees, thereby reducing the effects of winter injury, insect infestation, and drought. Less leaching of nutrients from soil, thereby improving the overall forest health.

Source: Adapted from *Emission Trends and Effects in the Eastern U.S.,* United States General Accounting Office, Report to Congressional Requesters, March, 2000.

Metals first. As we begin this discussion, remember that of the 100 or so elements on the periodic table, about 80% are metals. Metals typically are shiny and silvery in appearance; at least, they are shiny before they become tarnished or rusted by acid rain. Although acid rain (pH 3–5) does not affect all metals, unfortunately iron is one that it does.

Your Turn 6.15   Metals and Nonmetals

With the help of a periodic table, classify these elements as metals or nonmetals. Also give the chemical symbol for each.

a. iron   b. aluminum   c. fluorine
d. calcium   e. zinc   f. oxygen

As you can observe, iron is a major construction material. Bridges, railroads, and vehicles of all kinds depend on iron and the steel that is made from it. Rods of steel are used to strengthen concrete buildings and roadways. In many parts of the country, decorative iron fences and latticework both ornament and protect city and rural homes.

The problem with iron is that it rusts, as represented by this chemical equation.

$$4\ Fe(s) + 3\ O_2(g) \longrightarrow 2\ Fe_2O_3(s) \qquad [6.21]$$

Rusting is a slow process. Iron combines rapidly with oxygen only if you heat or ignite it, such as with a sparkler on the Fourth of July. But at room temperature, iron requires the presence of hydrogen ions to rust. Even pure water (pH = 7) has a sufficient concentration of H^+ to promote slow rusting. In the presence of acid, the rusting process is greatly accelerated. The role of H^+ is evident in equation 6.22, the first of a two-step process. In this step, iron metal dissolves.

$$4\ Fe(s) + 2\ O_2(g) + 8\ H^+(aq) \longrightarrow 4\ Fe^{2+}(aq) + 4\ H_2O(l) \qquad [6.22]$$

In the second step, the aqueous Fe^{2+} further reacts with oxygen.

$$4\ Fe^{2+}(aq) + O_2(g) + 4\ H_2O(l) \longrightarrow 2\ Fe_2O_3(s) + 8\ H^+(aq) \qquad [6.23]$$

The solid product, Fe_2O_3, is the familiar reddish brown material that we call rust.

Your Turn 6.16   Rust Adds Up

Show that rust formation, as represented in equation 6.21, is the sum of equations 6.22 and 6.23.

Your Turn 6.17   Careful with the Charges

On our planet, the element iron is found in several different chemical forms. This section just mentioned three: Fe, Fe^{2+}, and Fe^{3+}.

a. Which one of these is the familiar silvery iron metal?
b. With respect to iron metal, have any of these gained valence electrons? If so, which one(s) and how many electrons?
c. Have any lost valence electrons? If so, which one(s) and how many?

Answer
b. With respect to Fe, neither Fe^{2+} nor Fe^{3+} has gained valence electrons.

Because iron is inherently unstable when exposed to the natural environment, enormous sums of money are spent annually to protect exposed iron and steel in bridges, cars, buildings, and ships. Paint is the most common means of protection, but even paint degrades, especially when exposed to acidic rain and gases. Coating iron with a thin layer of a second metal such as chromium (Cr) or zinc (Zn) is another means of protection. Iron coated with zinc is called **galvanized iron.** Galvanized iron is still susceptible to the presence of acidic rain. Because of this, galvanized structures must be replaced more frequently than in the past.

Automobile paint can be spotted or pitted by acid deposition. To prevent this, automobile manufacturers now use acid-resistant paints. It is an irony that automobiles emit the very chemical that mars their paint. Follow the NO from your car's tailpipe and you may find that this chemical eventually ends up in droplets hitting the hood of your car.

Acidic rain also damages statues and monuments made of marble. For example, those in the Gettysburg National Battlefield have suffered irreparable damage. Figure 6.18 shows a recognizable, but much deteriorated statue of George Washington in New York City. Marble limestone, composed mainly of calcium carbonate, $CaCO_3$, slowly dissolves in the presence of H^+ ions.

$$CaCO_3(s) + 2\,H^+(aq) \longrightarrow Ca^{2+}(aq) + CO_2(g) + H_2O(l) \qquad [6.24]$$

Your Turn 6.18   Damage to Marble

Marble can contain both magnesium carbonate and calcium carbonate.

a. Analogous to equation 6.24, write the chemical equation for the reaction of acidic rain with magnesium carbonate.
b. Marble never contains sodium bicarbonate. Why?
Hint: See Section 5.9 on the solubilities of ionic compounds.

Your Turn 6.19   Damage from SO_2

Suppose that the acid represented in equation 6.24 by $H^+(aq)$ is sulfuric acid. Write the balanced chemical equation for the reaction of sulfuric acid with marble.

In 1944

At present

Figure 6.18

Acid rain damaged this limestone statue of George Washington. It was erected in New York City in 1944.

Figure 6.19

Acid rain knows no geographic or political boundaries. Acid rain has eroded Mayan ruins at Chichén Itzá, Mexico.

Visitors to the Lincoln Memorial in Washington, D.C., learn that the huge stalactites growing in chambers beneath the memorial are the result of acid rain eroding the marble, again a material containing either calcium carbonate or magnesium carbonate (or both). Other monuments in the eastern United States are suffering similar fates. Some limestone tombstones are no longer legible. Worldwide, many priceless and irreplaceable marble statues and buildings are being attacked by airborne acids (Figure 6.19). The Parthenon in Greece, the Taj Mahal in India, and the Mayan ruins at Chichén Itzá all show signs of acid erosion. Ironically, some of the acid deposition at these sites is due to the NO_x produced by the tour buses and vehicles with minimal emissions controls.

Consider This 6.20   **Deterioration and Damage**

Reexamine Figure 6.18. Although it may be tempting to blame acid rain for the damage, other agents may be at work. View possible other culprits for yourself by taking a photo tour of our nation's capitol, courtesy of a Web site on acid rain provided by the United States Geological Survey. A link is provided at the *Online Learning Center.* What kinds of damage do the photos show? What promotes damage by acid rain? What else has caused the deterioration?

Consider This 6.21   **Acid Rain Across the Globe**

The concerns of acid rain vary across the globe. Many countries in North America and Europe have Web sites dealing with acid rain. Search to locate one or use the links provided at the *Online Learning Center.* What are the issues in the country you selected? Does the acid deposition originate outside the borders of the country?

6.11   Acid Deposition, Haze, and Human Health

The more obvious effects of acid deposition often can be observed simply by looking out the window. Anyone living in the eastern half of the United States is familiar with the summer haze that may settle over the landscape. Ironically, you become

Visual range 20 miles

Visual range 100 miles

Figure 6.20

A hazy day and a clear day from Look Rock Tower in the Great Smoky Mountains National Park.

more aware of it on the occasional clear day when it really does seem that you can see forever. Airline passengers, as they peer down from 30,000 ft, may notice that the features and colors of the landscape are blurred. Just as the visitors to the Great Smoky Mountains National Park can view contrasting sets of photographs, so can you in Figure 6.20.

The causes of haze are well understood, but they differ from region to region. In the east, for example, coal-burning power plants in the Ohio Valley and elsewhere produce the smoke and particulate matter that in turn create the haze. In the west, a different set of particulates including soil dust and the soot of wood-burning stoves add to the haze.

East or west, power plants emit NO_x and SO_2. Although both contribute to haze, for the purposes of illustrating acid deposition we will focus on the latter. As we mentioned earlier, coal contains a few percent sulfur, and when the coal is burned, a steady stream of sulfur dioxide is released into the atmosphere. Since sulfur dioxide is colorless, this gas is not what we are peering through as "haze." Rather, SO_2 is the precursor to this haze.

> Aerosols consist of tiny particles that remain suspended in our atmosphere (Section 1.11).

Let's focus on a molecule of SO_2 as it exits the tall smokestack of a power plant. As it moves downwind, it can form an aerosol of sulfuric acid via a series of steps. The first is the reaction of SO_2 with oxygen to form SO_3, as we saw earlier in equation 6.17. Sulfur trioxide is also colorless gas, but it has the property of being **hygroscopic,** that is, it readily absorbs water from the atmosphere and retains it. As we saw in equation 6.11, a molecule of SO_3 can react rapidly with a water molecule to form sulfuric acid.

Your Turn 6.22   Droplets of Acid

As a review of the sulfur chemistry just described, write a set of chemical equations that start with elemental sulfur in coal and eventually produce sulfuric acid.

The tiny, tiny droplets of sulfuric acid then coagulate to produce larger droplets. These droplets form an aerosol with particles about a micrometer ($1\mu = 10^{-6}$ m) in size. These particles of sulfuric acid do not absorb sunlight. Rather, they scatter (reflect) sunlight, reducing visibility. The aerosols of sulfuric acid, which can persist for several days, can travel hundreds of miles downwind, which is why the haze can become so widespread. In addition, these fine particles of acid are stable enough that they enter our buildings and become part of the air that we breathe indoors.

You also may have heard of sulfate aerosols. Recall that sulfuric acid, H_2SO_4, ionizes to produce H^+, HSO_4^-, and SO_4^{2-}. The concentrations of each can be measured in an aerosol. But these acidic aerosols may react with bases to produce salts that contain the

sulfate ion. Typically this base is ammonia or in aqueous form, ammonium hydroxide. Thus, the particles in an aerosol may be a mixture of sulfuric acid, ammonium sulfate, $(NH_4)_2SO_4$, and ammonium hydrogen sulfate, NH_4HSO_4. Reporting the concentration of sulfate and hydrogen sulfate ions (rather than simply the pH) gives a better indication of how much sulfuric acid was initially present.

> Similarly, acidic aerosols of ammonium nitrate form from NO_x.

Your Turn 6.23   Sulfate Aerosols

As a review of the acid–base chemistry just described, write balanced chemical equations for these processes.

a. The reaction of sulfuric acid with ammonium hydroxide to form ammonium hydrogen sulfate and water.

b. The reaction of sulfuric acid with ammonium hydroxide to form ammonium sulfate and water.
Hint: This requires 2 mol of base.

Haze is most pronounced in summer when there is more sunlight to accelerate the photochemical reactions leading to sulfuric acid. As a result, the average visibility in the eastern United States is now about 20 miles and occasionally as low as 1 mile. By contrast, visibility in the western states is now lessened from the natural visual range of about 200 miles to 100 miles or less. Where you formerly might have been able to see the mountains 100 miles away, these mountains may now have disappeared into the haze. The visibility in many national parks has been affected, including Yellowstone, the Grand Canyon, Glacier, and Zion.

> Each summer, wild fires also contribute to the haze seen over parts of the western part of the United States.

The Clean Air Act of 1970 and its subsequent amendments included provisions to improve the visibility in our national parks. Although the standards set were by federal law, the states were charged with the implementation of these standards. Visibility continued to drop in the national parks. In the last few days of his presidency, Bill Clinton signed a bill authorizing the EPA to issue regulations to help clear the skies in national parks and wilderness areas. These regulations, called the Regional Haze Rule (1999), required the hundreds of older power plants that emitted vast quantities of SO_2, NO_x, and particulates to retrofit their operations with pollution controls. A final set of amendments, the Clean Air Visibility Rule, were issued on June 15, 2005 by President George W. Bush. These amendments limit SO_2 and NO_x emissions in western states and continue to be controversial.

Consider This 6.24   Hazy at Mount Rainier?

Web cams! Live on the Web, see for yourself the haze (or lack thereof) at Mount Rainer. Dozens of other places have Web cams as well, and the EPA posts a list of these. A link is provided at the *Online Learning Center*.

a. During daylight hours, look up several Web cams to see what's out there—or not.

b. Find the current air quality for a location of your choice. Some sites provide this information together with the Web cam photographs. For others, you can obtain the data from the EPA AIRNOW site.

c. How well does the air quality correlate with the visibility?

What you see is what you breathe. Once inhaled, the acidic droplets attack sensitive lung tissue. Those most susceptible include the elderly, the ill, and those with asthma, emphysema, and cardiovascular disease. People with preexisting conditions such as bronchitis

and pneumonia are likely to exhibit increased mortality rates. Those in good health feel the irritating effects of the acidic aerosols as well. Thus breathing air contaminated with aerosols of sulfate and sulfuric acid comes with a medical price tag.

Decreasing aerosol levels result in a huge savings, both in real dollars and in your health. The problem is that the costs and savings are not directly borne by the same groups. Industry must pay to clean up; people must pay medical bills. Government, of course, is involved in paying both.

The EPA estimates that the Clean Air Visibility Rule of 2005 will provide "substantial health benefits in the range of $8.4 to $9.8 billion each year—preventing an estimated 1,600 premature deaths, 2,200 non-fatal heart attacks, 960 hospital admissions, and more than 1 million lost school and work days." The cost-to-benefit ratio is thus *exceedingly* favorable. The EPA estimates the total annual costs for implementation are in the range of $1.5 billion.

Historically, polluted air has exacted a huge price. One of the worst recorded instances of pollution-related respiratory illness occurred in London in 1952. Periods of foggy bad air were nothing unusual to the British Isles, as factory chimneys had belched smoke into the air for several hundred years. But in December, 1952, the weather was colder than usual and people were burning large quantities of sulfur-rich coal in their home fireplaces. Due to unusual weather conditions, a deep layer of fog developed that trapped all the smoke and pollutants for five days, dropping visibility to practically zero. The deadly aerosol caused more than 4000 deaths, during its peak claiming 900 lives daily.

In 1948, a similar incident occurred in Donora, PA, a steel mill town south of Pittsburgh. Again a layer of fog trapped industrial pollutants close to the ground. By noon, the skies had darkened with a choking aerosol of fog and smoke (Figure 6.21). An 81-year-old fireman who took oxygen door-to-door to the victims reported, "It may sound dramatic or exaggerated, but you could barely see." High concentrations of sulfuric acid and other pollutants soon caused widespread illness. During the fog, 17 people died, to be followed by 4 more later. Although Donora and London were extreme and unusual incidents from the past, people still breathe highly polluted air today. The U.S. EPA and the World Health Organization currently estimate that 625 million people are still exposed to unhealthy levels of SO_2 released by the burning of fossil fuels.

Although acidic fogs can be immediately hazardous to one's health, public concern is growing over the indirect effects of acid deposition. For example, the solubilities of certain toxic metal ions, including lead, cadmium, and mercury are significantly increased in the presence of acids. These elements are naturally present in the environment, but normally are tightly bound in the minerals that make up soil and rock. Dissolved in acidified water and conveyed to the public water supply, these metals can pose serious health threats. Elevated concentrations of heavy metals already have been found in some of the major water reservoirs in Western Europe.

(a)

(b)

Figure 6.21

(a) A 1948 news headline from Donora, PA. (b) Donora at noon during the deadly smog of 1948.

Clearly there is a connection between burning fossil fuels, acidic precipitation, and human health. An article written in the journal *Science* in 2001 by an international team of authors bluntly assessed the situation, "For every day that policies to reduce fossil-fuel combustion emissions are postponed, deaths and illness related to air pollution will increase." Studies by the EPA have estimated that the reductions in SO_2 and associated acid aerosols pollution called for by the Clean Air Act Amendments of 1990 could result in saving billions of dollars in health care costs over time. The savings would come principally from reduced costs to treat pulmonary diseases such as asthma and bronchitis and from a decrease in premature deaths. But there is another connection between acidic precipitation and humans that may be less obvious. To find it, we need to return to NO_x.

6.12 NO_x—The Double Whammy

A slice of pizza? A glass of lemonade? A green salad with oil and vinegar? Rarely a day goes by that you don't ingest food in one form or another. Clearly, you need to eat in order to stay alive. As you are reading this, men and women across the globe are producing food by planting fields of grain, harvesting fruits and vegetables by the truckload, and perhaps even growing oregano or chives on a sunny windowsill. To their credit, humans have become quite expert in raising both plants and animals. However, a complication is that producing food such as a sausage pizza, just like driving a car (perhaps the one you used to pick up the pizza), adds to the acidity of the environment.

In earlier sections, we examined the link between energy production and the acidic emissions of SO_2 and NO_x. Here we will explore another link, this time between food production and NO_x emissions. The connection stems from a key difference between compounds of nitrogen and of sulfur in the environment; namely, that nitrates act as fertilizers and promote plant growth. Actually plants depend on sulfur as well, as they do on other elements such as carbon, hydrogen, phosphorus, and potassium. Except for nitrogen, however, these other elements tend to be readily available in the biosphere for uptake by plants. Since usable forms of nitrogen are in short supply, we need to add them in the form of fertilizers.

All living things (not just plants) require nitrogen.

The Sceptical Chymist might wonder how nitrogen levels can be low in soils when N_2 makes up so much of our atmosphere. Although abundant, the nitrogen molecule is *not* in a chemical form that most plants can use. As we have pointed out earlier, N_2 is far less reactive than O_2.

Your Turn 6.25   **Unreactive Nitrogen**

Review this information from earlier chapters.

a. Nitrogen is a major constituent of our atmosphere. Approximately what percent?
b. Draw the Lewis structure of N_2.
c. How does the bond energy of the triple bond in N_2 compare with other bond energies?
Hint: See Table 4.2.

In order to grow, plants need access to a more reactive form of nitrogen, such as the ammonium ion, ammonia, or the nitrate ion. These and other reactive forms are listed in Table 6.3. We refer to them collectively as **reactive nitrogen.** These compounds of nitrogen are biologically active, chemically active, or active with light in our atmosphere. As you might suspect, the air pollutants NO and NO_2 are among

Some scientists designate reactive nitrogen as Nr, where the r stands for reactive. We do not use this representation, as Nr resembles the chemical symbol for an element (and no element has this symbol).

Table 6.3	Some Reactive Forms of Nitrogen
Name	**Chemical Formula**
nitrogen monoxide	NO
nitrogen dioxide	NO_2
nitrous oxide	N_2O
nitrate ion	NO_3^-
nitrite ion	NO_2^-
nitric acid	HNO_3
ammonia	NH_3
ammonium ion	NH_4^+

Note: All are naturally occurring.

them. These forms of nitrogen all occur naturally and until recently were all present on our planet in *relatively small amounts*. Other forms of reactive nitrogen also exist, but we will introduce these when we need them later for our study of polymers, proteins, and DNA.

Your Turn 6.26   Reactive Nitrogen

From Table 6.3, select three forms of reactive nitrogen. For each one:
a. Write chemical reactions that illustrate the reactive nature of the chemical.
   Note: In the case of an ion, select a compound containing the ion.
b. Draw the Lewis structure, noting any unpaired electrons.
   Hint: Remember to add/subtract valence electron for negative/positive ions.

Figure 6.22
Nodules on the root of a soya plant that contain nitrogen-fixing bacteria.

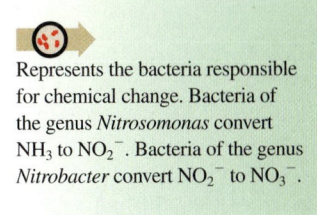
Represents the bacteria responsible for chemical change. Bacteria of the genus *Nitrosomonas* convert NH_3 to NO_2^-. Bacteria of the genus *Nitrobacter* convert NO_2^- to NO_3^-.

Of the oxides of nitrogen, N_2O is emitted naturally in the greatest amount. It is a potent greenhouse gas. See Section 3.8.

Although we categorized N_2 as unreactive, one reaction involving the nitrogen molecule is of utmost importance: biological nitrogen fixation. Plants such as alfalfa, beans, and peas remove, or "fix," N_2 from the atmosphere (Figure 6.22). To be more accurate, it is not the plants themselves, but rather the bacteria living on or near the roots of these plants that fix the nitrogen. As part of their metabolism, **nitrogen-fixing bacteria** remove nitrogen from the air and convert it to ammonia. When the ammonia dissolves in water, it releases the ammonium ion (see equation 6.4b). This ion is one of two forms of reactive nitrogen that most plants can absorb. Here is the pathway:

$$N_2 \longrightarrow NH_3 \xrightarrow{H_2O} NH_4^+ \qquad [6.25]$$
nitrogen fixation

The other form of reactive nitrogen that plants can absorb is the nitrate ion. **Nitrification** is the process of converting ammonia in the soil to the nitrate ion. Two types of bacteria are involved along this pathway.

$$NH_4^+ \longrightarrow NO_2^- \longrightarrow NO_3^- \qquad [6.26]$$
bacteria in the soil   bacteria in the soil

Finally, to come full circle, **denitrification** occurs, that is, the process of converting nitrates back to nitrogen gas. Again, bacteria accomplish this task. In so doing, these bacteria harness the energy released when the stable N_2 molecule forms. Depending on the soil conditions, the pathway may occur in steps that include NO and N_2O. Thus, these reactive forms of nitrogen also can be released from the soil.

$$NO_3^- \longrightarrow NO \longrightarrow N_2O \longrightarrow N_2 \qquad [6.27]$$
bacteria in the soil   bacteria in the soil   bacteria in the soil

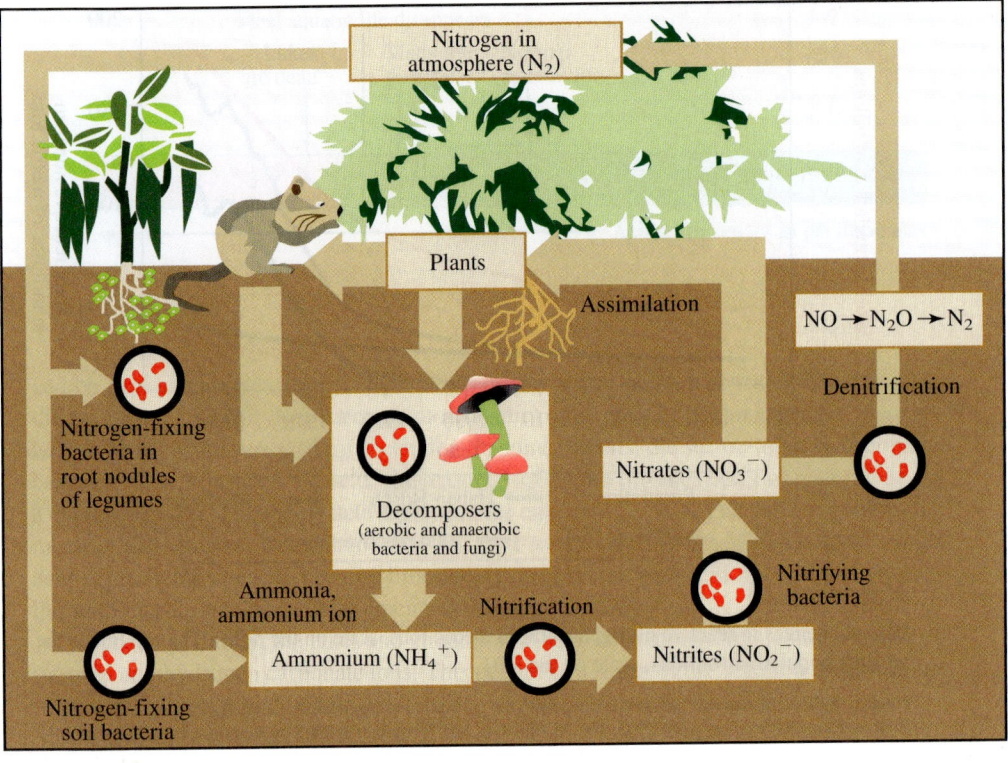

Figure 6.23

The nitrogen cycle (simplified).

All of these pathways are part of the **nitrogen cycle,** a set of chemical pathways whereby nitrogen moves through the biosphere. Figure 6.23 assembles pathways 6.25, 6.26, and 6.27 into a simplified version of the nitrogen cycle. In this cycle, all species are forms of reactive nitrogen except for N_2.

Returning now to the story of acidification, remember that reactive forms of nitrogen are needed for plant growth. Since bacteria in the soil cannot supply ammonia, ammonium ion, or nitrate ion in the amounts needed for the growth of crops, farmers use fertilizers. A few centuries ago, fertilizers were obtained by mining deposits of saltpeter (ammonium nitrate from the deserts of Chile) or by collecting guano, a nitrogen-rich deposit from bird and bat droppings in Peru. Neither source, however, was sufficient to meet the demand. An additional drain on the supply of nitrates was that they were used to make gunpowder and other explosives such as TNT. Thus, in the early 1900s, the search was on for a synthetic source of reactive nitrogen compounds.

How are fertilizers obtained in the large quantities needed for present-day agriculture? The answer lies in a second important reaction of N_2, one that literally captures it out of the air to synthesize ammonia:

$$N_2(g) + 3\,H_2(g) \longrightarrow 2\,NH_3(g) \qquad [6.28]$$

This famous chemical reaction is known as the Haber–Bosch process. It allows the economical production of ammonia, which in turn allows the large-scale production of fertilizers and nitrogen-based explosives. As a fertilizer, ammonia can be directly applied to the soil or can be applied as ammonium nitrate or ammonium phosphate. The green line that starts around 1910 in Figure 6.24 represents the large increase in reactive nitrogen from the Haber–Bosch process.

Also notice the gold line on this same graph. Clearly, the burning of fossil fuels is another large source of reactive nitrogen in our environment. At the high temperatures of combustion, N_2 reacts with O_2 to form NO. The top red line for population, of course, comes as no surprise. The increases in reactive nitrogen from burning fossil fuels (energy production) and fertilization (food production) parallel the growth in world population (people production).

Now we can understand the double whammy of NO_x emissions. The first problem is that they contribute to acid deposition that in turn forms haze and diminishes air quality.

In 1918, Fritz Haber received the Nobel Prize in chemistry for synthesizing NH_3 from N_2 and H_2. In 1931, Carl Bosch received the prize for using this synthesis commercially.

As it turns out, understanding the acidification of lakes is a good deal more compli-cated than simply measuring pH and acid-neutralizing capacities. One level of complex-ity is added by annual variations. Some years, for example, heavy winter snowfalls persist into the spring and then melt suddenly. As a result, the runoff may be more acidic than usual, because it contains all the acidic deposits locked away in the winter snows. A surge of acidity may enter the waterways at just the time when fish are spawning or hatching and are more vulnerable. In the Adirondacks, about 70% of the sensitive lakes are at risk for episodic acidification, in comparison with a far smaller percent that are chronically affected (19%). In the Appalachians, the number of episodically affected lakes (30%) is seven times those chronically affected.

Another level of complexity comes with the buildup of reactive nitrogen species such as the nitrate ion or the ammonium ion. **Nitrogen saturation** occurs when an area is over-loaded with "nitrogen," that is, when the reactive forms of nitrogen entering an ecosystem exceed the system's capacity to absorb the nitrogen. The patterns of nitrogen absorption depend on both the age of the vegetation (in general, younger, growing forests absorb nu-trients more than older ones) and the time of year (plant growth stops in the winter). But nitrogen absorption seems to have its limits. Once nitrogen saturation develops, the ni-trate ion accumulates with an accompanying rise in acidity. As a result, the soils have little ability to neutralize acidic precipitation before it runs off into the lakes and streams.

When, if ever, will the lakes recover? The good news is that the SO_2 emissions have been declining in recent years, and we have seen a corresponding decrease in the sulfate ion concentrations in the lakes of the Adirondacks. However, even though NO_x emissions have remained fairly constant, the amount of nitrates in the Adirondacks is increasing in more lakes than not. Thus, it appears that nitrogen saturation has occurred in the sur-rounding vegetation, with more of the acidity ending up in the lakes. The soil in the region of these lakes most likely has lost some of its acid-neutralizing capacity.

Recent findings are mixed. A March 2000 report to Congress puts it bluntly that "The lakes in the Adirondack Mountains are taking longer to recover than lakes located elsewhere and are likely to recover less or not recover, without further reductions of acid deposition." The Progress Report on Acid Rain issued by the EPA in 2004, however, reports some im-provements. For example, in comparison with earlier years when over 10% of the lakes in the Adirondacks were acidic, today the value is closer to 8%. Similar improvements are documented in the Midwest, where now only about 1% of the lakes are acidic. In contrast, the lakes in New England and the Blue Ridge Mountains remain stubbornly acidic.

6.14   Control Strategies

With the Clean Air Act Amendments of 1990, many hoped that the problems of acid rain would be solved. The Acid Rain Program that was established as part of the Clean Air Act Amendments of 1990 made reducing NO_x and SO_2 emissions a national priority. Although as a nation we have made significant reductions, we are still challenged to clean up regions of polluted and acidic air.

For NO_x, the Acid Rain Program set a target of reducing the annual emissions by 2 million tons by 2000. Phase I of the NO_x program applied to about 170 coal-fired boilers that produce electricity, specifying an emission rate of either 0.50 or 0.45 lb of NO_x per million Btu of heat input, depending on the type of boiler. Flexibility was built in, so that emission rates could be averaged over several units. Phase II began in 2000, tightening these emission standards and applying standards to still other types of boilers.

Btu stands for British thermal unit, the amount of heat needed to raise 1 pound of water 1 °F.

In spite of these efforts, the goal for NO_x emissions has not yet been achieved. Although emissions by electrical utilities (generating about a quarter of the NO_x) de-clined, NO_x emissions increased elsewhere, such as by the increasing number of trucks and automobiles on our highways. Reduction of nitrogen oxides from these vehicles is particularly challenging, because as sources they are small, individually owned, and by design mobile. There are more than 200 million motor vehicles in the United States and about 1 billion worldwide. Of these, the biggest contributors to NO_x pollution continue to be diesel engines, as noted in Figure 6.26.

Consider This 6.28   Less Dirty Diesels?

How successful are the current efforts to reduce diesel emissions of NO_x and particulate matter? In June 2006, the EPA released a progress report on the National Clean Diesel Campaign. The *Online Learning Center* has a link. Read the report, draw your own conclusions, and summarize them.

Figure 6.26
Sooty exhaust from a diesel truck. Note: NO emissions also are present but not visible.

Sceptical Chymist 6.29   Tractors and Cars

According to former EPA administrator Christine Todd Whitman, a large bulldozer produces 800 lb of pollution per year, the equivalent of 26 cars. From this, how many pounds of pollution is she crediting to each car per year? Which pollutants is a car producing and does her number seem to be within bounds? You may want to assume 10,000 miles driven per year at 20 miles per gallon as a basis for your calculations. *Hint:* As of January 2008, carbon dioxide was not considered a pollutant by the EPA.

To reduce NO_x, many techniques bearing a range of price tags are in use. It is chemically possible to reduce the NO emitted by cars and trucks by fitting them with catalytic converters and other emissions control devices. We already mentioned one of the functions of these catalysts: converting CO and unburned hydrocarbon fragments to CO_2. Other catalysts, typically in other parts of the catalytic converter, promote the reversal of the combination of nitrogen and oxygen that occurs in the engine at high temperatures. As the exhaust gases cool, the NO tends to decompose into its constituent elements.

$$2\,NO(g) \longrightarrow N_2(g) + O_2(g) \qquad [6.31]$$

Normally, this reaction proceeds slowly, but the appropriate catalyst can significantly increase its rate and thus decrease the amount of NO emitted. A current program funded by the EPA seeks to reduce emissions from school buses (Figure 6.27). Using catalysts is one of several strategies employed, and others include reducing engine idle time and using cleaner fuels.

Coal-fired utility plants, another major source of NO_x emissions, demonstrate other new technologies. For example, the Clean Coal Technology (CCT) Demonstration Program has developed and installed low-NO_x burners on numerous coal-fired plants. These burners decrease the amount of air during the combustion process, so that with less oxygen present, less NO_x is produced. In 2003, the U.S. Department of Energy reported that low-NO_x burners are now on 75% of the coal-burning power stations. Another CCT project involves "reburning" where additional fuel is injected into the combustion products to strip O out of NO_x. Both these new technologies are sufficiently complex that an artificial intelligence system may be needed to optimize the operating conditions. The successes in reducing NO_x emissions reported by one power station using new CCT technologies are shown in Figure 6.28. As we will describe shortly, lowering SO_2 emissions was more successfully achieved.

Figure 6.27
Clean School Bus Program of the EPA.

Clean Coal Technology was mentioned earlier in Section 4.6.

Consider This 6.30   CCT Demonstration Program

The CCT Program is funded both by government and industry and seeks technologies that better meet our environmental needs. Use the map at the Clean Coal Technology Compendium Web site to access a demonstration site of your choice. Describe the project you found. Links are provided at the *Online Learning Center.*

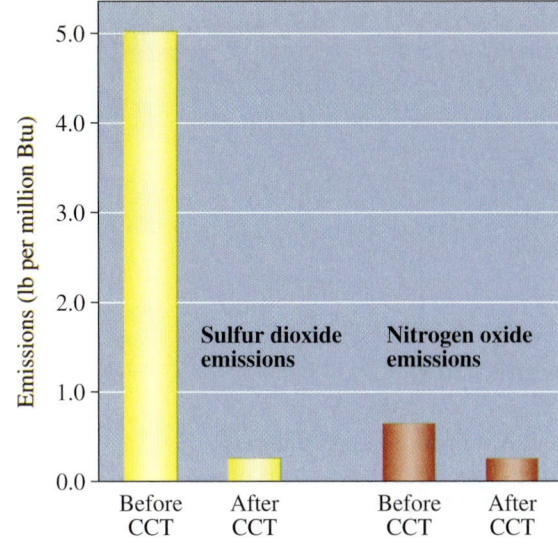

Figure 6.28

Changes in emissions at the Milliken Station power plant in Lansing, NY.

Note: CCT stands for Clean Coal Technology.
Source: U.S. Office of Fossil Energy. http://www.fossil.energy.gov/programs/powersystems/index.html

The Acid Rain Program also called for a 10-million-ton reduction of SO_2 emissions by the year 2000. Phase I, begun in 1995, required 263 mostly coal-burning boiler units at 110 electrical utility power plants (located in 21 different states) to reduce their emissions. Phase II, begun in 2000, further tightened the emissions on these plants. This phase also set further restrictions on power plants fired by natural gas and oil to encompass over 2000 boiler units. To date, the SO_2 emissions program has met with success. The fact that most anthropogenic SO_2 comes from a limited number of point sources (coal-burning power plants and factories) made the SO_2 problem easier to attack. As we already saw from Figure 6.17b, great strides have occurred in reducing U.S. SO_2 emissions.

Three major strategies have been employed to decrease SO_2 emissions: (1) switch to "clean coal" with lower sulfur content, (2) clean up the coal to remove the sulfur before use, and (3) use chemical means to neutralize the acidic sulfur dioxide in the power plant. We briefly consider the effectiveness and the cost of each of these.

Coal switching is an option because coals vary widely in sulfur content and their heat content. Anthracite, or "hard coal," is found mainly in Pennsylvania. It yields the greatest amount of energy and has the lowest percent of sulfur, but its supply is practically exhausted and more expensive. Bituminous, or "soft coal," is abundant in the Midwest. It has nearly the same heat content as anthracite but usually contains 3–5% sulfur. Western states have enormous deposits of low-sulfur sub-bituminous coal and lignite (brown coal); however, this coal has a low heat content and may contain up to 40% water.

Coal cleaning is relatively easy and the technology is available. The coal is crushed to a fine powder and washed with water so that the heavier sulfur-containing minerals sink to the bottom. But the process removes only about half of the sulfur, and it is expensive—from $500 to $1000 per ton of SO_2 eliminated.

An alternative to coal switching or coal cleaning is to chemically remove the SO_2 during or after combustion in the power plant. The chief method for doing this is called *scrubbing*. The stack gases are passed through a wet slurry of powdered limestone, $CaCO_3$. The limestone neutralizes the acidic SO_2 to form calcium sulfate, $CaSO_4$.

$$2\,SO_2(g) + O_2(g) + 2\,CaCO_3(s) \longrightarrow 2\,CaSO_4(s) + 2\,CO_2(g) \qquad [6.32]$$

Limestone is cheap and readily available. Although the process is highly efficient, installing scrubbers is expensive, so that the cost of this method has been estimated

at \$400–600 per ton of SO_2 removed. Part of the expense is associated with the disposal of the $CaSO_4$ formed. We simply cannot avoid the law of conservation of matter. The sulfur must end up somewhere; either it goes up the stack as SO_2 or gets trapped as $CaSO_4$.

Consider This 6.31   Emissions Close to Home

Thanks to the EPA, you now can find the acid rain emissions data for the power plants in your state. A direct link to an interactive map is provided at the *Online Learning Center.* Select a plant of your choice and use the Acid Rain Program (ARP) dataset to report:

a. the name of the plant.
b. the tons of SO_2 and NO_x emitted in a recent year.
c. the trend in emissions, by examining the data from previous years.

The principal reason compliance with the 1990 Clean Air Act Amendments regulations was achieved and even bettered was coal switching, in which high-sulfur coal was replaced by low-sulfur coal. By the early 1990s, the use of a new rail carrier and favorable railway tariffs made vast deposits of cheaper low-sulfur coal (even less than 1% S) in Montana and Wyoming available at costs lower than that for midwestern or eastern low-sulfur coal. In 1991, western low-sulfur coal averaged just \$1.30 per million Btu; eastern low-sulfur coal was \$1.60–1.70 per million Btu. High-sulfur eastern coal cost \$1.35–1.55 per million Btu. Given this price advantage, it is not surprising that nearly 60% of SO_2 reduction came from switching to low-sulfur western coal rather than using more expensive alternatives, such as scrubbing.

But this conversion to low-sulfur coal has hidden costs. It ignores the social and economic impact on the states that produce high-sulfur coal. Since 1990, it has been estimated that coal switching has caused a 30% decline in employment in areas where high-sulfur coal is mined. This includes regions of Pennsylvania, Kentucky, Illinois, Indiana, and Ohio, although half of the drop can be attributed to automation and other market factors. Western states now produce nearly 33% of the coal mined in the United States, up from only 6% in 1970.

The shift to low-sulfur western coal has another side to it. Because the coal produces less heat per gram than eastern coals, power plants must burn more of it to generate the same amount of electricity. Burning more coal may release more pollutants. For example, mercury and other trace metals are more prevalent in coal from western states. If more coal is burned, more metals are released unless steps are taken to remove them before they go up the smokestacks (a costly proposition).

6.15   The Politics of Acid Rain

The neutralization of acid rain will require more than chemistry. As we have noted throughout this textbook, industrial leaders, state officials, politicians, and citizens across the nation are all important players. We need workable solutions—both economically and in terms of human health.

One such solution lies in a unique feature of the Clean Air Act Amendments of 1990: a national "cap and trade" system. The SO_2 emissions were capped to meet goals that progressively are becoming lower. For example, in 2001 the release of SO_2 was set at 10.6 million tons from electrical utilities; in 2010 the releases will be lowered to 8.95 million tons. In order to reach these goals, each utility company operates with a *permit* that caps the pollution it can legally release per year. Exceeding this maximum carries fines of up to \$25,000 per day.

The Kyoto Protocol, now in effect (but not ratified by the United States), has a similar cap and trade system for CO_2 emissions.

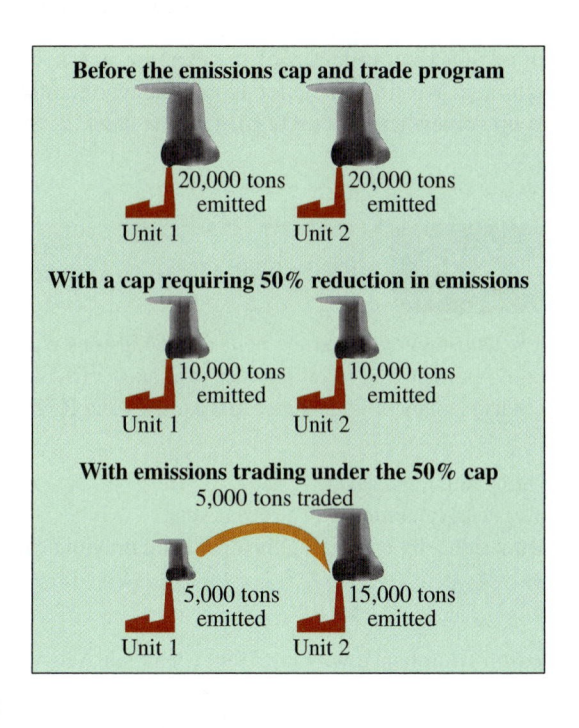

Figure 6.29

The emissions cap and trade concept.

Source: EPA, *Clearing the Air, The Facts About Capping and Trading Emissions,* 2002, page 3.
http://www.epa.gov/airmarkets/articles/clearingtheair.pdf

The "trade" part of the cap and trade system works through a system of allowances. Companies are assigned emission *allowances* that authorize the emission of 1 ton of SO_2, either during the current year or any year thereafter. At the end of a year, each company must have sufficient allowances to cover its actual emissions. If it has extra allowances, it can sell them or save them for a future year. If it has insufficient allowances, it must purchase them. Most of the allowance trading has taken place in the Ohio Valley.

An example of the cap and trade system is shown in Figure 6.29. With no controls, 20,000 tons is emitted from each of two units. Each one is capped at 10,000 tons, but one, through greater efficiency, performs better. The unit with emissions below its cap is assigned a credit for each ton of SO_2 saved. These credits can be sold to power plants that cannot efficiently meet their emission allowances. There is thus a financial incentive for power producers to achieve significant reductions of acidic oxide emissions. On the other hand, the purchase of credits by those who cannot yet meet the more stringent standards allows them to continue operation, at or below the permit level, while the plant works to reduce emissions.

The first official trade of emissions allowances under the provisions of the new law occurred in 1993. Since then, allowances have been bought and sold in private transactions and at public auctions. The Chicago Board of Trade even has a commodity trading market in emission allowances. Prices have ranged considerably, most less than the $1000 predicted by utility officials. At the 2005 acid rain allowance auction, the lowest successful bid for current year allowances was $300, and the highest at $750.

Consider This 6.32   Up for Auction

The year 2007 marked the 15th annual auction for sulfur dioxide allowances conducted now by the EPA (formerly by the Chicago Board of Trade). Be a detective on the Web to answer these questions.

   a. Are the allowances more costly or less this year than last?
   b. How many allowances were auctioned last year?
   c. Are most industries still achieving compliance without having to buy credits?
   Hint: Direct links to Web sites are provided at the *Online Learning Center.*

A national emissions trading program has not been finalized for NO_x. Meanwhile, a complex set of programs exist across the United States that shares a common goal: reducing tropospheric ozone. One such program, the NO_x Budget Trading Program, currently is operational in 11 northeastern states. Since 2003, this program has successfully achieved significant NO_x reductions in stationary sources such as coal-burning power plants. In turn, ozone levels have decreased in the neighboring areas.

Your Turn 6.33   **Summer and Winter NO_x**

The current NO_x emissions trading program in the northeast is aimed at reducing ozone levels.

a. Summer emissions of NO_x lead to ozone formation on hot sunny days. Explain why.
Hint: See Section 1.12.

b. Winter emissions of NO_x still need to be addressed as well. Explain why.
Hint: Review the episodic acidity of lakes described in the previous section.

In the long run, the most compelling argument for emissions reductions may be the benefits to human health. Not only does cleaner air add up to less illness and suffering, but also to real dollar savings for health care. The figures are compelling. A report to Congress in 2003 credited the Acid Rain Program as providing the "largest quantified annual human health benefits (over $70 billion) of any federal regulatory program implemented in the last 10 years." If you do the math, you will find this is a benefit-to-cost ratio of over 40 to 1. We hope the math that we cited at the opening of the chapter will change to reflect this. Namely, if somebody stops five people on the street and asks them about nitrogen (or sulfur) emissions, we hope they will report that reducing these is an *exceedingly* cost-effective investment.

Conclusion

If you have learned anything from this chapter, we hope it has been skepticism, prudence, and the recognition that complex problems cannot be solved by simple or simplistic strategies. "Acid rain" is not the dire plague once described by environmentalists and journalists. Nor is it a matter to be ignored. It is sufficiently serious that federal legislation, the Clean Air Act Amendments of 1990, have been enacted to reduce SO_2 and NO_x emissions, precursors to acid deposition. In addition, nitrogen emissions are tied to many issues other than acid rain. The release of reactive forms of nitrogen in our environment has caused a cascading set of problems.

Any failure to acknowledge the intertwined relationships involving the combustion of coal and gasoline, the production of sulfur and nitrogen oxides, and the reduced pH of fog and precipitation is to deny some fundamental facts of chemistry. Knowledge of ecology and biological systems is needed as well, so that acid deposition can be understood in the context of entire ecosystems, a task that requires that experts from several disciplines collaborate.

Public health also is at issue. Economic analyses reveal that allocating funds to reduce sulfur and nitrogen emissions will have a huge payoff in terms of lower mortality rates, fewer illnesses, and higher quality of living.

One response that we as individuals and as a society might make to the problems of acid precipitation has hardly been mentioned in this chapter, yet it is potentially one of the most powerful. It is to conserve energy. Sulfur dioxide and nitrogen oxides are by-products of our voracious demand for energy, especially for electricity and transportation. Carbon dioxide, of course, is an even more plentiful product. If our personal, national, and global

appetite for fossil fuels continues to grow unchecked, our environment may well become a good deal warmer and a good deal more acidic. Moreover, the problem may be intensified as petroleum and low-sulfur coals are consumed and we become even more reliant on high-sulfur coal.

There are other sources of energy—nuclear fission, water and wind, renewable biomass, and the Sun itself. All currently are being utilized, and this no doubt will increase. We explore nuclear fission in the next chapter. But we conclude this chapter with the modest suggestion that, for a multitude of reasons, the conservation of energy by industry and collectively by individuals could have profoundly beneficial effects on our environment.

Chapter Summary

Having studied this chapter, you should be able to:

- Define the terms *acid* and *base* and know how to use these definitions to distinguish acids from bases (6.1–6.3)
- Use chemical equations to represent the dissociation (ionization) of acids and bases (6.1–6.2)
- Write neutralization reactions for acids and bases (6.3)
- Describe solutions as acidic, basic, or neutral based on their pH or concentrations of H^+ and OH^- (6.3–6.4)
- Calculate pH values given hydrogen or hydroxide ion in whole-number concentrations (6.4)
- Describe the differences between the pH of water, the pH of ordinary rain, and the pH of acid rain, and locate on a map of the United States where the most acidic rain falls (6.4–6.5)
- Explain the role of sulfur oxides and nitrogen oxides in causing acid rain (6.7–6.8)
- List the different sources of NO_x and of SO_2 and explain the variations in the levels of these pollutants over the past 30 years (6.9)
- Explain the production of acidic aerosols and their effects on building materials and human health (6.10–6.11)

- Explain why N_2 is a relatively inert element. Describe different forms of reactive nitrogen and how they are produced both naturally and by humans. Use the nitrogen cycle to explain the cascading effects of reactive nitrogen (6.12)
- Describe how the industrial production of ammonia and the acidic deposition of nitrates both contribute to the buildup of reactive nitrogen on our planet (6.12)
- Describe nitrogen saturation and its consequences for lakes (6.13)
- Discuss the 1990 Clean Air Act Amendments and the cap and trade program. Describe the effect these continue to have on SO_2 emissions (6.13–6.14)
- Describe how NO_x emissions have been controlled differently from SO_2 emissions (6.13)
- Outline different ways to control acid rain, noting the cost–benefit considerations involved (6.13–6.14)
- Explain why acid rain control is an exceedingly wise investment in terms of the benefits to human health (6.15)

Questions

Emphasizing Essentials

1. **a.** Give names and chemical formulas for any five acids.

   b. Name three observable properties associated with acids.

   c. Give the Lewis structure for each species in equation 6.1.

2. Write a chemical equation that shows the release of one hydrogen ion from a molecule of each of these acids.

   a. HBr*(aq)*, hydrobromic acid

   b. H_2SO_3*(aq)*, sulfurous acid

   c. $HC_2H_3O_2$*(aq)*, acetic acid

3. **a.** Give names and chemical formulas for any five bases.

   b. Name three observable properties associated with bases.

   c. Give the Lewis structure for each species in equation 6.4c.

4. These bases dissolve in water. Write a chemical equation that shows the release of hydroxide ions as each dissolves.

   a. KOH*(s)*, potassium hydroxide

   b. $Ba(OH)_2$*(s)*, barium hydroxide

5. Consider these ions: nitrate, sulfate, carbonate, and ammonium.

   a. Give the chemical formula for each.

   b. Write a chemical equation in which the ion (in aqueous form) appears as a product.

6. Write a balanced chemical equation for each acid–base reaction.

   a. Potassium hydroxide is neutralized by nitric acid.

   b. Hydrochloric acid is neutralized by barium hydroxide.

   c. Sulfuric acid is neutralized by ammonium hydroxide.

7. Classify these aqueous solutions as acidic, neutral, or basic.

   a. HI(aq)

   b. NaCl(aq)

   c. NH$_4$OH(aq)

   d. $[H^+] = 1 \times 10^{-8}$ M

   e. $[OH^-] = 1 \times 10^{-2}$ M

   f. $[H^+] = 5 \times 10^{-7}$ M

   g. $[OH^-] = 1 \times 10^{-12}$ M

8. For parts **d** and **f** of question 7, calculate the $[OH^-]$ that corresponds to the given $[H^+]$. Similarly, for parts **e** and **g**, calculate the $[H^+]$.

9. Again referring back to question 7, calculate the pH for the concentrations given in parts **d**–**g**.

10. Give the difference in the $[H^+]$ between these pairs of solutions. Do your answers show that small changes in pH give rise to large changes in hydrogen ion concentration?

    a. pH = 6 and pH = 8

    b. pH = 5.5 and pH = 6.5

    c. $[H^+] = 1 \times 10^{-8}$ M and $[H^+] = 1 \times 10^{-6}$ M

    d. $[OH^-] = 1 \times 10^{-2}$ M and $[OH^-] = 1 \times 10^{-3}$ M

11. In terms of taste, pH, and amount of dissolved gas, how does carbonated water differ from rain that contains dissolved carbon dioxide?

12. Consult Figure 6.11 to find the data necessary to answer these questions.

    a. Of the cities Chicago, Atlanta, Seattle, and San Francisco, which would be likely to have the least acidic precipitation? The most?

    b. In terms of average pH, how does the rain in these cities compare with the rain where you live?

13. Suppose you have a new mountain bike and accidentally spilled a can of carbonated cola on the metallic handle bars and paint.

    a. Soft drinks are more acidic than acid rain. About how many times more acidic? *Hint:* Consult Figure 6.6.

    b. In spite of the higher acidity, this spill is unlikely to damage your handle bars and paint (although the sugar probably isn't great on your gears). Why is damage unlikely?

14. Write a balanced chemical equation for the chemical reaction involving sulfur shown in Figure 6.13.

15. The text states that the reaction of SO$_2$ with a free radical accounts for 20–25% of the sulfuric acid in the atmosphere.

    a. Write the balanced chemical equation.

    b. What is the source of this free radical in the atmosphere?

16. For each of these acids, write the chemical formula for the acid anhydride.

    a. carbonic acid, H$_2$CO$_3$

    b. sulfurous acid, H$_2$SO$_3$

17. Assume that coal can be represented by the formula C$_{135}$H$_{96}$O$_9$NS.

    a. What is the percent of nitrogen by mass in coal?

    b. If 3 tons of coal were burned completely, what mass of nitrogen in NO would be produced? Assume that all of the nitrogen in the coal is converted to NO.

    c. Actually more NO is produced than you just calculated. Explain.

18. In 2006, the United States burned about 1.1 billion tons of coal. Assuming that it was 2% sulfur by weight, calculate the tons of sulfur dioxide that were emitted.

19. Acid rain can damage marble statues and limestone building materials. Write the balanced chemical equation.

20. **a.** On the label of a shampoo bottle, what does the phrase "pH-balanced" imply?

    b. Does the phrase "pH-balanced" influence your decision to buy a particular shampoo? Explain.

21. Many gases are associated with exhaust from jet engines, including CO, CO$_2$, O$_3$, NO, NO$_2$, SO$_2$, and SO$_3$.

    a. Which of these gases do jet engines emit *directly*?

    b. Which ones form secondarily, that is, resulting from the emissions of part **a**?

22. Figure 6.14 offers information about SO$_2$ emissions from fuel combustion (mainly from electrical power production) and from transportation. Figure 6.16 offers information about NO$_x$ emissions from fuel combustion

(again mainly from power production) and from transportation. Relative to fuel combustion and transportation, how do the emissions of SO_2 and NO_x differ? Explain this difference.

23. Almost equal *masses* of SO_2 and NO_x are produced by human activities in the United States.

   a. How does their production compare based on a *mole* basis? Assume that all the NO_x is produced as NO_2.

   b. Suggest reasons why the U.S. percentage of global emissions is greater for NO_x than for SO_2.

24. Reactive nitrogen compounds affect the biosphere both directly and indirectly through other chemicals they help form.

   a. Name a direct effect of reactive nitrogen compounds that is a benefit.

   b. Name two direct effects of reactive nitrogen compounds that are harmful to human health.

   c. Ozone formation is a harmful *indirect* effect. Explain the connection between reactive nitrogen compounds and the formation of ozone.

25. Calculate the mass of $CaCO_3$ (in tons) necessary to react completely with 1.00 ton of SO_2 according to the reaction shown in equation 6.32.

26. A garden product called dolomite lime is composed of tiny chips of limestone that contain both calcium and magnesium carbonate. This product is "intended to help the gardener correct the pH of acid soils," as it is "a valuable source of calcium and magnesium."

   a. Write chemical formulas for magnesium carbonate and calcium carbonate. Is the calcium in the form of calcium ion or calcium metal?

   b. Write a chemical equation that shows why limestone "corrects" the pH of acidic soils.

   c. Will the addition of dolomite lime to soils cause the pH to rise or fall?

   d. Plants such as rhododendrons, azaleas, and camellias should not be given dolomite lime. Explain.

27. The Clean Air Act was discussed both in this chapter and in Chapter 1, the Montreal Protocol in Chapter 2, and the Kyoto Protocol in Chapter 3.

   a. What principal issue does each of these address?

   b. Place all three on a timeline.

Concentrating on Concepts

28. Professor James Galloway, an expert on acid rain, wrote "Human activity is not making the world acidic, rather it is making the world *more* acidic."

   a. Explain why the world is naturally acidic.

   b. Explain how humans are making the world more acidic.

   c. One large part of our planet is basic. Which one? *Hint:* Consult Figure 6.6.

29. Judging by the taste, do you think there are more hydrogen ions in a glass of orange juice or in a glass of milk? Explain your reasoning.

30. The formula for acetic acid, the acid present in vinegar, is commonly written as $HC_2H_3O_2$. Many chemists write the formula as CH_3COOH.

   a. Draw the Lewis structure for acetic acid.

   b. Show that both formulas represent acetic acid.

   c. What are the advantages and disadvantages of each formula?

   d. How many hydrogen atoms can be released as hydrogen ions per acetic acid molecule? Explain.

31. Television and magazine advertisements tout the benefits of antacids. A friend suggests that a good way to get rich quickly would be to market "antibase" tablets. Explain to your friend the purpose of antacids and offer some advice about the potential success of "antibase" tablets.

32. In Your Turn 6.7, you listed the ions present in aqueous solutions of acids, bases, and common salts. Now add water, a molecular species, to this list.

   a. List all molecular and ionic species in order of decreasing concentration in a 1.0 M aqueous solution of NaOH.

   b. List all molecular and ionic species in order of decreasing concentration in a 1.0 M aqueous solution of HCl.

33. Which of these has the *lowest* concentration of hydrogen ions: 0.1 M HCl, 0.1 M NaOH, 0.1 M H_2SO_4, pure water? Explain your answer.

34. Explain why rain is naturally acidic, but not all rain is classified as "acid rain."

35. As mentioned in Section 6.6, rain samples currently are being analyzed for acidity in the Central Analytical Laboratory in Illinois instead of out in the field.

   a. The pH values tended to be slightly higher in the lab. Did the acidity increase or decrease?

   b. Speculate on the causes of the pH increase.

36. Speaking of field measurements, in January 2006 those living in the metropolitan St. Louis area experienced a severe hail storm. Hail stones were still visible a day later, as shown in the photo. A chemistry professor took

samples of the hail, analyzed them in her laboratory and reported a pH of 4.8.

a. How does this pH compare with the normal precipitation in the St. Louis area? To acidity levels further east and further west of St. Louis?

b. What factors might lead to acid rain in St. Louis?

37. Mammoth Cave National Park in Kentucky is in close proximity to the coal-fired electric utility plants in the Ohio Valley. Noting this, the National Parks Conservation Association (NPCA) reported that this national park had the poorest visibility of any in the country.

a. What is the connection between coal-fired plants and poor visibility?

b. The NPCA reported "the average rainfall in Mammoth Cave National Park is 10 times more acidic than natural." From this information and that in your text, estimate the pH of rainfall in the park.

38. In the United States over the past few decades, emissions of ammonia have dramatically increased, although less so in the east than in the west.

a. Show with a chemical equation that ammonia dissolves in rain to form a basic solution.

b. Write the neutralization reaction for rain that contains both ammonia and nitric acid.

c. Ammonium sulfate also is found in rain. Write a chemical equation that demonstrates how it could have formed.

39. Ozone in the troposphere is an undesirable pollutant, but stratospheric ozone is beneficial. Does nitric oxide, NO, have a similar "dual personality" in these two atmospheric regions? Explain. *Hint:* Consult Chapter 2.

40. The mass of CO_2 emitted during combustion reactions is much greater than the mass of NO_x or SO_2, but there is less concern about the contributions of CO_2 to acid rain than from the other two oxides. Suggest two reasons for this apparent inconsistency.

41. The average pH of precipitation in New Hampshire and Vermont is low, even though these states have relatively fewer cars and virtually no industry that emits large quantities of air pollutants. How do you account for this low pH?

42. Admittedly, global sulfur emissions are difficult to estimate and you will find a range of values published in the literature. One set is shown in this figure for 1850–2000. These are estimates from a paper published in 2004 by David Stern at Rensselaer Polytechnic Institute. Gg stands for gigagrams, or 1×10^{12} grams.

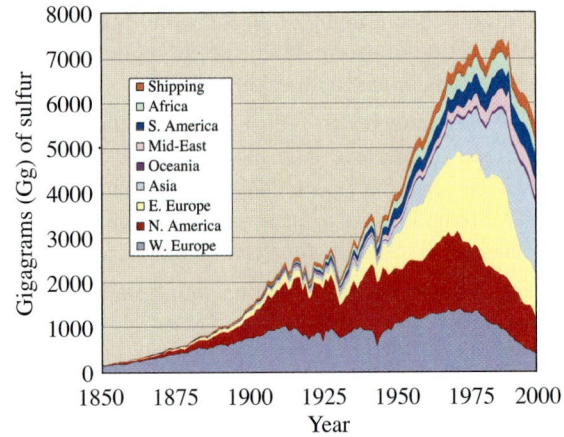

a. The figure shows total sulfur emissions. In what chemical form is this sulfur most likely to be?

b. According to the figure, in which years did the sulfur emissions peak?

c. List reasons for the decline in more recent years.

d. In 2000, which region of the world contributed the highest level of sulfur emissions?

e. Which regions were the largest contributors in the early 1970s?

43. The chemistry of NO in the atmosphere is complicated. NO can destroy ozone, as seen in Chapter 2. But remember from Chapter 1 that NO can react with O_2 to form NO_2. In turn, NO_2 can react in sunlight to produce ozone. Summarize these reactions, noting in which region of the atmosphere they each occur.

44. The chemical reaction in which NO reacts to form NO_2 in the atmosphere (equation 6.19) involves intermediate species A′ and A″. Here are possible structures.

a. In each, what does the dot (·) represent? For the atom with the dot, redraw to show all valence electrons, both bonding and nonbonding pairs. *Hint:* This atom does not have an octet of electrons.

b. Name a chemical property that A′ and A″ have in common.

45. a. Efforts to control air pollution by limiting the emission of particulates and dust can sometimes contribute to an increase in the acidity of rain. Offer a possible explanation for this observation. *Hint:* These particulates may contain basic compounds of calcium, magnesium, sodium, and potassium.

b. In Chapter 2, stratospheric ice crystals in the Antarctic were involved in the cycle leading to the destruction of ozone. Is this effect related to the observations in part **a**? Explain.

46. a. Several strategies to reduce SO_2 emissions are described in the text. The most effective ones in the last 10 years have been coal switching and stack gas scrubbing. Prepare a list of the advantages and disadvantages associated with each method.

b. Explain why coal cleaning has not been an effective strategy.

47. Discuss the validity of the statement, "Photochemical smog is a local issue, acid rain is a regional one, and the enhanced greenhouse effect is a global one." Describe the chemistry behind each issue. Do you agree that the magnitude of the problem is really so different in scope?

Exploring Extensions

48. Sometimes by trying to use a technological fix for one problem, we inadvertently create another.

a. How do the problems associated with the buildup of reactive nitrogen in the environment fit this statement? Explain.

b. Select another issue explored in Chapters 1–5. Again explain how your choice fits the statement as well.

49. Some local newspapers give forecasts for pollen, UV Index, and air quality. Why do you suppose that no forecast for acid rain is provided?

50. The compound $Al(OH)_3$ contains OH in its chemical formula. However we do not write a reaction analogous to equation 6.3. Explain. *Hint:* Consult a solubility table.

51. In Your Turn 6.7, you listed the ions present in aqueous solutions of acids, bases, and common salts. In question 32, you added molecular substances to the list. To quantify this list,

a. calculate the molar concentration of all molecular and ionic species in a 1.0 M solution of NaOH.

b. calculate the molar concentration of all molecular and ionic species in a 1.0 M solution of HCl.

52. A workshop was held to establish research priorities relating to acidic deposition. A presenter made this statement: "We have found a control strategy that is successful in cutting the acidity in half. However, an evil conspiracy of chemists will only allow the pH of precipitation to increase by 0.3." As a Sceptical Chymist in attendance, you realize that this statement should

be checked out. Do so and state your response to the presenter. *Hint:* See Appendix 3 on logarithms.

53. Equation 6.18 shows that energy (in the form of a hot engine or other source of heat) must be added to get N_2 and O_2 to react to form NO. A Sceptical Chymist wants to check this assertion and determine how much energy is required. Show the Sceptical Chymist how this can be done. *Hint:* Draw the Lewis structures for the reactants and products, noting that NO does not have an octet of electrons. The bond energy is 607 kJ/mol for the N-to-O double bond.

54. The text describes a green chemistry solution to reducing NO emissions for glass manufacturers.

a. Identify the strategy.

b. Use the Web to research what other industries might use this green chemistry strategy. Write a report to summarize your findings.

55. Here are examples of what an individual might do to reduce acid rain. For each, explain the connection to producing acid rain.

a. Hang your laundry to dry it.

b. Walk, bike, or take public transportation to work.

c. Avoid running dishwashers and washing machines with small loads.

d. Add additional insulation on hot water heaters and pipes.

e. Buy locally grown produce and locally produced food.

f. Fertilize lawns less frequently.

56. How do researchers determine whether the negative effects of acid deposition on aquatic life are a direct consequence of low pH or the result of Al^{3+} released from rocks and soil? Find one or more research articles. Prepare a summary of the experimental plan and the results.

57. One way to compare the acid-neutralizing capacity of different substances is to calculate the mass of the substance required to neutralize 1 mol of hydrogen ion, H^+.

a. Write a balanced equation for the reaction of $NaHCO_3$ with H^+. Use it to calculate the acid-neutralizing capacity for $NaHCO_3$.

b. If $NaHCO_3$ costs $9.50/kg, determine the cost to neutralize one mole of H^+.

58. Why are developing countries likely to emit an increasingly higher percentage of the global amount of SO_2? Pick a nation, research its current emissions of SO_2, and calculate its percent of global emissions. Are emissions likely to continue to increase in the future? Make a prediction and give your reasoning.

59. Like diesel trucks, sport utility vehicles (SUVs) emit more than their share of pollutants. Are these NO_x, SO_2, or both? Is legislation passed or pending to clean up their emissions? Research these questions using the Web or an owner's manual.

60.   Blue-baby syndrome (methemoglobinemia) can occur as a result of nitrate ion in the drinking water. Use the resources of the Web to answer these questions. *Hint*: Include "nitrate ion" in your search to avoid the congenital causes of methemoglobinemia.

a. What is the likely source of nitrate ion in the drinking water?

b. What happens to infants and young children who ingest too much nitrate?

c. What are the current guidelines for the nitrate ion in drinking water?

d. The nitrite ion is involved as well. How?

The Fires of Nuclear Fission

"It is difficult to see how we can reduce our dependence on fossil fuels without the help of nuclear power."

Lord Robert May, Oxford University, United Kingdom
Former President of the Royal Society
August 19, 2005, Science

With our continued dependence on fossil fuels comes a rising sense of urgency. Can we continue to burn them as we have in the past? Economic, environmental, and personal concerns accompany our use of fossil fuels. The price of crude oil continues to rise. Communities continue to suffer the tragedies of coal mining accidents. Coal-fired power plants contribute to acid rain and snow. And each gallon of gasoline we purchase at the pump to fuel our driving adds additional carbon dioxide to the atmosphere. Is now the time to play the nuclear piece on the energy game board?

Clearly we need clean and sustainable sources of energy. Without a decisive move to secure these, our consumption of fuels in the United States may quickly come to the point of checkmate. In the past 40 years, however, nothing about nuclear power has been either quick or decisive. As a nation, we continue to be divided about the desirability of using nuclear power. The nuclear waste from our aging reactors sits in short-term storage, because we have yet to agree on how we should store it for the long haul.

Can we overcome our paralysis? In part, our lack of decisive action is a result of the tremendous baggage that the word *nuclear* carries. The associations are disturbing: the bombing of Hiroshima, the radioactive fallout from atmospheric weapons testing, the tragedies of Chernobyl, the hazards of high-level radioactive waste, and the ultimate threat of nuclear annihilation. Probably no other topic in the physical sciences is more likely to provoke an emotional response.

And yet, people recognize the many benefits of nuclear science, including radiation therapy to treat cancer, nuclear diagnostic scans that bypass both anesthesia and surgery, and of course, the production of electricity by nuclear power plants. The applications of nuclear phenomena, harmful at one extreme and beneficial at the other, present us with both risks and benefits.

As we exploit the power of the atom, we face real and pressing questions. Are nuclear power plants safe to operate? Can they be kept safe from acts of terrorism? Can our communities deal with the wastes they produce? Can we prevent the diversion of nuclear materials to nuclear weapons? As has been the case in earlier chapters, science and societal issues are tightly connected. We will begin this chapter by examining the prospects for nuclear power in the years to come. But before we start, we ask you to consider your own position regarding nuclear power.

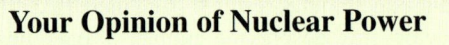

Consider This 7.1   Your Opinion of Nuclear Power

a. Given a choice between purchasing electricity generated by a nuclear plant or by a coal-burning plant, would you choose one over the other? Explain.

b. What circumstances, if any, would change your position on the use of nuclear power for generating electricity?

Save your answers to these questions, because you will revisit them at the end of the chapter.

7.1   A Comeback for Nuclear Energy?

Most people mindlessly switch on lights, giving no thought to the source of the energy that makes the bulbs glow. But for anybody who has lost electrical power because of a storm or repeated power blackouts, flipping a light switch may trigger a set of memories. Has the power come back on? Are we still without electricity? Can I brew my coffee in the morning (Figure 7.1)? The reliability of our power sources depends on the choices we make now and in the years to come.

Let's assume you have electrical power and can switch on your coffee maker. The odds are about 1 in 5 that the electricity to brew your coffee comes from a nuclear power plant. In 2007, about 20% of the electrical power in the United States was produced by the 103 nuclear reactors licensed by the Nuclear Regulatory Commission (NRC). These reactors are at 65 sites in 31 states. As you can see by Figure 7.2, the total electricity

Figure 7.1
We have a high demand for electricity (and caffeine).

283

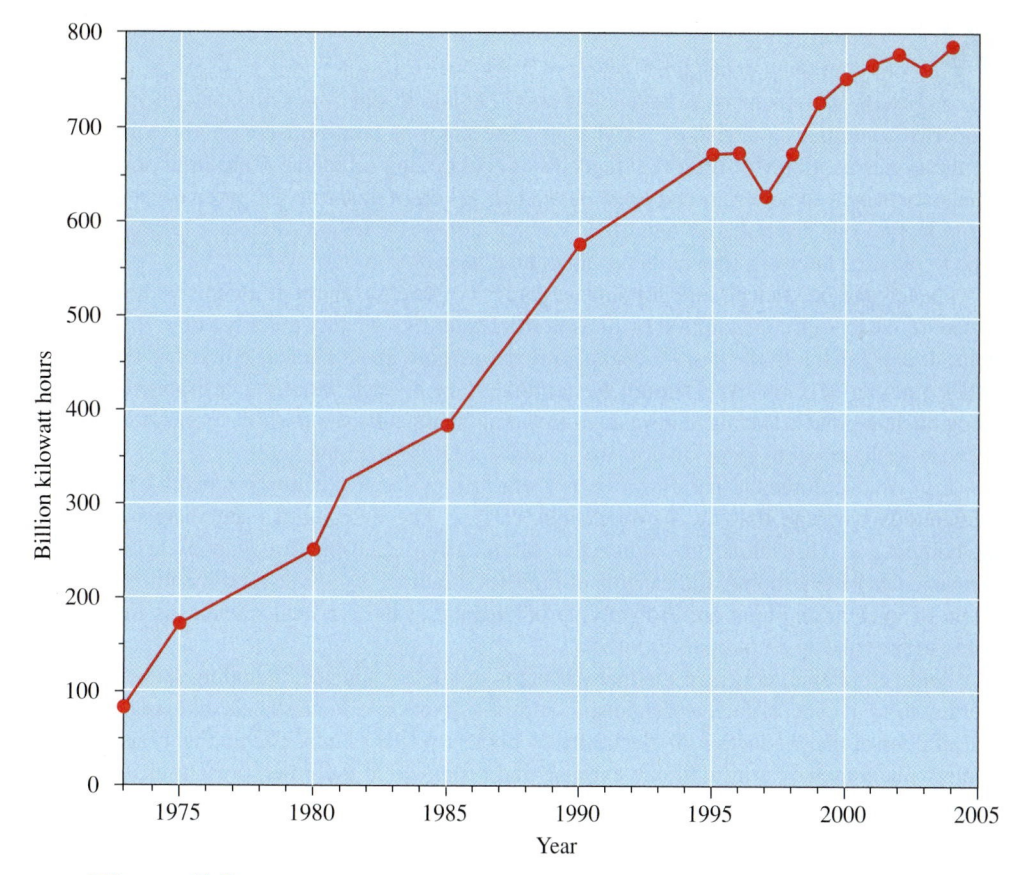

Figure 7.2

Nuclear power generation since 1973.

Source: Energy Information Administration.

generated by nuclear plants continues to increase slowly, though the overall percentage of total production in the United States has remained constant for the past 15 years. The increased production has resulted from increased efficiency and power upgrades at many of the reactor sites, despite the number of operating reactors dropping from its peak of 112 in 1990 to 103 in 2007.

When you brew your coffee a decade from now, from where will the electricity come? No new nuclear plants have been built since 1978; a moratorium was placed on their construction after the Three Mile Island accident in 1979. Furthermore, nine nuclear plants ceased their operations, some of them before their licenses even expired (Table 7.1). They include what was once the nation's largest nuclear plant, the Zion nuclear power station on the shores of Lake Michigan. Reasons cited for plant closings included the competition of natural gas and the competitive pressures of energy deregulation.

As of late 2007, about 20 new reactors were being considered by power utilities.

Consider This 7.2   Nuclear Power State-by-State

The Nuclear Energy Institute provides a map showing the locations of commercial nuclear power plants in the United States. A direct link is provided at the *Online Learning Center.*

a. Select a state with one or more nuclear power plants. What percent of this state's electrical energy comes from nuclear power?

b. Find a state in which more than half of the electrical energy is nuclear in origin.

c. Select a state with no nuclear power plants. How instead is the electricity generated?

Table 7.1	Nuclear Plant Closings Since 1990		
Nuclear Plant	**State**	**License Issued**	**Date Shut Down**
Millstone 1	Connecticut	1966	1998
Zion 1, Zion 2	Illinois	1973	1998
Big Rock Point	Michigan	1962	1997
Maine Yankee	Maine	1972	1997
Haddam Neck	Connecticut	1967	1996
San Onofre 1	California	1967	1992
Trojan	Oregon	1975	1992
Yankee–Rowe	Maine	1960	1991

Source: From Environmental Law & Policy Center, http://www.elpc.org/energy/nuclear_closings.html. Reprinted with permission.

It is hard to predict how long the nuclear plants near you (or in a neighboring state) will continue to operate. In the late 1990s, the decommissioning of a dozen or so plants seemed a foregone conclusion, given their age. Some people worried that so few plants would renew their licenses that the percent of electrical power generated by nuclear reactors would drop to less than 10%. Table 7.1 shows that eight plants did indeed close since 1990.

But, given the recent increased demands for electricity in the United States, a renaissance in nuclear power now is likely. For example, the Oyster Creek plant in New Jersey was scheduled for shutdown in 2000, but the plant is still in operation, having been purchased by an international power company. In Maryland, Calvert Cliffs Nuclear Power Station was the first to successfully complete the lengthy and costly process necessary to renew its operating licenses for another 20 years. Oconee Nuclear Power Station in South Carolina followed shortly thereafter. Ten reactor units have received extensions until 2033–2036. We will focus on the future of nuclear power in Section 7.12.

The construction and continued operation of nuclear plants is not only a matter of energy supply and demand, but also one of public acceptance. Depending on your age, you may have little recollection of the controversy that surrounded some nuclear power plants when they were proposed or being constructed. People have been lining up on one side or the other of the nuclear fence for quite some time.

> Decommissioning (shutting down) a nuclear plant is a complex operation. All parts must be analyzed and removed according to strict criteria.

Consider This 7.3	Take a Stand

As you can see in this photo taken in 1977 during the construction of the Seabrook nuclear power plant in New Hampshire, signs are one way to convey your position. If a nuclear plant were being built near your community today, what would your sign say? Prepare a list of three talking points for an interview with a reporter.

In the section that follows, we will examine the process of nuclear fission, thus taking the first step in explaining both the controversies and the hopes for nuclear energy as a power source that will lead us into the future.

7.2   How Fission Produces Energy

The key to answering this question is probably the most famous equation in all of the natural sciences, $E = mc^2$. This equation dates from the early years of the 20th century and is one of the many contributions of Albert Einstein (1879–1955). It summarizes the equivalence of energy, E, and matter, or mass, m. The symbol c represents the speed of light, 3.0×10^8 m/s, so c^2 is equal to 9.0×10^{16} m^2/s^2. The large value of c^2 means that

it should be possible to obtain a tremendous amount of energy from a small amount of matter, whether in a power plant or in a weapon.

For over 30 years, Einstein's equation was a curiosity. Scientists believed that it described the source of the Sun's energy, but as far as anyone knew, no one had ever observed on Earth a transformation of a substantial fraction of matter into energy. But in 1938, two German scientists, Otto Hahn (1879–1968) and Fritz Strassmann (1902–1980), discovered otherwise. When they bombarded uranium with neutrons, they found what appeared to be the element barium (Ba) among the products. The observation was unexpected because barium has an atomic number of 56 and an atomic mass of about 137. Comparable values for uranium are 92 and 238, respectively. At first, the scientists were tempted to conclude that the element was radium (Ra, atomic number 88), a member of the same group in the periodic table as barium. But Hahn and Strassmann were good chemical researchers, and the chemical evidence for barium was too compelling.

The German scientists were unsure of how barium could have been formed from uranium, so they sent a copy of their results to their colleague, Lise Meitner (1878–1968), for her opinion (Figure 7.3). Dr. Meitner had collaborated with Hahn and Strassmann on related research, but was forced to flee Germany in March 1938 because of the anti-Semitic policies of the Nazi government. When she received their letter, she was living in Sweden. She discussed the strange results with her physicist nephew, Otto Frisch (1904–1979), as the two of them went walking in the snow. In a flash of insight, she understood. Under the influence of the bombarding neutrons, the uranium atoms were splitting into smaller ones such as barium. The nuclei of the heavy atoms were dividing, like biological cells undergoing fission.

That word from biology is applied to a physical phenomenon in the letter that Meitner and Frisch published on February 11, 1939, in the British journal *Nature.* In the letter, entitled "Disintegration of Uranium by Neutrons: A New Type of Nuclear Reaction," the authors state the following:

"Hahn and Strassmann were forced to conclude that isotopes of barium are formed as a consequence of the bombardment of uranium with neutrons. At first sight, this result seems very hard to understand. . . . On the basis, however, of present ideas about the behavior of heavy nuclei, an entirely different . . . picture of these new disintegration processes suggests itself. . . . It seems therefore possible that the uranium nucleus . . . may, after neutron capture, divide itself into two nuclei of roughly equal size. . . . The whole "fission" process can thus be described in an essentially classical way."

Although just over a page long, this letter was immediately recognized for its significance. In fact, it would be difficult to think of a more important scientific communication.

Figure 7.3

Lise Meitner is pictured shortly after her arrival in New York in January, 1946.

Niels Bohr (1885–1962), an eminent Danish physicist, learned of the news directly from Frisch and brought it to the United States on an ocean liner several days before its publication. Within a few weeks of Meitner and Frisch's letter in *Nature*, scientists in a dozen laboratories in various countries confirmed that the energy released by the fission of uranium atoms was that predicted by Einstein's equation. Lise Meitner's contributions to the discovery of nuclear fission were honored by naming element 109 meitnerium. Earlier, element number 96, curium, had been named to honor Marie Curie, another woman who was a nuclear pioneer (Section 7.5).

Nuclear fission is the splitting of a large nucleus into smaller ones with the release of energy. Energy is released because the total mass of the products is slightly less than the total mass of the reactants. In spite of what you may have been taught, neither matter nor energy is *individually* conserved. Matter disappears and an equivalent quantity of energy appears. Alternatively, one can view matter as a very concentrated form of energy; nowhere is it more concentrated than in the atomic nucleus. Remember that an atom is mostly empty space. If a hydrogen nucleus were the size of a baseball, then its electron would be found in a sphere half a mile in diameter. Because almost all the mass of an atom is associated with its nucleus, the nucleus is incredibly dense. Indeed, a pocket-sized matchbox full of atomic nuclei would weigh over 2.5 billion tons! Given the energy–mass equivalence of Einstein's equation, the energy content of all nuclei is, relatively speaking, immense.

Only the nuclei of certain elements undergo fission and these only under certain conditions. The relative factors to consider are the size of the nucleus, the numbers of protons and neutrons it contains, and the energy of the neutrons used to initiate the fission. For example, relatively light and stable atoms such as oxygen, chlorine, and iron do not split. Extremely heavy nuclei may fission spontaneously. And the familiar heavy atoms, such as uranium and plutonium, will split if hit hard enough with a neutron. Some (but not all) isotopes of uranium will even fission with a more gentle nudge, such as in the conditions of a nuclear power plant.

Let's examine uranium more closely. *All* uranium atoms contain 92 protons. If these atoms are electrically neutral, these protons are accompanied by 92 electrons. In nature, though, uranium is found predominantly as two isotopes. The more abundant (99.3%) contains 146 neutrons. The mass number of this isotope of uranium is 238, that is, 92 protons plus 146 neutrons. We represent this isotope as uranium-238, or more simply as U-238. The less abundant isotope (0.7%) contains 143 neutrons plus 92 protons. We can represent this isotope as U-235.

To review the terms mass number and isotope, see Section 2.2.

Your Turn 7.4   Another Isotope of Uranium

A trace amount of U-234 also is found in nature. How many protons are in its nucleus? How many neutrons?

It can be useful to include both the mass number and the atomic number with an isotope. We usually write the mass number (as a superscript) and the atomic number (as a subscript) to the left of the chemical symbol. Using this convention, uranium-238 becomes:

$$\text{Mass number} = \text{number of protons} + \text{number of neutrons} \longrightarrow {}^{238}_{92}\text{U}$$
$$\text{Atomic number} = \text{number of protons} \longrightarrow$$

Similarly, U-235 can be written as ${}^{235}_{92}\text{U}$. The difference between these two isotopes is a mere three neutrons, but in nuclear terms this difference is significant. For example, under the conditions present in a nuclear reactor where the neutrons are of relatively low energy, U-238 does *not* undergo fission, yet U-235 does. Small differences in the nucleus can mean large differences in *nuclear* behavior.

The process of fission usually requires neutrons to initiate it and always releases neutrons, as can be seen by this nuclear equation with uranium-235.

$$_0^1n + _{92}^{235}U \longrightarrow [_{92}^{236}U] \longrightarrow _{56}^{141}Ba + _{36}^{92}Kr + 3\,_0^1n \qquad [7.1]$$

Nuclear equations are similar to, but not the same as the chemical equations that you have seen in earlier chapters. Let's look at the components, from left to right. Initially, a neutron hits the nucleus of U-235. This neutron, $_0^1n$, has a subscript of 0, indicating no charge; the superscript is 1 because the mass number of a neutron is 1. The nucleus of $_{92}^{235}U$ captures the neutron, forming a heavier isotope of uranium, $_{92}^{236}U$. This isotope is written in square brackets indicating that it exists only momentarily. Uranium-236 immediately splits into two smaller atoms (Ba and Kr) with the release of three more neutrons.

In a nuclear equation, the sum of the subscripts on the left must equal that of the subscripts on the right. Likewise, the sum of superscripts on each side of the equation must be equal. Coefficients in nuclear equations, such as the 3 preceding the $_0^1n$ in equation 7.1, are treated the same way as in chemical equations, multiplying the term that follows it. We demonstrate these features with nuclear equation 7.1.

Left	Right
Superscripts: $1 + 235 = 236$	$141 + 92 + (3 \times 1) = 236$
Subscripts: $0 + 92 = 92$	$56 + 36 + (3 \times 0) = 92$

A wide array of fission products can be formed when the nucleus of an atom of U-235 is struck with a neutron. Your Turn 7.5 gives other possibilities.

Your Turn 7.5   Other Examples of Fission

With the help of a periodic table, write these two nuclear equations. For both, a neutron of appropriate energy to initiate the fission process first hits U-235.

a. U-235 fissions to form Ba-138, Kr-95, and neutrons.
b. U-235 fissions to form an element (atomic number 52, mass number 137), another element (atomic number 40, mass number 97), and neutrons.

Answer

a. $_0^1n + _{92}^{235}U \longrightarrow _{56}^{138}Ba + _{36}^{95}Kr + 3\,_0^1n$

Look again at nuclear equation 7.1. Both sides contain neutrons. Why don't we cancel them out? Although you would do this in a mathematical equation, nuclear equations are not handled in the same manner. The neutrons on both sides of the equation are important: The one on the left side *initiates* fission and ones on the right side are *produced* from the fission process. The net production of neutrons allows a **chain reaction** to occur in which the fission reaction becomes self-sustaining. Each neutron produced can in turn strike another U-235 nucleus, cause it to split, and release a few more neutrons. The result is a rapidly branching chain reaction that spreads in a fraction of a second (Figure 7.4). With exactly this chain reaction, the first controlled nuclear fission reaction took place at the University of Chicago in 1942.

A **critical mass** is the amount of fissionable fuel required to sustain a chain reaction, providing that the fuel is held together long enough for the reaction to proceed. For example, the critical mass of U-235 is about 15 kg, or 33 lb. Were this mass of pure U-235 to be brought together in one place, fission would spontaneously occur. Nuclear weapons work on this principle. But as you will soon see, the uranium fuel in a nuclear power plant is far from pure U-235 and is unable to explode like a nuclear bomb.

We mentioned earlier that energy is given off during fission because the mass of the products is slightly less than that of the reactants. However, from the nuclear equations we have just written no mass loss is apparent, because the sum of the mass numbers is

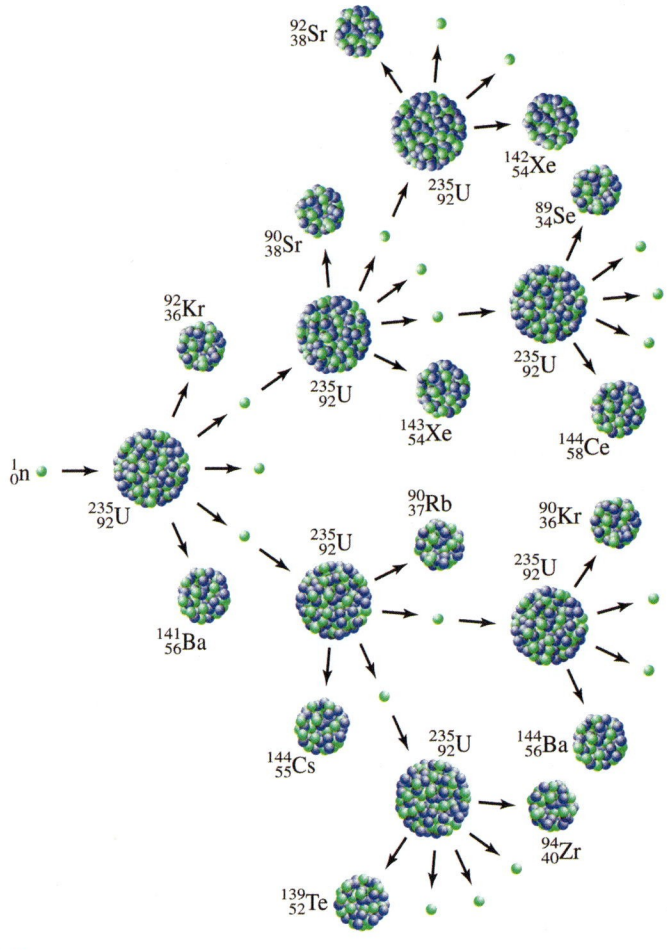

Figure 7.4

A chain reaction with U-235.

Figures Alive! Visit the *Online Learning Center* to learn more about nuclear fission and chain reactions.

the same on both sides. In fact, the actual mass does decrease slightly. To understand this, remember that the actual masses of the nuclei are not the mass numbers (the sum of the number of protons and neutrons); rather, they have measured values with many decimal places. For example, an atom of uranium-235 weighs 235.043924 atomic mass units. Were you to keep all six decimal places and compare the masses on both sides of the nuclear equation for the fission of U-235, you would find that the mass of the products is less by about 0.1%, or 1/1000th. As a consequence, the energy of the products is less than that of the reactants. This difference corresponds to the energy released.

How much energy would be released if all the nuclei in 1.0 kg (2.2 lb) of U-235 were to undergo fission? We can calculate an answer by using an equation closely related to $E = mc^2$, namely, $\Delta E = \Delta mc^2$. Here the Greek letter delta (Δ) means "the change in," so now with a change in mass we can calculate a change in energy. Since 1/1000 of this mass is lost, the value for Δm, the change in mass, is 1/1000 of 1.0 kg, which is 1.0 g or 1×10^{-3} kg. Now substitute this value and $c = 3.0 \times 10^8$ m/s into Einstein's equation.

$$\Delta E = \Delta mc^2 = (1.0 \times 10^{-3} \text{ kg}) \times (3.0 \times 10^8 \text{ m/s})^2$$
$$\Delta E = (1.0 \times 10^{-3} \text{ kg}) \times (9.0 \times 10^{16} \text{ m}^2/\text{s}^2)$$

Completing the calculation gives an energy change in what may appear to be unusual units.

$$\Delta E = (9.0 \times 10^{13} \text{ kg} \cdot \text{m}^2/\text{s}^2)$$

Atomic mass units are convenient for weighing atoms. Each is exactly 1/12 the mass of a C-12 atom or 1.66×10^{-27} kg.

The unit $kg \cdot m^2/s^2$ is identical to a joule (J). Therefore, the energy released from the fission of an entire kilogram of uranium-235 is a whopping 9.0×10^{13} J, or 9.0×10^{10} kJ.

To put things into perspective, 9.0×10^{13} J is the amount of energy released by the explosion of 33,000 tons (33 kilotons) of TNT. This is enough energy to raise about 700,000 cars 6 miles into the sky or to vaporize all the water in 37 Olympic-sized swimming pools! Yet, this massive amount of energy comes from the fission of a single kilogram of U-235, in which only 1 g (0.1% mass change) is actually transformed into energy.

> As described in Section 4.1, the joule (J) is a unit of energy.
> $$1 \text{ J} = 1 \text{ kg} \cdot m^2/s^2$$

Your Turn 7.6   Coal Equivalence

Select a grade of coal from Table 4.4. What mass of coal would be needed to produce the same amount of energy as would the fission of 1 kg of U-235?

As it turns out, one cannot fission a kilogram or two of pure U-235 in one fell swoop. In an atomic weapon, for example, the energy that is released blasts the fissionable fuel apart in a fraction of a second, thus halting the chain reaction before all the nuclei can undergo fission. Nonetheless, the energy released is enormous—on the order of 10 kilotons of TNT for the atomic bomb dropped on the city of Hiroshima in 1945. Figure 7.5 shows an atomic explosion at the Nevada Test Site. Code-named Priscilla, this test in 1957 had more than twice the explosive power of the bombs at Hiroshima and Nagasaki in 1945.

Recognize, though, that the energy of nuclear fission can be harnessed. This is exactly the objective of a nuclear power plant. Here, the energy is slowly and *continually* released under controlled conditions, as we shall see in the next section.

Figure 7.5

The nuclear test "Priscilla" was exploded on a dry lake bed northwest of Las Vegas on June 24, 1957.

7.3 How Nuclear Reactors Produce Electricity

Chapter 4 described how a conventional power plant burns coal, oil, or some other fuel to produce heat. The heat is then used to boil water, converting it into high-pressure steam that turns the blades of a turbine. The shaft of the spinning turbine is connected to large wire coils that rotate within a magnetic field, thus generating electrical energy. A nuclear power plant operates in much the same way, except that the water is heated not by combustion of a fuel, but by the energy released from the fission of nuclear "fuel" such as U-235. Like any power plant, a nuclear one is subject to the efficiency constraints imposed by the second law of thermodynamics. The theoretical efficiency for converting heat to work depends on the maximum and minimum temperatures between which the plant operates. This thermodynamic efficiency, typically 55–60%, is significantly reduced by other mechanical, thermal, and electrical inefficiencies.

A nuclear power station has parts that are both nuclear and nonnuclear (Figure 7.6). The nuclear reactor is the hot heart of the power station. The reactor, together with one or more steam generators and the primary cooling system, is housed in a special steel vessel within a separate reinforced concrete dome-shaped containment building. The nonnuclear portion contains the turbines that run the electrical generator. It also contains the secondary cooling system. In addition, the nonnuclear portion must be connected to some means of removing heat from the coolants. Accordingly, a nuclear power station will have one or more cooling towers or be located near a sizeable body of water (or both). Look back at Figure 4.3 that shows a diagram of a fossil-fuel power plant. This plant also requires a means of removing heat, as shown by the stream of cooling water.

The uranium fuel in the reactor core is in the form of uranium dioxide (UO_2) pellets, each comparable in height to a dime, as shown in Figure 7.7. These pellets are placed end to end in tubes made of a special metal alloy, which in turn are grouped into stainless-steel clad bundles (Figure 7.8). Each tube contains at least 200 pellets. Although a fission reaction, once started, can sustain itself by a chain reaction, neutrons are needed to induce

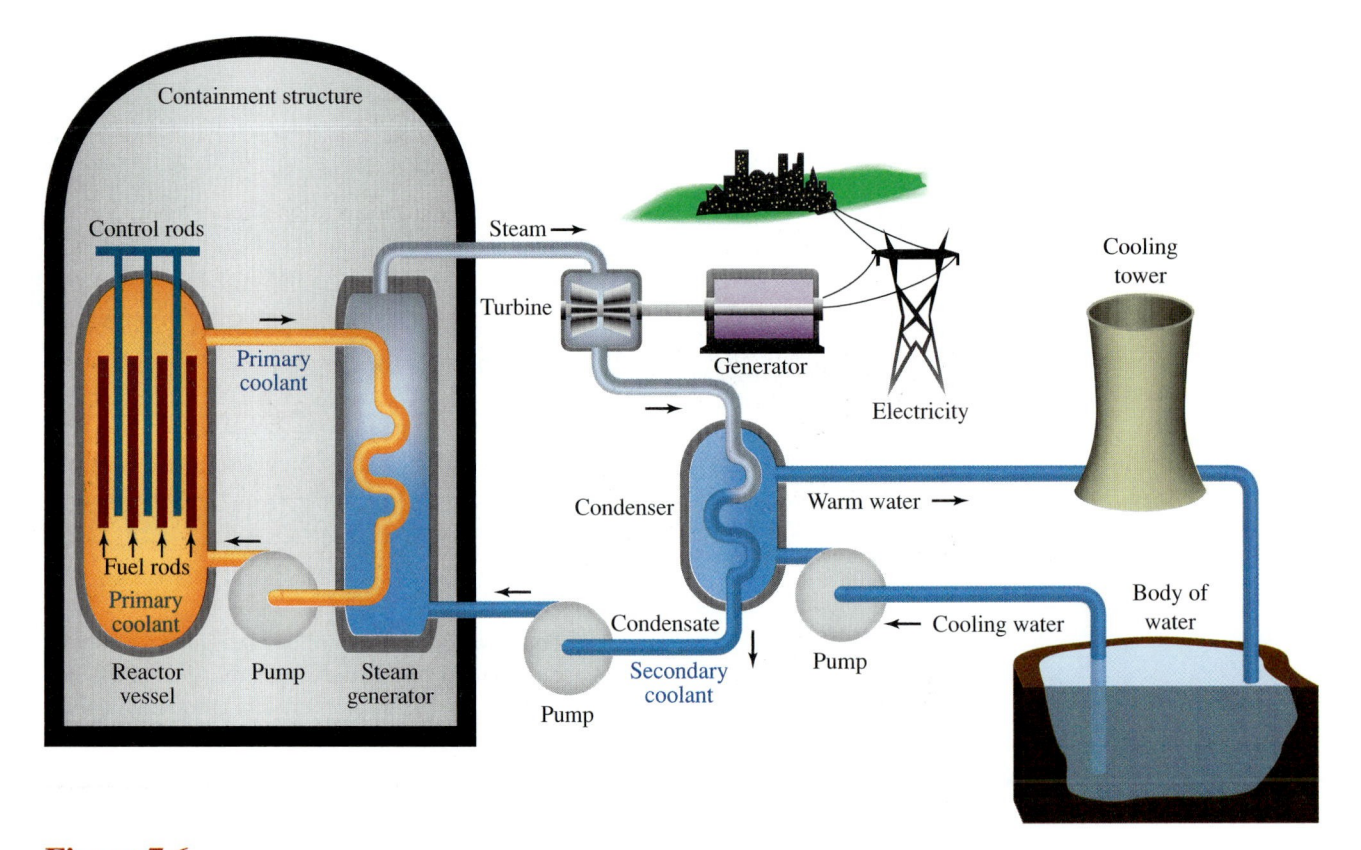

Figure 7.6

Diagram of a nuclear power plant.

Figure 7.7

Nuclear fuel pellets and a U.S. dime.

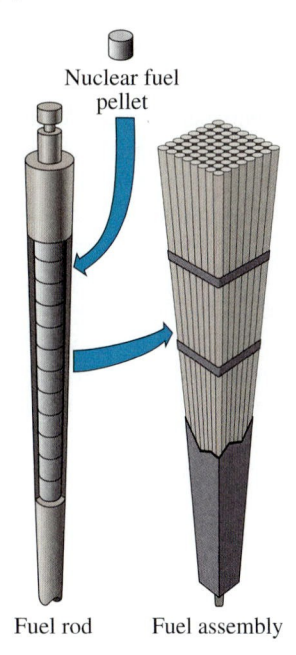

Nuclear fuel pellet

Fuel rod   Fuel assembly

Figure 7.8

Fuel pellet, fuel rod, and fuel assembly making up the core of a nuclear reactor (*left*). The fuel assembly submerged in an active reactor core (*right*).

the process (see equation 7.1, Figure 7.4). One means of generating neutrons is to use a combination of beryllium-9 and a heavier element such as plutonium. The heavier element releases alpha particles, $^4_2$He.

$$^{238}_{94}\text{Pu} \longrightarrow ^{234}_{92}\text{U} + ^4_2\text{He} \qquad [7.2]$$
$$\text{alpha particle}$$

These alpha particles in turn strike the beryllium, releasing neutrons, carbon-12, and gamma rays, $^0_0\gamma$. Here is the nuclear equation.

$$^4_2\text{He} + ^9_4\text{Be} \longrightarrow ^{12}_6\text{C} + ^1_0\text{n} + ^0_0\gamma \qquad [7.3]$$
$$\text{gamma ray}$$

The neutrons produced in this way can initiate the nuclear fission of uranium-235 in the reactor core.

Your Turn 7.7   Poo-Bee and Am-Bee

A neutron source constructed with Pu and Be is called a PuBe or "poo-bee" source. Similarly, the AmBe or "am-bee" source is constructed from americium and beryllium. Analogous to the PuBe source, write the set of reactions that produce neutrons from an AmBe source. Start with Am-241.

Remember—one fission event produces two or three neutrons. The trick is to "sponge up" these extra neutrons, but still leave enough to sustain the fission reaction. A delicate balance must be maintained. With extra neutrons, the reactor will run at too high a temperature; with too few neutrons, the chain reaction will halt and the reactor will cool. To achieve the needed balance, one neutron from each fission event should in turn cause another.

Metal rods interspersed among the fuel elements serve as the neutron "sponges." These **control rods,** composed primarily of an excellent neutron absorber such as cadmium or boron, can be slid up or down to absorb fewer or more neutrons. With the rods

fully inserted, the fission reaction is not self-sustaining. But as the rods are withdrawn, the reactor can "go critical"; that is, the fission chain reaction can become self-sustaining. This condition will not last. Over time, fission products that absorb neutrons build up in the fuel pellets. To compensate, the control rods are pulled further out. Eventually, the reactor fuel bundles must be replaced.

Your Turn 7.8   **Earthquake!**

Suppose a serious tremor were to occur. Automatically, any reactor near the epicenter should immediately be shut down. Should the software be programmed to fully insert the control rods into the reactor core, or should they be pulled out? Explain.

The fuel bundles and control rods are bathed in the **primary coolant,** a liquid that comes in direct contact with the nuclear reactor to carry away heat. In the Byron nuclear reactor (Figure 7.9) and in many others, the primary coolant is an aqueous solution of boric acid, H_3BO_3. The boron atoms absorb neutrons and thus control the rate of fission and the temperature. The solution also serves as a **moderator** for the reactor, slowing the speed of the neutrons and making them more effective in causing fission. Another major function of the primary coolant is to absorb the heat generated by the nuclear reaction. Because the primary coolant solution is at a pressure more than 150 times normal atmospheric pressure, it does not boil. It is heated far above its normal boiling point and circulates in a closed loop from the reaction vessel to the steam generators, and back again. This closed primary coolant loop thus forms the link between the nuclear reactor and the rest of the power plant (see Figure 7.6).

The heat from the primary coolant is transferred to what is sometimes referred to as the **secondary coolant,** the water in the steam generators that does not come in contact with the reactor. At the Byron nuclear plant, more than 30,000 gallons of water is converted to vapor each minute. The energy of this hot vapor turns the blades of turbines that are attached to an electrical generator. To continue the heat transfer cycle, the water vapor is then cooled and condensed back to a liquid and returned to the steam generator. In many nuclear facilities the cooling is done using large cooling towers that commonly are mistaken for the reactors. The reactor buildings are not as large (see Figure 7.9).

Cooling towers also are used in coal-fired plants.

Figure 7.9
The two cooling towers (with clouds of condensed water vapor) at the Byron nuclear power plant in Illinois. The reactors are located in the two cylindrical containment buildings in the foreground.

Your Turn 7.9   Clouds (not mushroom-shaped)

Some days you can see a cloud coming out of the cooling tower of a nuclear power plant, as shown in Figure 7.9. What causes the cloud? Does it contain radioisotopes produced from the fission of U-235?

Nuclear power plants also use water from lakes, rivers, or the ocean to cool the condenser. For example, at the Seabrook nuclear power plant in New Hampshire, every minute 398,000 gallons of ocean water flows through a huge tunnel (19 feet in diameter and 3 miles long) bored through rock 100 feet beneath the floor of the ocean. A similar tunnel from the plant carries the water, now 22 °C warmer, back to the ocean. Special nozzles distribute the hot water so that the observed temperature increase in the immediate area of the discharge is only about 2 °C. The ocean water is in a separate loop from the fission reaction and its products. The primary coolant (water with boric acid) circulates through the reactor core inside the containment building. However, this boric acid solution is kept isolated in a closed circulating system, which makes the transfer of radioactivity to the secondary coolant water in the steam generator highly unlikely. Similarly, the ocean water does not come in direct contact with the secondary system, so the ocean water is well protected from radioactive contamination. Clearly the electricity generated by a nuclear power plant is identical to the electricity generated by a fossil-fuel plant; the electricity is not radioactive, nor can it be.

Consider This 7.10   The Palo Verde Reactors

One of the most powerful nuclear plants in operation in the United States is the Palo Verde complex in Arizona. At maximum capacity, just one of its three reactors generates 1243 million joules of electrical energy every second. Calculate the total amount of electrical energy produced per day and the loss of mass of U-235 each day.

Hint: Start by calculating the quantity of energy generated not per second, but per day. Then use the equation $\Delta E = \Delta mc^2$ and solve for the change in mass, Δm. Report the mass loss in grams.

7.4 Nuclear Power Worldwide

Worldwide, just over 16% of the electricity produced and consumed is generated in roughly 440 nuclear power plants. Although the international reliance on nuclear energy is relatively low, nonetheless it is significant. For example, to replace this amount of energy would require the entire annual coal production of the United States!

Where does the United States rank in regard to nuclear power? Although the United States has more nuclear reactors than any other nation (Figure 7.10), nearly a third of these are over 30 years old. Furthermore, with only 20% of its electrical power generated by nuclear power reactors (Figure 7.11), the United States clearly does *not* lead the nuclear pack.

In contrast, France is a world leader in nuclear power. As of late 2007, the French have 59 nuclear power plants that generate 78% of their electricity. All of the countries that generate 40% or more of their electricity from nuclear power plants are in Europe. The Swiss generate 40% of their electricity with only five reactors.

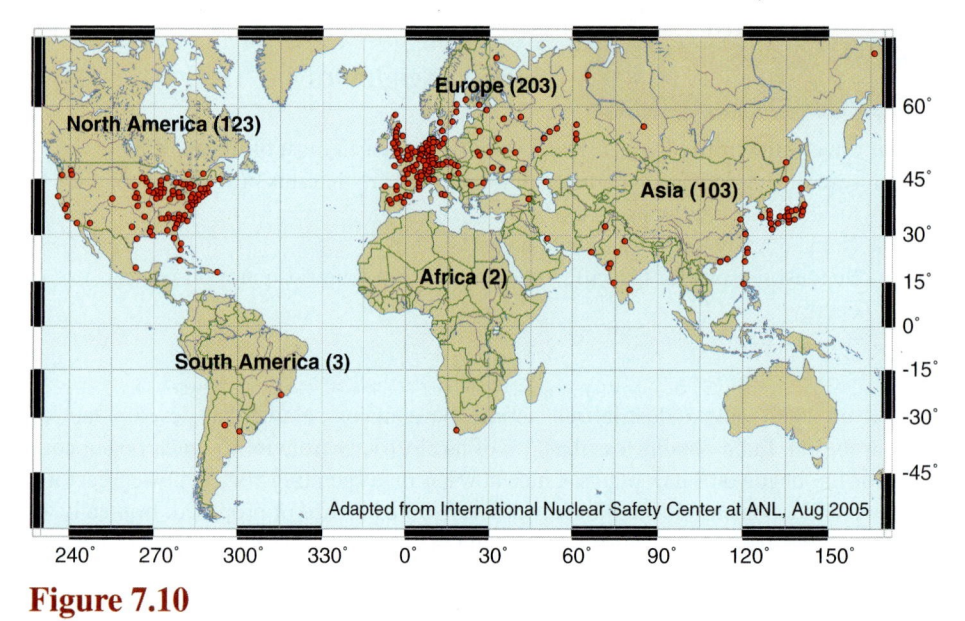

Figure 7.10

Number of reactors in operation worldwide, as of December 2005. Some sites have more than one reactor.

Source: http://www.insc.anl.gov/pwrmaps/map/world_map.php

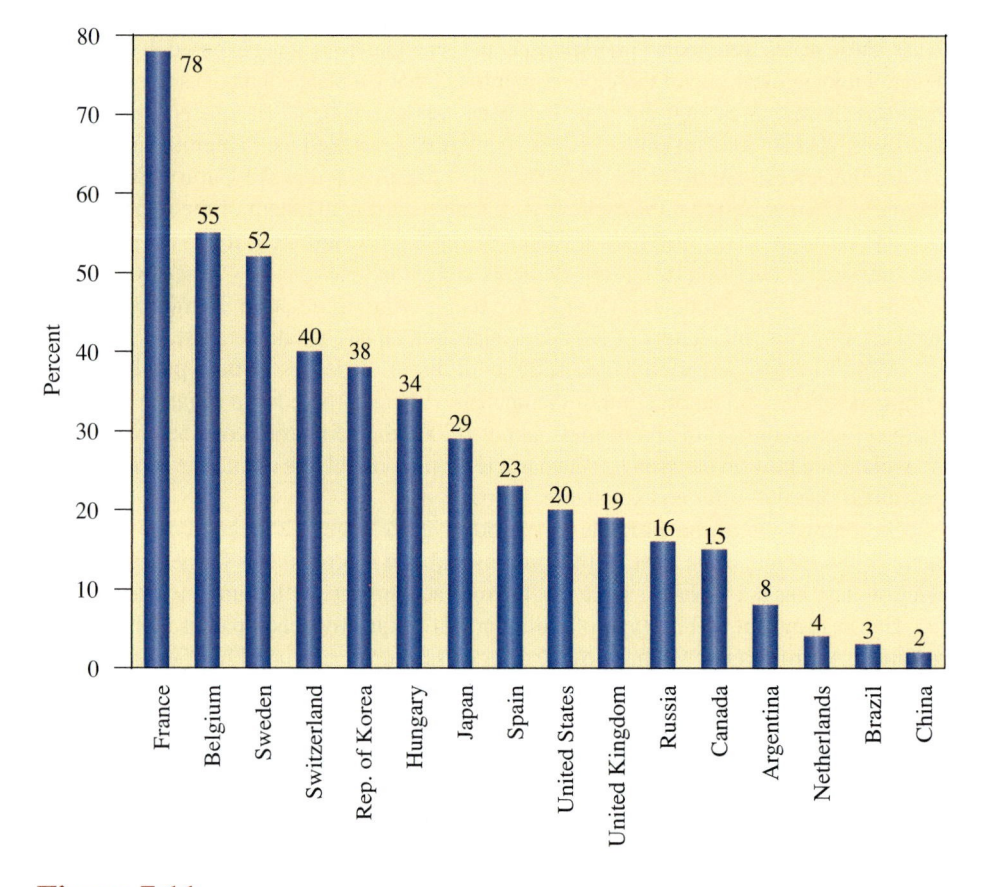

Figure 7.11

Percent of electrical power generated by nuclear power reactors in selected countries, November 2005.

Source: World Nuclear Association, http://www.world-nuclear.org/info/inf01.htm

Consider This 7.11   **Nuclear Neighbors**

a. Determine from Figure 7.10 the countries in which most of the nuclear reactors are located. Characterize these countries in terms of their geographic location and energy use.
b. Name three countries without nuclear reactors.
c. Suggest reasons for the different emphasis that countries place on nuclear energy.

As of 2007, only industrialized nations have major commercial development of nuclear fission. India obtains less than 3% of its electricity from its 15 reactors, but construction has begun on 8 new plants. China now has nine operating nuclear power reactors, with two under construction and over two dozen more planned or proposed. Ironically, in spite of the fact that much of the world's uranium comes from Africa, South Africa is the only nation on that continent that uses nuclear energy.

The topics we have been discussing—nuclear fission, uranium, nuclear fuel, nuclear weapons—all rest on an understanding of radioactivity. We now turn to this topic.

7.5 What Is Radioactivity?

Our knowledge of radioactive substances is just over 100 years old. In 1896, the French physicist Antoine Henri Becquerel (1852–1908) discovered radioactivity. At the time, his research involved using photographic plates; film, of course, had not yet been invented. Prior to use, these plates were sealed in black paper to keep them from being exposed. By accident he left a mineral near one of these plates and found that the plate's light-sensitive emulsion darkened. It was as though the plate had been exposed to light! Becquerel immediately recognized that the mineral emitted powerful rays that penetrated the lightproof paper.

Further investigation by the Polish scientist Marie Sklodowska Curie (1867–1934) (Figure 7.12) revealed that the rays were coming from a constituent of the mineral—the element uranium. In 1899, Marie Curie applied the term **radioactivity** to the spontaneous emission of radiation by certain elements. Subsequent research by Ernest Rutherford (1871–1937) led to the identification of two major types of radiation. Rutherford named them after the first two letters of the Greek alphabet, alpha (α) and beta (β).

Alpha and beta radiation have strikingly different properties. A **beta particle (β)** is a high-speed electron emitted out of the nucleus. A beta particle has a negative electrical charge (1−) and only a tiny bit of mass, about 1/2000 that of a proton or a neutron. If you are wondering how an electron (a beta particle) could possibly be emitted from a nucleus, stay tuned. We will offer an explanation shortly.

In contrast, an **alpha particle (α)** is positively charged (2+) and consists of the nucleus of a helium atom, that is, two protons and two neutrons. It is far heavier than an electron, and has a 2+ charge since no electrons accompany the helium nucleus.

Gamma rays are a third type of radiation that frequently accompanies alpha or beta radiation. A **gamma ray (γ)** has no charge or mass and is made up of high-energy, short-wavelength photons of energy. Just like infrared (IR), visible, and ultraviolet (UV) radiation, gamma rays are part of the electromagnetic spectrum. In terms of their energy, they are similar to X-rays. You will learn more about them in connection with food preservation in Chapter 11. Table 7.2 summarizes these three types of radiation.

The term *radiation* can be confusing. People say "radiation" and expect that the listener will understand from the context whether they mean electromagnetic or nuclear radiation. *Electromagnetic radiation* refers to all the different types of light: visible, infrared, ultraviolet, microwave, and, of course, gamma rays. For example, it is perfectly correct to say visible radiation instead of visible light. *Nuclear radiation,* however, refers to the radiation emitted by the nucleus, such as alpha, beta, or gamma radiation. Watch out for one more source of confusion. Gamma rays are *both* a type of electromagnetic

Figure 7.12
Marie Sklodowska Curie won two Nobel Prizes—one in chemistry, the other in physics—for her research on radioactive elements.

Rutherford was born in New Zealand and later worked first in England and then at McGill University in Canada.

Section 2.4 introduced gamma rays as part of the electromagnetic spectrum.

Table 7.2		Types of Nuclear Radiation		
Type	**Symbol**	**Composition**	**Charge**	**Change to the Nucleus That Emits It**
Alpha	$_2^4\text{He}$	2 protons 2 neutrons	2+	Mass number decreases by 4. Atomic number decreases by 2.
Beta	$_{-1}^{0}\text{e}$	an electron	1−	Mass number does not change. Atomic number increases by 1.
Gamma	$_0^0\gamma$	a photon	0	No change in either the mass number or the atomic number.

radiation and of nuclear radiation. When emitted from the nucleus of a radioactive substance, we refer to gamma rays as nuclear radiation. In contrast, when emitted from a galaxy far away, we call these gamma rays electromagnetic radiation.

Your Turn 7.12   "Radiation"

For each sentence, use the context to decipher whether the speaker is referring to nuclear or electromagnetic radiation.

a. Name a type of radiation that has a shorter wavelength than visible light.
b. Gamma radiation can penetrate right through the walls of your home.
c. Watch out for UV rays! If you have lightly pigmented skin, this type of radiation can give you a sunburn.
d. Rutherford detected the radiation emitted by uranium.

Answers
a. Electromagnetic radiation   **b.** Nuclear radiation

When either an alpha or beta particle is emitted, a remarkable transformation occurs—the atom that emitted the particle changes its identity. Earlier with the PuBe neutron source (see equation 7.2), you saw that alpha emission resulted in the nucleus of plutonium becoming that of uranium. Similarly, when uranium emits an alpha particle, it becomes the element thorium. This nuclear equation shows the process for uranium-238.

$$_{92}^{238}\text{U} \longrightarrow _{90}^{234}\text{Th} + _2^4\text{He} \qquad [7.4]$$

The mass number of the product, thorium-234, is 4 lower. The loss of the alpha particle (two protons, two neutrons) accounts for this. Also notice that the sum of the mass numbers on both sides of the nuclear equation is equal: 238 = 234 + 4. The same is true for the atomic numbers: 92 = 90 + 2.

The nucleus formed as the result of radioactive decay may still be radioactive. This is the case for thorium-234, the product of the alpha decay by uranium-238. Radioactive thorium-234 undergoes beta decay to form the new element protactinium (Pa).

$$_{90}^{234}\text{Th} \longrightarrow _{91}^{234}\text{Pa} + _{-1}^{0}\text{e} \qquad [7.5]$$

In contrast to alpha emission, with beta emission the atomic number *increases* by 1 and the mass number remains unchanged. One model that can help you make sense of this seemingly unusual set of changes is to regard a neutron as a combination of a proton and an electron. Beta emission can be thought of as breaking a neutron apart. Equation 7.6 shows this process, giving us an explanation of how an electron can be emitted from the nucleus.

$$_0^1\text{n} \longrightarrow _1^1\text{p} + _{-1}^{0}\text{e} \qquad [7.6]$$

Figure 7.15
An aerial view of the Chernobyl Unit 4 reactor after the chemical explosion.

$$2 \, H_2O(l) + C(graphite) \longrightarrow 2 \, H_2(g) + CO_2(g) \qquad [7.7]$$

$$2 \, H_2(g) + O_2(g) \longrightarrow 2 \, H_2O(g) \qquad [7.8]$$

The explosion blasted off the 4000-ton steel plate covering the reactor (Figure 7.15). Although a "nuclear" explosion never occurred, the fire and explosions of hydrogen blew vast quantities of radioactive material out of the reactor core and into the atmosphere.

Fires started in what remained of the building. In a short time, the plant lay in ruins. The head of the crew on duty at the time of the accident wrote: "It seemed as if the world was coming to an end . . . I could not believe my eyes; I saw the reactor ruined by the explosion. I was the first man in the world to see this. As a nuclear engineer I realized the consequences of what had happened. It was a nuclear hell. I was gripped with fear." (*Scientific American,* April 1996, p. 44)

The disaster continued. As the reactor burned, it continued to spew large quantities of radioactive fission products into the atmosphere for 10 days (see Figure 7.15). The release of radioactivity was estimated to be on the order of 100 of the atomic bombs dropped on Hiroshima and Nagasaki. People in nearby regions reported an odd, bitter, and metallic taste as they inhaled the invisible particles. The radioactive dust cut a swath across Ukraine, Belarus, and up into Scandinavia. Nearly 150,000 people living within 60 km of the power plant were permanently evacuated after the meltdown.

The human toll was immediate. Several people working at the plant were killed outright, and another 31 firefighters died in the cleanup process from acute radiation sickness, a topic we will examine in a later section. An estimated 250 million people were exposed to levels of radiation that may ultimately shorten their lives. Included in this figure are 200,000 "liquidators," people who buried the most hazardous wastes and constructed a 10-story concrete structure ("the sarcophagus") to surround the failed reactor.

One of the hazardous radioisotopes released was iodine-131. It decays by beta emission with an accompanying gamma ray.

The thyroid gland incorporates iodine in the form of iodide ion to manufacture thyroxin, a hormone essential for growth and metabolism.

$$^{131}_{53}I \longrightarrow {}^{131}_{54}Xe + {}^{0}_{-1}e \qquad [7.9]$$

If ingested, I-131 can cause thyroid cancer. In the contaminated area near Chernobyl, the incidence of thyroid cancer increased sharply, especially for those younger than age 15 (Figure 7.16). As of 2001, more than 700 children in Belarus, a neighboring country, were treated for thyroid cancer. Fortunately, with treatment, the survival rate for thyroid cancer is high and most have survived. As Dr. Akira Sugenoya, a Japanese physician who volunteered his expertise in Belarus to treat the children suffering from thyroid cancer, remarked "The last chapter of the terrible accident is far from written."

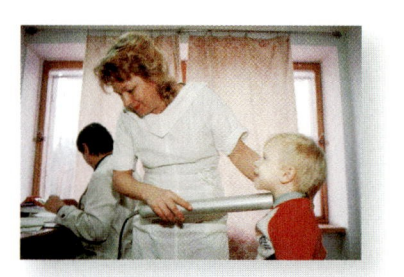

Figure 7.16

A child being checked for the level of radioactivity in his thyroid gland at a clinic north of Minsk, near the Chernobyl nuclear reactor.

Your Turn 7.14   **"Iodine"**

When people speak of iodine, depending on the context they may be referring to an iodine atom, an iodine molecule, or an iodide ion.

  a. Write Lewis structures to distinguish among these chemical forms.
  b. Which is the most chemically reactive and why?
  c. Which chemical form of iodine-131 is implicated in thyroid cancer?

Answer
  c. Iodide ion (I^-) is taken up by the thyroid gland.

Given the demonstrable problems with their design, the four reactors at Chernobyl have been shut down. On Friday, December 15, 2000, the control rods slid into the core at Unit 3, the last remaining reactor operating at Chernobyl, permanently shutting it down. Ukrainian President Leonid Kuchma reported, "This decision came from our experience of suffering. We understand that Chernobyl is a danger for all of humanity and we forsake a part of our national interests for the sake of global safety."

Today, most of the 1000 square miles of contaminated land in the vicinity of Chernobyl has not yet returned to farming. One small exception in Belarus is the land near Viduitsy, as shown in Figure 7.17. A report in 2005 indicated that none of the summer crop of rye and barley harvested in this area tested positive for radioisotopes.

Consider This 7.15   **Chernobyl's Legacy**

Twenty years after the accident, the report *Chernobyl's Legacy: Health, Environmental and Socio-economic Impacts* was issued. This document was an initiative of the International Atomic Energy Agency in cooperation with many other international agencies. *The Online Learning Center* provides a link.

  a. What diseases already have resulted from the radiation exposure? Describe the disease and which people were affected. When possible, cite the number of people affected.
  b. Why is it not possible to reliably assess the number of fatal cancers caused by the accident?

This recounting of the solemn facts of Chernobyl leads to an inevitable question: "Could it happen here?" America's closest brush with nuclear disaster occurred in March 1979, when the Three Mile Island power plant near Harrisburg, Pennsylvania, lost coolant and a partial meltdown occurred. Although some radioactive gases were released during the incident, no fatalities resulted. A 20-year study concluded in 2002 that the total cancer deaths among the exposed population were not higher than those of the general population. In spite of the initial failure, the system held and the

Figure 7.17

This area in Belarus is gradually being returned to agriculture.

damage was contained. Since then, refinements in design and safety were made to existing reactors and those under construction. Nuclear engineers agree that no commercial nuclear reactors in the United States have the design defects that led to the Chernobyl catastrophe.

Consider, for example, the Seabrook nuclear power plant in Massachusetts that was hailed as an example of state-of-the-art engineering when it was built. The energetic heart of the station is a 400-ton reaction vessel with 44-foot high walls made of 8-inch thick carbon steel. Unlike the Chernobyl plant, a reinforced concrete and a dome-shaped containment building must surround all reactors in the United States. As the name suggests, the containment structure is built to withstand accidents and prevent the release of radioactive material. The inner walls of the building are several feet thick and made of steel-reinforced concrete; the outer wall is 15 inches thick. The containment building is constructed to withstand hurricanes, earthquakes, and high winds.

Nonetheless, if you live near a nuclear power plant, you probably read about small reactor "incidents" in the daily news. For example, the *New York Times* regularly reports on the Indian Point reactor Units 2 and 3 just north of Manhattan on the Hudson River. Reactor 1 opened in 1962, was closed in 1974, but has not yet been decommissioned. Many of the incidents relate to the spent nuclear fuel, a topic that we will take up in Section 7.9.

Could a nuclear meltdown happen in some other region of the world? That possibility does exist, because such disasters result from the complex interplay of faulty plant design, human error, and political instability. Each of these factors must be minimized to keep a nuclear power plant operating safely. Although the nuclear units in many parts of the world get high rankings on all three factors, this is not the case everywhere. For example, in July 2001 the German government urged the closing of a Czech nuclear power plant near the German border because of safety concerns. Several reactors in Russia have long histories of safety violations and raise similar concerns. A plume of radioactive dust easily crosses international boundaries, and so the concerns of neighboring nations are well placed.

A related and far scarier question relates to acts of terrorism. Are nuclear reactors being considered as targets? This question must be taken seriously. Fortunately, as this book went to press, no incidents have occurred. Even before the threat of terrorists, nuclear reactors were built to withstand earthquakes since about a fifth are in regions of seismic activity, such as on the Pacific rim. These reactors are fitted with detectors that quickly can shut a reactor down if a tremor occurs. The impact of one or more fully fueled commercial jets on a containment dome, however, is another matter entirely. Were the dome to be breached, the results could truly be catastrophic, exceeding the radioactive releases of Chernobyl.

Sceptical Chymist 7.16   More About the Pacific Rim

We just stated that about a fifth of the world's reactors are in regions of seismic activity, such as the Pacific rim. Is this true? Use Figure 7.10, your knowledge of earthquake zones, and any other information you may need to look up on the Web to check the accuracy of this statement. See also if you can find the details of how reactors are built to withstand seismic shocks.

Today, nuclear plants and their past operations continue to be under intense scrutiny, hence the title of this section. We must look backward in order to gain the wisdom to move forward into our future which, undoubtedly, will involve nuclear energy.

7.7   Radioactivity and You

It would be a serious mistake to dismiss radioactivity as harmless. The evidence of the past makes this quite clear. Unfortunately, though, some of the scientists who first studied radioactive substances were not fully aware of their dangers. Marie Curie, for example, died of a blood disorder that most likely was induced by her exposure to radiation.

The dangers arise because alpha and beta particles have sufficient energy to ionize the molecules they strike. As you might expect, the same is true for gamma rays and X-rays. For this reason, these all are termed *ionizing radiation*. For example, when a beta particle penetrates your tissue, it is likely to hit a water molecule and can knock out an electron.

$$H_2O \xrightarrow{\text{ionizing radiation}} H_2O^+ + e^- \qquad [7.10]$$

The positively charged product, H_2O^+, is highly reactive because it has an unpaired electron. In your body, H_2O^+ will further react, often with another water molecule. The products in turn can react with still other molecules, including your DNA. This cascading set of radiation-induced molecular changes can range from being perfectly harmless to those causing the death of the cell.

Consider This 7.17   Free Radicals

Species with unpaired electrons (free radicals) are highly reactive. Below we have rewritten equation 7.10 to show the unpaired electron on H_2O^+.

$$H_2O \xrightarrow{\text{ionizing radiation}} [H_2O{\cdot}]^+ + e^-$$

In turn, the product may react with another water molecule.

$$[H_2O{\cdot}]^+ + H_2O \longrightarrow H_3O^+ + HO{\cdot}$$

In this equation, draw Lewis structures for all reactants and products.

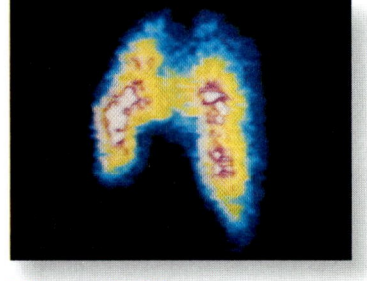

Figure 7.18
A thyroid image produced with I-131. Radioactive iodine has concentrated in the red and yellow areas.

End-of-chapter question #52 relates to taking potassium iodide tablets to treat exposure to I-131.

Rapidly dividing cells are particularly susceptible to damage by ionizing radiation. As a consequence, nuclear radiation can be used to *treat* certain kinds of cancer, such as prostate cancer and breast cancer. Radioactivity can treat other diseases as well. For example, in Graves' disease, the thyroid is hyperactive. Patients receive small amounts of radioactive I-131 orally in the form of potassium iodide. Since iodine concentrates in the thyroid gland, the same is true for radioactive iodine. The radiation it emits destroys the overactive tissue, in whole or in part (Figure 7.18). To restore normal thyroid function,

most patients need a supplement of a synthetic form of thyroxin, the iodine-containing hormone normally secreted by the thyroid.

But radiation also can damage healthy rapidly dividing cells, such as those in the bone marrow, the skin, hair follicles, stomach, and intestine. People who receive radiation treatments for cancer often experience a host of side effects that relate to the damage of *healthy* cells. Collectively, these side effects are termed **radiation sickness,** the illness characterized by early symptoms of anemia, nausea, malaise, and susceptibility to infection that are the result of a large exposure to radiation. Radiation sickness affected those near the Chernobyl accident, as well as the victims of Hiroshima and Nagasaki. Radiation-induced transformations of DNA also can produce genetic mutations, some that occasionally lead to cancer or birth defects.

Today, considerable care is taken to protect workers from nuclear radiation. This is accomplished in many ways, including by using shields made from a dense metal such as lead. Remember, though, that our world (and our bodies) naturally contains radioactive substances, so that radiation levels can never be reduced to zero. **Background radiation** is the radiation, on average, that exists at a particular location, usually due to natural sources (Figure 7.19). The level of background radiation depends primarily on where you live. It also depends on what type of dwelling you live in, because the Earth itself and the building materials quarried or manufactured from it contain tiny amounts of uranium and its decay products. About 80% of background radiation is natural in origin. The largest natural source is radon, a radioactive gas released in the decay series of uranium in the soils and rocks (see Figure 7.13).

See Section 1.13 for more about radon as an indoor air pollutant.

Consider This 7.18   Radon and You

Radon-222 is an alpha emitter.

a. Write the nuclear equation for the decay of radon-222.
b. The decay product is a solid. Would you expect it to be radioactive?
c. Given your previous answers, explain why radon can cause lung cancer.

Answer

a. $^{222}_{86}Rn \longrightarrow {}^{218}_{84}Po + {}^{4}_{2}He$

b. Polonium-218 is radioactive, as are all elements above atomic number 83.

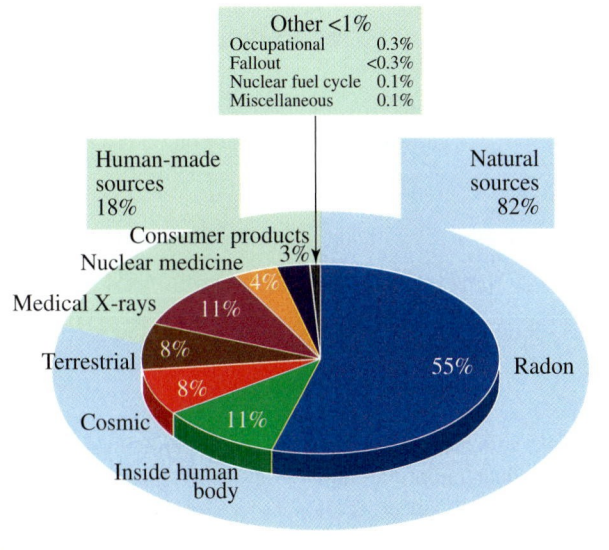

Figure 7.19

U.S. sources of background radiation.

Source: National Council on Radiation Protection Measures (NCRP) Report No. 93, *Ionizing Radiation Exposure of the Population of the United States, 1987.*

We use different units to measure the radioactivity of a sample and the damage that exposure to a dose of radiation can cause. In regard to the former, radioactivity is assessed by counting the number of disintegrations of a sample (alpha, beta, or gamma emissions) in a given time period. The **curie (Ci),** named in honor of Marie Curie, is a measure of radioactivity equivalent to the number of decays per second from one gram of radium.

$$1 \text{ curie (Ci)} = 3.7 \times 10^{10} \text{ disintegrations/s}$$

Since radium is highly radioactive, one curie is a huge amount of radiation. Accordingly, people typically measure radioactivity using *milli*curies (mCi), *micro*curies (µCi), *nano*curies (nCi), or even *pico*curies (pCi). For example, household radon measurements are quoted in picocuries, as you will see in Your Turn 7.24. Chemists working with radioisotopes in the lab typically use millicurie or microcurie amounts. If a laboratory worker spilled an amount as large as 100 mCi, serious cleanup procedures would be needed. In contrast, a spill of 100 µCi would not.

To put these values in perspective, the explosion at Chernobyl spewed 100–200 million curies into the atmosphere. In terms of the amount of radioactivity itself, this is the equivalent of dispersing 100–200 million grams of radium. At the time of the accident, the radioactivity levels near Chernobyl were measured from 5 to over 40 Ci per square kilometer. The amount of radiation released by the atomic bombs that exploded on Nagasaki and Hiroshima was lower by two orders of magnitude.

Pico:
1×10^{-12}
1/1,000,000,000,000

Nano:
1×10^{-9}
1/1,000,000,000

Micro:
1×10^{-6}
1/1,000,000

Milli:
1×10^{-3}
1/1000

Consider This 7.19   Assessing Radioactive Releases

It is not sufficient to just report the amount of a radioactive release. Rather, the *identity* of the radioisotopes also should be reported. Explain why, using the nuclear fission products I-131, Sr-90, and Cs-137 as examples.

Answer
Radioactive Cs-137 is particularly dangerous because of its uptake into living things (including humans) in the form of the cesium ion, Cs^+. Cs-137 also is dangerous because it has a half-life long enough to persist in the environment for decades (see next section).

When measuring the damage that exposure to a dose of radiation can cause, we use a different set of units. We also switch our focus from the *sample* to the *tissue* that is absorbing the dose. In part, the damage depends on the total amount of energy that the tissue absorbs. The **rad,** short for radiation absorbed dose, is defined as the absorption of 0.01 joule of radiant energy per kilogram of tissue. Thus, if a 70-kg person were to absorb 0.70 J of energy, a dose of 1 rad would be received. Although this is not very much energy, the energy is localized where the radiation hits and can ionize the molecules in its path.

The biological damage is more than simply a matter of the energy deposited. To estimate the physiological damage, another unit multiplies the number of rads by a factor Q determined by the type of radiation, the rate the radiation is delivered, and the type of tissue. Less damaging types, including beta, gamma, and X-rays, are arbitrarily assigned a Q of 1. Highly damaging radiation, such as alpha particles and high-energy neutrons, have a Q as high as 20. The **rem,** short for "roentgen equivalent man," is Q multiplied by the number of rads and is a measure of the damage caused to human tissue.

Q is called the relative biological effectiveness, sometimes referred to as RBE.

$$\text{Number of rems} = Q \times (\text{number of rads})$$

Thus for beta or gamma radiation ($Q = 1$), a dose of 10 rads is 10 rems. In contrast, a 10-rad dose of alpha radiation ($Q = 20$) may be as high as 200 rems. Thus, it would take approximately 20 times as many beta particles to do the same damage as a given number of alpha particles. This makes sense, as alpha particles are larger and deposit a greater amount of energy in the target tissue.

The scientific community in the United States is moving toward using the sievert rather than the rem. The **sievert (Sv)** is an international unit equal to 100 rem. Alternatively,

Table 7.3	Annual Radiation Dose (Sample Calculation)*	
Sources of Radiation		**(µSv/yr)**
1. Cosmic radiation		
a. Sea level (U.S. average)		260
b. Additional dose if you are above sea level		
up to 1000 m (3300 ft) add 20 µSv		20
1000 to 2000 m (6600 ft) add 50 µSv		
2000 to 3000 m (9900 ft) add 90 µSv		
3000 to 4000 m (13,200 ft) add 15 µSv		
4000 to 5000 m (16,500 ft) add 21 µSv		
2. Building material(s) used in your dwelling		
Stone, brick or concrete add 70 µSv		
Wood or other add 20 µSv		20
3. Rocks and soil		460
4. Food, water, and air (K and Rn)		2400
5. Fallout from nuclear weapons testing		10
6. Medical and dental X-rays		
a. Chest X-ray, add 100 µSv each		0
b. Gastrointestinal tract X-ray, add 5000 µSv each		0
c. Dental X-rays, add 100 µSv each		100
7. Airplane travel		
5-hour flight at 30,000 feet, add 30 µSv/flight		300
8. Other		
a. Live within 50 miles of a nuclear plant, add 0.09 µSv		0.09
b. Live within 50 miles of a coal-fired power plant, add 0.3 µSv		0.3
c. Use a computer terminal, add 1 µSv		1
d. Watch TV, add 10 µSv		10
e. Smoke one pack of cigarettes/day, add 10,000 µSv		0
Total Annual Radiation Dose		**3581**
U.S. annual average = 3600 µSv		

*Sample calculation is for an adult nonsmoker living in the Midwest.
Sources: U.S. Environmental Protection Agency, American Nuclear Society.

1 microsievert = 0.1 millirem

1 rem equals 0.0100 Sv. Because most doses of radiation are significantly less than a sievert (or a rem), smaller units such as microsieverts (µSv) and millirems (mrem) are employed.

$$1 \text{ microsievert (µSv)} = 1/1{,}000{,}000 \text{ of a sievert} = 1 \times 10^{-6} \text{ Sv}$$
$$1 \text{ millirem (mrem)} = 1/1000 \text{ of a rem} = 1 \times 10^{-3} \text{ rem}$$

Table 7.3 shows a sample calculation for an annual radiation dose. The values (in microsieverts) were selected for an adult nonsmoker living in the Midwest. In Your Turn 7.20 that follows, you can do your own calculation.

Your Turn 7.20   Your Personal Radiation Dose

a. Thanks to a Web page posted by the EPA, you can calculate the radiation dose that you receive in a year. Are you above or below the national average? *Hint:* Remember to convert from millirems (mrem) to microsieverts (µSv).

b. The Los Alamos National Laboratory also hosts a Web page that allows you to estimate your annual radiation dose. Use it to perform similar calculation. How do your results compare to the previous part? Explain any differences.

Links to both radiation dose calculators are provided at the *Online Learning Center.*

Nearly 3000 μSv, or about four fifths of the 3600 μSv absorbed per year by a typical (nonsmoking) U.S. resident, comes from natural background sources. As Table 7.3 shows, radon is the major source with additional natural contributions from cosmic rays, soil, and rock. The remainder of the annual dose comes from human sources, typically medical procedures such as diagnostic X-rays.

Table 7.3 also reveals that the radiation from a properly operating nuclear power plant is negligible. In fact, it is less than that of the naturally occurring radioisotopes in your own body. For example, about 0.01% of all the potassium ions (K^+) essential to your body are radioactive K-40. These K-40 ions give off about 200 μSv per year, approximately 1000 times more than the exposure from living within 20 miles of a nuclear power plant. Although bananas are rich in K^+, a steady diet of them will not increase your personal radioactivity because the potassium ion continuously moves in and out of your body.

In addition to K-40, our food contains other naturally occurring radioisotopes. One is carbon-14. This radioisotope is produced in our upper atmosphere by the interaction of nitrogen atoms and molecules with cosmic rays. Carbon-14 gets incorporated into carbon dioxide molecules that diffuse down into the troposphere where we live and breathe. The late Isaac Asimov, a prolific science writer, pointed out that a human body contains approximately 3.0×10^{26} carbon atoms, of which 3.5×10^{14} are C-14. Each breath you inhale contains carbon dioxide, including about 3.5 million carbon dioxide molecules that contain C-14 atoms. This number of atoms is so insignificant, that Table 7.3 contains no entry for radioactive carbon-14.

Your Turn 7.21   Radioactive Carbon and You

Assume that Isaac Asimov's figures are correct, and that 3.5×10^{14} of the 3.0×10^{26} carbon atoms in your body are radioactive. Calculate the percent that is C-14.

To further put your personal dose of radiation into perspective, the likely effects of a single dose of radiation are described in Table 7.4. Below a single dose of 0.25 Sv, or 250,000 μSv, no immediate physiological effects are observable. A dose of 0.25 Sv is nearly 70 times the average annual exposure!

The long-term effects of low doses of radiation, however, are still under debate. The issue is how to extrapolate from the known high-dose data to lower doses. Extrapolation is necessary because, of course, we cannot do experiments on humans to make reliable measurements. Additionally, the effects of low doses would be small and would only show up over a long time span.

Two radiation dose-response models are illustrated in Figure 7.20. The first, the more conservative of the two, is the **linear, nonthreshold model**. This model assumes a linear relationship between the adverse effects and the radiation dose, with radiation being harmful at all doses, even low ones. Thus, if the adverse effect is to get cancer, doubling the

---

Table 7.4	Physiological Effects of a Single Dose of Radiation	
Dose (rem)	**Dose (Sv)**	**Likely Effect**
0–25	0–0.25	No observable effect
25–50	0.25–0.50	White blood cell count decreases slightly
50–100	0.50–1.00	Significant drop in white blood cell count, lesions
100–200	1.00–2.00	Nausea, vomiting, loss of hair
200–500	2.00–5.00	Hemorrhaging, ulcers, possible death
>500	>5.00	Death

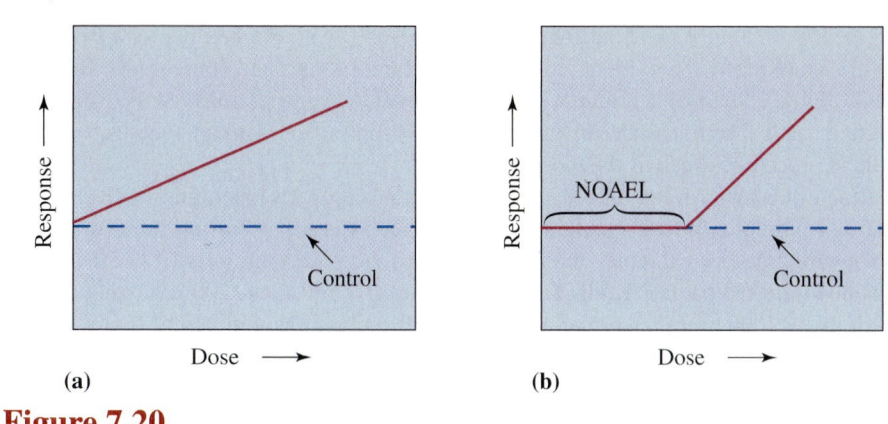

Figure 7.20

Two dose-response curves for radiation. **(a)** Linear, nonthreshold model. **(b)** Threshold model.

Note: NOAEL means no observed adverse effect level.
Source: *Environmental Science & Technology*, "Redrawing the Dose-Response Curve," Vol. 38, No. 5, March 1, 2004, p. 90A.

End-of-chapter question #56 allows you to further investigate the hormesis phenomenon.

radiation dose doubles the incidence of cancer, and tripling it causes three times as much. No cellular repair of the damage caused by radiation is assumed to take place, even at low doses. Although not illustrated here, other models also exist. For example, **hormesis** is the concept that low doses of a harmful substance (such as radiation) may actually be beneficial.

The model represented by Figure 7.20(b) makes a different assumption about low doses. Here, no observable adverse effects occur until a certain threshold is reached. Presumably, cellular repair can take place so that the response curve initially remains flat. By this model, low doses of radiation are safe.

Currently, the linear, nonthreshold model is being used by the EPA and other federal agencies in setting exposure standards. An exhaustive report compiled in 2002 by the National Council on Radiation Protection and Measurement states that there is "no conclusive evidence" on which to reject the linear, nonthreshold model, also noting that it may never be possible to "prove or disprove the validity." Steve Page, the director of EPA's Office of Air Quality Planning and Standards, adds that with the nonthreshold model "the risk from radiation is within the allowable range from toxic chemicals, 1-in-10,000 to 1-in-a-million chances of developing cancer." As a conservative model, it is more costly to implement and hence more controversial.

> **Consider This 7.22** **Standards and Your Job**
>
> The ramifications of adopting a specific radiation dose-response model are both biological and economic. By using the nonthreshold model, are we being "better safe than sorry" or are we wasting a lot of money protecting ourselves needlessly? As a health professional who must operate under the stricter federal limits for radiation safety, draw up a list of points that you could use with the public to support the stricter standards.

7.8 Nuclear Waste: Here Today, Here Tomorrow

It is hard to exaggerate the problems of nuclear waste. Indeed, the storage of spent reactor fuel presents us with formidable challenges. This waste contains radioactive fission products that will remain hazardous for thousands of years. The nuclear decay process cannot be hastened; we must store waste safely while we wait for the nuclear clock to tick.

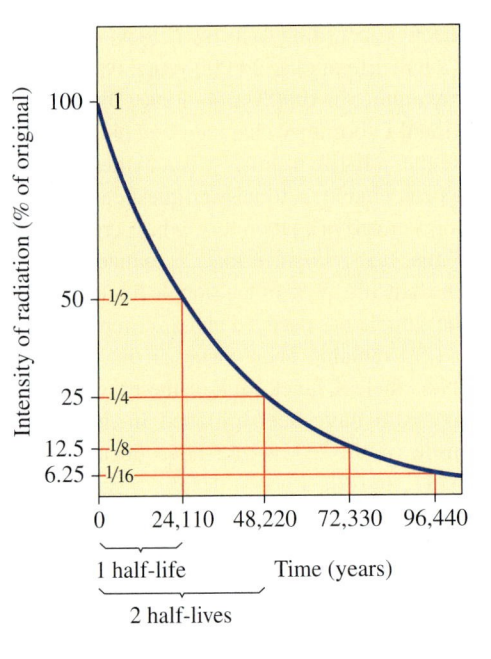

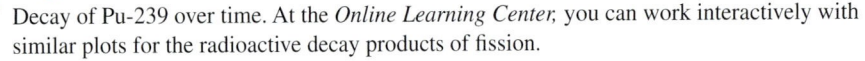

Figure 7.21

Decay of Pu-239 over time. At the *Online Learning Center,* you can work interactively with similar plots for the radioactive decay products of fission.

The radioactivity of a sample can decrease quickly over a short period of time or decrease more slowly over a much longer one. Each radioisotope has its own **half-life** ($t_{1/2}$), the time required for the level of radioactivity to fall to one half of its initial value. For example, plutonium-239, an alpha-emitter formed in nuclear reactors fueled with uranium, has a half-life of 24,110 years. Accordingly, it will take 24,110 years for the radioactivity of a sample of Pu-239 to halve. After a second half-life (another 24,110 years), the level of radioactivity will be one fourth of the original amount. And in three half-lives (72,330 years), the level will be one eighth (Figure 7.21). From these times, you can see that plutonium decay takes a very long time!

Other radioisotopes decay even more slowly. For example, the half-life of U-238 is 4.5 billion years. Coincidentally, this is approximately the age of the oldest rocks on Earth, a determination made by measuring their uranium content. The half-life for each particular isotope is a constant and is independent of the physical or chemical form in which the element is found. Moreover, the rate of radioactive decay is essentially unaltered by changes in temperature and pressure. From Table 7.5 you can see that half-lives range from milliseconds to millennia.

See nuclear equations 7.12 and 7.13 for the production of plutonium.

Table 7.5	Half-Lives for Selected Radioisotopes
Radioisotope	**Half-life ($t_{1/2}$)**
uranium-238	4.5×10^9 years
potassium-40	1.3×10^9 years
plutonium-239	24,110 years
carbon-14	5715 years
cesium-137	30.2 years
strontium-90	29.1 years
thorium-234	24.1 days
iodine-131	8.04 days
radon-222	3.82 days
plutonium-231	8.5 minutes
polonium-214	0.00016 seconds

Each radioisotope decays according to its own clock. As we just saw, plutonium-239, discovered in 1941, has a half-life of over 24,000 years. Other isotopes of plutonium have different half-lives. For example, in 1999 Carola Laue, Darleane Hoffman, and a team at Lawrence Berkeley National Laboratory characterized plutonium-231. These researchers had to work fast because the half-life of Pu-231 is a matter of mere minutes!

Half-life values ($t_{1/2}$) can enable us to answer questions of several different types. For example, once Pu-231 is generated in a laboratory, what percent of the sample remains after 25 minutes? To answer this, first recognize that 25 minutes is roughly three half-lives or 3×8.5 minutes. After one half-life, 50% of the sample has decayed and 50% remains. After two half-lives, 75% of the sample has decayed and 25% remains. And after three half-lives, 87.5% has decayed and 12.5% remains. These values are not exact, because 25 minutes is not exactly three half-lives. Nonetheless, quick back-of-the-envelope calculations can be useful.

This question also could have been phrased in this way: "After 25.5 minutes what percent of a sample of Pu-231 would have decayed?" This type of question requires one more step. To find the amount decayed, simply subtract the percent that remains from 100%. If 12.5% remains, then 87.5% has decayed. Check the math: (87.5% decayed) + (12.5% remaining) = 100%.

Speaking of back-of-the-envelope type calculations, let's do the same calculation with a different radioisotope. For example, if you had a sample of U-238 ($t_{1/2} = 4.5 \times 10^9$ years), what percent of it would remain after 25 minutes? To answer this, recognize that minutes, days, or even months would be a mere instant in the span of a 4.5-billion-year half-life. Thus, essentially all of the uranium-238 would remain. The next two exercises offer you more practice with half-life calculations.

Your Turn 7.23   Tritium Calculation

Hydrogen-3 (tritium, H-3) is sometimes formed in the primary coolant water of a nuclear reactor. Tritium is a beta-emitter with $t_{1/2} = 12.3$ years. For a given sample containing tritium, after how many years will only about 12% of the radioactivity remain?

Your Turn 7.24   Radon Calculation

Radon-222 is a radioactive gas produced from the decay of radium, a radioisotope naturally present in many rocks.

a. Where did the radium in rocks come from?
Hint: See Figure 7.13.
b. Radon activity is usually measured in picocuries (pCi). Suppose that the radioactivity from Rn-222 in your basement were measured at 16 pCi, a high value. If no additional radon entered the basement, how much time would pass before the level dropped to 0.50 pCi?
Hint: In dropping from 16 to 1 pCi, the level of radioactivity is reduced four times by half: 16 to 8 to 4 to 2 to 1. This corresponds to four $t_{1/2}$, each 3.82 days.
c. Why is it incorrect to assume that no more radon will enter your basement?

One final difficulty with reactor waste is that the fission products, if released, may enter and accumulate in your body, with potentially fatal consequences. One culprit is strontium-90, a radioactive fission product that entered the biosphere in the 1950s from the atmospheric testing of nuclear weapons. Strontium ions are chemically similar to calcium ions; both elements are in Group 2A of the periodic table. Hence, like Ca^{2+}, Sr^{2+} concentrates in milk and bones. Once ingested, radioactive strontium poses a lifelong threat because of its half-life of 29 years. Like I-131, Sr-90 was among the harmful fission products released in the vicinity of the Chernobyl reactor.

Your Turn 7.25   **Strontium-90**

Sr-90 is formed from the fission of U-235 in a reaction that produces three neutrons and another element. Write the nuclear equation.
Hint: Remember to include the neutron that induces the fission.

On a cheerier note, we end this section with carbon-14, a radioisotope mentioned in Section 7.7. With a $t_{1/2}$ of 5715 years, carbon-14 decays to nitrogen-14 through the process of beta decay. Our atmospheric carbon dioxide contains a constant steady-state ratio of one radioactive C-14 atom for every 10^{12} atoms of nonradioactive C-12. Living plants and animals incorporate the isotopes in that same ratio. However, when the organism dies, exchange of CO_2 with the environment ceases. Thus, no new carbon is introduced to replace the C-14. As a consequence, the concentration of C-14 in any material that once was alive decreases with time, halving every 5715 years.

In the 1950s, W. Frank Libby (1908–1980) first recognized this decrease by experimentally measuring the C-14/C-12 ratio in a sample. The ratio provided an estimate of when the organism died. Human remains and many human artifacts contain carbon, and fortunately, the rate of decay of C-14 is a convenient one for measuring human activities. Charcoal from prehistoric caves, ancient papyri, mummified human remains, and suspected art forgeries have all revealed their ages by this technique. The C-14 technique provides ages that agree to within 10% of those obtained from historical records, thus validating the legitimacy of the radiocarbon-dating technique.

Carbon-14 dating was used to establish the age of the famous Shroud of Turin.

Sceptical Chymist 7.26   **Ancient Shroud**

Using carbon-14 dating, a controversial burial cloth was dated at over 100,000 years ago. Does this age seem reasonable to you, given that the half-life of C-14 is 5715 years?
Hint: After more than 10 half-lives have passed, the amount of the radioisotope left is vanishingly small.

7.9   Options for Dealing with Nuclear Waste

Of the issues surrounding nuclear power, the safe disposal of nuclear waste is the most pressing. There is no apparent "silver bullet" (or silver waste canister). In a June 1997 *Physics Today* article, John Ahearne, past chair of the U.S. Nuclear Regulatory Commission, reminds us that, ". . . Like death and taxes, radioactive waste is with us—it cannot be wished away. . . ." Before we discuss some possible options, we need to define the types of nuclear waste materials.

High-level radioactive waste (HLW), as the name implies, has high levels of radioactivity and, because of the long half-lives of the radioisotopes involved, requires essentially permanent isolation from the biosphere. HLW comes in a variety of chemical forms, including ones that are highly acidic or basic. It also contains toxic metals. Thus, HLW is sometimes labeled as a "mixed waste" in that it is hazardous *both* because of the chemicals it contains *and* because of their radioactivity. Furthermore, this waste also presents a national security risk because it contains fissionable plutonium that could be extracted and used to construct nuclear weapons (see Section 7.10).

In contrast, **low-level radioactive waste (LLW)** is waste contaminated with smaller quantities of radioactive materials than HLW and specifically excludes spent nuclear fuel. LLW includes a wide range of materials. Items in LLW include contaminated laboratory clothing, gloves, and cleaning tools from medical procedures using radioisotopes and even discarded smoke detectors. For some LLW, the radioactivity levels are quite low. As

you might suspect, the hazards associated with LLW are significantly less than those from HLW. Nearly 90% of the volume of all nuclear waste is low level rather than high level.

High-level radioactive waste largely comes from nuclear power plants, both commercial and military. For example, each of the 103 commercial nuclear reactors in the United States produces about 20 tons of spent fuel annually. **Spent nuclear fuel (SNF)** is the radioactive material remaining in fuel rods after they have been used to generate power in a nuclear reactor and is regulated as HLW. After removal from the reactor, the spent fuel rods are still "hot," both in temperature and their radioactivity. They contain isotopes of uranium, plutonium-239 formed by the capture of neutrons by U-238, together with a wide variety of highly radioactive fission products such as iodine-131, cesium-137, and strontium-90. Huge quantities of HLW also were created during the Cold War because reactor fuel was reprocessed to produce plutonium for military uses. In 1996, the U.S. Department of Energy reported that the accumulated HLW generated by the Department of Defense occupied a volume of approximately 350,000 m^3 with a radioactivity of about 900 million curies. To help conceptualize this, think in terms of football fields. This volume corresponds to nine of them covered to a depth of 30 feet! Such military waste is in the inconvenient form of solutions, suspensions, slurries, and salt cake stored in barrels, bins, and tanks.

Approximately 30% of the fuel rods in every reactor are replaced annually, on a rotating schedule. After the spent rods are removed from the reactor, they are transferred to deep pools in which they are cooled by water that contains a neutron absorber. As of 2005, over 52,000 metric tons of SNF have been discharged from the nation's nuclear reactors at the sites where it was created (Figure 7.22). Typical reactor sites were designed to hold about 25 years' worth of nuclear waste.

Today, almost all reactor waste is being stored on site where it was generated. The storage facilities, not built for the long term, are hardly ideal. In the 1950s and early 1960s, the plan had been to reprocess the spent fuel to extract plutonium and uranium from it and recycle these elements as nuclear fuel to produce additional energy. Storage capacity for spent fuel rods on site was designed with such reprocessing in mind. However, only one of several planned reprocessing plants ever went into operation and then only briefly (1967–1975). Thus, reprocessing never was capable of keeping up with the rate of spent fuel production, about 2000 tons every year. In 1977, then President Jimmy Carter, a nuclear engineer, declared a moratorium on commercial nuclear fuel reprocessing that continues to this day.

Such restrictions are not worldwide. France, the United Kingdom, Germany, and Japan all reprocess some of their SNF. An option that is becoming more and more attractive is the use of a **breeder reactor,** a nuclear reactor that can produce more fissionable fuel (usually Pu-239) than it consumes (usually U-235). This seems like a dream come true to an energy-hungry planet. Imagine if your car synthesized gasoline as you drove! Scientists in the United States, as well as in the countries mentioned earlier, know how to recover plutonium from the spent fuel of breeder reactors. In the 1970s, several factors led

Figure 7.22

Thousands of canisters containing spent fuel rods in an underwater storage pool.

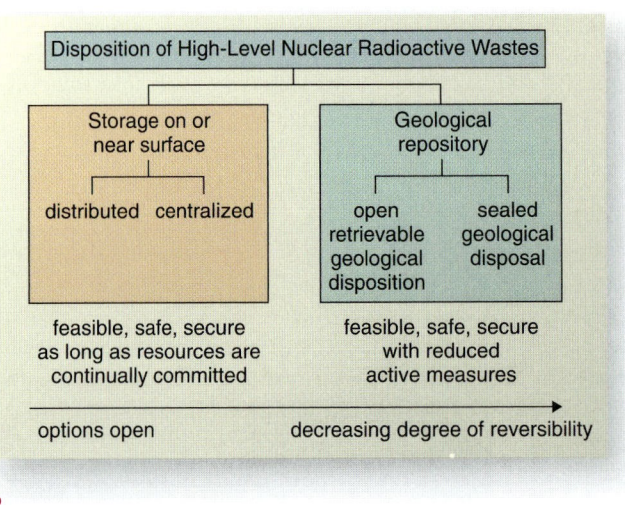

Figure 7.23

Methods of high-level nuclear waste deposition.

Source: *Disposition of High-Level Wastes and Spent Nuclear Fuel,* National Academy Press, 2000.

the United States to stop pursuing breeder reactor technology, including the objection to plutonium reprocessing and the more complicated reactor design required. We will take a more detailed look at the possible future of reprocessing in Section 7.12.

In the absence of reprocessing, two feasible options for the storage of HLW are now under consideration: monitored storage on or near the surface and storage in geological repositories deep underground. These differ in a key variable: *active management* (Figure 7.23). In surface storage, human societies over thousands of years must commit resources to maintain the integrity of the wastes. In geological repository storage, the wastes may be accessible and retrievable (although less easily) or sealed "forever," requiring minimal human vigilance. In a report published in 2000 by the National Academy of Sciences (NAS), the latter option of deep-underground storage was favored, noting that it was not prudent to assume that future societies on Earth would be able to maintain surface storage facilities. Long-term geological storage of HLW, first proposed in 1957 by NAS, became the alternative option.

Because of the long half-lives of the radioisotopes present, HLW must remain isolated from the groundwater for at least 10,000 years to allow the high levels of radioactivity to decrease significantly. Most plans employ a method known as **vitrification,** in which the spent fuel elements or other mixed waste are encased in ceramic or glass. First, the waste is dried, pulverized, and then mixed with finely ground glass and melted at about 1150 °C. Then the molten glass and wastes are poured into stainless-steel canisters, cooled, and capped for on-site storage. More than 1 million pounds of waste already have been treated in this way and await the development of a long-term underground repository (Figure 7.24). The radioactivity remains, but the nuclear materials are trapped in solid glass.

Figure 7.24

Encapsulating reprocessed HLW in glass canisters (vitrification).

After 10 half-lives, the radioactivity of a sample drops essentially to background level.

Consider This 7.27   **Nuclear Waste Warning Markers**

On February 15, 2007, the International Atomic Energy Agency (IAEA) unveiled this new symbol to warn the public about the dangers of radiation. You can read the details at the IAEA Web site.

a. Describe what this new symbol conveys to you.

b. Suppose you were asked to design markers to be installed near an underground nuclear waste repository. These markers must warn future generations of the existence of nuclear waste and must last for at least 10,000 years (more than four times the age of the pyramids of Egypt). The message must be intelligible to Earthlings of the future. Try your hand at designing these warning markers, keeping in mind the changes that have occurred in *Homo sapiens* during the past 10,000 years and those that might occur in the next 10 millennia.

(a)

(b)

Figure 7.25

(a) Map of Yucca Mountain and state of Nevada. **(b)** Yucca Mountain, looking south into the desert.

Source: **(a)** U.S. Department of Energy.

Whatever physical form it may take, the ultimate resting place for the waste remains in question. In 1997, the Nuclear Waste Policy Amendments Act designated Yucca Mountain (Figure 7.25) in Nevada as the sole site to be studied as an underground long-term, high-level nuclear waste repository. To date, utility companies have paid over $14 billion to fund the development of such a repository. However, it is not certain that the Yucca Mountain depository will ever become operational. Difficult political, legal, and technical barriers remain, and solutions do not appear near at hand. To establish deep geological storage of HLW, the federal government must deal with state legislatures and tribal governments whose land rights are involved.

Significant opposition to the proposal centers on the need to transport the waste to Yucca Mountain from reactor sites all over the country. Critics have dubbed the plan a "mobile Chernobyl." Roughly 36,000 combined truck and rail shipments taking over 30 years will be required to relocate all of the HLW to Yucca Mountain. Current plans call for shipping the spent fuel and high-level waste through 43 states within half a mile of 50 million Americans before it reached the proposed Nevada Test Site interim repository. Former Speaker of the House Dennis Hastert of Illinois, a supporter, declared that the bill "assures that another 15 years will not pass before the federal government lives up to its responsibility of accepting spent fuel." His position was not surprising given that Illinois has more commercial nuclear reactors (11) than any other state.

Since the terrorist attacks of September 11, 2001, additional concerns have surfaced over the issue of having HLW stored around the country, rather than in a central and presumably more secure location. Currently, many perceive the HLW housed in numerous temporary storage facilities as more dangerous. Others argue that the massive number of truck and rail shipments will offer easier terrorist targets.

In early 2002, President George W. Bush sent a letter to both houses of Congress stating, "I consider the Yucca Mountain site qualified for application for a construction authorization for a repository. Therefore, I recommend the Yucca Mountain site for this purpose." His recommendation was based on an earlier one to him from the Secretary of

Energy, which in turn was based on thousands of pages of documentation. But in April, 2002, Nevada Governor Kenny C. Guinn countered with an official Notice of Disapproval to the Senate. His accompanying statement proclaimed, "As a matter of science and the law, and in the interests of state comity and sound national policy, Yucca Mountain should not be developed as a high-level nuclear waste repository."

Consider This 7.28   **To Centralize or Not to Centralize?**

Would you favor a large, centralized stockpile of nuclear waste? Or would you favor the continued storage of waste right at the plant where it is generated? Prepare a list of advantages and disadvantages of each approach.

In July 2002, the Senate gave final approval for the site at Yucca Mountain, essentially overriding the wishes of the State of Nevada. Accordingly, the 2004 spending bill proposed by President George W. Bush set a high priority in completing the Yucca Mountain project. In 2004 however, the U.S. appellate court handed the project a major legal setback when it ruled that the Environmental Protection Agency (EPA) did not set adequate guidelines for acceptable future radiation levels at the site. The judges rejected the 10,000-year compliance period, citing the National Academy of Sciences study that concluded peak exposure could occur 300,000 years after the site is filled and sealed. Officials from both EPA and DOE admit that it will be a very difficult task to design any repository to meet such far-ranging safety requirements.

Meanwhile, on the technological side, tunnels are being dug 1400 feet beneath the surface of Yucca Mountain (Figure 7.26). A complex rail system and canister-carrying cars are being designed, and engineers are studying different methods for ensuring that no water can reach the waste containers, if and when the repository opens. The federal government has already spent an estimated $54 billion on the project, and money continues to be allocated. If completed, the site will be the largest radioactive storage facility in the world, with a capacity of over 70,000 metric tons of spent fuel and 8000 tons of high-level military waste. The nuclear power plants currently in operation add about 2000 metric tons of SNF every year to the over 54,000 tons of HLW already in existence. Therefore, you can surmise that when the repository opens, if ever, it possesses only the capacity to receive waste produced until about 2012.

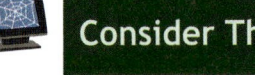

Consider This 7.29   **Yucca Mountain**

What is the current status of Yucca Mountain as a HLW long-term repository? What positions are being taken by the players involved? Use the resources of the Web to find out. Be sure to cite your sources.

Figure 7.26

A worker inspects one of the tunnels beneath Yucca Mountain.

Congress (or its successor) may still be debating the issue 24,110 years from now, when the plutonium-239 we create today completes its *first* half-life. Other disposal methods seem even less promising. Disposal in deep-sea clay sediments beneath 3000–5000 m of water was investigated. Proposals to bury the radioactive waste under the Antarctic ice sheet or to rocket it into space have largely been discredited. But one thing is sure. Whatever disposal methods are ultimately adopted, they must be effective over an extremely long time. Nuclear engineers William Kastenberg and Luca Gratton conclude their *Physics Today* article (June 1997) with the sobering thought:

"For a high-level waste depository of the type proposed for Yucca Mountain, it is clear that natural processes will eventually redistribute the waste materials. Present design efforts are directed toward ensuring that, at worst, the degraded waste configurations will eventually resemble stable, natural ore deposits, preferably for periods exceeding the lifetimes of the more hazardous radionuclides. Perhaps that's the best we can hope for."

7.10   The Nuclear Weapons Connection

> Most commercial reactors worldwide use enriched uranium as fuel. However, some British and Canadian reactors are designed to run on natural (unenriched) uranium.

Although both nuclear power plants and nuclear bombs derive energy from nuclear fission, each requires a different rate of reaction. A nuclear power plant needs a slow, controlled energy release; in contrast, a nuclear weapon requires one that is rapid and uncontrolled. In either case, the fission reaction is essentially the same. Both are fueled by **enriched uranium,** that is, uranium that has a higher percent of U-235 than its natural abundance of about 0.7%. The difference lies in the *extent* of the enrichment. Commercial nuclear power plants typically operate with 3–5% U-235, whereas atomic weapons use fuel that may be as high as 90% U-235. The latter is sometimes referred to as highly enriched or weapons-grade uranium.

Your Turn 7.30   Enriched Uranium

The fuel pellets in a nuclear power plant are enriched to 3–5% uranium-235.

a. Which other isotope of uranium is present in the pellets? Is this isotope fissionable under the conditions in a nuclear reactor?

b. After use in a reactor, spent fuel pellets contain radioisotopes of many different elements, including strontium, barium, krypton, and iodine. Explain the origin of these radioisotopes.

Hint: See Figures Alive! for more about fuel pellets.

In a nuclear reactor, the concentration of fissionable U-235 is low. Most of the neutrons given off during fission of U-235 are absorbed by U-238 nuclei in the fuel pellets and by other elements such as cadmium and boron in the control rods. Consequently, the neutron stream cannot build up sufficiently to establish an explosive chain reaction. In contrast, atomic weapons use highly enriched uranium in which neutrons are likely to encounter another U-235 nucleus. As we noted earlier, a spontaneously explosive nuclear fission reaction (that is, the explosion of an atomic bomb) will occur only if a critical mass of U-235 (about 33 lb) is quickly assembled in one place.

The isotopes U-235 and U-238 behave essentially the same in all chemical reactions, so the separation of these two isotopes is *extremely* difficult and relies on advanced technology that is not readily available. Uranium-235 differs from U-238 only by three neutrons; nonetheless, this mass difference can be exploited to achieve a separation. How? On average, lighter gas molecules move faster than heavier ones. Therefore gas molecules containing U-235 should travel slightly more rapidly than their analogs containing U-238. One way to separate molecules of different masses is by **gaseous diffusion,** a process used to separate gases with different molecular weight by forcing them through a series of permeable membranes. Lighter gas molecules diffuse more rapidly through the membrane than heavier ones.

But uranium ore clearly is not a gas; rather, it is a mineral that contains UO_3 and UO_2. Most other uranium compounds are solids as well. However, the compound uranium hexafluoride (UF_6) has a notable property. Known as "hex," UF_6 is a solid at room temperature but readily vaporizes when heated to 56 °C (about 135 °F). To produce hex, the uranium ore is converted to UF_4, which in turn is reacted with more fluorine gas.

$$UF_4(g) + F_2(g) \longrightarrow UF_6(g) \qquad [7.11]$$

On average, a $^{235}UF_6$ molecule travels about 0.4% faster than a $^{238}UF_6$ molecule. If the gaseous diffusion process is allowed to occur repeatedly through a long series of permeable membranes, significant amounts of $^{235}UF_6$ and $^{238}UF_6$ can be separated. For more than four decades, uranium isotopes were separated by gaseous diffusion at the Oak Ridge National Laboratory in Tennessee.

Currently, large commercial enrichment plants are operating in the United Kingdom, the Netherlands, France, Germany, and the former Soviet Union with smaller ones elsewhere. In the United States, the only currently operating commercial enrichment plant is in Paducah, Kentucky. It carries out uranium enrichment based on gaseous diffusion (Figure 7.27). In 2006, the NRC issued a license to a consortium of U.S. and European energy companies to build a state-of-the-art gas centrifuge facility in New Mexico. That plant was scheduled to begin operations in 2008 but not reach full capacity until 2013. In a centrifuge, UF_6 enters a rotor that spins at high speed inside an evacuated chamber. The heavier $^{238}UF_6$ molecules move closer to the wall of the rotor due to the centrifugal force, thus producing partial separation of the isotopes. If the mixed isotopes are introduced to the middle of the rotor, enriched and depleted streams are removed from the ends. Several centrifuges must be used consecutively to achieve the required enrichment level, but the process requires about 95% less energy than diffusion for the same amount of enrichment.

Regardless of the enrichment method, once the U-235 has been separated from U-238, depleted uranium remains. Nicknamed DU, **depleted uranium** contains almost entirely U-238 (99.8%) and has been depleted of most of the U-235 that it once naturally contained. Estimates indicate that over 1 billion metric tons of DU is stored currently in the United States. The military also has deployed DU as casings for armor-piercing munitions.

> Equation 7.11 is a *chemical* equation, not a *nuclear* equation.

Figure 7.27

The transportation of "hex" at Paducah, Kentucky. A cylinder of enriched uranium product from the Paducah plant is being loaded into an autoclave of the transfer and shipping facility.

Consider This 7.31   Depleted Uranium

Depleted uranium is used to tip antitank shells. These first were used in the Gulf War in 1991 and later in other armed conflicts, including Kuwait, Bosnia, Afghanistan, and Iraq. For example, the Department of Defense estimates that over 120,000 kg of DU was employed in the first year of the Iraq War (March 2003–March 2004). Research the properties of DU to learn why it is used in this way. Also summarize the controversies involved.

At enrichment levels of 3–5%, nuclear fuel rods cannot be incorporated into functional atomic bombs. However, the technology to transform the uranium ore into weapons-grade uranium (about 90% U-235) is essentially identical to that used to produce reactor-grade fuel. In recognition of this fact, only certain countries are authorized to produce enriched uranium according to the Nuclear Non-Proliferation Treaty. That agreement bestows on signatory sovereign nations the right to pursue nuclear power, and hence uranium enrichment, for peaceful purposes. Iran, a signer of the treaty, restarted its uranium enrichment program in the summer of 2005, despite protests from the United States and other countries.

A more likely scenario for clandestine weapon manufacturing would be to use the plutonium-239 formed from U-238 in a conventional reactor. Analogous to U-235 in equation 7.1, U-238 absorbs a neutron and forms the unstable species U-239. In this case, fission does *not* occur, but rather in a matter of hours U-239 undergoes beta decay.

$$^{1}_{0}n + ^{238}_{92}U \longrightarrow [^{239}_{92}U] \longrightarrow ^{239}_{93}Np + ^{0}_{-1}e \qquad [7.12]$$

The new element formed, neptunium-239, also is a beta-emitter and decays to form plutonium-239.

$$^{239}_{93}\text{Np} \longrightarrow \; ^{239}_{94}\text{Pu} + \; ^{0}_{-1}\text{e} \qquad\qquad [7.13]$$

This transformation was discovered early in 1940. The chemical and physical properties of plutonium were determined with an almost invisible sample of the element on the stage of a microscope. The chemical processes devised on such minute samples were scaled up a billionfold and used to extract plutonium from the spent fuel pellets from a reactor built on the Columbia River at Hanford, Washington. The plutonium was chemically separated from the uranium and used in the first test explosion of a nuclear device on July 16, 1945, near Alamogordo, New Mexico. The bomb dropped on Nagasaki a little less than a month later also was fueled by plutonium.

Plutonium-239 poses an international security problem because the plutonium produced in nuclear power reactors could possibly be incorporated into nuclear bombs. It has been widely speculated that the 1981 bombing of a nuclear facility in Iraq by war planes from Israel was done to prevent Iraq from being able to produce plutonium-containing nuclear weapons. A more recent (and ongoing) international crisis involves efforts to dissuade North Korea from reprocessing plutonium. Beginning in the early 1990s, international suspicions arose over North Korea and its possible nuclear weapons program. These suspicions were intensified by North Korea's withdrawal from the Nuclear Non-Proliferation Treaty. Tests by the International Atomic Energy Agency (IAEA) on used fuel rods from the 5 MW research reactor at Yongbyon indicated that indeed reprocessing had been occurring (Figure 7.28). In 1994, negotiations between North Korea and the United States resulted in an agreement in which North Korea would freeze its nuclear weapons program in exchange for shipments of fuel oil and an easing of economic sanctions.

Unfortunately, the sanctity of this agreement didn't last long. In 2002, North Korean officials publicly acknowledged the existence of a clandestine uranium enrichment program, expelled the international inspectors, and resumed construction of two larger reactors, all in violation of the 1994 agreements. During negotiations with China and the United States in 2003, North Korean officials admitted for the first time that they possessed nuclear weapons. In 2005, the North Korean government declared that plutonium extraction from 8000 fuel rods had been completed. Unequivocal confirmation of North Korea's reprocessing efforts came in October 2006 when a successful detonation of a nuclear device was announced. Analysis of the air near the test site indicated that plutonium was the nuclear material used. As worrisome from a security perspective is the threat made by North Korean officials of their intentions to export nuclear materials if certain concessions are not made. Given the risks associated with Pu-239 and U-235, it is essential that both

Figure 7.28

A recent satellite view of the plutonium reprocessing plant in Yongbyon, North Korea.

national and international organizations carefully monitor the supplies and distribution of these isotopes throughout the world.

Safeguarding existing nuclear materials has taken on a new meaning since the end of the Cold War and the demise of the former Soviet Union. One part of the problem is the plutonium and highly enriched uranium in Russia's nuclear arsenal (about 20,000 warheads). At present, though, these warheads are stored with relatively good security. Furthermore, any thief would find it difficult to remove the plutonium and uranium from the warheads. In contrast, Russia's legacy from the Cold War, a stockpile of highly enriched uranium and plutonium (about 600 metric tons), is far more accessible and hence far more threatening to world security (Figure 7.29). The fissionable materials stored in labs, research centers, and shipyards across the former Soviet Union are vulnerable to theft. These 600 tons of fissionable material translate into the capacity to construct approximately 40,000 new nuclear weapons.

Nuclear weapons are not the only threat. Consider also the "**dirty bomb**," a device that employs a conventional explosive to disperse a radioactive substance. If used in a city, such a device would create havoc both from the explosion and from the dispersal of the radioactive substance. Cobalt-60 and strontium-90 are candidates for use in dirty bombs, as both these radioisotopes are relatively easy to obtain and have long enough half-lives to persist in the environment. Again no fission is involved with a dirty bomb; only a conventional explosive.

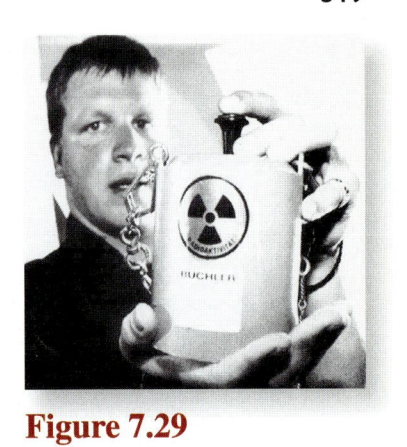

Figure 7.29
A smuggled canister of military grade Pu-239 captured in Germany.

Sceptical Chymist 7.32 — Radioactivity Levels

A brochure on nuclear terrorism makes this assertion: "A nuclear weapon, if exploded, would create more radioactive substances than originally present in the weapon. In contrast, if a dirty bomb were to be exploded, the amount of radioactivity would be the same before, during, and right after the explosion." Are these statements accurate? Explain.

The world community clearly recognizes the dangers of nuclear trafficking and the need for effective safeguards. Since September 11, 2001, the threat has increased substantially, as it appears that terrorists will use any weapons available. In recognition of the global threat, the Nobel Peace Prize for 2005 was shared equally between the International Atomic Energy Agency (IAEA) and its Director General, Mohamed ElBaradei. The Nobel committee cited ". . . their efforts to prevent nuclear energy from being used for military purposes and to ensure that nuclear energy for peaceful purposes is used in the safest possible way." The future safety of nations, if not of our planet, may depend on our ability to safeguard (and ultimately recycle) plutonium and highly enriched uranium.

Consider This 7.33 — The Nobel Peace Prize

The Nobel Peace Prize, awarded since 1901, has honored people who have worked on many issues relating to peace in the world, including those relating to nuclear energy and nuclear warfare.

a. Visit the official Web site of the Nobel Foundation (link provided at the *Online Learning Center*). Learn more about the circumstances surrounding the 2005 Nobel Peace Prize by reading the press release. Summarize what you learned in four to six bullet points.

b. From a list of other recipients of the Peace Prize, find at least one other person or group who worked on nuclear issues. Again summarize what you learned in a series of bullet points.

c. On the same Web site, access the materials under the educational link. Play the Peace Dove Game, one that has you match a region of the world with peace doves bearing a message intended for it. What new information did you gain from this game?

7.11   Risks and Benefits of Nuclear Power

In earlier sections, we described both the benefits and risks of nuclear power. We also noted the markedly different extents to which countries around the world use nuclear power to generate electricity (see Section 7.4). Regardless of whether a country contains many nuclear-powered electrical generators or only a few, the associated risks and benefits must be weighed by all involved.

The 19th century poet William Wordsworth spoke of technological risks and benefits as "… Weighing the mischief with the promised gain …" He was speaking, in this case, about the railroad, a technology new in his time.

Such risk–benefit analyses are never easy, though in a sense we carry them out every day. We commonly regard risk as the probability of being injured or losing something, but there are many types of risk. They can be voluntary, such as those associated with wind surfing or bungee jumping, or involuntary, such as inhaling someone else's cigarette smoke. When we drive a car we control the risks (at least to some extent), but we have no control over the increased risk of radiation exposure at high altitudes or of a commercial plane crash.

Counterbalancing risks are benefits such as the improvement of health, increased personal comfort or satisfaction, saving money, or reducing fatalities. Everyday living inevitably involves risks and their related benefits: crossing a street, riding a motorcycle or in a car, cooking a meal or eating one, and even the simple act of getting up in the morning. Because there is some element of risk in everything we do, we almost automatically make judgments about what level of risk we consider acceptable. Most people do not intentionally put themselves at high risk, even when the potential benefit is also high, such as going into a burning building to save a child. On the other hand, there is an alarming increase in the number of people who expect "zero risk" in whatever they do or whatever surrounds them, although it is impossible to achieve. *There is no such thing as zero risk.*

In the case of nuclear energy, we are dealing with social benefits in relation to technological risks, but we must not make the mistake of only considering the risks and benefits that relate directly to nuclear fission. We also must weigh the risks associated with the alternatives—especially the coal-fired power plants that nuclear reactors are designed to replace. Recall, for example, the estimate in Chapter 4 that over 100,000 workers have been killed in American coal mines since 1900, most prior to the 1950s when higher safety standards were instituted. Table 7.6 summarizes the risk of fatalities from the annual operation of a 1000-megawatt power plant using either coal or nuclear power. The conclusion is that, at least for the hazards identified here, the risks associated with energy produced by nuclear power are considerably less than those from coal-burning plants.

Table 7.6	Risks from Coal and Nuclear-Powered Electricity Generators	
Hazard Type	**Coal**	**Nuclear**
Routine occupational hazards	Coal mining accidents and black lung disease constitute a uniquely high risk.	Risks from sources not involving radioactivity dominate.
Deaths*	2.7	0.3–0.6
Routine population hazards	Air pollution produces a relatively high, though uncertain, risk of respiratory injury.	Low-level radioactive emissions are more benign than the corresponding risks from coal.
Deaths*	1.2–50	0.03
Catastrophic hazards (excluding occupational)	Acute air pollution episodes with hundreds of deaths are not uncommon. Long-term climatic change, induced by CO_2, is conceivable.	Risks of reactor accidents are small compared with other quantified catastrophic risks.
Deaths*	0.5	0.04
General environmental degradation	Strip mining and acid runoff; acid rainfall with possible effect on nitrogen cycle.	Long-term contamination with radioactivity.

*Deaths are the number expected per year for a 1000-megawatt power plant. In all cases, 6000 person-days lost are assumed to equal one death.

Source: Modified from *Perilous Progress: Managing the Hazards of Technology*, by Robert W. Kates, Ed., 1985, Westview Press, Boulder, Colorado.

Coal-fired power plants produce huge amounts of carbon dioxide, a waste product of combustion for which currently we have no large-scale technology for remediation. Annually, one 1000-MW coal-fired electric power plant releases about 4.5 million tons of CO_2 into the atmosphere. In a year, such a plant also generates about 3.5 million cubic feet of waste ash, a substantial volume. By comparison, a 1000-MW nuclear power reactor produces about 70 ft^3 of high-level waste per year. Thus, the risks and benefits for power stack up differently, depending on how this power is generated. Because nuclear power plants do not burn fossil fuels, they release no carbon dioxide. Nuclear power also has been touted as a way to reduce acid rain, because nuclear fission releases no acidic oxides of sulfur and nitrogen. In contrast, a typical 1000-MW coal-fired power plant burns over 10,000 tons of coal and could easily release 300 tons of SO_2 and perhaps 100 tons of NO_x *every day!*

In addition to C, H, O, N, and S, coal contains other elements. Uranium and thorium are two radioactive impurities in coal. According to W. Alex Gabbard, a physicist at the Oak Ridge National Laboratory, trace quantities of uranium in coal can be as high as 10 ppm and the amount of thorium is usually more than twice that of uranium. Current U.S. coal consumption is just over 1100 million tons annually. At that level, over 1300 tons of uranium and 2600 tons of thorium are being released into the environment every year, exceeding the amount of uranium consumed in nuclear plants. Ironically, therefore, a coal-fired power plant releases more radioactivity on a daily basis than a properly functioning nuclear plant does. Although much of the radioactive metals are collected in the fly ash residue, that too becomes a more difficult waste disposal problem.

As we have noted, nuclear energy carries tremendous emotional overtones. In part, these stem from mystery, misunderstandings, and the powerful imagery of mushroom clouds. The risks of radiation and the possibility of a major disaster, however remote, loom large in human consciousness. The accidents at Three Mile Island and Chernobyl, though hardly equivalent, have made the public wary. We have limited trust in technology and perhaps even less in people. We are apprehensive about human error in the design, construction, and management of nuclear power plants. After all, human errors and technicians' responses to them were the weak points in the prescribed safety procedures that caused the accidents at Three Mile Island and Chernobyl.

> During the useful lifetime of the nuclear fuel rods, a conventional power plant would need 200 train cars each carrying 15 tons of coal **every day** to produce the same amount of electricity.

Consider This 7.34   Informed Citizens

In this chapter, we have touched on many topics that relate to nuclear power. To convey what you have learned to others, draft a set of FAQ (frequently asked questions) that relate to the use of nuclear reactors to generate electricity. Write at least five questions and provide the answers. For example, you might begin with "Can a nuclear power plant explode like an atomic bomb?" and then use the information provided in Section 7.2 to write your answer.

Hint: Check out the fact sheets and brochures posted at the Web site of the U.S. Nuclear Regulatory Commission. Here you can learn more about what is of interest to the public. A direct link is provided at the *Online Learning Center.*

7.12 What Is the Future for Nuclear Power?

Beginning in the early 1960s, the United States experienced a dramatic growth in the nuclear power industry that lasted until 1979. The accident at Three Mile Island and the fear that accompanied it certainly contributed to the end of that growth phase. More important, however, were the economics of nuclear energy. With the retreat of fossil-fuel prices and the added costs of nuclear safety and oversight imposed in the 1980s, it simply was not economically feasible for utilities to construct new nuclear plants.

Figure 7.30

An artist's impression of the new Westinghouse AP1000 light-water reactor.

A second growth phase of nuclear power may be upon us. The comprehensive energy bill passed by Congress and signed by President George W. Bush in 2005, contained several provisions that may pave the way for an expansion of nuclear power in the United States. Included in the bill are tax incentives for electricity produced by new nuclear plants and loan guarantees to utility companies to defray the enormous costs of such construction projects. The bill seems to be having the desired effect. According to a press release from Senator Pete Domenici (R, New Mexico), in the first three months after the bill was signed, eight utility companies announced plans to construct 13 new reactors in the next 15 years. Central to this effort is the approval of new reactor designs. In 2005, the NRC approved the Westinghouse AP1000 (Figure 7.30) pressurized water design, and a simplified boiling water design was approved in early 2007. Both of the new designs are simpler, faster to construct than existing plants, and rely on more passive safety systems that have fewer mechanical pumps and valves. The accident at Three Mile Island was blamed on a valve malfunction.

Expansion of nuclear power is underway in other countries as well, though opponents are offering varying degrees of resistance. For example, in 2005 British Prime Minister Tony Blair proposed new reactor construction. His proposal faces opposition from his own Labor Party, in part stemming from the charge that these new reactors will require government subsidies. As of late 2006, nearly 30 reactors were under construction worldwide, including 8 in India and 4 in Russia. Both China and India are scheduled to build 30 nuclear plants by 2020. The Iranian Parliament has also approved plans for increased use of nuclear power. As this book went to press, a Russian company was nearing completion of Iran's first 1000-MW reactor. This reactor has the potential to generate more than just electricity; we discussed the international concern over possible weapons development in Iran in Section 7.10.

To address both the issue of spent nuclear fuel disposal and nuclear nonproliferation, many countries are implementing the use of mixed oxide, or MOX, fuel technology. Conventional nuclear reactor fuel can be made by combining fissionable plutonium oxide (about 10%) with nonfissionable depleted uranium oxide. However, plutonium from both spent nuclear fuel and from nuclear warheads can be incorporated into MOX fuel. In 2000, the United States–Russian Federation Plutonium Disposition Agreement was signed requiring "the safe, transparent, and irreversible disposition of 68 metric tons of weapons-grade plutonium—enough plutonium to make thousands of nuclear weapons." Incorporating the plutonium from warheads into MOX fuel pellets will eliminate over 10% of the nuclear weapons constructed during the Cold War. Furthermore, should the United States decide to end the moratorium on fuel reprocessing, MOX fuel would significantly extend our reserves of uranium, as well as lessen the amount of HLW requiring long-term storage.

The future of nuclear power does not stand in isolation from other international issues, the most pressing of which are environmental protection and the proliferation of nuclear weapons, as we have discussed. Whether the risks associated with nuclear power

outweigh those of global warming, acid rain, and nuclear terrorism is a difficult question for which there is no clear-cut or easy answer, in spite of extended study and debate. Hans Blix, former director general of the International Atomic Energy Agency, in describing the international dimensions of the problem, noted that developing nations will not likely build nuclear plants in the near future:

"If nuclear power is to be relied on to alleviate our burdening of the atmosphere with carbon dioxide, it is therefore to the industrialized countries that we must first look. They are in the position to use these advanced technologies—and they are also the greatest emitters of carbon dioxide."

Consider This 7.35　　The Nuclear Greenhouse?

Although no greenhouse gases are produced in the actual nuclear fission process, other sources of CO_2 emissions are associated with the nuclear industry. List as many as you can and indicate the significance of each.

Is nuclear the answer? Proponents line up on each side of the argument. Scott Peterson, vice president of the Nuclear Energy Institute, remarked in December 2005:

"We really think that there's no more important time than right now for the world and the United States to consider nuclear energy as part of the solution to reducing carbon emissions and improving our air quality, and, at the same time, meeting the incredible appetite that the world is going to have for electricity."

Others conclude that nuclear power can reduce global warming only slightly and suggest that it would be much less expensive to invest in enhanced energy efficiency. Bill Keepin and Gregory Kats of the Rocky Mountain Institute have estimated that reducing carbon dioxide emissions significantly with nuclear power would require the completion of a new nuclear plant every 2 days for the next 38 years. Furthermore, physics professor at the Massachusetts Institute of Technology and former Under Secretary of Energy, Ernest J. Moniz, said that if the world built enough reactors to provide energy without contributing to global warming, a new Yucca Mountain would be needed every 3.5 years. Oak Ridge National Laboratory staff members reported that before any massive replacement of fossil fuel with nuclear power could occur, several new techniques would have to be developed. These include commercial-scale recycling of nuclear fuels, breeder reactors to extend existing fuels, and possibly uranium recovery from seawater.

Sceptical Chymist 7.36　　The Effect of Nuclear Power

One side claims that increased reliance on nuclear power cannot have a significant effect in reducing greenhouse gas emissions. What assumptions are being made in this statement? Do you think they are valid? Explain.

So where does that leave us? In the United States and around the world, the public remains divided on the issues of nuclear power. Our need for energy expands daily, as does the mass of radioactive waste with which we must cope. The era of cheap oil is over and the era of climate change is dawning, yet hazards (both real and perceived) associated with radioactivity and nuclear weapons remain. It is a classic risk–benefit situation, and the final compromise has not been reached. For now, it is clear that nuclear power is not the cure-all for the world's energy woes, but it will likely remain a piece of the remedy.

Consider This 7.37   Risks and Benefits

We have examined nuclear fission as a source of electrical power in some detail. Now we ask you to list the risks and benefits associated with currently operating nuclear fission reactors. Using this list, take a stand on the question of the future use of nuclear fission-powered plants. Write an editorial proposing your view of a viable 20-year national policy on the issues.

Consider This 7.38   Second Opinion Survey

Now that you are near the end of your study of nuclear power, return to the personal opinion survey of Consider This 7.1 and answer the questions one more time. After completing the survey for a second time, compare your answers in the second survey with those from the first. Are there any striking differences in your opinions of nuclear power? If so, which of your opinions about nuclear power changed the most? What was responsible for this shift?

Conclusion

Over 50 years have passed since the first commercial nuclear power plant began producing electricity in the United States. The glittering promise of boundless, unmetered electricity, drawn from the nuclei of uranium atoms, has proved illusory. But the needs of our nation and our world for safe, abundant, and inexpensive energy are far greater today than they were in 1957. Therefore, scientists and engineers continue their atomic quest. Where the search will lead is uncertain, but it is clear that people and politics will have a major say in ultimately making the decision. Reason, together with a regard for those who will inhabit our planet in both the near and far future, must govern our actions. Maybe Homer Simpson was right when he proclaimed, ". . . Lord, we are especially thankful for nuclear power, the cleanest, safest energy source there is. Except for solar, which is just a pipe dream." As it happens, we explore a little of the rationality behind that pipe dream in the following chapter.

Chapter Summary

Having studied this chapter, you should be able to:

- Give an overview of our past and current use of nuclear power in the United States (7.1)

- Explain the process of nuclear fission, the role of neutrons in sustaining a chain reaction, and the source of the energy it produces (7.2)

- Compare and contrast how electricity is produced in a conventional power plant and in a nuclear power plant (7.3)

- Report on the use of nuclear power for electricity generation worldwide (7.4)

- Compare the processes of alpha, beta, and gamma decay in terms of what happens in the nucleus (7.5)

- Interpret the meaning of the word *radiation*, depending on the context (7.5)

- Explain how the radioactive decay of uranium-238 leads to the production of a series of radioisotopes. Also explain why naturally occurring radioisotopes such as carbon-14 and hydrogen-3 are *not* part of this series (7.5)

- Describe the accident at Chernobyl and explain why radioactive iodine was released and was hazardous to people (7.6)

- Rank the sources that contribute to your annual dose of radiation, both natural and human-made (7.7)

- Explain why nuclear radiation is also termed *ionizing radiation*. In your body, explain the connection between ionizing radiation and the production of free radicals (7.7)

- Some units describe the radioactive sample; others describe the damage it does to tissue. Use the curie, the rad, and the rem to illustrate this (7.7)
- Apply the concept of half-life to radiocarbon dating and to the storage of nuclear waste (7.8)
- Evaluate radioisotopes in terms of their health hazards, discussing factors such as half-life, type of radioactive decay, effect once in the body, and route of entry into the body. For example, compare radon-222, iodine-131, and strontium-90 (7.8)
- Describe the issues associated with the production and storage of high-level radioactive waste, including spent nuclear fuel (7.9)

- Take an informed stand on how high-level radioactive wastes should be handled and stored (7.9)
- Evaluate news articles on nuclear power and nuclear waste with confidence in your ability to understand the scientific principles involved (7.9–7.11)
- Describe the connections between nuclear power and nuclear weapons proliferation (7.10)
- Assess the risks and benefits in regard to the use of nuclear power (7.11)
- Take an informed stand on the use of nuclear power for electricity production (7.12)
- Outline the factors that will favor or oppose the growth of nuclear energy in the next decade (7.12)

Questions

Emphasizing Essentials

1. Give two ways in which an atom of carbon can differ from another atom of carbon. Then give three ways in which *all* carbon atoms differ from *all* uranium atoms.

2. Representations such as ^{14}N or ^{15}N give more information than simply the atomic symbol N. Explain.

3. **a.** How many protons does an atom of $^{239}_{94}Pu$ contain?

   b. What element contains one more proton than uranium? Two more?

   c. How many protons does radon-222 contain?

4. Determine the number of protons and neutrons in each of these nuclei.

   a. C-14 (radioactive)

   b. C-12 (stable)

   c. H-3 (tritium, a radioisotope of hydrogen)

   d. Tc-99 (a radioisotope used in medicine)

5. $E = mc^2$ is one of the most famous equations of the 20th century. What do the symbols in the equation represent?

6. Give an example of a nuclear equation and of a chemical equation. In what ways are the two equations alike? Different?

7. This nuclear equation represents a plutonium target being hit by an alpha particle. Show that the sum of the subscripts on the left is equal to the sum of the subscripts on the right. Then do the same for the superscripts.

$$^{239}_{94}Pu + {}^{4}_{2}He \longrightarrow [^{243}_{96}Cm] \longrightarrow {}^{242}_{96}Cm + {}^{1}_{0}n$$

8. For the nuclear equation shown in the previous question,

   a. Suggest how the $^{4}_{2}He$ was produced.

   b. $^{1}_{0}n$ is a product. What does this symbol represent?

   c. Explain why curium-243 is written in square brackets. *Hint:* See equation 7.1.

9. Californium, element number 98, was first synthesized by bombarding an element with alpha particles. The products were californium-245 and a neutron. What was the target isotope used in this nuclear synthesis?

10. Explain the significance of neutrons in initiating and sustaining the process of nuclear fission. In your answer, define and use the term *chain reaction*.

11. Nuclear fission occurs through many different pathways. For the fission of U-235 induced by a neutron, write a nuclear equation to form:

    a. bromine-87, lanthanum-146, and more neutrons

    b. a nucleus with 56 protons, a second with a total of 94 neutrons and protons, and 2 additional neutrons

12. This schematic diagram represents the reactor core of a nuclear power plant.

Match each letter with one of these terms.

   fuel rods

   cooling water into the core

   cooling water out of the core

   control rod assembly

   control rods

13. Identify the segments of the nuclear power plant diagrammed in Figure 7.6 that contain radioactive materials and those that do not.

14. Explain the difference between the primary coolant and the secondary coolant. The secondary coolant is not housed in the containment dome. Why not?

15. Boron can absorb neutrons.

    a. Write the nuclear equation in which boron-10 absorbs a neutron to produce lithium-7 and an alpha particle.

    b. Boron, like cadmium, can be used in control rods. Explain.

16. What is an alpha particle? How is it represented? Answer these same questions for a beta particle and for a gamma ray.

17. Plutonium-239 decays by alpha emission.

    a. Write the nuclear equation.

    b. Plutonium is most hazardous when inhaled in particulate form. Explain.

    c. Would you expect a sample of Pu-239 to decrease to background levels in hours, days, or years? Explain.

18. Iodine-131 decays by beta emission.

    a. Write the nuclear equation.

    b. Iodine, radioactive or not, accumulates in the body. Where?

    c. Would you expect a sample of I-131 to decay in hours, days, or years? Explain.

19. Radioactive decay is accompanied by a change in the mass number, a change in the atomic number, a change in both, or a change in neither. For the following types of radioactive decay, which change(s) do you expect?

    a. alpha emission

    b. beta emission

    c. gamma emission

20. In a fashion similar to U-238 (see Figure 7.13), U-235 goes through a series of alpha and beta decays before reaching a stable isotope. For practice, write the first six, which will bring you to an isotope of radon. In order, the steps in the full radioactive decay series are α, β, α, β, α, α, α, β, α, β, α, ending in stable Pb-207. Some steps have accompanying γ radiation, but you may omit this.

21. Given that the average U.S. citizen receives 3600 μSv of radiation exposure per year, use the data in Table 7.3 to calculate the percentage of radiation exposure the average U.S. citizen receives from each of these sources.

    a. food, water, and air

    b. a dental X-ray twice a year

    c. the nuclear power industry

22. What percent of a radioactive isotope would remain after two half-lives, four half-lives, and six half-lives? What percent would have decayed after each period?

23. Estimate the half-life of radioisotope X from this graph.

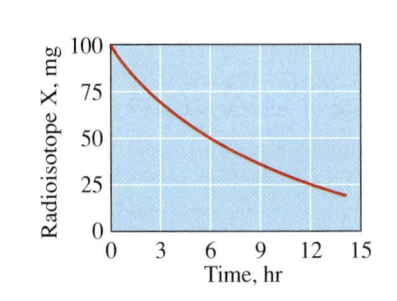

Concentrating on Concepts

24. The opening lines of this chapter connect acid rain and global warming to burning fossil fuels. Explain the connection.

25. In Consider This 7.1, you were asked to answer several questions about nuclear power. Ask the same questions of someone at least one generation older than you and someone younger. In comparison with your answer, what similarities and differences do you find?

26. The isotopes U-235 and U-238 are alike in that they are both radioactive. However, these two isotopes have very different abundances in nature. List their natural abundances and explain the significance of the difference.

27. Consider the uranium fuel pellets used in commercial nuclear power plants.

    a. Describe one way in which U-235 and U-238 can be separated.

    b. Why is it necessary to enrich the uranium for use in the fuel pellets?

    c. The fuel pellets are enriched only to a few percent, rather than to 80–90%. Give three reasons why.

    d. Explain why it is not possible to separate the isotopes of uranium by chemical means.

28. a. Why must the fuel rods in a reactor be replaced every couple of years?

    b. What happens to the fuel rods after they are taken out of the reactor?

29. At full capacity, each reactor in the Palo Verde power plant uses only a few pounds of uranium to generate 1243 megawatts of power. To produce the same amount of energy would require about 2 million gallons of oil or about 10,000 tons of coal in a conventional power plant. How is energy produced in the Palo Verde plant, compared with conventional power plants?

30. One important distinction between the Chernobyl reactors and those in the United States is that those in Chernobyl used graphite as a moderator to slow neutrons, whereas U.S. reactors use water. In terms of safety, give two reasons why water is a better choice.

31. If you look at nuclear equations in sources other than this textbook, you may find that the subscripts have been

omitted. For example, you may see an equation for a fission reaction written this way.

$$^{235}U + {}^1n \longrightarrow [{}^{236}U] \longrightarrow {}^{87}Br + {}^{146}La + 3\ {}^1n$$

a. How do you know what the subscripts should be? Why can they be omitted?

b. Why are the superscripts *not* omitted?

32. Using the model of a neutron presented in equation 7.6, explain how a high-speed electron can be ejected from the nucleus in beta decay.

33. Coal can contain trace amounts of uranium. Explain why thorium must be found in coal as well.

34. Suppose somebody tells you that a radioisotope is "gone" after about seven half-lives. Critique this statement, explaining both why it could be a reasonable assumption and why it might not be.

35. A Web site describing an X-ray procedure reports, "Despite its negative connotations, people are exposed to more radiation on a daily basis than they may realize. For example, infrared radiation is released whenever there is extreme heat. The Sun generates ultraviolet radiation, and a little exposure to it will tan a lighter skinned person. In addition, the body contains naturally radioactive elements." Examine the three examples given in this explanation. Do they refer to nuclear or electromagnetic radiation?

36. Consider this representation of a Geiger–Müller counter (also called a Geiger counter), a device commonly used to detect ionizing radiation. The probe contains a gas under low pressure.

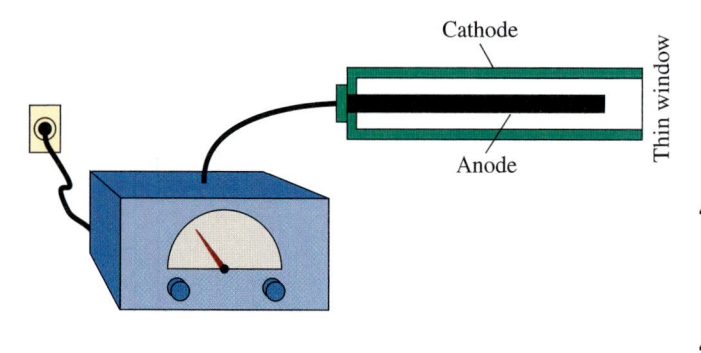

a. How does radiation enter the Geiger–Müller counter?

b. Why does this device only detect radiation that is capable of ionizing the gas contained in the probe?

c. What are other methods for detecting the presence of ionizing radiation?

37. Rapidly dividing cells are present in several places in the adult body. These include the skin, the hair follicles, the stomach and intestines, the lining of your mouth, and your bone marrow. Match the symptoms listed in Table 7.4 with the type of cell that was affected by the radiation.

38. Exposure to ionizing radiation can cause cancer. A beam of ionizing radiation also can be used to cure certain types of cancer. Explain.

39. Fluorine only has one naturally occurring radioisotope, F-19. If fluorine also occurred in nature as F-18, would this necessarily complicate the separation of $^{238}UF_6$ and $^{235}UF_6$? Explain.

40. **a.** Is depleted uranium (DU) still radioactive? Explain.

b. Is spent nuclear fuel (SNF) still radioactive? Explain.

41. It is generally believed that terrorists would be more likely to construct a nuclear bomb using Pu-239 reclaimed from breeder reactors than using U-235. Use your knowledge of chemistry to offer reasons for this.

42. Weapons-grade plutonium is almost completely Pu-239. In contrast, the plutonium produced in the normal operation of a water-cooled power reactor (reactor-grade plutonium) generally has a higher concentration of heavier isotopes such as Pu-240 and Pu-241. Propose an explanation for this observation.

43. **a.** What are the characteristics of high-level radioactive waste (HLW)?

b. Explain how low-level waste (LLW) differs from HLW.

Exploring Extensions

44. Alchemists in the Middle Ages dreamed of converting base metals, such as lead, into precious metals—gold and silver. Why could they never succeed? Today could we convert lead to gold? Explain.

45. Make a time line of nuclear history, putting at least a dozen dates on your line. For example, start with Becquerel's discovery of radioactivity in 1896. Other candidates for inclusion are Chernobyl, Hiroshima, the opening of the first commercial reactor, the discovery of various medical isotopes, the use of uranium glazes in Fiesta ware, and the Nuclear Test Ban Treaty.

46. The Tennessee Valley Authority's nuclear reactor, Browns Ferry 1, did not operate between 1985 and 2007. As this book went to print, it was back on line. What are the current news reports?

47. Explain the term *decommission*, as in "decommissioning a nuclear power plant." What technical challenges are involved? You might want to start by learning more about the decommissioning of the Yankee–Rowe facility (see Table 7.1). The resources of the Web can help you.

48. Einstein's equation, $\Delta E = \Delta mc^2$ applies to chemical changes as well as to nuclear reactions. An important chemical change studied in Chapter 4 was the combustion of methane, which releases 50.1 kJ of energy for each gram of methane burned.

a. What mass loss corresponds to the release of 50.1 kJ of energy?

b. To produce the same amount of energy, what is the ratio of the mass of methane burned in a chemical

reaction to the mass loss converted into energy according to the equation $\Delta E = \Delta mc^2$?

c. Use your results in parts **a** and **b** to comment on why Einstein's equation, although correct for both chemical and nuclear changes, usually is only applied to nuclear changes.

49. When 4.00 g of hydrogen nuclei undergoes fusion to form helium in the Sun, the change in mass is 0.0265 g and energy is released. Use Einstein's equation, $\Delta E = \Delta mc^2$, to calculate the energy equivalent of this change in mass.

50. Under conditions like those on the Sun, hydrogen can fuse with helium to form lithium, which in turn can form different isotopes of helium and of hydrogen.

$$^2_1\text{H} + {}^3_2\text{He} \longrightarrow [{}^5_3\text{Li}\ ] \longrightarrow {}^4_2\text{He} + {}^1_1\text{H}$$
2.01345 g 3.01493 g   4.00150 g 1.00728 g

a. What is the mass difference between a mole of the reactant and the product isotopes? Molar masses are given below each isotope.

b. How much energy (in joules) is released in this reaction?

51. Lise Meitner and Marie Curie were both pioneers in developing an understanding of radioactive substances. You likely have heard of Marie Curie and her work, but may not have heard of Lise Meitner. How are these two women related in time and in their scientific work?

52. Taking potassium iodide tablets can protect your thyroid from exposure to radioactive iodine, thus reducing your risk of thyroid cancer.

a. Give the chemical formula for potassium iodide.

b. By what mechanism does potassium iodide protect you?

c. How long does the protection last?

d. Are the tablets expensive? *Hint:* The FDA Web site is a good source of information for parts **b** and **c**.

53. A stockpile of approximately 50 metric tons of plutonium exists in the United States as a result of disassembling warheads from the nuclear arms race. What should the fate of this plutonium be? *Hint:* A search for "plutonium disposal" on the Web will bring up references. Try also including United States and DOE as search terms.

a. Some propose that the plutonium be sent to local nuclear power plants to "burn" as fissionable fuel. What are the advantages and disadvantages of such a course of action?

b. Others propose that it be stored permanently in a repository. Again, list the advantages and disadvantages.

54. Advertisements for Swiss Army watches stress their use of tritium. One ad states that the "... hands and numerals are illuminated by self-powered tritium

gas, 10 times brighter than ordinary luminous dials. . . ." Another advertisement boasts that the "... tritium hands and markers glow brightly making checking your time a breeze, even at night. . . ." Evaluate these statements and, after doing some Web research, discuss the chemical form of tritium in these watches, and what its role is.

55. The amount of radiation you receive from medical X-rays varies considerably, depending on the procedure. For example, dental X-rays may be taken as "bitewings," full mouth, or panoramic, each involving a different dose. Select a type of medical X-ray and research the amount of radiation involved. Report your data in both rems and sieverts (using mrem, mSv, or μSv as appropriate).

56. The hormesis phenomenon, defined in Section 7.7, is that toxic substances in small amounts can increase one's resistance to the same substance in large amounts. Analogous to the dose-response curves of Figure 7.20, the figure here indicates the zone of hormesis where the curve dips *below* the control line into a therapeutic region. Use the resources of the Web to investigate hormesis. Prepare a summary of your findings. One starting point is an article, "Is Radiation Good for You?" at the science Web site The Why? Files. *Hint:* As in Figure 7.20, NOAEL means no observed adverse effect level.

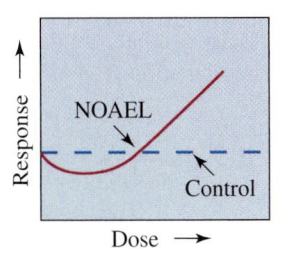

57. According to Table 7.3, smoking 1.5 packs of cigarettes a day adds 15,000 μSv to your annual radiation dose.

a. Which radioactive isotopes are responsible for this dose in cigarette smoke?

b. A nonsmoker, living with a 1.5-pack-a-day smoker, may receive the equivalent of 12 chest X-rays per year as a result of the secondhand smoke. How many microsieverts would this add to the nonsmoker's annual dose?

58. MRI, or magnetic resonance imaging, is an important tool for some types of medical diagnoses.

a. What is the scientific basis for this technique?

b. What information can an MRI give a physician that cannot be obtained through direct examination of a patient?

c. This MRI method used to be called NMR, nuclear magnetic resonance. Why do you think the name was changed?

59. Deciding where to locate a nuclear power plant requires

analysis of both risks and benefits associated with the plant. If you were to play the role of a CEO of a major electrical utility considering whether to pursue permits for the construction of a nuclear power plant in your area, what risks and benefits would you cite?

60. As this book went to press, final plans were being drafted for the 2010 construction of a new containment dome for Chernobyl. Accomplishing this feat will be difficult, as radiation levels on site are still high, and the melted core of the reactor is still thermally hot. Use the resources of the Web to find out the current status of the project, including the safety measures proposed to protect the workers.

Energy from Electron Transfer

"Our digital future may offer do-it-all devices that are nonetheless simple to use; batteries that can be recharged on the fly, perhaps wirelessly; and tech gadgets that resemble nothing available today."

Liane Cassavoy, *PC World*
Symposium at the Massachusetts Institute of Technology, 2004.

"I'd put my money on the sun and solar energy. What a source of power! I hope we don't have to wait 'til oil and coal run out before we tackle that."

Thomas Edison, Inventor
(1847–1931)

Our way of life demands long-lasting, reliable, and conveniently sized personal power sources, better known as batteries. Without them, the portable electronic gadgetry we depend on would be nothing more than extra weight. From small to large, batteries are needed to operate hearing aids, cell phones, digital cameras, and cars. This chapter reveals that the fundamental process powering many of our personal power sources is the transfer of electrons. But are all batteries the same? Do they all "run down"? What makes a battery rechargeable?

Increasingly, our way of life also demands that we find cleaner ways to transport our goods and ourselves. To this end, fuel cells, hydrogen, and gasoline–electric hybrid power sources are gaining prominence. What are fuel cells, and how can they be used to transfer electrons and generate power? Will hydrogen, with its potential to provide clean, reliable, and affordable energy, be the answer to all of our energy needs? What are the advantages of hybrid technology in which the combustion of gasoline combines with electron-transfer methods to produce energy? Will the promise of obtaining endless power from sunlight be realized? Each of these questions about energy from electron transfer will be considered in this chapter.

Consider This 8.1	Take a Battery Inventory

To help understand your personal dependence on electron transfer in consumer products, make a list of things that run on batteries.

a. Which items on your list use a battery as the main source of energy and which as a backup source?
b. What type of battery is used in each case?
c. Which batteries are rechargeable and which are not?
d. How do you dispose of batteries that are not rechargeable?

This is not the first time in this text that we have discussed our need for energy. For example, Chapter 4 emphasized the energy obtained from burning fossil fuels to produce electrical energy and to power transportation. Combustion processes also involve the transfer of electrons, but the widespread use of fossil-fuel combustion gave us reason to study them in a separate chapter. Chapter 7 focused on energy released by nuclear fission, a process that does not involve the transfer of electrons. Centralized power plants, whether fueled by coal or fissionable isotopes, distribute electricity regionally through vast power networks to offices, classrooms, and residences. But supplies of fossil fuels will not last forever, and fissionable isotopes, though available, are difficult to obtain. Moreover, both fuels have huge environmental costs. The combustion of coal and petroleum releases vast quantities of carbon dioxide, a significant contributor to global warming. Sulfur dioxide and nitrogen oxides are also released, leading to decreases in air quality and increases in acid precipitation. Spent nuclear fuel must be safely stored for generations to come. The conclusion seems obvious. If we are to continue to inhabit this planet and achieve the quality of life we desire, we must develop and depend on other sources of energy.

8.1 Electrons, Cells, and Batteries: The Basics

The famous inventor Thomas Edison was convinced in the late 19th century that batteries were doomed to failure. He branded such devices as ". . . a sensation, a mechanism for swindling the public by stock companies, . . ." He went on to say that, although such batteries were scientifically all right, their commercial success would be ". . . as absolute a failure as one can imagine." He clearly was better at inventing than at judging the future success of batteries! If you are carrying a cellular phone, listening to your iPod, taking

Edison's view might have been clouded by his ownership of a large, municipal electrical power company.

a digital photograph, or using a laptop computer, you know that you depend on batteries every day.

Although the term *battery* is in common use, a standard flashlight "battery" is more correctly called a galvanic cell, or simply a cell. A **galvanic cell** is a device that converts the energy released in a spontaneous chemical reaction into electrical energy. A collection of several galvanic cells wired together constitutes a true battery, such as the battery in your automobile. A galvanic cell is the opposite of an **electrolytic cell,** one in which electrical energy is converted to chemical energy.

A **battery** is a device consisting of one or more cells that can produce a direct current by converting chemical energy to electrical energy. Batteries are found everywhere in today's society because they are convenient, transportable sources of stored energy. Batteries are big business in the United States, with consumers spending over $100 billion on electronics in 2005 according to the Consumer Electronics Association. Many of these consumer products require batteries, spurring continued growth in the battery industry.

All galvanic cells produce useful energy from reactions involving the transfer of electrons from one substance to another. The transfer of electrons involves two changes, each represented by a separate equation. A **half-reaction** is a type of chemical equation that shows the electrons either lost or gained. The **oxidation half-reaction** shows the reactant that loses electrons, and the electrons appear on the product side of the equation. The **reduction half-reaction** shows the reactant that gains electrons, and the electrons appear on the reactant side of the equation. A galvanic cell always involves two half-reactions: one oxidation and the other reduction.

Let us start by looking at a simplified version of the reaction that takes place in a nickel–cadmium (NiCad) battery:

Oxidation half-reaction:	$Cd \longrightarrow Cd^{2+} + 2\,e^-$	[8.1]
Reduction half-reaction:	$2\,Ni^{3+} + 2\,e^- \longrightarrow 2\,Ni^{2+}$	[8.2]
Overall cell reaction:	$2\,Ni^{3+} + Cd \longrightarrow 2\,Ni^{2+} + Cd^{2+}$	[8.3]

> Oxidation = Loss of electrons
> Reduction = Gain of electrons

> NiCad is an abbreviation, not a chemical formula for the nickel–cadmium battery.

In this case, two electrons are given off, or "lost," in the oxidation half-reaction (equation 8.1). Where do they go? They are transferred to the ion being reduced. The number of electrons given off during oxidation must equal the number of electrons gained through reduction for the overall equation to balance. For this reason, the coefficient "2" appears in the reduction half-reaction (equation 8.2). Electrons do *not* appear in the overall cell reaction shown in equation 8.3, obtained by adding equations 8.1 and 8.2.

Your Turn 8.2 Electrons in Half-Reactions

Which of these represents an oxidation half-reaction? Which represents a reduction half-reaction? What is the basis for your decision?

a. $Al^{3+} + 3\,e^- \longrightarrow Al$

b. $Zn \longrightarrow Zn^{2+} + 2\,e^-$

c. $Mn^{7+} + 3\,e^- \longrightarrow Mn^{4+}$

d. $2\,H_2O \longrightarrow 4\,H^+ + O_2 + 4\,e^-$

e. $2\,H^+ + 2\,e^- \longrightarrow H_2$

The transfer of electrons through an external circuit produces **electricity,** the flow of electrons from one region to another that is driven by a difference in potential energy. The reaction provides the energy needed to drive a cordless razor, a power tool, or countless other battery-operated devices. The chemical species oxidized in the cell and the species reduced must be connected in a way such that the electrons released during oxidation are transferred to the species being reduced. To enable this transfer, **electrodes,** electrical conductors placed in the cell as sites for chemical reactions, are used. Different

processes take place at each electrode, and the electrodes are given different names. The **anode** is the electrode where oxidation takes place. The **cathode** receives the electrons sent from the anode through the external circuit. At the cathode, the electrons are used in the reduction half-reaction. The electrical energy is the result of the spontaneous reaction that occurs in the cell. The resulting **voltage,** the difference in electrochemical potential between the two electrodes, is expressed in the unit called volt (V). The greater the difference in potential between two electrodes, the higher the voltage will be and the greater the energy associated with the transfer. In the case of a NiCad cell, the maximum difference in electrochemical potential is measured as 1.46 V under specified conditions. Several cells are connected together in series to produce greater potential differences necessary to power larger devices (Figure 8.1).

The reaction in a NiCad battery is actually more complicated than represented in equations 8.1–8.3. Cadmium metal, the anode, contains atoms of cadmium that are oxidized to Cd^{2+} ions. These in turn combine with OH^- ions to form $Cd(OH)_2$. Simultaneously, Ni^{3+} ions (in hydrated $NiO(OH)$) on a nickel cathode are reduced to Ni^{2+} ions (in $Ni(OH)_2$). A water-based paste containing $NaOH$ or KOH acts as the electrolyte separating the electrodes.

Anode (oxidation half-reaction):

$$Cd(s) + 2\,OH^-(aq) \longrightarrow Cd(OH)_2(s) + 2\,e^- \quad [8.4]$$

Cathode (reduction half-reaction):

$$2\,NiO(OH)(s) + 2\,H_2O(l) + 2\,e^- \longrightarrow 2\,Ni(OH)_2(s) + 2\,OH^-(aq) \quad [8.5]$$

Overall cell reaction:

$$Cd(s) + 2\,NiO(OH)(s) + 2\,H_2O(l) \longrightarrow 2\,Ni(OH)_2(s) + Cd(OH)_2(s) \quad [8.6]$$

These three equations show exactly the same transfer of electrons represented in equations 8.1–8.3, but now all the different states and chemical forms are indicated. Note that because oxidation always accompanies reduction, both electrode materials undergo a chemical change as the battery discharges (and charges). Figure 8.2 illustrates a common use for NiCad batteries.

The unit name volt honors the Italian physicist Alessandro Volta (1745–1827). He is credited with inventing the first electrochemical battery in 1800.

Electrolytes were defined in Section 5.7.

Figure 8.1

This 7.2-V Ryobi portable power drill comes with two NiCad battery packs and a recharging unit.

Your Turn 8.3   Checking Balance and Charge

Consider equations 8.1–8.6.

a. From the standpoint of number and type of atoms, is each equation balanced?
b. From the standpoint of number and type of atoms, is the total charge zero for both reactants and products? Does it have to be?
c. Name a quick way to tell an overall cell reaction from a half-reaction.

The NiCad battery is rechargeable, an important characteristic for consumer use. The starting materials that undergo oxidation and reduction are solids, as are the products. The solids formed by the forward (discharging) reaction cling to a stainless-steel grid within the battery. Thus, they are still available and permit the reaction to be reversed during the recharging process. No gases are produced during either the recharging or the discharging, so the battery can be totally sealed. Transfer of electrons takes place during both the discharging and recharging processes, just in opposite directions (equation 8.7). Although this rechargeable battery can be discharged and recharged many times, eventually the accumulation of impurities or breakdown of the separators will end the useful life of the battery.

$$Cd(s) + 2\,NiO(OH)(s) + 2\,H_2O(l) \underset{recharging}{\overset{discharging}{\rightleftharpoons}} 2\,Ni(OH)_2(s) + Cd(OH)_2(s) \quad [8.7]$$

Commercially successful batteries must be affordable, last a reasonable length of time, and be safe to use and discard or recharge. In some applications, the size and weight of the cell is of paramount importance. Solids, pastes, gels, or thick slurries act as electrolytes to

Cathode, $NiO(OH)(s)$ | Anode, $Cd(s)$

Separator, $KOH(aq)$ paste

Figure 8.2

A NiCad galvanic cell in which Cd is oxidized at the anode, and Ni^{3+} is reduced at the cathode.

carry ions between electrodes. Aqueous solutions are often too hazardous to use in commercial cells because of the possibility for leakage. However, aqueous solutions are often used in simple laboratory cells. Electrons are transferred through the external circuit from anode to cathode, but positive and negative ions are often transported internally through a conducting salt bridge to complete the circuit. Ask if you can set up the galvanic cell in Your Turn 8.4 in the laboratory, or if you can see it as a demonstration in class.

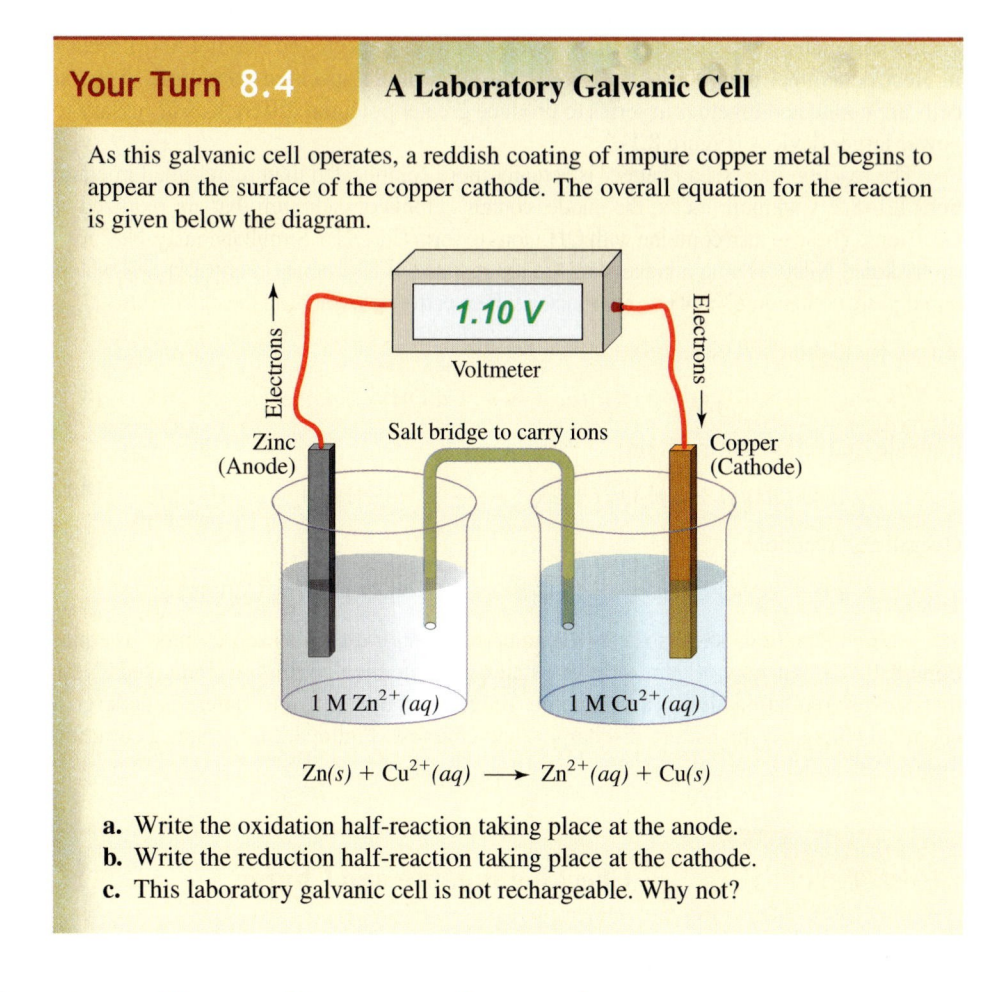

Your Turn 8.4   **A Laboratory Galvanic Cell**

As this galvanic cell operates, a reddish coating of impure copper metal begins to appear on the surface of the copper cathode. The overall equation for the reaction is given below the diagram.

$$Zn(s) + Cu^{2+}(aq) \longrightarrow Zn^{2+}(aq) + Cu(s)$$

a. Write the oxidation half-reaction taking place at the anode.
b. Write the reduction half-reaction taking place at the cathode.
c. This laboratory galvanic cell is not rechargeable. Why not?

8.2 Some Common Batteries

Almost everyone has used an alkaline battery such as one of those shown in Figure 8.3. How does the electron transfer work? Figure 8.4 details the composition of a typical alkaline battery.

These are the half-reactions for the alkaline cell illustrated in Figure 8.4.

Anode (oxidation half-reaction):

$$Zn(s) + 2\,OH^-(aq) \longrightarrow Zn(OH)_2(s) + 2\,e^- \qquad [8.8]$$

Cathode (reduction half-reaction):

$$2\,MnO_2(s) + H_2O(l) + 2\,e^- \longrightarrow Mn_2O_3(s) + 2\,OH^-(aq) \qquad [8.9]$$

The overall cell reaction is the sum of the two half-reactions.

$$Zn(s) + 2\,MnO_2(s) + H_2O(l) \longrightarrow Zn(OH)_2(s) + Mn_2O_3(s) \qquad [8.10]$$

The cell voltage depends primarily on the chemicals that participate in the reaction. The voltage does not depend on factors such as the overall size of the cell, the amount of material it contains, or the size of the electrodes. Thus all alkaline cells, from the tiny AAA size to the large D cells, have the same voltage, 1.54 V. The oxidation half-reaction at the anode sends electrons through the external circuit to power your flashlight, radio,

Alkaline (basic) solutions were discussed in Section 6.4.

Figure 8.3
These AAA to D alkaline cells all produce 1.54 V, but the larger cells can sustain the transfer of electrons through the external circuit for a longer time.

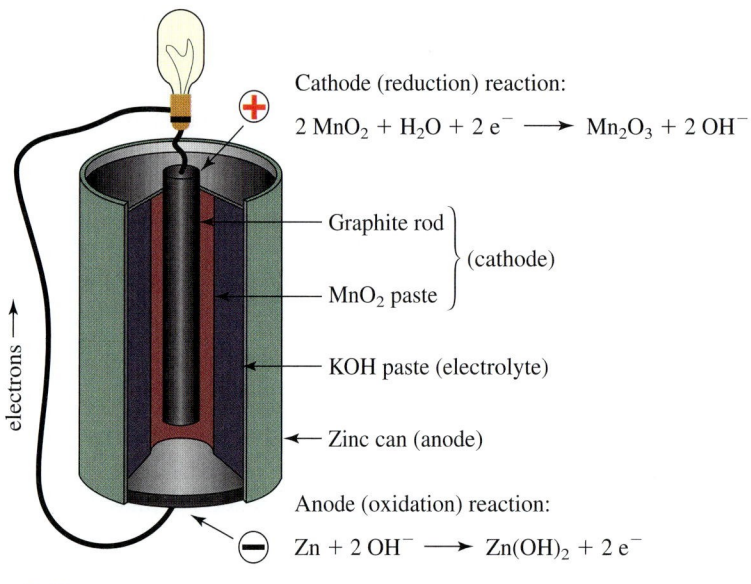

Cathode (reduction) reaction:

$$2\,MnO_2 + H_2O + 2\,e^- \longrightarrow Mn_2O_3 + 2\,OH^-$$

Graphite rod ⎤
⎥ (cathode)
MnO_2 paste ⎦

KOH paste (electrolyte)

Zinc can (anode)

Anode (oxidation) reaction:

$$Zn + 2\,OH^- \longrightarrow Zn(OH)_2 + 2\,e^-$$

electrons →

Figure 8.4

Diagram of an alkaline cell (battery).

or other device, before returning the electrons to the cathode to participate in the reduction half-reaction. The **current**, the rate of electron flow through the external circuit, is measured in amperes (amps, A) or very likely in milliamps (mA) for small cells. Larger cells contain more materials and can sustain the transfer of electrons over a longer period, making them useful for many applications. Many different galvanic cells have been developed, each for a specific purpose. Examples are listed in Table 8.1.

Current is measured in amperes (amps, A) to honor André Ampère (1775–1836). He was a largely self-taught French mathematician who devoted himself to the study of electricity and magnetism.

Consider This 8.5 Cell Phone Batteries

Many cell phones use lithium ion batteries. Use the resources of the Web to learn more.
Hint: Search for "lithium ion battery chemistry."

a. Why is this battery suited for use in portable devices?
b. What materials form the anode and cathode?
c. How does the lithium ion battery differ from a lithium–iodine battery?

Table 8.1 Some Common Galvanic Cells

Type	Voltage	Rechargeable?	Examples of Uses
Alkaline	1.54	No	Flashlights, small appliances
Lithium–iodine	2.8	No	Camera batteries, pacemakers
Lithium ion	3.7	Yes	Laptop computers, cell phones, digital music players
Lead–acid (storage battery)	2.0	Yes	Automobiles
Nickel–cadmium (NiCad)	1.25	Yes	Portable consumer electronics
Nickel metal hydride (NiMH)	1.25	Yes	Replacing NiCad for many uses; hybrid vehicles
Mercury	1.3	No	Formerly used widely in cameras, other appliances

Compact, long-lasting batteries play a special role in medical applications. The widespread use of cardiac pacemakers has been due, in large part, to the improvements made in the batteries used to power them, rather than in the pacemakers themselves. All lithium batteries take advantage of the low density of lithium metal to make a lightweight battery, but lithium–iodine cells are so reliable and long-lived that they are often the battery of choice for this application. A lithium–iodine pacemaker battery implanted in the chest can last as long as 10 years before it needs to be replaced. Persons with pacemakers are advised to avoid electromagnetic fields that can interfere with the electronic components of the battery-powered device, such as holding a cell phone close to the location of the pacemaker or going through an airport security arch.

Consider This 8.6   Oxyride Batteries

Use the resources of the Web to learn more about the recently introduced Oxyride battery. This was developed in Japan for Panasonic.

a. What advantages does the Oxyride battery have over a typical alkaline battery?
b. This battery differs in two ways from a typical alkaline battery. Describe the differences.
c. Is this battery rechargeable? Explain.

Most batteries convert chemical energy into electrical energy with an efficiency of about 90%. Compare this with the much lower efficiencies that typically characterize the conversion of heat to work (30–40%) in electricity-generating plants. However, remember that considerable energy is required to manufacture galvanic cells. Ores and minerals must be mined and processed, and the various components manufactured and assembled. Moreover, a battery has a finite life. Even rechargeable batteries will eventually have to be replaced. Sooner or later, the voltage will drop below usable levels, and electrons will no longer flow. At this point, the battery is "dead" and ready for disposal.

8.3 Lead-Acid (Storage) Batteries

The best known rechargeable battery is the lead–acid battery. Lead–acid batteries are called **storage batteries** because they "store" electrical energy. Until very recently, such batteries were used in every automobile. The lead–acid storage battery is a true battery because it consists of six cells, each generating 2.0 V, for a total of 12.0 V. Here is the overall cell reaction.

$$\text{Pb}(s) + \text{PbO}_2(s) + 2\,\text{H}_2\text{SO}_4(aq) \underset{\text{recharging}}{\overset{\text{discharging}}{\rightleftharpoons}} 2\,\text{PbSO}_4(s) + 2\,\text{H}_2\text{O}(l) \qquad [8.11]$$

lead   lead dioxide   sulfuric acid   lead sulfate   water

The anode is made of metallic lead and the cathode of lead dioxide, PbO_2. The electrolyte is a sulfuric acid solution (Figure 8.5). Although the weight of the lead and the corrosive properties of the acid are disadvantages, the lead–acid storage battery is dependable and long-lasting.

The key to the success of the lead–acid storage battery is the reversibility of equation 8.11. This reaction produces the energy necessary to power a car's starter, headlights, and various devices. As the reaction proceeds spontaneously to the right, the battery "discharges." But the electrical demands of a modern car are so great that in a short time most of the reactants would be converted to products, significantly reducing the voltage and the current. To counter this, the battery is attached to an alternator turned by the engine. The alternator produces direct current electricity that is run back through the battery. This input of energy reverses equation 8.11 and thus recharges the battery. In a high-quality

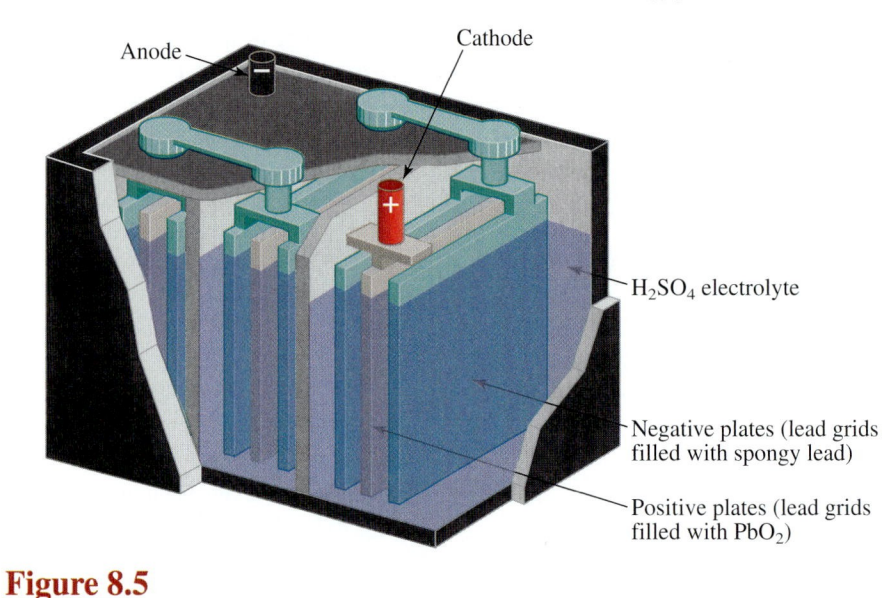

Figure 8.5

Cutaway view of a lead–acid storage battery.

Older lead–acid batteries required addition of water and therefore were designed with easily removable screw caps.

lead–acid storage battery, these processes of discharging and recharging can continue for five years or more.

In environments where the emissions from internal combustion engines cannot be tolerated, sealed lead–acid storage batteries can provide energy. Thus, forklifts in warehouses, passenger carts in airport terminals, and wheelchairs are typically powered by lead–acid storage batteries. The weight of the lead–acid batteries becomes an advantage in helping to stabilize such transportation devices. Because of the dependability of lead–acid batteries, they are sometimes used together with wind turbine electrical generators. The generator recharges the batteries when the wind is blowing, and the batteries provide electricity when the wind stops.

Your Turn 8.7   **Another Look at the Storage Battery**

Reexamine equation 8.11 for the reaction that takes place in a lead–acid storage battery.

a. When the battery is discharging, what is being oxidized? Reduced?

b. When the battery is recharging, what is being oxidized? Reduced?

A group of Israeli chemists led by Doron Aurbach and his colleagues at Bar-Ilan University are developing a more environmentally friendly storage battery. Their strategy has been to replace lead with magnesium, a metal that is less toxic, lighter in weight, and less expensive. The new magnesium–acid battery, like the lead–acid storage battery, would be rechargeable. Prototypes current in 2007 produce up to 1.3 V, but the chemists hope to improve that to 1.7 V in the near future with further refinements to the materials. Magnesium is a far more reactive metal than lead, and it is a challenge to prevent unwanted side reactions between magnesium and the electrolytes used in the battery. Earlier these chemists explored a solid-state battery to avoid this problem, but research published in 2006 provides new insights into electrolyte solutions that could be used. One of the first applications envisioned for the new type of storage battery would be to provide uninterrupted power to computer networks during power outages. This battery might also become useful to power gasoline–electric hybrid vehicles as we describe in the next section.

8.4   Hybrid Vehicles

Prius in Latin means "to go before."

As concerns grow about the cost and supply of gasoline and about vehicle emissions, more and more automobile owners are considering **hybrid vehicles**. These vehicles combine conventional gasoline engines with battery technology. Honda and Toyota have led the way in developing such vehicles. The Honda Insight, a small two-seater was the first hybrid sold in the United States in 1999. The Toyota Prius (Figure 8.6), a hybrid about the size of a Toyota Corolla, was first available in Japan in 1997 and in 2000 in the United States. Many other manufacturers now produce models of hybrid cars, SUVs, and trucks.

(a)

(b)

Figure 8.6

(a) The Toyota Prius gasoline–electric hybrid automobile.
(b) The "engine" of the Prius. The NiMH batteries are not shown because they are under the back seat and trunk areas.

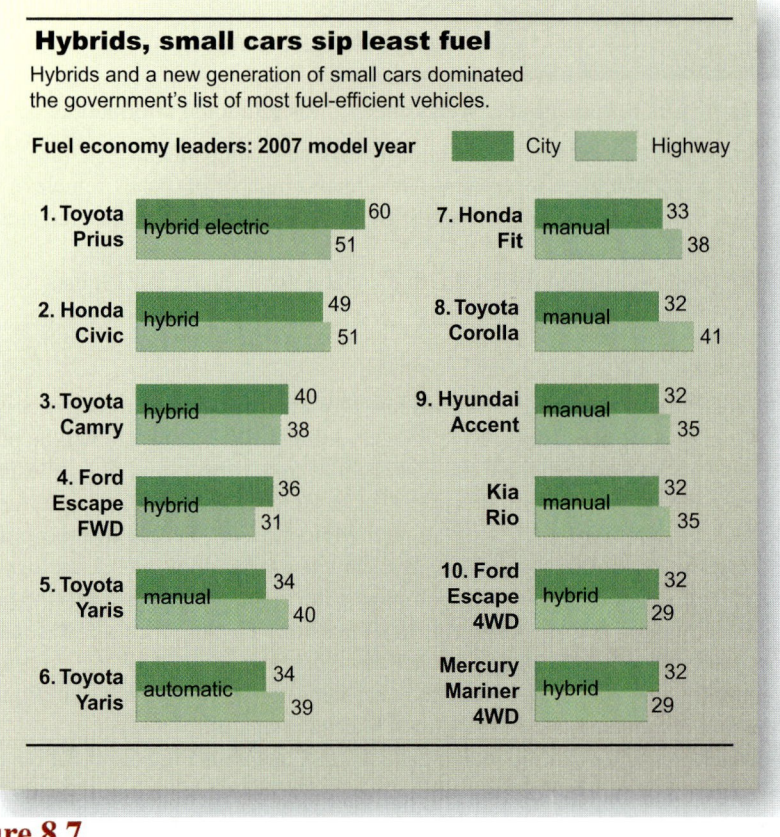

Figure 8.7

News report of fuel economy leaders, 2007 model year.

Source: Data from the U.S. Environmental Protection Agency.

With a 1.5-L gasoline engine sitting side-by-side with nickel metal hydride batteries, an electric motor, and an electric generator, the Prius needs no recharging. It consumes only about half the gasoline, emits 50% less carbon dioxide and far less nitrogen oxides than a conventional car, while delivering 52 miles per gallon (mpg) of gasoline around town and 45 mpg out on the highway. The electric motor draws power from the batteries to start the car moving or to power it at low speeds. Using a process called regenerative braking, the kinetic energy of the car is transferred to the alternator, which charges the batteries during deceleration and braking. The gasoline engine assists the electric motor during normal driving, with the batteries boosting power when extra acceleration is needed. The newer midsize Toyota Prius, introduced in the United States in 2003, is rated at 60 mpg in the city and 51 mpg on the highway. Figure 8.7 shows the 2007 rankings for the most fuel-efficient automobiles.

The fuel economy leaders in 2007 differed markedly from those in 2006. For example, since its introduction in model year 2000, the Honda Insight hybrid has topped the EPA's fuel-efficiency list, with 60 mpg city and 66 mpg highway. However, Honda discontinued the Insight after model year 2006, and now the Toyota Prius has captured the top spot. In model year 2007, several new small cars were introduced by Toyota, Honda, Hyundai, and Kia. These vehicles broke into the top 10, displacing the fuel-efficient diesels made by Volkswagen.

The fuel economy leaders in the United States soon may include more diesels. Although diesels nearly disappeared from the U.S. car market in the 1980s because they were judged to be dirty, dull, and unreliable, they never left the stage in Europe or Japan. One factor reviving interest in the United States is the availability of cleaner, low-sulfur fuel for diesels. Another is a more effective trap for the tiny but dangerous soot particles that diesels emit. According to J. D. Power, an automotive market research firm, the market share of diesel cars will reach 15% by 2015, a considerable increase from the 2005 value of 3.2%.

Many hybrids, unlike conventional gasoline-powered cars, deliver better mileage in city driving than at highway speeds.

Consider This 8.10   Fuel-Efficient Cars

Consider Figure 8.7 and use the Web as needed for additional information.

a. How many cars on this list are built by U.S. auto companies?
b. Which manufacturer accounts for the greatest number of top 10 vehicles?
c. Even though they are not in the top 10, which manufacturer makes the most fuel-efficient diesels?
d. The Hummer H2 and the Ford Excursion are not included in EPA lists for fuel-efficiency. Why is that?

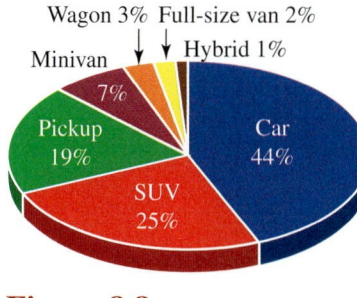

Figure 8.8

2005 U.S. auto market. The values do not total 100% due to rounding.

Source: From Keith Naughton, "Hybrid Nation? Nope," *Newsweek,* October 10, 2005, p. 46. Copyright 2005 Newsweek, Inc. All rights reserved. Reprinted by permission.

With such advantages, have we turned into a hybrid nation? The short answer is "no." In spite of rising gas prices, automobile industry leaders agree that there will be no mass market for alternative energy vehicles—battery-powered, hybrid cars, or ones using fuel cells—unless the performance and price of such vehicles match those of conventional models. Jack Smith, former CEO of General Motors, said: "People are too practical. There's a certain element who will fall in love with new technology, but technology won't survive unless it's cost effective." The common strategy for companies now is to provide more than one option for fuel-efficient, low-emission vehicles, then allow consumers to choose the best option for their personal transportation needs. Although hybrid sales only account for 1.1% of the U.S. auto market, their sales are rising each year. Figure 8.8 shows the mix of vehicles sold in the United States during 2005.

Passenger cars will continue to dominate the U.S. auto market, but one of every four vehicles sold is a SUV. Despite pain at the pump, many Americans refuse to downsize. SUVs and trucks remain hot items, and they can employ the same hybrid technology proven successful in cars. Reluctant consumers have to be convinced that their desire for muscle and size will not prevent them from enjoying the fuel savings and environmental benefits possible with hybrid technology. The auto researcher J. D. Power estimated in 2007 that by 2010, the number of SUVs on the market would increase 24% to 109 models, but only 44 of these would be hybrids. Hybrid technology is also entering the luxury car market, including the Lexus RX 400h that is rated at 30 mpg in the city and 26 mpg on the highway.

Consider This 8.11   Hybrids and the Drive for Power

a. Why are some of the newer hybrid car models less fuel efficient than the first ones introduced, such as the Honda Insight or the Toyota Prius?
b. The federal government, some states, and even some cities encourage the purchase of hybrid cars by offering benefits. List some of these.
   Hint: Use the terms *tax incentives* and *hybrid cars* in your Web search.

8.5 Fuel Cells: The Basics

Suppose someone were to suggest a way to combine H_2 and O_2 to form H_2O without the hazards of combustion. Suppose that this individual also claimed that the reaction could be carried out with no direct contact between the hydrogen and the oxygen. The Sceptical Chymist might well dismiss such assertions as sheer nonsense—an outright impossibility. And yet, sometimes what appears to be completely contrary to common sense can happen in the natural world. The operation of a fuel cell is a case in point. A **fuel cell** is a galvanic cell that produces electricity by converting the chemical energy of a fuel directly into electricity without burning the fuel. Fuel cells are sometimes called "flow" batteries, because both fuel and oxidizer must constantly flow into the cell to continue the chemical reaction. After we consider this important type of battery, we can better understand new approaches to powering vehicles and generating electricity.

William Grove, an English physicist, invented fuel cells in 1839. However, they remained a mere curiosity until the advent of the U.S. space program in the 1960s. The Space Shuttle, for example, carries three sets of 32 cells fueled with hydrogen. The electricity they generate illuminates bulbs, powers small motors, and operates computers.

A fuel cell functions somewhat like a conventional flashlight battery. But, unlike batteries, fuel cells use a constant external supply of fuel (such as hydrogen) and oxidant (such as the oxygen in air). Therefore, they produce electricity as long as fuel and oxidant are provided and consequently do not need to be recharged in the same sense as traditional batteries. As in a battery, the overall chemical reaction is physically divided into two parts: the oxidation half-reaction and the reduction half-reaction. But instead of the anode *itself* being the source of electrons, the anode is the electrode at which the oxidation of the fuel takes place. The other electrode behaves as the cathode, where reduction of the oxidant (usually oxygen) takes place.

The electrolyte that separates the anode from the cathode in a fuel cell serves the same purpose as in a traditional galvanic cell, allowing the flow of ions but not electrons. Fuel cells using aqueous phosphoric acid (H_3PO_4) as the electrolyte were the first type to be commercially available and continue to be used on the Space Shuttle. But such strongly acidic solutions are very corrosive and require that the liquid be fully contained, adding complexity and cost to these types of fuel cells. Perhaps this does not seem like much of a problem. After all, flashlight batteries and car batteries also contain corrosive electrolytes and must be sealed. The added difficulty lies in the fact that a fuel cell, unlike a battery is not a closed system. To function, oxidant and fuel must be constantly supplied from outside the cell, so complete isolation is not possible.

Over the past several decades, other types of fuel cells based on different electrolyte materials have been developed for a variety of applications. One type incorporates a solid polymer electrolyte separating the reactants. We will use this type to explain the general operation of fuel cells before describing other types in the next section. The polymer electrolyte membrane, also called a proton exchange membrane (PEM), is permeable to H^+ ions and is coated on both sides with a platinum-based catalyst. These electrolytes can operate at reasonably low temperatures, typically from 70 °C to 90 °C, and can start transferring electrons to provide electrical power quickly. As a result, PEM fuel cells are currently very popular among automakers for new fuel cell prototype vehicles and for personal consumer applications. A typical design is shown of Figure 8.9.

Polymers are discussed in Chapter 9.

The H^+ ion is a proton, the simplest possible cation. See Section 5.7.

In fuel cells, both oxidation (a loss of electrons) and reduction (a gain of electrons) always occur. Typically, hydrogen gas (H_2) is the fuel used in conjunction with oxygen. If so, the oxidation and reduction half-reactions are represented by equations 8.12 and 8.13. As a molecule of hydrogen passes through the membrane, it is oxidized and loses two electrons to form two H^+ ions.

Anode (oxidation half-reaction):

$$H_2(g) \longrightarrow 2\,H^+(aq) + 2\,e^- \qquad [8.12]$$

These H^+ ions flow through the proton exchange membrane and combine with oxygen (O_2) and two electrons to form water.

Cathode (reduction half-reaction):

$$\tfrac{1}{2}\,O_2(g) + 2\,H^+(aq) + 2\,e^- \longrightarrow H_2O(l) \qquad [8.13]$$

The overall cell reaction is the sum of the two half-reactions.

$$H_2(g) + \tfrac{1}{2}\,O_2(g) + \cancel{2\,H^+(aq)} + \cancel{2\,e^-} \longrightarrow \cancel{2\,H^+(aq)} + H_2O(l) + \cancel{2\,e^-} \quad [8.14]$$

The electrons lost by H_2 through oxidation (equation 8.12) are gained by O_2 in the reduction process (equation 8.13). The $2\,e^-$ and $2\,H^+$ that appear on each side of the arrow in equation 8.14 can be canceled to give the net equation shown in equation 8.15.

$$H_2(g) + \tfrac{1}{2}\,O_2(g) \longrightarrow H_2O(l) \qquad [8.15]$$

This same equation appears in Chapters 1, 4, 5, and 7, an indication of its significance.

Thus, in a fuel cell, there is a transfer of electrons from H_2 to O_2. Oxidation cannot occur alone; that would be rather like one hand clapping. Oxidation (electron loss) must always

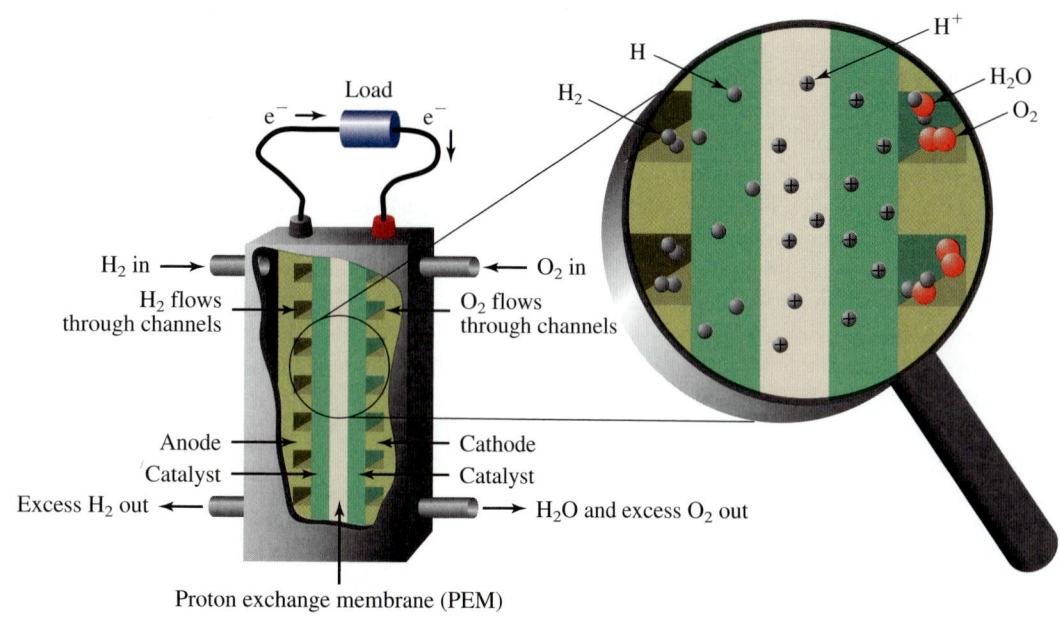

Figure 8.9

A PEM fuel cell in which H_2 and O_2 combine to form water without combustion.

Figures Alive! Visit the *Online Learning Center* to learn more about the chemistry of a PEM fuel cell.

be paired with reduction (electron gain). The net reaction is a combination of the oxidation and the reduction half-reactions, taking into account gain and loss of electrons.

To produce electricity, fuel cells (and batteries) are constructed in such a way that the electrons released during the oxidation half-reaction are transferred to the species being reduced (see Figure 8.9). The electrodes provide the sites for the chemical reactions; oxidation occurs at the anode, and reduction takes place at the cathode. The electrons flowing from the anode to the cathode of a fuel cell move through an external circuit to do work, which is the whole point of the device.

The electrical energy produced is the result of the familiar reaction (see equation 8.15) that occurs in the fuel cell. But in a fuel cell, this reaction takes place without a flame and with relatively little heat and no light being produced. Therefore, it is not classed as a combustion reaction. Water is the only chemical product if hydrogen is the fuel. No carbon-containing greenhouse gases or spent nuclear fuel are produced, and no air pollutants are emitted during the electron exchange in a fuel cell. If only the power-producing step is considered, hydrogen fuel cells are judged to be a more environmentally friendly way to produce electricity than are coal-fired or nuclear power plants. A comparison of the combustion of H_2 with O_2 and fuel cell technology is given in Table 8.2.

> Remember that it *requires* energy to produce H_2 from compounds containing hydrogen.

Table **8.2**	**Comparison of Combustion with Fuel Cell Technology**			
Process	**Fuel**	**Oxidant**	**Products**	**Other Considerations**
Combustion	Hydrocarbons, alcohols, H_2, wood, etc.	O_2 from air	H_2O, CO/CO_2, heat, light, and sound	Rapid process, flame present, lower efficiency, most useful for producing heat
Fuel cell	H_2*	O_2 from air	H_2O, electricity, some heat	Slower process, no flame, quiet, higher efficiency, most useful for generating electricity

*Compounds containing hydrogen, such as natural gas or alcohols, can be used in some fuel cells. Since these compounds contain carbon as well, CO or CO_2 (or both) are products.

The overall cell reaction represented by equation 8.15 releases 286 kJ of energy per mole of water formed. But instead of liberating most of this energy in the form of heat, the fuel cell converts 45–55% or more of it into electrical energy. This direct production of electricity eliminates the inefficiencies associated with using heat to do work to produce electricity. Internal combustion engines are only 20–30% efficient in deriving energy from fossil fuels.

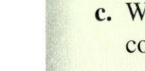

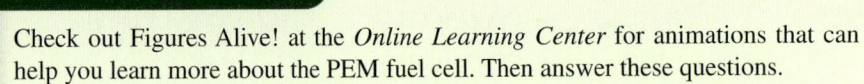

Consider This 8.12   Revisiting the PEM Fuel Cell

Check out Figures Alive! at the *Online Learning Center* for animations that can help you learn more about the PEM fuel cell. Then answer these questions.

    a. How is a fuel cell different from other batteries discussed earlier in this chapter?
    b. Can a PEM fuel cell be recharged? Explain.
    c. Why is the combination of H_2 and O_2 in a fuel cell not classified as combustion? Explain.

8.6   Fuel Cells: One Type Doesn't Fit All

Just as batteries, motors, and electrical generators come in different sizes and types, so do fuel cells. Though the fuels and principles of operation are essentially the same, different electrolytes give each type of fuel cell unique characteristics useful for a specific application.

For automobiles, PEM fuel cells currently are the best option, due mainly to their high **power density**, the energy capacity per unit of fuel cell mass. Furthermore, the PEM electrolytes can operate at reasonably low temperatures, typically from 70 °C to 90 °C and can start transferring electrons to provide electrical power quickly, a property required for automobile applications. Several companies manufacture fuel cell vehicles on a limited scale (Figure 8.10).

Ironically, the low operating temperatures are related to one major drawback of PEM fuel cells. They must remain moist in order to function. The species that moves through the membrane is not just a proton, but rather a water molecule plus a proton, called the **hydronium ion, H_3O^+**. Water produced at the cathode is recycled back to the anode to maintain the required moisture. If the membrane dries out, the PEM fuel cell abruptly stops working. Because constant moisture levels are essential, PEM fuel cells are extremely sensitive to heat. If the temperature rises above 100 °C, the water may evaporate from the electrolyte. Unfortunately, this relatively low operating temperature requires expensive reforming catalysts, very pure hydrogen, and complex construction techniques to ensure the fuel cell's reliability over the long term.

Researchers at Cal Tech have demonstrated that solid acid electrolyte fuel cells offer a possible alternative. Solid acid electrolytes transport bare protons (H^+, as opposed

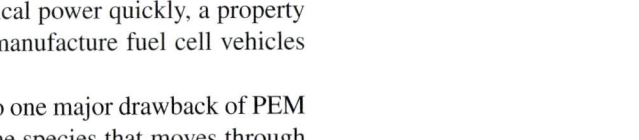

The attraction between a proton and polar water molecules was discussed in Section 5.6, and the hydronium ion was introduced in Section 6.1.

Figure 8.10
The Honda FCX is powered by hydrogen fuel cells.

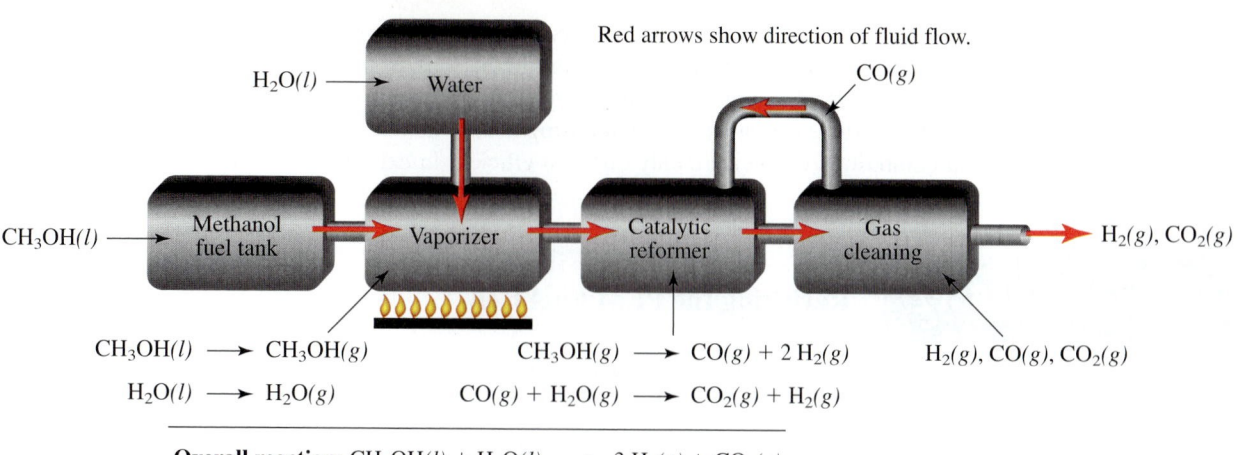

Figure 8.11

Hydrogen obtained from methanol by a reforming process.

to hydronium ions, H_3O^+) through their structure, and therefore can operate at higher temperatures that would "dry out" PEM fuel cells. In addition, the solid is less porous than the polymer membranes. The decreased gas leakage between the anode and cathode increases power output. Furthermore, they can be used with various fuels, not just hydrogen gas.

For some applications, having to store gaseous hydrogen is inconvenient. In such cases, hydrogen can be formed from the onboard decomposition of liquid methanol, CH_3OH, producing H_2 as needed. The catalytic decomposition of methanol into hydrogen and carbon dioxide is an example of **reforming**, a process using heat, pressure, and catalysts to rearrange the atoms within molecules (Figure 8.11). Reforming processes also allow gasoline or even diesel fuel to be used as the source of hydrogen.

Section 4.8 discussed reforming *n*-octane to produce isooctane for gasoline.

Sceptical Chymist 8.13   "Zero" Emissions from Fuel Cells

Hydrogen–oxygen fuel cells are considered to be environmentally friendly "zero-emissions" power sources. Does the Sceptical Chymist agree with this statement? What if the source of hydrogen were methanol or natural gas, which is largely methane? Are the same emissions expected and are they still just as environmentally friendly?

A newer type of fuel cell, the direct methanol fuel cell (DMFC), uses a methanol–water solution as the fuel. Increased portability and decreased size are two advantages to getting rid of the fuel reformer. The methanol fuel is oxidized at the anode.

Anode (oxidation half-reaction):

$$H_2O(l) + CH_3OH(aq) \longrightarrow CO_2(g) + 6\,H^+(aq) + 6\,e^- \qquad [8.16]$$

Air or oxygen is blown into the cathode compartment where the oxygen picks up electrons and reacts with hydrogen ions to produce water.

Cathode (reduction half-reaction):

$$3/2\,O_2(g) + 6\,H^+(aq) + 6\,e^- \longrightarrow 3\,H_2O(l) \qquad [8.17]$$

So the net reaction is:

$$CH_3OH(aq) + 3/2\,O_2(g) \longrightarrow CO_2(g) + 2\,H_2O(l) \qquad [8.18]$$

The key to the DMFC is the metal catalyst electrode. Because of the low operating temperatures and CO_2 produced, the catalysts have to be more active and more resistant

to poisoning. Interestingly, incorporation of some less expensive metals (for instance, dissolving some nickel or cobalt into the platinum catalyst) improves the efficiencies dramatically. Power densities of DMFCs are 20 times higher than they were just 15 years ago.

In addition to removing the need for a reformer, the liquid fuel technology is much easier to reduce in size. It appears that **microcells**, literally very tiny fuel cells, will become reality before larger scale applications such as electrical generation and powering automobiles (Figure 8.12).

In 2005, Toshiba announced the fabrication of a 300-mW prototype DMFC. It is smaller than a deck of cards and delivers enough power to keep an MP3 player running for approximately 60 hours. The company expects commercial products powering laptops and cell phones by 2007. This may be just the beginning. Allied Business Intelligence (ABI) predicts that by 2011, over 20% of handheld electronic devices with rechargeable power packs will be powered by DMFCs. One disadvantage inherent in these microcells is the toxicity of methanol. Researchers are currently investigating methods of substituting ethanol as the fuel. Although efficiencies of so-called DEFCs are low, one day it may be possible to run your cell phone with a beer!

> *Poisoning* occurs when impurities react with a catalyst and decrease its ability to promote reactions.

> Methanol is also called wood alcohol. Drinking it can cause blindness.

Consider This 8.14   Alcohol Fuel Cells

For the proposed ethanol–oxygen fuel cell under development, the overall chemical reaction is:

$$C_2H_5OH(l) + 3\ O_2(g) \longrightarrow 2\ CO_2(g) + 3\ H_2O(l)$$

a. Identify the chemicals that are oxidized and reduced.
b. Unlike batteries, fuel cells do not store chemical energy. Explain the significance of this statement for the future of fuel cells and batteries.
c. List several reasons why fuel cells have not been the energy source of choice in the past, but may become viable in the future.

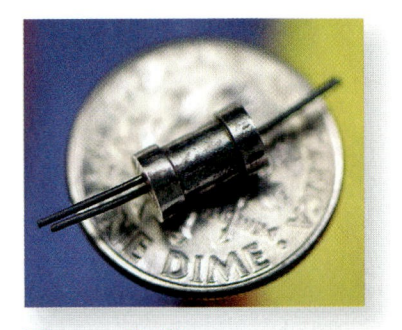

Figure 8.12
A micro fuel cell and a U. S. dime.

Interest also exists in using fuel cells to generate power at stationary locations that may or may not be connected to a power grid. The energy may be used locally or for standby or backup power on the grid. This practice is called **distributed generation,** placing many power-generating modules of 30 MW or less near the end user, as opposed to giant centralized power plants supplying many thousands of users. Worldwide, over 150 demonstration fuel cell plants have been installed for power generation. Typical sizes range from institutional or factory units capable of producing several megawatts down to 1-kilowatt systems for an individual home (the average U.S. house normally uses about 1–2 kW and up to 15 kW during peak usage). Nearly 75% of these installations are in Japan, about 15% are in North America, and 9% are in Europe.

Solid oxide fuel cells (SOFCs) are an attractive alternative for stationary power generation. A unique feature of SOFCs is the identity of the ions that move through the membrane. Instead of hydronium ions, oxide ions (O^{2-}) are transported from the cathode to the anode. These are the reactions that occur in an SOFC.

Anode (oxidation half-reaction):

$$H_2(g) + O^{2-} \longrightarrow H_2O(g) + 2\ e^- \qquad [8.19]$$

Cathode (reduction half-reaction):

$$\tfrac{1}{2}\ O_2(g) + 2\ e^- \longrightarrow O^{2-} \qquad [8.20]$$

Overall cell reaction:

$$H_2(g) + \tfrac{1}{2}\ O_2(g) \longrightarrow H_2O(g) \qquad [8.21]$$

The electrolyte has to be made from a different material than the proton exchange membranes discussed earlier. The best materials for oxide ion transport are hard ceramic

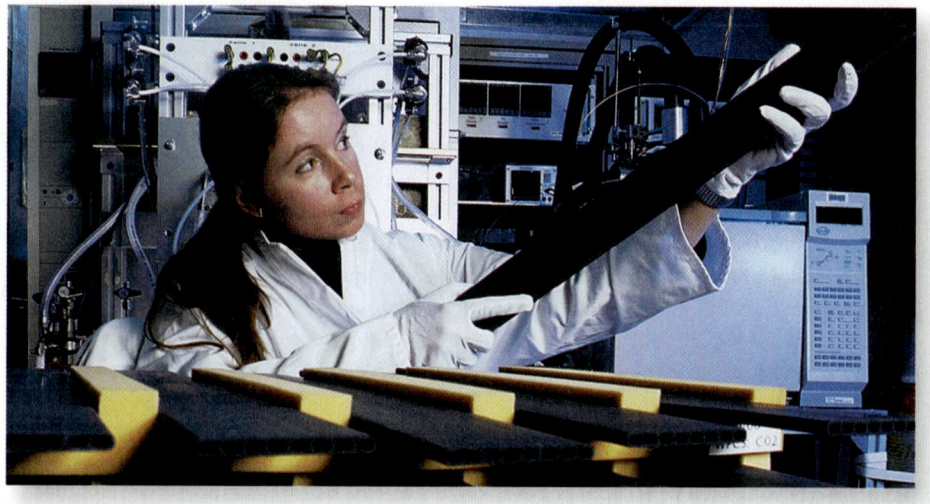

Figure 8.13
A Siemens researcher checks new, extremely flat solid oxide fuel cells.

oxides. Zirconium oxide with a small amount of yttrium oxide works very well. The surface of the ceramic is coated on both sides with specialized porous metallic electrode materials. This design of fuel cell operates at higher temperatures than PEM fuel cells, typically in the range of 650 °C to 1000 °C.

The high operating temperatures of SOFCs offer distinct advantages. Most importantly, SOFCs are "fuel-flexible." They are tolerant to impurities because they can at least partially reform hydrocarbon fuels internally. They have efficiencies around 60%, and are reliable over long periods. The H_2O produced from a solid oxide fuel cell is in the form of high-temperature steam. It can be used to spin a gas turbine (as in fossil-fuel and nuclear power plants), generating more electricity. Such hybrid, or cogeneration, schemes have theoretical efficiencies approaching 70%. Furthermore, the waste thermal energy from SOFCs can be harnessed and bring the efficiencies to over 80% (Figure 8.13).

Consider This 8.15   Military Electrons

Fuel cells find many applications in the military. To find out about military fuel cell demonstration sites, go to the Department of Defense (DoD) Web site. Check out several of the demonstration sites across the country to learn the different uses of these fuel cells and any cost savings reported.

> Unlike batteries, fuel cells cannot deliver large bursts of power. Because of this, some portable power designs combine both battery and fuel cell technology.

Table 8.3 gives a comparison of the types of fuel cells discussed in this chapter. In all of their forms, the future for fuel cells looks promising. Improvements in portability and efficiency are predicted to drive fuel cell sales worldwide beyond $2.6 billion by 2009. PEM cells will continue their market dominance, but both SOFCs (predicted to be the second largest market segment in 2009) and direct methanol fuel cells (due to their ability to power portable devices) are predicted to lead this growth.

Consider This 8.16   New Fuel Cell Technologies

Search the Web to learn about the latest developments in fuel cell technologies. Write a brief summary of one advancement and comment on its possible impact on any future applications.

Hint: Use the term *fuel cell applications* in your Web search.

Table 8.3	Comparison of Different Fuel Cell Systems						
Fuel Cell Type	Common Electrolyte	Operating Temperature	System Output	Efficiency	Applications	Advantages	Disadvantages
Phosphoric acid	H_3PO_4 soaked in a solid matrix	150–200 °C	50 kW–1 MW	36–42% 80–85%*	Distributed generation Outer space	High efficiency	Corrosive, expensive catalyst
PEM	Solid organic polymer	50–100 °C	<1 kW–250 kW	50–60%	Transportation Backup power	Noncorrosive Quick start-up	Highly sensitive to fuel impurities Expensive catalyst
DMFC	Solid organic polymer	60–90 °C	<1 W–100 W	50–60%	Small, portable power supply	No reformer Liquid fuel	Toxicity of methanol
SOFC	ZrO_2/Y_2O_3 ceramic	650–1000 °C	5 kW–3 MW	60–85%*	Distributed generation	Fuel Flexible High efficiency	Slow start-up

* Combined heat and power.

8.7 Splitting Water: Fact or Fantasy?

An increase in the demand for fuel cells will be accompanied by an increased demand for hydrogen. Where is it all going to come from? On one hand, things look promising because hydrogen is the most plentiful element in the universe. Over 93% of all atoms are hydrogen atoms. Although hydrogen is not nearly that abundant on Earth, there is still an immense supply of the element. On the other hand, essentially all of it is tied up in compounds. Hydrogen is too reactive to exist for long in its diatomic form, H_2, in the presence of the other elements and compounds that make up the atmosphere and Earth's crust. Therefore, to obtain hydrogen for use as a fuel, we must extract it from hydrogen-containing compounds, a process that requires energy.

Methane, the major component of natural gas, is currently the chief source of hydrogen. It can be produced via an endothermic reaction with steam.

$$165 \text{ kJ} + CH_4(g) + 2 \, H_2O(g) \longrightarrow 4 \, H_2(g) + CO_2(g) \qquad [8.22]$$

Methane also is produced in landfills, and a new process tested in Saskatchewan, Canada, exploits this source of methane. The hydrogen can be produced via a reaction with carbon dioxide.

$$247 \text{ kJ} + CO_2(g) + CH_4(g) \longrightarrow 2 \, H_2(g) + 2 \, CO(g) \qquad [8.23]$$

You can see why this reaction has not received much attention. It is highly endothermic. However, the Solar Hydrogen Energy Corporation has developed a solar mirror array that can focus the Sun's energy to heat the reactants, CO_2 and CH_4. Not only can this technology produce hydrogen, but also it can do so from a waste material found in a landfill.

Your Turn 8.17   Back to Bond Energies

a. Use the average bond energy values in Table 4.2 to check the energy required by the reaction in equations 8.22 and 8.23. Show your work clearly.
b. Are the reactions endothermic or exothermic?
c. Did your calculated value match the values given in the equation? Explain.

Still, each of the reactions just described contains a major flaw; carbon dioxide is produced. Is there another source of hydrogen around? In Jules Verne's 1874 novel, *Mysterious Island,* a shipwrecked engineer speculates about the energy resource that will be used when the world's coal supply has been used up. "Water," the engineer declares, "I believe

that water will one day be employed as fuel, that hydrogen and oxygen which constitute it, used singly or together, will furnish an inexhaustible source of heat and light."

Is this simply science fiction, or is it energetically and economically feasible to break water into its component elements? Can it be done economically? And what does this have to do with advanced vehicles and light from the Sun? To answer these questions and assess the credibility of the claim by Verne's engineer, a Sceptical Chymist needs to examine the energy cost of this reaction.

$$2 H_2(g) + O_2(g) \longrightarrow 2 H_2O(l) + 572 \text{ kJ} \qquad [8.24]$$

Experiment shows that the reaction, as written, gives off 572 kJ when 2 mol of liquid water is formed from the combination of 2 mol of hydrogen and 1 mol of oxygen. Burning of 1 mol of H_2 will yield half of this, or 1 mol of H_2O and $1/2$ (572) kJ of energy. Here is the chemical equation.

$$H_2(g) + 1/2 O_2(g) \longrightarrow H_2O(l) + 286 \text{ kJ} \qquad [8.25]$$

Sceptical Chymist 8.18   Checking Bond Energies

Use the bond energy values in Table 4.2 to check the energy released by the reaction in equation 8.25. Are the values the same? Explain. To be convincing, clearly show your reasoning.

Because energy is *evolved,* the energy change in this combustion reaction is −286 kJ for each mole of H_2 burned to form liquid water. This is equivalent to releasing 143 kJ per gram of H. In comparison, the heat of combustion of coal is 30 kJ/g, octane (a major component in gasoline) is 46 kJ/g, and methane (natural gas) is 54 kJ/g when the products of combustion are $CO_2(g)$ and $H_2O(l)$. Clearly, hydrogen has the capability of being a powerful energy source. In fact, on a per-gram basis, hydrogen has the highest heat of combustion of any known substance. The extraordinary energy per gram of hydrogen when it burns raises a tantalizing prospect—the practical use of hydrogen as a fuel to power motor vehicles that would produce only water vapor. No pollutants such as the carbon monoxide and nitrogen oxides would form, unlike what happens when burning fossil fuels.

Your Turn 8.19   Back to Basics

Calculate the number of moles and grams of H_2 that would have to be burned to yield an American's daily energy share of 260,000 kcal.
Hint: 1 kcal = 4.18 kJ

Answer
3800 mol, 7600 g

Because the formation of 1 mol of water from hydrogen and oxygen releases 286 kJ (see equation 8.25), an identical quantity of energy must be absorbed to reverse the reaction to produce hydrogen. Figure 8.14 summarizes the processes.

To bring about this reaction, all that is needed is a source of 286 kJ. The most convenient method of decomposing water into hydrogen and oxygen is by **electrolysis, the process of passing a direct current of electricity of sufficient voltage through water to decompose it into H_2 and O_2** (Figure 8.15). When water is electrolyzed, the volume of hydrogen generated is twice that of the oxygen. This suggests that a water molecule contains twice as many hydrogen atoms as oxygen atoms, testimony to the formula H_2O.

$$286 \text{ kJ} + H_2O(l) \longrightarrow H_2(g) + 1/2 O_2(g) \qquad [8.26]$$

Of course, the question remains: How will the electricity for large-scale electrolysis be generated? Most electricity in the United States is produced by burning fossil fuels in

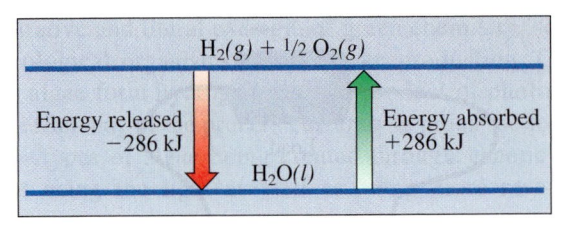

Figure 8.14

Energy differences in the hydrogen–oxygen–water system.

conventional power plants. If we only had to deal with the first law of thermodynamics, the best we could possibly achieve would be to burn an amount of fossil fuel equal in energy content to the hydrogen produced in electrolysis. But we must also deal with the consequences of the second law of thermodynamics. Because of the inherent and inescapable inefficiency associated with transforming heat into work, the maximum possible efficiency of an electrical power plant is 63%. When we add the additional energy losses caused by friction, incomplete heat transfer, and transmission over power lines, it would require at least twice as much energy to produce the hydrogen than we could obtain from its combustion. This is comparable to buying eggs for 10 cents each and selling them for 5, which is no way to do business.

<aside>See Section 4.2 for a discussion of the second law of thermodynamics.</aside>

Considerations such as these lead some observers to say that the pollution from producing hydrogen could offset the benefits. Furthermore, most methods of generating electricity have a variety of negative environmental effects. At one time it was thought that the "cheap" extra electricity from nuclear fission could be used to produce hydrogen to fuel the economy, but that energy utopia has hardly been realized. It should be apparent, therefore, that electricity generated from fossil fuels or nuclear fission is not a currently practical method to split water to produce hydrogen for use as a fuel.

A second possibility would be to use heat energy to decompose water. Simply heating water to decompose it thermally into H_2 and O_2 is not commercially promising. To obtain reasonable yields of hydrogen and oxygen, temperatures of over 5000 °C would be required. The attainment of such temperatures is not only extremely difficult, but also it would consume enormous amounts of energy—at least as much as would be released when the hydrogen was burned. Thus, we have again reached a point where we would be investing a great deal of time, effort, money, and energy to generate a quantity of hydrogen that would, at best, return only as much energy as we invested. In practice, a good deal less energy would result.

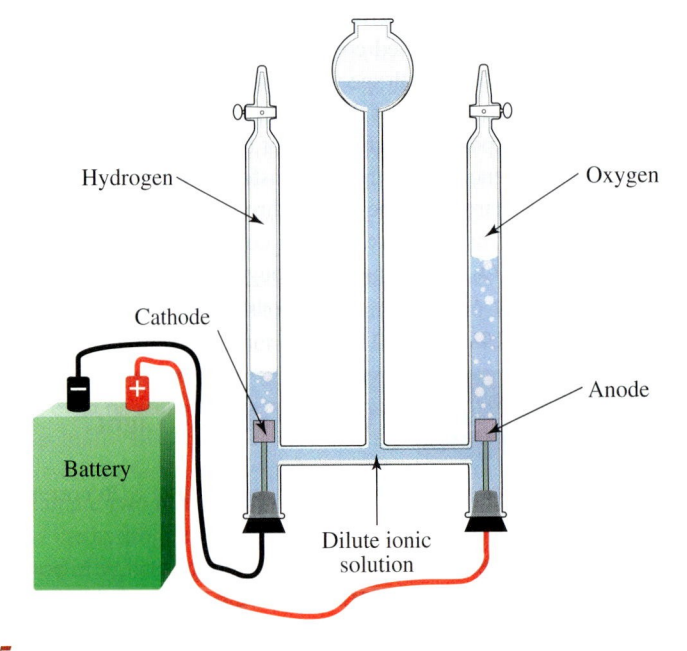

Figure 8.15

Electrolysis of water.

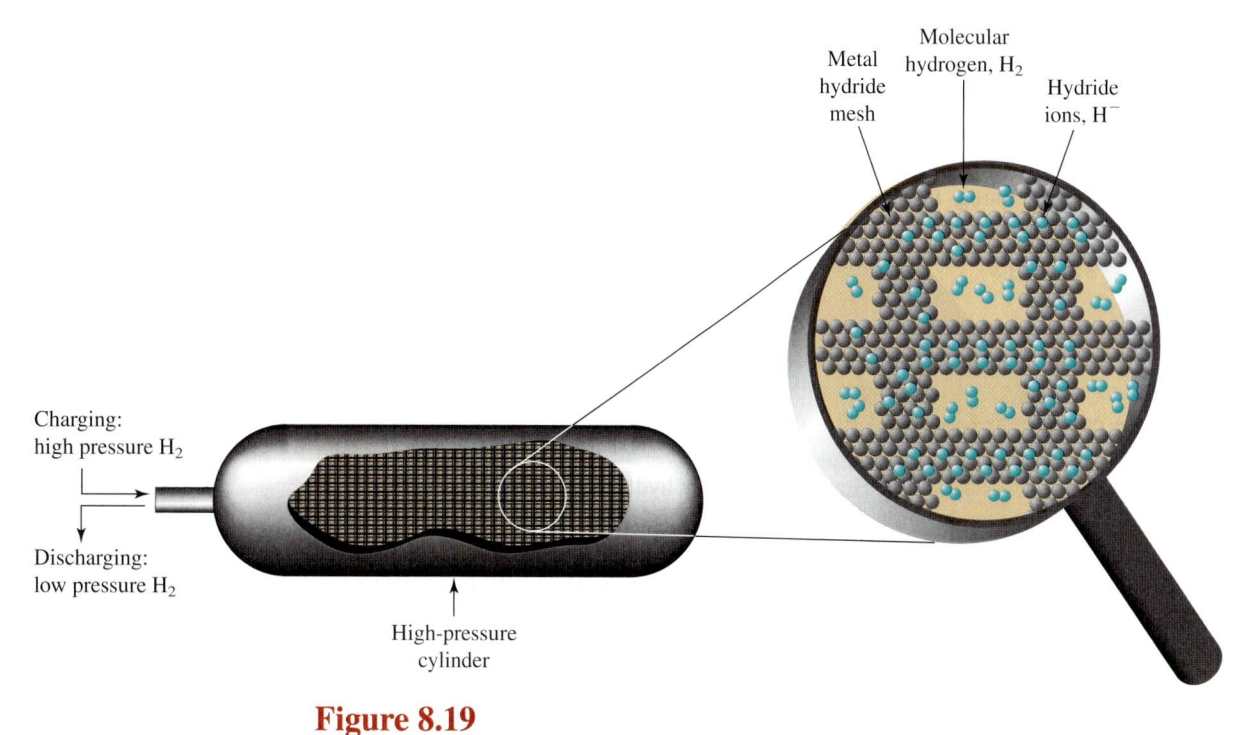

Figure 8.19

Absorption and release of hydrogen from a metal hydride.

Your Turn 8.21   Active Metal Hydrides

a. Write chemical reactions for the charging and discharging of metal hydrides shown in Figure 8.19. Use M as the symbol for a generic metal.

b. Compare the role of the metals in NiMH batteries (see Consider This 8.9) with those used for hydrogen storage. List some similarities and differences.

Clearly, such approaches would greatly improve the safety and convenience of handling H_2 and perhaps would be the decisive factor in determining the extent of its acceptance as a fuel. As we discussed earlier in this chapter, a hydrogen-fueled car would produce only water vapor and none of the carbon monoxide or nitrogen oxides emitted from using gasoline-fueled internal combustion engines.

Even if we were to manage to solve the production, storage, and transport problems just identified, we must consider how best to extract energy from our hydrogen. The most obvious way would be to burn it in power plants, vehicles, and homes. A stream of pure hydrogen burns smoothly, quietly, and safely in air, delivering 143 kJ per gram and forming only nonpolluting water as the end product. But when hydrogen is mixed directly with oxygen, a spark can be sufficient to produce a devastating explosion, limiting the usefulness of hydrogen as a direct fuel. However, the promise of using hydrogen in fuel cells is creating renewed interest in the hydrogen economy.

In his State of the Union address in 2003, President George W. Bush pledged $1.2 billion for a new Hydrogen Fuel Initiative. Since that announcement, over $900 million has been allocated, with an additional $300 million requested in the 2008 budget. The major goal is to resolve many of the technical and economic barriers to widespread use of hydrogen-powered vehicles by 2020.

Iceland's Hydrogen Economy

The small country of Iceland is taking bold steps to cut its ties to fossil fuels by 2050. Part of the plan, first announced in 1999, is to demonstrate that the country can produce, store, and distribute hydrogen to power both public and private transportation.

a. Name three factors that motivate Iceland to cut its ties to fossil fuels.
b. What tangible outcomes have resulted to date?
c. Do you think the lessons learned in Iceland will be relevant for the United States? Explain.

8.9 Photovoltaics: The Basics

All life on Earth depends on the Sun's energy. It warms the planet, making life possible. Plants harness that radiant energy, using it to convert carbon dioxide to carbohydrates and oxygen. Indirect conversion of solar energy into electricity is not new. Looking back through time, the fossil fuels we burn today once depended on solar energy that enabled plant material to grow. Now we burn those fuels for energy, much of it in the form of electricity. Electricity supplies about 35% of U.S. energy needs, and two thirds of that electricity is used in residential and commercial buildings.

Every hour of every day, enough energy from the Sun reaches Earth to meet the world's energy demand for an entire year. But, currently, less than 0.5% of the power generated in the United States comes directly from the Sun. The solar alternative seems like a utopian solution to the tremendous energy demands of the 21st century. So why does solar energy account for less than 1% of U.S. energy production? Although prodigious amounts of sunshine hit the Earth daily, the rays do not strike any specific spot on our planet for 24 hours a day, 365 days a year. A gallon of gasoline is a lot of energy in a relatively small volume, whereas sunlight hits the Earth everywhere and is very difficult to concentrate. The challenge then, is to convert solar radiation *directly* into electricity in a practical and economically feasible way.

Photovoltaic cells (solar cells) convert radiant energy directly to electrical energy, without the intermediary of hydrogen or some other fuel. The photovoltaic (PV) technology industry considers research results on photovoltaic cells and develops practical energy sources that help to meet our energy needs. It takes only a few PV cells to produce enough electricity for your calculator or digital watch. If more power is required, PV cells can be connected together to form arrays capable of generating electricity for large-scale use. PV devices have already demonstrated their practical utility for satellites, highway signs, security and safety lighting (Figure 8.20), automobile recharging stations, and navigational buoys, just to mention a few common uses. Cost savings can be substantial. For example, using solar cells rather than batteries in navigational buoys saves the U.S. Coast Guard an estimated $6 million annually through reduced maintenance and repair.

PV cells are made from a class of materials called **semiconductors,** materials that do not normally conduct electricity well, but can do so under certain conditions, such as exposure to sunlight. Semiconductors are solids with structures closely resembling metals. We know that the structure of metals, such as copper or aluminum, enables them to be good conductors of electricity. **Metallic bonding** can be described with the "electron sea" model, in which outermost (valence) electrons are shared among all the atoms in the substance. There is a highly regular array of positively charged nuclei, surrounded by a "sea" of electrons. The outermost electron on each metal atom is loosely attracted to its nucleus. When a large number of atoms form a tightly packed array, the valence electrons can act like a liquid that is spread over all the nuclei. The solid is held together by the mutual attraction of the metal cations for the mobile, highly delocalized electrons.

It is easy to induce electrons to move in this metallic electron sea, since the bonding of an electron to any single individual nucleus is relatively weak. The motion of electrons is what constitutes a current, what we think of as conducting electricity. In semiconductors, the bonding is very similar to that in metals. The nuclei are arranged in regular

Figure 8.20
Photovoltaic (solar) cells are used to improve security, enhance safety, and direct pedestrians and vehicles.

Conductivity in aqueous ionic solution was discussed in Section 5.7.

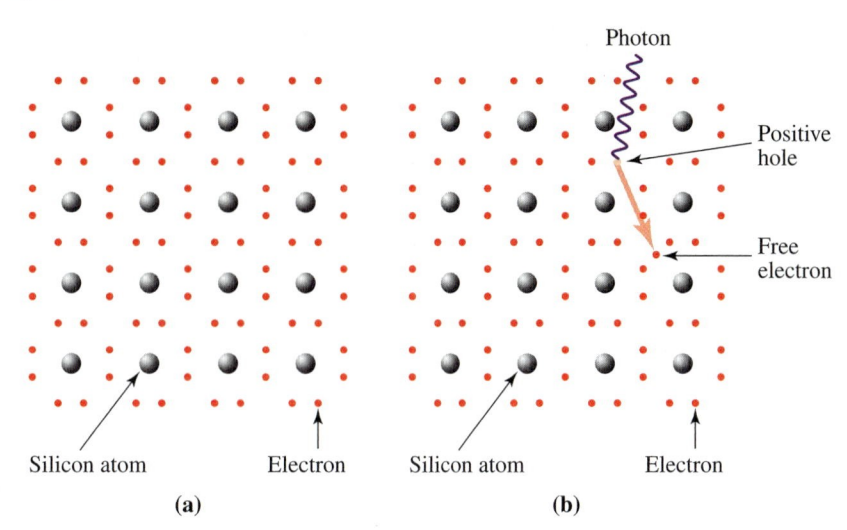

Figure 8.21
(a) Schematic of bonding in silicon.
(b) Photon-induced release of a bonding electron in a silicon semiconductor.

arrays. However, the electrons around each nucleus have a stronger attraction to the nuclei that they surround. Therefore, before the electrons are free to move around and conduct electricity, energy must be added, in the form of heat or light.

The element silicon was one of the first semiconducting materials developed for use in PV cells. A crystal of silicon consists of an array of silicon atoms, each bonded to four other atoms by means of shared pairs of electrons (Figure 8.21a). These shared electrons are normally fixed in the bonds and unable to move through the crystal. Consequently, silicon is not a very good electrical conductor under ordinary circumstances. However, if a valence electron absorbs sufficient energy, it can be excited and released from its bonding position (Figure 8.21b). Once freed, the electron can move throughout the crystal lattice, making the silicon an electrical conductor.

For a PV cell to generate electricity, light must induce such a movement of electrons in the cell. This electron movement depends on the interaction of matter and photons of radiant energy—a topic already treated in considerable detail in Chapters 2 and 3. In those chapters, we pointed out that the portion of the Sun's radiation reaching Earth's surface is mainly in the visible and infrared regions of the spectrum, with a maximum intensity near a wavelength of 500 nm. Light of this wavelength is in the visible range and has energy of about 3.6×10^{-19} J per photon, corresponding to 220 kJ per mole of photons. The energy required for silicon to release an electron from a bond is 1.8×10^{-19} J per photon, which is equivalent to radiation with a wavelength of 1100 nm. Visible light has a wavelength range of 400–700 nm. Recall that the shorter the wavelength of radiation, the greater the energy per photon. Therefore, photons of visible sunlight have more than enough energy to excite electrons in silicon semiconductors.

Consider This 8.23   PV Efficiency

a. What region of the electromagnetic spectrum corresponds to the 1100 nm light required to release an electron in silicon?
b. What do you think happens to the *excess* energy when shorter wavelength light strikes the silicon?
c. Comment on the effect this would have on the efficiency of such a photovoltaic device.

In order to make PV cells that can capture and store energy from sunlight, the electric currents generated must have certain predictable and controllable properties. Therefore, it became necessary to fabricate the PV cells not from a single pure substance, but from

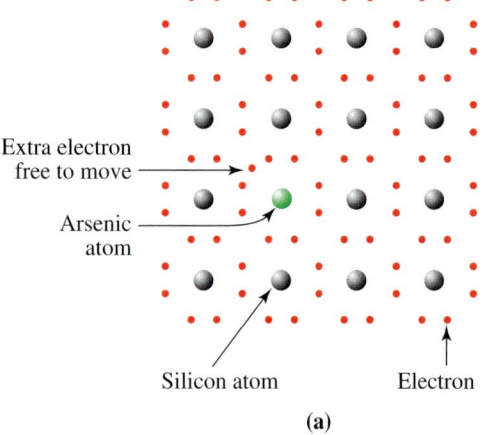

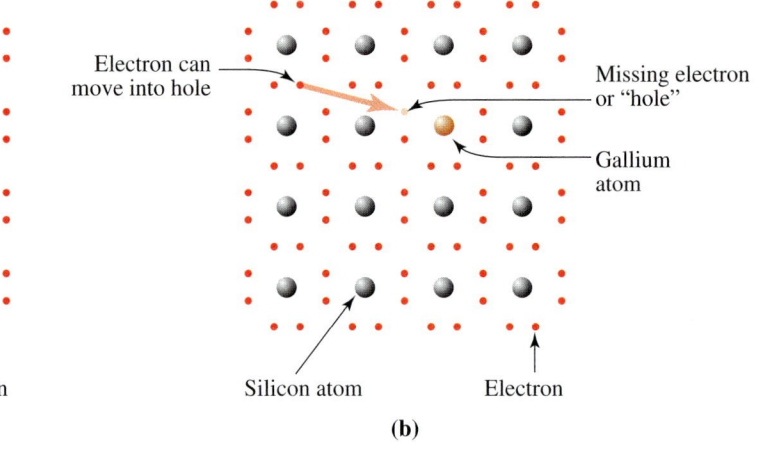

(a) (b)

Figure 8.22

(a) An arsenic-doped *n*-type silicon semiconductor.
(b) A gallium-doped *p*-type silicon semiconductor.

a combination of materials with different characteristics. A very common method of adjusting the properties of a pure semiconductor material is through **"doping,"** a process of intentionally adding small amounts of other elements to pure silicon. The doping materials are chosen for their ability to facilitate the transfer of electrons. For example, about 1 ppm of gallium (Ga) or arsenic (As) is often introduced into the silicon. These two elements and others from the same periodic groups are used because their atoms differ from silicon by a single outer electron. Silicon has four electrons in its outer energy level, gallium has three, and arsenic has five. Thus, when an atom of As is introduced in place of Si in the silicon lattice, an extra electron is added. The replacement of a Si atom with a Ga atom means that the crystal is now one electron "short."

Ga is in Group 3A.
Si is in Group 4A.
As is in Group 5A.

The extra electrons in arsenic-doped silicon are not confined to bonds between atoms. Rather, electrons move easily through the lattice, increasing the electrical conductivity of the material over that of pure silicon. Silicon doped in this manner is called an ***n*-type semiconductor** in which there are freely moving negative charges, the electrons. On the other hand, for each silicon atom replaced with a gallium ion, an electronic vacancy, or "hole," is introduced into what is normally a two-electron bond. When an electron moves into this vacancy, a new hole appears where the mobile electron formerly was located and the positive charge has developed in a new location. Gallium-doped silicon is therefore a ***p*-type semiconductor** in which there are freely moving positive charges, or holes. Negatively charged electrons and positively charged holes move in opposite directions. Figure 8.22 illustrates both *n*- and *p*-type semiconductors. Both types of doping increase the conductivity of the silicon because less energy is needed to get extra electrons or holes moving. This means that photons of lower energy (longer wavelength) can induce electron release and transport in doped crystals.

Your Turn 8.24   Other Doping Materials

Some solar cell designs use phosphorus and boron to dope silicon crystals.

a. Which will form an *n*-type semiconductor? Explain your reasoning.
b. Which will form a *p*-type semiconductor? Explain your reasoning.

"Sandwiches" of *n*- and *p*-type semiconductors are used in transistors and other miniaturized electronic devices that have revolutionized communication and computing. Similar sandwich structures are central to the direct conversion of sunlight to electricity. A photovoltaic cell typically includes sheets of *n*- and *p*-type silicon in close contact (Figure 8.23). The *n*-type semiconductor is rich in electrons and the *p*-type is rich in posi-

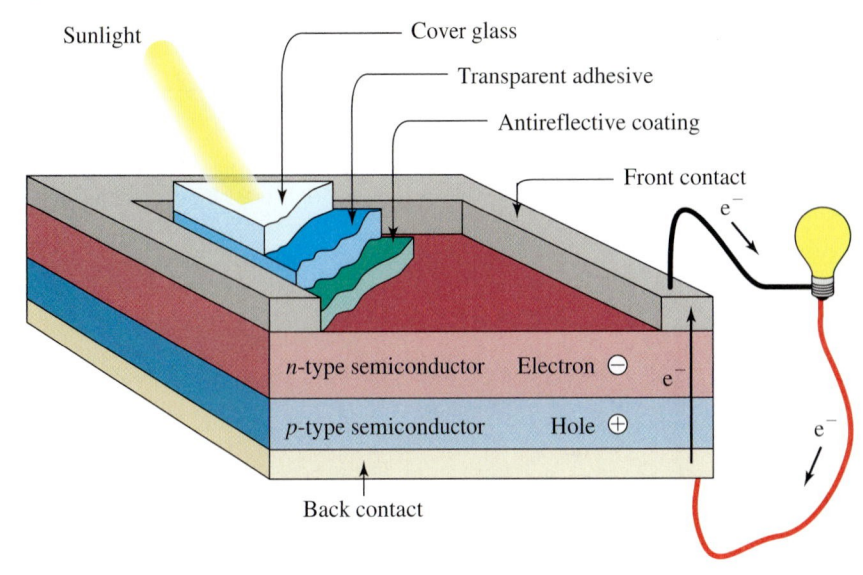

Figure 8.23

Schematic diagram of a photovoltaic (solar) cell.

tive holes. When they are placed in contact, electrons tend to diffuse from the *n*-region into the *p*-region. Likewise, the positive holes tend to be displaced from the *p*-region to the *n*-region. This generates a voltage, or potential difference, at the junction between the semiconductors. The voltage difference accelerates the electrons released when sunlight strikes the doped silicon. If a conducting wire connects the two layers, electrons flow through the external circuit from the *n*-semiconductor, where their concentration is higher, to the *p*-semiconductor, where it is lower.

Not only does the use of a *p-n* junction facilitate the conduction of electricity, but also it ensures that the current will flow in a specific direction through the PV cell. The transfer of electrons generates a direct current of electricity that can be intercepted to do essentially all the things electricity does, including being stored in batteries for later use. As long as the cell is exposed to light, the current will continue to flow, powered only by solar energy.

The fabrication of photovoltaic cells poses some significant challenges. The first is that although silicon is the second most abundant element in Earth's crust, it is found combined as silicon dioxide, SiO_2. You know this material by its common name, sand, or more correctly as quartz sand. The good news is that the starting material from which silicon is extracted is cheap and abundant. The not so good news is that processes to extract and purify silicon are expensive. Many of the early PV cell designs require ultrapure 99.999% silicon (Figure 8.24).

A second challenge is that the direct conversion of sunlight into electricity is not very efficient. A photovoltaic cell could, in principle, transform up to 31% of the radiant energy to which it is sensitive into electricity. However, some of the radiant energy is reflected by the cell or absorbed to produce heat instead of an electric current. Typically, a commercial solar cell now has an efficiency of only 15%, but even this is a significant increase from the first solar cells built in the 1950s, which had efficiencies of less than 4%. In Chapter 4, we lamented the 63% maximum efficiency of converting heat to work in a conventional power plant. It might seem that we should be even more distressed at the lower limits that can be achieved by photovoltaics. Remember, however, that the first use of solar cells was to provide electricity in NASA spacecraft. For that application, the intensity of radiation was so high that low efficiency was not a serious limitation and costs were not of paramount concern. For commercial use on Earth, costs and efficiency are issues. Our Sun is an essentially unlimited energy source, and converting it to electricity even inefficiently is free from many of the environmental problems associated with burning fossil fuels or with storage of spent fuel from nuclear fission. These considerations add impetus to research and development of solar cells.

In addition to making an effort to improve the performance of silicon semiconductors by doping, scientists have been searching for other substances that exhibit the

An individual solar cell produces at most 0.5 V.

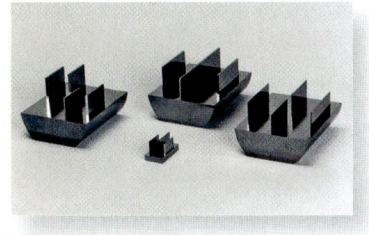

Figure 8.24

Ultrapure silicon is sliced into thin wafers to make solar cells.

same or better semiconductor properties. Promising substitutes include germanium, an element found in the same group of the periodic table as silicon, and compounds in which the elements have the same number of outer electrons as Sn or Ge. Included in this latter list are gallium arsenide (GaAs), indium arsenide (InAs), cadmium selenide (CdSe), and cadmium telluride (CdTe). There are also some very new combinations of indium, gallium, and nitrogen. Some of these new semiconductors have enhanced the efficiency of sunlight-to-electricity conversion and made possible photovoltaic cells that are responsive to wider regions of the spectrum. Siemens Solar Industries has developed a copper indium selenide semiconducting thin film that has significant advantages over amorphous silicon and cadmium telluride.

Your Turn 8.25	**Two-Element Semiconductors**

Using the periodic table as a guide, show why the electron arrangement and bonding in GaAs and CdSe would be very similar to that in pure Ge.

Replacing crystalline silicon with the noncrystalline form of the element is another approach to commercial viability. Photons are more efficiently absorbed by less highly ordered Si atoms, a phenomenon that permits reducing the thickness of the silicon semiconductor to 1/60th or more of its former value. The cost of materials is thus significantly reduced.

Other researchers are developing multilayer solar cells. By alternating thin layers of *p*-type and *n*-type doped silicon, each electron has only a short distance to travel to reach the next *p-n* junction. This lowers the internal resistance within the cell and raises the efficiency of the cells at an increasingly competitive price. Maximum theoretically predicted efficiencies could improve to 50% for 2 junctions, to 56% for 3 junctions, and to 72% with 36 junctions. As of 2007, the maximum efficiency actually demonstrated with a multijunction solar cell was 40.7%. Figure 8.25 gives a sense of just how thin these layers actually are. Multilayer technology, compared with single-cell technology, uses smaller quantities of silicon and the production process can become highly automated.

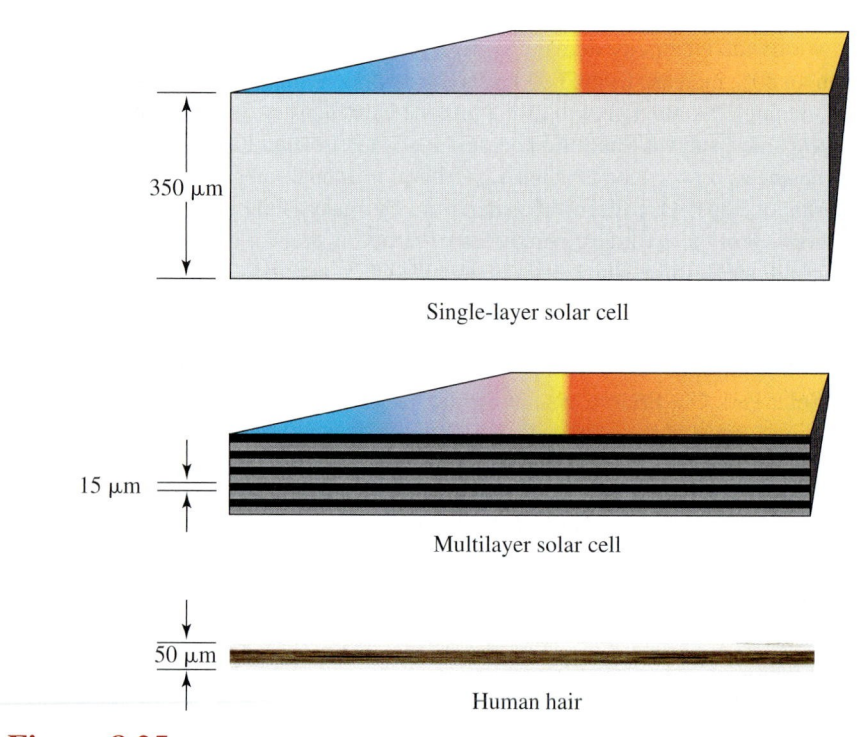

350 μm

Single-layer solar cell

15 μm

Multilayer solar cell

50 μm

Human hair

Figure 8.25

A comparison of the relative thickness of a solar cell layer, either in a single or multilayer cell, to the diameter of an average human hair. Note: 1 μm = 10^{-6} m.

8.10　Photovoltaics: Plugging into the Sun

At currently attainable levels of operating efficiency, all the electricity needs of the United States could be supplied by a photovoltaic generating station covering an area of 85 miles by 85 miles, roughly the area of New Jersey. Long-range prospects for photovoltaic solar energy are encouraging. Its cost is decreasing while the cost of electricity generated from fossil fuels is increasing. Limited and uncertain supply and the expense of pollution controls are driving the cost of fossil-fuel electricity still higher. Given this situation and the continued improvements in the performance and decreases in the cost of solar cells, electricity from photovoltaic systems could become even more competitive with that from fossil fuels early in the 21st century.

Sceptical Chymist 8.26　Checking Geographic Claims

The text claims that a solar farm the size of New Jersey would be adequate to supply the electricity needs of the entire United States. Are you surprised by this claim? Did you expect the required area to be larger or smaller? What assumptions are implicit in this statement? Do you think they are valid?

Because of advances in photovoltaic technology and economies of scale, the cost of producing electricity in this manner has declined by about 4% per year in the last 15 years. As a result, electricity can now be generated for somewhere between $0.22 and $0.40 per kilowatt-hour. If government incentive programs are in place, as was true for the nuclear industry during its formative years, the cost can drop to $0.10–0.12 per kWh. By comparison, the average U.S. residential price of electricity is $0.085 per kWh.

The decrease in cost of PV electricity has been accompanied by a dramatic increase in usage. Although photovoltaic power has grown at an average rate of 16% per year since 1990, it still represents a minute fraction of global power supplies. At the end of 2005, there were over 1090 kW of installed PV capacity for those countries reporting to the International Energy Agency (Figure 8.26). According to the U.S. Department of Energy, worldwide electricity demand is expected to jump to 26 trillion kWh by 2025, more than an 80% increase from 2005 levels. Many believe that the finite supply of fossil fuels will dictate continued growth of the photovoltaic industry. As a result, most major energy companies, such as Shell, Amoco, and British Petroleum have invested heavily in the solar business, now valued at about $2.7 billion in annual sales worldwide.

In terms of large-scale electrical generation, Germany is the global leader and currently operates 8 of the 10 largest photovoltaic "farms" in the world. The biggest of these was the Solar Park Bavaria, which utilizes over 57,000 solar panels and occupies a total of 67 acres. At peak capacity, it can generate 10 MW, enough to power about 2000 homes in the surrounding area. The replacement of fossil-fuel combustion will decrease emissions of CO_2 by 6200 tons per year over the Solar Park's 30-year lifetime. An even larger facility is the Solar Park Gut Erlasee, also in Bavaria (Figure 8.27), which was inaugurated in 2006. A unique aspect of the solar cells in both facilities is their ability to tilt to follow the Sun's path across the sky, thereby increasing their overall efficiency.

Germany may not boast the largest solar farm for long. In April 2007, construction began on an 18-MW solar plant outside Las Vegas. When completed, the plant will provide Nellis Air Force Base, and the 12,000 military and civilian personnel who live there, with over 25% of their electricity.

Because of the diffuse nature of sunlight, photovoltaic technology is better suited to distributed generation (Section 8.6). Photovoltaics produce needed electricity from sunlight and solar thermal systems produce heat that can be used for domestic water and space heating, or even to heat swimming pools. Although solar energy is both clean and plentiful, only about 500,000 homeowners worldwide use solar cells to generate their own electricity. Due to the high initial costs, various local and federal government incentive programs have been created to encourage photovoltaic use. For example, in 1989, the Sacramento

A typical nuclear power plant generates 1000 MW of electricity.

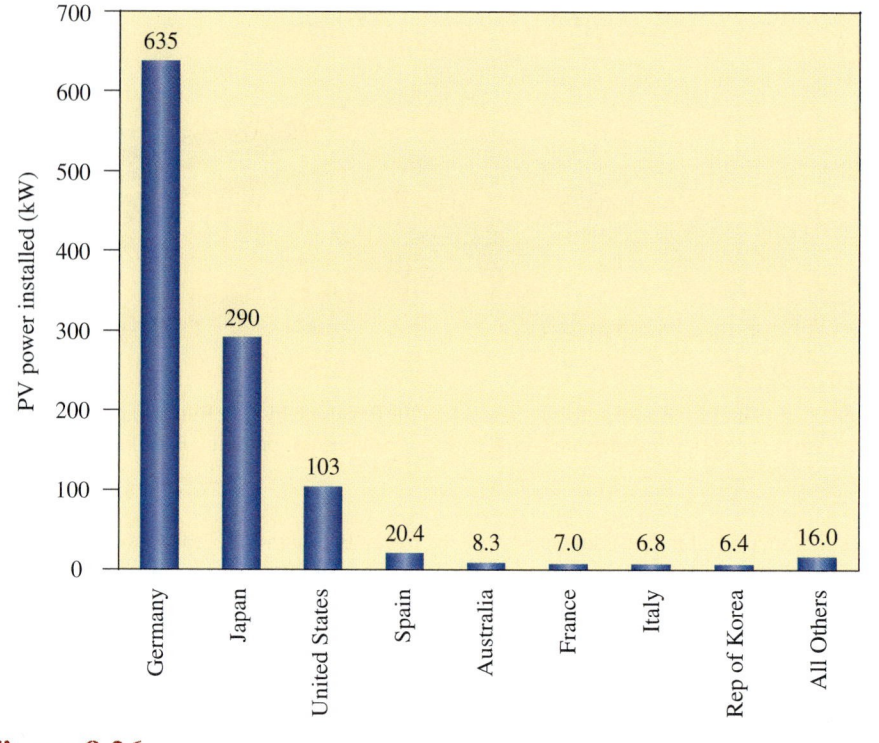

Figure 8.26

Installed PV power around the world, 2005.

Source: International Energy Agency.

Municipal Utility District voted to close its nuclear reactors in favor of using photovoltaics and other "cleaner" energy technologies. They now offer cash rebates to those homeowners and businesses that install PV systems for power generation. In a touch of irony, the small photovoltaic plant is adjacent to the now-closed nuclear power plant.

The Million Solar Roofs (MSR) program was designed to facilitate deployment of a million rooftop photovoltaic systems in the United States by 2010. The U.S. program started in 1997, and was folded into the Solar America Initiative (SAI) in 2007. The $148 million program establishes a two-pronged approach. Funding for research and development of new PV systems as well as thermal energy concentrators is the main technological

Figure 8.27

An aerial view of the Solar Park Gut Erlasee in Bavaria. At peak capacity, it can generate 12 MW.

Figure 8.28

Two model solar Habitat for Humanity homes built in Tennessee by Oak Ridge National Laboratories (ORNL) in partnership with the Million Solar Roofs program. Each has a 2-kW solar electric system.

thrust. A second element addresses ways to lower the economic barriers impeding the expansion of the PV market. Both programs use tax credits in combination with private, local, and state partnerships to help meet the goal. Habitat for Humanity (Figure 8.28) has been one of the most active partners, helping to promote energy-efficient homes and increased use of solar energy in many different regions of the country.

Such a transformation in energy generation and delivery is not without some technological obstacles. To make such a distributed generation network feasible (whether with fuel cells or photovoltaics), the transmission lines will have to move power in both directions: to the users when demand is high, but also from users producing surplus power when their demand is low.

Consider This 8.27   A Million and One Solar Roofs

Use the Web to learn more about recent solar energy projects. Then propose a solar energy project in a community of your choice. List at least five factors that are important to consider before proceeding with the project.

More than a third of Earth's population is not hooked into an electrical network because of the costs associated with constructing and maintaining equipment, and supplying the fuel to generate the electricity. Because photovoltaic installations are essentially maintenance-free and can be used almost anywhere, they are particularly attractive for electrical generation in remote regions and are already affecting the lives of millions of people across the globe (Figure 8.29). For example, the highway traffic lights in certain parts of Alaska, far from power lines, operate on solar energy. A similar, but more significant application of photovoltaic cells may be to bring electricity to isolated villages in developing countries. In recent years, more than 200,000 solar lighting units have been installed in residential units in Colombia, the Dominican Republic, Mexico, Sri Lanka, South Africa, China, and India.

An exemplary application of the use of photovoltaic technology is in Indonesia, an archipelago of more than 13,000 islands. About 70% of all households there do not have access to electrical lines. Therefore, installing photovoltaic cells is an attractive alternative. In the village of Leback, solar electric units have been installed in 500 homes. Electricity has also been supplied to public buildings, shops, streetlights, and a satellite antenna system for the 11 public television units. Before the photovoltaic systems, Leback villagers used kerosene for lighting and batteries for radios. Kerosene costs $6–12 per month, depending on availability. Under a loan–purchase agreement, villagers pay $4–5 per month for their home solar electric system.

The transfer of electrons in solar cells can even power vehicles. A series of long-distance races, originally called Sunrayce, has been held since 1990. Student teams design, build, test, and drive cars powered by photovoltaic cells and battery packs. Now referred to as the North American Solar Challenge, the 2005 race was an 11-day, 2500-mile solar car race sponsored by the U.S. Department of Energy (DOE), its National Renewable Energy

Figure 8.29

Photovoltaics can power water pumps in remote areas of the world where there is no access to electricity.

Figure 8.30

The University of Michigan solar car Sunrunner crosses the finish line amid about 10,000 spectators at the 2005 North American Solar Challenge.

Laboratory (NREL), Natural Resources Canada, and several private corporations. The 2005 race was distinguished by being the longest (2500 miles) and by being the first international race in the series. It was won by the University of Michigan solar car, setting a record by averaging a speed of 46.2 mph in covering the distance from Austin, Texas, to Calgary in Alberta, Canada in just under 54 hours (Figure 8.30). Because of their avant-garde designs, the cars are not ready to be put on the roads for everyday use. In fact, one solar car was ticketed during testing procedures for failing to have proper head and tail-lights and failing to have a current license plate displayed!

> North Amercian Solar Challenge races are scheduled every 2 to 3 years.

Consider This 8.28   **Solar Cars Near You**

Check to see if your college or a college near you has participated in building and racing a solar car. If so, interview one or more of the students involved and write up a brief press statement about their efforts. If not, what are the barriers to participation?

Although prodigious amounts of sunshine hit the Earth daily, the rays do not strike any specific spot on our planet for 24 hours a day, 365 days a year. This means that the electricity generated by photovoltaic cells during the day must be stored using batteries for use at night. Nevertheless, the direct conversion of sunlight to electricity has many advantages. In addition to relieving some of our dependence on fossil fuels, an economy based on solar electricity would reduce the environmental damage of extracting and transporting these fuels. Furthermore, it would help to lower the levels of air pollutants such as sulfur oxides and nitrogen oxides. It would also help avert the dangers of global warming by decreasing the amount of carbon dioxide released into the atmosphere. Fossil fuels will certainly remain the preferred form of energy for certain applications. On balance, however, the future looks sunny for solar-based energy.

In the long run, sustainable energy will likely come from a variety of sources within an integrated system. These include photovoltaics and hydrogen, as well as other renewables such as wind, tidal, and geothermal energy. Using a variety of sources to generate electricity and a combination of collectors, electricity can be passed to the electrical grid or to an electrolysis device for hydrogen generation. Systems are under development, but there are many challenges before they are in widespread use.

Conclusion

We look to many different forms of electron transfer for our energy needs, and the future promises more developments. Batteries can store chemical energy and convert it to a flow of electrons that is useful for thousands of applications. Hybrid vehicles use new battery technology with internal combustion engines to improve fuel efficiency. Fuel cells are one of the most efficient new strategies for energy transformation and may become a major energy source for future personal power use, transportation, and perhaps even large-scale electricity production. Advances in research and changes in global economies may make it fiscally and energetically feasible to use solar radiation to extract hydrogen from water or some other hydrogen source. The hydrogen can either be burned directly as a clean fuel or combined with oxygen in a fuel cell that generates electricity rather than heat. Photovoltaic cells convert sunlight directly into electricity. There is no need for intermediate steps in which heat energy is transformed first into mechanical and then into electrical energy, with accompanying power losses along the way.

We hope that the discussions of energy in this book have begun to give you some background and perspective on the depth and importance of this issue. Now, you are in a position to take overall stock of the situation and maybe look ahead to the future. A few facts seem beyond doubt. The world's thirst for energy will not abate; it will almost assuredly continue to increase. In his 2006 State of the Union address, President George W. Bush proclaimed "... we are addicted to oil," which came as no surprise. Moreover, that addiction most definitely is not sustainable. The coal, petroleum, and natural gas from which we derive the vast majority of our power are not renewable resources and are destined to become scarce in the not-too-distant future. Nuclear power, though not directly responsible for emission of greenhouse and acid-rain-causing gases, is not without risks and seems unlikely at present to shoulder much more of the energy load that it does currently. A transformation is required.

But the laws of thermodynamics and human nature are such that these transformations will not occur spontaneously. Energy alternatives cannot be developed without hard work and the investment of intellect, time, and money. Yet, in the United States, the amount of effort and money devoted to research on new energy sources sometimes appears to be directly proportional to the cost of oil. When international crises or natural disasters drive up the price of petroleum or when regional energy shortages occur, there is a sudden flurry of official interest in energy conservation and the development of alternative technologies. When oil supplies are plentiful and prices are low at the gasoline pump, few seem to care about preparing for the time when fossil fuels will be depleted or they become much too polluting or too expensive to burn. As recent years have clearly shown, the cycles of cheap and expensive oil may be gone for good. Paraphrasing the words of Winston Churchill, this is not the end of oil, but maybe the beginning of the end. The sacrifices and compromises that "kicking our oil habit" will require, as a nation and a world, depend heavily on the choices we make today. We need to establish national and personal priorities and act on them. We have been the beneficiaries of a bountiful nature, but in turn, we have an obligation to ensure energy sources for generations yet to come.

Chapter Summary

Having studied this chapter, you should be able to:

- Discuss the principles governing the transfer of electrons in galvanic cells, including the processes of oxidation and reduction (8.1)
- Describe the design, operation, applications, and advantages of several different types of batteries (8.1–8.3)
- Compare and contrast the principles, advantages, and challenges of producing and using hybrid vehicles (8.4)
- Describe the design, operation, applications, and advantages of typical fuel cells (8.5–8.6)
- Understand why one type of fuel cell does not meet all needs (8.6)
- Explain the energy costs and gains of producing hydrogen and using it as a fuel (8.7)
- Discuss issues related to developing a hydrogen economy (8.8)
- Describe the principles governing the operation of photovoltaic (solar) cells and their current and future uses (8.9)
- Express informed opinions about the future development of all types of electron transfer technology for producing electrical energy on personal, regional, national, and global scales (8.1–8.9)

Questions

Emphasizing Essentials

1. **a.** Define the terms *oxidation* and *reduction*.

   b. Why must these processes take place together?

2. Which of these half-reactions represent oxidation and which reduction? Explain your reasoning.

   a. $Fe \longrightarrow Fe^{2+} + 2\,e^-$

   b. $Ni^{4+} + 2\,e^- \longrightarrow Ni^{2+}$

   c. $2\,H_2O + 2\,e^- \longrightarrow H_2 + 2\,OH^-$

3. You have seen several examples of oxidation–reduction reactions in this chapter. Now examine these equations and decide which are oxidation–reduction reactions and which are not. Explain your decisions.

   a. $Zn(s) + 2\,MnO_2(s) + H_2O(l) \longrightarrow$
   $$Zn(OH)_2(s) + Mn_2O_3(s)$$

   b. $HCl(aq) + NaOH(aq) \longrightarrow NaCl(aq) + H_2O(l)$

   c. $CH_4(g) + 2\,O_2(g) \longrightarrow CO_2(g) + 2\,H_2O(g)$

4. Comment on the statement: Every combustion reaction is an oxidation–reduction reaction.

5. Two common units associated with electricity are the volt and the amp. What does each unit measure?

6. Consider this galvanic cell. A coating of impure silver metal begins to appear on the surface of the silver electrode as the cell discharges.

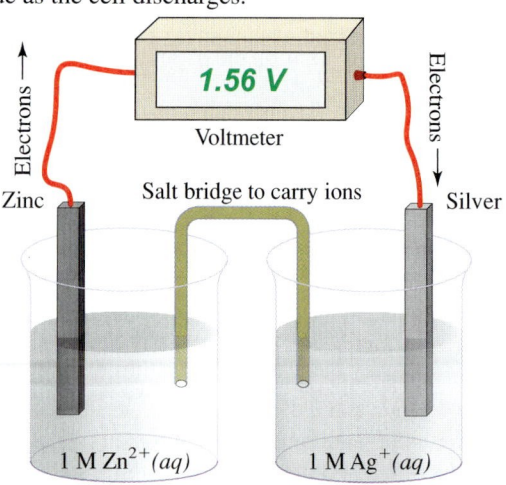

a. Identify the anode and write the oxidation half-reaction.

b. Identify the cathode and write the reduction half-reaction.

7. In the lithium–iodine cell, Li is oxidized to Li^+; I_2 is reduced to $2\,I^-$.

   a. Write equations for the two half-reactions that take place in this cell, labeling one as oxidation and the other as reduction.

   b. Write an equation for the overall reaction in this cell.

   c. Identify the half-reaction that occurs at the anode and the half-reaction that occurs at the cathode.

8. **a.** Is the voltage from a tiny AAA-size alkaline cell the same as that from a large D alkaline cell? Explain.

   b. Will both batteries sustain the flow of electrons for the same amount of time? Why or why not?

9. Identify the type of battery commonly used in each of these consumer electronic products. Assume none uses solar cells.

   a. battery-powered watch   **c.** digital camera

   b. MP3 player   **d.** handheld calculator

10. The mercury battery has been used extensively in medicine and industry. Its overall reaction can be represented by this equation.

$$HgO(l) + Zn(s) \longrightarrow ZnO(s) + Hg(l)$$

   a. Write the oxidation half-reaction.

   b. Write the reduction half-reaction.

   c. Why is the mercury battery no longer in common use?

11. **a.** What is the function of the electrolyte in a galvanic cell?

   b. What is the electrolyte in an alkaline cell?

   c. What is the electrolyte in a lead–acid storage battery?

12. These are the *incomplete* equations for the half-reactions in a lead storage battery. They do not show the electrons either lost or gained.

$$Pb(s) + SO_4^{2-}(aq) \longrightarrow PbSO_4(s)$$
$$PbO_2(s) + 4\,H^+(aq) + SO_4^{2-}(aq) \longrightarrow$$
$$PbSO_4(s) + 2\,H_2O(l)$$

  a. Balance both equations with respect to charge by adding electrons on either side of the equations, as needed.

  b. Which half-reaction represents oxidation and which represents reduction?

  c. One of the electrodes is made of lead, the other of lead dioxide. Which is the anode and which is the cathode?

13. What is meant by the term *hybrid car?*

14. a. What is the role of the electrolyte in a fuel cell?

  b. List two advantages fuel cells have over internal combustion engines.

15. Is the conversion of $O_2(g)$ to $H_2O(l)$ in a fuel cell an example of oxidation or reduction? Use electron loss or gain to support your answer.

16. Consider this diagram of a hydrogen–oxygen fuel cell used in earlier space missions.

  a. How does the reaction between hydrogen and oxygen in a fuel cell differ from the combustion of hydrogen with oxygen?

  b. Write the half-reaction that takes place at the anode in this fuel cell.

  c. Write the half-reaction that takes place at the cathode in this fuel cell.

17. a. What is a PEM fuel cell? How does it differ from the fuel cell represented in question #16?

  b. What is an SOFC? How does it differ from the fuel cell represented in Question #16?

18. In addition to hydrogen, methane also has been studied for use in PEM fuel cells. Balance the given oxidation and reduction half-reactions and write the overall equation for a methane-based fuel cell.

Oxidation half-reaction:

$$\_\_\ CH_4(g) + \_\_\ OH^-(aq) \longrightarrow$$
$$\_\_\ CO_2(g) + \_\_\ H_2O(l) + \_\_\ e^-$$

Reduction half-reaction:

$$\_\_\ O_2(g) + \_\_\ H_2O(l) + \_\_\ e^- \longrightarrow \_\_\ OH^-(aq)$$

19. The reactions in a hydrogen-fueled solid oxide fuel cell (SOFC) are shown in equations 8.19–8.21. This is the skeleton equation for the oxidation half-reaction if CO, rather than H_2, is the fuel.

$$CO(g) + O^{2-} \longrightarrow CO_2(g)$$

  a. Balance by adding electrons as needed.

  b. Combine the balanced equation for oxidation with that for reduction.

  c. Write the overall equation for a carbon monoxide-based SOFC.

20. a. Potassium, a Group 1A metal, reacts with H_2 to form potassium hydride, KH. Write the chemical equation for the reaction.

  b. Potassium hydride reacts with water to release H_2 and form potassium hydroxide. Write the chemical equation.

  c. Offer a possible reason that potassium is not used to store H_2 for use in fuel cells.

21. a. What is meant by "the hydrogen economy"?

  b. Even if methods for producing hydrogen cheaply and in large quantities were to become available, what problems would still remain for the hydrogen economy?

22. a. How are equations 8.24 and 8.25 the same and how are they different?

  b. How will the energy released in the reaction shown in equation 8.24 compare with the energy released in the reaction represented by equation 8.25? Explain your reasoning.

23. Given that 286 kJ of energy is released per mole of H_2 burned, what is the maximum amount of energy that can be released when 370 kg of H_2 is burned?

24. a. Use bond energies from Table 4.2 to calculate the energy released when 1 mol of H_2 burns.

  b. Compare your result with the stated value of 286 kJ. Account for any difference.

25. Every year, 5.6×10^{21} kJ of energy comes to Earth from the Sun. Why can't this energy be used to meet all of our energy needs?

26. This *unbalanced* equation represents the last step in the production of pure silicon for use in solar cells.

$$\_\_\ Mg(s) + \_\_\ SiCl_4(l) \longrightarrow \_\_\ MgCl_2(l) + \_\_\ Si(s)$$

  a. How many electrons are transferred per atom of pure silicon formed?

  b. Is the production of pure silicon an oxidation or a reduction reaction? Why do you think so?

27. The symbol . represents an electron and the symbol ⬤ represents a silicon atom. The darker sphere in the center of the diagram represents either a gallium or an arsenic atom. Does this diagram represent a gallium-doped *p*-type silicon semiconductor, or does it represent an arsenic-doped *n*-type silicon semiconductor? Explain your answer.

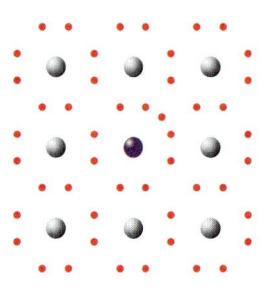

28. Describe the main reasons why solar cells have solar energy conversion efficiencies significantly less than the theoretical value of 31%.

Concentrating on Concepts

29. Explain the significance of the title of this chapter, "Energy from Electron Transfer."

30. Consider these three sources of light: a candle, a battery-powered flashlight, and an electric light bulb. For each source, provide:

    a. the origin of the light

    b. the immediate source of the energy that appears as light

    c. the original source of the energy that appears as light. *Hint:* Trace this back stepwise as far as possible.

    d. the end-products and by-products of using each

    e. the environmental costs associated with each

    f. the advantages and disadvantages of each light source

31. Explain the difference between a rechargeable battery and one that must be discarded. Use a NiCad battery and an alkaline battery as examples.

32. Is there a difference between a galvanic cell and an electrochemical cell? Explain, giving examples to support your answer.

33. What is the difference between a storage battery and a fuel cell?

34. Why are electric cars powered by lead–acid storage batteries alone only a short-term solution to the problem of air pollution emissions from automobiles? Outline your reasoning.

35. AgZn batteries are replacing lead–acid batteries in small airplanes, such as Cessna 172s.

    a. Why are these batteries, although more expensive, preferable to the lead–acid batteries used previously?

    b. Write the half-reaction of oxidation and of reduction.

36. The battery of a cell phone discharges when the phone is in use. A manufacturer, while testing a new "power boost" system, reported these data.

Time, min.sec	Voltage, V
0.00	6.56
1.00	6.31
2.00	6.24
3.00	6.18
4.00	6.12
5.00	6.07
6.35	6.03
8.35	6.00
11.05	5.90
13.50	5.80
16.00	5.70
16.50	5.60

    a. Prepare a graph of these data.

    b. The manufacturer's goal was to retain 90% of its initial voltage after 15 minutes of continuous use. Has that goal been achieved? Justify your answer using your graph.

37. Assuming that hybrid cars are available in your area, what questions would you ask the car dealer before deciding to buy or lease a hybrid? Which of these questions do you consider most important? Offer reasons for your choices.

38. You never need to plug in Toyota's gasoline–battery hybrid car to recharge the batteries. Explain.

39. Prepare a list of the environmental costs and benefits associated with hybrid vehicles. Compare that list with the environmental costs and benefits of vehicles powered by gasoline. On balance, which energy source do you favor, and why?

40. William C. Ford, Jr., chairman of the board of Ford Motor Company, is quoted as saying that going "totally green" with zero-emissions vehicles will be a real challenge. Regular drivers won't buy high-tech clean cars, Ford admits, until the industry has a "no-trade-off" vehicle widely available. What do you think he means by a no-trade-off vehicle? Do you think he is justified in this opinion?

41. Fuel cells were invented in 1839, but never developed into practical devices for producing electrical energy until the U.S. space program in the 1960s. What advantages did fuel cells have over previous power sources?

42. Hydrogen, H_2, and methane, CH_4, can each be used with oxygen in a fuel cell. Hydrogen and methane also can be burned directly. Which has greater heat content when burned, 1.00 g of H_2 or 1.00 g of CH_4? *Hint:* Write the balanced chemical equation for each reaction and use the bond energies in Table 4.2 to help answer this question.

43. Engineers have developed a prototype fuel cell that converts gasoline to hydrogen and carbon monoxide. The carbon monoxide, in contact with a catalyst, then reacts with steam to produce carbon dioxide and more hydrogen.

   a. Write a set of reactions that describes this prototype fuel cell, using octane (C_8H_{18}) to represent the hydrocarbons in gasoline.

   b. Speculate as to the future economic success of this prototype fuel cell.

44. At this time, the U.S. Department of Transportation (DOT) prohibits passengers from carrying flammable fluids aboard aircraft. Explain how this might affect the development of microfuel cells for use in consumer electronics such as portable computers.

45. Consider this representation of two water molecules in the liquid state.

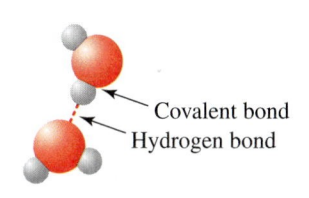

   Covalent bond
   Hydrogen bond

   a. What bonds are broken when water boils? Are these intermolecular or intramolecular bonds? (*Hint*: See Chapter 5 for definitions.)

   b. What bonds are broken when water is electrolyzed? Are these intermolecular or intramolecular bonds?

46. Although hydrogen gas can be produced by the electrolysis of water, this reaction is usually not carried out on a large scale. Suggest a reason for this fact.

47. Small quantities of hydrogen gas can be prepared in the lab by reacting metallic sodium with water, as shown in this equation.

$$2\,Na(s) + 2\,H_2O(l) \longrightarrow H_2(g) + 2\,NaOH(aq)$$

   a. Calculate the grams of sodium needed to produce 1.0 mol of hydrogen gas.

   b. Calculate the grams of sodium needed to produce sufficient hydrogen to meet an American's daily energy requirement of 1.1×10^6 kJ.

   c. If the price of sodium were $94/kg, what would be the cost of producing 1.0 mol of hydrogen? Assume the cost of water is negligible.

48. a. As a fuel, hydrogen has both advantages and disadvantages. Set up parallel lists for the advantages and the disadvantages of using hydrogen as the fuel for transportation and for producing electricity.

   b. Do you advocate the use of hydrogen as a fuel for transportation or for the production of electricity? Explain your position in a short article for your student newspaper.

49. Fossil fuels have been called ". . . Sun's ancient investment on Earth." Explain this statement to a friend who is not enrolled in your course.

50. The cost of electricity generated by solar thermal power plants currently is greater than that of electricity produced by burning fossil fuels. Given this economic fact, suggest some strategies that might be used to promote the use of environmentally cleaner electricity from photovoltaics.

51. Name some of the current applications of photovoltaic cells *other* than the production of electricity in remote areas.

Exploring Extensions

52. The aluminum–air battery is being explored for use in automobiles. In this battery, aluminum metal undergoes oxidation to Al^{3+} and forms $Al(OH)_3$. Oxygen from the air undergoes reduction to OH^- ions.

   a. Write equations for the oxidation and reduction half-reactions. Use H_2O as needed to balance the number of hydrogen atoms present, and add electrons as needed to balance the charge.

   b. Add the half-reactions to obtain the equation for the overall reaction in this cell.

   c. Specify which half-reaction occurs at the anode and which occurs at the cathode in the battery.

   d. What are the benefits of the widespread use of the aluminum–air battery? What are some of the limitations? Write a brief summary of your findings.

   e. What is the current state of development of this battery? Is it in use in any vehicles at the present time? What is its projected future use?

53. An iron-based "superbattery" is a promising alternative for delivering more power with fewer environmental effects than alkaline batteries. Find out how the superbattery is designed and its state of commercial acceptance.

54. Although Alessandro Volta is credited with the invention of the first electric battery in 1800, some feel this is a reinvention. Research the "Baghdad battery" to evaluate the merit of this claim.

55. If all of today's technology presently based on fossil-fuel combustion were replaced by H_2–O_2 fuel cells, significantly more H_2O would be released into the environment. Is this effect a concern? Find out what other effects might be anticipated from switching to a hydrogen economy.

56. a. Hydrogen is generally considered an environmentally friendly fuel, only producing water after reacting with oxygen. What effect could the widespread use of hydrogen have on urban air quality?

   b. Some scientists are reporting concerns that leakage of hydrogen gas from cars, hydrogen production plants, and fuel transportation could cause problems in the Earth's ozone layer. How significant are these concerns? What is the mechanism by which hydrogen could destroy ozone?

57. At the cutting edge of technology the line between science and science fiction often blurs. Investigate the "futuristic" idea of putting mirrors in orbit around the Earth to focus and concentrate solar energy for use in generating electricity.

58. Although silicon used to make solar cells is one of the most abundant elements in the Earth's crust, extracting it from minerals is costly. The increased demand for solar cells has some companies worried about a "silicon shortage." Use the resources of the Web to find out how silicon is purified and how the PV industry is coping with the rising prices.

59. Figure 8.27 shows an array of photovoltaic cells installed at Solar Park Gut Erlasee in Germany. Where in the United States is the largest photovoltaic power plant? Use the Web to learn of other large-scale photovoltaic cell installations. What factors help to influence this approach, one that uses a centralized array rather than individual rooftop solar units?

60. The Solar America Initiative program receives so little publicity that most people in the United States are unaware of its existence. Design a poster to explain and promote some part of this program to the general public.

Chapter 9

The World of Polymers and Plastics

On the trail.

"Hey, I'm out of here this weekend. I sure need a break from studying. I've got a date with my road bike before the cold weather really sets in. It doesn't matter to me if it rains. Check out the new biking gear I just bought. I have my Thinsulate-lined Gore-Tex jacket that will keep me dry. And if it gets cold? No problem. I have a C-Tech polyester microfiber jersey and new Porelle Drys socks to keep me warm. My Spandura tights are 18 times tougher than my older nylon/Lycra ones. And I can stash the rest of my gear in my lightweight nylon backpack with polyurethane padded straps for comfort. Like I said, I'm outta here. It will feel great to get away from the synthetic world and commune with Ma Nature for a change. Catch ya later."

Our biker is right. When she hits the road she can get away from school, exams, and the stress that may accompany these. But she cannot get away from the world of synthetics —synthetic polymers, that is. Polymers are everywhere. They are present in synthetic forms, such as Gore-Tex and nylon. They also are present in natural forms, such as proteins and cellulose, the structural material in plants. While both are important, synthetic polymers are the focus of this chapter.

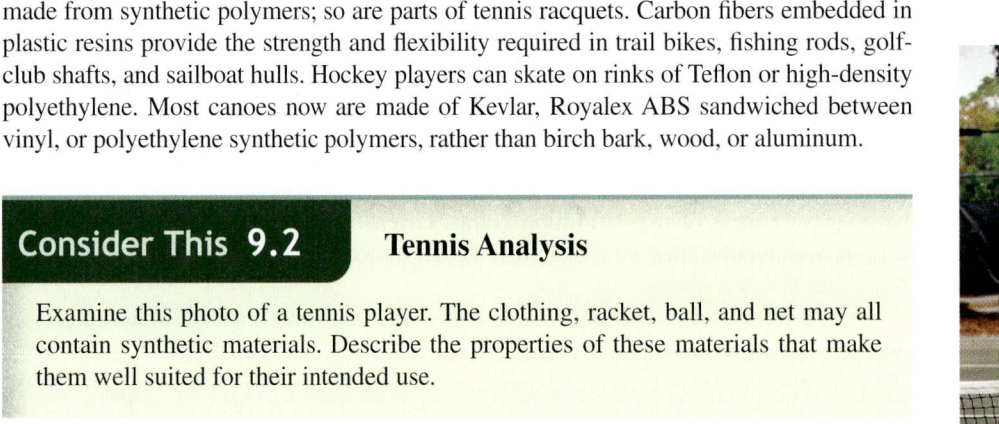

Consider This 9.1 — Polymers for Fun

Choose a favorite activity—on the land, water, or in the air. Search the Web to find a company that manufactures or sells equipment for this activity. Which polymers are mentioned on the Web site? Make a table of their names, the equipment in which they are used, and any desirable properties cited as advertising points.
Hint: Some polymer names start with *poly,* such as polyester or polypropylene. Others have trade names, such as Gore-Tex, Orlon, or Styrofoam. Still other polymers are coatings and resins and may be mentioned as epoxides or acrylics.

Recreation has been revolutionized by the introduction of synthetic polymers and plastics. Football is played on artificial turf by players wearing plastic helmets. Tennis balls are made from synthetic polymers; so are parts of tennis racquets. Carbon fibers embedded in plastic resins provide the strength and flexibility required in trail bikes, fishing rods, golf-club shafts, and sailboat hulls. Hockey players can skate on rinks of Teflon or high-density polyethylene. Most canoes now are made of Kevlar, Royalex ABS sandwiched between vinyl, or polyethylene synthetic polymers, rather than birch bark, wood, or aluminum.

Consider This 9.2 — Tennis Analysis

Examine this photo of a tennis player. The clothing, racket, ball, and net may all contain synthetic materials. Describe the properties of these materials that make them well suited for their intended use.

At this moment, you probably are wearing or carrying at least a dozen materials that did not exist 75 years ago, with some new in the last decade. Your running shoes alone contain a variety of polymers: the sole, the trim, the foam padding, the upper, the laces, and even the lace tips. These polymers add cushioning, support, and shock absorption as you walk, jog, or run. Your shirt and pants may contain synthetic fibers as well. These can add a bit more stretch, minimize wrinkles, repel rain or stains, or provide additional strength. Your clothing also may be made of microfibers, that is, polymers made into small-diameter threads. These fine threads can add desirable properties such as the ability to insulate from the cold and breathability.

Also look for polymers in the world around you. For example, think about your pen and cell phone. Most likely both have plastic cases. The same is true for calculators and computers. And of course the CDs and DVDs you play are made of polymers.

Polymers such as these come primarily from a single raw material: petroleum. As you learned earlier, our planet's supply of petroleum is limited and most of it is refined to produce fuels. Only a small percent is used to manufacture polymers and other important chemicals. In principle, polymers can be made from any carbon-containing starting material. Crude oil, however, remains the most convenient and economical.

Chemical companies now can make "eco-friendly" polymers from renewable materials such as wood, cotton fibers, straw, starch, and sugar. This work has implications for land use, crop productivity, and no doubt much more. For example, the NatureWorks unit of the company Cargill has introduced a plastic made from corn glucose that can be

Recall Dmitri Mendeleev's words (Section 4.11) that burning petroleum as a fuel "would be akin to firing up a kitchen stove with bank notes."

used for clothing, food packaging, and even the plastic parts of automobiles. In 2006, spokeswoman Ann Tucker reported "triple-digit sales growth for the past two years." Another area of growth is "eco-dinnerware" that is designed with an eye toward minimizing use of petroleum and maximizing disposal options. For example, the company Earth-Shell produces plates and bowls from renewable materials such as potato starch that are more readily composted. These now are on the shelves at some large food stores. In August 2006, EarthShell CEO Vincent Truant was quoted in *USA Today* as saying, "People prefer a product from a Midwest cornfield than a Middle East oil field."

Consider This 9.3	Eco-Friendly Plates

Do you want to eat your tofu off something other than a Styrofoam plate? Search on the Web for NatureWorks from Cargill, Sorona from DuPont, eco-dinnerware from EarthShell, or for other bio-based polymers. What do these polymers promise? Which products are now on the market? Are they biodegradable?

To understand the complexities surrounding the sources and uses of polymers, we first must understand their structure and how they are synthesized. We will focus on these in the next few sections and then will return to the issues of polymer use in our society.

9.1 Polymers: Long, Long Chains

Rayon, nylon, Lycra, polyurethane, Teflon, Saran, Styrofoam, Formica! These seemingly different materials all are synthetic polymers. What they have in common is evident at the molecular level. **Polymers** are large molecules consisting of a long chain or chains of atoms covalently bonded together. A polymer molecule can contain thousands of atoms and have a molar mass of over a million grams. Given their size, polymers are referred to as **macromolecules**, that is, molecules of high molecular mass that have characteristic properties because of their large size.

Monomers (*mono* meaning "one"; *meros* meaning "unit") are the small molecules used to synthesize the larger polymeric chain. Each monomer is analogous to a link of the chain. The polymers (*poly* means "many") can be formed from the same type of monomer or from a combination of monomers. The long chain shown in Figure 9.1 may help you to imagine a polymer made from identical monomers, that is, identical links in a chain.

Keep in mind that chemists did not invent polymers. Natural polymers are found both in plants and animals. For example, wood, wool, cotton, starch, natural rubber, skin, and

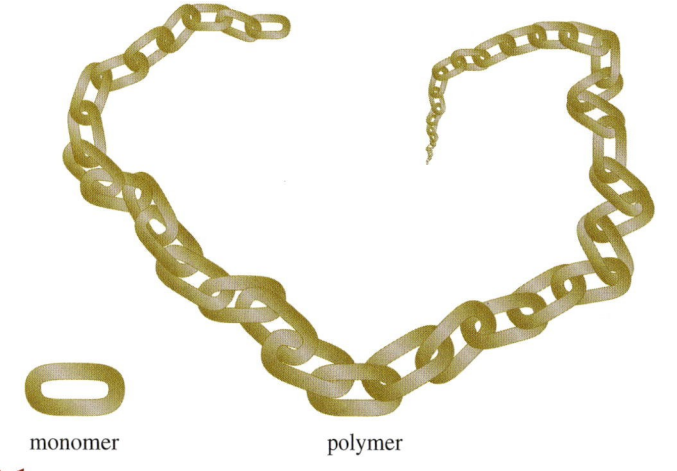

monomer   polymer

Figure 9.1

Representations of a monomer (*single link*) and a polymer (*long chain*) made from one type of monomer.

Figure 9.2

Oak logs and grass have in common the natural polymer cellulose. The monomer is glucose.

hair are all natural polymers. Like synthetic polymers, natural ones exhibit a stunning variety of properties. They give strength to an oak tree, delicacy to a spider's web, softness to goose down, and flexibility to a blade of grass (Figure 9.2).

When chemists first created polymers, they used natural ones such as cotton and rubber as models. Indeed, many synthetic polymers originally were substitutes for expensive or rare materials that occurred naturally. As the number of new polymers grew, their uses expanded dramatically. Many applications make use of the fact that plastics deliver comparable strength at a lower weight. For example, polymers generally have a density of 1–2 g/cm^3 compared with that of approximately 8 g/cm^3 for steel. Thus an automobile body built from polymers would weigh far less than its steel counterpart and would require less energy (and therefore less fuel) to move. Similarly, plastic packaging reduces weight, eliminates breakage, and helps save fuel during shipping.

> Density was discussed in Section 5.6.

Synthetic polymers sometimes are called plastics, a term that applies to materials with a broad range of properties and applications. The word *plastic* is both an adjective, "capable of being molded," and a noun, "something capable of being molded." More specifically, the *Merriam-Webster Collegiate Dictionary,* 11th edition, refers to plastics as "any of numerous organic synthetic or processed materials that are mostly . . . polymers of high molecular weight and that can be molded, cast, extruded, drawn, or laminated into objects, films, or filaments." As it turns out, some metals have plastic-like properties because they can be "cast, extruded, and drawn." Therefore, the word *plastic* has many applications beyond that of describing synthetic polymers. We will generally use the word *polymer* in this chapter. Either way, plastic or polymer, we are talking about a large molecule that has been synthesized from smaller ones.

9.2 Adding up the Monomers

How do monomers combine to make a polymer? In the previous section, we used a chain to represent a polymer, but made no mention of how the chain was formed. In this section, we will provide the details. As you may surmise, a polymer has no chain links as such. Rather, covalent chemical bonds connect the monomers.

Polyethylene will serve as our first example. As the name indicates, polyethylene is synthesized from the monomer ethylene, $H_2C{=}CH_2$. Ethylene is the common name for ethene, the smallest member in the family of hydrocarbons containing a C-to-C double bond. In the polymerization reaction, n molecules of ethylene combine to form polyethylene.

> Polyethylene is called polythene in the United Kingdom.

$$n \begin{array}{c} H \\ \diagdown \\ C \\ \diagup \\ H \end{array}{=}\begin{array}{c} H \\ \diagup \\ C \\ \diagdown \\ H \end{array} \xrightarrow{\;R\cdot\;} \left[\begin{array}{c} H \quad H \\ | \quad\; | \\ C{-}C \\ | \quad\; | \\ H \quad H \end{array} \right]_n \qquad [9.1]$$

Closely examine each part of Equation 9.1. On the left is the ethylene monomer. The coefficient n in front of it specifies the number of molecules that react. In turn, this specifies the length and the molecular mass of the polymer. Molecular masses of polyethylene generally fall between 10,000 and 100,000 g/mol, but can run into the millions. On the right, the n also appears as a subscript in the product, indicating that each monomer becomes part of the long chain. The large square brackets enclose the repeating unit of the polymer.

Polyethylene is the sole product. It contains exactly the same number and kinds of atoms as did the monomers. Literally, the monomers add to one another to form a long chain of n units. As a result, we call this **addition polymerization**: the monomers add to the growing polymer chain in such a way that the product contains all the atoms of the starting material. No other products are formed.

(a)

(b)

Figure 9.3

(a) Bottles made from polyethylene. **(b)** A sign posted on a railway tank car that transports liquefied ethylene. The 1038 identifies it as ethylene, the red diamond indicates high flammability, and the 2 indicates moderate reactivity.

In equation 9.1, also notice the R• over the arrow. To appreciate its significance, you need to know a bit more about the ethylene monomer. Often produced at oil refineries, this compound is a colorless gas with a faint gasoline-like odor. It is distinctly different from the odorless solid polyethylene (Figure 9.3a). As hydrocarbons, however, ethylene and polyethylene share the property of being flammable, with ethylene being much more highly so. Although not classified as an air pollutant, ethylene still is a VOC, that is, a volatile organic compound. As you learned in Chapter 1, any VOC released into the atmosphere can contribute to the buildup of photochemical smog. For several reasons, then, safety precautions are needed when transporting ethylene from the refineries to the sites where polyethylene is manufactured. Since it is inconvenient to transport ethylene as a gas (think how large the container would have to be), it first is subjected to pressure and low temperature. The resulting liquid then can be transported in tank cars that bear a label like the one shown in Figure 9.3b. But under these conditions will the ethylene polymerize in the tank car? Fortunately, no. Clearly the end user of the ethylene would be distressed to receive a tank car full of solid polyethylene! Perhaps you now can better appreciate the R• that appears over the arrow in equation 9.1. In the absence of a catalyst such as R•, ethylene does not polymerize.

The catalyst R• warrants further examination. As a free radical, it has an unpaired electron and is highly reactive. As shown in Figure 9.4, R• attaches to a $H_2C{=}CH_2$ molecule to initiate the chain-forming process. Also recall that the double bond in ethylene contains *four* electrons. After an ethylene molecule reacts with R•, *two* of these electrons remain, leaving a single bond. The other two electrons move (as the red arrows indicate) to form two new single bonds, one to R and the other where the ethylene molecule adds to the chain. Thus the chain grows.

The reactive hydroxyl radical, •OH, was discussed earlier in Sections 1.11, 2.8, and 6.7.

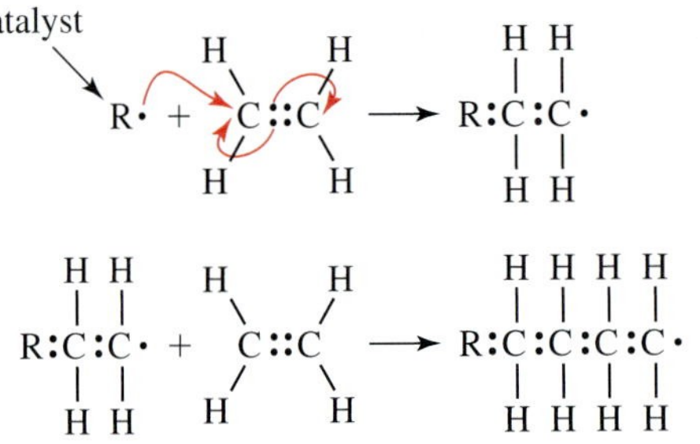

Initiating
free-radical
catalyst

$$R\cdot + \quad \overset{H}{\underset{H}{C}}::\overset{H}{\underset{H}{C}} \longrightarrow R:\overset{H}{\underset{H}{C}}:\overset{H}{\underset{H}{C}}\cdot$$

$$R:\overset{H}{\underset{H}{C}}:\overset{H}{\underset{H}{C}}\cdot + \quad \overset{H}{\underset{H}{C}}::\overset{H}{\underset{H}{C}} \longrightarrow R:\overset{H}{\underset{H}{C}}:\overset{H}{\underset{H}{C}}:\overset{H}{\underset{H}{C}}:\overset{H}{\underset{H}{C}}\cdot$$

Figure 9.4

The polymerization of ethylene.

 Figures Alive! Visit the *Online Learning Center* to see an animated version of this figure. Look for the Figures Alive! icon in this chapter as a guide to related activities.

As each monomer of ethylene adds, two things happen: the double bond between the carbons converts to a single bond, and a new bond forms between the monomer and the growing chain. This process is repeated many times over in many chains simultaneously. Occasionally, the ends of two polymer chains join and stop the chain growth. The process also can be stopped by adding specific compounds to the reaction that "cap" the reactive chain end. And of course the process stops if you run out of monomers. The result of all this chemistry is that gaseous ethylene is converted to solid polyethylene.

Although we placed R• over the arrow in equation 9.1, in truth, we also could have represented the reaction like this.

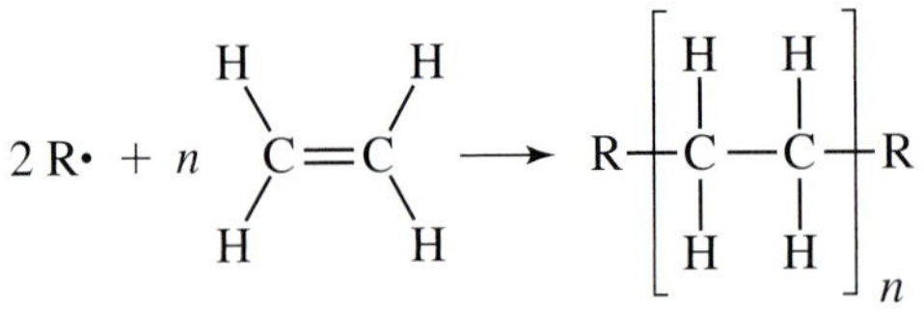

$$2\,R\cdot + n \quad \overset{H}{\underset{H}{C}}=\overset{H}{\underset{H}{C}} \longrightarrow R\left[\overset{H}{\underset{H}{\overset{|}{C}}}-\overset{H}{\underset{H}{\overset{|}{C}}}\right]_n R \qquad [9.2]$$

However, we will continue with our practice of omitting R• as a reactant as we did in equation 9.1. The R group that "caps" the end of the molecule can be considered chemically insignificant with respect to the long chain.

The numerical value of n and hence the length of the chain can vary. During the manufacturing process, n will be adjusted in order to create specific properties for the polymer. Moreover, within a single sample the individual polymer molecules can have varying lengths. In every case, however, the molecules contain a chain of carbon atoms that are attached to one another by single bonds. In essence, any given polyethylene molecule is like an octane molecule, but much, much longer.

> Industrial chemists use several synthetic routes to produce polyethylene. The most common employ a metal catalyst and mild temperatures.

Your Turn 9.4   Polymerization of Ethylene

In equation 9.1 for the polymerization of ethylene, suppose that $n = 4$.

   a. Rewrite equation 9.1 to indicate this.
   b. Draw the structural formula of the product without using brackets. Remember to put R groups at the ends of the chain.
   c. In terms of its molecular structure, how does the product differ from octane?

Answer

   c. Octane is C_8H_{18}. Although the product molecule similarly has eight carbon atoms, it has two fewer hydrogen atoms and two R groups at the ends of the molecule.

Figure 9.5

Packaging made from polyethylene.

9.3 Polyethylene: A Closer Look

Polyethylene has a wide variety of uses. For example, it is found in plastic milk jugs, detergent containers, baggies, and packing materials (Figure 9.5). Yet, as we have seen in the previous section, all polyethylene is made from the same starting material, $H_2C{=}CH_2$. How can this one monomer form polymers that can be used in so many different ways?

Your Turn 9.5   **Polyethylene Hunt**

Take time to explore the properties of polyethylene. As we will see in the next section, polyethylene containers, baggies, and packaging materials are marked either as low density (LDPE) or high density (HDPE). Using this code as your guide, locate some items made of polyethylene and note their properties. Do LDPE and HDPE differ in flexibility? Is one more translucent? Is one more often colored with a pigment than the other?

LDPE

HDPE

The different properties of polyethylene largely stem from differences in the long molecules that compose it. Relatively speaking, these molecules are very long indeed. Imagine a polyethylene molecule to be as wide as a piece of spaghetti. If this were the case, the molecule would be almost a half mile long! To continue the analogy, the polyethylene used to make plastic bags contains molecular chains arranged somewhat like cooked spaghetti on a plate. The strands are jumbled up and not very well aligned, although in some regions the molecular chains run in parallel. Moreover, the polyethylene chains, like the spaghetti strands, are not bonded to one another.

Recall that we used hydrogen bonding to describe the attractive force *between* water molecules in the liquid phase. The hydrogen bond is not a covalent bond. Rather, it is an **intermolecular attractive force,** that is, an attraction between two molecules resulting from the interactions of the electron clouds and nuclei. These attractive forces are different from the covalent bonds that exist *within* each molecule in which electrons are truly shared between two atoms. Intermolecular attractive forces are much weaker than covalent bonds, but they still have an effect on the behavior of a group of atoms. In water, these forces help to keep the molecules very close together, sliding past and around one another as the liquid flows, unlike in the gas phase where the molecules are far apart and bump into one another less frequently.

Polymers that contain only hydrogen and carbon, such as HDPE, LDPE, and polypropylene (PP), do not have hydrogen bonds. Rather, there is another type of intermolecular attractive force that keeps the molecules close to one another. Each atom in the long polymeric chain contains its own electrons. But these electrons can be attracted to

> Hydrogen bonding was explained in Section 5.6. The unusual properties of water also were mentioned.

the atoms on neighboring molecular chains, and the magnitude of the attraction between strands of polyethylene is a direct result of the large number of atoms involved. The attraction is a bit like that between the two halves of Velcro. The bigger the surface area of one Velcro strip, the better it will hold to the other. The intermolecular forces holding polyethylene together are called **dispersion forces**, and they are attractions between molecules that result from a distortion of the electron cloud that causes an uneven distribution of the negative charge. Dispersion forces can be quite significant in very large molecules such as polymers.

Evidence of the molecular arrangement of polyethylene can be obtained by doing a short experiment. Cut a strip from a heavy-duty transparent polyethylene bag, grab the two ends of the strip, and pull. A fairly strong pull is required to start the plastic stretching, but once it begins, less force is needed to keep it going. The length of the plastic strip increases dramatically as the width and thickness decrease (Figure 9.6a). A little shoulder forms on the wider part of the strip and a narrow neck almost seems to flow from it in a process called "necking." Unlike the stretching of a rubber band, the necking effect is not reversible, and eventually the plastic thins to the point where it tears.

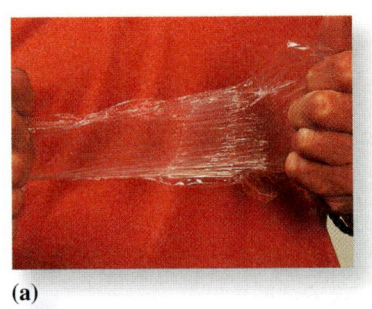

(a)

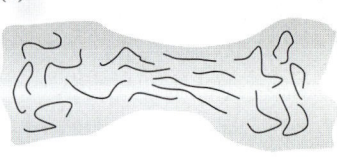

(b)

Figure 9.6
(a) A plastic bag stretched until it "necks." **(b)** A representation of "necking" at the molecular level.

Consider This 9.6   "Necking" Polyethylene

Necking permanently changes the properties of a piece of polyethylene.

a. Does necking affect the number of monomer units, n, in the average polymer?

b. Does necking affect the bonding between the monomer units within the polymer chain?

Figure 9.6b represents the necking of polyethylene from a molecular point of view. As the strip narrows down, the previously mixed-up molecular chains shift, slide, and align parallel to one another in the direction of pull. In some plastics, such stretching, or "cold drawing," is carried out as part of the manufacturing process to alter the three-dimensional arrangement of the chains in the solid. Of course, as the force and stretching continue, the polymer eventually reaches a point at which the strands can no longer realign, and the plastic breaks. Paper, a natural polymeric material, tears when pulled because the strands (fibers) in paper are rigidly held in place and are not free to slip like the long molecules in polyethylene.

A strategy to control the molecular structure and physical properties of polymers is to regulate the branching of the polymer chain. This approach is used to produce high-density polyethylene (HDPE) and low-density polyethylene (LDPE) (Figure 9.7).

As you may have discovered in Your Turn 9.5, the plastic bags dispensed in the produce aisles of supermarkets are often LDPE. These bags are stretchy, transparent, and not very strong. This low-density form was the first type of polyethylene to be manufactured. A study of its structure reveals that the molecules consist of about 500 monomeric units, and that the central polymer chain has many side branches, like the limbs radiating from a central tree trunk.

About 20 years after the discovery of LDPE, chemists were able to adjust reaction conditions to prevent branching. Thus HDPE was born. In their Nobel Prize-winning research, Karl Ziegler (1898–1973) and Giulio Natta (1903–1979) developed new catalysts that enabled them to make linear (unbranched) polyethylene chains consisting of about 10,000 monomer units. Having no side branches, these long chains can arrange parallel to one another (see Figure 9.7a). The structure of HDPE is thus more crystalline than the irregular tangle of the polymer chains in LDPE. The more highly ordered structure of HDPE gives it greater density, rigidity, strength, and a higher melting point than LDPE. Furthermore, the high-density form is opaque; the low-density form tends to be transparent.

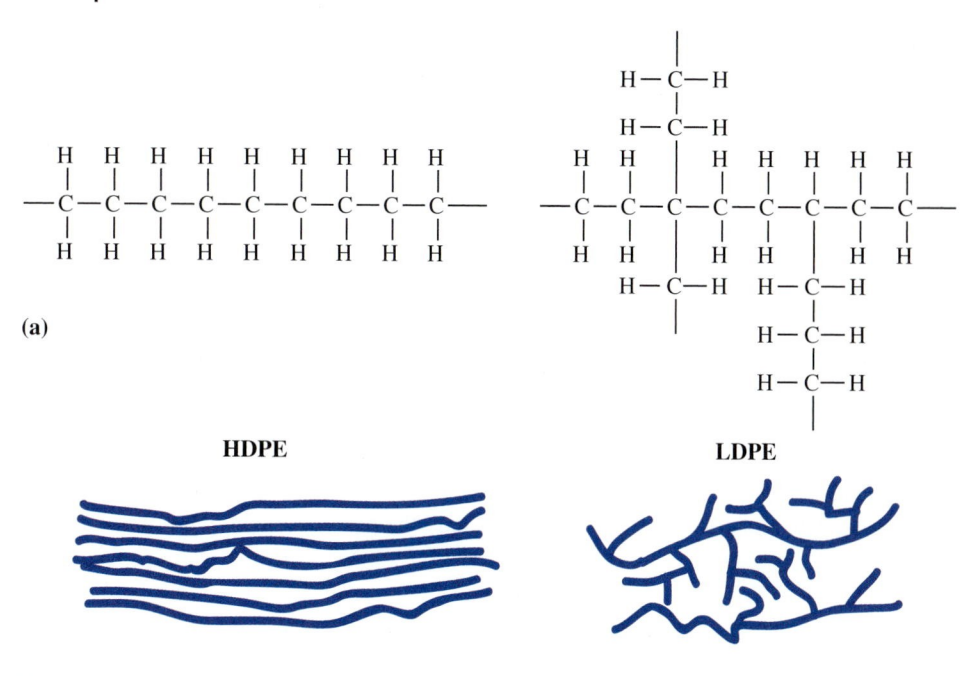

HDPE   **LDPE**

(b)

Figure 9.7

High-density (linear) polyethylene and low-density (branched) polyethylene.
(a) Details of bonding. **(b)** Schematic representations.

Consider This 9.7   HDPE and LDPE

The densities of HDPE and LDPE are actually quite close: 0.96 g/cm³ and 0.93 g/cm³, respectively. Nonetheless, this density difference results in different properties. Use Figure 9.7 to rationalize why the density of HDPE is slightly greater than that of LDPE.

Given these differences in properties, HDPE and LDPE have different applications. High-density polyethylene is used to make toys, containers, stiff or "crinkly" plastic bags, and heavy-duty pipes. A newer use of HDPE was spurred by surgery patients who have HIV/AIDS. Without suitable protection, surgeons would run the risk of acquiring the disease through contact with a patient's blood. Allied-Signal Corporation has produced a linear polyethylene fiber called Spectra that can be fabricated into liners for surgical gloves. Spectra gloves are said to have 15 times more resistance to cuts than medium-weight leather work gloves, but are so thin that a surgeon can retain a keen sense of touch. A sharp scalpel can be drawn across the glove with no damage to either the fabric or the hand inside. Such strength is in marked contrast to the properties of the common polyethylene plastic grocery bag.

Consider This 9.8   Shopping for Polymers

The *Macrogalleria* is an intriguing Web site that contains a virtual shopping mall with stores selling items made from polymers. Visit the stores to find at least six items made of LDPE or HDPE. Make your selections from several different shops. A link to the site is provided at the *Online Learning Center*.

It would be a mistake to conclude that polyethylene is restricted to the extremes represented by highly branched or strictly linear forms. By modifying the extent and location of branching in LDPE, its properties can be varied from the soft and wax-like coatings on milk cartons to stretchy plastic food wrap. Because of its linearity and close packing, HDPE is sufficiently rigid to be used for plastic milk bottles. Unfortunately, consumers sometimes are unaware of the consequences of such structural tinkering. For example, the higher melting point of HDPE (130 °C) means that it can go through the dishwasher safely. In contrast, objects made of LDPE (melting point of 120 °C) may melt if near the heating element. The hot water of a dishwasher will melt neither HDPE nor LDPE.

Finally, one of the first and most important uses of polyethylene was a consequence of its being a good electrical insulator. During World War II, polyethylene was used by the Allies to coat electrical cables in aircraft radar installations. Sir Robert Watt, who discovered radar, described polyethylene's critical importance.

"The availability of polythene [polyethylene] transformed the design, production, installation, and maintenance problems of airborne radar from the almost insoluble to the comfortably manageable . . . A whole range of aerial and feeder designs otherwise unattainable was made possible, a whole crop of intolerable air maintenance problems was removed. And so polythene played an indispensable part in the long series of victories in the air, on the sea, and on land, which were made possible by radar."

Quoted by J.C. Swallow in "The History of Polythene" from
Polythene—The Technology and Uses of Ethylene Polymers (2nd ed.)
A. Renfrew, Editor. London: Iliffe and Sons, 1960.

Consider This 9.9   **Other Types of Polyethylene**

In addition to LDPE and HDPE, polyethylene is manufactured as MDPE and LLDPE. Use the Web to find out about these and other types. How do the properties of these differ?

9.4 The "Big Six": Theme and Variations

Today, more than 60,000 synthetic polymers are known. Since 1976, the United States has produced a larger volume of synthetic polymers than the volumes of steel, copper, and aluminum combined. Although many polymers were developed for specialized uses, six account for roughly 75% of those used in the United States. We refer to these everyday polymers as the "Big Six." These six polymers, listed in Table 9.1, are polyethylene (low-density and high-density), polypropylene, polystyrene, polyvinyl chloride, and polyethylene terephthalate. Pronounce this last word as "ter-eh-THAL-ate" (the "ph" is silent).

Table 9.1 also lists some of the important properties of these six polymers. All are solids that can be colored with pigments. Although all are insoluble in water, some dissolve or soften in the presence of hydrocarbons, fats, and oils. These six polymers are **thermoplastic,** meaning that with heat they can be melted and reshaped over and over again. However, they exhibit a range of melting points depending on the route by which they were manufactured. In general, polyethylene has a relatively low melting point, with LDPE and HDPE melting at about 120 °C and 130 °C, respectively. Polypropylene (PP) has a higher melting point of 160–170 °C.

Depending on the arrangement of their molecules, polymers have varying degrees of toughness and strength. At the microscopic level, in some parts of a polymer the molecules may have a very orderly and repeating pattern, such as one would find in a crystalline solid. In these **crystalline regions,** the long polymer molecules are arranged neatly and tightly in a regular pattern. In other parts of the same polymer, you can find **amorphous regions.** Here, the long polymer molecules are found in a random, disordered

Table 9.1	The Big Six		
Polymer	**Monomer**	**Properties of Polymer**	**Uses of Polymer**
Polyethylene (LDPE) △ 4 LDPE	Ethylene $H_2C=CH_2$	Translucent if not pigmented. Soft, and flexible. Unreactive to acids and bases. Strong and tough.	Bags, films, sheets, bubble wrap, toys, wire insulation.
Polyethylene (HDPE) △ 2 HDPE	Ethylene $H_2C=CH_2$	Similar to LDPE. More rigid, tougher, slightly more dense.	Opaque milk, juice, detergent, and shampoo bottles. Buckets, crates, and fencing.
Polyvinyl chloride △ 3 PVC, or V	Vinyl chloride $H_2C=CHCl$	Variable. Rigid if not softened with a plasticizer. Clear and shiny, but often pigmented. Resistant to oils, acids, bases, and most chemicals.	Rigid: Plumbing pipe, house siding, charge cards, hotel room keys. Softened: Garden hoses, waterproof boots, shower curtains, IV tubing.
Polystyrene △ 6 PS	Styrene $H_2C=CH(C_6H_5)$	Variable. "Crystal" form transparent, sparkling, somewhat brittle. "Expandable" form lightweight foam. Both forms rigid and dissolve in many organic solvents.	"Crystal" form: Food wrap, CD cases, transparent cups. "Expandable" form: Foam cups, insulated containers, food packaging trays, egg cartons, packaging peanuts.
Polypropylene △ 5 PP	Propylene $H_2C=CH(CH_3)$	Opaque, very tough, good weatherability. High melting point. Resistant to oils.	Bottle caps. Yogurt, cream, and margarine containers. Carpeting, casual furniture, luggage.
Polyethylene terephthalate △ 1 PETE, or PET	Ethylene glycol $HO-CH_2CH_2-OH$ Terephthalic acid $HOOC-C_6H_4-COOH$	Transparent, strong, shatter-resistant. Impervious to acids and atmospheric gases. Most costly of the six.	Soft-drink bottles, clear food containers, beverage glasses, fleece fabrics, carpet yarns, fiber-fill insulation.

Note: The structures of the first five monomers differ only by the atoms shown in blue.

arrangement and their packing is much looser. Because of their structural regularity, the crystalline regions impart toughness and resistance to abrasion. This accounts for the strength of HDPE and PP. Although some polymers are highly crystalline, most still include amorphous regions as well. These regions impart flexibility. For example, the amorphous regions in PP give it the ability to be bent without breaking. The range of properties among polymers means that they are differently suited for specific applications. The next exercise provides an opportunity to match polymers with their uses.

Consider This 9.10　　Uses of the "Big Six"

Use the information in Table 9.1 about the Big Six polymers to determine:

a. Which polymer would not be suitable for margarine tubs because it softens with oil.

b. Which ones are transparent and which one is used in soft-drink bottles.

c. Which one is tough and used for bottle caps. Name another application in which toughness counts.

d. Which ones are listed as unreactive/impervious to acids and thus could serve as orange juice containers.

e. Which one has the lowest melting point. What are the implications for leaving items made from this polymer in a car on a sunny day?

Table 9.1 also shows that six monomers are used to make six different polymers, but perhaps not in the way you would expect. Two of the polymers use the same monomer, ethylene. Three others each use a monomer closely related to ethylene: styrene, vinyl chloride, and propylene. The sixth polymer, polyethylene terephthalate, is the odd one of the six being made using two different monomers. We will return to polyethylene terephthalate, but after first spending time with the other five members of this sextet.

In this section, we will more closely examine the three monomers that structurally are related to ethylene. Here they are together with ethylene.

Visit the *Online Learning Center* to learn more about these monomers.

$$\begin{array}{cccc}
\underset{H}{\overset{H}{\diagup}}C=C\underset{H}{\overset{H}{\diagup}} &
\underset{H}{\overset{H}{\diagup}}C=C\underset{Cl}{\overset{H}{\diagup}} &
\underset{H}{\overset{H}{\diagup}}C=C\underset{CH_3}{\overset{H}{\diagup}} &
\underset{H}{\overset{H}{\diagup}}C=C\underset{C_6H_5}{\overset{H}{\diagup}} \\
\text{ethylene} & \text{vinyl chloride} & \text{propylene} & \text{styrene}
\end{array}$$

In the case of vinyl chloride, one of the hydrogen atoms in ethylene is replaced by a chlorine atom. In propylene, a methyl group ($-CH_3$) substitutes for a hydrogen atom. And in styrene, the substituent is $-C_6H_5$, known as a phenyl group.

As you might suspect, all three undergo addition polymerization just like ethylene. But the substituent introduces variety. To see this, let us examine the polymerization of n molecules of vinyl chloride to form polyvinyl chloride (PVC).

$$n\ \underset{H}{\overset{H}{\diagup}}C=C\underset{Cl}{\overset{H}{\diagup}} \xrightarrow{R\cdot}
\left[\begin{array}{cc} \overset{H}{\underset{H}{\mid}}{C} & \overset{H}{\underset{Cl}{\mid}}{C} \end{array}\right]_n$$

[9.3]

The chlorine atom creates a type of asymmetry in the monomer. Visualize this by arbitrarily thinking of the carbon atom bearing two hydrogen atoms as the "head" and the carbon with the chlorine atom as the "tail."

When vinyl chloride molecules add to form polyvinyl chloride, the molecules can orient in one of three ways: head-to-tail, alternating head-to-head/tail-to-tail, and randomly (Figure 9.8). In the repeating head-to-tail arrangement, chlorine atoms are on alternate carbons. In a head-to-head/tail-to-tail arrangement, chlorine atoms are on adjacent carbons. In the random arrangement, an irregular mixture of the previous two occurs. The head-to-tail arrangement is the usual product for polyvinyl chloride.

The arrangement of monomers in the chain is one factor that affects the flexibility of the polymer. Thus each arrangement of PVC has somewhat different properties, with the most regular one, repeating head-to-tail, being the stiffest because the molecules pack more easily together to form crystalline regions. Stiff PVC finds use in drain and sewer pipes, credit cards, house siding, furniture, and various automobile parts. The random arrangement is still stiff, but somewhat less so.

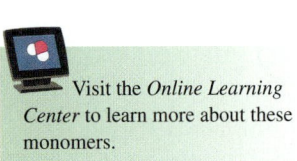

tail　head

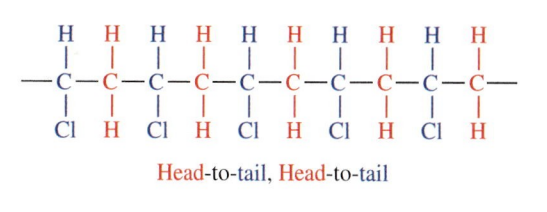

Head-to-tail, Head-to-tail

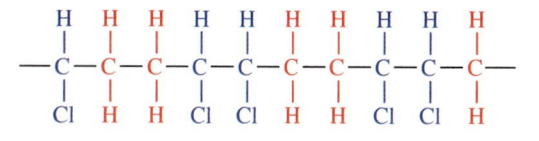

Head-to-head, tail-to-tail

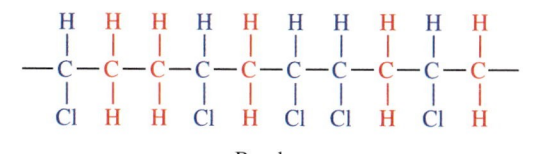

Random

Figure 9.8

Three possible arrangements of the monomers in PVC.

PVC can be further softened with **plasticizers**, compounds that are added in small amounts to polymers to make them softer and more pliable. Plasticizers work by fitting in between the large polymer molecules, thus disrupting the regular packing of the molecules. Flexible PVC that contains plasticizers is familiar in shower curtains, "rubber" boots, garden hoses, clear IV bags for blood transfusions, artificial leather ("patent leather"), and flexible insulation coatings on electrical wires.

Next, let us examine the polymerization of *n* molecules of propylene to form polypropylene. Again, we have addition polymerization, and again several arrangements of the monomer are possible. A particularly useful form of polypropylene is the repeating head-to-tail, head-to-tail arrangement. This regularity imparts a high degree of crystallinity and makes the polymer strong, tough, and able to withstand high temperatures. These properties are reflected in the uses. For example, polypropylene is found as the strong fibers of indoor–outdoor carpeting as well as in tough videocassette cases. Strength and chemical resistance make polypropylene a good choice for applications in which structural ruggedness is required.

> Just as ethylene also is called ethene, propylene also is called propene.

Your Turn 9.11   From Propylene to Polypropylene

a. Analogous to equation 9.3, write the polymerization reaction of *n* monomers of propylene to form polypropylene.
Note: To get the repeating head-to-tail, head-to-tail arrangement, a special catalyst is used, so just put "catalyst" over the arrow.

b. Analogous to Figure 9.8, show a random arrangement of the monomers in a segment of polypropylene.

Consider This 9.12   Polypropylene, "the Tough One"

Polypropylene, one of the Big Six, is used to construct items for which toughness counts. This polymer may not be as familiar to you as polyethylene and PET, because polypropylene items sometimes do not carry a recycling symbol and are not collected curbside. Search the Web to find five items manufactured from polypropylene.

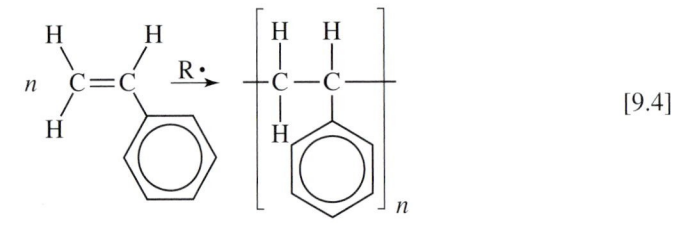

Random

(a)

(b)

Figure 9.9

(a) The random arrangement of polystyrene.
(b) Partyware made from "crystal" (general purpose) polystyrene.

Finally, let us examine the polymerization of *n* molecules of styrene to form polystyrene. Styrene usually polymerizes to polystyrene with the head-to-tail arrangement. Here is the addition polymerization, and in this case *n* equals about 5000.

$$ n \ \overset{H}{\underset{H}{C}} = \overset{H}{\underset{}{C}} \overset{R \cdot}{\longrightarrow} \left[\overset{H}{\underset{H}{C}} - \overset{H}{\underset{}{C}} \right]_n $$

[9.4]

Polystyrene is a low-cost plastic with several familiar uses. Transparent cases for CDs and clear rigid plastic party glasses are both made from polystyrene. In this form, sometimes referred to as general purpose or "crystal" polystyrene, the polymer is hard and brittle. Have you ever squeezed too hard on your clear plastic party glass causing it to split down the side? It probably was polystyrene. "Crystal" polystyrene is in the random arrangement shown in Figure 9.9.

The familiar foam hot beverage cup, foam egg carton, and foam meat packaging tray are examples of another type of polystyrene (PS), sometimes referred to as "expandable." Such items are made from "expandable" polystyrene beads. These beads contain 4-7% of a **blowing agent**, that is, either a gas or a substance capable of producing a gas to manufacture a foamed plastic. For PS, the blowing agent is typically a low-boiling liquid such as pentane. If the beads are placed in a mold and heated with steam or hot air, the pentane vaporizes. In turn, the expanding gas expands the polymer. The expanded particles are fused together into the shape determined by the mold. Because it contains so many bubbles, this plastic foam not only is light, but also is an excellent thermal insulator. The hard, transparent version of polystyrene is made by molding the melted polymer *without* the blowing agent. It is used to fabricate the cases of computers and TVs, in addition to CD cases and partyware.

At one time, chlorofluorocarbons were included in the list of compounds used as blowing agents. Concern over the involvement of CFCs in the destruction of stratospheric ozone led to their phaseout by 1990. Gaseous pentane (C_5H_{12}) and carbon dioxide are now frequently used for this purpose. The Dow Chemical Company developed a new process that uses pure carbon dioxide as a blowing agent to produce Styrofoam for packaging material. Using the Dow 100% CO_2 technology eliminates the use of 3.5 million pounds of CFC-12 or HCFC-22 as blowing agents. The CO_2 used in the process is a by-product from existing commercial and natural sources, such as ammonia plants and natural gas wells. Thus, it does not contribute additional CO_2, a greenhouse gas. Because it is nonflammable, carbon dioxide is preferred as a blowing agent over pentane, which is flammable.

Visit the *Online Learning Center* to learn more about other addition polymers.

Styrofoam is a brand name of polystyrene foam insulation, produced by the Dow Chemical Company.

Chapter 10 discusses functional groups in more depth.

> **Your Turn 9.13**   **Polystyrene Possibilities**
>
> Show the arrangement of atoms in a polystyrene chain in the repeating head-to-tail arrangement. Why do you think this arrangement is favored rather than the head-to-head arrangement?

9.5   Condensing the Monomers

Monomers make the polymer! As we just saw, structural changes in the monomer lead to changes in the properties of the polymer. In this section, we will expand on this theme and offer another variation.

To understand different monomers, it is helpful if you have a way to categorize the common groups of atoms that they may contain. These common groups are known as **functional groups.** They are distinctive arrangements of groups of atoms that impart characteristic chemical properties to the molecules that contain them (Table 9.2).

Earlier you met the phenyl group, $-C_6H_5$, in the styrene monomer. This group consists of six carbon atoms arranged to form a hexagon. Typically, in representations, the C and H atoms in the phenyl group are omitted, and the entire structure is represented by a hexagon with a circle inside. Here are three representations of the phenyl group.

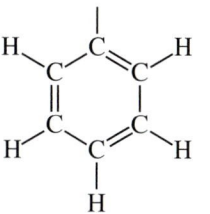 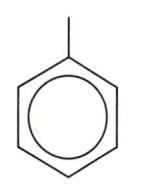 

> **Your Turn 9.14**   **Benzene and Phenyl**
>
> The difference between a phenyl group, $-C_6H_5$, and the compound benzene, C_6H_6, is simply one hydrogen atom.
>
> **a.** Analogous to the three structures just given for the phenyl group, sketch these for benzene.
> **b.** Both the phenyl group and benzene have two resonance structures. Draw them.
> **c.** Given these resonance structures, why is the shorthand symbol of a circle within a hexagon a particularly good representation of both benzene and the phenyl group? *Hint:* Resonance was introduced in Section 2.3.
>
> **Answer**
> **c.** The two equivalent resonance structures indicate that in benzene the electrons are uniformly distributed around the ring. All the C-to-C bonds are of equal strength and length. The circle inside indicates the uniformity of these six bonds.

Another example of a functional group is the hydroxyl group, $-OH$. This group is found in all alcohol molecules. The carboxylic acid group, $-COOH$, also has this $-OH$ group within it:

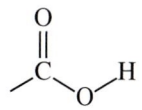

Table 9.2	Selected Functional Groups	
Name	**Chemical Formula**	**Structural Formula**
hydroxyl	—OH	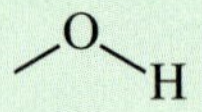
carboxylic acid	—COOH	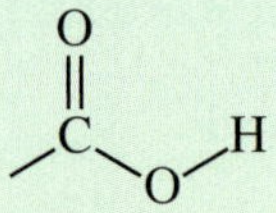
ester	—COOC—	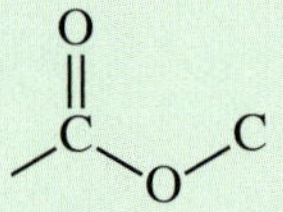
amine	—NH$_2$	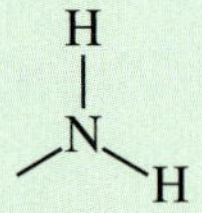
amide	—CONH$_2$	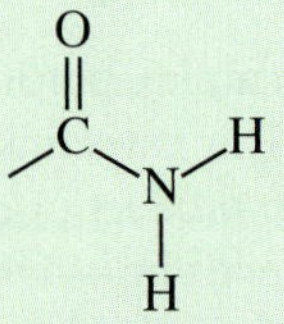
phenyl	—C$_6$H$_5$	

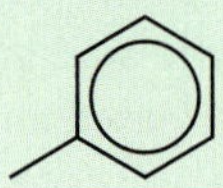

However, the —OH within a —COOH group is *not* an alcohol. Think of the entire carboxylic acid group, —COOH, as a unit.

The ester functional group is closely related to the carboxylic acid group. The general structure is:

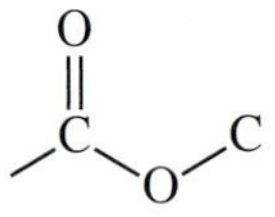

These functional groups and others will be discussed in greater detail later. This set will be sufficient for you to understand the basics of condensation polymers, the topic of this section.

Polyethylene terephthalate (PET or PETE) is formed by **condensation polymerization**, a process in which the monomers join by eliminating (splitting out) a small molecule such as water. Thus, a condensation polymerization has two products: the polymer itself plus molecules released during the polymer's formation. Many polymers are formed by condensation reactions. Natural ones include cellulose, starch, wool, silk, and proteins; synthetics include nylon, Dacron, Kevlar, ABS, and Lexan.

Polyethylene terephthalate is a **copolymer**, a combination of two or more different monomers—ethylene glycol and terephthalic acid (see Table 9.1). With two monomers, we have double the trouble or double the fun, depending on how you wish to approach this. One monomer, ethylene glycol, HOCH$_2$CH$_2$OH, contains two hydroxyl groups, one on each carbon atom. The other monomer, terephthalic acid, contains two carboxylic acid groups, one on each side of the benzene ring. Thus each monomer is armed with two functional groups.

To understand how copolymers can form, let us first consider just one molecule of each monomer. In the equation that follows, keep your eye on the OH in the carboxylic acid and the H in the alcohol. These are released to form a water molecule.

terephthalic acid   ethylene glycol

[9.5]

The remaining portions of the alcohol and the acid connect through the ester functional group. This ester linkage is highlighted in blue.

The product of equation 9.5 still has groups remaining that can react: —COOH on the left end and —OH on the right. The carboxylic acid group can react with the hydroxyl group of another ethylene glycol molecule; likewise, the hydroxyl group can react with a carboxylic acid group of another terephthalic acid molecule. Each time, a molecule of water is released and an ester is formed. This process, represented in Figure 9.10, occurs many times over to yield polyethylene terephthalate. This polymer is a type of polyester, as all the monomers are joined with ester functional groups.

Consider This 9.15   Esters and Polyesters

You have seen that terephthalic acid and ethylene glycol can react. Now consider acetic acid and ethyl alcohol (ethanol):

acetic acid   ethanol

a. Show how this carboxylic acid and alcohol can react to form an ester.
Hint: Remember the water molecule is formed as a product.
b. Could acetic acid and ethanol react to form a polyester? Explain your reasoning.

The most common use for PET is in beverage bottles because PET is semi-rigid, colorless, and gas-tight (Figure 9.11). Narrow, thin ribbons of it (under the trade name Mylar) are coated with metal oxides and magnetized to make audiotapes and videotapes. Artificial hearts contain parts made of PET. Photographic and X-ray film are made from PET, and containers are made from it for medical supplies to be sterilized by irradiation.

PET is not the only polyester in town. By varying the number and type of carbon atoms in the monomers, chemists have synthesized different polyester polymers. Trade names include Dacron, Polartec, Fortrel, and Polarguard. Polyester can easily be spun into fibers that are easy to wash and dry quickly. Since their introduction, polyester fibers have found many uses in fabrics. Dacron and other polyesters are frequently mixed with

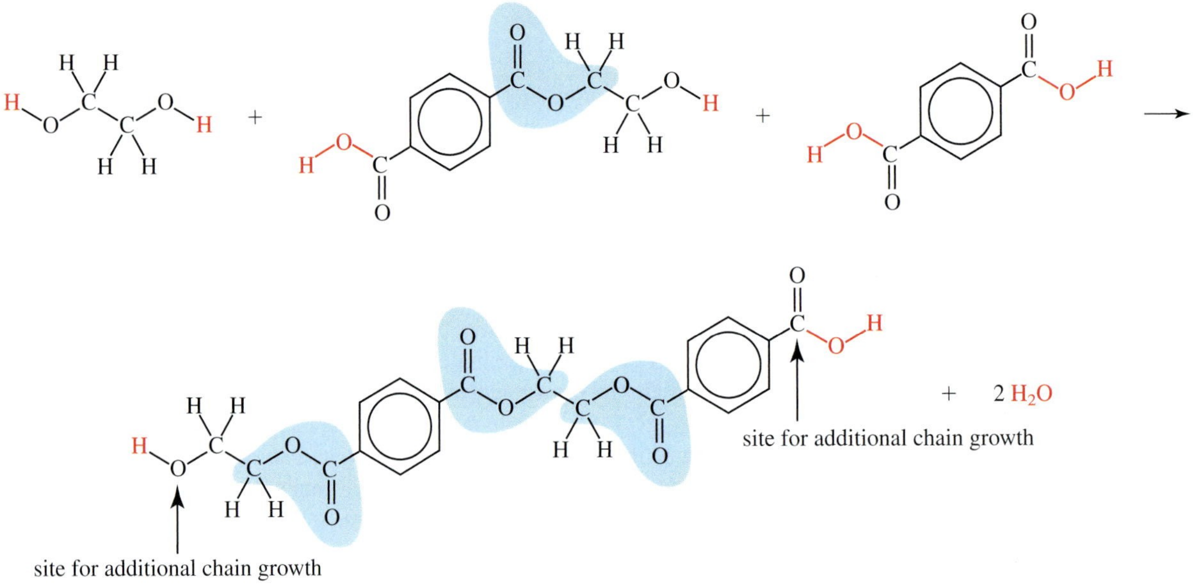

Figure 9.10

Two different monomers are used to build PET, a polyester. The ester functional group is highlighted in blue.

cotton or wool to make fabric blends. Polartec and Polartec fleece are commonly used in outer vests, jackets, hats, blankets, scarves, and gloves. The next activity shows the two monomers for polyethylene naphthalate (PEN), a polymer similar to PET, but with better temperature resistance.

Consider This 9.16 **From PET to PEN**

In both PET and PEN, the alcohol monomer is ethylene glycol, but the organic acid monomers differ slightly. Here is the organic acid monomer in PEN.

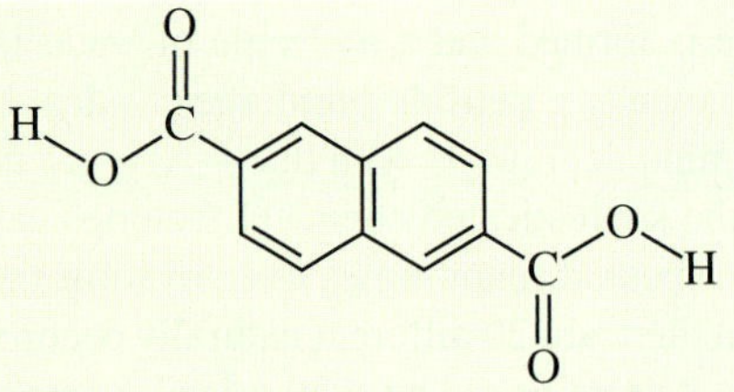

Use structural formulas to show the reaction of two molecules of naphthalic acid with two molecules of ethylene glycol.

Figure 9.11

Two-liter soft-drink bottles made from PET.

9.6 Polyamides: Natural and Nylon

No discussion of condensation polymerization can be complete without examining two other classes of polymers. One is proteins, which are natural polymers; the other is nylon, a synthetic substitute that brilliantly duplicates some of the properties of silk, a naturally occurring protein. Other natural proteins can be found in our skin, fingernails, hair, and muscle tissue.

Amino acids are the monomers from which our body builds proteins. Each amino acid molecule contains two functional groups: an amine group ($-NH_2$) and a carboxylic acid group ($-COOH$). Twenty different amino acids occur naturally, each differing in one of the groups attached to the central carbon atom. This side chain is represented with an R, as shown in this general structural formula for an amino acid.

See Sections 11.4, 12.3, and 12.4 for more about amino acids and proteins.

In some amino acids, R consists of only carbon and hydrogen atoms; in others, R may include additional atoms, such as oxygen, nitrogen, and even sulfur. Some R groups have acidic properties; others are basic.

As monomers, amino acids join to form a long chain via condensation polymerization. However, there are three key differences between the condensation polymer PET and any given protein: (1) PET monomers are linked by the ester functional group. In contrast, both proteins and nylon are **polyamides;** that is, condensation polymers that contain the amide functional group. (2) PET contains just two monomers, ethylene glycol and terephthalic acid, that must be in a 1:1 ratio. In contrast, proteins can contain up to 20 different amino acid monomers that can be in any ratio. And (3) as monomers, each amino acid has two *different* functional groups, an amine and a carboxylic acid. In PET, each monomer contained two of the *same* functional group: an alcohol or a carboxylic acid.

Let's see how these three differences play out. Here is a representation of the reaction between two amino acids, one with the side chain R and the other with R'.

[9.6]

peptide bond

In this reaction, an amide is formed and a molecule of water is eliminated. This amide contains a C-to-N bond known as a **peptide bond,** the covalent bond that forms when the —COOH group of one amino acid reacts with the —NH_2 group of another, thus joining the two amino acids. In the sophisticated chemical factories called biological cells, this condensation reaction is repeated many times over to form the long polymeric chains called proteins. Given that there are 20 different naturally occurring amino acids as building blocks, a great variety of proteins can be synthesized. Some contain hundreds of these 20 amino acids, whereas others contain only a few.

Chemists often attempt to replicate the chemistry of nature. For example, a brilliant chemist working for the DuPont Company, Wallace Carothers (1896–1937) (Figure 9.12) was studying many polymerization reactions, including the formation of peptide bonds (see equation 9.6). Instead of using amino acids, Carothers tried combining adipic acid and hexamethylenediamine, also known as 1,6-diaminohexane.

adipic acid   hexamethylenediamine

Figure 9.12

Wallace Carothers, the inventor of nylon.

Note that the molecule of adipic acid has two carboxylic acid groups, one on each end; similarly, hexamethylenediamine has two amine functional groups. As in the case of protein synthesis, the acid and amine groups react to eliminate water and form an amide. But in this instance, the polymer consisted of two alternating monomers. This polymer is known as nylon.

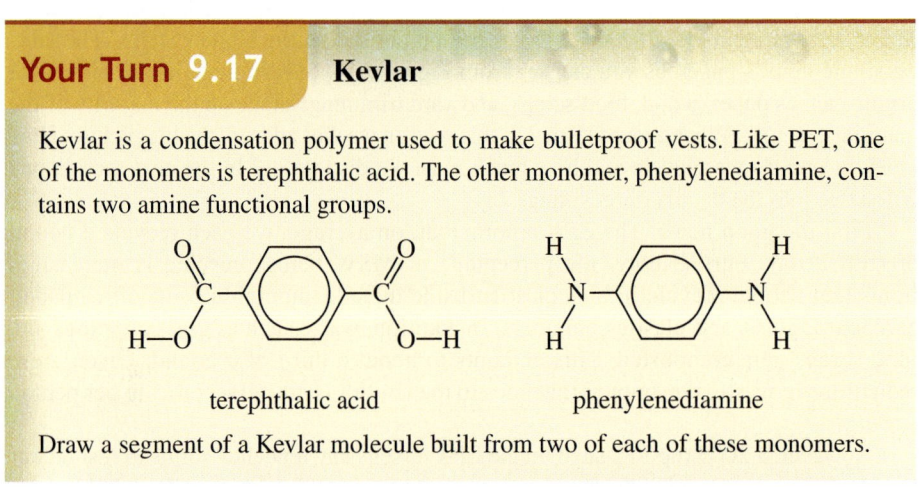

adipic acid   hexamethylenediamine

$$+ \ H_2O \qquad [9.7]$$

DuPont executives decided the new polymer had promise, especially after company scientists learned to draw it into thin filaments. These filaments were strong and smooth and very much like the protein spun by silkworms. Therefore, nylon was first introduced to the world as a substitute for silk. The world greeted it with bare legs and open pocketbooks. Four million pairs of nylon stockings were sold in New York City on May 15, 1940, the first day they became available (Figure 9.13). But, in spite of consumer passion for "nylons," the civilian supply soon dried up, as the polymer was diverted from hosiery to parachutes, ropes, clothing, and hundreds of other wartime uses. By the time World War II ended in 1945, nylon had repeatedly demonstrated that it was superior to silk in strength, stability, and rot resistance. Today this polymer, in its many modifications, continues to find wide applications in clothing, sportswear, camping equipment, the workroom, the kitchen, and the laboratory.

Your Turn 9.17   Kevlar

Kevlar is a condensation polymer used to make bulletproof vests. Like PET, one of the monomers is terephthalic acid. The other monomer, phenylenediamine, contains two amine functional groups.

terephthalic acid   phenylenediamine

Draw a segment of a Kevlar molecule built from two of each of these monomers.

Figure 9.13

Customers eagerly lined up to buy nylon stockings in 1940, when they were first available commercially.

9.7   Recycling: The Big Picture

This chapter opened noting that most polymers are synthesized from petroleum feed-stocks. But on a day-to-day basis, people seem to be a good deal more concerned about where plastics go than where they came from. The next activity will enable you to do your own tally.

Consider This 9.18   Plastic You Toss

Keep a journal of all the plastic you throw away (not recycle) in one week. Include plastic packaging from food and other products that you purchase. Estimate the weight of this plastic—is it a few ounces, a pound, or more? Keep the journal handy because you will be asked to review it later.

Let's put the plastic you discard in perspective. Fortunately, the EPA has been keeping statistics about municipal solid waste—better known as garbage—for over 30 years. **Municipal solid waste** (MSW) includes everything you discard or throw into your trash, including food scraps, grass clippings, and old appliances. MSW does not include industrial waste or waste from construction sites. For the past few decades, the amount of MSW has averaged just under 4.5 lb per person per day in the United States. The largest single item is paper, as you can see from Figure 9.14. In fact, materials of biological origin such as paper, wood, food scraps, and yard trimmings make up the majority of the materials we toss. What about plastic? In this section we will address the big picture: How much plastic do we recycle and how much must we dispose of? In the next section, we will delve into the details of recycling.

First the good news. The EPA reports that, on average, we each recycle a pound of stuff a day. Furthermore, the percentage of MSW being recycled is increasing. Items that people recycle in bins or at curbside include aluminum cans, office paper, cardboard, glass, and plastic containers. In addition, waste such as grass clippings and food scraps gets composted. This amounts to about a third of a pound. Given these reductions in waste, the amount that goes to the landfill is now closer to 3 lb per person per day.

However, a 2005 report issued by the EPA lessens any tendency for us to rest on our laurels. It states: "Despite sustained improvements in waste reduction, household waste remains a constant concern because trends indicate that the overall tonnage we create continues to increase." The tonnage is indeed large. Our current MSW per year is just short of 240 million tons and growing.

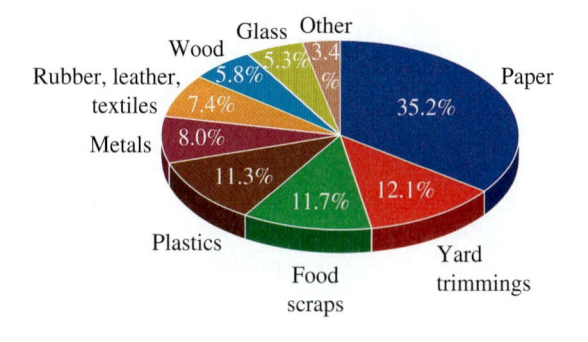

Figure 9.14

What's in your garbage? Composition by weight of municipal solid waste before recycling in 2003.

Source: U.S. Environmental Protection Agency, EPA 530 F-05-003, April 2005.

How much of this is plastic? Consult Figure 9.14 to see that roughly 11% of what we discard is plastic. EPA tallies the data for three kinds of plastics. The first is durable items, such as plastic furniture and garden hoses. The second is nondurable items such as textiles, plastic pens, and safety razors. The third is containers and packaging, such as beverage bottles and food containers. Doing the math, we as a nation generate about 11% of 240 million tons of plastic a year, or about 27 million tons. Depending on the type of plastic, our efficiency in recycling may be low or high, as the next activity will reveal.

Consider This 9.19 | Recycling Scorecard

According to the EPA, here is our recycling scorecard for 2003.

Type of Plastic	Weight Generated (millions of tons)	Weight Recovered (millions of tons)
Durable goods	8.39	0.33
Nondurable goods	6.35	Negligible
Containers/Packaging	11.9	1.06

a. For each type of plastic, calculate the plastic recovered as a percent of the waste generated.

b. Nondurable goods have very low recycling rates. Name some nondurable items and suggest reasons why.

If you did the calculations, you saw that we recycle plastics at a surprisingly low rate. This may seem at odds with all the milk jugs and plastic bottles you see being recycled in your own community. Indeed, you are correct. Certain plastics are recycled much more consistently than others. For example, polyethylene milk containers are recycled at a rate of 32% and clear PET soft drink bottles at 25%. Pitching these plastics into recycling bins adds up!

Sceptical Chymist 9.20 | Your Personal Tally

Earlier, in Consider This 9.18, you kept a journal of all the plastic you discarded in a week. Do the figures just cited by the EPA ring true? That is, do you throw out far more plastic than you recycle? In your deliberations, remember that you may not have discarded any durable goods (garden hoses and the like) on a particular day. Factor in your use of polymers such as nylon, polyester, and polyethylene as part of the bigger picture.

The plastic that we do not recycle or reuse eventually ends up in one of two places: a landfill (the usual "out of sight, out of mind" approach) or an incinerator. Both are problematic. Although landfill space is still available and in some areas even plentiful, landfills nonetheless have drawbacks. They take up space in congested areas, they have costs associated with their construction and upkeep, they may leak, and they emit methane, a greenhouse gas.

Another complication is that the majority of plastics do not readily biodegrade in the landfill (or anywhere else). Most bacteria and fungi lack the enzymes necessary to break down synthetic polymers. Some microbes, however, possess the enzymes to break down

Figure 9.15

Logo awarded by the Biodegradable Products Institute.

naturally occurring polymers such as cellulose. For example, in Chapter 3 you read about the release of methane by cattle. Actually, the methane is produced when bacteria obtain energy by decomposing cellulose in the cow's rumen. In the same chapter, you also saw that methane is generated by natural decomposition of organic materials in landfills, another result of bacterial activity.

Chemists can engineer biodegradability into some polymers by introducing certain bonds or groups into the chain. For example, research scientists at DuPont recently developed a polymer called Biomax that decomposes in about eight weeks in a landfill. This new polymer is a chemical relative of PET, but differs in that smaller amounts of other monomers are included with the usual ones, terephthalic acid and ethylene glycol. These additional monomers create weak spots in the polymers that are susceptible to degradation by moisture. Once water does its job of breaking the polymer into smaller chains, microorganisms feed off the smaller chains to produce CO_2 and water. Biomax, not yet in wide use, nonetheless has many potential applications: fast-food packaging, lawn bags, and the liners of disposable diapers. In 2003, Biomax was awarded the Compostable Logo for products "designed to compost quickly, completely, and safely" (Figure 9.15).

Achieving biodegradability in polymers raises some concerns, as an EPA report cautions.

"Before the application of these technologies can be promoted, the uncertainties surrounding degradable plastics must be addressed. First, the effect of different environmental settings on the performance (e.g., degradation rate) of degradables is not well understood. Second, the environmental products or residues of degrading plastics and the environmental impacts of degradables on plastic recycling are unclear."

Part of the difficulty is that even natural polymers do not decompose completely in landfills. Modern waste disposal facilities are covered and lined to prevent leaching of waste and waste by-products into the surrounding ground. Landfill linings and coverings also create anaerobic (oxygen-free) conditions that impede bacterial and fungal action. As a result, many supposedly biodegradable substances decompose slowly or not at all. Excavation of old landfills has unearthed old newspapers that are still readable (Figure 9.16) and five-year-old hot dogs that, while hardly edible, are at least recognizable.

Figure 9.16

Some buried wastes can remain intact for a long time. This newspaper from 1952 was excavated 37 years later.

Consider This 9.21   **Landfill Liners**

Landfill liners include natural clay and human-made plastics. For example, thick sheets of high-density polyethylene may be employed. Even the best HDPE liners, however, can crack and degrade.

  a. From Table 9.1, which types of chemicals soften HDPE?
  b. Name five substances sent to the landfill that over time could degrade a HDPE liner.

Answer

  b. Cooking oil, shoe polish, and alcohol, to name a few.

What about incineration? Because the Big Six and most other polymers are primarily hydrocarbons, incineration is an excellent way to dispose of them. The chief products of combustion are carbon dioxide, water, and a good deal of energy. In fact, pound for pound, plastics have a higher energy content than does coal. Although plastics account for only 11% of the weight of municipal solid waste, they have approximately 30% of its energy content.

But incineration of plastics is not without its drawbacks. The repeated message of Chapters 1–4, that burning does not destroy matter, applies here as well. The gases produced by combustion may be "out of sight," but they had best not be "out of mind." Burning plastics produces CO_2, a greenhouse gas. Of special concern in incineration are chlorine-containing polymers such as polyvinyl chloride that release hydrogen chloride during combustion. Because HCl dissolves in water to form hydrochloric acid, such smokestack exhaust could make a serious contribution to acid rain. Chlorine-containing plastics sometimes burn to produce toxic gases. Moreover, some plastic products have inks containing lead and cadmium. These toxic metals concentrate in the ash left after incineration and thus contribute to a secondary disposal problem. However, if carefully monitored and controlled, incineration can lead to a large reduction in plastic waste, generate much-needed energy, and have only a small negative effect on the environment.

Refer back to Section 4.10 to see how garbage safely can be burned to produce electricity.

Your Turn 9.22   **Burning a Plastic**

Polypropylene burns completely to produce carbon dioxide and water. Write a balanced chemical equation. Assume an average chain length of 2500 monomers.

Given the problems associated with landfill disposal and incineration of natural and synthetic polymers, attention has logically turned to *recycling* both. Although recycling polymers does not literally dispose of them as does incineration or biodegradation, it helps to reduce the amount of new plastic entering the waste stream. However, in contrast to incineration, recycling polymers requires an input of energy. Furthermore, if the waste plastic is dirty or of low quality, more energy may be needed to recycle it than to produce a comparable quantity of new virgin plastic. Nonetheless, recycling is one of several ways to divert plastic from landfills and incinerators. In the next section, we examine the details.

9.8 Recycling Plastics: The Details

In the United States, each year we generate more and more tons of municipal solid waste. The overall rate of recycling in recent years, however, has been just over 30%. The graph in Figure 9.17 shows the values over the years for *total* recycling, not just for plastics. To see how the recycling of *plastics* fared over time, we must examine not just one plastic,

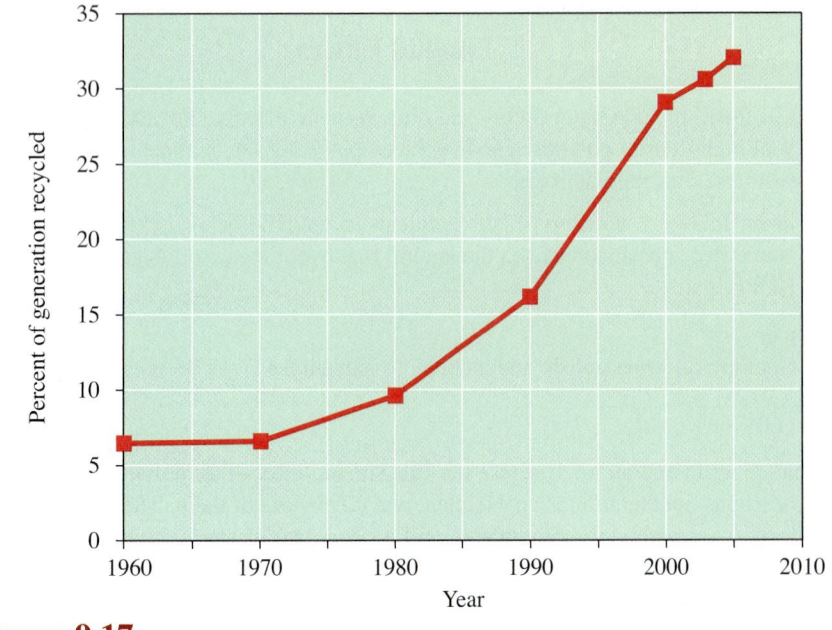

Figure 9.17

Municipal solid waste (MSW) recycling from 1960 to 2005.

Source: U.S. Environmental Protection Agency, EPA 530 F-05-003, April 2005.

but a whole host of them. Some plastics can be easily recycled; others present nothing short of a logistical nightmare. The devil is in the details!

According to the American Plastics Council, the amount of plastic we recycle is steadily increasing. For example, examine Table 9.3 to see the success we are achieving with PET and HDPE. It shows that we are adept in dropping PET pop bottles into recycling bins and in setting out HDPE milk jugs at the curb. The American Plastics Council attributes the success with PET and HDPE to several factors, including increased curb-side pick-ups, the rising recycled value of plastic, and successful recycling legislation in California and New York City. In contrast, LDPE and polyvinyl chloride are barely on the radar screen, as these plastics are far more difficult to reuse. Many factors must work synergistically to reduce the plastic waste stream.

Sceptical Chymist 9.23   Pounds or Tons Recycled

Examine the values in Table 9.3 from American Plastics Council (APC).

a. Are these values comparable to those quoted by the EPA in Consider This 9.19? Assume that the tons quoted were short tons, that is, 2000 lb per ton.

b. Why do you think the EPA reported in one unit, and the APC reported in another?

Answer

a. 1003 + 904 + 0.9 + 6 = 1914 million pounds recycled, or 0.96 million tons. This is in the ballpark of the 1.06 million tons quoted by the EPA, especially considering that one set of data is for 2003 and the other for 2004. Also, data for polystyrene is missing in the APC set.

But counting individual pop bottles begs the bigger question relating to the overall recycling rate: Are *proportionately* more (or fewer) pop bottles landing in the recycling bin? To answer this, we need data that compare the amount recycled with the total plastic waste stream. According to the Container Recycling Institute, the news is not good.

Table 9.3	Recycled Plastics in 2004	
Plastic	Amount Recycled in 2004 (Million pounds)	Change from 2003 (Million pounds)
PET	1003	+165
HDPE	904	83
PVC	0.9	+0.7
PP	6	+0.3
LDPE	0.3	+0

Source: American Plastics Council, 2004 National Post-Consumer Plastics Recycling Report.

In 1995, one out of every three containers was recycled; today the ratio stands closer to one in five. Thus while *more* containers are being recycled, this is offset by the fact that *more* containers are being manufactured. Figure 9.18 shows the bad news, at least for the United States. The values shown on these graphs translate to billions of plastic soda bottles in the landfills, in the incinerators, or more likely, in both. Not just Sweden, but many European countries are more successful.

Can bottle bills help? Certainly in some regards, such as in reducing the amount of litter and municipal solid waste. These bills require consumers to pay a deposit for their beverages at the time of purchase that is later refunded on return of the container. In essence, a bottle bill is a "container deposit law." Such refunds for beer and soft-drink bottles are nothing new. Older people may remember as children the incentive to collect glass bottles in order to claim 1- or 2-cent refunds at the local grocery store. Back in the 1950s and 1960s, of course, there were no plastic beverage bottles. The aluminum can was introduced gradually during this time. According to the Container Recycling Institute, "By 1970, cans and one-way bottles had increased to 60% of beer market share, and one-way containers had grown from just 5% in 1960 to 47% of the soft-drink market. British Columbia enacted the first beverage container recovery system in North America in 1970." So dealing with bottle waste, plastic or otherwise, is nothing new.

Today, several states have bottle bills that apply to some or all beverage containers. If you live in Oregon, your state was the first to enact such a bill in 1970. As of 2006, eleven states have bottle bills, and four others have legislation pending (Figure 9.19). In addition, eight provinces in Canada have bottle bills. In the United States, only one bill requiring a deposit ever has been repealed. This was a local ordinance in Columbia, Missouri, that was instituted in 1977 and repealed in 2002. Bottle bills are not without controversy, and you can explore the origins in the next activity.

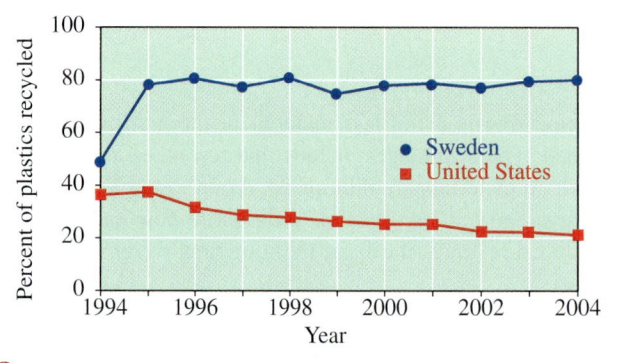

Figure 9.18

Recycling of all plastics in United States and Sweden, 1994–2004.

Source: Container Recycling Institute, including data derived from the American Plastics Council, the National Association for PET Container Resources, and AB Svenska Returpack.

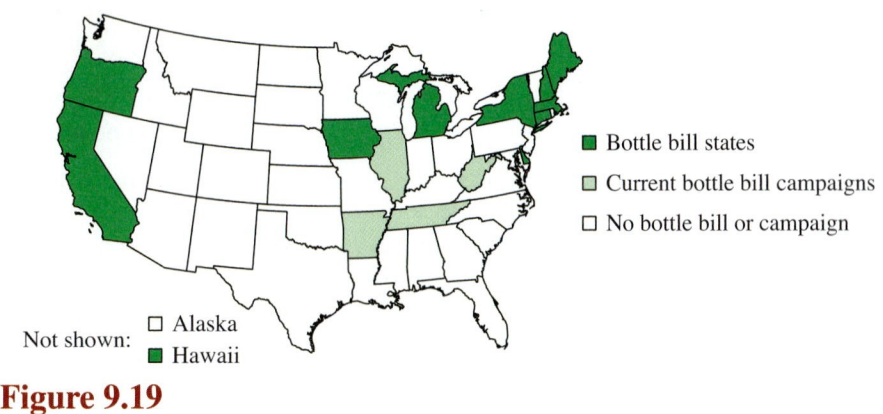

Figure 9.19

States with bottle bill or current bottle bill campaigns.

Source: Container Recycling Institute, 2007.

Consider This 9.24   Bottle Bill Controversies

Cartoons by Mark Wilson appear regularly in newspapers in up-state New York. In this one from June 2, 2005, what position is he taking? Explain why some grocers and bottle associations stand strongly against bottle bills, yet some consumer groups argue strongly for them. To get you started, links are provided at the *Online Learning Center.*

Source: Mark Wilson (MARQUIL)
http://www.empire.com. Reprinted with permission.

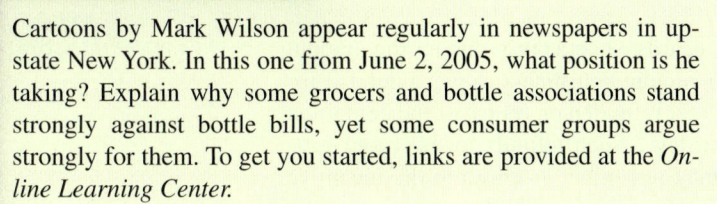

Figure 9.20

Solid waste management hierarchy.

Source: Adapted from http://www.epa.gov/epaoswer/non_hw/muncpl/fag.htm.

What other than legislation can help reduce the plastic waste stream? Source reduction and reuse also are of key importance. In fact, these are the preferred ways to manage solid waste, as indicated by the hierarchy from the EPA (Figure 9.20).

Your Turn 9.25   Management Hierarchy

As just mentioned, some states have deposit-refund systems. Where do these fall on the solid waste management hierarchy?
Hint: Your answer may vary, depending on whether the deposits are on glass or plastic bottles.

Source reduction means using less material in the first place and thus generating less waste in the end. The advantages include that resources are conserved, pollution is reduced, and any toxic materials are minimized. As an example, consider the reductions now possible in beverage bottles. The 2-L soda bottle now uses 25% less plastic than when it was introduced in 1975; the 1-gal milk jug now weighs 30% less than a decade ago.

Another reduction has occurred with polystyrene (PS). This plastic is used extensively in food packaging and beverage cups. As a result, it tends to be contaminated by food residues, making it difficult to clean and recycle. Enter source reduction. The Polystyrene Packaging Council reports that since 1974, 9% less of the plastic is now needed to manufacture the products. All told, according to the Council this reduction has "eliminated the need for more than 2900 billion pounds of polystyrene."

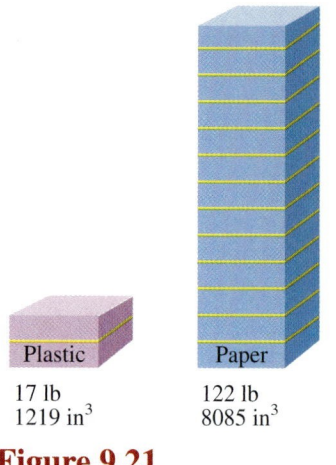

Figure 9.21

Relative volumes occupied by 1000 plastic bags and by 1000 paper bags.

Polystyrene also can be reused and with good reason. Packing "peanuts," while a tiny part of the waste stream, are a huge nuisance. These peanuts seem to end up just about everywhere, including storm drains, sewers, and waterways. Reuse is the option of choice, and in some parts of the country, the rate is as high as 30%. Similarly, different types of insulated polystyrene foam containers are reused multiple times.

The reuse option also is an important factor in the paper versus plastic debate. As we pointed out in the previous section, only just over a tenth of municipal solid waste is plastic. In contrast, paper and cardboard make up the largest percentage of municipal solid waste (35%). This raises a question that is often discussed: Which constitutes the lesser environmental burden, paper or plastic? One issue to consider is that 1000 plastic bags weigh 17 lb and have a volume of 1219 in^3 (about 2/3 of a cubic foot). The same number of paper grocery bags weighs 122 lb and takes up 8085 in^3, or about 4.5 ft^3. Figure 9.21 shows the volume relationship to scale.

Consider This 9.26   Paper or Plastic

When you are in a supermarket, which do you usually request, plastic or paper grocery bags? List the advantages and disadvantages of each. How often do you bring your own bags, thereby using neither?

Clearly when it comes to paper or plastic, the answer is neither. Rather, the best method is source reduction by bringing your own sacks to the grocery store.

Returning to our discussion of the solid waste management hierarchy (see Figure 9.20), we come to recycling. Observe that recycling is not the preferred option; rather, it occupies a middle position. For recycling to be successful and self-sustaining, a number of factors must be coordinated. These involve not only science and technology, but also economics and sometimes politics. True recycling involves a closed loop (Figure 9.22) in which plastics are collected, sorted, and then converted into products that consumers buy, use, and later recycle.

In order to recycle, it is necessary to collect the plastic. We already have mentioned several options: collecting at curbside, at local drop-off centers, and through bottle bill programs involving a deposit and refund. For recycling to be successful, a dependable supply of used plastic must be consistently available at designated locations.

Once collected, the plastic needs to be transported to a facility at which it can be sorted and prepared for some marketable commodity. The codes that appear on plastic objects (see Table 9.1) help facilitate the sorting process. Because of the large volume of material, automated sorting methods have been developed. Once sorted, almost any polymer that is not extensively cross-linked can be melted. The molten polymer can be used directly in the manufacturing of new products. Alternatively, it can be solidified, pelletized, and stored for future use.

If a mixture of various polymers is melted, the product tends to be darkly colored and have different properties depending on the nature of the mixture. This type of reprocessed material is generally good enough for lower grade uses such as parking lot bumpers, disposable plastic flower pots, and cheap plastic lumber. Such mixed material is obviously not as valuable as the pure, homogeneous recycled polymer. This underscores the importance of sorting plastics. For similar reasons, manufacturers prefer to use only a single polymer in a product to avoid the need to separate.

Successful recycling requires visionaries. The chemist Nathaniel Wyeth (Figure 9.23) used his creativity to develop the plastic soda bottle in contrast with his artist brother Andrew Wyeth, who expresses his creativity on canvas. Nathaniel Wyeth was quoted as saying: "One of my dreams is that we're going to be able to melt the returned bottles down, mix them with reinforcing fibers, and make car bodies out of them. Then, once the car has served its purpose, rather than put it in the junk pile, melt the car down and make bottles out of it."

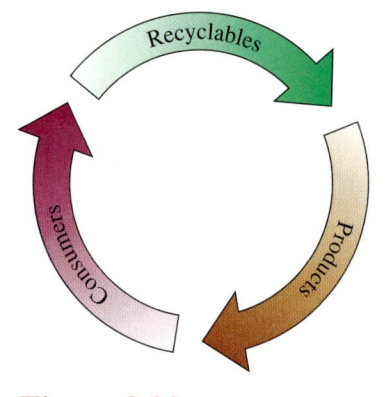

Figure 9.22

True recycling involves a never-ending loop.

Source: Reprinted by permission of the National Association for PET Container Resources.

Figure 9.23

Nathaniel Wyeth (1911–1990), with his invention of the PET soda bottle still in use today.

A variant on this method of reprocessing plastics is to decompose the polymers into simpler molecules, in some cases back to the actual original monomers. Fanciful as Nathaniel Wyeth's bottles-to-cars-to-bottles idea might seem, it is nonetheless a reality. DuPont chemists were nominated for a Presidential Green Chemistry Challenge Award for improving on Wyeth's dream. They developed a proprietary process for treating post-consumer PET that, like pulling beads apart from a necklace, unlocks (depolymerizes) the polymer back to its original monomers. These can then be reused to make new PET for other products. Mary Johnson, a DuPont employee, says "Because these monomers retain their original properties, they can be reused over and over again in any first-quality application. A popcorn bag can become an overhead transparency, then a polyester peanut butter jar, then a snack food wrapper, then a roll of polyester film, then a popcorn bag again."

Given a vision and a supply of plastic (ideally clean and sorted), the manufacturers can get to work. The items produced contain varying percentages and types of recycled materials. The terminology is confusing. **Recycled-content products** are those produced from materials that otherwise would have been in the waste stream. These include items manufactured from discarded plastic as well as rebuilt items, such as plastic toner cartridges that are refilled. Trash bags, laundry detergent bottles and carpeting are common plastic items that may qualify as recycled-content products. Some playground equipment and park benches also are made from discarded plastic.

"Recycled" products are now beginning to provide the origin of the recycled material. **Postconsumer content** is material that previously was used individually or industrially that otherwise would have been discarded as waste. **Preconsumer content** is waste left over from the manufacturing process itself, such as scraps and clipping. Preconsumer plastic, for example, could have been sent to the landfill instead of being "recycled." Finally **recyclable products** are simply ones that you can recycle. They do not necessarily contain any recycled materials.

Your Turn 9.27   Recycled and Recyclable

Give three examples of recyclable items that you might purchase. Also give three examples of recycled-content products. Can an item fall into both categories?

To complete the cycle shown in Figure 9.22, the recycled items are marketed and (ideally) purchased by consumers. A company or a city would be well advised, however, to determine or create the demand for recycled polymers before completing all the other steps. Without a product and buyers, recycling programs are doomed to fail. In fact, recycling laws in a number of cities have not been implemented and enforced because one of the links in this polymeric chain of supply, collecting, sorting, processing, manufacturing, and marketing was missing. Without all the activities needed for recycling, the system will not work, unless it is heavily subsidized. More municipalities are becoming willing to provide the necessary funds as the market for post-consumer plastics grows stronger and prices increase enough to justify significant recycling activity.

Since PET is more successfully recycled than most plastics, we should take a closer look. Approximately 1 billion pounds of PET is recycled in the United States alone! But PET soft-drink bottles need special handling before they can be melted and reused. The bottles usually are sorted to remove tinted PET containers and PVC containers. The latter, if left in the batch, can weaken the final product. Any labels, bottle caps, or food that adhered to the plastic also must be separated or scrubbed off. Bottle caps, for example, are usually made out of the tougher polypropylene. The next exercise shows how PET can be separated from other polymers by density. This is helpful in the case of PET mixed with PVC, as these both can be transparent and identical to the eye.

Consider This 9.28 — Sink or Float?

When placed in a liquid, a plastic will float if its density is less than that of the liquid and sink if it is greater. Here is the density for PET and three other plastics that are likely to be found with it in a recycling bin.

Plastic	Density (g/cm³)
PET	1.38–1.39
HDPE	0.95–0.97
PP	0.90–0.91
PVC	1.18–1.65

The densities of six liquids at the same temperature are:

Liquid	Density (g/mL)
methanol	0.79
an ethanol/water mixture	0.92
a different ethanol/water mixture	0.94
water	1.00
saturated solution of $MgCl_2$	1.34
saturated solution of $ZnCl_2$	2.01

Given a mixed sample containing PET, propose a way to separate it from the other three plastics.

Much of recycled PET is converted into polyester fabrics, including carpeting, T-shirts, the popular "fleece" used for jackets and pullovers (Figure 9.24), and the fabric uppers in jogging shoes. Five recycled 2-L bottles can be converted into a T-shirt or the insulation for a ski jacket; it takes just about 450 such bottles to make polyester carpeting for a 9 × 12 foot room. Currently only a small percentage of beverage bottles are made from recycled PET, but the amount is expected to increase in the next decade.

In this section we have explored the complexities of recycling. The technologies are new and rapidly developing as the markets shift. And recycling is not the only game in town. There is no single, best solution to the problems posed by plastic waste, or more generally by *all* solid waste. Incineration, biodegradation, reuse, recycling, and source reduction all provide benefits, and all have associated costs. Therefore, it is likely that the most effective response will be the development of an integrated waste management system that will employ all of these strategies. The goal of such an integrated system would be to optimize efficiency, conserve energy and material, and minimize cost and environmental damage.

Figure 9.24
Activewear made from recycled PET.

Consider This 9.29 — Responsible Consumerism

Unless consumers buy products made from recycled plastics, manufacturers will have little financial incentive to produce such products, and this may threaten the viability of recycling. Use the Web to find five or more products from recycled plastics, other than those mentioned in this chapter. Identify the polymer(s) used in each product.

Conclusion

Synthetic polymers are at the very center of modern living, yet their existence depends on a precious resource that rapidly is being consumed—crude oil. We have come to not only depend on synthetic polymers, but in many cases to take them for granted to the point of being wasteful. Once more we revisit the issue of lifestyle. Over time, chemists have created an amazing array of polymers and plastics—new materials that have made our lives more comfortable and more convenient. Many of these plastics represent a significant improvement over the natural polymers they replace. Furthermore, many products we use today would be impossible without synthetic polymers and plastics. There would be no DVDs, no cell phones, no breathable contact lenses, no Polartec clothing, no kidney dialysis apparatus, and no artificial hearts. We have become dependent on polymers, and it would be difficult if not impossible to abandon their use. The chemical industry has given consumers what they want. But there now appears to be more of it than we would like or perhaps than we can deal with responsibly. We must learn to cope with this glut of plastic waste while saving matter and energy for tomorrow. To create a new world of plastics and polymers will require the intelligence and efforts of policy planners, legislators, economists, manufacturers, consumers, and of course, chemists.

Chapter Summary

Having studied this chapter, you should be able to:

- Give examples of both natural and synthetic polymers (9.1)

- Understand at the molecular level the relationship between polymers and the monomers from which they are synthesized (9.1)

- Understand the molecular mechanism of addition polymerization (9.2)

- Compare and contrast low-density polyethylene and high-density polyethylene, both at the molecular level and in terms of their properties (9.3)

- Recognize the molecular structures for each of the Big Six polymers and be able to draw the structures of the monomers from which they were made (9.4)

- Match the properties of the Big Six polymers with their uses (9.4):

    low-density polyethylene (LDPE) and high-density polyethylene (HDPE) (9.3)

    polyvinyl chloride (PVC) (9.4)

    polystyrene (PS) (9.4)

    polypropylene (PP) (9.4)

    polyethylene terephthalate (PET) (9.5)

- Compare and contrast condensation polymerization with addition polymerization (9.5)

- Be able to name and draw structural formulas for different functional groups (9.5)

- Explain the relationship between amino acids and proteins (9.6)

- Use structural formulas to write the chemical equation for the synthesis of nylon (9.6)

- Explain and interpret trends in plastics recycling over the past decade (9.7)

- Relate the technical, economic, and political issues in methods for disposing of waste plastic: incineration, biodegradation, reuse, recycling, and source reduction (9.7–9.8)

- Discuss the different activities involved in recycling and their inherent complexities (9.8)

- Explain the waste reduction hierarchy, and why source reduction and reuse are preferred (9.8)

Questions

Emphasizing Essentials

1. Give two examples each of natural and of synthetic polymers.

2. Polymers sometimes are referred to as macromolecules. Explain.

3. Equation 9.1 contains an n on both sides of the equation. The one on the left is a coefficient; the one on the right is a subscript. Explain.

4. In equation 9.1, explain the function of the R• over the arrow.

5. Describe how each of these strategies would be expected to affect the properties of polyethylene. Also provide an explanation at the molecular level for each effect.

    a. increasing the length of the polymer chain

    b. aligning the polymer chains with one another

    c. increasing the degree of branching in the polymer chain

6. Figure 9.3a shows two bottles made from polyethylene. How do the two bottles differ at the molecular level?

7. Ethylene (ethene) is a hydrocarbon. Give the names and structural formulas of two other hydrocarbons that, like ethylene, can serve as monomers.

8. Why is a repeating head-to-tail arrangement not possible for ethylene?

9. Determine the number of $H_2C{=}CH_2$ monomeric units, *n*, in one molecule of polyethylene with a molar mass of 40,000 g. How many carbon atoms are in this molecule?

10. A structural formula for styrene is given in Table 9.1.

   a. Redraw it to show all of the atoms present.

   b. Give the chemical formula for styrene.

   c. Calculate the molar mass of a polystyrene molecule consisting of 5000 monomers.

11. Vinyl chloride polymerizes to form PVC in several different arrangements, as shown in Figure 9.8. Which is shown here?

12. Here are two segments of a larger PVC molecule. Do these two structures represent the same arrangement? Explain your answer by identifying the orientation in each arrangement. *Hint:* See Figure 9.8.

and

13. Butadiene, $H_2C{=}CH{-}HC{=}CH_2$, can be polymerized to make Buna rubber. Would this be by addition or condensation polymerization?

14. Which of the "Big Six" most likely would be used for these applications?

   a. clear soda bottles

   b. opaque laundry detergent bottles

   c. clear, shiny shower curtains

   d. tough indoor–outdoor carpet

   e. plastic baggies for food

   f. packaging "peanuts"

   g. containers for milk

15. Polyethylene is the most widely used synthetic polymer, but it is not the plastic of choice for margarine containers. Similarly it is not used for soft-drink bottles. Explain.

16. Plastics are widely used as containers. Check the recycling code on 10 containers of your choice (see Table 9.1). In your sample, which polymer was the most widely used?

17. Name the functional group(s) in each of these monomers.

   a. styrene

   b. ethylene glycol

   c. terephthalic acid

   d. the amino acid where R=H

   e. hexamethylenediamine

   f. adipic acid

18. Circle and identify all the functional groups in this molecule:

19. Kevlar is a type of nylon called an *aramid*. It contains rings similar to that of benzene. Because of its great mechanical strength, Kevlar is used in radial tires and in bulletproof vests. Your Turn 9.17 gives the structures for the two monomers, terephthalic acid and phenylenediamine. Name the functional groups in both the monomers and in the polymer.

20. Explain how a copolymer is a subset of the more general term *polymer.* Give an example of a polymer that is a copolymer, and one that is not.

21. Silk is an example of a natural polymer. Name three properties that make silk desirable. Which synthetic polymer has a chemical structure modeled after silk?

22. The Dow Chemical Company has developed a process that uses CO_2 as the blowing agent to produce Styrofoam packaging material.

   a. What is a blowing agent?

   b. What compound does CO_2 likely replace in the process, and why is this substitution environmentally beneficial?

23. Consider these data:

Year	U.S. Population (millions)	Plastics Produced in the United States (billions of pounds)
1997	269	89
2003	290	107

In the United States,

a. How many pounds of plastic were produced in 1997? In 2003?

b. How many pounds of plastic were produced per person in these same years?

c. Between 1997 and 2003, what is the percent change in the number of pounds of plastic produced per person?

Concentrating on Concepts

24. You were asked in Consider This 9.18 to keep a journal of all the plastic products you *throw away* in one week. Now consider all of the plastic items that you *recycle* in one week. Are there any from your first list of items thrown away that could be on your second list? Explain.

25. Draw a diagram to show the relationships among these terms: *natural, synthetic, polymer, nylon, protein.* Add other terms as needed.

26. Celluloid was the first commercial plastic, developed in response to the need to replace ivory for billiard balls and piano keys. Speculate on the properties of celluloid that made it a successful substitute for ivory in these products.

27. Glucose from corn is the source of some new bio-based polymer materials. Glucose also is the monomer in cellulose. Earlier in this text you encountered glucose in the chemical reaction of photosynthesis. What is photosynthesis and from what is glucose produced?

28. The properties of a plastic are a consequence of more than just its chemical composition. Name two other features of a polymer chain that can influence its properties.

29. Many monomers contain a C-to-C double bond. Select one and draw its structural formula together with the corresponding polymer. Describe the similarities and differences between the monomer and the polymer.

30. What structural features must a monomer possess to undergo addition polymerization? Explain, giving an example. Do the same for condensation polymerization.

31. This equation represents the polymerization of vinyl chloride. At the molecular level as the reaction takes place, how does the Cl-to-C-to-H bond angle change?

$$n \; \overset{\text{H}}{\underset{\text{H}}{\text{C}}} = \overset{\text{H}}{\underset{\text{Cl}}{\text{C}}} \quad \xrightarrow{\;R\cdot\;} \quad \left[\overset{\text{H}}{\underset{\text{H}}{\overset{|}{\underset{|}{\text{C}}}}} - \overset{\text{H}}{\underset{\text{Cl}}{\overset{|}{\underset{|}{\text{C}}}}} \right]_n$$

32. Polyacrylonitrile is a polymer made from the monomer acrylonitrile, CH_2CHCN.

a. Draw a Lewis structure of this monomer.

b. Polyacrylonitrile is used in making Acrilan fibers used widely in rugs and upholstery fabric. What danger do rugs or upholstery made of this polymer create in the case of house fires?

33. Roy Plunkett, a DuPont chemist, discovered Teflon while experimenting with gaseous tetrafluoroethylene. Here is the monomer.

$$\overset{\text{F}}{\underset{\text{F}}{\text{C}}} = \overset{\text{F}}{\underset{\text{F}}{\text{C}}}$$

a. Analogous to equation 9.1, write the chemical reaction for the polymerization of *n* molecules of tetrafluoroethylene to form Teflon.

b. Why is a repeating head-to-tail arrangement not possible for this polymer?

34. Equation 9.1 shows the polymerization of ethylene. From the bond energies of Table 4.2, is this reaction endothermic or exothermic?

35. Would your answer from the previous question differ if tetrafluoroethylene were used as the monomer? See Question 33 for the monomer.

36. Do you expect the heat of combustion of polyethylene, as reported in kilojoules per gram (kJ/g), to be more similar to that of hydrogen, coal, or octane, C_8H_{18}? Explain your prediction.

37. Recycling is not the same as waste prevention. Explain.

38. This graph shows U.S. production of plastics from 1977 through 2003.

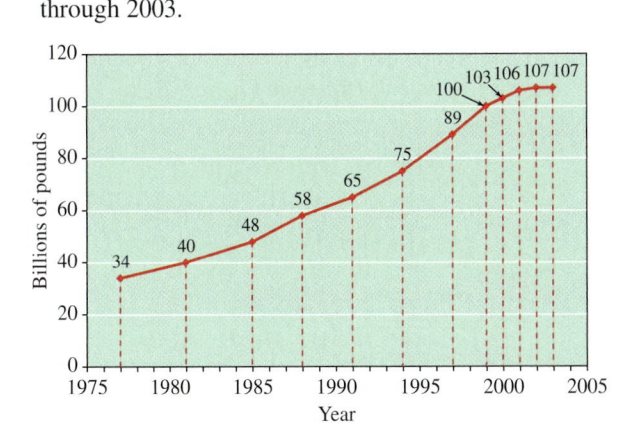

a. What is the approximate increase in plastic production for any five-year period before 2003?

b. How many years were required for the production in 1977 to double?

c. Redraw this as a bar graph that shows the relationship between the year and the pounds of plastics produced. Discuss whether the bar graph is easier than the line graph to establish the doubling time.

d. Suggest factors that may have contributed to changes in production from 2001 to 2003.

39. Consider the polymerization of 1000 ethylene molecules to form a large segment of polyethylene.

$$1000 \; CH_2 = CH_2 \quad \xrightarrow{\;R\cdot\;} \quad \left(CH_2CH_2 \right)_{1000}$$

a. Calculate the energy change for this reaction. *Hint:* Use Table 4.2 of bond energies.

b. To carry out this reaction, must heat be supplied or removed from the polymerization vessel? Explain.

40. Here is the structural formula for Dacron, a condensation polyester:

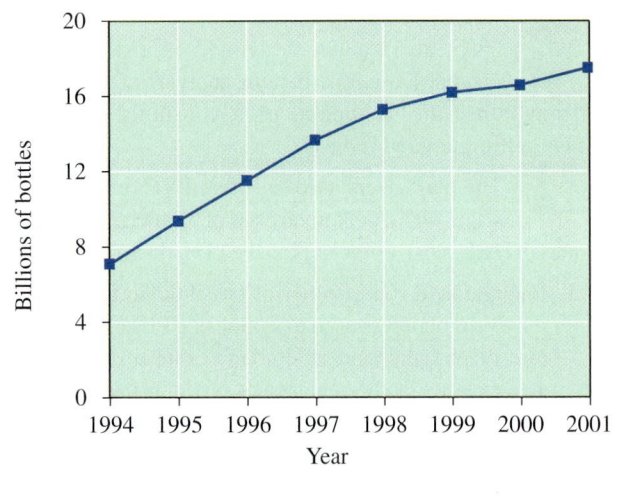

Dacron is formed from two monomers, one with two hydroxyl groups (−OH) and the other with two carboxylic acids (−COOH). Draw the structural formulas for the alcohol and the acid monomers used to produce Dacron.

41. Catalysts are used to help control the average molar mass of polyethylene, an important strategy to control polymer chain length. During World War II, low-pressure polyethylene production used varying mixtures of triethylaluminum, $Al(C_2H_5)_3$, and titanium tetrachloride, $TiCl_4$, as a catalyst. Here are some data showing how the molar ratio of the two components of the catalyst affects the average molar mass of the polymer produced.

Moles $Al(C_2H_5)_3$	Moles $TiCl_4$	Average Molar Mass of Polymer, g
12	1	272,000
6	1	292,000
3	1	298,000
1	1	284,000
0.63	1	160,000
0.53	1	40,000
0.50	1	21,000
0.20	1	31,000

a. Prepare a graph to show how the molar mass of the polymer varies with the mole ratio of $Al(C_2H_5)_3/ TiCl_4$.

b. What conclusion can be drawn about the relationship between the molar mass of the polymer and the mole ratio of $Al(C_2H_5)_3/TiCl_4$?

c. Use the graph to predict the molar mass of the polymer if an 8:1 ratio of $Al(C_2H_5)_3$ to $TiCl_4$ were used.

d. What ratio of $Al(C_2H_5)_3$ to $TiCl_4$ would be used to produce a polymer with a molar mass of 200,000?

e. Can this graph be used to predict the molar mass of a polymer if either pure $Al(C_2H_5)_3$ or pure $TiCl_4$ were used as the catalyst? Explain.

42. When you try to stretch a piece of plastic bag, the length of the piece of plastic being pulled increases dramatically and the thickness decreases. Does the same thing happen when you pull on a piece of paper? Why or why not? Explain on a molecular level.

43. Consider Spectra, Allied-Signal Corporation's HDPE fiber, used as liners for surgical gloves. Although the Spectra liner has a very high resistance to being cut, the polymer allows a surgeon to maintain a delicate sense of touch. The interesting thing is that Spectra is *linear* HDPE, which is usually associated with being rigid and not very flexible.

a. Suggest a reason why branched LDPE cannot be used in this application.

b. Offer a molecular level reason for why linear HDPE is successful in this application.

44. One limitation of the Big Six is the relatively low temperatures, 90–170 °C, at which they melt (see Table 9.1). Suggest ways to raise the upper temperature limits while maintaining the other desirable properties of these substances.

45. All the Big Six polymers are insoluble in water, but some dissolve or at least soften in hydrocarbons (see Table 9.1). Use your knowledge of molecular structure and solubility to explain this behavior.

46. When Styrofoam packing peanuts are immersed in acetone (the primary component in some nail-polish removers), they dissolve. If the acetone is allowed to evaporate, a solid remains. The solid still consists of Styrofoam, but now it is solid and much denser. Explain. *Hint:* Remember that Styrofoam is made with foaming agents.

47. This figure, entitled "Plastic Soda Bottles Wasted," was adapted from one shown in *Beverage World* magazine. The *y*-axis is in units of billions of beverage bottles.

a. Which polymer is used in clear plastic beverage bottles?

b. If recycled, to what uses can this polymer be put?

c. More bottles are recycled each year, yet this graph still shows an increase in bottles wasted. Explain.

Exploring Extensions

48. a. Name two functional groups not discussed in this chapter. Give an example of a molecule containing each one. *Hint:* You might want to look ahead to Chapter 10.

    b. Find the structural formula for the acetone molecule. What functional group does it contain?

49. 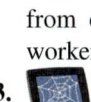 Cotton, rubber, silk, and wool are natural polymers. Consult other sources to identify the monomer in each of these polymers. Which are addition polymers and which are condensation polymers?

50. 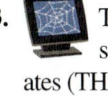 Until recently, most PET beverage bottles were colorless. Now, however, Dasani water is sold in blue PET, as are other beverages. Similarly, Sprite is sold in green PET. Find out what effect, if any, the blue and green PET is having on recycling.

51. A Teflon ear bone, fallopian tube, or heart valve? A Gore-Tex implant for the face or to repair a hernia? Some polymers are biocompatible and now used to replace or repair body parts.

    a. List four properties that would be desirable for polymers used *within* the human body

    b. Other polymers may be used outside your body, but in close contact with it. For example, no surgeon is needed for you to use your contact lenses—you insert, remove, clean, and store them yourself. From which polymers are contact lenses made? What properties are desirable in these materials? Either a call to an optometrist or a search on the Web may provide some answers.

    c. What is the difference in the material used in "hard" and "soft" contact lenses? How do the differences in properties affect the ease of wearing the contact lenses?

52. PVC, also known as "vinyl," is a controversial plastic. Do a risk–benefit analysis of using PVC from either the standpoint of a consumer or from a worker in the vinyl industry.

53. The plasticizers used to soften PVC are controversial as well. A common class of plasticizers is phthalates (THAL-ates), esters of phthalic (THAL-ic) acid.

    a. Phthalic acid is an isomer of terephthalic acid, one of the two monomers used to synthesize PETE. The structure of phthalic acid is similar to that of terephthalic acid except that the acid groups on the benzene ring are adjacent to each other. Draw the structural formula for phthalic acid.

    b. Write the chemical equation in which phthalic acid reacts with two molecules of ethanol to form a double ester.

    c. The plasticizers DINP and DEHP (here the P stands for phthalate) use a longer chain alcohol than ethanol, resulting in an ester with longer side-chains. Given the role that plasticizers play, why do you think a longer chain alcohol is needed?

    d. Use the resources of the Web to research why plasticizers such as DINP and DEHP are controversial.

54. Who first synthesized Kevlar? What was the background and academic training of these scientists? Was the potential for using this polymer in radial tires immediately understood? What are other applications of Kevlar? Write a short report on the results of your findings. Be sure to cite your sources.

55. Isoprene is the monomer that forms natural rubber. Here is its structural formula, with the carbons numbered.

$$CH_2 \overset{2}{\underset{1}{=}} \underset{|}{C} \overset{CH}{\underset{3}{\diagup}} \underset{4}{\diagdown} CH_2$$
$$CH_3$$

When isoprene monomers add to form polyisoprene (natural rubber), the polymer has a C-to-C double bond between carbon atoms 2 and 3. How does this double bond form? *Hint:* Each double bond has four electrons in it. When a new single bond is formed between two monomers, that single bond only needs two electrons in it, one from each of the monomers joined by that new bond.

56. a. What are the structures of the monomers used in SBR synthetic rubber?

    b. How are natural and synthetic rubber alike, and how do they differ?

57. Synthetic rubber is usually formed through addition polymerization. An important exception is silicone rubber, which is made by the condensation polymerization of dimethylsilanediol. This is a representation of the reaction.

$$n\ HO-\underset{\underset{CH_3}{|}}{\overset{\overset{CH_3}{|}}{Si}}-OH \longrightarrow \left[O-\underset{\underset{CH_3}{|}}{\overset{\overset{CH_3}{|}}{Si}}-O \right]_n + n\ H_2O$$

    a. Predict some of the properties of this polymer. Explain the basis for your predictions.

    b. Silly Putty is a popular form of silicone rubber. What are some of the properties of Silly Putty?

58. A few decades ago, recycling personal computers was not a concern because not enough of them were around to matter. Today, however, there is good

reason to keep keyboards, monitors, and "mice" out of the landfill.

a. Which polymers do your computer and its accessories contain?

b. What are the options for recycling the plastics in computers?

59. Free-radical peroxides promote the polymerization of ethylene into polyethylene. They also play a key role in tropospheric smog formation. Use the Web to learn more about how the peroxides promote ethylene polymerization and how peroxides are involved with photochemical smog formation in the troposphere. Write a brief report comparing the types of peroxides important with each of these cases. Cite all sources.

60. In 2007, Cargill won a green chemistry award for using soybeans instead of petroleum to produce polyols. What is a polyol, and how are polyols used to produce "soybean plastics"?

Chapter 10

Manipulating Molecules and Designing Drugs

Medications range from prescription drugs. . .

. . .to over the-counter-medicines. . .

. . . to herbal alternatives . . .

. . .to illicit drugs.

Drugs. This word elicits hope, relief, fear, intrigue, outrage, or maybe simply disdain. Pharmaceuticals (drugs) are substances intended to prevent, moderate, or cure illnesses. Medicinal chemistry is the science that deals with the discovery or design of new therapeutic chemicals and their development into useful medicines.

Modern pharmacology has its origins in folklore, and the history of medicine is full of herbal and folk remedies. The use of herbs, roots, berries, and barks for relief from illness can be traced to antiquity as illustrated in documents recorded by ancient Chinese, Indian, and Near East civilizations. The Rig-Veda (compiled in India between 4500 and 1600 BC), one of the oldest repositories of human learning, refers to the use of medicinal plants. The Chinese Emperor Shen Nung prepared a book of herbs over 5000 years ago. In it, he described a plant called *ma huang* (now called *Ephedra sinica*), used as a heart stimulant. This plant contains ephedrine, a drug we will consider later in this chapter.

More recently, chemists have designed, synthesized, and characterized a vast array of prescription and over-the-counter drugs. Today, drugs help patients regulate their blood sugar, blood pressure, cholesterol, and allergies. They help AIDS patients stay alive while scientists search for a cure. Effective anticancer drugs and powerful analgesics now exist. Other drugs can even manage mental disorders that once were thought to be untreatable.

People have long taken drugs for the purpose of altering their perceptions and moods. The famed philosopher Nietzsche said that no art could exist without intoxication. Many writers and artists have found that drugs act as a creative and destructive force in their lives and work. People abuse drugs primarily because of the promise of instant relief or pleasure and the possibility of heightened awareness. It is a common misconception that today's problem with drug abuse is a recent phenomenon. The reality is that human history has been marked with drug use and abuse.

In discussing drugs, we will consider these questions: Where do the ideas and resources to develop new drugs come from? What is the process by which pharmaceuticals make it to market? Why does a drug have a certain effect, and which features of its molecular structure contribute to the biological activity? How do drugs move from being available by prescription only to being sold over the counter? What are the merits and pitfalls of herbal medicines, and are naturally occurring drugs "safer" than synthetic ones? Which drugs are most commonly abused? Chemistry concepts key to answering these questions will be presented in this chapter. Understanding some basic chemistry can go a long way toward staying healthy in today's complicated world.

Consider This 10.1 **Today's Drugs**

   a. Consider the modern pharmaceuticals prescribed in the United States today. List what you think are the top five most frequently prescribed drugs.

   b. List what you think are the top five most frequently abused drugs.

   c. Share your lists with a small group of students. Do some of the drugs that you or others listed appear in both parts **a** and **b**?

10.1   A Classic Wonder Drug

In the fourth century BC, Hippocrates, perhaps the most famous physician of all time, described a "tea" made by boiling willow bark in water. The concoction was said to be effective against fevers. Over the centuries, that folk remedy, common to many different cultures, ultimately led to the synthesis of a true "wonder drug," one that has aided millions of people.

One of the first systematic investigators of willow bark (Figure 10.1) was Edmund Stone, an English clergyman. His report to the Royal Society (1763) set the stage for a series of further chemical and medical investigations. Chemists were subsequently able

Figure 10.1

The white willow tree, *Salix alba,* source of a miracle drug.

to isolate small amounts of yellow, needle-shaped crystals of a substance from the willow bark extract. Because the tree species was *Salix alba,* this new substance was named salicin. Experiments showed that salicin could be chemically separated into two compounds. Clinical tests provided evidence that only one of these components reduced fevers and inflammation. It also was demonstrated that the active component was converted to an acid in the body. Unfortunately, the clinical testing revealed some troubling side effects. The active component not only had a very unpleasant taste, but also its acidity led to acute stomach irritation in some individuals.

The active acid was used as a treatment for pain, fever, and inflammation. But because of its serious side effects, chemists set out to modify the structure of the active acid to form a related compound that still would be effective, but without the undesirable taste or stomach distress. The first modification attempt took a very simple approach. The acid was neutralized with a base, either sodium hydroxide or calcium hydroxide, to form a salt of the acid. It turned out that the resulting salts had fewer side effects than the parent compound. Based on this finding, chemists correctly concluded that the acidic part of the molecule was responsible for the undesirable properties. Consequently, the next step was to seek a structural modification that would lessen the acidity of the compound without destroying its medicinal effectiveness.

One of the chemists working on the problem was Felix Hoffmann, an employee of a major German chemical firm. Hoffmann's motivation was more than just scientific curiosity or assigned task. His father regularly took the acidic compound as treatment for arthritis. It worked, but he suffered nausea. The younger Hoffmann succeeded in converting the original compound into a different substance, a solid that reverted back to the active acid once it was in the bloodstream. This molecular modification greatly reduced nausea and other adverse reactions; a new drug had been discovered (1898).

Extensive hospital testing of Hoffmann's compound began along with simultaneous preparation for its large-scale manufacture by a well-known pharmaceutical company. The new drug itself could not be patented because it was already described in the chemical literature. However, the company hoped to recoup its investment by patenting the manufacturing process. Clinical trials showed the drug to be nonaddicting. Its toxicity is classified as low, but 20–30 g ingested at one time may be lethal. At the suggested dose of 325–650 mg every 4 hours, it is a remarkably effective antipyretic (fever-reducing), analgesic (anti-pain medication), and anti-inflammatory agent. Data from clinical tests uncovered the side effects noted in Table 10.1. The drug was also found to inhibit blood clotting and to cause at least some small, almost always medically insignificant, amounts of stomach bleeding in about 70% of users.

> Acid–base neutralization reactions were discussed in Section 6.3.

Table 10.1   Side Effects of the "Wonder Drug"

Symptoms	Frequency	Severity*
Drowsiness	Rare	4
Rash, hives, itch	Rare	3
Diminished vision	Rare	3
Ringing in the ears	Common	5
Nausea, vomiting, abdominal pain	Common	2
Heartburn	Common	4
Black or bloody vomit	Rare	1
Blood in the urine	Rare	1
Jaundice	Rare	3
Shortness of breath	Rare	3

*The severity scale ranges from 1, life-threatening, seek emergency treatment immediately to 5, continue the medication and tell the physician at the next visit.

Source: H. W. Griffith, *The Complete Guide to Prescription and Non-Prescription Drugs,* 1983, HP Books, Tucson, Arizona.

> ## Consider This 10.2   Miracle Drug
>
> In the United States, the final step for approval of a drug is the submission of all of its clinical test results to the Food and Drug Administration (FDA) for a license to market the product.
>
> **a.** If you were an FDA panel member presented with the information in Table 10.1, would you vote to approve this drug that treats pain, fever, and inflammation?
>
> **b.** If approved, should this drug be released as an over-the-counter drug or should its availability be restricted as a prescription drug? Write a one-page position paper.

Perhaps you have already guessed the identity of the miracle drug related to willow bark tea. Its chemical name, 2-(acetyloxy)-benzoic acid or (more commonly) acetylsalicylic acid, may not help much. But the power of advertising is such that, had we revealed that the firm that originally marketed the drug was the Bayer division of I. G. Farben, we would have let the tablet out of the bottle. The compound in question is the world's most widely used drug, even a century after its discovery. People in the United States annually consume nearly 80 billion tablets of this miracle medicine. You know it as aspirin.

Admittedly, we have only given the highlights in the history of aspirin. Most of the development, testing, and design of aspirin occurred in the 18th and 19th centuries. Stone's letter to the Royal Society was written in 1763, and Felix Hoffmann's modification of salicylic acid to yield aspirin was done in 1898. Furthermore, the clinical testing of aspirin was somewhat less systematic than our account implies. But the basic facts and the steps that led to aspirin's full development are essentially correct. We must also add one more very important fact. Aspirin did not have to receive drug approval before being put on the market; no such certifying process was in place at the time. Had approval based on clinical test results been necessary, it is quite likely that aspirin would have been available only on a prescription basis.

> ## Consider This 10.3   What Should a Drug Be Like?
>
> Make a list of the properties you think a drug should have. Then compare your list with those of your classmates. Note similarities and differences.
>
> **a.** Are any items missing from your list that you now think you should include?
>
> **b.** Are any items present on your list that you now think you should delete?

10.2   The Study of Carbon-Containing Molecules

One of carbon's interesting properties is its ability to form a wide variety of molecules. This element is so ubiquitous in nature that one of the largest subdisciplines of chemistry, **organic chemistry,** is devoted to the study of carbon compounds. The name *organic* is historical and suggests a biological origin for the substances under investigation, but this is not necessarily true. In practice, most organic chemists confine themselves to compounds in which carbon is combined with a relatively small number of other elements: hydrogen, oxygen, nitrogen, sulfur, chlorine, phosphorus, and bromine. Even with this restriction, over 12 million of the 27 million total known compounds are considered organic. The chemical behavior (i.e., properties and reactivity) of organic compounds enables us to organize them into a relatively small number of categories. As a result, in this chapter we concentrate on only a few and stress their important roles within living things.

To specify an organic compound from among the myriad of possibilities, you must be able to name it correctly. An international committee called the International Union of Pure and Applied Chemists (IUPAC) established and periodically updates a formal set of nomenclature rules so each of the known compounds can be uniquely named. However, many of these compounds have been known for a long time by common names such as alcohol, sugar, or morphine. When a headache strikes, even chemists do not call out for 2-(acetyloxy)-benzoic acid; they simply say "Give me some aspirin!" Likewise, prescriptions specify penicillin-N rather than 6[(5-amino-5-carboxy-1-oxopentyl)amino] 3,3-dimethyl-7-oxopentyl-4-thia-1-azabicyclo[3.2.0]heptane-2-carboxylic acid. A mouthful like this is the cause of great merriment to those who like to satirize chemists. Nonetheless, chemical names are important and unambiguous to those who know the system. You can rest easy because in this chapter, we will use common names in almost all cases.

An incredible variety of organic compounds exists because of the remarkable ability of carbon atoms to bond in multiple ways both to other carbon atoms and to atoms of other elements. To better understand such possibilities, we need a few basic rules for bonding in organic molecules. You used one of these in Chapter 2, the octet rule. When bonded, each carbon atom has a share in eight electrons, an octet. Eight electrons can be arranged to form four bonds, with a pair of shared electrons in each covalent bond. The most common bonding arrangements for these four bonds around a carbon atom are (a) four single bonds, (b) two single bonds and one double bond, and (c) one single bond and one triple bond. These arrangements are illustrated in Figure 10.2.

Other elements exhibit different bonding behavior in organic compounds. A hydrogen atom is always attached to another atom by a single covalent bond. An oxygen atom typically attaches either with two single bonds (to two different atoms) or one double bond (to a single atom). A nitrogen atom commonly forms three single bonds (to three different atoms), but also can form either a triple bond (to one other atom), or a single and a double bond.

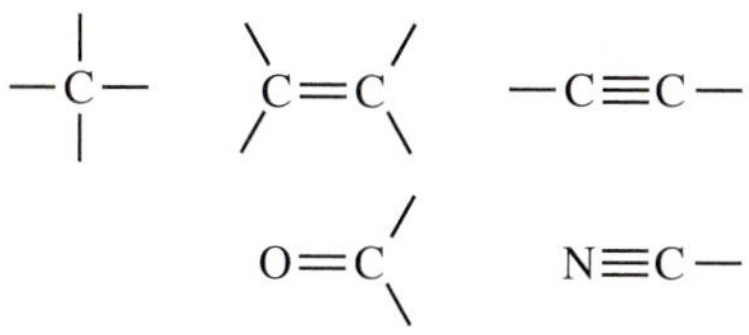

> The octet rule was discussed in Section 2.3.

(a)   (b)   (c)

Figure 10.2

Common bonding arrangements for carbon.

Your Turn 10.4   Satisfying the Octet Rule

Examine each carbon atom in Figure 10.2. Does each atom follow the octet rule?

Chemical formulas such as C_4H_{10} indicate the kinds and numbers of atoms present in a molecule, but do not show how the atoms are arranged or connected. To get that higher level of detail, **structural formulas** are used to show the atoms and their arrangement with respect to one another in a molecule. Here is the structural formula for normal butane, or *n*-butane (C_4H_{10}), a hydrocarbon fuel used in cigarette lighters and camp stoves.

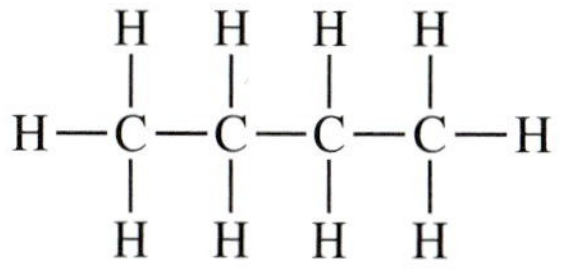

A drawback to writing structural formulas, at least in a textbook, is that they take up considerable space. To convey the same information in a format that is easily typeset into a single line, we use **condensed structural formulas** where carbon-to-hydrogen bonds are not drawn out explicitly, but simply understood to be single bonds. Here are two condensed structural formulas for *n*-butane.

$$CH_3-CH_2-CH_2-CH_3 \quad \text{or} \quad CH_3CH_2CH_2CH_3$$

Note that even though these condensed structural formulas give the opposite impression, the carbons are really bonded directly to other carbon atoms, and that the hydrogen atoms do not intervene in the chain. Rather, two or three hydrogens are attached to each carbon atom, depending on its position in the molecule.

The same number and kinds of atoms can be arranged in different ways, helping to explain why there are so many different organic compounds. **Isomers** are molecules

> Although a CH₃ group at the left end of a chain could be drawn as H₃C—, for ease we often reverse the order to CH₃— with the understanding that the bond is to the C atom, not to the H atoms.

with the same chemical formula (same number and kinds of atoms), but with different structures and properties. You already encountered isomers with the chemical formula C_8H_{18} in Chapter 4.

Isomers were also described in Section 4.9.

Here we illustrate the concept of isomers with C_4H_{10}. One way to arrange the carbon atoms is in a chain to form *n*-butane. Another arrangement is possible in which the four carbon atoms are not all in a line. This isomer is known as *iso*butane. The linear *n*-butane is shown for comparison, now represented in a more realistic zigzag form.

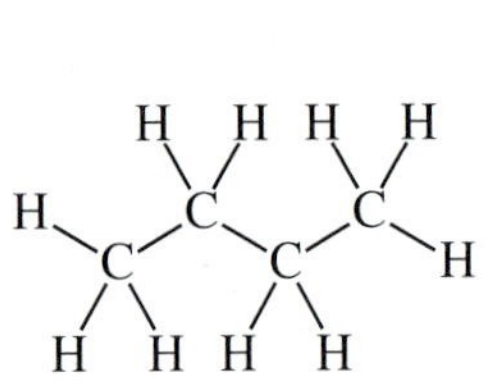

n-butane 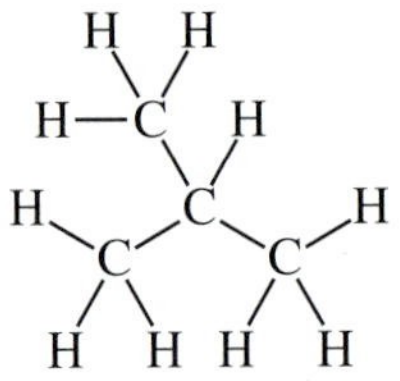   *iso*butane

The chemical formulas of these two isomers are the same; the way the atoms are connected is different. Note that the central carbon atom in *iso*butane has three carbon atoms connected to it; *n*-butane has no such carbon atom. You cannot rotate the bonds of one of these representations to make it look like the other; rotating the bonds won't change how the atoms are connected.

Just like its linear isomer, *iso*butane can be represented with a condensed structural formula.

$$CH_3 - CH - CH_3 \quad \text{or} \quad CH_3CH(CH_3)CH_3 \quad \text{or} \quad CH_3CH(CH_3)_2$$

Here, the parentheses around the CH_3 groups indicate that they are attached to the carbon atom to their left. Note that the CH_3 attached to the central CH carbon atom introduces a "branch" into the molecule.

Figure 10.3 shows three representations of *n*-butane and *iso*butane. The first column shows the simple structural formula, the second a ball-and-stick model. The third column shows a space-filling model that presents a more realistic view of the molecular shape.

Only two isomers of C_4H_{10} exist. As the number of atoms in a hydrocarbon increases, so does the number of possible isomers. Thus, C_8H_{18} has 18 isomers and $C_{10}H_{22}$ has 75. Given a chemical formula, no simple calculation can be performed to obtain the number of isomers.

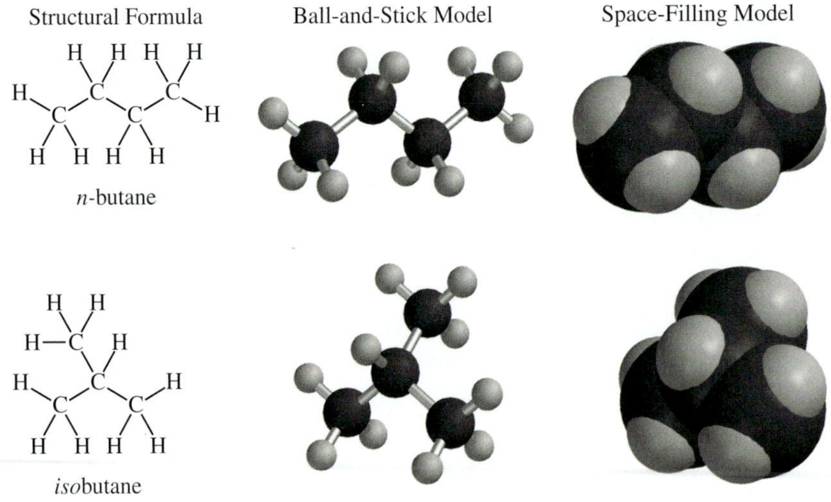

Structural Formula   Ball-and-Stick Model   Space-Filling Model

n-butane

*iso*butane

Figure 10.3

Various representations of the isomers *n*-butane and *iso*butane.

Your Turn 10.5   Drawing Structures

Draw a structural formula for each of these condensed structural formulas.

a. $CH_3CH_2CH_2CH(CH_3)_2$ **b.** $CH_3CH(CH_3)CH_2CH_3$
c. $CH_3CH_2C(CH_3)_3$ **d.** $CH_3CH(CH_2CH_3)CH_3$

Answers

Chemists also routinely use a **line-angle drawing** to represent the structure of a molecule. This is a simplified version of a structural formula that is most useful for representing larger molecules. A line-angle drawing helps you to focus on the backbone of carbon atoms. One carbon atom is assumed to occupy each vertex position. Any line extending from the backbone signifies another carbon atom (actually a —CH_3 group), unless the symbol for another element is given. Hydrogen atoms are not indicated in the line-angle drawing, but are implied as required by the octet rule. Remember that each carbon atom will have four bonds to it, sharing a total of eight electrons. Line-angle drawings for *n*-butane, *iso*butane, and two other simple molecules are shown in Table 10.2.

Table 10.2   Molecular Representations

Compound	Chemical Formula	Structural Formula	Line-Angle Drawing
n-butane	C_4H_{10}		
*iso*butane	C_4H_{10}		
n-hexane	C_6H_{14}		
cyclohexane	C_6H_{12}		

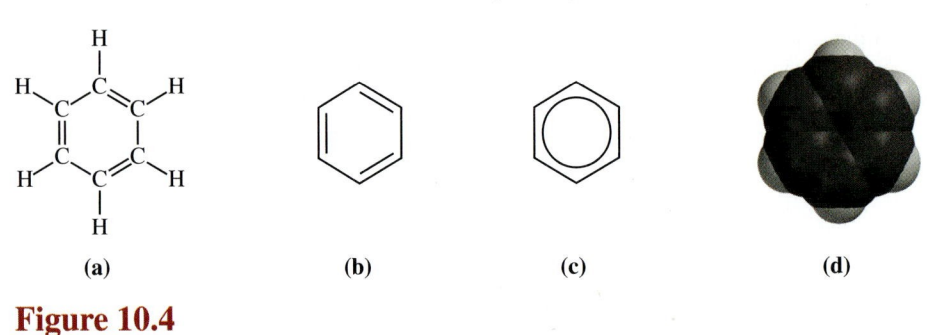

Figure 10.4
Representations of benzene, C_6H_6.

Your Turn **10.6**   **Structural Isomers**

a. Are *n*-butane and *iso*butane isomers? Explain.
b. Are *n*-hexane and cyclohexane isomers? Explain.

Your Turn **10.7**   **Isomers of C_5H_{12}**

Three isomers have the formula C_5H_{12}. For each, draw a structural formula, a condensed structural formula, and a line-angle drawing.

Many molecules, including aspirin, have carbon atoms arranged in a ring. For example, examine the structure of cyclohexane, C_6H_{12}, in Table 10.2. The ring in cyclohexane has six carbons, and rings most commonly contain five or six carbon atoms. In aspirin, however, the six-membered ring is based on benzene, C_6H_6, rather than on cyclohexane. The structural formula for benzene is shown in Figure 10.4a.

Structure (b) in Figure 10.4 is a line-angle drawing for benzene. Although this structure has alternating single and double bonds, the actual structure of benzene, based on experimental evidence indicates that all C-to-C bonds have the same length. Since C-to-C single bonds are longer than C-to-C double bonds, benzene cannot have alternating single and double bonds. Consequently, the electrons must be uniformly distributed around the ring. The circle within the hexagon in structure (c) is an effort to convey this. Both structures (c) and (d) represent the uniformly distributed electrons around the ring, as described by resonance theory (see Section 2.3). This same hexagonal structure is found in the —C_6H_5 phenyl group that is part of many molecules, including styrene and polystyrene (Section 9.4).

> A C-to-C single bond length is 0.154 nm, a double bond length is 0.134 nm, and the bond lengths in benzene are 0.139 nm.

10.3   Functional Groups

Central to the study of drug discovery and interactions are functional groups. **Functional groups** are distinctive arrangements of groups of atoms that impart characteristic physical and chemical properties to the molecules that contain them. Indeed, these groups are so important that we often show them in structural formulas and represent the remainder of the molecule with an "R." The R is generally assumed to include at least one carbon or hydrogen atom. You already encountered some functional groups in Chapter 9. The generic formula for an alcohol is ROH, as in methanol, CH_3OH (an alcohol derived from degradation of wood), and ethanol, CH_3CH_2OH (alcohol derived from fermentation of grains and sugar). The presence of the —OH group attached to a carbon makes the compound an alcohol.

> Section 9.5 discussed alcohols in polymerization reactions.

> An alcohol has an —OH group *covalently* bound to the rest of the molecule. This is different from the hydroxide ion, OH⁻, which is *ionically* bonded to a cation.

Similarly, a carboxylic acid group, commonly written as [structure showing $\overset{O}{\underset{O}{\overset{\|}{C}}}$ H], —COOH, or —CO_2H, confers acidic properties. In aqueous solution, a H⁺ ion (a proton) is transferred

from the —COOH group to an H_2O molecule to form a hydronium ion, H_3O^+. We represent an organic acid with the general formula RCOOH, or RCO_2H. In acetic acid (CH_3COOH), the acid in vinegar, the R group is —CH_3, the methyl group.

Table 10.3 lists eight functional groups found in drugs and other organic compounds. Each functional group is characteristic of an important class of compounds.

Table 10.3　Some Important Organic Functional Groups

Functional Group	Generic Formula	Specific Examples Name*	Specific Examples Structural Formula	Specific Examples Condensed Structural Formula
hydroxyl	—O—H	ethanol (ethyl alcohol)		CH_3CH_2OH
ether	C—O—C	dimethyl ether		CH_3—O—CH_3 or CH_3OCH_3
aldehyde	O‖ —C—H	propanal		O‖ CH_3CH_2—C—H or CH_3CH_2CHO
ketone	O‖ C—C—C	2-propanone (dimethyl ketone, acetone)		O‖ CH_3—C—CH_3 or CH_3COCH_3
carboxylic acid	O‖ —C—O—H	ethanoic acid (acetic acid)		O‖ CH_3—C—OH or CH_3CO_2H or CH_3COOH
ester	O‖ —C—O—C	methyl ethanoate (methyl acetate)		O‖ CH_3—C—OCH_3 or CH_3COOCH_3
amine	H| —N—H	ethylamine		$CH_3CH_2NH_2$
amide	O‖ —C—N—H | H	propanamide		O‖ CH_3CH_2—C—NH_2 or $CH_3CH_2CONH_2$

*IUPAC names, common names in parentheses.

Your Turn 10.8   Line-Angle Drawings

For each of these condensed structural formulas, make a line-angle drawing. Name the functional group in each one.

a. $CH_3CH_2CH_2COCH_3$ **b.** $CH_3CH_2CH(CH_3)CH_2OH$
c. $CH_3CH(NH_2)CH_2CH_3$ **d.** $CH_3COOCH_2CH_3$
e. CH_3CH_2CHO

Answers

a. (ketone)   **b.** (hydroxyl)

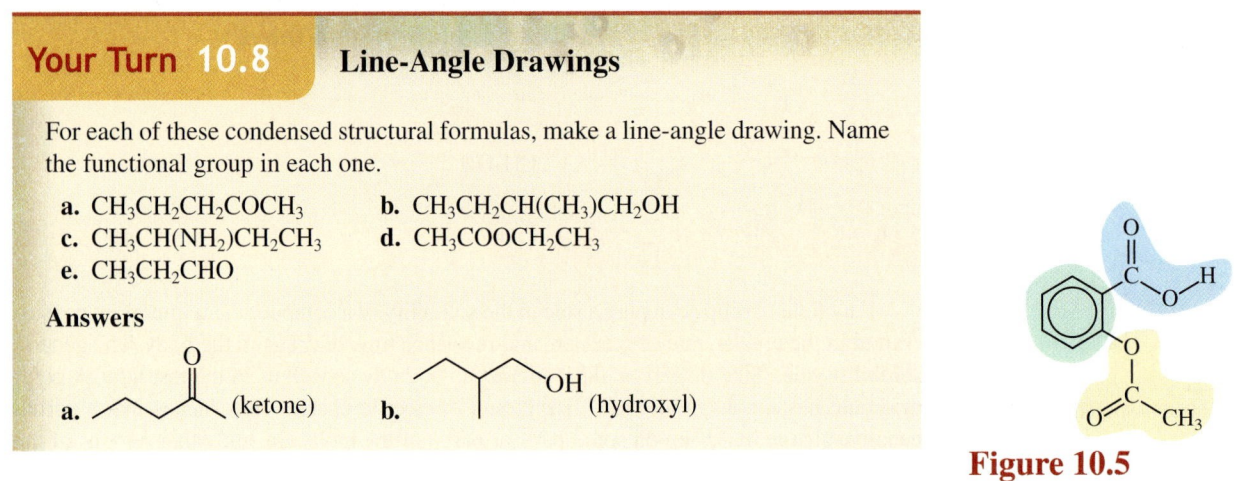

Figure 10.5
Structural formula of aspirin.

The presence and properties of functional groups are responsible for the action of all drugs. Aspirin has three functional groups, which are shown in Figure 10.5. You will recognize that the green area encloses a benzene ring. Its presence makes aspirin soluble in fatty compounds that are important cell membrane components. The other two functional groups are responsible for the drug's activity. You have just been reminded that the —COOH group indicates a carboxylic acid (blue area). The remaining functional group (yellow area) is an ester. An ester may be formed by reacting an acid and an alcohol; a water molecule is eliminated in the process. A catalytic amount of a stronger acid like H_2SO_4 speeds the reaction up; this is denoted in the reaction equation by placing an H^+ over the arrow (equation 10.1).

Felix Hoffmann prepared aspirin by modifying the structure of salicylic acid. But note that he did not modify the carboxylic acid group on the molecule. Salicylic acid also contains an —OH group that Hoffmann reacted with acetic acid as shown in equation 10.1. The product was an ester of acetic acid and salicylic acid, which accounts for one of aspirin's names: acetylsalicylic acid.

> Formation of an ester was shown for condensation polymers in Section 9.5.

$$\text{salicylic acid} + \text{acetic acid} \xrightarrow{H^+} \text{acetylsalicylic acid} + \text{water} \qquad [10.1]$$

salicylic acid   acetic acid   acetylsalicylic acid   water

Because aspirin retains the —COOH group of the original salicylic acid, it still has some of the undesirable acidic properties of the parent compound. However, the ester group (yellow area in Figure 10.5) makes the compound more palatable and less irritating to the stomach lining. Once aspirin is ingested and reaches the site of its action, equation 10.1 is reversed. The ester splits into acetic acid and salicylic acid, and the latter compound exerts its antipyretic (fever-reducing) and analgesic (pain-reducing) properties.

Your Turn 10.9   Ester Formation

Draw structural formulas for the esters that form when these acid and alcohol pairs react.

a. CH_3CH_2OH + $H{-}O{-}\overset{\overset{O}{\|}}{C}{-}CH_3$ $\xrightarrow{H^+}$

(continued)

Your Turn 10.9　　**Ester Formation (*continued*)**

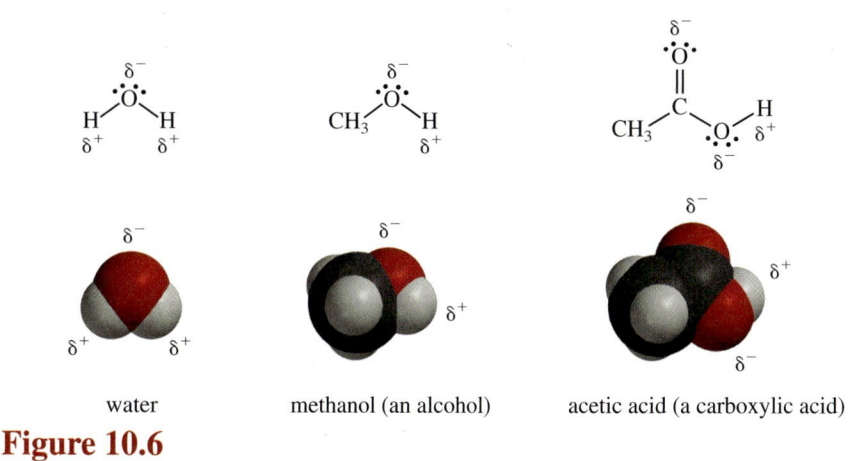

b.

Functional groups can play a role in the solubility of a compound, an important consideration in the uptake, rate of reaction, and residence time of drugs in the body. The general solubility rule, "like dissolves like," applies in the body as well as in the test tube. A polar molecule has a nonsymmetrical distribution of electric charge. This means that a partial negative charge builds up on some part (or parts) of the molecule, and other regions of the molecule bear a partial positive charge. Water is an excellent example of a polar molecule. Relatively speaking, the oxygen atom is slightly negatively charged and the hydrogen atoms are slightly positive. Because the molecule is bent, it has a nonsymmetrical charge distribution. This is represented in Figure 10.6; the δ^+ and δ^- symbols represent partial charges. Functional groups containing oxygen and nitrogen atoms (for example, —OH, —COOH, and —NH$_2$) usually increase the polarity of a molecule. This in turn enhances its solubility in a polar substance such as water, which is advantageous for drug molecules.

> The concept of *like dissolves like* was introduced in Section 5.9.

water　　　　　methanol (an alcohol)　　　　acetic acid (a carboxylic acid)

Figure 10.6

Examples of polar molecules.

By contrast, hydrocarbons that do not contain such functional groups are typically nonpolar and will not dissolve in polar solvents. For example, *n*-octane, C_8H_{18} is nonpolar and insoluble in water. However, it does dissolve in nonpolar solvents such as hexane (C_6H_{14}) and dichloromethane (CH_2Cl_2). For the same reasons, drugs with significant nonpolar character tend to accumulate in cell membranes and fatty tissues that are largely hydrocarbon and nonpolar.

The water solubility of a drug that is either acidic or basic can be improved by neutralizing it and forming a salt. For example, many drugs contain nitrogen and are basic. When such a drug is neutralized with HCl or H_2SO_4, the nitrogen accepts an H^+ from the acid and is protonated. As a result, the nitrogen becomes positively charged and paired with the negative charge on the chloride or hydrogen sulfate (HSO_4^-) ion. Prior to being protonated, the compound is electronically neutral and is said to be in its freebase form. A **freebase** is a nitrogen-containing molecule in which the nitrogen is in possession of its lone pair of electrons.

Consider the drug pseudoephedrine, a common decongestant used in over-the-counter remedies for the common cold:

[10.2]

pseudoephedrine (freebase)　　　　　　　　pseudoephedrine hydrochloride salt

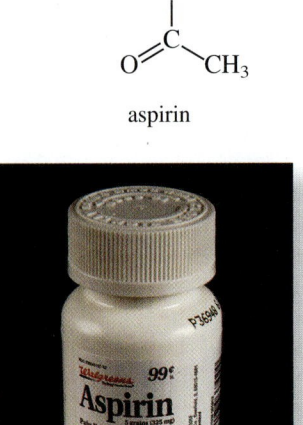
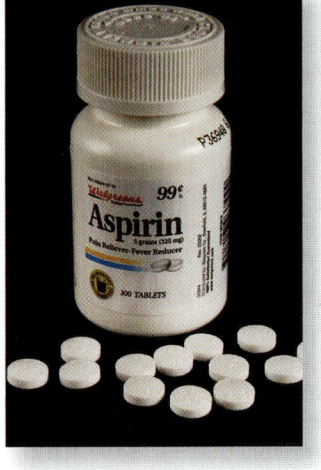
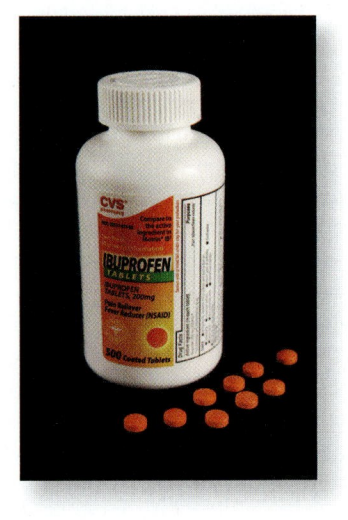
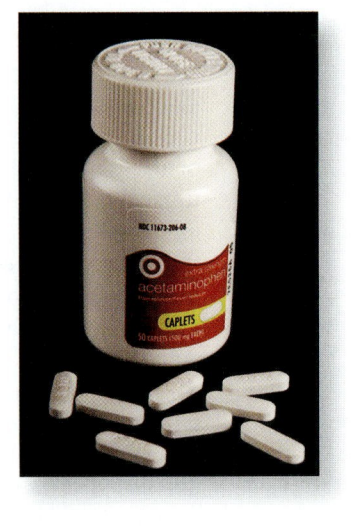

aspirin   ibuprofen   acetaminophen

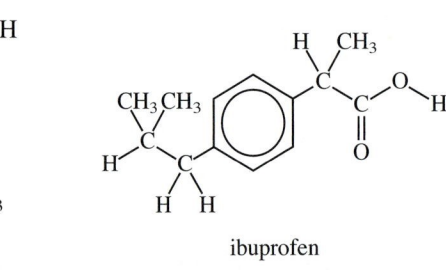

Figure 10.7

Structural formulas and samples of some common analgesics.

The nitrogen of the amino group reacts as a base when treated with hydrochloric acid. Pseudoephedrine can thus be converted to its hydrochloride salt (an ionic compound) in which the nitrogen bears a positive charge and the chloride ion a negative charge. The salt form of pseudoephedrine is preferable as a drug because it is more stable, has less of an odor, and is water-soluble. An estimated half of all drug molecules used in medicine are administered as salts that improve their water-solubility and stability, which in turn increase their shelf life. Conversion of a salt back into its freebase form may be accomplished by treating the salt with a base such as NaOH.

Drugs with similar physiological properties often have similar molecular structures and include some of the same functional groups. Of the approximately 40 alternatives to aspirin that have been produced, ibuprofen and acetaminophen are the most familiar. Figure 10.7 gives the structural formulas of the three leading analgesics. All are based on a benzene ring with two **substituents,** an atom or functional group that has been substituted for a hydrogen atom, but these substituents differ. In Your Turn 10.10, you have an opportunity to identify the structural similarities and differences of these analgesics.

Your Turn 10.10   **Common Structural Features of Analgesics**

Look at the structural formulas in Figure 10.7. Identify the structural features and functional groups that aspirin, ibuprofen, and acetaminophen have in common.

The current commercial method for producing ibuprofen is a stunning application of green chemistry. Previous methods of ibuprofen production required six steps, used large amounts of solvents, and generated significant quantities of waste. By using a

catalyst that also serves as a solvent, BHC Company, a 1997 Presidential Green Chemistry Challenge Award winner, makes ibuprofen in just three steps with a minimum of solvents and waste. In the BHC process, virtually all the reactants are converted to ibuprofen or another usable by-product; any unreacted starting materials are recovered and recycled. Nearly 8 million pounds of ibuprofen, enough to make 18 billion 200-mg pills, is produced annually in Bishop, Texas, at the BHC facility, built specifically for the commercial production of the drug.

10.4   How Aspirin Works: Function Follows Form

To understand the action of aspirin, you need to know something about the body's chemical communication system. We normally think of internal communication as consisting of electrical impulses traveling along a network of nerves. This is true for the system that triggers movement, breathing, heartbeats, and reflex actions. Most of the body's messages, however, are conveyed not by electrical impulses, but through chemical processes. In fact, your very first communication with your mother was a chemical signal saying "I'm here; better get your body ready for me." It is much more efficient to release chemical messengers into the bloodstream where they can be circulated to appropriate body cells, than to "hardwire" each individual cell with nerve endings.

The chemical messengers produced by the body's endocrine glands are called **hormones.** Figure 10.8 is a representation of such chemical communication. Hormones encompass a wide range of functions and a similarly wide range of chemical composition and structure. Thyroxine, a hormone secreted by the thyroid gland, is essential for regulating metabolism. The ability of the body to use glucose (blood sugar) for energy depends on insulin. This hormone, a small protein built from only 51 polymerized amino acids, is secreted by the pancreas. Persons who suffer from diabetes are often required to take daily injections of insulin. Yet another well-known hormone is adrenaline (epinephrine), a small molecule that prepares the body to "fight or flight" in the face of danger.

Aspirin and other drugs that are physiologically active, but not anti-infectious agents, are almost always involved in altering the chemical communication system of the body. A significant problem is that this system is very complex, allowing many

Section 12.6 describes the use of genetic engineering to obtain human insulin from bacteria.

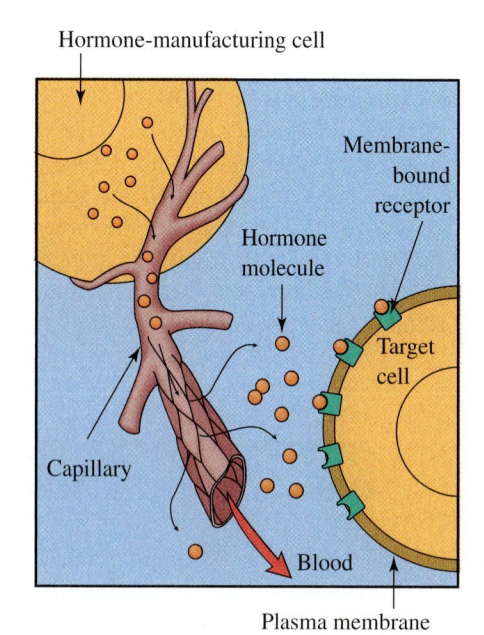

Figure 10.8

Chemical communication in the body. Hormone molecules travel through the bloodstream from the cell where they are made to the target cell.

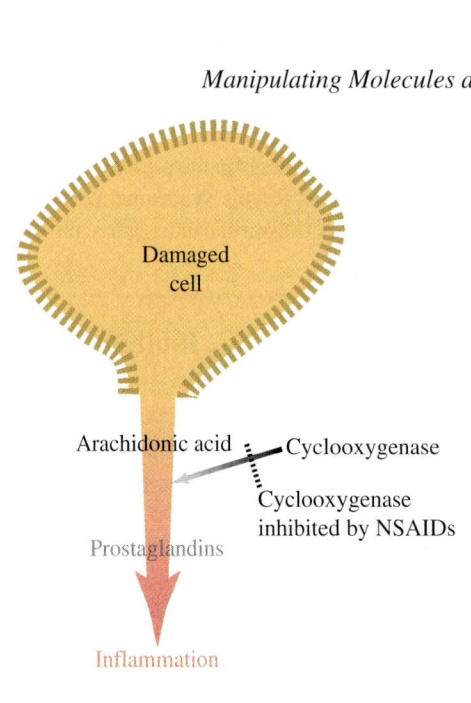

Figure 10.9
Aspirin's mode of action.

compounds to be used to send more than one message simultaneously. The wide range of aspirin's therapeutic properties, as well as its side effects, is clear evidence that the drug is involved in several chemical communication systems. It works in the brain to reduce fever and pain, it relieves inflammation in muscles and joints, and it appears to decrease the chances of stroke and heart attack. It may even lessen the likelihood of colon, stomach, and rectal cancer.

In large measure, the versatility of aspirin and similar nonsteroidal anti-inflammatory drugs (NSAIDs; steroids are covered in Section 10.7) is related to their remarkable ability to block the actions of other molecules. Research on the activity of aspirin indicates that one of its modes of action involves blocking cyclooxygenase (COX) enzymes. **Enzymes are proteins that act as biochemical catalysts, influencing the rates of chemical reactions.** Most enzymes speed up reactions and channel them so that only one product (or a set of related products) is formed. Cyclooxygenases catalyze the synthesis of a series of hormone-like compounds called prostaglandins from arachidonic acid (Figure 10.9).

Prostaglandins cause a variety of effects. They produce fever and swelling, increase sensitivity of pain receptors, inhibit blood vessel dilation, regulate the production of acid and mucus in the stomach, and assist kidney functions. By preventing prostaglandin production, aspirin reduces fever and swelling. It also suppresses pain receptors and so functions as a painkiller. Because the benzene ring conveys high fat solubility, aspirin is also taken up into cell membranes. In certain specialized cells, the drug blocks the transmission of chemical signals that trigger inflammation. This process also appears to be related to aspirin's effectiveness as a pain reliever.

The NSAIDs exhibit these same properties in varying degrees. For example, because acetaminophen blocks COX enzymes, but does not affect the specialized cells, it reduces fever but has little anti-inflammatory action. On the other hand, ibuprofen is a better enzyme blocker and specialized cell inhibitor. Consequently, ibuprofen is both a better pain reliever and fever reducer than aspirin. Ibuprofen has fewer functional groups than aspirin, which may be the reason why ibuprofen has fewer side effects. With fewer polar functional groups, ibuprofen is more fat-soluble than aspirin. Its anti-inflammatory activity is five to 50 times that of aspirin.

Interestingly, aspirin is unique among these three compounds in its ability to inhibit blood clotting. This property has led to the suggestion that low regular doses of aspirin can help prevent strokes or heart attacks. Of course, these anticoagulation characteristics also mean that aspirin is not the painkiller of choice for surgical patients, or those suffering from ulcers or blood-clotting problems. That is the rationale behind the ancient advertising of

You encountered catalysts in several other contexts, including automobile emissions control (Section 1.11), petroleum refining (Section 4.8), and addition polymerization (Section 9.3).

"more hospitals use Tylenol"; doctors don't want their patients to bleed unnecessarily. Another drawback of aspirin is that in rare cases in the presence of certain viruses, it can trigger a sometimes fatal response known as Reye's syndrome, particularly in children younger than 15. Furthermore, in some patients aspirin can trigger acute episodes of asthma.

Consider This 10.11   COX-2 Inhibitors

In 1992 researchers discovered that there were two separate types of COX enzymes, COX-1 and COX-2. The COX-2 enzyme is responsible for the production of the prostaglandins that regulate inflammation, pain, and fever, thereby reducing these symptoms. The COX-1 enzyme catalyzes the production of the prostaglandins responsible for maintaining proper kidney function and for keeping the stomach lining intact. Aspirin blocks the activity of both COX enzymes so it has the unwanted side effects of producing stomach aches and bleeding.

After this discovery, new "super aspirins" that would affect only the COX-1 enzymes were developed. Touted as drugs that would cause fewer gastrointestinal problems, they entered the market with great fanfare. They became instant blockbuster drugs, selling in the billions of dollars per year. Use the Web to find the names of two of these drugs. Why have they disappeared from the pharmaceutical radar screen?

A few final comments about NSAIDS seem appropriate. Because it is a specific chemical compound, aspirin is aspirin—acetylsalicylic acid, regardless of its brand. Indeed, about 70% of all acetylsalicylic acid produced in the United States is made by a single manufacturer. But, although all aspirin molecules are identical, not all aspirin tablets are the same. The products are mixtures of various components, including inert fillers and bonding agents that hold the tablet together. Buffered aspirin tablets also include weak bases that counteract the natural acidity of the aspirin. Some coated aspirins keep the tablet intact until it leaves the stomach and enters the intestine. These differences in formulation can influence the rate of uptake of the drug and hence, how fast it acts, and the extent of stomach irritation it produces. Furthermore, although standards for quality control are high, it is conceivable that individual lots of aspirin may vary slightly in purity. Aspirin also decomposes with time, and the smell of vinegar can signify that such a process has begun. Fortunately, none of this poses a significant threat to health, and the benefits of aspirin far outweigh the risks for the great majority of people.

Consider This 10.12   Supersize My Aspirin

A friend who suffers from heart disease has been told by the doctor to take one aspirin tablet a day. To save money, your friend often buys the large 300-tablet bottle of aspirin. You, on the other hand, rarely take aspirin, but cannot pass up a good bargain. You also buy the large bottle.

a. Why is the "giant economy size" bottle of aspirin not as good a deal for you as it is for your friend?
b. What chemical evidence supports your opinion?

10.5   Modern Drug Design

The evolution of "willow bark tea" to aspirin and further modifications to this painkiller's structure to enhance its beneficial effects and decrease its side effects represent stages in historical drug design. Penicillin is another example of a miracle drug

whose origin lies in "natural" sources. Molds had been used for treating infections for 2500 years, although their effects were unpredictable and sometimes toxic. The penicillin story includes an accidental discovery by the British bacteriologist Alexander Fleming in 1928. Fleming's curiosity was aroused by the chance observation that in a container of bacterial colonies, the area contaminated by the mold *Penicillium notatum* was free of bacteria (Figure 10.10). He correctly concluded that the mold produced a substance that inhibited bacterial growth, and he named this biologically active material penicillin.

A careful reconstruction has indicated that a series of critical, but fortuitous events had to occur for the discovery to be made. Spores from the mold, part of an experiment in a nearby lab, drifted into Fleming's laboratory and accidentally contaminated some Petri dishes containing *Staphylococcus* (bacteria) growing on a nutrient medium. Then came a series of chance incidents involving poor laboratory housekeeping, a vacation, and the effects of weather. Fleming fortunately noticed among a pile of dirty glassware the dish in which the *Staphylococcus* had been killed. His experience allowed him to interpret the phenomenon, recognizing that some unknown substance produced by the *Penicillium* was a potential antibacterial agent. "The story of penicillin," Fleming wrote, "has a certain romance in it and helps to illustrate the amount of chance, of fortune, of fate, or destiny, call it what you will, in anybody's career." But of course, the discovery would not have happened without Fleming's powers of observation and insight. The episode illustrates the often misquoted maxim of the great French scientist, Louis Pasteur: "In the fields of observation, chance favors only the prepared mind." Most versions of this famous aphorism neglect the "only." It was only because Fleming's mind was prepared that he was able to capitalize on this chain of unlikely events.

The process of taking penicillin from the Petri dish to the pharmacy was not much different from what is done today. The first step was a systematic effort to isolate the active agent produced by *Penicillium notatum*. Once identified, the substance had to be purified and concentrated by sophisticated techniques. Also, the efficacy of penicillin in treating humans had to be demonstrated. World War II gave increased impetus to this research and to the development of new methods for preparing large quantities of penicillin. Because the scientists were successful in doing so, thousands of lives were saved during the war, and millions since then (Figure 10.11).

Treatments of infections once considered incurable—pneumonia, scarlet fever, tetanus, gangrene, and syphilis—were revolutionized by Fleming's discovery. The discovery of penicillin may have been serendipitous, but the development of the next several "generations" of antibiotics in this class involved systematic and careful research. Small changes are made to a drug and the resulting substances are tested with the goal of optimizing desired activity and decreasing side effects. More than a dozen different penicillins are currently in clinical use including: penicillin G (the original discovered by Fleming and the form that causes an allergic reaction in about 20% of the population), ampicillin, oxacillin, cloxacillin, penicillin O, and amoxicillin (the pink, bubble-gum-flavored concoction you might have been given as a child). Amoxicillin is still available in capsule form; it is commonly prescribed for being effective against a broad spectrum of bacteria and is usually well tolerated.

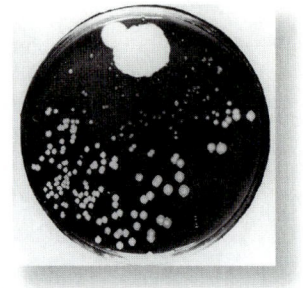

Figure 10.10

Photograph of the original culture plate of the fungus *Penicillium notatum*. This image was photographed by Sir Fleming for his 1929 paper on penicillin. The large white area at 12 o'clock is the mold *Penicillium notatum;* the smaller white spots are areas of bacterial growth.

Figure 10.11

A sign pasted at the entrance of a new facility for penicillin production during World War II.

Consider This 10.13 **Drugs by Chance**

Modern methods of drug discovery involve structure–activity relationship studies, computer modeling, among other techniques. Sometimes side effects of a drug may open the door for its usefulness in treating other illnesses. Screening programs where chance is built into the methodology. There are many examples where a new drug was discovered by "chance." Use the resources of the Web to find an example of a drug that was discovered by unusual circumstances.

The effectiveness of penicillin has unfortunately led to extreme overuse. As a result, cunning bacterial bugs have developed mechanisms for rendering penicillin (along with other antibiotics) useless. We are now witnessing strains of resistant bacteria or "super-bugs," a phenomenon Fleming predicted back in 1945. Bacteria develop resistance to penicillin by secreting an enzyme that attacks the penicillin molecule before it can act. Some of the newer penicillins differ in their effectiveness at killing certain bacteria and their susceptibility to the enzymes the organisms produce. Closely related to the penicillins are the cephalosporins (cephalexin, or Keflex) that are particularly effective against resistant strains of bacteria. Careful research on structural modifications has led to other important medicines like cyclosporine, a drug that prevents tissue rejection. Its development made possible the revolutionary success of organ transplant surgery.

So how do chemists know which structural features are important to a drug's function? The modern approach to chemotherapy and drug design probably began early in the 20th century with Paul Ehrlich's search for an arsenic compound that would cure syphilis without doing serious damage to the patient. His quest was for a "magic bullet," a drug that would affect only the diseased site and nothing else. He systematically varied the structure of many arsenic compounds, simultaneously testing each new compound for activity and toxicity on experimental animals. He finally achieved success with arsphenamine (Salvarsan 606), so named because it was the 606th compound investigated. Since then, medicinal chemists have adopted Ehrlich's strategy of carefully relating chemical structure and drug activity. Systematic changes made to a drug molecule and assessment of the resulting changes in activity is known as a **structure–activity relationship (SAR) study.**

Drugs can be broadly classified into two groups: those that produce a physiological response in the body and those that inhibit the growth of substances that cause infections. You already learned that aspirin falls in the first group. So do synthetic hormones and psychologically active drugs. These compounds typically initiate or block a chemical action that generates a cellular response, such as a nerve impulse or the synthesis of a protein. Antibiotics exemplify drugs that prevent the reproduction of foreign invaders. They do so by inhibiting an essential chemical process in the infecting organism. Thus, they are particularly effective against bacteria.

Consider This 10.14   Friend or Foe?

Make two lists of drugs for each of the two broadly classified groups: those that bring about a desired physiological response and those that kill foreign invaders. Propose three drugs for each list, using examples not given in this section.

Although drugs vary in their versatility, many of them act only against particular diseases or infections. This specificity is consistent with the relationship that exists between the chemical structure of a drug and its therapeutic properties. Both the general shape of the molecule and the nature and location of its functional groups are important factors in determining its physiological efficacy. This correlation between form and function can be explained in terms of the interaction between biologically important molecules. Although many of these molecules are very large, consisting of hundreds of atoms, each molecule often contains a relatively small active site or receptor site that is of crucial importance in the biochemical function of the molecule. A drug is often designed to either initiate or inhibit this function by interacting with the receptor site.

An example is provided by a receptor site that controls whether a cell membrane is permeable to certain chemicals. In effect, such a site acts as a lock on a cellular door. The key to this lock may be a hormone or drug molecule. The drug or hormone bonds to the receptor site, opening or closing a channel through the cell membrane. Whether

the channel is open or closed can significantly influence the chemistry that occurs in the cell. In fact, under some circumstances, the cell may be killed, which may or may not be beneficial to the organism.

This lock-and-key analogy is often used to describe the interaction of drugs and receptor sites. Just as specific keys fit only specific locks, a molecular match between a drug and its receptor site is required for physiological function. The process is illustrated in Figure 10.12. But if a perfect lock-and-key match were required in the body, it would mean that each of the millions of physiological functions would have a unique receptor site and a specific molecular segment to fit it. Simple logic suggests that such rigid demands would not be very efficient. Consequently, the lock-and-key model, although a good starting point that works in a limited number of cases, must be modified.

Using another analogy, a receptor site is like a size 9 right footprint in the sand. Only one foot will fit it exactly, and many feet (all left feet and any right feet larger than size 9) will not fit. But many other right feet can fit into the print reasonably well. So it is with receptor sites and the molecules (or their functional groups) that bind to them. Some active sites can accommodate a variety of molecules including drugs. Indeed, the way most drugs function is by replacing a normal protein, hormone, or other substance in the invading organism. The general term **substrate** refers to the substances whose reactions are catalyzed by an enzyme. In the substrate inhibition model of enzyme activity, the presence of the drug molecule prevents the enzyme from carrying out its required chemistry. As a result, the growth of an invading bacterium is inhibited, or the synthesis of a particular molecule is turned off (Figure 10.13).

Generally speaking, the drug that best fits the receptor site has the highest therapeutic activity. In some cases, however, a drug molecule does not need to fit the receptor site particularly well. The bonding of functional groups of the drug to the receptor site may even alter the shape of the drug, the site, or both. Often what counts is for the drug to have functional groups of the proper polarity in the right places. Therefore, one important strategy in designing drugs is to determine its **pharmacophore,** the three-dimensional arrangement of atoms or groups of atoms responsible for the biological activity of a drug molecule. Medicinal chemists then synthesize a molecule having that specific active portion, but with a much simpler, nonactive remainder. These researchers custom design the molecule to meet the requirements of the receptor site. In effect, they design feet to fit footprints.

An outstanding example of this approach is provided by opiate drugs such as morphine. Morphine, a complex molecule, is difficult to synthesize. However, the pharmacophore responsible for opiate activity has been identified and is highlighted in Figure 10.14. The flat benzene ring fits into a corresponding flat area of the receptor, and the nitrogen atom binds the drug molecule to the site. Incorporating this particular portion into other less complex molecules, such as meperidine (Demerol), creates opiate activity. Demerol is much less addictive than morphine but also less potent.

The lock-and-key analogy was first proposed in 1894 by the biochemist Emil Fischer.

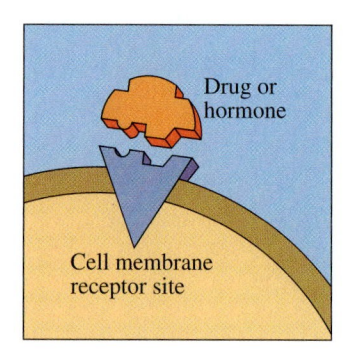

Figure 10.12

Lock-and-key model of biological interactions.

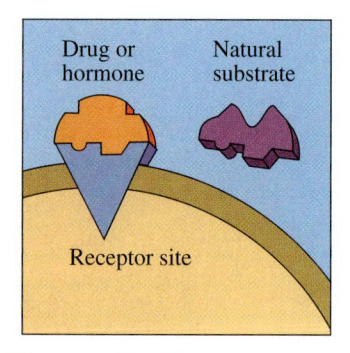

Figure 10.13

Drug molecule displacing a natural substrate on receptor site.

The term *pharmacophore* was originally described by Paul Ehrlich more than 100 years ago. Ehrlich was a German physician and biochemist who won a Nobel Prize in medicine in 1908 for his work on immunization.

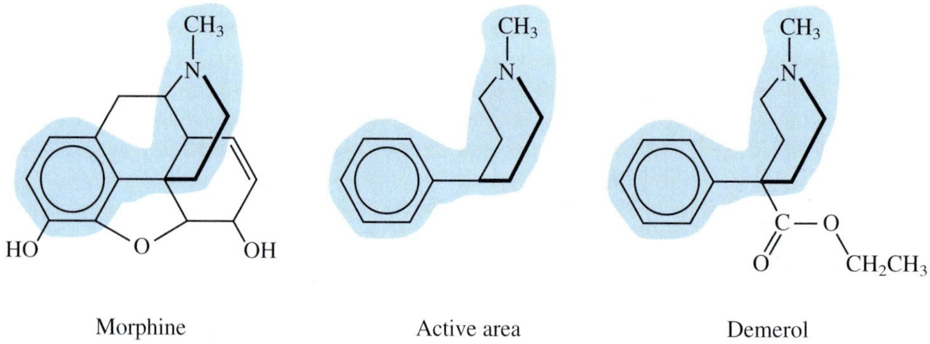

Morphine   Active area   Demerol

Figure 10.14

Molecular structures of morphine and Demerol. The highlighted "active areas," or pharmacophores, are the portions of the molecule that interact with the receptor. The darker lines indicate that these bonds are in front of the rings, or coming out in front of the plane of the page.

Consider This 10.15   3D Drugs

See for yourself how drug molecules appear in three dimensions by visiting the *On-line Learning Center,* where you will find a collection of biologically active drugs.

a. Select several drugs and examine their three-dimensional structure. How do these computer representations differ from the structural formulas of drugs shown in this chapter?

b. What advantages do the computer representations have over two-dimensional drawings? What are their limitations compared with "real" molecules? Are there any disadvantages?

The discovery that only certain functional groups are responsible for the therapeutic properties of pharmaceutical molecules was an important breakthrough in drug design. Sophisticated computer graphics are now used to model potential drugs and receptor sites. Thanks to these representations with their three-dimensional character, medicinal chemists can "see" how drugs interact with a receptor site. Computers can then be used to search for compounds that have structures similar to that of an active drug. Chemists can also modify structures in computer models and visualize how the new compounds will function.

Such techniques help to minimize the cost and time it takes to prepare a so-called **lead compound,** a drug (or a modified version of that drug) that shows high promise for becoming an approved drug. An important new methodology is **combinatorial chemistry,** the systematic creation of large numbers of molecules in "libraries" that can be rapidly screened in the lab for biological activity and the potential for becoming new drugs. Drug companies have created large populations of molecules, or libraries, with a sort of a "shotgun" approach. The sheer volume of compounds in the libraries increase the likelihood that lead compounds will be discovered. Advances in automation, robotics, and computer programming have refined the technique, making it one that no drug company can afford to ignore (Figure 10.15).

Although complex protocols for combinatorial chemistry exist, let's look at the concept in its simplest form. Consider a molecule with three different functional groups as shown in Figure 10.16. When a sample of this compound is reacted with some reagent (step 1), a set of products is formed. After the addition of a second reagent (step 2) it is easy to see that many compounds can be formed rapidly. Again, different protocols may involve splitting products at various points, but simple statistics show that a great number of new compounds can be made in short order. The process can be repeated numerous times, with the products

The term *lead compound,* pronounced differently, would refer to a compound containing the element Pb.

Figure 10.15
A scientist working with a combinatorial synthesis instrument.

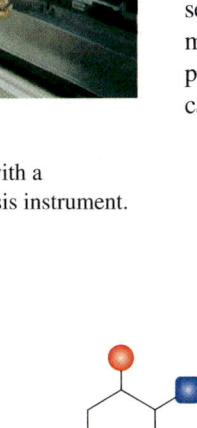

Figure 10.16
Illustration of a combinatorial synthesis process. Different colored shapes represent different functional groups.

being screened for desired activity at each step. Unpromising reactions can be screened out quickly. Used in conjunction with computers, combinatorial chemistry can minimize the trial-and-error aspects and expense, thus speeding up drug design and development. Using traditional methods, a medicinal chemist could prepare perhaps four lead compounds per month at an estimated cost of $7000 each. With combinatorial chemical methods, the chemist can prepare nearly 3300 compounds in that same time for about $12 each.

10.6   Give These Molecules a Hand!

Drug design is further complicated when drug–receptor interaction involves a common but subtle phenomenon called optical isomerism, or chirality. **Chiral,** or **optical, isomers** have the same chemical formula, but they differ in their three-dimensional molecular structure and their interaction with polarized light. Chirality most frequently arises when four different atoms or groups of atoms are attached to a carbon atom. A compound having such a carbon atom can exist in two different molecular forms that are nonsuperimposable mirror images of each other. One optical isomer will rotate polarized light in a clockwise manner, and this is called the dextro or (+) isomer. The other isomer is called the levo or (−) isomer, and it rotates polarized light in a counterclockwise manner.

Nonsuperimposable mirror images should be familiar to you. You carry two of them around with you all the time—your hands. If you hold them palms up, you can recognize them as mirror images. For example, the thumb is on the left side of the left hand and on the right side of the right hand. Your left hand looks like the reflection of your right hand in a mirror. But your two hands are not identical. Figure 10.17 illustrates this relationship for both hands and molecules.

> Polarized light waves move in a single plane; nonpolarized light waves move in many planes.

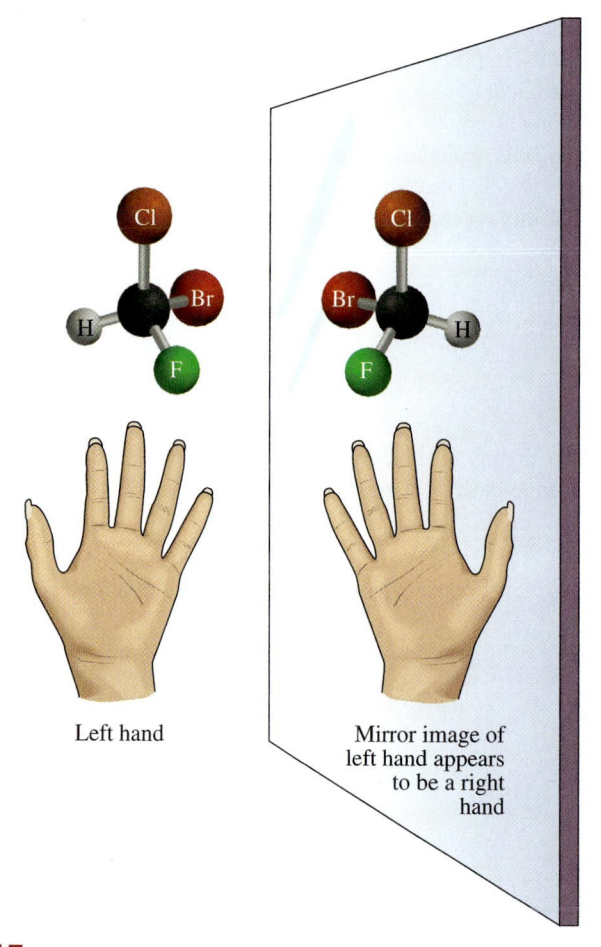

Left hand

Mirror image of left hand appears to be a right hand

Figure 10.17

Mirror image of a molecular model and a hand. The molecule CHBrClF is chiral and its shape is tetrahedral.

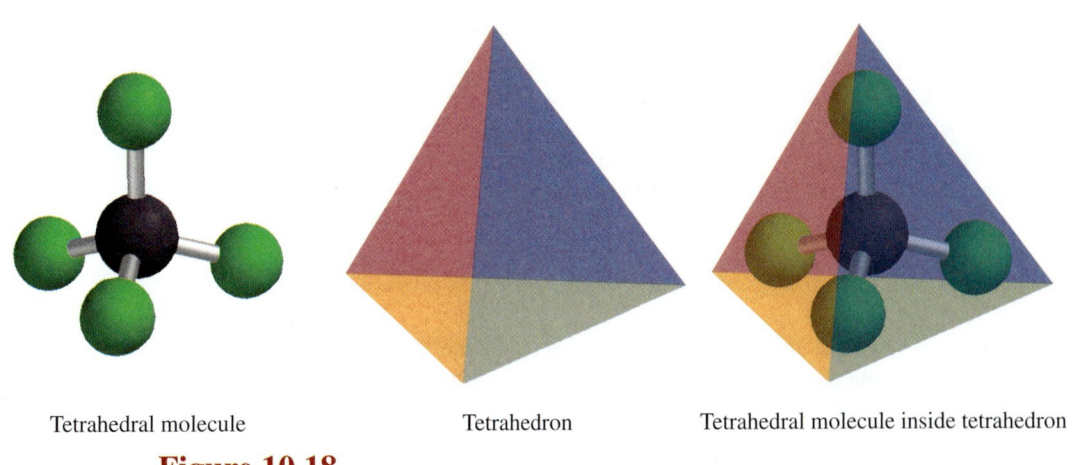

Tetrahedral molecule   Tetrahedron   Tetrahedral molecule inside tetrahedron

Figure 10.18

A tetrahedron has four equilateral triangular faces.

Note that the four atoms or groups of atoms bonded to the central carbon atom are in a tetrahedral arrangement (Figure 10.18). The positions of these four atoms correspond to the corners of a three-dimensional figure with equal triangular faces. The "handedness" of these molecules gives rise to the term *chiral,* from the Greek word for hand.

Chemists often use a formalism called a wedge–dash drawing when representing a central chiral carbon atom. For example, the molecule in Figure 10.17 could be drawn as the figure on the left. Here, the Cl and the Br are in the plane of the page. The dashed line to the H indicates that the H atom is behind the page, extending away from the viewer. The solid wedge going to the F indicates that the F atom is in front of the page, oriented toward the viewer. Similar to a line-angle drawing, the central carbon atom may be implied but not drawn.

Many biologically important molecules, including sugars and amino acids, exhibit chirality. This is significant because, although most chemical and physical properties of a pair of optical isomers are very nearly identical, their biological behavior can differ markedly. Generally, the explanation for this difference is related to the necessity of a good molecular fit between a molecule and its receptor site. Maybe Lewis Carroll's Alice had some inkling of this when, in *Through the Looking Glass,* she remarked to her cat, "Perhaps looking-glass milk isn't good to drink."

Wedge–dash drawing of one of the molecules in Figure 10.17.

Sugars and amino acids are discussed in Sections 11.2 and 11.6.

Your Turn 10.16   Chiral Molecules

Use an asterisk to identify the chiral carbon in these molecules. Then draw both chiral isomers using wedge-dash drawings. Place the chiral carbon as the central atom.

a. phenylalanine, an essential amino acid

b. $CH_3CH(OH)CH_2CH_3$ 2-butanol, an alcohol

c. $CHClFCH_3$ 1-chloro-1-fluoroethane, a hydrochlorofluorocarbon

d. methamphetamine, a notoriously dangerous street drug

Answer

a.

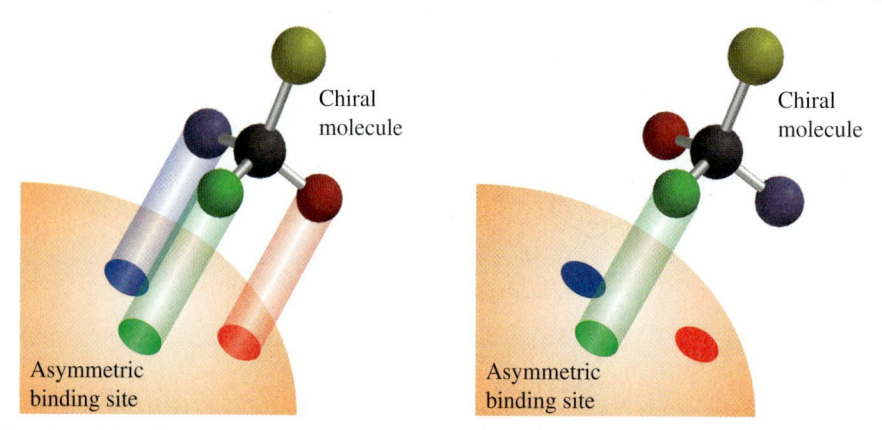

Figure 10.19

A chiral molecule binding (*left*) or not binding (*right*) to an asymmetrical site.

You can illustrate the relationship between chirality and biological activity by taking things in your own hands. Your right hand fits only a right-handed glove, not a left-handed one. Similarly, a right-handed drug molecule fits only a receptor site that complements and accommodates it. Any drug containing a carbon atom with four different atoms or groups attached to it will exist in chiral isomers, only one of which usually fits into a particular asymmetrical receptor site (Figure 10.19).

The extreme molecular specificity created by chirality complicates the medicinal chemist's task of synthesizing drugs. A drug molecule must include the appropriate functional groups, and these groups must have the three-dimensional configuration that gives the molecule its desired biological activity. In many chemical reactions, the "right" and "left" optical isomers are produced simultaneously. Such a situation results in a **racemic mixture** (±) consisting of equal amounts of each optical isomer. But frequently, only one optical isomer is pharmaceutically active. For example, many opiate drugs exist as optical isomers, only one of which may have opiate activity. In Figure 10.20, levomethorphan, the left-handed (levo, or −) isomer of methorphan, is an addictive opiate. On the other hand, its right-handed (dextro, or +) mirror image is a nonaddictive cough suppressant. This permits the use of dextromethorphan in many over-the-counter cough remedies, but the right-handed isomer must either be synthesized in pure dextro form or separated from a mixture with its levo isomer. The latter task can prove to be very difficult since the physical properties of the isomers are often identical.

Many other drugs exhibit chirality and are active only in one of the isomeric forms. This is true for some antibiotics and hormones and for certain drugs used to treat a wide range of conditions: inflammation, cardiovascular disease, central nervous system disorders, cancer, high cholesterol levels, and attention deficit disorder. Among the widely used chiral drugs are ibuprofen, cyclosporine (the drug used to prevent rejection in organ transplants), and the lipid-reducing drug atorvastatin (Lipitor). Ibuprofen is sold as a racemic mixture of (+) and (−) isomers (Figure 10.21 shows the (+) isomer). Only (−)-ibuprofen

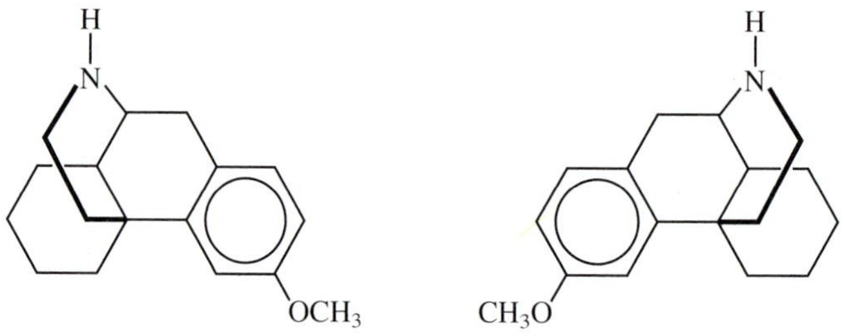

> The vitamin E sold in stores is generally a racemic mixture of + and − isomers. The + isomer is the physiologically active one that can be purchased in pure form at a significantly higher price.

levomethorphan　　　　　　　　　　dextromethorphan

Figure 10.20

Levo- and dextromethorphan.

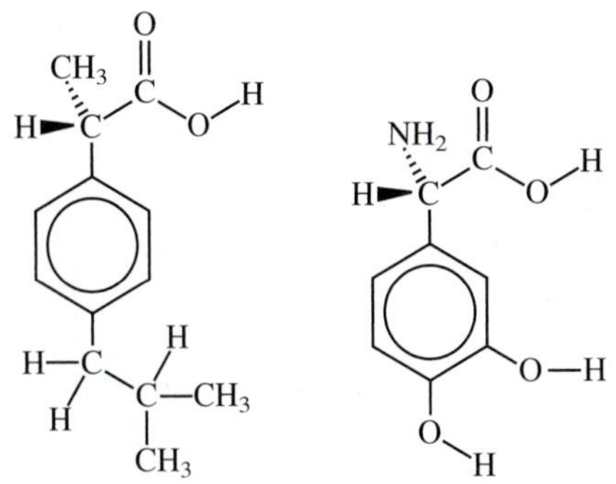

Figure 10.21

Two chiral drug molecules.

acts as a pain reliever; (+)-ibuprofen does not. However, in the body the (+) isomer is converted to the (−) form. Therefore, it is likely that someone taking ibuprofen is just as well off taking the racemic mixture rather than the more expensive (−)-ibuprofen.

Naproxen, a common pain reliever, is one example of many in which one isomer is preferred, even required. One form of naproxen relieves pain; the other causes liver damage. One last example involves a treatment for Parkinson's disease. The initial use of racemic dopa for treatment of this disease brought on adverse effects such as anorexia, nausea, and vomiting. The use of the single isomer (−)-dopa greatly reduced these side effects and the desired effect was achieved using half the initial doses.

Your Turn 10.17   Examining Ibuprofen and (−)-Dopa

Carefully examine the structural formulas for (+)-ibuprofen and (−)-dopa given in Figure 10.21.

a. For each drug, which is the chiral carbon atom?
b. Identify all of the functional groups present in both drugs.
c. Draw the structural formula of (−)-ibuprofen and (+)-dopa.

William Knowles, Barry Sharpless, and Ryoji Noyori shared the 2001 Nobel Prize in chemistry for their research that developed new catalytic methods for synthesizing chiral drugs.

Consequently, drug companies have active research programs designed to create chirally "pure" drugs, those having only the beneficial isomer of a drug as a single chiral form. Although making the proper, single isomer might seem like an exercise of interest only to chemists, it is big business. The majority of the most successful prescription drugs sold worldwide are single-isomer drugs, with total annual sales around $50 billion and rising, a trend that will no doubt continue into the future. These include the chiral "blockbusters" drugs Lipitor, simvastatin (Zocor), esomeprazole (Nexium), and sertraline (Zoloft), with global sales over $1 billion per year.

Lipitor, a chiral drug classified as a statin, lowers cholesterol by preventing its synthesis in the liver. It is a phenomenal bestseller with annual worldwide sales topping $10 billion. Producing Lipitor requires the synthesis of hydroxynitrile (HN), another chiral molecule. Until recently, HN had been produced only as a racemic mixture, thus requiring the separation of the two isomers. Worse still, this process involved large quantities of hydrogen bromide and cyanide. These chemicals are toxic to say the least, and the process produced enormous quantities of unwanted side products.

Enter Codexis, a company that won one of the 2006 Presidential Green Chemistry Challenge Awards. Chemists at Codexis developed an elegantly green route to the HN intermediate using enzymes to carry out highly specific reactions. Their green process increases yields, reduces the formation of by-products, reduces the generation of waste and use of solvents, reduces the use of purification equipment, and increases worker safety.

The widespread use of Lipitor creates an annual demand for the HN intermediate of about 200 metric tons. This indeed is an invention worthy of a Presidential award!

New chiral drugs are big business, and drug companies have recently devised a novel strategy for increasing profits in this area. A number of proven medicines formerly sold as racemates (± mixtures) are being reevaluated and remarketed as single isomers, a strategy known as a "chiral switch." The economic reasons for this are obvious: pharmaceutical companies are able to extend patent protection on their bestseller drugs and give them a hedge against generic competition. The therapeutic rationale for the switch is that the single isomer may provide benefits such as a wider margin of safety, fewer side effects, and simpler interactions in the body. Methylphenidate, sold under the trade name Ritalin, is a drug used for treating attention-deficit hyperactivity disorder (ADHD). Formerly prescribed as a racemate, it has made the chiral switch and is now marketed as a single isomer; it is reported to be equally effective at half the dose compared with the racemate, and has an improved side effect profile.

Consider This 10.18   Chiral Switch

In principle, the strategy of chiral switching should afford a therapeutic advantage for the single-isomer drug over the racemic mixture. But is this always the case? Use the Web to identify one drug (other than Ritalin) that has made the chiral switch and report on the relative merits of the single isomer versus the racemic mixture.

10.7   Steroids

Consider contraceptives, muscle-mass enhancers, and abortive agents. What do these chemically have in common? The surprising answer is that they are all **steroids,** a class of naturally occurring or synthetic fat-soluble organic compounds that share a common carbon skeleton arranged in four rings.

As a family of compounds, steroids arguably best illustrate the relationship of form and function. Certainly no other group of chemicals is more controversial because their uses range from contraception to vanity promoters. The naturally occurring members of this ubiquitous group of substances include structural cell components, metabolic regulators, and the hormones responsible for secondary sexual characteristics and reproduction. Among the synthetic steroids are drugs for birth control, abortion, and bodybuilding. Table 10.4 lists some of the functions.

In spite of their tremendous range of physiological functions, all steroids are built on the same molecular skeleton. Thus, these compounds also provide a marvelous example of the economy with which living systems use and reuse certain fundamental structural units for many different purposes. The common characteristic of steroids is a molecular

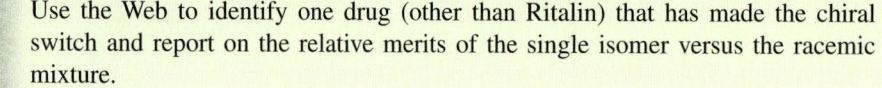

Table 10.4   Steroid Functions

Function	Examples
Regulation of secondary sexual characteristics	Estradiol (an estrogen); testosterone (an androgen)
Regulation of the female reproductive cycle	Progesterone RU-486 (the "abortion pill")
Regulation of metabolism	Cortisol; cortisone derivatives
Digestion of fat	Cholic acid
Component of cell membranes	Cholesterol
Stimulation of muscle and bone growth	Gestrinone, trenbolone

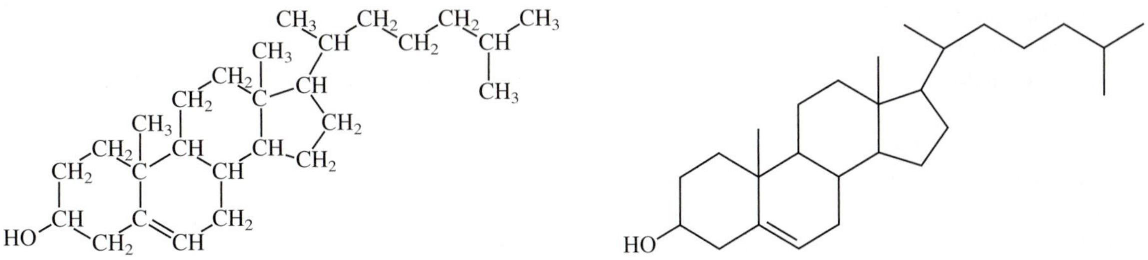

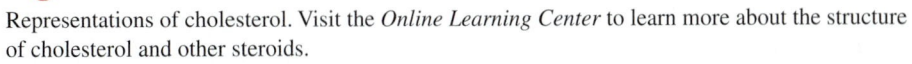

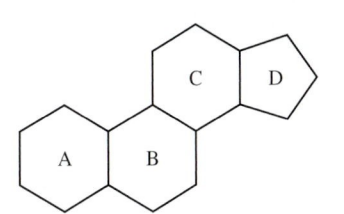

Figure 10.22

Representations of cholesterol. Visit the *Online Learning Center* to learn more about the structure of cholesterol and other steroids.

framework (nucleus) consisting of 17 carbon atoms arranged in four rings. The steroid nucleus is illustrated here.

Recall that in such a representation (a line-angle drawing), carbon atoms are assumed to occupy the vertices of the rings but are not explicitly drawn. The three six-membered carbon rings of the steroid nucleus are designated A, B, and C, and the five-membered ring is designated D. Although the steroid nucleus appears flat as drawn, it actually is three-dimensional in shape. The dozens of natural and synthetic steroids are all variations on this theme. Some differ only slightly in structural detail, but have radically different physiological function. Extra carbon atoms or functional groups at critical positions on the rings are responsible for this variation.

The steroid cholesterol is a major component of cell membranes and is shown in Figure 10.22. The figure on the left includes all the atoms in the molecule; the one on the right gives the skeletal representation using a line-angle drawing.

Careful examination of Figure 10.23 illustrates how subtle molecular differences can result in profoundly altered physiological properties. The difference between a molecule of estradiol and one of testosterone lies only in one of the rings. Are the only differences between men and women due to a carbon atom and a few hydrogen atoms? You be the judge.

Some other common natural steroids are the hormones cortisone and corticosterone (Figure 10.24). You may have used some form of cortisone preparation for treating rashes and other minor skin disorders. Corticosteroids are produced in the adrenal cortex and play a wide range of roles such as regulation of inflammation, immune response, stress response, and carbohydrate metabolism. Steroidal anti-inflammatory drugs are the most often used treatment option for people with asthma; one synthetic choice is prednisone.

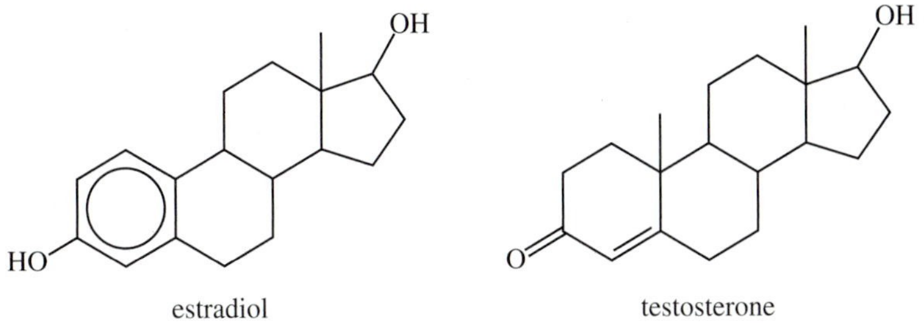

estradiol   testosterone

Figure 10.23

Estradiol and testosterone.

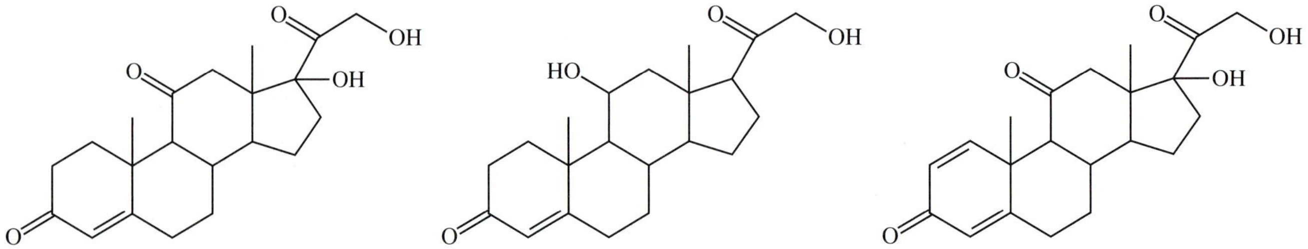

cortisone   corticosterone   prednisone

Figure 10.24
Cortisone, corticosterone, and prednisone.

Your Turn 10.19   **Structural Similarities of Steroids**

Here are pairs of steroids. Their structures are given either here in the text or in Figures Alive! at the *Online Learning Center.*

a. Identify the structural similarities in each pair.

estradiol and progesterone   corticosterone and cortisone
cholic acid and cholesterol   prednisone and cortisone
estradiol and testosterone

b. Write the chemical formula for five of the drugs in part **a.**

10.8   When Drug Researchers Think Small

No foray into the realm of pharmaceutical design and discovery would be complete without mentioning the latest innovations in the nanoscale arena. When describing nanochemistry, we are focusing on a size domain from roughly 1 to 100 nm. **Nanomedicine** is the union of nanoscale technology and medical treatment. An application relevant to this discussion is drug delivery systems, one use that promises to become reality in the very near future.

Current methods of drug delivery are limited in their efficiency, because the drug has to make its way through the body before it reaches the desired site. Nanotechnology may pave the way for more precise drug delivery, and one of the most exciting new products are called nanotubes. **Nanotubes,** or nanocapsules, are thin single-walled tubes that may be synthetic or partially synthetic (bio-nanotubes). In 2005, a collaborative effort between biologists and material scientists at the University of California, Santa Barbara, reported the development of "smart" bio-nanotubes for drug delivery vehicles. What makes them so smart? They have been designed to encapsulate a drug and then open up when it arrives in a particular location in the body, delivering the drug exactly where it is needed. Pretty smart.

How do researchers make bio-nanotubes? They combine **microtubules,** nano-sized hollow cylinders with outer diameters between 20 nm and 30 nm, with solid lipid membranes. Microtubules are involved in several basic cellular processes such as intracellular transport, maintenance of cell shape, and formation of the spindle structure in cell division. The scientists discovered that as they manipulated electric charges on the lipid membranes and the microtubules, it was possible to form either open-ended or closed bio-nanotubes. The latter were formed as, under certain electrical conditions, the lipid membrane flattened out to cover the whole cylindrical surface of the microtubule, forming a sort of nanoscale capsule with lipid caps. Under different electrical conditions, the lipid membrane will bead up on the surface of the microtubule, opening the ends of the capsule. The manipulation of electric charge makes it possible to toggle between the two states (open and closed) of bio-nanotubes forming the basis for controlled drug delivery.

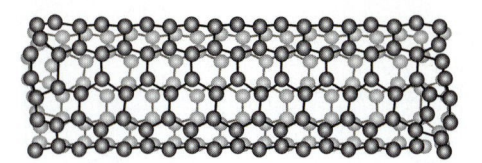

Figure 10.25

A hollow nanotube of pure carbon. It can be modified to increase solubility.

Carbon nanotubes, nano-sized tubes of pure carbon with wall thicknesses as thin as a single atom, have found applications in engineering projects, but none have materialized yet in the world of pharmaceuticals (Figure 10.25). Critics of the idea of using carbon nanotubes in the human body point to the obvious problem of water solubility, a desirable property for drug therapies. Carbon-based nanostructures typically are not very soluble in water. Can you imagine trying to dissolve soot from your fireplace in water? But a 2006 article in the *Journal of the American Chemical Society* describes the production of water-soluble carbon nanotubes. Researchers reported a quick and easy method for the production of new types of nanotubes, claiming that they are over 125 times more water-soluble than other known ones.

And quick and easy the procedure is; just add carbon nanotubes to a mixture of nitric and sulfuric acids and pop the mixture in the microwave oven. After just 3 minutes, pull it out of the oven, and voilà—water-solubility! Under these conditions, the nanotubes become covered with negatively charged sulfonate and carboxylate groups. These are the functional groups found in the salts of sulfuric and acetic acid. About 30% of the carbon atoms in the nanotubes have a carboxylate group attached, and about 10% of the carbon atoms have become bound to a sulfonate group. These highly polar groups increase the water-solubility of the nanotubes from about 0.08 mg/mL to 10 mg/mL.

Consider another benefit: with two different functional groups present, it is possible to simultaneously attach two different drugs to the nanotubes. This means the possibility of having a drug delivery system that can carry two different drugs to a targeted area. For example, the carboxylate group may bind to an antibiotic drug while the sulfonate group may be attached to a drug that lowers resistance to antibiotics.

Other 2007 news in the nanomedicine arena includes reports from several companies that are testing ceramic microstructures for controlled release of drugs. These hollow, porous spheres made of titanium and zirconium can be coated inside and out with an active drug. The sizes of the spheres are customizable, allowing controlled release of drugs. The hope is that this technology can be used to extend the patents for drugs that are soon to be available as generic equivalents.

10.9   Drug Testing and Approval

At one time, all drugs and medicines were available without a prescription. Before the Food and Drug Administration regulations existed, virtually anything could be sold as a remedy. In fact, alcohol, cocaine, and opium were included in some early products with no warnings to users. The Food, Drug, and Cosmetic Act of 1938 and its amended versions in 1951 and 1962 defined prescription drugs as ones that could be habit-forming, toxic, or unsafe for use except under medical supervision. Virtually everything else is available for sale over the counter (i.e, without a prescription).

Prescription drugs are manufactured on a colossal scale to meet patient demands. Each year, over 3 billion prescriptions are filled in the United States, accounting for hundreds of billions of dollars at the cash register. But the pathway for a new drug from a laboratory to a pharmacy shelf is long and complicated. All proposed new drugs, whether extracted from natural sources or synthesized in the laboratory, are subjected to an exacting series of tests before they obtain FDA approval. Current law requires evidence that new drugs are safe as well as effective before approval is granted.

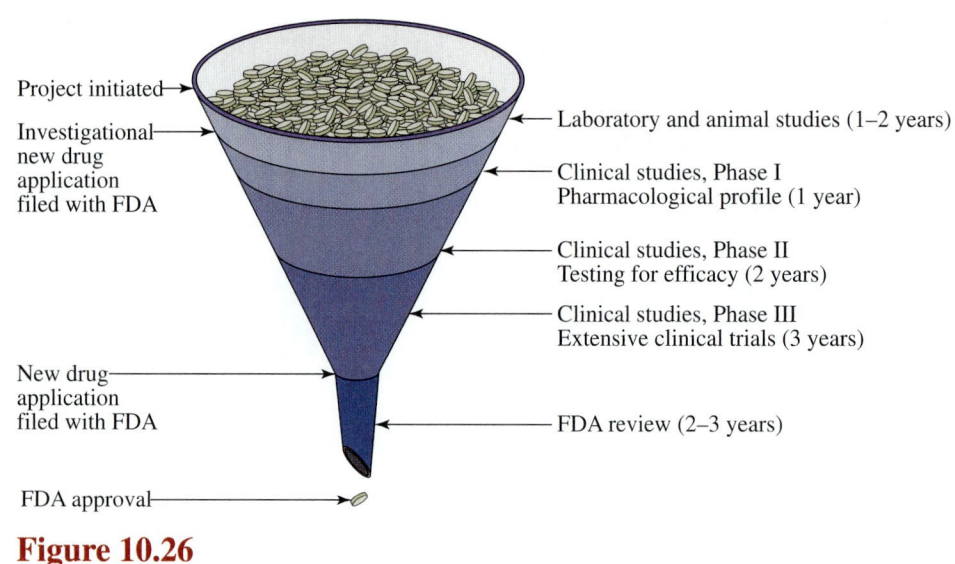

Project initiated → Laboratory and animal studies (1–2 years)

Investigational new drug application filed with FDA → Clinical studies, Phase I — Pharmacological profile (1 year)

Clinical studies, Phase II — Testing for efficacy (2 years)

Clinical studies, Phase III — Extensive clinical trials (3 years)

New drug application filed with FDA → FDA review (2–3 years)

FDA approval →

Figure 10.26

Schematic of the drug approval process in the United States.

From discovery to approval, the development of a new drug takes, on average, nearly 8-10 years and about $500 million—over three times the cost of a decade ago. The expenses are principally for the various stages of drug testing, probably the most complicated and thorough premarketing process ever developed for any product. Although the number of pills getting through the funnel of Figure 10.26 gets progressively smaller with time, the diagram does not begin to convey the high rejection rate of proposed drugs. The odds of getting a candidate drug from identification to approval are 1 in 10,000. For every 10,000 trial compounds that begin the process, 20 make it to the level of animal studies, half that many get clearance for use in clinical testing with humans, and finally 1 gets FDA approval.

Once the promising candidates have been identified, they are subject to in vitro studies, those carried out in laboratory flasks. Simultaneously, a wide range of activity is undertaken by the pharmaceutical company. Chemists and chemical engineers investigate whether the compound can be produced in large volume with consistent quality control. Chemists and pharmacists together carry out studies of the most effective way to formulate the drug for administration—as capsules, pills, injection, syrup, or perhaps something more unusual such as a nasal spray, skin patch, or implant. Chemical stability and shelf life are evaluated. Economists, accountants, patent attorneys, and market analysts conduct research on the likelihood of deriving a profit from the product.

In vitro means "in glass."

Only a small fraction of compounds survive this scrutiny to move on to animal testing. Such in vivo tests are designed to determine the drug's efficacy, safety, dosage, and side effects. At this stage, pharmacologists typically determine the drug's mode of action, how it is metabolized, and its rate of absorption and excretion. The tests are carefully controlled, requiring the collection of very specific kinds of data. For example, drugs are evaluated for their short- and long-term effects on particular organs (for example, the liver or kidneys) and on more general systems (for example, the nervous or reproductive system). Perhaps the most controversial toxicity testing involves the determination of the lethal dose-50 (LD_{50}), the minimum dose that kills 50% of the test animals (Figure 10.27).

In vivo means "in life."

Consider This 10.20 **Animals and Drug Testing**

Animal rights groups often target the LD_{50} standard as an example of callous indifference to animal welfare. Other groups argue that standards such as LD_{50} are necessary to ensure drug safety and effectiveness. Take a position on the issue and prepare a statement.

Figure 10.27

Animal testing draws passionate responses from both sides.

Results of animal tests must be submitted to the FDA for evaluation before permission is granted to proceed to the next stage—clinical testing of the drug on humans. In addition, approval must be obtained from local agencies and authorities such as a hospital's ethics panel or medical board. The FDA must establish whether the drug appears to be effective and safe before it can be sold to the public. What goes on the label regarding use, side effects, and warnings must also be determined.

Typically, clinical studies involve the three phases identified in Figure 10.26. First there is phase I, developing a pharmacological profile. Then comes phase II that tests the efficacy of the drug. Phase III is carrying out the actual clinical tests. Sometimes phase IV trials are conducted after a product is approved and on the market to find out more about the treatment's long-term risks and benefits. Sometimes the drug will be tested in different populations of people, such as children. Most of the safety tests of phase I are done with healthy volunteers who are given single and repeat doses of the drug in various amounts. It is also at this stage that researchers look for interactions with other drugs. A kind of test called a double-blind, placebo-controlled test is administered to small patient groups in phase II to test the drug's effectiveness on patients having the condition that the drug is designed to affect. In this protocol, neither the patient nor the physician knows which patients are receiving the drug and which are receiving a placebo, an inactive imitation that looks like the "real thing." Such tests are designed to eliminate bias in the interpretation of the results.

Long-term toxicity studies are also initiated during phase II. The clinical trials are expanded in phase III, while manufacturing processes are scaled up and tests are carried out on the stability of the drug. The entire process often requires six years or more. In the early 1980s, an average of 30 clinical trials were done on drugs that were ultimately approved. Over the past 20 years, that number has risen to over 70, the trials have become more complex, and the number of patients treated per trial (about 4000) has more than doubled.

Large-scale clinical trials are desirable because a large pool will more likely include a wide range of subjects. Variety is important because the drug in question may have markedly different effects on the young and the old, on men and women, on pregnant or lactating women, on infants, nursing infants, and unborn fetuses, and on persons suffering from diabetes, poor circulation, kidney problems, high blood pressure, heart conditions, or a host of other maladies.

Once clinical trials have been completed successfully (typically by only 10 drugs out of an original pool of 10,000 compounds), the test data are submitted to the FDA as part of a new drug application. This document can easily exceed 3500 pages. On review, the agency may require the repetition of experiments or the inclusion of new ones, thus adding years to the approval process. Of the drugs submitted to clinical testing, only about 1 in 10 is finally approved.

Consider This 10.21 Double-Blind Testing

Double-blind protocols have other uses than testing for the effectiveness of a drug. For example, physicians may use a double-blind test for diagnosing food allergies. In such tests, the physician administers a series of foods and placebos in disguised form. The test substances are labeled in code known only to a third person.

a. Why do you think double-blind tests for the effectiveness of a drug or for establishing a food allergy are necessary?

b. Compared with the single-blind tests in which only the patient is unaware of the drug or food being administered, how do double-blind tests affect the reliability of the information gained?

Once it receives the FDA's permission, a drug can be sold in the United States. Nevertheless, it still remains under scrutiny, monitored through reports from physicians. Drugs are removed from the market if serious problems occur. Some side effects show up only when large numbers of users are involved. A new drug application, for example, typically includes safety data on several hundred to several thousand patients. An adverse event occurring in 1 in 15,000 or even 1 in 1000 users could be missed in clinical trials, but it could pose a serious safety problem when the drug is used by many times that number of patients. Such an example is rofecoxib (Vioxx), a drug that represented a billion dollars a year in revenue to its manufacturer, the pharmaceutical company Merck. A study in 2000 showed it increased the risk for stroke and heart attacks. Merck withdrew it from the market in 2004. As of mid-April 2006 there had been 11,500 civil law suits filed against Merck.

The lengthy process for drug testing and approval is not without controversy. Most people probably favor thorough screening of any proposed drug, but the price of such protection is high. The most obvious costs are monetary. Bringing a new drug to market is incredibly expensive, and the numbers of new drug approvals are not keeping pace with rising research and development costs (Figure 10.28). Of course much of the expense is passed on to the consumer. For example, in a hospital a single dose of tirofiban (Aggrastat), a medication that dramatically increases the likelihood of surviving a heart attack (and eliminates subsequent health costs), has a price tag of $1100. Such prices have

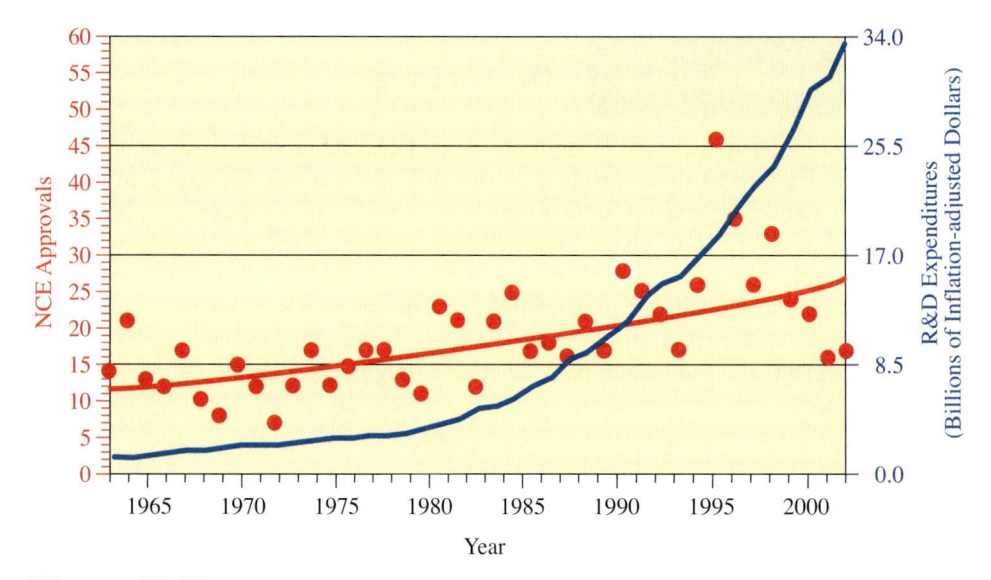

Figure 10.28

Drug development statistics: New certified approvals (NCE or new chemical entity) and research development costs (R&D expenditures).

Source: Tufts CSDD Approved NCE Database, PhRMA, 2004.

driven the costs of medical care and medical insurance in the United States to astronomical levels. One issue in the debate over health care reform is who will pay for the research and development that ultimately leads to new medication.

Consider This 10.22   Who Should Pay?

Who should pay for Aggrastat, a life-prolonging drug: individuals, medical insurance, pharmaceutical companies, or the government? Take a stand and defend your position.

But more than money is at stake. In some cases, the costs of the protracted drug approval process may be human lives. When a patient is suffering from an almost certainly fatal disease such as some cancers, the risk–benefit equation changes. When there is nothing to lose, people are willing to take great risks, possibly taking imperfectly tested drugs. Some mortally ill patients have smuggled drugs from countries where the approval process is less stringent than in the United States. Some have grasped at the straw of largely unproven remedies. And within the system, some advocates have urged that the FDA approval process be short-circuited to permit the use of experimental drugs on patients who have no other options.

Within the limits of its legal responsibilities to balance benefits with risks, the FDA has responded appropriately. Ten years ago, an FDA review for a new drug required nearly three years, whereas today it is less than a year. A new "fast-track" system has been instituted for priority drugs—those that address life-threatening ailments or new drug therapies for conditions that had no such therapies. The fast-track policy promises to have priority drugs, if found to be acceptable, approved within six months of application. Action on nonpriority drugs is to be taken within 10 months, down from the initial target of 12 months.

People suffering from rare diseases, however, may not be able to purchase appropriate medication at any price because it may not exist. There is a significant financial disincentive for a pharmaceutical company to invest heavily in developing a drug that will be used by only a small fraction of the population: such medications are called "orphan drugs."

Consider This 10.23   Orphan Drugs

Antibiotics, analgesics, and other drugs have a large and profitable market around the world. However, development and marketing of essential drugs needed by only a small number of people suffering from rare diseases can be a drain on a pharmaceutical company. If the pharmaceutical companies decide not to make and market these "orphan drugs" because of their low economic return, how will people who need these drugs obtain them? Should the government step in and require successful drug companies to contribute a percentage of their profits to a fund for research, development, and production of these orphan drugs? Take a position and outline your reasons.

To such considerations, one must add the objections some have to standard test protocols. Animal rights advocates are highly critical of the use of any animal subjects in drug screening. The sacrificing of test animals in establishing LD_{50} values is especially controversial. For others, the generally accepted methods of human testing are at issue. The argument is that because the drug may have some benefit and will probably do no significant harm, it is unethical to withhold it from a control population. Some terminally ill AIDS patients, by mixing and sharing test drugs and placebos, have intentionally refused to cooperate with double-blind clinical studies.

Consider This 10.24 **The High Price of Drugs**

In some cases, American prescription drugs are sold in foreign countries, at prices lower than those in the United States. As a result, the press has reported that senior citizens travel to Canada or Mexico to buy prescription drugs at reduced prices. In other cases, successful drugs used in treating AIDS patients are manufactured in violation of existing patents. Brazil has decided to disregard patents on the anti-AIDS drug nelfinavir (Viracept) and allow its manufacture in government laboratories. Pick one of these two cases and compile a list of arguments supporting each side. Which side do you support?

10.10 Prescription, Generic, and Over-the-Counter Medicines

Enter customer. The pharmacist asks "Brand name or generic?" This scenario is played out daily in thousands of pharmacies across the country. How is the person to decide? For millions of Americans, the cheaper generic version can mean the difference between getting the necessary medication and not being able to afford it, although not all approved drugs are available in generic form.

The two forms can be differentiated rather simply. A pioneer drug is the first version of a drug that is marketed under a brand name, such as Xanax, an antianxiety, or sedative drug. A **generic drug** is chemically equivalent to the pioneer drug, but cannot be marketed until the patent protection on the pioneer drug has run out after 20 years. The lower priced drug is commonly marketed under its generic name, in this case alprazolam instead of Xanax. The 20-year patent protection on the pioneer drug begins when it is patented, not when it's first put on the market. In cases requiring a long preapproval time, the actual marketing period can be relatively short, even less than six years. In such a situation, a drug company has very little time to recoup its research and development costs before a generic competitor can be manufactured. However, almost 80% of generic drugs are produced in brand-name equivalents (Figure 10.29). Like pioneer drugs, generic drugs must also be approved by the FDA.

In 1984, Congress passed the Drug Price Competition and Patent Restoration Act that greatly expanded the number of drugs eligible for generic status. This act eliminated the need for generics to duplicate the efficacy and safety testing done on counterpart pioneer drugs. Doing so saves drug manufacturers considerable time and money. The FDA also issued specific guidelines for a generic drug's comparability to the pioneer drug. By FDA mandate, the generic and pioneer versions must be bioequivalent in dosage

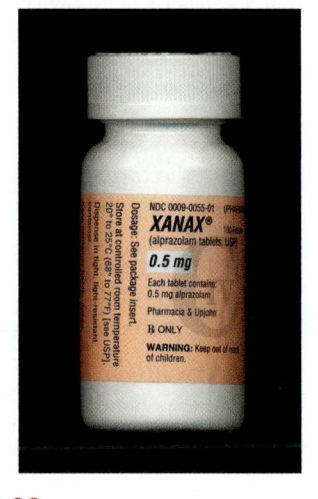
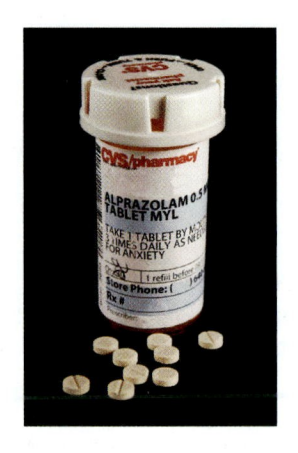

Figure 10.29

A brand-name drug (Xanax) and its generic counterpart (alprazolam).

form, safety, strength, route of administration, quality, performance characteristics, and intended use. In other words, it must deliver the same amount of active ingredient into a patient's bloodstream at the same rate. Once proven, companies can begin to drive the high price of a brand-name drug down through competition.

But it is not always that easy. In 2006, Congressional leaders were encouraged to take a look at a procedure called a "citizen petition" that was designed to alert the FDA of any safety issues surrounding a drug. Members of the generic drug industry and people at the FDA were complaining that this procedure was being misused by brand-name drug companies. When anyone, an ordinary citizen, group, or a drug company files one of these petitions, time-consuming reviews of the generic drug in question by the FDA must be carried out. Meanwhile, the brand-name drug company continues to enjoy the benefit of being the only company to market the drug, even after the patent has expired. An article in the July 3, 2006, *Washington Post* quoted Senator Debbie Stabenow (D, Michigan): "The brand-name drug industry has found a major new loophole. The way things stand now, even if the FDA finds that a petition was frivolous and rejects it, [the drug companies] can get hundreds of millions of dollars of profit from the delay." Senator Stabenow said that 20 of the 21 brand-name petitions settled by the FDA between 2003 and 2006 were ultimately rejected.

Consider This 10.25   Citizen Petitions

The *Washington Post* article mentioned in this section listed the hot-selling antidepressant drug bupropion (Wellbutrin XL) as an example of the misuse of a citizen petition. Search the Web using keywords such as citizen petition, generic drug, pharmaceutical company, and FDA. Find one petition and research it. Draw up a report explaining the nature of the petition, who filed it, and why. Then state your opinion. Was the petition filed because of a bona fide safety issue? Or was it designed to stave off generic competition?

Over-the-counter (OTC) drugs allow people to relieve many annoying symptoms and cure some ailments without the need to see a physician. Nonprescription medications now account for about 60% of all medications used in the United States. More than 80 therapeutic categories of OTC drugs exist, ranging from acne products to weight control products. Table 10.5 contains several major categories of OTC products and their chief components. In accordance with the laws and regulations prevailing in this country, drugs including OTC drugs are subjected to an intensive, extensive, and expensive screening process before they can be approved for sale and public use. Just as in the case of prescription drugs, the ultimate question to be answered with over-the-counter drugs is "Do the benefits outweigh the risks?" The answer for an OTC drug depends on whether a consumer is using it properly. Therefore, the FDA and pharmaceutical manufacturers must try to balance OTC safety and efficacy.

Table 10.5   Examples of Over-the-Counter Drugs

Analgesics and Anti-Inflammatory Drugs	Cough Remedies
aspirin	*Expectorant*
ibuprofen (Advil)*	guaifenesin
naproxen (Aleve)	
acetaminophen (Tylenol)	*Cough suppressants*
	codeine (only in some states)
Antacids and Indigestion Aids	dextromethorphan
aluminum and magnesium salts (Maalox)	
calcium carbonate (Tums)	*Antihistamines*
calcium and magnesium salts (Rolaids)	brompheniramine
	chlorpheniramine
	diphenhydramine

*Names are shown as chemical name followed by common or brand name in parentheses.

Earlier we described OTC pain relievers and NSAIDs and their mode of action. Their side effects include increased stomach bleeding (aspirin), gastrointestinal upset (aspirin, ibuprofen), aggravating asthma (aspirin), and kidney or liver (acetaminophen) damage at high dosage or with chronic use.

More than 100 viruses are responsible for the misery accompanying the common cold. A whole host of cold remedies, most with multiple components, are designed to help the sufferer. Decongestants reduce swelling when viruses invade the mucous membranes, but the adverse effects include nervousness and insomnia. Nasal sprays relieve the swollen nasal tissues, but their use beyond a three-day limit often leads to a rebound effect, or a return of the runny nose.

Antihistamines relieve the runny nose and sneezing associated with allergies, but cause drowsiness and often lightheadedness. Because they induce drowsiness, it is not surprising that approved sleep aids are often antihistamines. However, children often experience insomnia and hyperactivity after taking them.

Coughing is a natural way to rid the lungs of excess secretions. Expectorants make the phlegm thinner and therefore easier to cough up, while suppressants provide relief and restful sleep. The presence of both in most cough remedy preparations seems senseless! Codeine and dextromethorphan have equally good cough-suppressing potential, but the former has a reputation for being habit-forming, a property that limits its over-the-counter availability in some states.

Heartburn, indigestion, and "acid" stomach are targets of antacids and related drugs. Antacids are basic compounds containing aluminum, magnesium, or calcium hydroxides (or combinations of them) that neutralize excess stomach acid. The popularity of the calcium-containing alternatives has grown with the promotion of the need for younger and older adults to maintain a regular supply of dietary calcium to prevent degradation of bones (osteoporosis).

Consider This 10.26   Using Common Sense

Many people are of the opinion that it is best to only take the medicines that are indicated for a particular illness. This makes plain sense. Yet some companies selling OTC preparations seem to throw in everything but "the kitchen sink," hoping to give their product some sort of "value-added" feature. Check the contents of several cold remedies found at your local drug or grocery store (or go online to find the same information). In particular, look at several cough remedy preparations. Do you find the use of both guaifenesin and dextromethorphan in one product? List the names of the products that do, and then explain why a product containing both of these medicines makes no sense.

The self-care revolution of the last several decades has encouraged the availability of safe and effective OTC drugs and has provided additional pressure for the reclassification of many prescription drugs to OTC status. According to the Consumer Healthcare Products Association, about 80 ingredients or reduced dosages of drugs have made the OTC switch since 1976. Recent prescription to OTC switches include loratidine (Claritin, antihistamine, 2002), and omeprazole (Prilosec, acid reducer, 2003), and lovastatin (cholesterol control, 2005). Right now, more than 700 OTC products have ingredients that were once only available by prescription. Additional pressure to have the switch occur comes from the health insurance industry. Changing widely used prescription drugs to OTC status greatly diminishes the insurance companies' share of payments. The conditions for which the drugs are prescribed must be common, non-life-threatening, and self-diagnosable by the average consumer. The FDA can change the status of an OTC drug back to prescription status if significant safety problems are uncovered.

From the drug manufacturer's standpoint, changing the status of a product from prescription to OTC often allows the manufacturer to market the product for several more

years without generic competition. Sales volume also increases when a product is reclassified to OTC status. For the consumer, out-of-pocket expense in some cases may actually increase when going with OTC therapeutics because few third-party health insurance payers provide reimbursement for OTC products. But overall it is estimated that prescription to OTC switches save the American public over $20 million each year.

10.11 Herbal Medicine

Worldwide, a growing number of people are using herbal products for preventive and therapeutic purposes. Herbal remedies and folk medicines abound in most cultures. This should not be surprising or astounding; nature is a very good chemist. Some (but not all) compounds found in plants and simple organisms are likely to have positive physiological effects in humans. U.S. sales of such popular herbal remedies as ginkgo biloba, St. John's wort, echinacea, ginseng, garlic, and kava kava have steadily risen over the past decade to $4 billion in 2003. Table 10.6 lists some herbs and plants and reasons for ingesting them.

   Many people have been taking St. John's wort for its reported mood-elevating activity. The Herb Research Foundation reports data from over 2000 patients in 23 clinical studies that have consistently found that a preparation of St. John's wort, a plant, is just as effective against mild to moderate depression as standard antidepressant drugs (Figure 10.30). In April 2001, however, a study published in the *Journal of the American Medical Association* reported that St. John's wort is ineffective against severe depression. The herb worked no better than the placebo in over 200 adults diagnosed with severe depression. The study was funded partly by the National Institutes of Mental Health and partly by Pfizer Incorporated, which makes sertraline (Zoloft), the most commonly prescribed multibillion dollar a year antidepression drug in the United States. Dr. Richard Skelton from Vanderbilt University, a coauthor of the new study, recommended further studies on the use of the herb for mild depression, saying: "I would like to see people with mild depression studied, and see if it works in those folks. If it works, that would be great." He recommended against using the herbal medicine until further studies are done.

   Consider ephedra (Figure 10.31), a naturally occurring substance derived from the Chinese herbal *ma huang* as well as from other plant sources. Although ephedra has long been used to treat certain respiratory symptoms in traditional Chinese medicine, in recent years it has been heavily promoted and used for the purposes of aiding weight loss, enhancing sports performance, and increasing energy.

   Ephedra contains six amphetamine-like alkaloids including ephedrine and pseudoephedrine. Ephedrine, the main constituent, is a bronchodilator (opens the airways) and stimulates the sympathetic nervous system. It has valuable antispasmodic properties, acting on the air passages by relieving swelling of the mucous membranes. Pseudoephedrine

Table 10.6	Common Herbs and Their Possible Benefits
Herb or Plant	**Symptoms to Be Relieved**
Valerian, passion flower	Anxiety
Licorice, wild cherry bark, thyme	Coughs
Echinacea, garlic, goldenseal root	Colds, flu
St. John's wort	Depression
Chamomile, peppermint, ginger	Nausea, digestive problems
Valerian, passion flower, hops, lemon balm	Insomnia
Ginkgo biloba	Memory loss
Valerian, passion flower, kava kava, Siberian ginseng	Stress, tension

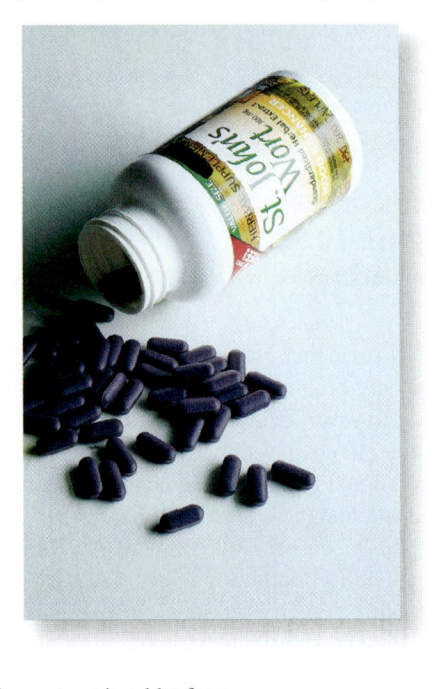

Figure 10.30

St. John's wort plant (*Hypericum perforatum*) and an extract in tablet form.

(Sudafed) is a nasal decongestant and has less stimulating effect on the heart and blood pressure.

In their synthetic form, these drugs were regulated as OTC drugs and used as a decongestant for the short-term treatment of runny nose, asthma, bronchitis, and allergic reactions by opening the air passages in the lungs. Figure 10.32 shows three related structures; methamphetamine is presented so you can see the structural similarities between ephedra drugs and this potent and dangerous stimulant. Ephedra does not contain methamphetamine.

Dietary supplements that contained ephedra made big headlines in 2003. The deaths of well-known athletes were linked to rare but serious side effects of ephedra use. Side effects reported by ephedra users included nausea and vomiting, psychiatric disturbances such as agitation and anxiety, high blood pressure, irregular heartbeat, and, more rarely, seizures, heart attack, stroke, and even death.

Figure 10.31

Ephedra's source and a common formulation.

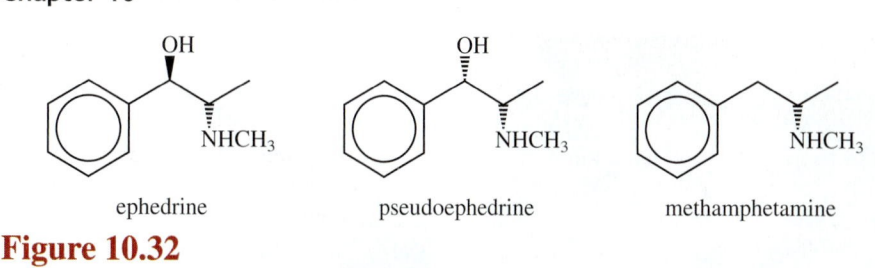

ephedrine   pseudoephedrine   methamphetamine

Figure 10.32

Chemical structures of ephedrine and two related drugs.

A solid wedge indicates a group coming out toward you. A dashed wedge indicates that the group is pointing away from you. These conventions were first introduced in Section 3.3.

No evidence currently supports the claim that ephedra enhances athletic performance. And only preliminary evidence suggests that ephedra aids in modest, temporary weight loss. However, evidence does indicate that ephedra is associated with an increased risk of side effects, possibly even fatal ones. In 2003, the International Olympic Committee, the National Football League, the National Collegiate Athletic Association, minor league baseball, and the U.S. Armed Forces banned the use of ephedra.

In December 2003, the FDA issued a consumer alert on the safety of dietary supplements containing ephedra. Consumers were advised to immediately stop buying and using ephedra products. In February 2004, the FDA published a final rule stating that dietary supplements containing ephedra present an unreasonable risk of illness or injury. The rule effectively banned the sale of these products, which took effect 60 days after its publication. But in April 2005, a federal judge in Utah struck down the FDA ban, allowing supplements with no more than 10 mg of ephedra per daily dose. Yet it seems that the last chapter on ephedra has yet to be written. In January 2006, the FDA, in conjunction with the U.S. Attorney General's Office in Pennsylvania, authorized the seizure of the ephedra-containing dietary supplement Lipodrene from an Oakmont, PA business.

The St. John's wort and ephedra examples illustrate important concerns about herbal medicines. Are they effective, and are they safe? Psychiatrists report that many of their patients with depression have tried St. John's wort before coming for medical help. One estimate suggests that over a million people in the United States alone have tried it or are using St. John's wort. Irrespective of the accuracy of the estimate, large numbers of Americans are apparently using herbal medicines under circumstances with little or no medical supervision.

Herbal remedies are only loosely regulated by the FDA. The Dietary Supplement Health and Education Act (DSHEA) of 1994 changed their classification from "food or drug" to "dietary supplement." **Dietary supplements** by definition include vitamins, minerals, amino acids, enzymes, and herbs and other botanicals. Many dietary supplements have shown no adverse effects and in fact have proven to be beneficial to good health. Others, like ephedra, have been shown to be problematic.

Under the DSHEA, the FDA does not review dietary supplements for safety and effectiveness before they are marketed. Rather, the law allows the FDA to prohibit sale of a dietary supplement if it "presents a significant or unreasonable risk of injury." When the FDA seeks to take regulatory action against a supplement (such as in the ephedra case), the burden of proof for establishing harm falls on the government. Unlike prescription or OTC drugs, there is no assessment of purity of preparations or concentrations of active ingredients, set amounts, or delivery protocols. Furthermore, there are no requirements for studies of interactions among herbal medicines or between them and traditional medicines. Since the manufacturers of herbal remedies are not required to submit proof of safety and efficacy to the FDA before marketing, information regarding the interactions between herbal remedies and other drugs are largely unknown.

In the spring of 2001, representatives from the American Society of Anesthesiologists reported concerns that patients undergoing surgery may risk unexpected bleeding when they take certain herbs within two weeks before surgery. To this point, no scientific studies linking the bleeding and a specific herb have been published. According to Dr. John Neeldt, president of the American Society of Anesthesiologists, the familiar question "Are you taking any medications?" should be augmented with, "Are you taking any herbal remedies?"

Consider This 10.27 **Does Natural Mean Safer?**

The legal standard of "significant or unreasonable risk" implies a risk–benefit calculation based on the best available scientific evidence. This suggests that the FDA must determine if a product's known or supposed risks outweigh any known or suspected benefits, based on the available scientific evidence. This must be done in light of the claims the manufacturer makes and with the understanding that the product is being sold directly to consumers without medical supervision.

When deciding to take any medication a consumer makes a risk–benefit analysis, sometimes unconsciously. One element of such an analysis is the *perceived* risk. Do you think that the general population perceives naturally occurring drugs as safer than synthetic ones? Explain using examples of your choice.

10.12 Drugs of Abuse

Before concluding our foray into the world of medicinal drugs, an examination of the abuse of drugs is warranted. Since the beginning of history, people have taken nonmedically indicated drugs for a variety of reasons. In this section we will examine a few of the most commonly abused drugs.

According to the 2004 Substance Abuse and Mental Health Services Administration (SAMHSA) survey an estimated 19.1 million Americans, or 7.9% of the population age 12 or older, were current illicit drug users. Current means use of an illicit drug during the month prior to the survey interview. Notable statistics from the SAMHSA survey include the following.

- Marijuana is the most commonly used illicit drug. Of the 14.6 million users of marijuana in 2004, about one third, or 4.8 million persons, used it on 20 or more days in the past month.

- In 2004, an estimated 2.0 million persons (0.9%) were current cocaine users, 467,000 of whom used crack. Hallucinogens were used by 929,000 persons, including 450,000 users of "ecstasy." There were an estimated 166,000 current heroin users.

- An estimated 6.0 million persons, or 2.5% of the population age 12 or older, were current users of psychotherapeutic drugs taken nonmedically. An estimated 4.4 million used pain relievers, 1.6 million used tranquilizers, 1.2 million used stimulants, and 0.3 million used sedatives.

- For oxycodone (OxyContin) there was a significant increase in the lifetime prevalence of use from 2003 to 2004 among persons age 18 to 25: from 8.9 to 10.1%.

- An estimated 121 million Americans age 12 or older reported being current drinkers of alcohol in the 2004 survey (50.3%). About 55 million (22.8%) participated in binge drinking, and 16.7 million (6.9%) were heavy drinkers (binge drinking 5 or more days in the past month).

- An estimated 70.3 million Americans (29.2% of the population age 12 or older) reported current use of a tobacco product in 2002.

Consider This 10.28 **Which Drug Is Most Harmful?**

Of all the drugs that humans abuse, which do you think causes the most harm? That is, can you point a finger at one drug that is the most disruptive in the home and workplace, causing more health problems and death than all of the others? Make your choice and be prepared to explain your choice in a group discussion.

Table 10.7	Drug Schedules			
Class	**Has Current Accepted Medical Uses**	**Potential for Abuse**	**Examples**	
Schedule I	No	High	heroin LSD marijuana, hashish mescaline MDMA ("ecstasy")	
Schedule II	Yes	High	oxycodone (in OxyContin and Percocet) morphine, opium methadone cocaine (as a topical anesthetic) methamphetamine	
Schedule III	Yes	Medium	hydrocodone with acetaminophen (Vicodin) codeine with acetaminophen anabolic steroids	
Schedule IV	Yes	Low	alprazolam (Xanax) propoxyphene and acetaminophen (Darvocet) diazepam (Valium)	
Schedule V	Yes	Lowest	cough suppressants with small amounts of codeine diphenoxylate and atropine (Lomotil) promethazine (Phenergan)	

In 1970, the Comprehensive Drug Abuse Prevention and Control Act was passed into law. Title II of this law, the Controlled Substances Act, is the legal foundation of narcotics enforcement in the United States. The Controlled Substances Act regulates the manufacture and distribution of drugs, and places all drugs into one of five schedules. The criteria for the schedules and examples of each are shown in Table 10.7.

 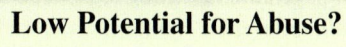

Consider This 10.29   Low Potential for Abuse?

The sedatives diazepam (Valium) and alprazolam (Xanax) are currently listed as schedule IV drugs, indicating that they have a low potential for abuse. Search the Web for information on addiction to these powerful sedative drugs. Do you agree with the current scheduling? Explain.

Cannabis sativa means useful (*sativa*) hemp (*cannabis*).

Marijuana, a schedule I drug, is the most commonly used illicit drug in the United States. Some slang names for marijuana include pot, weed, grass, Mary Jane, and chronic. It is a mixture of the dried leaves, stems, seeds, and flowers of the hemp plant, *Cannabis sativa* (Figure 10.33). The drug is usually smoked and occasionally eaten. The first known record of marijuana use dates back to the time of the Chinese Emperor Shen Nung (ca. 2737 BC), who prescribed use of the plant for the treatment of malaria, gas pains, and absentmindedness.

Hemp is the plant whose botanical name is *Cannabis sativa*. Other plants are called hemp, but *Cannabis* hemp is the most useful of these plants. Since prehistory hemp has been used for many purposes. Fiber is its most well-known product, and the word *hemp* can mean the rope or twine made from the hemp plant, as well as just the stalk of the plant that produced it. The major psychoactive drug in *Cannabis sativa* is concentrated in the

leaves and flowers of the hemp plant, so one cannot get high from smoking hemp rope or wearing clothing woven from hemp fibers.

The Chinese routinely used the marijuana plant for clothing fibers as well as for medicinal purposes. Much later, hemp was planted in the Jamestown area in 1611 for the purpose of making rope, but there is no recorded evidence of its medicinal use by the early settlers. George Washington kept a field of hemp at Mt. Vernon, and it is believed that he used the plant for both rope and medicine.

Extracts of marijuana were employed by physicians in the early 1800s for a tonic and euphoriant. But in 1937, the Marijuana Tax Act prohibited its use as an intoxicant and its medical use was regulated as national concern of its use emerged. The Marijuana Tax Act required anyone producing, distributing, or using marijuana for medical purposes to register and pay a tax that effectively prohibited nonmedical use of the drug. Although the act did not make medical use of marijuana illegal, it did make it expensive and inconvenient.

In 1942, marijuana was removed from the U.S. Pharmacopoeia, the government's official compendium of medicines, because it was believed to be a harmful and addictive drug that caused psychoses, mental deterioration, and violent behavior. The current legal status of marijuana was established in 1970 with the passage of the Controlled Substances Act.

The major psychoactive chemical in marijuana is Δ^9-tetrahydrocannabinol, or THC (Figure 10.34). The concentration of THC varies depending on how the hemp is grown: temperature, amount of sunlight, and soil moisture and fertility. High-potency varieties of marijuana are grown across the United States, with THC levels as high as 7%. In hashish, or hash, the dried and pressed flowers and resin of the plant, THC concentrations can be as high as 12%.

When marijuana smoke is inhaled, THC rapidly passes from the lungs into the bloodstream, which carries the chemical to organs throughout the body, including the brain. In the brain, THC connects to specific sites called cannabinoid receptors on nerve cells and influences the activity of those cells. Some brain areas have many cannabinoid receptors; others have few or none. Many cannabinoid receptors are found in the parts of the brain that influence pleasure, memory, thought, concentration, sensory and time perception, and coordinated movement. THC leaves the blood rapidly through metabolism and uptake into the tissues. The chemical may remain stored in body fat for long periods; research has indicated that a single dose can take up to 30 days for complete elimination. The short-term effects of marijuana use can include problems with memory and learning, distorted perception, difficulty in thinking and problem solving, loss of coordination, decreased blood pressure, and increased heart rate.

Medicinal marijuana may be indicated for treatment of nausea, glaucoma, pain management, and appetite stimulation. Such treatment may be a last resort when all other medications have failed, such as with the unrelenting nausea and vomiting that may accompany weeks of chemotherapy in treating diseases such as leukemia and AIDS. Clinical studies on the usefulness of marijuana are difficult to conduct as many barriers discourage researchers. For example, the scarcity of funding combined with complicated regulations enforced by both federal and state agencies make research in this area daunting.

The current debate over the medical use of marijuana is basically a debate over the value of its medicinal properties relative to the risk posed by its use. The debate is colored by complex moral and social judgments that underlie current drug control policy in the United States. The 1996 California referendum known as Proposition 215 allowed

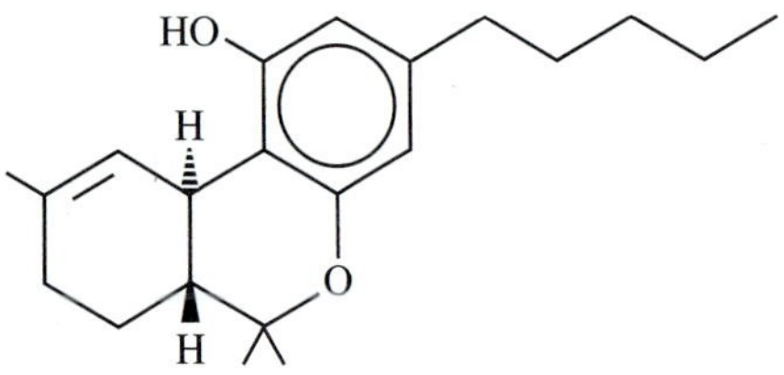

Figure 10.33

Pot plants.

Figure 10.34

Molecular structure of THC.

seriously ill Californians to obtain and use marijuana for medical purposes without criminal prosecution or sanction, with the stipulation that a physician's recommendation is required. The Institute of Medicine, a part of the National Academy of Sciences, published a review in 1999 that found marijuana to be "moderately well suited for particular conditions, such as chemotherapy-induced nausea and vomiting, and AIDS wasting."

By 2005, ten states (Alaska, California, Colorado, Hawaii, Maine, Montana, Nevada, Oregon, Vermont, and Washington) had laws that permit physicians to prescribe marijuana for medical purposes. In 2006, Rhode Island became the 11th state to pass a medical marijuana law for the seriously ill. The significance of this was that Rhode Island was the first state to move on the issue since the U.S. Supreme Court ruled in June 2006 that patients who use the drug can still be prosecuted under federal law. This ruling came on the heels of an FDA report that stated "no sound scientific studies supported the medical use of marijuana," which was in direct contradiction to the Institute of Medicine's 1999 report. Dr. Jerry Avorn, a medical professor at Harvard Medical School was quoted saying "Unfortunately, this is yet another example of the FDA making pronouncements that seem to be driven more by ideology than by science."

Consider This 10.30   Is Your State Going to Pot?

The legality of medical marijuana seems to shift like dunes of sand. Use the Web to determine your home state's current position on the use of marijuana for medicinal purposes. Is this position in conflict with that of the U.S. federal government? Be sure to have up-to-date information on the federal government's position. Do you agree or disagree with both your home state and the U.S. government on this issue? List the sources of your information.

> Oxycodone is a schedule II drug under the Controlled Substances Act because of its high propensity to cause dependence and abuse.

OxyContin is a schedule II drug (Figure 10.35). Known on the street as "oxy," OxyContin is the tradename for the drug oxycodone hydrochloride, a morphine-like narcotic. Other analgesics like Percocet and Percodan contain the same chemical, but in much smaller amounts (5–7.5 mg/tablet). OxyContin tablets contain up to 80 mg of oxycodone with a time release mechanism that allows the drug to be delivered slowly over a longer period. For those enduring acute and long-term pain, such as some elderly patients, and those suffering from some terminal forms of cancer, this drug delivers much-needed relief.

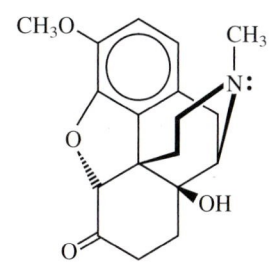
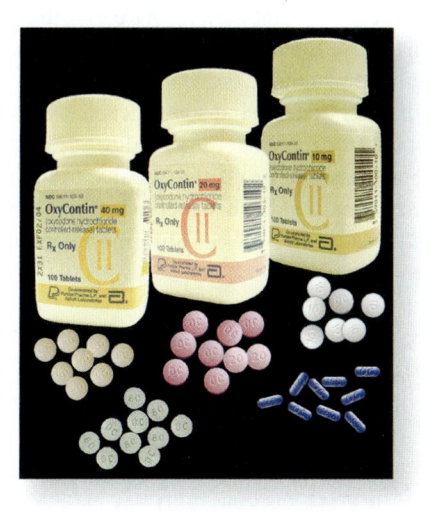

Figure 10.35
Oxycodone molecular structure and OxyContin pills.

Source: Photo © 2004, Publishers Group, http://www.streetdrugs.org.

The drug is supposed to be taken orally, but abusers of OxyContin are able to get around the time release mechanism by crushing the tablet and either snorting it or dissolving it in water and injecting the solution. The effect is similar to that achieved from using heroin. Obviously, absorbing this much of the powerful narcotic can have dramatic consequences. Since the drug arrived on the market, emergency room overdoses linked to this narcotic increased dramatically.

The problem was first observed in rural areas of Kentucky, Virginia, West Virginia, and Maine and the drug OxyContin was given the derogatory slang names of "hillbilly heroin" and "poor man's heroin." But it has since spread to other areas in the United States and even globally. In March 2002, an 18-year-old female became the first U.K. fatality attributed to OxyContin.

Purdue Pharma, the manufacturer of the product, has come under fire recently for allegedly turning a blind eye to the mounting reports of abuse of the drug. As of 2006 thousands of lawsuits were pending against Purdue Pharma for overpromoting the prescription painkiller while playing down its addictive side. In an effort to educate health care providers about these risks, the drug manufacturer issued a warning in the form of a "Dear Healthcare Professional" letter. The company's other efforts have included discontinuing the most powerful pill, a 160-mg form of the drug; stamping pills from Mexico and Canada to help authorities trace illicit supplies; passing out tamper-proof prescription pads to doctors; and instituting educational programs aimed at doctors as well as potential abusers, with a particular focus in the Appalachian region.

Your Turn 10.31   Oxycodone Formulation

The oxycodone in an OxyContin pill is not a freebase, but rather is formulated as a salt. What acid is used to form this salt? Draw a structural formula for the drug in its salt form.

Consider This 10.32   Who Is Right? Wrong?

Use the Web to investigate several sites that discuss the abuse of OxyContin. Also find sites that describe the legal woes that Purdue Pharma has encountered. Do you think that the drug manufacturer bears some responsibility in this matter? Write an essay taking one side or another or present your stance in a class discussion.

Morphine, oxycodone, hydrocodone (found in Vicodin, Lortab, and Lorcet), and codeine belong to a class of drugs known as opiates. Morphine and codeine are extracted from *Papaver somniferum,* the opium poppy. The other opiates listed are synthesized from morphine. They vary in their strength and thus their drug scheduling. But they all have the same mode of action. These drugs bind to a chemical receptor called *mu,* that interrupts the transmission of pain in the spinal cord. Opiates also stimulate areas of the brain involved in pleasure, called the reward, or endorphin, pathways. In a normal brain, the chemical dopamine crosses between brain cells in the reward pathway, producing pleasurable feelings. Opiates stimulate the release of higher levels of dopamine, strengthening reward signals, and producing intense euphoria. Repeated use of opiates causes the brain of a user to become accustomed to an overstimulated reward pathway. This in turn brings about the phenomenon of tolerance—greater amounts of opiates are required to achieve the euphoria the user once experienced.

The drugs just described are but a few of today's commonly abused substances. There are many more, including alcohol, which is possibly the most damaging of all drugs when taken to excess. Just as for prescribed medicines, the choice to take illicit drugs involves a risk–benefit analysis. All drug users should have this basic information before making this choice.

Conclusion

Molecular modifications by chemists have created a vast new pharmacopoeia of wonder drugs that have significantly increased the number and quality of our days. Prior to penicillin, an infection could mean a death sentence. Now thanks to penicillin, sulfa drugs, and antibiotics, the great majority of bacterial infections are easily controlled. Dreaded killers such as typhoid, cholera, and pneumonia have been largely eliminated—at least in many nations of the world.

New methods of drug discovery are allowing vast libraries of novel compounds to be built. These can be stored for future testing by assays that have not yet been dreamed of. Nanomedicine opens new horizons for drug delivery systems, in which selective drugs will be directed right where they are needed.

But no drug can be completely safe, and almost any drug can be misused. These issues become the focus when the FDA is asked to change the status of a drug from prescription to over-the-counter. Taking any medication (generic or brand name) is a conscious choice between the benefits derived from the drug and the risks associated with its side effects and limits of safety. Because most drugs have very wide, carefully established margins of safety, their benefits far outweigh their risks for the general population. For some drugs, however, the trade-off between effectiveness and safety involves a different balance. A drug with severe side effects may be the only treatment available for a life-threatening disease. Someone suffering from HIV/AIDS or advanced, inoperable cancer understandably has a different perspective on drug risks and benefits than a person with a cold. And the impersonal anonymity of averages involved in clinical drug trials takes on new meaning at the bedside of a loved one.

Herbal and alternative medicines raise new questions. Who is responsible for defining their efficacy, monitoring their purity, and developing contraindications to their use with other drugs? Doctors must be advised of all medicines their patients are taking, including herbal remedies. The abuse of herbal and synthetic drugs continues to cause major problems throughout our society. Advances are being made in all of these areas. When chemistry is applied to medicine, science must be guided by morality, and reason must be tempered with compassion.

Chapter Summary

Having studied this chapter, you should be able to:

- Describe the discovery, development, and physiological properties of aspirin (10.1)

- Understand bonding in carbon-containing (organic) compounds (10.2)

- Apply the concept of isomerism to organic molecules (10.2)

- Convert chemical formulas of carbon-containing compounds to structural formulas, condensed structural formulas, and line-angle drawings (10.2)

- Recognize functional groups and the classes of organic compounds that contain them; draw structural formulas for organic molecules containing various functional groups (10.3)

- Understand that functional groups may be chemically modified to change a molecule's properties (10.3)

- Predict the products of ester formation reactions and describe how amines may be converted to their salt forms (10.3)

- Relate the molecular structure of aspirin to other analgesics (10.3)

- Understand the mode of action of aspirin and other analgesics (10.4)

- Describe the discovery of penicillin (10.5)

- Explain the lock-and-key mechanism of drug action (10.5)

- Describe how combinatorial synthesis can be employed in the creation of large collections of new drugs at lower costs than previous methods (10.5)

- Understand differences in molecular structure between a pair of chiral (optical) isomers (10.6)

- Appreciate the economic effect of chiral drugs (10.6)

- Identify the basic carbon skeleton arrangement of steroids (10.7)

- Recognize that minor changes in steroid structure may result in large changes in bioactivity (10.7)

- Discuss the emerging field of nanotechnology in terms of drug delivery systems (10.8)

- Explain the procedure for drug testing and approval and the associated benefits and costs (10.9)

- Compare and contrast brand-name and generic drugs (10.10)

- Identify some of the over-the-counter drug categories and their uses (10.10)

- Understand the process of a drug going from prescription to OTC (10.10)

- Describe some of the potential benefits and risks of herbal medicines (10.11)

- Explain the scheduling of prescription drugs (10.12)

- Discuss the use of marijuana and oxycodone in terms of their physiological and social effects (10.12)

Questions

Emphasizing Essentials

1. Give the intended effect of each. Can one drug exhibit all of these effects?

   a. an antipyretic drug

   b. an analgesic drug

   c. an anti-inflammatory drug

2. The field of chemistry has many subdisciplines. What do organic chemists study?

3. Write condensed structural formulas and line-angle drawings for the three isomers of C_5H_{12} assigned in Your Turn 10.7.

4. Write the structural formula and line-angle drawings for each different isomer of C_6H_{14}. *Hint:* Watch out for duplicate structures.

5. Consider the isomers of C_4H_{10}. How many different isomers could be formed by replacing a single hydrogen atom with an —OH group? How many different alcohols have the formula C_4H_9OH? Draw the structural formula for each.

6. For each compound, identify the functional group present.

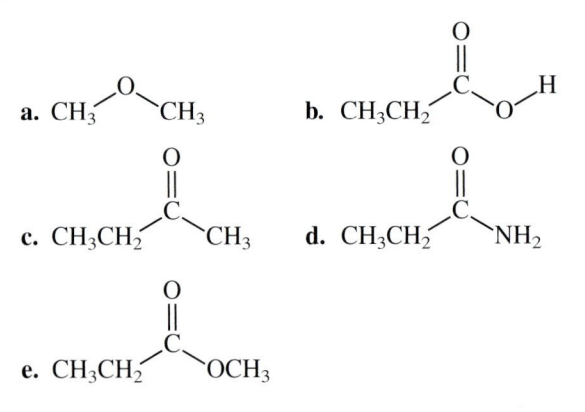

7. Draw the simplest compound that can contain each of these functional groups. In some cases, only one carbon atom is required; in other cases two.

   a. an alcohol   **d.** an ester

   b. an aldehyde   **e.** an ether

   c. a carboxylic acid   **f.** a ketone

8. For each of these, identify the functional group. Then, draw an isomer that contains a different functional group.

   a. CH_3CH_2—OH   **b.** CH_3CH_2 C(=O)H

   c. CH_3CH_2 C(=O)OCH_3

9. In allergy sufferers, histamine causes runny noses, red eyes, and other symptoms. Here is its structural formula.

   a. Give the chemical formula for this compound.

   b. Circle the amine functional groups in histamine.

   c. Which part (or parts) of the molecule make the compound water-soluble?

10. Figure 10.7 shows a somewhat condensed structural formula for acetaminophen, the active ingredient in Tylenol.

    a. Draw the structural formula for acetaminophen, showing all atoms and all bonds.

    b. Give the chemical formula for this compound.

    c. Children's Tylenol is a flavored aqueous solution of acetaminophen. Predict what part (or parts) of the molecule make acetaminophen water-soluble.

11. Identify the functional groups in each of these.

    a. Barbital (a sedative)

b. Penicillin-G (an antibiotic)

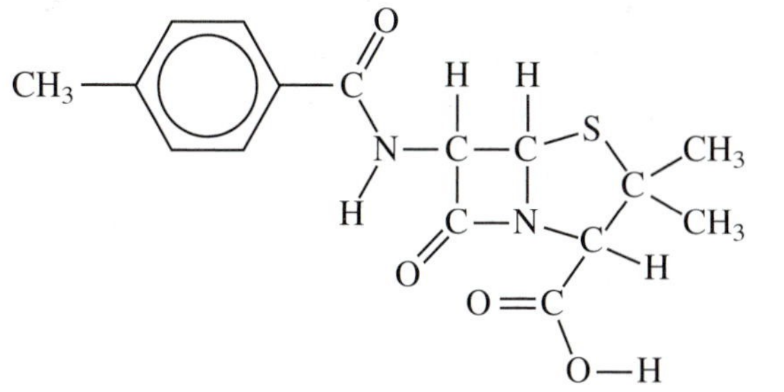

c. Amyl dimethylaminobenzoate (an ingredient in sunscreens)

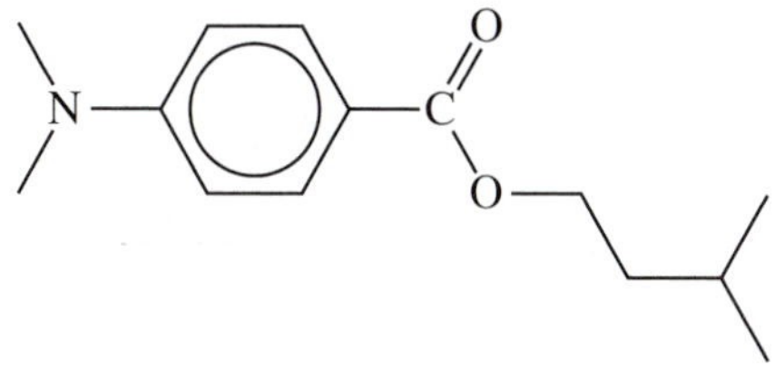

12. Ibuprofen is relatively insoluble in water but readily soluble in most organic solvents. Explain this solubility behavior based on its structural formula.
Hint: See Figures 10.6 and 10.7.

13. Here is the structural formula for diazepam, the sedative found in Valium.

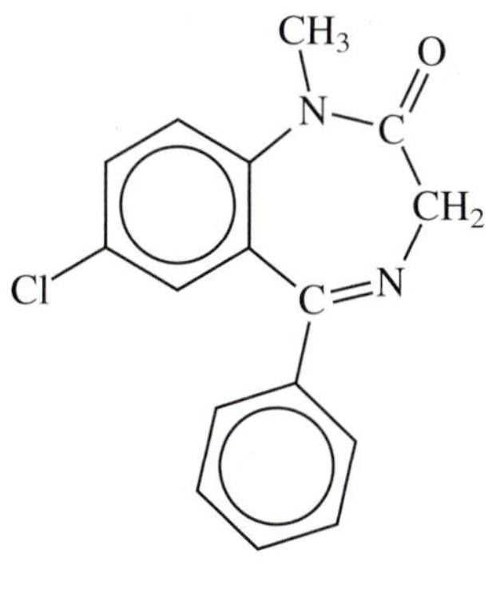

Judging from its structure, do you expect it to be more soluble in fats or in aqueous solutions? Explain.

14. Draw structural formulas for the esters formed when acetic acid reacts with these alcohols:

a. *n*-propanol, $CH_3CH_2CH_2OH$

b. *iso*propanol, $(CH_3)_2CHOH$

c. *t*-butanol, $(CH_3)_3COH$

15. Interpret this sentence by giving the meaning of each acronym and explaining the effect. "NSAIDs have an effect on COX enzymes."

16. Usually carbon forms four covalent bonds, nitrogen three, oxygen two, and hydrogen only one bond. Use this information to draw structural formulas for:

a. A compound that contains one carbon atom, one nitrogen atom, and as many hydrogen atoms as needed.

b. A compound that contains one carbon atom, one oxygen atom, and as many hydrogen atoms as needed.

17. Would aspirin be more active if it were to interact with prostaglandins directly, rather than by blocking the activity of COX enzymes? Explain your reasoning.

18. Examine the combinatorial synthesis process diagrammed in Figure 10.16. If you started with a molecule that had two active functional groups, how many products could form after two synthetic steps, where each step adds a molecule with one functional group? How many products could be formed if the reagent in the first step had two reactive functional groups itself?

19. The text states that 80 billion tablets of aspirin a year are consumed in the United States. If the average tablet contains 500 mg of aspirin, how many pounds of aspirin does this consumption represent?

20. Identify the functional groups in morphine and meperidine (Demerol). Can these molecules be assigned to a particular class of compound (i.e., an alcohol, ketone, or amine)? Explain. *Hint*: See Figure 10.14 for structural formulas.

21. What is meant by the term *pharmacophore*?

22. Sulfanilamide is the simplest sulfa drug, a type of antibiotic. It appears to act against bacteria by replacing *para*-aminobenzoic acid, an essential nutrient for bacteria, with sulfanilamide. Use these structural formulas to explain why this substitution is likely to occur.

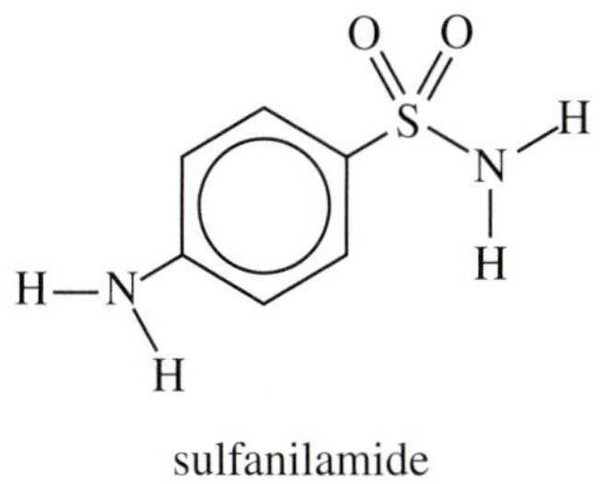

sulfanilamide

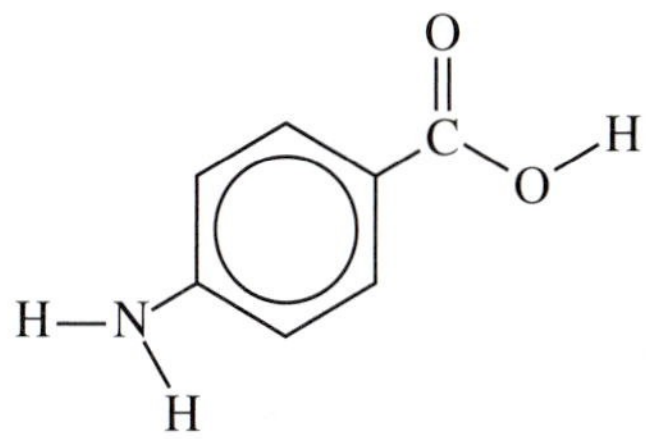

para-aminobenzoic acid

23. Which of these molecules is chiral?

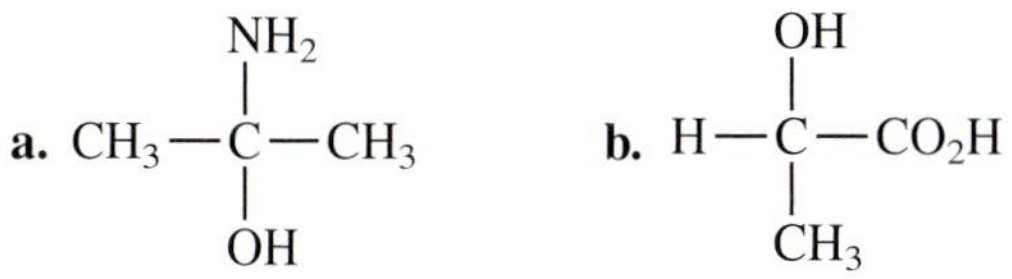

a. $CH_3-\underset{\underset{OH}{|}}{\overset{\overset{NH_2}{|}}{C}}-CH_3$

b. $H-\underset{\underset{CH_3}{|}}{\overset{\overset{OH}{|}}{C}}-CO_2H$

c. $CH_3 \overset{\overset{\displaystyle NH_2}{|}}{\underset{\underset{\displaystyle N}{\overset{\displaystyle |||}{C}}}{C}} CO_2H$ d. $CH_3 \overset{\overset{\displaystyle OH}{|}}{\underset{\underset{\displaystyle CH_3}{|}}{C}} CO_2H$

24. Which of these molecules has chiral forms?

a. $CH_3 \overset{\overset{\displaystyle NH_2}{|}}{\underset{\underset{\displaystyle OH}{|}}{C}} CH_2CH_3$ b. $H \overset{\overset{\displaystyle OH}{|}}{\underset{\underset{\displaystyle H}{|}}{C}} C_2H_5$

c. $CH_3 \overset{\overset{\displaystyle NH_2}{|}}{\underset{\underset{\displaystyle CH_2OH}{|}}{C}} CO_2H$ d. $CH_3 \overset{\overset{\displaystyle OH}{|}}{\underset{\underset{\displaystyle CH_2SH}{|}}{C}} CO_2H$

25. Methamphetamine hydrochloride is a powerful stimulant that is dangerous and highly addictive. This drug also goes by the street names "crystal," "crank," or "meth." This structural formula shows its salt form.

The freebase form of this drug, called "ice," is also abused. What does the term *freebase* mean, and how might the drug be converted to this form?

26. Examine the structures of levo- and dextromethorphan (see Figure 10.20). Identify any chiral carbons and list any functional groups present.

27. Molecules as diverse as cholesterol, sex hormones, and cortisone contain common structural elements. Use a line-angle drawing to show the structure they share.

Concentrating on Concepts

28. The text states that some remedies based on the medications of earlier cultures contain chemicals that are effective against disease, others are ineffective but harmless, and still others are potentially harmful. How might it be determined into which of these three categories a recently discovered substance fits?

29. Draw structural formulas for each of these molecules and determine the number and type of bonds (single, double, or triple) for each carbon atom.

    a. H_3CCN (acetonitrile, used to make a type of plastic)

    b. $H_2NC(O)NH_2$ (urea, an important fertilizer)

    c. C_6H_5COOH (benzoic acid, a food preservative)

30. Compare the physiological effects of aspirin with those of acetaminophen and ibuprofen. Relate differences to the nature of each compound at the molecular and cellular levels.

31. In Your Turn 10.7, you were asked to draw structural formulas for the three isomers of C_5H_{12}. One student submitted this set, with a note saying that six isomers had been found. (*Note:* The hydrogen atoms have been omitted for clarity.) Help this student see why some of the answers are incorrect.

32. Styrene, $C_6H_5CH{=}CH_2$, the monomer for polystyrene described in Chapter 9, contains the phenyl group, C_6H_5. Draw structural formulas to show that this molecule, like benzene, has resonance structures.

33. Aspirin is a specific compound, so what justifies the claims for the superiority of one brand of aspirin over another?

34. Figure 10.8 represents chemical communication within the body. Write a paragraph explaining what this figure means to you in helping to explain chemical communication.

35. Consider this statement. "Drugs can be broadly classed into two groups: those that produce a physiological response in the body and those that inhibit the growth of substances that cause infections." Into which class does each of these drugs fall?

    a. aspirin   c. (Keflex) antibiotic   e. amphetamine

    b. morphine   d. estrogen   f. penicillin

36. Consider the structure of morphine in Figure 10.14. Codeine, another strong analgesic with narcotic action, has a very similar structure in which the —OH group attached to the benzene ring is replaced by an —OCH₃ group.

    a. Draw the structural formula for codeine and label its functional groups.

b. The analgesic action of codeine is only about 20% as effective as morphine. However, codeine is less addictive than morphine. Is this enough evidence to conclude that replacement of —OH groups with —OCH₃ groups in this class of drugs will always change the properties in this way? Explain.

37. Dopamine is found naturally in the brain. The drug (−)-dopa is found to be effective against the tremors and muscular rigidity associated with Parkinson's disease. Identify the chiral carbon in (−)-dopa, and comment on why (−)-dopa is effective, whereas (+)-dopa is not.

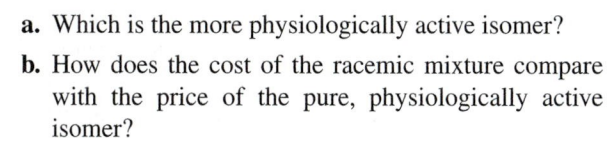

(−)-dopa

38. Vitamin E is often sold as a racemic mixture of (+) and (−) isomers. Use the Web to find answers to these questions.

a. Which is the more physiologically active isomer?

b. How does the cost of the racemic mixture compare with the price of the pure, physiologically active isomer?

39. Consider the fact that levomethorphan is an addictive opiate, but dextromethorphan is safe enough to be sold in many over-the-counter cough remedies. From a molecular point of view, how is this possible?

40. Describe the lock-and-key analogy for the interaction between drugs and receptor sites. Use the analogy in a discussion as if you were explaining this to a friend.

41. Why are many projects to isolate or synthesize new drugs started in this country, but few actually receive FDA approval for general use?

42. Until the early 19th century, it was believed that organic compounds had some sort of "life force" and could only be produced by living organisms. This was the basis of a concept called vitalism. This view was dispelled in the late 1800s. Use the Web to find out what changed that opinion.

Exploring Extensions

43. One avenue for successful drug discovery is to use the initial drug as a prototype for the development of other similar compounds called analogs. The text states that cyclosporine, a major antirejection drug used in organ transplant surgery, is an example of a drug discovered in this way. Research the discovery of this drug to verify this statement. Write a brief report describing your findings, citing your sources.

44. Dorothy Crowfoot Hodgkin first determined the structure of a naturally occurring penicillin

compound. What in her background prepared her to make this discovery? Write a short report on the results of your findings, citing your sources.

45. Before the cyclic structure of benzene was determined (see Figure 10.4), there was a great deal of controversy about how the atoms in this compound were arranged.

a. Count outer electrons for C and H in C₆H₆. Then draw the structural formula for a possible linear isomer.

b. Give the condensed structural formula for your answer in part **a**.

c. Compare your structure with those drawn by classmates. Are they all the same? Why or why not?

46. Antihistamines are widely used drugs for treating symptoms of allergies caused by reactions to histamine compounds. This class of drug competes with histamine, occupying receptor sites on cells normally occupied by histamine. Here is the structure for a particular antihistamine.

a. Give the chemical formula for this compound.

b. What similarities do you see between this structure and that of histamine (shown in Question 9) that would allow the antihistamine to compete with histamine?

47. Over the next few years, the FDA may consider deregulating more than a dozen drugs, nearly as many as have already been approved for over-the-counter sales during the past decade. The products that have led this trend have been the widely advertised drugs for heartburn.

a. Which questions need to be answered before a drug is deregulated?

b. Will these questions change if you are considering this need from the viewpoint of the FDA, a pharmaceutical company, or as a consumer?

48. Find out more about the new process for the manufacture of hydroxynitrile (HN), a precursor to Lipitor that won a 2006 Presidential Green Chemistry Challenge Award. How does this process differ from the earlier one for manufacturing HN? Write a brief report on your research, citing your sources.

49. The steroids testosterone and estrone were first isolated from animal tissue. One ton of bull testicles was needed to obtain 5 mg of testosterone and 4 tons of pig ovaries was processed to yield 12 mg of estrone.

a. Assuming complete isolation of the hormones was achieved, calculate the mass percentage of each steroid in the original tissue.

b. Explain why the calculated result very likely is incorrect.

50. Herbal remedies are prominently displayed in supermarkets, drug stores, and discount stores.

    a. What influences your decision to buy one of these remedies?

    b. Choose a remedy and carefully examine its label for information about the active ingredients, inert ingredients, anticipated side effects, the suggested dosage, and the cost per dose.

    c. How confident are you that the safety and efficacy of these remedies are ensured? Explain your answer.

51. Habitrol was the most successful smoking-cessation prescription until the introduction of bupropion (Zyban) in 1997, which quickly claimed 50% of the market. How are these two approaches to drug therapy different?

52. Drug approval laws in other nations are not the same as those in the United States. Choose a country and find out how its drug approval process works compared with the process in the United States. Construct relative time lines that reveal what steps must be taken and approximately the length of time each step may require. Also, find out if specific drugs are available in that country that are not available in the United States and comment on what factors may be influencing the policies of each country.

53. Over-the-counter drugs allow consumers to treat a myriad of symptoms and ailments. An advantage is that the user can purchase and administer the treatment without the effort or the expense of consulting a physician. To provide a wider margin of safety for these circumstances, the OTC versions of drugs are often administered in lower doses. See if this is true by looking at information for prescription and OTC versions of painkillers like ibuprofen (Motrin) and heartburn treatments like nizatidine (Axid) and famotidine (Pepcid). Report on your findings.

54. Herbal or alternative medicines are not regulated in the same way as prescription or OTC medicines. In particular, the issues of concern are identification and quantification of the active ingredient, quality control in manufacture, and side effects when the herbal remedy is used in conjunction with another alternative or prescription medicine. Look for evidence from herbal supplement manufacturers that address these issues, and write a report documenting your findings, giving your references.

55. Danco Laboratories, the U.S. company that produces RU-486 (mifeprex, the "abortion pill") makes claims about the safety of this steroidal drug including a comparison to aspirin. Do some research on RU-486, and write a short report on the drug. Include its structure, mode of action, and safety record.

56. The antibiotic ciprofloxacin hydrochloride (Cipro) treats bacterial infections in many different parts of the body. This drug made headlines in 2001 for use in patients who had been exposed to the inhaled form of anthrax. Use the Web or another source to obtain the structure of Cipro. Draw its structure and identify the functional groups.

57. Direct-to-consumer advertising of prescription drugs has proved to be a successful marketing tool for pharmaceutical companies. Twenty percent of consumers say that advertisements prompted them to call or visit their doctor to discuss the drug, according to PharmTrends, a patient-level syndicated tracking study of consumer behavior by market research organization Ipsos-NPD. Make a list of the pros and cons of this type of marketing from both the patient's and physician's point of view.

58. In 2003, a series of spot examinations of mail shipments of foreign drugs to U.S. consumers conducted by the FDA and U.S. Customs and Border Protection revealed that these shipments often contain unapproved or counterfeit drugs that pose serious safety problems. Although many drugs obtained from foreign sources purport to be, and may even appear to be, the same as FDA-approved medications, these examinations showed that many are of unknown quality or origin. Of the 1153 imported drug products examined, the overwhelming majority, 1019 (88%), were illegal because they contained unapproved drugs. Many of these imported drugs could pose clear safety problems. Use the FDA Web site to determine which drugs were most commonly counterfeited and their countries of origin.

59. Thalidomide was first marketed in Europe in the late 1950s. It was used as a sleeping pill and to treat morning sickness during pregnancy. At that time it was not known to cause any adverse effects. By the late 1960s, however, the drug was banned after it was found to be a teratogen, causing deformed limbs in the children of women who took it early in pregnancy. Use the Web for information to write a short paper that describes the optical isomers of thalidomide and why the FDA did not approve thalidomide for use in the United States until recently. For what purpose has the FDA recently approved the use of thalidomide?

60. Conventional acne treatments are maintenance therapies. Antibiotics, such as tetracycline and erythromycin, are not expected to result in long-term improvement once they are stopped. Some individuals may not respond to any of the conventional medications for acne. A possible solution for those patients is the vitamin A derivative known as isotretinoin (Accutane). However, Accutane is by no means a frontline therapy. Using the Web, find the serious side effects that may accompany the use of Accutane.

Chapter

11

Nutrition: Food for Thought

"Study Finds Low-Fat Diet Won't Stop Cancer or Heart Disease," ran the headline in the New York Times. *"Big Study Finds No Clear Benefit of Calcium Pills," the* Times *declared a week later."*

Source: Reprinted with permission.

If complicated scientific findings get translated into sound bites, the message often gets twisted. Consumers need to look behind the headlines.

Nutrition Action Health Letter, April 2006

Have you found yourself confused by conflicting dietary advice in the news? What happened to the low-carb craze of a few years ago? How are good carbs different from bad carbs? And what about all those terms related to fats—low fat, fat-free, and trans fats? Aren't saturated fats the only unhealthy part of our daily diet? Has chocolate really become a health food? Are artificial sweeteners safe? Given all these questions, it is no wonder that consumers often just give up trying to follow dietary recommendations found in the news headlines. Surrounded by food choices and perhaps partially because of uncertainty as to what constitutes a proper diet, the U.S. population now finds that obesity and its related adverse health effects are overtaking smoking as the No. 1 cause of death in the United States.

To add scientific understanding to our dietary choices, we start with the categories of food—the nature and structures of the family of fats, carbohydrates, proteins, vitamins, and minerals. We also know that too many calories combined with too little exercise accounts for part of the increase in obesity, and this is a tip off that our discussion will also include energy. We will consider both the energy content of different nutrients and how the body handles the various structural forms that store the energy from food. The chapter ends with a review of some old and new methods of food preservation and a more global perspective on nutrition.

Consider This 11.1   The Changing American Thirst

This graph presents data from the U.S. Department of Agriculture for per capita consumption of common beverages.

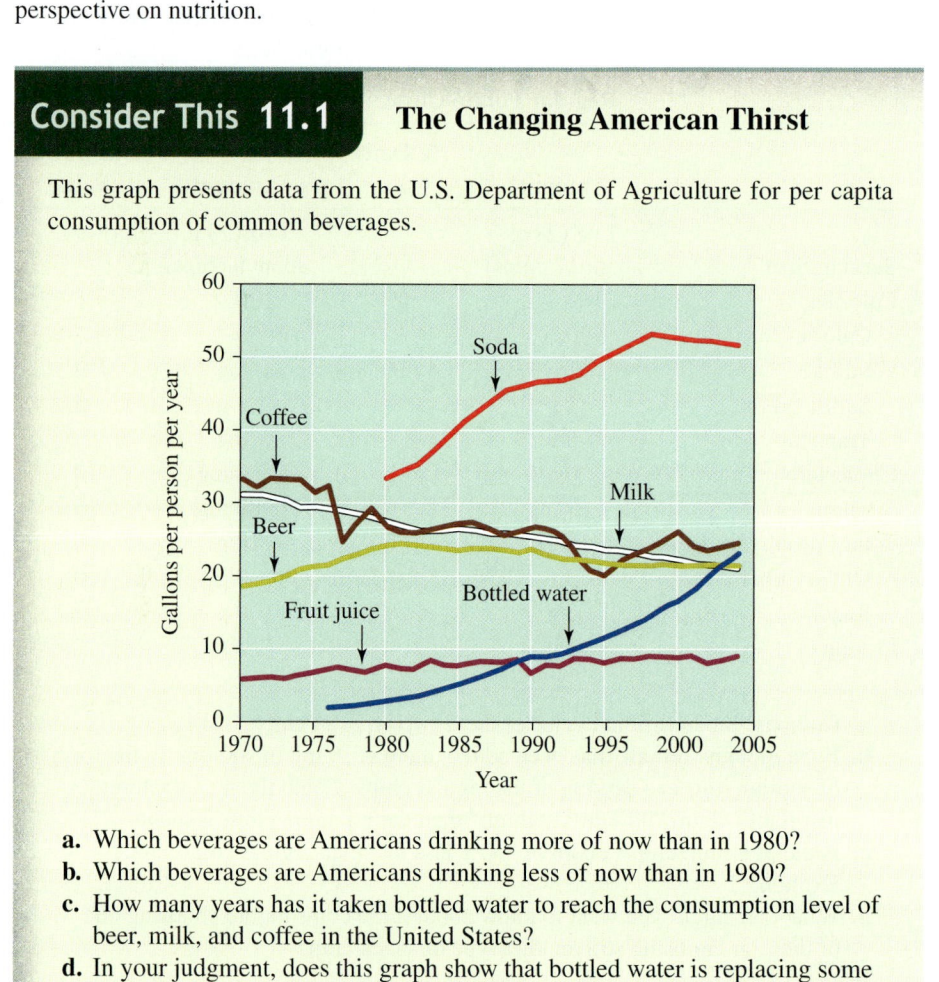

a. Which beverages are Americans drinking more of now than in 1980?
b. Which beverages are Americans drinking less of now than in 1980?
c. How many years has it taken bottled water to reach the consumption level of beer, milk, and coffee in the United States?
d. In your judgment, does this graph show that bottled water is replacing some other beverage in American diets? Explain.

Source: From *Nutrition Action Health Letter,* "The changing American Diet," by Bonnie Liebman and David Schardt, April 2006. Copyright©2006 by Center for Science in the Public Interest via Copyright Clearance Center.

11.1   You Are What You Eat

Whether we eat delicious culinary creations or some junk food gobbled on the run, all of us eat because food provides four fundamental types of materials required to keep our bodies functioning: water, energy sources, raw materials, and metabolic regulators.

Many of these important properties of water were discussed in Chapter 5.

Water serves as both a reactant and a product in metabolic reactions, as a coolant and thermal regulator, and as a solvent for the countless substances that are essential for life. Our bodies are approximately 60% water, but H_2O cannot be burned in the body or elsewhere. Therefore, we need food as a source of energy to power processes as diverse as muscle action, brain and nerve impulses, and the movement of molecules and ions in suitable ways at appropriate times and places. We also eat because we need raw materials for the syntheses of new bone, blood, enzymes, muscles, hair, and for the replacement and repair of cellular materials. And finally, food supplies some of the enzymes and hormones that function to control the biochemical reactions associated with metabolism and all other vital processes.

1 dietary calorie = 1 Calorie
= 1 kcal
= 1000 calories

Consider This 11.2 | Chip Choices

The composition of two types of potato chips from the same manufacturer is given here: one a regular potato chip advertising "no salt added" and the second promoting itself as a premium potato chip with 40% less fat than regular potato chips. The percent of daily values given are based on a 2000-Calorie diet.

	Regular Potato Chips		Premium Potato Chips	
	Amount	% Daily Value	Amount	% Daily Value
Serving Size	1 oz (28 g, about 17 chips)		1 oz (28 g, about 9 chips)	
Calories	160		130	
Calories from fat	90		50	
Total Fat	11 g	16	6 g	9
Saturated fat	1 g	5	1 g	5
Trans fat	0 g		0 g	
Cholesterol	0 mg	0	0 mg	0
Sodium	10 mg	0	80 mg	3
Total Carbohydrates	15 g	5	18 g	6
Dietary fiber	3 g	8	0 g	0
Sugars	3 g		0 g	
Protein	1 g		2 g	

a. Compare the nutritional value of these two types of chips.

b. Even though "no salt" has been added, there is 10 mg of sodium (in the form of sodium ion) in a serving of the regular chips. Could the manufacturer justifiably claim "90% less sodium" than the premium chip? Explain.

c. Is the premium chip's claim of "40% less fat" than regular chips valid? Explain.

d. What else might you want to know about these chips before choosing one of these or choosing still another type of snack chip?

Eating properly is a matter of consuming a proper variety of foods, not just eating sufficient amounts of food. It is possible to consume food regularly, even to the point of being overweight, and still be malnourished. The meaning of this commonly used term is important. **Malnutrition** is caused by a diet lacking in the proper mix of nutrients, even though the energy content of the food eaten may be adequate. Contrast malnutrition with **undernourishment,** a condition in which a person's daily caloric intake is insufficient to meet metabolic needs. Significant numbers of people in this country are malnourished and undernourished, while even greater proportions of Americans are overweight

than ever before. In data reported by the Centers for Disease Control and Prevention in the *Journal of the American Medical Association* in 2006, 66% of all adults are classified as overweight with nearly half of that population classified as obese. Even children are getting fat, with more than one third now overweight.

Sceptical Chymist 11.3   A Lifetime of Food

During a lifetime, you will eat a truly prodigious amount of food, estimated to be about 700 times your adult body weight. This statement is itself quite an extraordinary assertion. Do calculations to check that the statement is in the ballpark. State all of your assumptions clearly.

Hint: Start by assuming a life span of approximately 78 years and that your present weight is your adult weight. Estimate the weight of food eaten daily at present, and use these data to project your lifetime consumption of food.

Processed foods are required to provide nutritional information. A label, such as that shown in Figure 11.1, first displays the amounts of fats, carbohydrates, and proteins. These are the **macronutrients** that provide essentially all of the energy and most of the raw material for body repair and synthesis. Sodium and potassium ions (not the metals) are present in much lower concentrations, but these ions are essential for the proper electrolyte balance in the body. Several other minerals (Section 11.5) and an alphabet soup of vitamins (Section 11.5) are listed in terms of the percent of recommended daily requirements supplied by a single serving of the product. It should be evident that all these substances, whether naturally occurring or added during processing, are chemicals. Unfortunately, this fundamental fact is apparently lost on those who pursue the impossible dream of a "chemical-free" diet. *All* food is inescapably and intrinsically chemical, even food claiming to be "organic" or "natural."

Table 11.1 indicates the mass percentages (grams of component per 100 g of food item) of water, fats, carbohydrates, and proteins in several familiar foods. For this particular selection of foods, the variation in composition is considerable. But in every case, these four components account for almost all of the matter present. Water ranges from a high of 89% in 2% milk to a low of 1% in peanut butter. Peanut butter is comparable to steak and fish in percent of protein and also leads these foods in fat content. Chocolate chip cookies have the highest percentage of carbohydrate because of their high sugar and refined flour content.

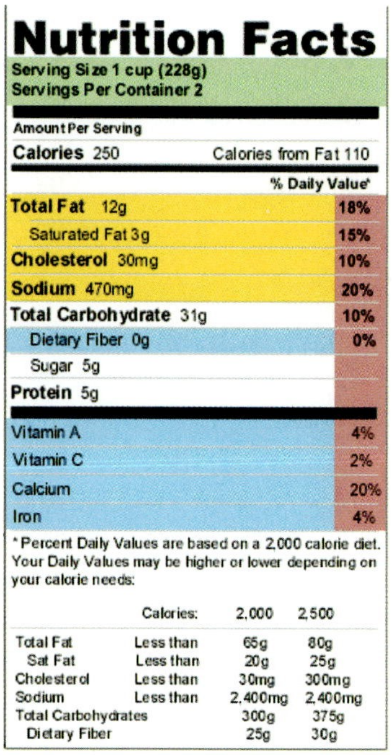

Nutrition Facts
Serving Size 1 cup (228g)
Servings Per Container 2

Amount Per Serving	
Calories 250	Calories from Fat 110

	% Daily Value*
Total Fat 12g	**18%**
Saturated Fat 3g	**15%**
Cholesterol 30mg	**10%**
Sodium 470mg	**20%**
Total Carbohydrate 31g	**10%**
Dietary Fiber 0g	**0%**
Sugar 5g	
Protein 5g	

Vitamin A	4%
Vitamin C	2%
Calcium	20%
Iron	4%

* Percent Daily Values are based on a 2,000 calorie diet. Your Daily Values may be higher or lower depending on your calorie needs:

		Calories:	2,000	2,500
Total Fat	Less than		65g	80g
Sat Fat	Less than		20g	25g
Cholesterol	Less than		30mg	300mg
Sodium	Less than		2,400mg	2,400mg
Total Carbohydrates			300g	375g
Dietary Fiber			25g	30g

Figure 11.1
A nutrition facts label from a package of nuts.

Table 11.1	Percentage of Water, Fats, Carbohydrates, and Proteins			
Food	**Water**	**Fats**	**Carbohydrates**	**Proteins**
White bread	37	4	48	8
2% Milk	89	2	5	3
Chocolate chip cookies	3	23	69	4
Peanut butter	1	50	19	25
Sirloin steak	57	15	0	28
Tuna fish	63	2	0	30
Black beans (cooked)	66	<1	23	9

Source: U.S. Department of Agriculture, Agricultural Research Service, *Home and Garden Bulletin* 72.

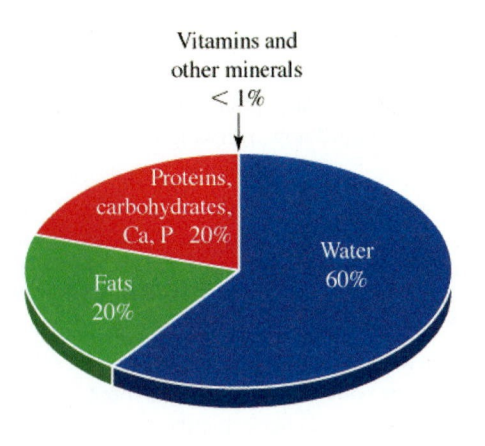

Figure 11.2

Composition of the human body.

Now compare Table 11.1 with Figure 11.2, which presents similar data for the human body. It is not surprising that the human body is composed of the same basic types of molecules that make up the food we eat. We are wetter and fatter than bread and contain more protein than milk; we are more like steak than chocolate chip cookies. From the data in Figure 11.2, we can calculate that a 150-lb person consists of 90 lb of water (150 lb × 60-lb water/100-lb body) and 30 lb of fat. The remaining 30 lb is almost all composed of various proteins and carbohydrates plus the calcium and phosphorus in the bones. The other minerals and the vitamins weigh less than 1 lb. This indicates that a little bit of each of them goes a long way, a point discussed in Section 11.5.

Of the nearly 90 naturally occurring elements, just 11 make up over 99% of the mass of your body. Table 11.2 lists the mass percentages (grams of element/100 g body mass) of these 11 elements in the human body and gives their relative atomic abundances. Figure 11.3 shows their location on the periodic table.

Just four elements—oxygen, carbon, hydrogen, and nitrogen—account for 95.7% of our body mass. Not surprisingly, they also are the major elements in what we eat and drink. Hydrogen and oxygen are, of course, the elements of water. Moreover, along with carbon atoms, they constitute all fats (Section 11.2) and carbohydrates (Section 11.3). Finally, these three elements plus nitrogen are found in all proteins (Section 11.4). Thus, nature uses very simple aggregates of oxygen, carbon, hydrogen, and nitrogen atoms in

Table 11.2	Major Elements of the Human Body		
Element	**Symbol**	**Grams per 100 g of Body Mass**	**Relative Abundance per Million Atoms in Body**
oxygen	O	64.6	255,000
carbon	C	18.0	94,500
hydrogen	H	10.0	630,000
nitrogen	N	3.1	13,500
calcium	Ca	1.9	3,100
phosphorus	P	1.1	2,200
chlorine	Cl	0.40	570
potassium	K	0.36	580
sulfur	S	0.25	490
sodium	Na	0.11	300
magnesium	Mg	0.03	130

Your Turn 11.4	**How Much and How Many?**

Table 11.2 gives both the mass of the major elements per 100 g of body mass and the relative abundance in the number of atoms per million atoms in the body. Why is oxygen the most abundant element when measured in grams per 100 g of body mass, but hydrogen is the most abundant when measured in terms of relative abundance per million atoms in the body?

Hint: Consider the atomic masses of each element.

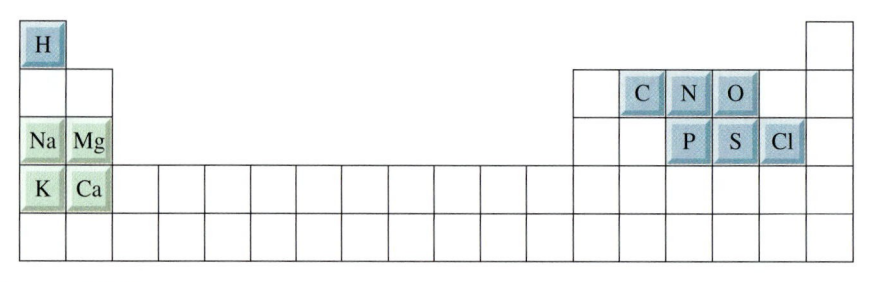

Figure 11.3
Periodic table showing the 11 most abundant elements in the human body.

a myriad of elegantly functional combinations to produce the major constituents of a healthy body and a healthful diet.

In the next three sections, we will take a look at these three macronutrient classes of compounds—fats, carbohydrates, and proteins—before turning to minerals and vitamins. Once we have considered all the main food chemical groups, it will be easier to understand how to make sound nutritional choices based on chemistry.

11.2   Fats and Oils: Part of the Lipid Family

Everyone knows the properties of fats from personal experience. They are greasy, slippery, soft, low-melting solids that are not particularly soluble in water. Butter, cheese, cream, whole milk, and certain meats and some fish are loaded with them. All of these products are of animal origin. But margarine and cooking oils are of vegetable origin. Oils, such as those obtained from olives, corn, or nuts, exhibit many of the properties of animal-based fats. Unlike fats, oils are liquids. Whether solid or liquid, fats and oils belong to a family of compounds known as **lipids,** a class of compounds that includes not only the edible fats and oils but also diverse materials such as cholesterol and other steroids.

Petroleum-based greases and oils share similar properties of being greasy and slippery, suggesting a chemical and structural similarity to edible fats and oils. But there are some important differences between edible fats and oils and petroleum-based greases and oils. You are well aware that petroleum is made up almost exclusively of hydrocarbons. In these compounds, carbon atoms are bonded to one another (often in chains) and to hydrogen atoms. There are no polar bonds present and therefore hydrocarbon molecules are nonpolar. Unlike some edible fats and oils, petroleum-based greases and oils do not mix at all well with water or other polar substances.

Naturally occurring **fatty acid molecules** are characterized by a long nonpolar hydrocarbon chain generally containing an even number of carbon atoms (typically 12–24) and a polar carboxylic acid group at the end of the chain. For example, consider the common fatty acid found in beef fat, stearic acid, $C_{17}H_{35}COOH$ (Figure 11.4). The oxygen

Hydrocarbons in petroleum were discussed in Section 4.8. Reasons why nonpolar substances do not dissolve in water were examined in Section 5.12.

$$CH_3(CH_2)_{16}COOH$$

Condensed structural formula

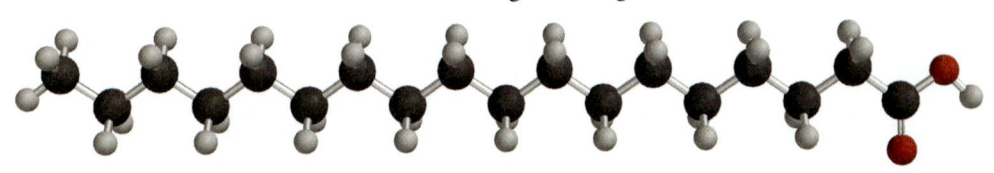

Semi-expanded structural formula

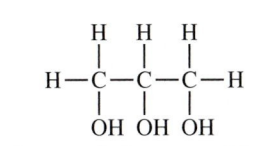

Line-angle drawing

Line-angle drawings were
introduced in Section 10.2.

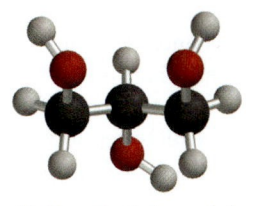

Ball-and-stick model

Figure 11.4

Representations of stearic acid, a fatty acid.

Figures Alive! Visit the *Online Learning Center* to learn more about fats and fatty acids.

is in the carboxylic acid group (—COOH) found at the end of a long hydrocarbon chain. This functional group is what adds the term *acid* to fatty acid because —COOH can release a hydrogen ion (H⁺). The long hydrocarbon chains, on the other hand, give fats most of their characteristic physical properties.

A second player in forming fats and oils is the compound glycerol, also called by the common name glycerine. Glycerol is a sticky, syrupy liquid that is sometimes added to soaps and hand lotions. The representations in Figure 11.5 show that a glycerol molecule contains three —OH groups, which classifies it as an alcohol.

If fatty acids react with glycerol, one product is a **triglyceride, an ester of three fatty acid molecules and one glycerol molecule.** Three molecules of water are also formed. The formation of a triglyceride can be represented by a word equation:

3 fatty acid molecules + 1 glycerol molecule ⟶ 1 triglyceride molecule + 3 water molecules   [11.1]

Moving from the general statement to a specific example, equation 11.2 represents the combination of three stearic acid molecules with a glycerol molecule to form a triglyceride. Note that the triglyceride contains three ester functional groups, explaining why a triglyceride is also known as a triester.

organic acid + alcohol ⟶
ester + water

This group is characteristic
of an ester:

The ester group was discussed
in Sections 9.2 and 10.3.

$$CH_2(OH)CH(OH)CH_2OH$$

Condensed structural formula

Semi-expanded structural formula

Line-angle drawing

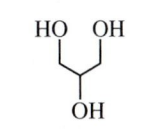

Ball-and-stick model

Figure 11.5

Representations of glycerol, an alcohol.

$$3 \ CH_3(CH_2)_{16}-C{\overset{O}{\underset{O-H}{}}} \ + \ H-O-\overset{\overset{H}{|}}{\underset{\underset{H}{|}}{C}}-\overset{\overset{O-H}{|}}{\underset{\underset{H}{|}}{C}}-\overset{\overset{H}{|}}{\underset{\underset{O-H}{|}}{C}}-H \ \longrightarrow \ \text{(triglyceride)} \ + \ 3 \ H_2O$$

[11.2]

This is the process involved in the formation of most animal and vegetable fats and oils. Variety is introduced by having different fatty acids incorporated into the same triglyceride rather than just one, as was the case in equation 11.2. The overwhelming majority of fatty acids in the body, almost 95%, are transported and stored in the form of triglycerides. Using this term, we can now define **fats** as triglycerides that are solid at room temperature, whereas **oils** are triglycerides that are liquid at room temperature.

Your Turn 11.5   **Triglyceride Formation**

Stearic acid (see Figure 11.4) and palmitic acid ($CH_3(CH_2)_{14}COOH$) are fatty acids found in animal fats and in butter.

   a. Draw the line-angle representation for palmitic acid.
   b. Name the functional group responsible for the acidic properties of both fatty acids.
   c. Write an equation analogous to equation 11.2 that uses structural formulas to show the reaction of 3 molecules of palmitic acid with glycerol to form a triglyceride and 3 molecules of water.
   d. In your answer to part **c,** circle the structural features responsible for classifying the product triglyceride as a triester.

Answer
   a. (line-angle structure of palmitic acid with OH and O)

   b. carboxylic acid group, —COOH

   Fats and oils differ in properties not only because of the length of the hydrocarbon chain in the fatty acids forming the triglyceride, but also because of the bonding within that chain. A fatty acid is **saturated** if the hydrocarbon chain contains only single bonds between the carbon atoms. In a saturated hydrocarbon chain, the C atoms contain the maximum number of H atoms that can be accommodated and therefore is saturated in hydrogen. This is the case with stearic acid. However, other fatty acids contain one or more C-to-C double bonds. In this case, the fatty acid is **unsaturated** if the molecule contains one or more double bonds between carbon atoms. Oleic acid, with only one double bond between carbon atoms per molecule, is classified as **monounsaturated**. **Polyunsaturated** fatty acids contain more than one double bond between carbon atoms per molecule. Linoleic acid, which contains two C-to-C double bonds per molecule, and linolenic acid with three C-to-C double bonds per molecule, are both examples of polyunsaturated fatty acids. Note in Figure 11.6 that each of these three different unsaturated fatty acids contains 18 carbon atoms.

Oleic acid, a **monounsaturated** fatty acid

$$CH_3(CH_2)_7CH=CH(CH_2)_7COOH$$

Linoleic acid, a **polyunsaturated** fatty acid

$$CH_3(CH_2)_4CH=CHCH_2CH=CH(CH_2)_7COOH$$

Linolenic acid, a **polyunsaturated** fatty acid

$$CH_3CH_2CH=CHCH_2CH=CHCH_2CH=CH(CH_2)_7COOH$$

Figure 11.6
Unsaturated fatty acids.

Your Turn 11.6   Unsaturated Fatty Acids

a. What structural feature identifies oleic, linoleic, and linolenic acids as unsaturated fatty acids?
b. Lauric acid, $CH_3(CH_2)_{10}COOH$, is one of the components of palm oil. Draw the line-angle representation for lauric acid and classify it as saturated or unsaturated.
c. Write a structural equation similar to equation 11.2 showing the reaction of 2 molecules of oleic acid and 1 molecule of lauric acid with glycerol to form a triglyceride and 3 molecules of water.
d. In your answer to part **c,** circle the structural features that identify the product as a triester.

Answer
a. Oleic, linoleic, and linolenic fatty acids all have at least one C-to-C double bond.

b.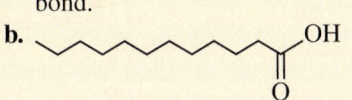

It is a saturated fatty acid with all C-to-C single bonds.

The three fatty acids in a single triglyceride molecule can be identical, two can be the same, or all three can be different. Moreover, the fatty acids in a triglyceride molecule can exhibit varying degrees of unsaturation. The fatty acids can also be sequenced differently, all factors contributing to the variety of fats and oils. It is the characteristics of the fatty acids that a given fat or oil contains that govern its overall extent of unsaturation.

The physical properties of fats also depend on their fatty acid content. Table 11.3 indicates trends within a given family of fatty acids. In saturated fatty acids, for example, the melting points increase as the number of carbon atoms per molecule (and the molecular mass) increase. On the other hand, in a series of fatty acids with a similar number of carbon atoms, increasing the number of C-to-C double bonds decreases the melting

Table 11.3	Comparing Fatty Acids		
Name	Number of C Atoms per Molecule	Number of C-to-C Double Bonds per Molecule	Melting Point, °C
Saturated Fatty Acids			
Capric acid	10	0	32
Lauric acid	12	0	44
Myristic acid	14	0	54
Palmitic acid	16	0	63
Stearic acid	18	0	70
Unsaturated Fatty Acids			
Oleic acid	18	1	16
Linoleic acid	18	2	−5
Linolenic acid	18	3	−11

point. Thus, when the melting points of the 18 carbon fatty acids are compared, saturated stearic acid (no C-to-C double bonds) is found to melt at 70 °C, oleic acid (one C-to-C double bond per molecule) melts at 16 °C, and linoleic acid (two C-to-C double bonds per molecule) melts at −5 °C. These trends carry over to the triglycerides containing the fatty acids and explain why fats rich in saturated fatty acids are solids at room or body temperature, whereas highly unsaturated ones are liquids.

Figure 11.7 gives evidence of this generalization. All naturally occurring lipids are mixtures of various triglycerides formed from different fatty acids. In general,

> Stearic acid is a solid at body temperature, whereas oleic and linoleic acids are liquids.

> Normal body temperature is 37 °C; room temperature is approximately 20 °C.

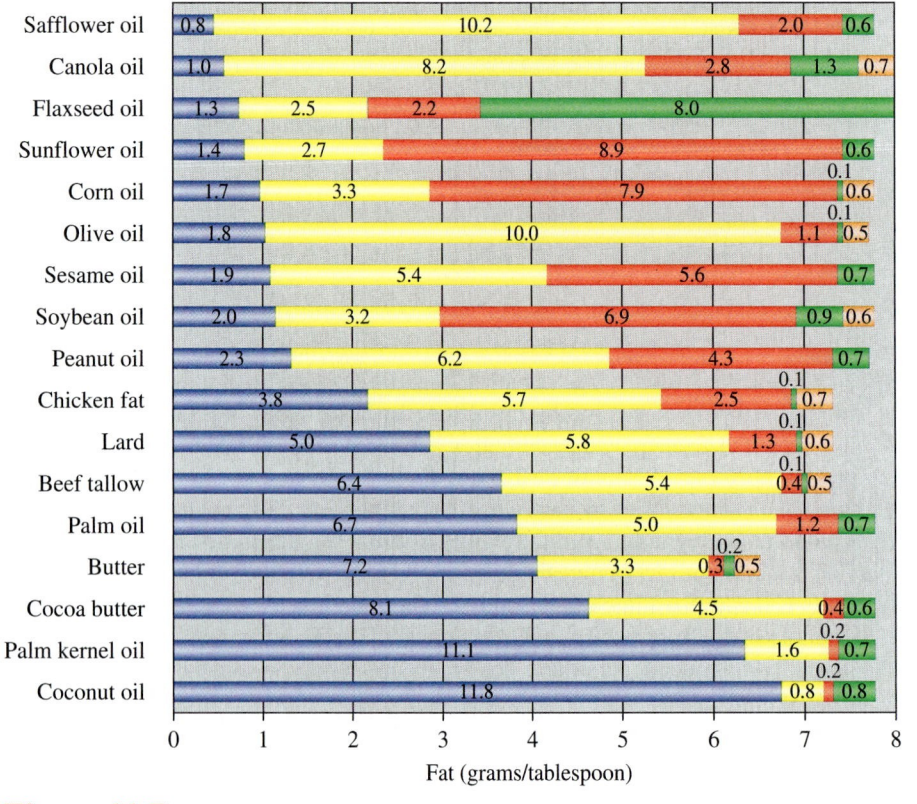
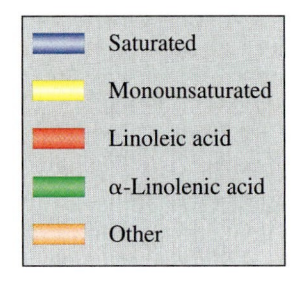

Figure 11.7
Saturated and unsaturated oils and fats.

solid or semisolid animal fats, such as lard and beef tallow, are high in saturated fats. In contrast, olive, safflower, and other vegetable oils consist mostly of unsaturated triglycerides. However, the figure reveals some surprising differences in the composition of oils. For example, flaxseed oil is particularly rich in alpha-linolenic acid (α-linolenic acid, or ALA), a polyunsaturated acid that is being studied for its health benefits. Palm kernel and coconut oils contain much more saturated fat than corn and canola oil. Ironically, the coconut oil used in some nondairy creamers is 87% saturated fat, far more than the percentage found in the cream it replaces. In fact, coconut oil contains more saturated fat than pure butterfat. Concern over the high degree of saturation in coconut and palm oil accounts for the statement sometimes printed on food labels: "Contains no tropical oils."

The solid form of coconut oil is called coconut butter. It melts to form an oil around room temperature.

Consider This 11.7   **Cooking Oil Chemistry**

Nutrition Facts
Serving Size 1 Tbsp (15 mL)
Servings Per Container about 63

Amount Per Serving		
Calories 120	Cal. from fat 120	
		% Daily Value*
Total Fat 14g		21%
Saturated Fat 1g		6%
Trans Fat 0g		
Polyunsaturated 11g		
Monounsaturated 2g		
Cholesterol 0g		0%
Sodium 0g		0%
Total Carbohydrate 0g		
Protein 0g		
Vitamin E 20%		

Not a significant source of dietary fiber, sugars, vitamin A, vitamin C, calcium, and iron
*Percent Daily Values are based on a 2,000 Calorie diet.

a. Consider this label from a popular brand of cooking oil. Is the major component likely to be safflower oil, canola oil, or soybean oil? Explain.

b. This brand of cooking oil has one unusual ingredient, vitamin E. Do you think this is a part of the oil itself or an added component?
Hint: You may want to look ahead to Section 11.5.

Another category of fats has been much in the news lately. It is the **trans fats,** fats that have been transformed by the addition of hydrogen to unsaturated vegetable oils. For example, unless you eat "natural" peanut butter, the jar on your shelf probably is labeled something like, "oil modified by partial hydrogenation." The oil extracted from peanuts is rich in mono- and polyunsaturated fatty acids. You can see it floating on top of the butter in jars of "natural" peanut butter. For creamy-style peanut butter, the oil can be treated chemically and converted into a semisolid that does not separate from peanut butter. This is an example of **hydrogenation,** a process in which hydrogen gas, in the presence of a metallic catalyst, is added to a double bond and converts it to a single bond. Hydrogenation of compounds makes them more saturated, most commonly converting one or more C-to-C double bonds into C-to-C single bonds. Equation 11.3 shows how this works with linoleic acid in peanut oil.

[11.3]

Notice that in this case, the hydrogenation was partial and only one of the double bonds in linoleic acid was changed to a single bond. As a result of this partial hydrogenation, the oil is changed to a semisolid fat. The extent of hydrogenation can be carefully controlled by temperature and pressure conditions to yield products of desired saturation and therefore melting point, softness, and spreadability. Such customized fats and oils are in many products, including margarines, cookies, and candy bars.

In most natural unsaturated fatty acids, the hydrogen atoms attached to the carbon atoms are on the *same* side of the C-to-C double bond. Hydrogenation, in addition to replacing double bonds with single bonds, can reconfigure the geometry around a double bond so that the hydrogen atoms are on the *opposite* side of the double bond. For example, oleic acid and elaidic acid have identical chemical formulas, $CH_3(CH_2)_7CH=CH(CH_2)_7COOH$. Both are monounsaturated, but their properties, uses, and possible health effects are different. Oleic acid, a natural unsaturated fatty acid, is a major component of olive oil but the elaidic acid, found in some types of soft margarines, is a trans fatty acid. The two structures are shown in Figure 11.8. Be sure to note the relative location of the hydrogen atoms around the C-to-C double bond for each fatty acid.

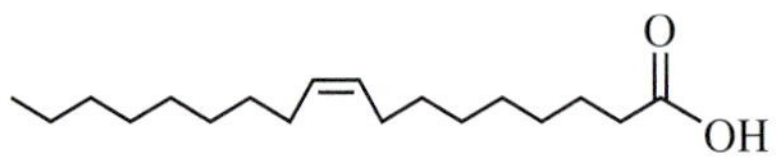

Oleic acid, a natural fatty acid 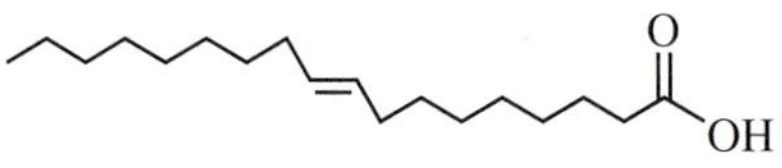   Elaidic acid, a trans fatty acid

Figure 11.8

Structure of natural and trans fatty acids with the same chemical formula.

H atoms are on the *same* side of the double bond:

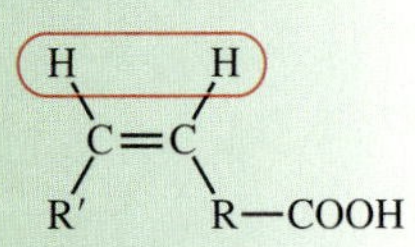

H atoms are on the *opposite* side of the double bond:

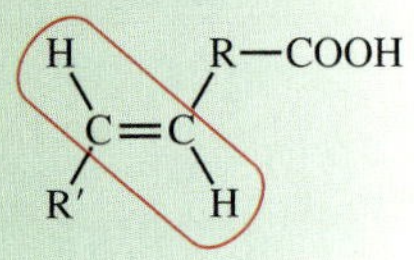

Scientific studies now reveal that trans fats raise the level of triglycerides and "bad" cholesterol in the blood. This finding came as somewhat of a surprise because partially hydrogenated fats still contain at least some C-to-C double bonds and unsaturation is definitely a plus in a healthy diet. However, comparing molecular structures can help us understand why trans fats behave as saturated fats. Saturated fats with their long "straight" hydrocarbon chains, tend to pack together very nicely, one reason that they are solids at room temperature. The molecules of unsaturated edible oils with their "bends" do not pack as well, one reason they are liquids at room temperature. Natural oils all have these bends, such as in the unsaturated fatty acids shown in Figure 11.6. The surprise is that when partial hydrogenation takes place, the trans fatty acids in fats more closely resemble the shape of saturated fatty acids in fats and therefore behave in a similar manner in the body.

LDL (low-density lipoprotein) cholesterol is referred to as "bad" because of its tendency to build up on artery walls, causing heart attacks and strokes.

Starting in January 2006, the FDA required that foods include trans fat information on the labels. In December 2006, New York City banned restaurants from selling foods with anything but trace amounts of trans fats by July 1, 2007. Any food containing more than 0.5 g of trans fat must be removed from the menu by July 1, 2008. In March 2003, Denmark became the first country to strictly regulate foods containing trans fats. Canada followed suit in 2004 and the European Food Safety Authority was preparing a scientific study due in 2007 before setting regulations.

Most nutritionists now recommend that consumers go easy on any product with either trans fats or saturated fats. Manufacturers are responding by looking for substitutes for the trans fats. Using tropical oils such as palm or coconut oil would not be acceptable because of their high percentage of saturated fats. Some manufacturers are adding other oils such as sunflower oil or flaxseed oil to their products. Both oils are high in polyunsaturated fatty acids. Food manufacturers also are responding by encouraging farmers to produce new soybean varieties. New types of soybean oil reduce the need for partial hydrogenation and do not alter taste, texture, or shelf life in testing by the Kellogg company. Both Monsanto and the Hi-Bred International unit of DuPont are currently planting hundreds of thousands of acres of new soybean varieties in an attempt to meet the growing demand.

Genetic engineering of crops is discussed in Section 12.10.

Consider This 11.8   Spreadables and Fat Content

This table lists the fat content for Crisco and three soft, butter substitutes. You will need to visit your local supermarket or use the Web to answer parts **c–e**.

	Crisco	Brummel & Brown	I Can't Believe It's Not Butter	Benecol
Serving Size	1 tbsp = 12 g	1 tbsp = 14 g	1 tbsp = 14 g	1 tbsp = 14 g
Total fat (g)	12	5	8	8
Saturated	3	1	2	1
Trans	1.5	0	0	0
Polyunsaturated	3	2.5	4	2
Monounsaturated	4	1	2	4.5

a. Of the three butter substitutes, which has the highest percentage of saturated fats? How does it compare with Crisco and butter?
Hint: See Figure 11.7 for values for butter.
b. What percentage of the total fat in Crisco is polyunsaturated? How does this compare with the three butter substitutes?
c. How does the fat content of the stick formulation of I Can't Believe It's Not Butter compare with that of the spreadable formulation given above? Explain any differences.
d. How does the fat content of the new formulation of Crisco (look for the green label, rather than the familiar blue label) compare with that given here? Explain any differences.
e. Conduct a minisurvey of butter and margarine products in a supermarket in your area. List the fat content (in the same categories as above) for five different products.

Cholesterol is a lipid, but it has a very different structure from the rest of the fats and oils that we have been discussing. The cholesterol molecule has a four-ring structure with a hydrocarbon chain attached to the five-carbon ring. This structure is similar to that found in other hormones such as progesterone and testosterone. The class of compounds is known as steroids, and cholesterol is the most abundant steroid in the human body, being both synthesized internally and ingested from food.

Two compounds present in the body help carry cholesterol and triglycerides through the bloodstream: **HDL** (high-density lipoprotein) and **LDL** (low-density lipoprotein). The HDLs are the "good" lipoproteins, more effective in transporting cholesterol than LDLs. It appears that people with high values of LDL relative to HDL concentration are particularly susceptible to heart disease. Restricting dietary intake of cholesterol, combined with regular exercise increases HDL at the expense of LDL and also serves to burn calories, a process necessary to prevent weight gain. Still, perhaps the most important factor influencing cholesterol levels in our blood is genetics.

In our preoccupation with dietary fat, realize that fats often enhance our enjoyment of food. They improve "mouth feel" and intensify certain flavors. Of prime significance, however, is the fact that fats are essential for life. They are the most concentrated source of energy in the body (Section 11.6), and they provide insulation that retains body heat and cushions internal organs. Moreover, triglycerides and other lipids, including cholesterol, are the primary components of cell membranes and nerve sheaths. Although "fathead" is hardly a compliment, in fact our brains are rich in lipids. Because fats play many important roles in our bodies, a variety of triglycerides are required to make them, incorporating a wide range of fatty acids—saturated, monounsaturated, and polyunsaturated. Fortunately, our bodies can synthesize almost all of the necessary fatty acids from the starting materials provided by a normal diet. The exceptions are

Structures of steroids were given in Section 10.7 and in Figures Alive! for Chapter 10.

linoleic and linolenic acids. These two essential fatty acids must be obtained directly from the foods we eat; our body cannot produce them. Generally this does not create a problem because linoleic and linolenic acids are found in many foods including plant oils, fish, and leafy vegetables.

11.3   Carbohydrates: Sweet and Starchy

Carbohydrates have the job of providing all the cells in the body with the energy they need. The best known dietary carbohydrates are sugars—simple carbohydrates such as those found in fruits and processed foods. Simple carbohydrates are recognized by their sweet taste and are easily digested by the body. The best known complex carbohydrate is likely starch, found in nearly all plant-based foods such as grains, potatoes, and rice. Complex carbs can be pleasant to the taste buds, but they do not taste sweet and take a bit longer to digest.

Chemically, **carbohydrates** are compounds containing carbon, hydrogen, and oxygen, the last two elements in the same 2:1 atomic ratio as found in water. Glucose, for example, has the formula $C_6H_{12}O_6$. This composition gives rise to the name, "carbohydrate," which implies "carbon plus water." But the hydrogen and oxygen atoms are not bonded together to form water molecules. Rather, carbohydrate molecules are built of rings containing carbon atoms and an oxygen atom. The hydrogen atoms and —OH groups are attached to the carbon atoms. There are many opportunities for differences in molecular structure. For example, 32 distinct isomers (including chiral isomers) have the formula $C_6H_{12}O_6$. The isomers differ slightly in their properties, including intensity of sweetness. Figure 11.9 shows the structures of three simple sugars. These *monosaccharides,* or "single sugars" of fructose and glucose, both have the formula $C_6H_{12}O_6$.

From Figure 11.9, you can see that each monosaccharide contains a single ring consisting of four or five carbon atoms and one oxygen atom. The best way to visualize its three-dimensional structure is to imagine that the ring is perpendicular to the plane of the paper, with the bold print edges facing you. The H atoms and —OH groups are thus either above or below the plane of the ring. This results in two forms of glucose: alpha (α) and beta (β). In α-glucose, the —OH group on carbon 1 is on the opposite side of the ring from the —CH_2OH group attached to carbon 5. As shown in Figure 11.9, the β-glucose form has the two groups on the *same* side of the ring. This is also the case for β-fructose, with the —OH on carbon 2 and the —CH_2OH group at carbon 5 being on the same side of the ring.

Ordinary table sugar, sucrose, is an example of a **disaccharide,** a "double sugar," formed by joining two monosaccharide units. In forming a sucrose molecule, an α-glucose and a β-fructose unit are connected by a C-to-O-to-C linkage created when an H and an OH are split from the monosaccharides to form a water molecule. This is a condensation reaction. Water produced in this way is used for other reactions in the body. This reaction and the structure of the sucrose molecule are also shown in Figure 11.10.

Chirality and optical isomerism were discussed in Section 10.6.

The ending *-ose* is typical for sugars.

Condensation reactions were also described in Sections 9.5 and 10.3.

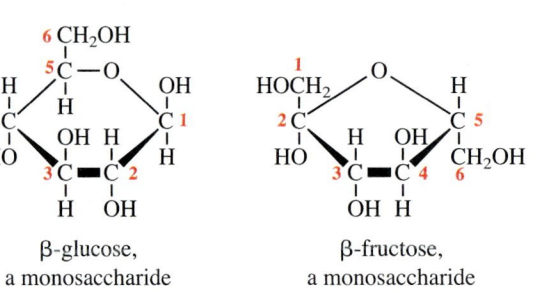

α-glucose, a monosaccharide   β-glucose, a monosaccharide   β-fructose, a monosaccharide

Figure 11.9

Molecular structures of three monosaccharides.

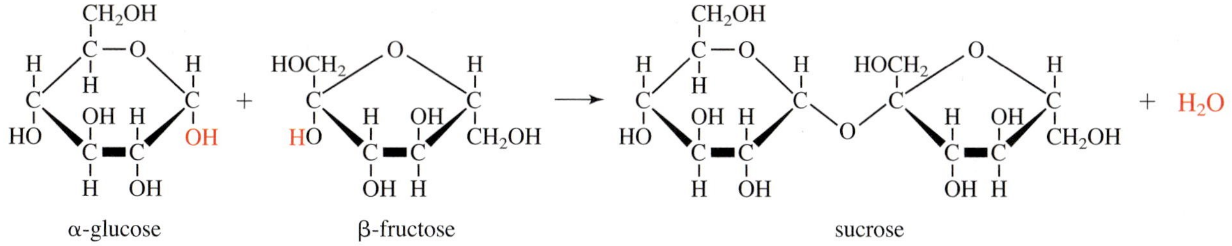

Figure 11.10

Formation of sucrose, a disaccharide.

The linking of monosaccharide molecules is by no means restricted to the formation of disaccharides. Some of the most common and abundant carbohydrates are **poly**saccharides, polymers made up of thousands of glucose units. As the name implies, these macromolecules consist of "many sugar units." Monosaccharide monomers combine to form polysaccharides, releasing a water molecule each time a monomer is incorporated into the chain. Starch, cellulose, and glycogen are three familiar examples of polysaccharides, also called **complex carbohydrates**.

Like starch, cellulose is a polymer of glucose, but these two polysaccharides behave very differently in the body. Humans are able to digest starch by breaking it down into individual glucose units; in contrast, we cannot digest cellulose. Consequently, we depend on starchy foods such as potatoes or pasta as carbohydrate sources rather than devouring toothpicks or textbooks. The reason is a subtle difference in how the glucose units are joined in starch and in cellulose. In the alpha (α) linkage in starch, the bonds connecting the glucose units have a particular orientation, whereas the beta (β) linkage between glucose units in cellulose is in a different orientation (Figure 11.11).

Our enzymes and the enzymes of many mammals are unable to catalyze the breaking of beta linkages in cellulose. Consequently, we can't dine on grass or trees. Cows, goats, sheep, and other ruminants manage to break down cellulose with a little help. Their digestive tracts contain bacteria that decompose cellulose into glucose monomers. The animals' own metabolic systems then take over. Similarly, the fact that termites

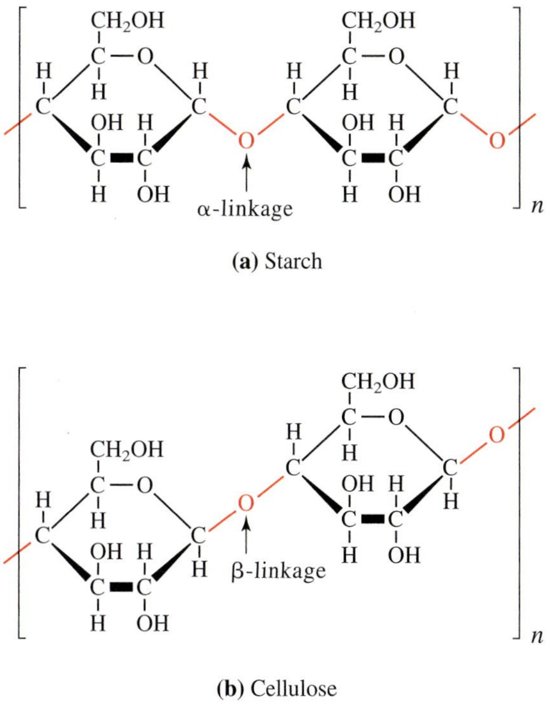

(a) Starch

(b) Cellulose

Figure 11.11

The bonding between glucose units in **(a)** starch and **(b)** cellulose.

contain cellulose-hungry bacteria means that wooden structures are sometimes at risk. The methane released by all of these busy termites may even be contributing to global warming.

See Section 3.8 for more about methane and other greenhouse gases.

Consider This 11.9   Differences in Molecular Structure

Speculate why the slight difference in molecular structure between starch and cellulose is enough to make the latter polysaccharide indigestible to human beings. *Hint:* Section 10.5 can be helpful.

Glycogen is the form in which polysaccharides are stored in our bodies and has a molecular structure similar to that of starch. The chains of glucose units in glycogen, however, are longer and more branched than those in starch. Glycogen is vitally important because it stores energy for use in our bodies. It accumulates in muscles and especially in the liver, where it is available as a quick source of internal energy.

A healthful diet derives more of its carbohydrates from polysaccharides than from simple sugars, such as mono- and disaccharides. Enzymes in our saliva initiate the process of breaking down the long polysaccharide chains into glucose molecules, an important first step in metabolism. But our body also synthesizes glucose from a variety of precursors, including other sugars.

Your Turn 11.10   Tasty Crackers

Sustained chewing of an unsweetened, unsalted cracker results in a sweet taste. What is the molecular explanation for this phenomenon?

Consider This 11.11   Sucralose: A Sweet Story

A yellow packet of Splenda sugar substitute contains sucralose. Since its debut on the market in 1998, its retail sales have soared to $188 million per year—more than its competitors of Equal and Sweet'N Low combined. Use the resources of the Web to answer these questions.

  a. How many Calories does a packet of Splenda contain?
  b. Equal advertises that it is 200 times sweeter than sugar; Sweet'N Low is 300 times sweeter. What is the comparable value for Splenda?
  c. Retail sales of sugar substitutes are just one market. What other types of products may contain sugar substitutes?
  d. Splenda's slogan is "Made from Sugar, So It Tastes Like Sugar." Is that a correct statement or false advertising? Explain.

Lactose intolerance, a common metabolic dysfunction, is somewhat related to the difference in digestibility between starch and cellulose. This condition is shared by at least 80% of the world's population. Although most Northern Europeans, Scandinavians, and people of similar ethnic background are able to consume milk, cheese, and ice cream with no ill effects, they are the exception. Excluding very young children, lactose intolerance affects the majority of people, making it difficult for them to digest dairy products. Consumption of these foods is often followed by diarrhea and excess gas. The symptoms result from the inability to break down lactose (milk sugar) into its component monosaccharides, glucose and galactose. The linkage between the two monosaccharides in

lactose is a beta form, similar to that in cellulose. People who are lactose intolerant have a lack of or a low concentration of lactase, the enzyme that catalyzes the breaking of this bond. In such individuals, the intact lactose is instead fermented by their own intestinal bacteria. This process generates carbon dioxide and hydrogen gases, and lactic acid, the principal cause of the diarrhea. Given milk's importance for growing bones and teeth, it is significant that infants of all ethnic groups generally produce sufficient lactase to digest a milk-rich diet. But, as we age, this production decreases. By adulthood, most people in the world do not have enough of the enzyme to accommodate a diet heavy in dairy products.

11.4   Proteins: First Among Equals

The word *protein* derives from *protos,* Greek for "first." The name is misleading. Life depends on the interaction of thousands of chemicals, and to assign primary importance to any single compound or class of compounds is simplistic. Nevertheless, proteins are an essential part of every living cell. They are also major components in hair, skin, and muscle; and they transport oxygen, nutrients, and minerals through the bloodstream. Many of the hormones that act as chemical messengers are proteins, as are all the enzymes that catalyze the chemistry of life.

Proteins are polyamides or polypeptides, polymers built from amino acid monomers. The great majority of proteins are made from various combinations of the 20 different naturally occurring amino acids. Molecules of all amino acids share a common structural pattern. Four chemical species are attached to a carbon atom: a carboxylic acid group, an amine group, a hydrogen atom, and a side chain designated R in the structure shown in Figure 11.12.

Variations in the R side chain group differentiate the individual amino acids (Figure 11.13). In glycine, the simplest amino acid, R is a hydrogen atom. In alanine, R is a —CH_3 group; in aspartic acid (found in asparagus), it is —CH_2COOH; and in phenylalanine, —$CH_2(C_6H_5)$.

Two of the 20 naturally occurring amino acids have R groups that bear a second —COOH functional group, three have R groups containing amine groups, and two others contain sulfur atoms. Because all amino acids except glycine involve four different units bonded to a central carbon atom, they all exhibit chirality. All the naturally occurring amino acids that are incorporated into proteins are in the left-handed isomeric form.

Combining amino acids to form proteins takes place through a reaction between an amine group and a carboxylic acid group. Equation 11.4 represents the reaction of glycine with alanine to form a **dipeptide**, a compound formed from two amino acids. Here, the acidic —COOH group of a glycine molecule reacts with the —NH_2 group of alanine, and an H_2O molecule is eliminated. In the process, the two amino acids become linked by a peptide bond (indicated in the shaded area). Once incorporated into the peptide chain, the amino acids are known as **amino acid residues.**

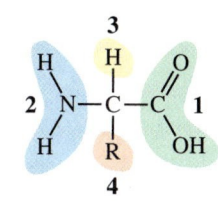

Figure 11.12

General structure for an amino acid. (1) a carboxylic acid group; (2) an amine group; (3) a hydrogen atom; and (4) a side chain designated R.

glycine   alanine   aspartic acid   phenylalanine

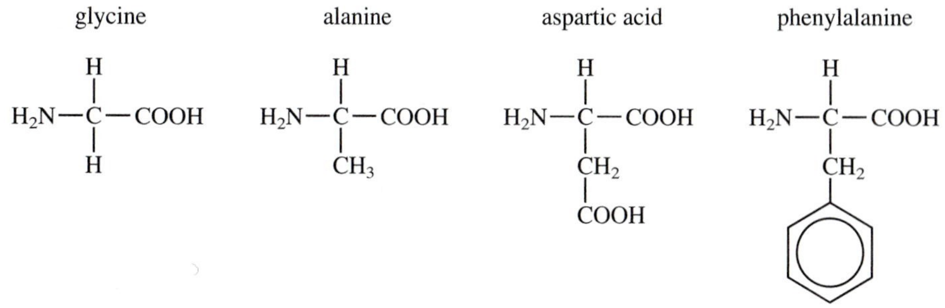

Figure 11.13

Examples of amino acids with different side chains.

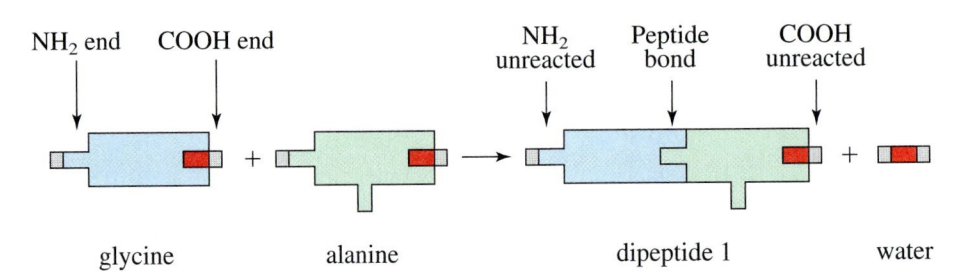

glycine   alanine   dipeptide   water   [11.4]

Because each amino acid bears an amine group and an acid group, the amino acids can join in two ways. Hence, two dipeptides are possible. We illustrate the options with simple block diagrams for the amino acids. The first case is shown in equation 11.4 in which glycine provides the acid and alanine provides the amine. This is another example of condensation polymerization, already encountered in Section 11.3 in the formation of polysaccharides.

NH$_2$ end COOH end		NH$_2$ unreacted Peptide bond COOH unreacted

glycine   alanine   dipeptide 1   water

In the second case, the amino acids reverse roles; alanine provides the —COOH and glycine the —NH$_2$.

NH$_2$ end COOH end		NH$_2$ unreacted Peptide bond COOH unreacted

alanine   glycine   dipeptide 2   water

Examination of the molecular structures of the two dipeptides indicates that they are not identical. In dipeptide 1, the unreacted amine group is on the glycine residue and the unreacted acid group is on the alanine residue; in dipeptide 2, the —NH$_2$ is on the alanine residue and the —COOH on the glycine residue.

The point of all this is that the order of amino acid residues in a peptide makes a difference. The particular protein formed depends not only on what amino acids are present, but also on their sequence in the protein chain. Assembling the correct amino acid sequence to make a particular protein is like putting letters in a word; if they are in a different order, a completely new meaning results. Thus, a tripeptide consisting of three different amino acids is like a three-letter word containing the letters *a, e,* and *t.* There are six possible combinations of these letters. Three of them—*ate, eat,* and *tea*—form recognizable English words; the other three—*aet, eta,* and *tae*—do not. Similarly, some sequences of amino acids may be biological nonsense.

Still restricting ourselves to three-letter words and only the letters *a, e,* and *t,* but allowing the duplication of letters, we can make perfectly good words such as *tee* and *tat,* and lots of meaningless combinations such as *aaa* and *tte.* There are, in fact, a total of 27 possibilities, including the 6 identified earlier. Just as many words use letters more than once, most proteins contain specific amino acids incorporated more than once.

Putting the amino acids of a protein into their proper order is like assembling a train correctly by placing each car in the right sequence.

See Section 12.4 for more information about the structure and synthesis of proteins.

Your Turn 11.12   **Making Tripeptides**

You have seen from the block diagrams in this section that one glycine (Gly) and one alanine (Ala) molecule can combine to form two dipeptides: GlyAla and AlaGly. If one permits multiple use of each of the two amino acids, two other dipeptides are possible: GlyGly and AlaAla. Thus, a total of four different dipeptides can be made from two amino acids if each amino acid can be used more than once. Eight different tripeptides can be made from supplies of two different amino acids, assuming that each amino acid can be used once, twice, three times, or not at all. Use the symbols Gly and Ala to write down representations of the amino acid sequence in all eight of these tripeptides.
Hint: Start with GlyGlyGly.

Consider This 11.13   **3-D Amino Acids**

Structural features of amino acids are more readily apparent if you view their three-dimensional representations. You will find links to do this at the *Online Learning Center*.

a. View a 3-D representation of glycine and compare it with the 2-D structure shown in Figure 11.13. List the advantages and drawbacks of each representation.

b. View a 3-D representation of leucine. Draw structures for the four groups bonded to the central carbon atom and explain why this molecule is optically active.

c. Now examine any three amino acid structures of your choice. Which ones did you select? Describe two ways in which their structures are more complex than that of glycine.

The body does not normally store a reserve supply of protein, so foods containing protein must be eaten regularly. As the principal source of nitrogen for the body, proteins are constantly being broken down and reconstructed. A healthy adult on a balanced diet is in nitrogen balance, excreting as much nitrogen (primarily as urea in the urine) as she or he ingests. Growing children, pregnant women, and persons recovering from long-term debilitating illness or burns have a positive nitrogen balance. This means that they consume more nitrogen than they excrete because they are using the element to synthesize additional protein. A negative nitrogen balance exists when more protein is being decomposed than is being made. This occurs in starvation, when the energy needs of the body are unmet from the diet, and muscle is metabolized to maintain physiological functions. In effect, the body feeds on itself.

Another cause of a negative nitrogen balance may be a diet that does not include enough of the **essential amino acids,** those required for protein synthesis but that must be obtained from the diet because the body cannot synthesize them. Of the 20 natural amino acids that make up our proteins, we can synthesize 11 from simpler molecules, but 9 must be ingested directly. If any of the nine essential amino acids identified in Table 11.4 are

The *–ine* ending is used for naming most amino acids.

Table 11.4	The Essential Amino Acids	
histidine	lysine	threonine
isoleucine	methionine	tryptophan
leucine	phenylalanine	valine

missing from the diet, the result can be severe malnutrition because some of the amino acids ingested cannot be utilized and are thus excreted.

Good nutrition thus requires protein in sufficient quantity and suitable quality. Beef, fish, poultry, and other meats contain all the essential amino acids in approximately the same proportions found in the human body. Therefore, meat is termed a complete protein. However, most people of the world depend on grains and other vegetable crops rather than meat as their major sources of protein. If such a diet is not sufficiently diversified, some essential amino acids may be lacking. For example, Mexican and Latin American diets are rich in corn and corn products, a protein source that is *incomplete* because corn is low in tryptophan, an essential amino acid. A person may eat enough corn to meet the total protein requirement, but still be malnourished because of insufficient tryptophan.

Fortunately for millions of vegetarians, a reliance on vegetable protein does not doom them to malnutrition. The trick is to apply a principle that nutritionists call **protein complementarity,** combining foods that complement essential amino acid content so that the total diet provides a complete supply of amino acids for protein synthesis. You do this, likely unknowingly, every time you eat a peanut butter sandwich. Bread is deficient in lysine and isoleucine, but peanut butter supplies these amino acids. On the other hand, peanut butter is low in methionine, a compound provided by bread. The traditional diets in many countries meet protein requirements through nutritional complementarity. In Latin America, beans are used to complement corn tortillas; soy foods are eaten with rice in parts of Southeast Asia and Japan. People in the Middle East combine bulgur wheat with chickpeas or eat hummus, a paste made from sesame seeds, and chickpeas, with pita bread. In India, lentils and yogurt are eaten with unleavened bread.

Consider This 11.14   Protein Complementarity

Use the Web to find at least two additional examples of protein complementarity.

a. What essential amino acids are involved in the combination?
b. Are the foods containing each combination ones that are common to a particular country or region, much as peanut butter and bread are common in the United States?

Livestock, especially beef cattle, also benefit from protein complementarity. They are fed a variety of grains with a complete set of amino acids to incorporate ultimately into steaks and hamburger. However, the second law of thermodynamics applies to beef cattle as well as to everything else. There is a loss of efficiency with each step of energy transfer, whether in electrical power plants or in cells during metabolism. Cattle are notoriously inefficient in converting the energy in their feed into meat on the hoof. It takes about 7 lb of grain to produce 1 lb of beef. Put into human terms, the 1.75 lb of grain used to produce a "quarter-pounder" can provide two days of food for someone on a vegetarian diet. Other animals are more efficient than cattle in converting grain to meat. Hogs require 6 lb of grain per pound of meat, turkeys need 4 lb, and chickens even less, only 3 lb. It is obviously much more efficient to get food energy directly from grains, rather than through secondary or tertiary sources further along the food chain. On the other hand, one should keep in mind that pasture land used to graze cattle is often unsuitable for growing crops. Moreover, much of the food consumed by animals would be indigestible or unpalatable to humans.

A postscript to the protein story is provided by the unusual case of aspartame, a sweet dipeptide. Because of the great American preoccupation and battle with excess Calories and excess pounds, artificial sweeteners have become a $2 billion-plus business. Gram for gram, these compounds are much sweeter than sugar, but they have little if any nutritive value. Hence, they are nonfattening. The principal use of artificial sweeteners is in soft drinks and there the most widely used artificial sweetener is aspartame, the principal ingredient in NutraSweet and Equal. Somewhat surprisingly, the compound is related to proteins and not to the sweet simple carbohydrates. Aspartame is a dipeptide made from

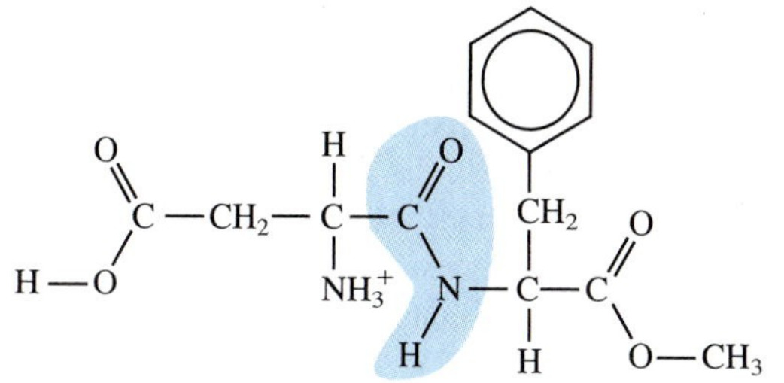

Figure 11.14

The structural formula of aspartame.

Figure 11.15

A warning: Phenylketonurics: Contains Phenylalanine.

The ketone functional group was shown in Table 10.3.

aspartic acid and a slightly modified phenylalanine (Figure 11.14). Notice how different this structure is from that of sucrose shown in Figure 11.10.

Alone, neither of the two amino acids in aspartame tastes sweet. Yet, the compound that results from their chemical combination is about 200 times sweeter than sucrose. The fact that sucrose and aspartame are chemically and structurally very different invites speculation about the molecular features that convey sweetness. For whatever reason, aspartame is sufficiently sweet to be used by millions of people worldwide. A few cases of adverse side effects have been attributed to aspartame, but exhaustive reviews have failed to show an unequivocal and direct connection between the symptoms and the sweetener. A study from the National Cancer Institute looked at aspartame consumption in over 500,000 people. Results released in April 2006 again confirmed that there was no cause for concern, even among heavy users of aspartame.

For the vast majority of consumers, aspartame is a safe alternative to sugar. One group of people, however, definitely should not use aspartame. The warning on packets of artificial sweeteners and products containing aspartame is explicit: "Phenylketonurics: Contains Phenylalanine" (Figure 11.15).

This is a case where one person's meat is another's poison. Phenylalanine is an essential amino acid converted in the body to tyrosine, another amino acid. Individuals with phenylketonuria, a genetically transmitted disease, lack the enzyme that catalyzes this transformation. Consequently, the conversion of dietary phenylalanine to tyrosine is blocked and the phenylalanine concentration rises. To compensate for the elevated phenylalanine, the body converts it to phenylpyruvic acid, excreting large quantities of this acid in the urine. Phenylpyruvic acid is termed a "keto" acid because of its molecular structure; hence, the disease is known as phenyl*keto*nuria or PKU. People with the disease are called phenylketonurics.

Excess phenylpyruvic acid causes severe mental retardation. Therefore, the urine of newborn babies is tested for this compound, using special test paper placed in the diaper. Infants diagnosed with PKU must be put on a diet severely limited in phenylalanine. This means avoiding excess phenylalanine from milk, meats, and other sources rich in protein. Commercial food products are available for such diets, their composition adjusted to the age of the user. Because phenylalanine is an essential amino acid, a minimum amount of it must still be available, even in phenylketonurics. Supplemental tyrosine may also be needed to compensate for the absence of the normal conversion of phenylalanine to tyrosine. A phenylalanine restricted diet is recommended for phenylketonurics at least through adolescence. Adult phenylketonurics also must limit their phenylalanine intake and hence curtail their use of aspartame.

11.5   Vitamins and Minerals: The Other Essentials

Your daily diet not only should supply an adequate number of Calories and macronutrients, but also must provide vitamins and minerals. These are **micronutrients**, substances needed only in miniscule amounts, but essential for the body to produce enzymes, hormones, and other substances needed for proper growth and development.

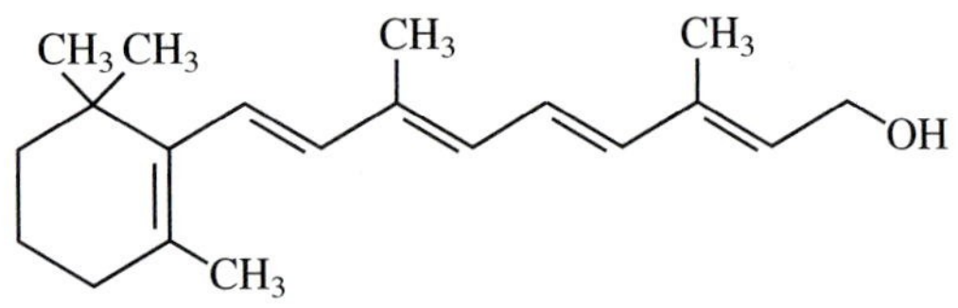

Vitamin A, a lipid-soluble vitamin 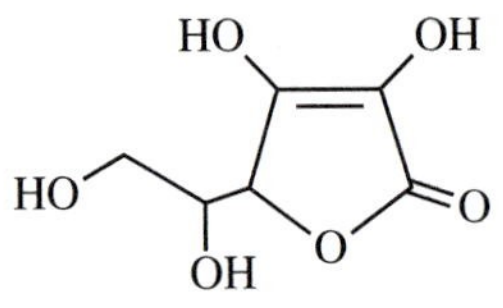   Vitamin C, a water-soluble vitamin

Figure 11.16

Examples of lipid-soluble and water-soluble vitamins.

Nearly everyone in the United States knows that vitamins and minerals are important, but a thriving multimillion-dollar supplement industry will remind us if we forget. Unfortunately, many popular processed foods that are high in sugars and fats lack these essential micronutrients.

A detailed understanding of the role of vitamins and minerals is of relatively recent origin. Over the ages, humans learned that they became ill if certain foods were lacking. However, the correlation between diet and health was often accidental and anecdotal. More systematic studies began early in the 20th century, with the discovery of "Vitamine B_1" (thiamine). The particular designation, B_1, was the label on the test tube in which the sample was collected. The general term *vitamin* was chosen because the compound, which is vital for life, is chemically classed as an amine. The final "e" disappeared with the discovery that not all vitamins are amines. **Vitamins** are organic molecules with a wide range of physiological functions. Although only small amounts are needed in the diet, vitamins are essential for good health, proper metabolic functioning, and disease prevention. Vitamins generally are not used as a source of energy, although some of them help break down macronutrients.

Vitamins often are classified on the basis of solubilities; they either are lipid-soluble or water-soluble. For example, the structural formula of vitamin A, shown in Figure 11.16, contains carbon and hydrogen atoms almost exclusively. Thus, it is similar to the hydrocarbons derived from petroleum. Vitamins that are not lipid-soluble are soluble in water because these polar molecules contain several —OH groups that form hydrogen bonds with water molecules. Vitamin C is a case in point.

Hydrogen bonds were discussed in Section 5.6. The relationship between molecular structure and solubility was explored in Section 5.9.

Consider This 11.15   **Classifying Vitamins**

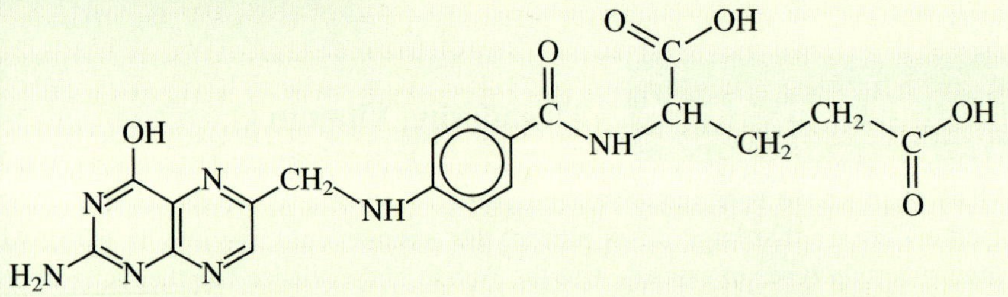

This is the structure for folic acid, a vitamin that helps prevent certain types of anemia and aids in nucleic acid synthesis. It is particularly important for pregnant women. Is this a lipid-soluble or a water-soluble vitamin? Explain.

These solubility differences among vitamins have significant implications for nutrition and health. Because of their fat-solubility, vitamins A, D, E, and K are stored in cells rich in lipids, where they are available on biological demand. This means that the fat-soluble vitamins need not be taken daily. It also means that these vitamins can build up to toxic levels if taken too far in excess of normal requirements. For example, high doses of vitamin A can result in fatigue, headache, dizziness, blurred vision, dry skin, nausea,

and liver damage. Vitamin D toxicity occurs at just four to five times its Recommended Daily Intake (RDI), making vitamin D the most toxic vitamin. Cardiac and kidney damage can result. Such high levels of the vitamin are reached using vitamin supplements, not through a normal diet.

Water-soluble vitamins, by contrast, are not generally stored; any unused excess is excreted in urine. Thus, they must be consumed frequently and in small doses. Unfortunately, when taken in extremely large doses, even water-soluble vitamins can accumulate at toxic levels, although such cases are rare. For example, there are reports that vitamin B_6, taken at 10–30 times the recommended dose per day for extended periods, results in nerve damage, including paralysis. Even higher doses of vitamin B_6 supplements, up to 1000 times the recommended dosage, have been consumed to alleviate the symptoms of premenstrual syndrome (PMS), again causing abnormal neurological symptoms. For most people, a balanced diet should provide all the necessary vitamins and minerals in appropriate amounts, making vitamin supplements unnecessary.

Many of the water-soluble vitamins serve as **coenzymes**, molecules that work in conjunction with enzymes to enhance the enzyme's activity. Lipid, protein, and carbohydrate synthesis and metabolism all depend on this important function. Members of the vitamin B family are particularly adept in acting as coenzymes. Niacin plays an essential role in energy transfer during glucose and fat metabolism. The synthesis of niacin in the body requires the essential amino acid tryptophan. Thus, a diet deficient in tryptophan may lead to niacin deficiency. Such a deficiency causes pellagra, a condition involving a darkening and flaking of the skin, as well as behavioral aberrations.

Some vitamins were discovered when observers correlated diseases with the lack of specific foods. For example, vitamin C (ascorbic acid) must be supplied in the diet, typically via citrus fruits and green vegetables. An insufficient supply of the vitamin leads to scurvy, a disease in which collagen, an important structural protein, is broken down. The link between citrus fruits and scurvy was discovered more than 200 years ago when it was found that feeding British sailors limes or lime juice on long sea voyages prevented the disease. Ascorbic acid also is required for the uptake, use, and storage of iron, important in the prevention of anemia.

> This practice also led to British sailors being called "limeys."

The last vitamin in this brief overview is vitamin E, important in the maintenance of cell membranes and as protection against high concentrations of oxygen, such as those that occur in the lungs. Vitamin E is so widely distributed in foods that it is difficult to create a diet deficient in it, although people who eat very little fat may need supplements. Vitamin E deficiency in humans has been linked with nocturnal cramping in the calves and fibrocystic breast disease.

Consider This 11.16   Megadosing Vitamin C

Many claims have been made that megadoses of vitamin C are therapeutic. Two such claims are that large doses prevent the common cold and may be effective against certain types of cancers. Use the Web to find evidence to either support or refute the claims.

Hint: "Linus Pauling" and "vitamin C" are useful search terms.

a. What range in daily amount of vitamin C constitutes a megadose?
b. Select one of the claims and further investigate it. List the potential therapeutic benefits.
c. What research studies back up this claim? Do any refute it? Give the citations.

An adequate supply of dietary minerals (ions or inorganic compounds) is also essential for good health. Table 11.2 listed the major elements in the body. There are seven

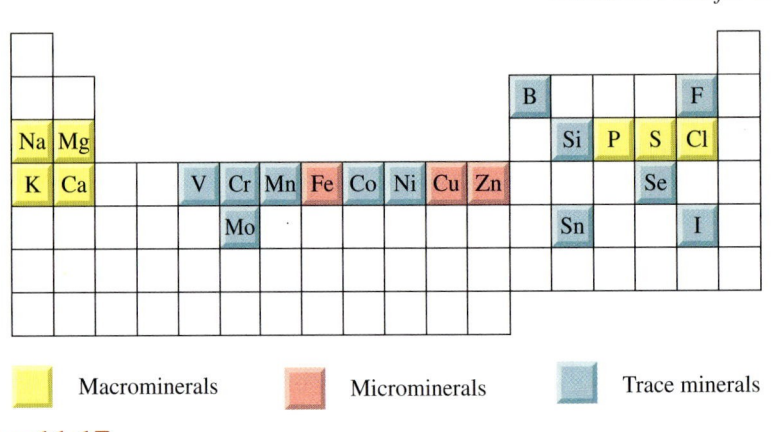

Figure 11.17

Periodic table indicating dietary minerals necessary for human life.

macrominerals—Ca, P, Cl, K, S, Na, and Mg—elements that are necessary for life but are not nearly as abundant as O, C, H, or N. Macrominerals are only required in amounts greater than 100 mg (0.100 g) or are present in the body in amounts greater than 0.01% of body weight. The adult RDAs for these macrominerals typically range from 1 to 2 g. The body requires lesser amounts of Fe, Cu, and Zn—the so-called **microminerals**. **Trace minerals**, including I, F, Se, V, Cr, Mn, Co, Ni, Mo, B, Si, and Sn, are usually measured in micrograms (1×10^{-6} g). Although the total amount of trace elements in the body is only about 25–30 g, their slight amounts belie the disproportionate importance they have in good health.

Only the essential dietary minerals are shown in the periodic table in Figure 11.17. The metallic elements exist in the body as cations, for example, Ca^{2+} (calcium ion), Mg^{2+} (magnesium ion), K^+ (potassium ion), and Na^+ (sodium ion). The nonmetals typically are present as anions, thus chlorine is found as Cl^- (chloride ion) and phosphorus appears in PO_4^{3-} (phosphate ion).

Other minerals ingested at dietary levels below a microgram have been promoted as beneficial, even some we consider dangerous for human health such as As, Cd, or Pb. Often the initial evidence comes from animal studies, suggesting the results might also apply to humans. However, convincing evidence that ultra-low trace element deficiencies are responsible for human diseases has been elusive. Setting a human RDA for a dietary mineral that is present in extremely low amounts is a near impossible task, although still an active field of research.

There can be danger in exceeding the RDA values set for minerals. The reason is that there is sometimes a very narrow margin between beneficial and harmful doses. This is not because of lipid-solubility, as was the case for some vitamins, but rather because of competition between trace elements. For example, high doses of zinc can interfere with the metabolism of copper, an element essential to making blood and tissue. In extreme cases, anemia is the result. Many such interactions among essential minerals are still being investigated.

The physiological functions of minerals are widely diverse. Calcium is the most abundant mineral in the body. Along with phosphorus and smaller amounts of fluorine, it is a major constituent of bones and teeth. Blood clotting, muscle contraction, and transmission of nerve impulses also require Ca^{2+} ions.

Sodium is also essential for life, but not in the relatively excessive amounts supplied by the diets of most Americans. Most salt we eat comes from processed foods and fast foods, not necessarily from the saltshaker. The labels of processed foods list "sodium" content, meaning the number of mg of Na^+ per serving. For example, different brands of tomato soup may have between 700 and 1260 mg of Na^+ per serving, providing 29–53% of the recommended daily value of no more than 2400 mg of Na^+ per day. The major concern with excess dietary sodium is its correlation with high blood pressure (hypertension) in susceptible individuals. Some people are advised by their doctors to ingest no more than 1500 mg of Na^+ per day.

Remember that macrominerals are still micronutrients.

Ions were defined and discussed in Section 5.7.

The Latin word for salt is *sal.* Salt was so highly valued in Roman times that soldiers were paid in *sal,* thereby forming the root for the modern word *salary.*

Your Turn 11.17   Sodium in Your Diet

Compare sodium content for foods in the same category, such as different brands of pretzels, bread, frozen pizza, salad dressing, or even tomato soup. Have your findings surprised you or influenced your future choices? Recall that 1 g = 1000 mg.

Oranges, bananas, tomatoes, and potatoes help supply the recommended daily requirement of 2 g of potassium (in the form of K^+ ion), a mineral that is essential for the transmission of nerve impulses and intracellular enzyme activity. Because sodium and potassium ions have similar chemical properties, their physiological functions are also closely related. In intracellular fluid (the liquid within cells), the concentration of potassium ions is considerably greater than that of sodium ions. The reverse situation holds in the lymph and blood serum outside the cells. There the concentration of potassium ions is low and that of sodium ions is high. The relative concentrations of K^+ and Na^+ are especially important for the rhythmic beating of the heart. Individuals who take diuretics to control high blood pressure may also take potassium supplements to replace potassium excreted in the urine. However, such supplements should be taken only under a physician's directions because of the potential danger that they could dramatically alter the potassium–sodium balance and lead to cardiac complications.

In most instances, microminerals and trace elements have very specific biological functions and are incorporated in relatively few biomolecules. Iodine is an example. Most of the body's iodine, in the form of the iodide ion, I^-, is concentrated in the thyroid gland. There it is incorporated into thyroxine, a hormone that regulates metabolism. Excess thyroxine is associated with hyperthyroidism, or Graves' disease, in which basal metabolism is accelerated to an unhealthy level, rather like a racing engine. On the other hand, a thyroxine deficiency, sometimes caused by insufficient dietary iodine, slows metabolism and results in tiredness and listlessness. Both hyper- and hypothyroidism can lead to goiter, an enlargement of the thyroid gland. One way to help prevent goiter is by consuming adequate amounts of the iodide ion. Seafood is a rich source of the element. Another source is iodized salt, sodium chloride (NaCl) to which 0.02% of potassium iodide (KI) has been added. The tendency of the thyroid gland to concentrate iodine is key to the use of radioactive I-131 as a treatment for an overactive thyroid and for the use of nonradioactive KI tablets to counter accidental exposure to any radioactive isotopes of iodine.

> Section 7.6 discussed the continuing need for monitoring thyroid function in children exposed after Chernobyl.

Consider This 11.18   Getting Well Using Radioactive Iodine

Individuals with hyperthyroidism have an overactive thyroid gland. They may experience symptoms associated with an accelerated metabolism such as heat intolerance, agitation, and excessive sweating. Use the resources of the Web to answer these questions.

  a. Radioactive I-131 is used to treat hyperthyroidism. What role does it play?
  b. List some risks and benefits for an individual treated with I-131 for hyperthyroidism.
  c. After I-131 therapy, what further treatment does the individual require?

11.6   Energy from the Metabolism of Food

All the energy needed to run the complex chemical, mechanical, and electrical systems called the human body comes from fats, carbohydrates, and proteins. This energy initially arrives on Earth in the form of sunlight, which is absorbed by green plants during photosynthesis. Under the influence of a catalyst called chlorophyll, carbon dioxide and

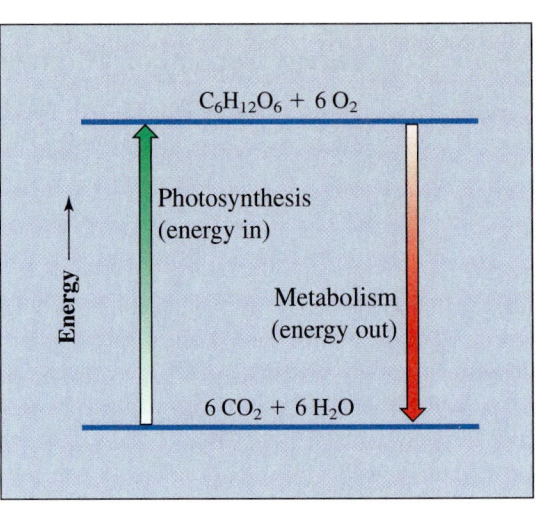

Figure 11.18

Energy balance for photosynthesis and metabolism of glucose.

water are combined to form glucose, $C_6H_{12}O_6$. In the process, the Sun's energy is stored in chemical bonds of the sugar.

$$\text{Energy (from sunshine)} + 6\,CO_2 + 6\,H_2O \xrightarrow{\text{chlorophyll}} C_6H_{12}O_6 + 6\,O_2 \quad [11.5]$$

During metabolism, the photosynthetic process is reversed, the food is converted into simpler substances and the stored energy is released.

$$C_6H_{12}O_6 + 6\,O_2 \longrightarrow 6\,CO_2 + 6\,H_2O + \text{Energy (from metabolism)} \quad [11.6]$$

The energy balance between equations 11.5 and 11.6 is schematically represented in Figure 11.18.

The food we eat provides the energy to drive the chemical reactions that constitute the processes of life. One example of an energy-requiring process is muscular motion, including the beating of the heart. But most of the energy released by metabolism goes to maintain differences in ionic concentrations across cell membranes. The natural tendency is for osmosis to move substances from regions of higher concentration to those of lower concentration. Energy is required to prevent this from happening. The proper concentration differences essential for nerve action and other physiological functions are maintained at great energetic expense. In short, spontaneous reactions furnish the energy for nonspontaneous reactions to occur. As an analogy, consider an automobile storage battery. The battery produces electrical energy because of chemical reactions in the battery. These spontaneous processes provide energy that can be used to drive nonspontaneous processes such as starting the car or making the headlights and horn work.

In addition to having a supply of sufficient energy, the body must have some way of regulating the rate at which the energy is released. Without such control, wild temperature fluctuations and high inefficiency could result. Again, the automobile provides an analogy. Dropping a lighted match into the fuel tank would burn all the gasoline all at once (and the car as well). This is a drastic and not particularly effective way to move a car. Under normal operating conditions, just enough fuel is delivered to the ignition system to supply the automobile with the energy it needs without raising the temperature of the car and its occupants beyond reason. In this way, by releasing a little energy at a time, the efficiency of the process is enhanced. So it is with the body. The conversion of foods ultimately into carbon dioxide and water occurs over many small steps, each one involving enzymes, enzyme regulators, and hormones. As a result, energy is released gradually, as needed, and body temperature is maintained within normal limits. The energy associated with each macronutrient is given in Table 11.5.

The major source of energy in a well-balanced diet is fat. On a gram per gram basis, fats provide about 2.5 times as much energy as the other macronutrients. This observation

If you need a refresher on energy changes at the molecular level, see Section 4.4.

Each heartbeat uses about one joule (1 J) of energy.

Osmosis was discussed in Section 5.15.

Table 11.5	Average Energy Content of Macronutrients	
Fats		9 Cal/g
Carbohydrates		4 Cal/g
Proteins		4 Cal/g

1 dietary calorie = 1 Cal
= 1 kcal
= 1000 calories

makes it easy to understand the popularity of low-fat diets for losing weight. Although proteins, like carbohydrates, yield about 4 Cal/g if metabolized, proteins are not used in the body primarily as an energy source. Rather proteins store molecular parts for building skin, muscles, tendons, ligaments, blood, and enzymes.

The reason for the dramatic difference in energy between fats and carbohydrates is evident from the chemical composition of these two types of material. Compare the formula of a fatty acid, lauric acid, $C_{12}H_{24}O_2$, with that of sucrose (table sugar), $C_{12}H_{22}O_{11}$. Both compounds have the same number of carbon atoms per molecule and very nearly the same number of hydrogen atoms. When the fatty acid or the sugar burns, its carbon and hydrogen atoms combine with added oxygen to form CO_2 and H_2O, respectively. But more oxygen is required to burn a gram of lauric acid, $C_{12}H_{24}O_2$, than a gram of sucrose, $C_{12}H_{22}O_{11}$. Examine these two reactions.

The structure of lauric acid is revealed by writing its formula as $CH_3(CH_2)_{10}COOH$.

$$C_{12}H_{24}O_2 + 17\,O_2 \longrightarrow 12\,CO_2 + 12\,H_2O + 8.8\ \text{Cal/g} \qquad [11.7]$$
lauric acid

$$C_{12}H_{22}O_{11} + 12\,O_2 \longrightarrow 12\,CO_2 + 11\,H_2O + 3.8\ \text{Cal/g} \qquad [11.8]$$
sucrose

Oxygenated fuels were discussed in Section 4.9.

In the language of chemistry, the sugar is already more "oxygenated" or more "oxidized" than the fatty acid. Weaker C-to-H bonds (416 kJ/mol) have already been replaced by stronger O-to-H bonds (467 kJ/mol) in the sucrose. The result is that even though fewer O-to-O double bonds (498 kJ/mol) must be broken for sucrose to combine with O_2, less energy overall is released than is the case for the combustion of lauric acid.

Given how many tasty foods contain fat, it is easy to get an unhealthy percentage of our daily Calories from fats. The problem is illustrated by considering Sceptical Chymist 11.19. In accordance with the *Dietary Guidelines for Americans* released by the U.S. Department of Agriculture (USDA) and the U.S. Department of Health and Human Services (HHS) in January 2005, no more than 20–35% of Calories should come from total fat intake. Furthermore, the guidelines recommend that less than 10% of Calories come from saturated fatty acids and that trans fat consumption be kept as low as possible.

Sceptical Chymist 11.19   Low-Fat Cheese

A popular brand of low-fat shredded cheddar cheese advertises that it provides 1.5 g of fat with 15 Cal from total fat per serving. There are 50 Cal per serving and of the total fat, 1.0 g is saturated fat. A serving is defined as 1/4 cup, or 28 g. Is this a "low-fat" cheese? Support your decision with some numbers. Remember that the dietary recommendation is that no more than 30% of Calories should come from fat.

So how many Calories does a person need? The answer is: "It depends." The number of Calories your diet should supply each day depends on your level of activity, the state of your health, your sex, age, body size, and a few other factors. You can start by finding the category that best describes you in Table 11.6, which summarizes the daily food

Table 11.6	Estimated Calorie Requirements (United States)		
Gender by Age (yr)	**Activity Level**		
	Sedentary[a]	**Moderately Active**[b]	**Active**[c]
Females			
14–18	1800	2000	2400
19–30	2000	2000–2200	2400
31–50	1800	2000	2200
51+	1600	1800	2000–2200
Males			
14–18	2200	2400–2800	2800–3200
19–30	2400	2600–2800	3000
31–50	2200	2400–2600	2800–3000
51+	2000	2200–2400	2400–2800

a. *Sedentary* means a lifestyle that includes only the light physical activity associated with typical day-to-day life.

b. *Moderately active* means a lifestyle that includes physical activity equivalent to walking about 1–3 miles per day at 3–4 miles per hour, in addition to the light physical activity associated with typical day-to-day life.

c. *Active* means a lifestyle that includes physical activity equivalent to walking more than 3 miles per day at 3–4 miles per hour, in addition to the light physical activity associated with typical day-to-day life.

Source: *Dietary Guidelines for Americans,* 2005, USDA.

energy intakes that have been recommended for Americans. The estimated Calorie requirements are presented by gender and age groups at three different levels of physical activity. Although not included in this table, growing children need a proportionally large energy intake to fuel their high level of activity and provide raw material for building muscle and bone. Therefore, children are particularly susceptible to undernourishment and malnutrition. Indeed, mortality rates among infants and young children are disproportionately high in famine-stricken countries.

Consider This 11.20   Calories by Gender and Age

Consider the information in Table 11.6 and the *Dietary Guidelines for Americans,* 2005 (link provided at the *Online Learning Center*) to answer these questions.

  a. Do males and females of the same age require the same number of Calories for the same level of activity? Explain.

  b. As an active male or female ages, how does the estimated Calorie requirement change?

  c. Do the *Dietary Guidelines for Americans,* 2005 offer practical advice for meeting your estimated Calorie requirements? Explain.

Where does all this food energy go? The first call on the Calories consumed is to keep the heart beating, the lungs pumping, the brain active, the blood circulating, all major organs working, and the body temperature at 37 °C. These requirements define the **basal metabolism rate (BMR),** the minimum amount of energy required daily to support basic body functions. This corresponds to approximately one Calorie per kilogram (2.2 lb) of body weight per hour, although it varies with size and age. The BMR is experimentally determined in a resting state, and the quantity of energy used in digestion is eliminated by having the subject fast for 12 hours before the measurement is made. To put this on a personal basis, consider a 20-year-old female weighing 55 kg (121 lb). If her

Your basal metabolism rate is approximately 1 Cal/kg body mass per hour.

Table 11.7	Energy Expenditure (Cal/Hr) for Common Physical Activities*			
Moderate Physical Activity		Cal/hr	Vigorous Physical Activity	Cal/hr
Hiking		370	Running/jogging (5 mph)	590
Light gardening/yard work		330	Heavy yard work (chopping wood)	440
Dancing		330	Swimming (freestyle laps)	510
Golf (walking, carrying clubs)		330	Aerobics	480
Bicycling (<10 mph)		290	Bicycling (>10 mph)	590
Walking (3.5 mph)		280	Walking (4.5 mph)	460
Weight lifting (light workout)		220	Weight lifting (vigorous workout)	440
Stretching		180	Basketball (vigorous)	440

* Values include both resting metabolic rate and activity expenditure for a 70-kg (154-lb) person. Calories burned per hour will be higher for persons heavier than 154 lb and lower for persons who weigh less.

$$\frac{1300 \text{ Cal}}{2200 \text{ Cal}} \times 100 = 59\%$$

body has a minimum requirement of 1 Cal/(kg·h), her daily basal metabolism rate will be 1 Cal/(kg · h) 55 kg × 24 h/day or about 1300 Cal/day. According to Table 11.6, the recommended daily energy intake for a woman of this age and weight is a maximum of 2200 Cal if she is moderately active. This means that 59% of the energy derived from this food goes just to keep her body systems going. Dieters must beware not to cut their caloric intake so low that essential bodily functions are affected.

Where the rest of the Calories go depends on what she does. The law of conservation of energy decrees that the energy must go somewhere. If she "burns off" the extra Calories in exercise and activity, none will be stored as added fat and glycogen. But, if the excess energy is not expended, it will accumulate in chemical form. Putting it more crassly, "those who indulge, bulge," unless they work and play hard.

Some indication of how hard and how long we have to work or play to use up dietary Calories is given in Table 11.7. This table reports the energy expenditures for various activities as a function of body weight. Table 11.8 quantifies exercise in readily recognizable units such as hamburgers, potato chips, and beer. Of course, by combining the information in this section with the information in earlier parts of this chapter about the *types* of nutrients in food, it should be clear that a healthful diet cannot be achieved simply by consuming the correct number of Calories. A 2000 Cal diet of only potato chips and beer would leave a person malnourished. Proper nutrition is not simply a matter of how much, but also of what kind of food a person consumes.

Table 11.8	How Much Exercise Must I Do if I Eat This Cookie?*		
Food	Calories	Time if Walking at 3.5 mph, (min)	Time if Running at 5 mph, (min)
Apple	125	27	13
Beer (regular) 8 oz	100	21	10
Chocolate chip cookie	50	11	5
Hamburger	350	75	59
Ice cream, 4 oz	175	38	18
Pizza, cheese, 1 slice	180	39	18
Potato chips, 1 oz	108	23	11

* Values include both resting metabolic rate and activity expenditure for a 70-kg (154-lb) person.

Your Turn 11.21   **Playing off the Calories**

A 70-kg person consumes a meal consisting of two hamburgers, 3 oz of potato chips, 8 oz of ice cream, and a 12-oz beer. Calculate the number of Calories in the meal and the number of minutes the person would have to vigorously play basketball in order to "work off" the meal.

Answer
1524 Cal, 208 min

11.7 Quality Versus Quantity: Dietary Advice

Despite great advances in improving healthcare and modern medicine in the last century, obesity is now on the rise in the United States. In the last 30 years scientific research on nutrition has made tremendous progress. However, with advertising and the mass media, people seem more frustrated than ever in trying to understand dietary advice. One day the "experts" say one thing; the next they seem to say the opposite. Because of an emphasis on short but often sensational news pieces, the media generally only reports the results of single studies. Often these are only reported when they are at odds with the currently accepted standards. But nutritional studies are very complex and the scientific results cannot be summarized into simple sound bites while still being fully explained. Because human beings are complex biochemical machines, a great deal of care must be taken to understand the limitations and true implications of any single data set. Figure 11.19 gives a glimpse of changing dietary advice from the U.S. Department of Agriculture in the past.

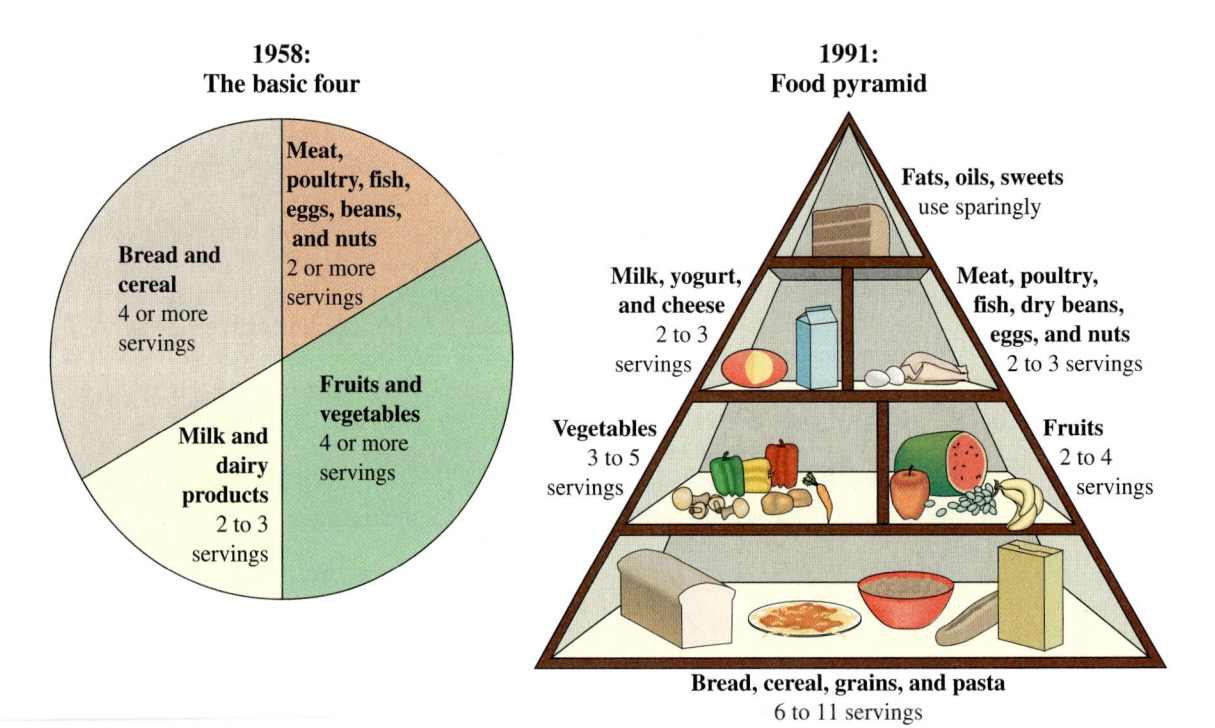

Figure 11.19

The USDA basic four food groups and the food pyramid.

Source: United States Department of Agriculture.

Consider This 11.22   Is Chocolate a Health Food?

Chocoholics have reason to rejoice that certain types of chocolate are reported to have health benefits. Critics point out, however, that chocolate is full of fat, is addictive, and has empty calories. Use these questions to explore the controversies.

a. Do the reported health benefits apply to all types of chocolate?
b. Is the main fat found in chocolate saturated or unsaturated?
c. Is there any scientific evidence that chocolate is addictive?
d. What components of chocolate have been shown to have health benefits?

An even greater change took place in April 2005 when the U.S. Department of Agriculture introduced a different style of pyramid (Figure 11.20). The 1991 pyramid has been turned on its side, stairs have been added, and a person is climbing the stairs. Extensive interactive resources accompany "Steps to a Healthier You," allowing you to complete an in-depth assessment of your diet quality and physical activity status. Incorporating recommendations from the January 2005 *Dietary Guidelines for Americans*, the accompanying Web site offers personalized advice for different types of diets. There are many ideas and resources to help the computer-savvy consumer stay on a sound dietary plan, with tips ranging from cutting down on saturated fats to increasing exercise or eating out without busting a diet. You will have a chance to explore this resource in Consider This 11.23.

Consider This 11.23   Step Up the Pyramid

The USDA's 2005 symbol and interactive food guidance system can be found at the USDA Web site (available at the *Online Learning Center*). Explore MyPyramid and then consider these questions.

a. Enter your age, sex, and physical activity level into MyPyramid Plan feature. What did you find out about what you need to eat?
b. For a more detailed assessment of your food intake and physical activity level, click on MyPyramid tracker. What did you find out?
c. What features of the interactive Web site are most helpful to you? Explain.
d. When unveiling the MyPyramid symbol, U.S. Agriculture Secretary Mike Johanns said that the symbol was "deliberately simple" to encourage consumers to get more in-depth information about healthier food choices. Do you think the symbol accomplishes that goal? Why is a person walking up the steps of the pyramid?
e. The USDA has no budget to promote the Dietary Guidelines or the pyramid itself. How will the information be distributed to the public?

History can help us understand the proliferation of diets. In the 1960s, experiments based on controlled feeding of particular food items to participants for several weeks showed that saturated fat increased cholesterol levels. But these studies also showed that polyunsaturated fats—found in vegetables and fish—reduced cholesterol. Advice in the following decades was to replace saturated fats with unsaturated ones rather than reducing total fat. The "sat fat is bad" movement led to the greatly expanded use of vegetable oils. Unfortunately, as was noted in Section 11.2, the partial hydrogenation of these oils to

Figure 11.20

MyPyramid: Steps to a Healthier You.

Source: U.S. Department of Agriculture, 2005.

create products such as margarine gave rise to trans fats, a category of fat that may prove to be worse than saturated forms.

Two of the largest and longest running studies of diet are the Harvard-based Nurses' Health Study and Health Professionals Follow Up Study, that followed over 90,000 women and 50,000 men, respectively, over several decades. These studies provided information that often has led to suggestions for dietary changes (including development of the 1991 USDA food pyramid), and sometimes fueled diet trends. The Harvard studies have shown that a participant's risk of heart disease was strongly influenced by the type of dietary fat consumed. Eating trans fat increases the risk substantially while saturated fat increases it slightly. Unsaturated fats decrease the risk. Therefore, the total fat intake alone is not associated with heart disease risk. Furthermore, epidemiological studies have shown little evidence that total fat or specific fats affect the risks of several cancers. The rise in obesity has been blamed on fat but American consumption of calories from fat has decreased since the 1980s while the rate of obesity still continues to grow. All of this would seem to imply that fat is not the culprit it was once believed to be. Or, at least, that the solution to American dietary woes is not quite so clear-cut.

So what about the "carbs"? Diet books addressing the issue of carbohydrates have spanned the spectrum from banning all carbs to a "good carb versus bad carb" differentiation. The claim is that "bad" carbohydrates cause a quick rise in blood sugar, followed by a spike in blood insulin level and a weight gain caused by making the body store fat or making the person feel hungry again from the low blood sugar. Insulin is a hormone secreted by the pancreas that allows the cells in your body to absorb and store sugar that is in the blood. The sugar that is not immediately burned for energy is converted to and stored as fat in your cells. Furthermore, glucagon also is a hormone secreted by the pancreas that, essentially, has the opposite effect of insulin: it promotes the use of stored glucose in cells. The release of glucagon will decrease after a "glucose spike," such as would result from eating "bad carbs," but it is known to increase after consumption of proteins. Therefore, many of the "low-carb" diets promote the consumption of increased amounts of protein as a way to use up stored calories. The long-term health effects of this approach have yet to be seen.

Clearly, nutrition and dieting are complex. However, research in this area increasingly is contributing to our understanding of these issues. In the United States, a concern is how to keep people from dying of obesity. But in the United States and many parts of the world as well, another concern has been how to keep people from dying of starvation. This highlights, once again, the distinction between undernourishment and malnourishment. We turn next to the issue of hunger on our planet.

The growth in girth of Americans has been accompanied by a proliferation of diet plans: Weight Watchers, the Atkins diet (phase I and II), the South Beach diet, and the Dean Ornish diet (very low fat), among others. The American Heart Association also promotes healthful eating.

a. Pick two of these diets and compare their features. How close are these to your own diet?
b. What scientific arguments lie behind the diets you selected?
c. Report on another diet that is different from those above.

11.8   Feeding a Hungry World

The Food and Agriculture Organization of the United Nations (FAO), in their December 2005 report, estimated that more than 852 million people worldwide are undernourished. Of those, 808 million (95%) live in developing countries, 34 million (4%) in countries in transition, and 10 million (1%) in developed countries. Translating these sterile statistics into terms of human misery provides evidence of the magnitude and tenacity of hunger in the world. Clearly, the world's food supply is not equally distributed among its inhabitants. Piles of corn and wheat rot in the American Midwest or on docks around the world, while half a billion men, women, and children go to bed hungry. Figure 11.21 maps the location of the 38 countries facing serious food shortages.

Food shortages have multiple and interrelated causes. Some are related to geography and climate, as certain areas of the world simply do not have enough arable land and adequate soil to produce sufficient food for their people. Prolonged droughts can reduce crop yields, and episodic floods can wash away soil and crops. Other causes are economic. For example, the fertilizers needed to supplement mineral-depleted soils may not be available because of their high cost. Furthermore, animals for working the fields, to say nothing of tractors, may be too expensive for farmers in some regions. Still other causes are political

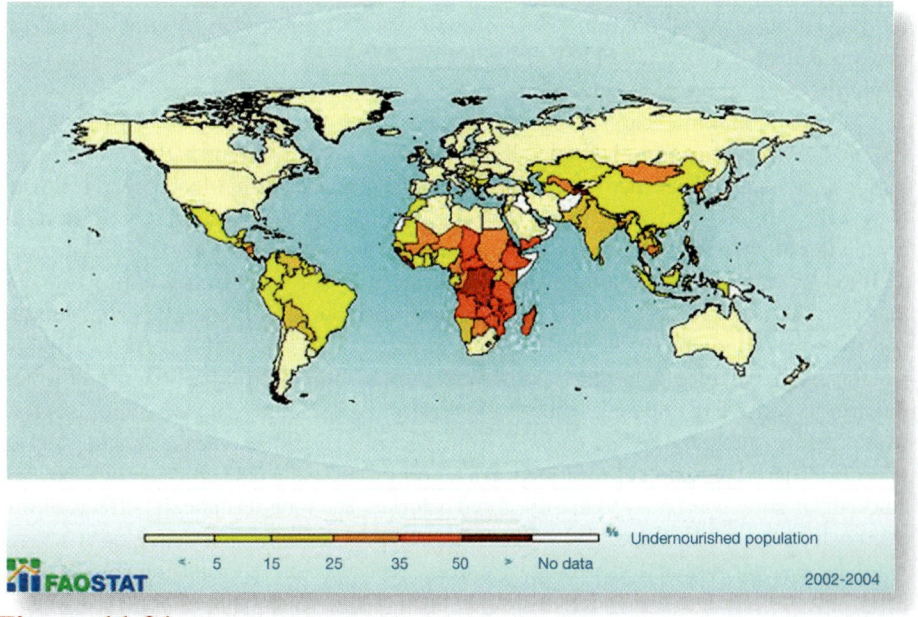

Figure 11.21

Countries facing undernourished population.

Source: Reprinted with permission of The Food and Agriculture Organization of the United Nations, from *The State of Food and Agriculture*, 2002–2004.

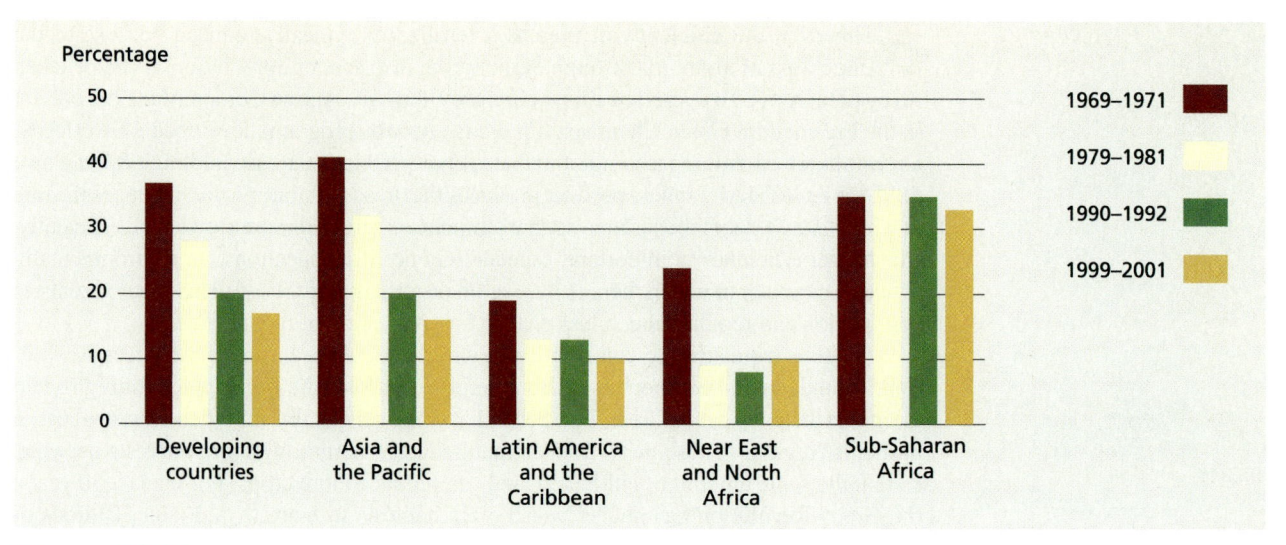

Figure 11.22

Percentage of population that is undernourished in developing countries, by region.

Source: Reprinted with permission of The Food and Agriculture Organization of the United Nations, 2002–2004, from *www.foa.org.*

or military. Civil strife in some countries can block the flow of food and other agricultural products. Finally, food shortages are compounded by disease. For example, the HIV-AIDS pandemic has affected food production, marketing, and transportation.

The situation is changing in many parts of the world, as seen in Figure 11.22. To meet growing populations, global grain and cereal production has doubled over the past quarter century, and supplies of vegetables, fruits, milk, meat, and fish have also increased. In many developing countries, food production has generally not kept pace with growing populations. The exception has been Asia, where the rate of increase in per capita food production has been greater than that even in the developed world. In stark contrast, sub-Saharan Africa has experienced a long-term, continuing decline in per capita food production. One major reason for the increase in Asian crop yields has been greater use of fertilizers and pesticides. The application of both has often been criticized as being harmful to the environment, but the fact remains that millions have been saved from starvation, with fertilizers and pesticides playing a role.

| Consider This 11.25 | Worldwide Trends in Undernourishment |

Over the past two decades, progress has been made in many regions in reducing the percentage of the population that is undernourished. Use the information in Figure 11.22 and the resources of the Web to answer these questions.

a. How has the percentage of population in developing countries that is undernourished changed from 1969–1971 to 1999–2001?

b. Have any regions shown an increase in the percent undernourished? Discuss possible reasons.

c. In developing countries, has the decrease in the percentage undernourished been accompanied by a decrease in absolute numbers? Explain.

d. Which region of the world has experienced the greatest decrease in the percentage of undernourished people? Offer some possible reasons.

Urea, $(NH_2)_2CO$, is a major fertilizer used worldwide. It provides nitrogen because enzymes found in soil decompose it to ammonia and CO_2. The ammonia is then taken up by plants.

$$(NH_2)_2CO + H_2O \xrightarrow{\text{urease}} 2\,NH_3 + CO_2 \qquad\qquad [11.9]$$

However, the efficiency of urea as a fertilizer is typically reduced because of the direct loss of ammonia through evaporation, in excess of 30%, before it can be taken up by plant roots. To overcome this inefficiency, the IMC Agrico Company, a 1997 entrant in the Presidential Green Chemistry Challenge Awards program, developed AGROTAIN, a formulation containing a compound that is converted into a urease inhibitor. Spread on a field, the AGROTAIN linked product produces the urease inhibitor, which reduces the rate at which urease decomposes urea so that ammonia is released more slowly and efficiently. The higher efficiency is important, especially in no-till applications, an environmentally friendly approach in which there is little or no disturbance of topsoil. This method reduces soil erosion and requires much less energy for application of the fertilizer.

An even more striking contribution to world agriculture has been the Green Revolution. A fundamental component of this enterprise has been the development of high-yield grains, principally wheat, rice, and corn, that were genetically modified to grow best in particular regions. These new varieties mature faster, permitting more harvests per year, so that the same amount of cultivated land can produce more crops. For the last 50 years, the Green Revolution has helped world grain harvests to more than double. Billions of people in India, Asia, and Africa have benefited from the practice. But the Green Revolution is neither a panacea nor the ultimate answer. In spite of its successes, the Green Revolution is not universally applicable and has not been without costs. It works best in areas where water for irrigation is abundant; where money is available for supplemental fertilizers such as ammonia, urea, or nitrates; and where technological understanding and application exist.

Researchers estimate that, within 20 years, global demand for the world's three most important crops—rice, maize (a type of corn), and wheat—will have increased by 40%, simply to keep pace with global food requirements. Genetic engineering and other applications of biotechnology now hold out promise for a second Green Revolution to meet such demands.

Genetic engineering for food production will be discussed in Section 12.10.

11.9   Food Preservation

Food poisoning is a serious matter. If you ever have experienced even a mild case, you know its discomforts and sometimes embarrassments. In severe cases, food poisoning can claim the lives of loved ones. Thus over the centuries, people in every culture have found ways to preserve foods.

Two time-tested food preservatives are sugar and salt. As you may know first-hand, jellies, jams, and pickles make use of them. These preservatives work by creating a high salt or sugar concentration that is unfavorable to the growth of bacteria, yeasts, and molds. By the process of osmosis, water will flow out of the cells of these organisms, thus rupturing their cell membranes. The drawback, however, is that sugar and salt greatly alter the taste of the food. Although changes in taste may be considered desirable (pickled herring is a delicacy to some), fortunately we have other options for preserving food.

Another time-tested method of preservation is to heat or chill the food. The former, if done properly, kills microorganisms. Home canning and the pasteurization of milk or juice are examples. The latter, refrigeration, retards but does not ultimately prevent spoilage. You may have observed this for yourself with a container that remained too long on a refrigerator shelf.

Newer food additives also are used to preserve food for freshness, especially in packaged "convenience" foods. People want their cookies and potato chips to be fresh and free of any tinge of rancidity. For foods such as these, oxygen is the enemy. Although essential for our breathing, oxygen slowly reacts with fats and oils to make them go rancid. Fortunately, food additives can mitigate the effects of oxygen. For example, **antioxidants** are compounds added to foods, drugs, and cosmetics to minimize the oxidation of unsaturated oils and fats that can cause rancidity, color loss, and flavor changes.

If you examine the labels of foods that contain fats or oils, you may find that BHT or BHA has been added either to the food or the cardboard packaging. These compounds are *b*utylated *h*ydroxy*t*oluene and *b*utylated *h*ydroxy*a*nisole, respectively, two antioxidants

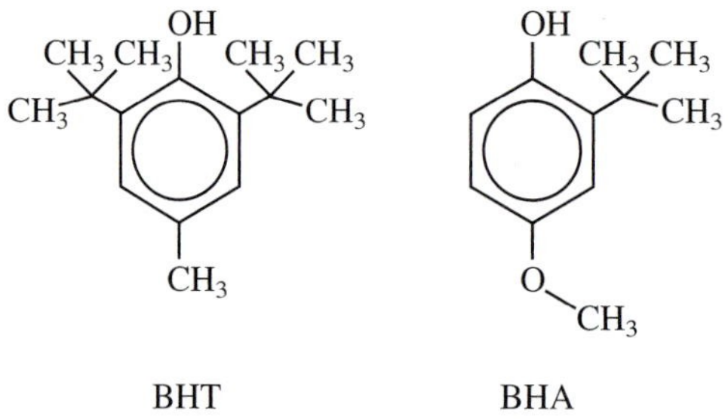

Figure 11.23
Structural formulas for BHT and BHA antioxidants.

(Figure 11.23). BHT and BHA act by preventing the buildup of free radicals that form when fats and oils react with oxygen from the air.

$$\text{fat (or oil)} + \text{oxygen} \longrightarrow \text{free radicals} + \text{other products}$$

With their unpaired electrons, free radicals are highly reactive. BHT, BHA, and other antioxidants scavenge the unpaired electron from the free radical to form a stable species. This prevents further oxidation of the fat, thus keeping your cereal, cookies, and potato chips more tasty and free from rancid oils.

Free radicals were discussed in Sections 1.11, 2.11 and 9.4 in relation to air quality, ozone depletion, and polymerization, respectively.

Your Turn 11.26   Antioxidants in Cereal

Find a supply of cereal boxes, either in a kitchen or grocery store. Select any 10 and examine the labels on these boxes carefully.

a. What percent of the cereals list a preservative for freshness? Of these, how many use BHT or BHA?
b. What percent list BHT or BHA as part of the packaging?
c. Name any other compounds mentioned as preservatives.

As the previous activity demonstrates, we now have alternative ways of preserving food freshness. Some cereals are packed in a nitrogen atmosphere, others use other antioxidants such as vitamin E, and still others choose to use no preservative. Although the risks of BHT and BHA appear to be low, these additives may not be necessary.

A more recent and far more controversial method of food preservation is irradiation. Food, of course, is subjected to radiation for many different purposes, with the outcome dependent on the part of the electromagnetic spectrum used. Visible light from the Sun drives photosynthesis. We cook food with infrared radiation (heat), and we use microwave radiation to warm up leftovers. Entirely different parts of the spectrum are used for **food irradiation,** a process of subjecting food to high-energy ionizing radiation to kill or reduce the levels of undesirable contaminants such as bacteria, spores, and insects. Gamma rays, X-rays, and beta particles are three types of ionizing radiation in use for food irradiation. On impact, they have sufficient energy to remove electrons to form ions or free radicals. In turn, these species interact with the organisms in food and kill them.

Classified as a food additive by Congress in 1958, food irradiation was approved by the FDA in 1963. The irradiation procedure is relatively straightforward and carried out at over 160 facilities worldwide. The material to be irradiated is placed on a conveyer belt that moves past a beam of high-energy radiation that is enclosed and shielded. The sources of the radiation can be a Co-60 or Cs-137 gamma source, an X-ray generator, or an electron-beam generator that produces high-energy electrons.

Only a small number of irradiated foods, including potatoes, fish, shrimp, grapefruit for export, and strawberries for domestic consumption, have been approved by the FDA

The electromagnetic spectrum was discussed in Section 2.4. Gamma radiation was introduced in Section 7.5.

The U.S. Army has used spent fuel rods for irradiation.

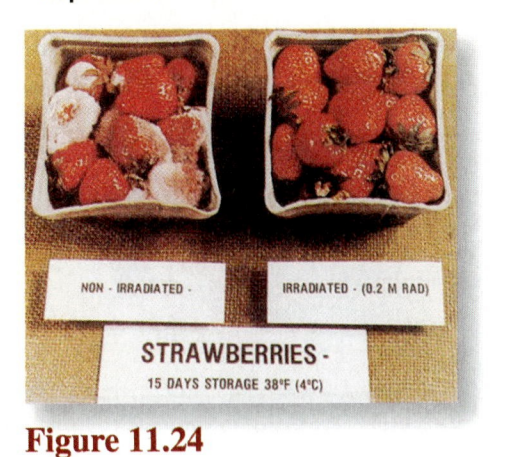

Figure 11.24

Strawberries preserved by irradiation.

Figure 11.25

The international label for irradiated food.

for sale in the United States (Figure 11.24). Irradiation of fresh or frozen poultry was approved in 1992. In 1999, the irradiation of beef, pork, and lamb was approved as a way to reduce or eliminate disease-causing organisms. The World Health Organization, the Food and Agricultural Organization of the United Nations, and the U.S. FDA all have concluded that food irradiation is safe when proper procedures and practices are used. Irradiated foods even have their own international logo (Figure 11.25). Yet irradiated foods are not widely accepted. Why the controversy?

Those opposed to food irradiation question whether irradiated foods are safe to eat. The most serious charge concerns the by-products (sometimes called radiolytic products) generated when chemical bonds break. For example, if the food contains a high percentage of water, high-energy ionizing radiation can cause this reaction to occur.

$$H_2O \xrightarrow{\text{ionizing radiation}} H_2O^+ + e^- \qquad [11.10]$$

> This reaction also was shown in equation 7.10.

The product, H_2O^+, is highly reactive and can react with additional water molecules to produce other reactive species. These, in turn, can react with the molecules in the food to produce undesirable by-products. Cooking, of course, similarly produces chemical changes in the food, often many times greater than those from ionizing irradiation. Decades of research suggest that the by-products of irradiation mostly are those found normally in food; nonetheless, a few are of concern. One is benzene, C_6H_6. It appears, however, that the amounts are so tiny that the benefits of irradiation for food preservation far outweigh the risks.

> Section 4.9 introduced you to the benzene molecule. Its structural formula was given in Section 10.3.

The very need for this nuclear technology has itself been called into question. One issue is whether the radioactive materials used as gamma ray sources would be desirable for terrorist activities. Critics also express concern that food irradiation will be used to cover up improper food handling processes, giving consumers misplaced confidence in the safety of food. Even so, critics and proponents agree that irradiating foods to preserve them does not make them radioactive beyond the normal background radiation that all foods naturally possess.

Consider This 11.27   Cobalt-60 Suitability

Cobalt-60, a radioisotope that emits both beta and gamma radiation, has a half-life of roughly 5 years. Rods of Co-60 are extremely useful as gamma sources, both for cancer treatment and for food irradiation. Comment on the fate of a person who acquired a rod of Co-60, planning on dispersing it as an instrument of terror such as a dirty bomb. Assume this person transported the rod or attempted to grind it up.

The need for food preservation is serious and should be kept in perspective. Worldwide, food spoilage and contamination is a significant problem, claiming up to half of all food crops in some parts of the world. Closer to home, we are not immune to such contamination in our food supply. Outbreaks of food poisoning in the United States occur periodically from *Salmonella,* due to inadequate treatment in chicken- and beef-processing plants. Food contaminated with *Salmonella* has been linked to thousands of deaths in the United States alone. The symptoms of food poisoning—abdominal pain, diarrhea, nausea, and vomiting—mimic those of short-term gastroenteritis (stomach flu). Therefore, food poisoning is often misdiagnosed. The more widespread irradiation of chicken meat, for example, could lower the threat of accidental poisoning by *Salmonella.*

Like many issues we have examined, food irradiation can be viewed in terms of its risks and benefits. For example, does the benefit of irradiating strawberries to keep them fresh for a few days longer outweigh the risks? People are not likely to become ill or die from eating strawberries that are a bit past their peak. On the other hand, trichinosis is a serious disease that can occur by eating pork contaminated with the *Trichinella spiralis* parasite. Low-level irradiation of pork kills the parasite, making the pork safe to eat. Although irradiated beef, pork, and chicken have been approved by the FDA, firms that process these meats are wary that consumers will not buy the irradiated products. This is in spite of the fact that in countries where humans have consumed irradiated foods for years, including poultry and seafood, no adverse effects have been observed. Apparently the U.S. chicken-processing companies feel that the costs do not outweigh the benefits (at least to them).

Consider This 11.28   **Food Irradiation . . . Thanks or No Thanks?**

Food irradiation remains a controversial topic. Pick one food item that is irradiated and prepare a position paper on it. Feel free to compose your arguments from the standpoint of a food company executive, a manufacturer of irradiation equipment, a government official, an organic grower, or simply a hungry citizen. Be sure to cite all sources. The *Online Learning Center* offers some helpful links.

Conclusion

Nutrition has become a national issue—in medicine and health, in business and advertising, and in our daily lives. From television advertisements to the neighborhood newsstand, information abounds about using diet and exercise to lose weight and prevent disease. This chapter began with the conflicting information consumers receive and how the American diet is changing to meet the rising epidemic of obesity. It ends with the grim reality of trying to find enough food worldwide to feed the growing population and ways to store and protect that food. Even though our individual tastes vary, our biological needs are much the same. We need carbohydrates and fats as our energy sources; fats for cell membranes, synthesis, and lubrication; proteins to build muscle and create the enzymes that catalyze the wonderful chemistry of life; and vitamins and minerals to help make that chemistry happen. Nutrition, like water quality, is a global issue that affects the health of all human beings, regardless of where they live. People with too much to eat, like most Americans, seem preoccupied with food, although generally with too little regard for what they eat. The hungry and the starving think of little else beyond how to feed themselves. Chemistry is only part of the solution to one of the great challenges of our time—how to meet all individual dietary needs, regardless of region or wealth.

Chapter Summary

Having studied this chapter, you should be able to:

- Differentiate between malnutrition and undernourishment (11.1)
- Understand the physiological functions of food (11.1)
- Describe the distribution of water, fats, carbohydrates, and proteins in the human body and some typical foods (11.1)
- Identify the major elements found in the human body (11.1)
- Recognize and use the chemical composition and molecular structure of fats and oils (11.2)
- Identify sources of saturated and unsaturated fats and their significance in the diet (11.2)
- Show how fatty acids and glycerol can combine to form a triglyceride (11.2)
- Understand how hydrogenation leads to the formation of trans fats (11.2)
- Discuss sources of cholesterol and its significance in the diet (11.2)
- Recognize and use the chemical composition and molecular structure of carbohydrates (11.3)
- Differentiate among the structures and properties of sugars, starch, and cellulose (11.3)
- Describe the symptoms and cause of lactose intolerance (11.3)
- Give the general molecular structure of an amino acid (11.4)
- Identify and use the chemical composition and molecular structure of proteins (11.4)
- Discuss the importance of essential amino acids and their dietary significance (11.4)

- Explain the principal of protein complementarity (11.4)
- Describe the symptoms and cause of phenylketonuria (11.4)
- Discuss the effects of selected vitamins on human health (11.5)
- Differentiate chemically between fat-soluble and water-soluble vitamins (11.5)
- Describe the effects of selected minerals on human health (11.5)
- Discuss the necessity of macrominerals, microminerals, and trace minerals for human health (11.5)
- Explain carbohydrates, fats, and proteins as energy sources (11.6)
- Discuss typical recommended daily energy intakes (11.6)
- Relate energy expenditures in various activities (11.6)
- Identify and use basal metabolism rate (BMR) (11.6)
- Know appropriate resources for obtaining up-to-date dietary advice (11.7)
- Use resources to determine personal diet plans (11.7)
- Discuss the problems of undernourishment in the world (11.8)
- Identify some contributing factors to observed trends in undernourishment (11.8)
- Describe various strategies for feeding the world's growing population (11.8)
- Discuss various methods of food preservation, including use of antioxidants (11.9)
- Weigh the risks and benefits of food irradiation and take an informed stand (11.9)

Questions

Emphasizing Essentials

1. Food provides four essentials to keep our bodies functioning. What are these and what are their roles?

2. Can a person be malnourished even when eating enough Calories every day to meet metabolic needs? Explain.

3. **a.** What are macronutrients and what role do they play in keeping us healthy?

   b. Name the three major classes of macronutrients.

4. Water is not considered a macronutrient, but it clearly is essential in maintaining health. What are some of the roles that water plays in our bodies? *Hint:* You may want to refer to Chapter 5.

5. Consider this chart.

Based on the relative percentages of protein, carbohydrate, water, and fat given, is this graph more likely a representation of steak, peanut butter, or chocolate chip cookies? Justify your choice.

6. Answer these questions using Table 11.1.

   a. Identify the top three foods that are good sources of carbohydrates and arrange them in order of decreasing percentage of carbohydrates.

   b. Identify the top three foods that are good sources of protein and arrange them in order of decreasing percentage of protein.

   c. Which of these foods should be avoided if you are controlling dietary intake of fat? Identify the top three and arrange them in order of decreasing percentage of fat.

7. An 18-oz steak is the manager's special at a local restaurant. Use the information in Table 11.1 to calculate the ounces of protein, fat, and water that the customer eating this entire steak would consume.

8. Examine the data in Table 11.2 and explain why hydrogen ranks first in atomic abundance in the human body, but third behind oxygen and carbon in terms of mass percent.

9. Use the information in Table 11.2 to answer these questions.

   a. What is the ratio of the relative abundance of potassium to sodium in the human body?

   b. What is the ratio of grams of potassium to grams of sodium in the human body?

   c. Are these elements included in the composition of the human body shown in Figure 11.2? Why or why not?

10. a. Consider the composition of the human body shown in Figure 11.2. What are the principal elements that make up water, proteins, carbohydrates, and fats? Which elements are in common among these major components of the human body?

   b. Compare your answers for part **a** with the relative abundance of the elements in the body given in Table 11.2. Is there a correlation? Explain the correlation between your lists and the relative abundances of the elements in the body.

11. What are the similarities between fats and oils? The differences?

12. Lactic acid is $CH_3CH(OH)COOH$.

   a. Draw a structural formula for lactic acid and indicate the geometry around each C atom.

   b. Is lactic acid saturated or unsaturated? Explain.

   c. Is lactic acid a fatty acid? Explain.

13. From the entries in Figure 11.7, identify the fat or oil with the highest percentage of:

   a. polyunsaturated fat   **c.** total unsaturated fat

   b. monounsaturated fat   **d.** saturated fat

14. Identify the fatty acids likely to be present in each of these. *Hint*: Use Figure 11.7 and the Figures Alive! for Chapter 11 at the *Online Learning Center.*

   a. canola oil   **b.** olive oil   **c.** lard

15. The label of a popular brand of soft margarine lists "partially hydrogenated soybean oil" as an ingredient. What does "partially hydrogenated" mean? Why does the label not simply say soybean oil, rather than partially hydrogenated soybean oil?

16. Your friend wants to cut food costs and has learned that peanut butter is a good protein source. What additional information should your friend consider before making the decision to make peanut butter the major dietary protein source? *Hint:* See Table 11.1.

17. Fructose, $C_6H_{12}O_6$, is a carbohydrate.

   a. Rewrite the formula for fructose to emphasize the original meaning of the term *carbohydrate*.

   b. Draw a structural formula for one of the isomers of fructose.

   c. Do you expect different isomers of fructose to have the same sweetness? Explain.

18. Fructose and glucose both have the formula, $C_6H_{12}O_6$. How do their structural formulas differ?

19. State what is meant by each term, and give an example.

   a. monosaccharide

   b. disaccharide

   c. polysaccharide

20. What problems can arise from regularly consuming excess dietary servings of carbohydrates?

21. What is Splenda? How is it chemically similar to and different from sucrose?

22. Use the lock-and-key model discussed in Section 10.5 to offer a possible explanation why individuals who suffer from lactose intolerance can digest other sugars such as sucrose and maltose, but not lactose.

23. Why should phenylketonurics not drink a soft drink containing aspartame but can drink one sweetened with sucralose?

24. What is the nutritional significance of the elements shaded on this periodic table?

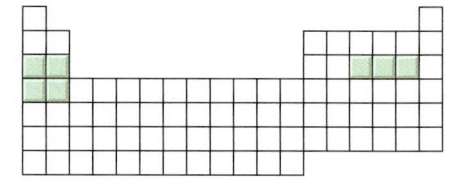

25. Why is it safer to take large doses of vitamin C than vitamin D?

26. Here is a schematic diagram of the 1991 food pyramid.

```
        F
      D   E
      B   C
        A
```

  a. What foods are in each section of the pyramid, and how many servings of each should you consume daily?

  b. [icon] Correlate each of these lettered regions on the 1991 food pyramid with the colors of the wedges on the 2005 "Steps to a Healthier You" logo.

27. A piece of sausage pizza contains items from several food groups. Identify each group and name the part of the pizza that is responsible for representing that particular food group.

28. Use the information in Figures 11.21 and 11.22 to answer these questions.

  a. Which areas of the world are experiencing food supply shortfalls and require exceptional assistance?

  b. Give reasons why these areas do not include North America.

29. What is the purpose of adding BHT or BHA to processed food?

30. What is a gamma ray? How are gamma rays useful for food preservation?

Concentrating on Concepts

31. Explain to a friend why it is impossible to go on a highly advertised "all organic, chemical-free" diet.

32. a. What percentage of the elements in the periodic table is found in proteins, carbohydrates, and fats?

  b. Give the relationship between the type of bonds these elements can form and what makes them so prevalent in the human body.

33. The data in this table are excerpted from a 2006 report by the Centers for Disease Control and Prevention.

Percentage of People Who Are Obese and Overweight in the United States

Years	1999–2000		2003–2004	
	Obese	Overweight	Obese	Overweight
All adults	30.5	64.5	32	66
Men	27.5	67	31	71
Women	33	62	33	62
Children	14	28	17	34

  a. When comparing the 1999–2000 with the 2003–2004 data, which group (men, women, or children) has shown the greatest percentage increase in being overweight? Explain.

  b. One news source reported these data with the headline "U.S. obesity epidemic may be leveling off." Do you think these data support such a headline? Explain.

  c. These data were gathered through in-person examinations that include actual weight measurements. Offer some possible reasons why this approach is considered more reliable than telephone surveys.

34. For each statement, indicate whether it is always true, may be true, or cannot be true. Justify your answers by explaining your reasoning.

  a. Plant oils are lower in saturated fat than are animal fats.

  b. Lard is more healthful than butterfat.

  c. There is no need to include fats in our diets because our bodies can manufacture fats from other substances we eat.

35. [icon] Americans eat about 22 lb of snack foods per capita every year. Olean, a nonfattening, nonmetabolizable fat developed by the Procter & Gamble Company, was approved by the FDA in 1996 for use in salty snack foods such as potato chips and tortilla chips. In spite of having FDA approval, Olean remains controversial, with supporters and detractors. Use the Web to locate the Olean Web site as well as other sites that present contrasting viewpoints.

  a. How does Olean work? Why is it not digested?

  b. Why is the use of Olean in snack foods controversial?

  c. In August 2003, the FDA removed the requirement for a warning label for products containing Olean. Why did the FDA take this action? Give examples of opposing viewpoints regarding the use of the Olean warning labels.

36. Experimental evidence suggests that some physiological effects of saturated fats, compared with unsaturated fats, may be caused by differences in packing the molecules. The hydrocarbon chains in saturated fatty acids can pack more tightly than those of unsaturated or polyunsaturated fatty acids.

  a. Explain why saturated fatty acid molecules are able to pack more tightly than molecules of unsaturated or polyunsaturated fatty acids. *Hint:* You could use molecular models to help you see the effect single or double bonds can have on the ease of packing.

  b. Explain why the extent of molecular packing influences the melting points of stearic, oleic, linoleic, and linolenic acids. See Table 11.3 for melting point values.

  c. How is structural packing related to the harmful effects of trans fats?

37. Some people prefer to use nondairy creamer rather than real cream or milk. Some, but not all nondairy creamers, use coconut oil derivatives to replace the butterfat in cream. Is a person trying to reduce dietary saturated fats wise to use nondairy creamers such as these? Explain.

38. An avocado is a tropical fruit, but is unlike the tropical coconut or palm in its type of fat/oil composition. What is the primary type of fat found in avocados?

39. Why is it more difficult for a person to control her or his cholesterol level than to control her or his fat intake? What steps are effective in minimizing cholesterol in the blood?

40. **a.** Which are the "good" lipoproteins, LDLs or HDLs?

    b. What function do the "good" lipoproteins perform?

41. Substitutes have been developed for fat ("fake fats" such as Olean) and sugar (sucralose and aspartame, for example). Why have there not been attempts to develop a comparable substitute for protein?

42. In people who exhibit lactose intolerance, the enzyme lactase that normally catalyzes the lactose breakdown is either missing or is present at levels too low to support normal enzymatic activity. How does this inability to break down lactose parallel our ability to metabolize starch, but not cellulose?

43. Here is information about the sugar content of different foods.

Food Product	Sugar	Calories	Serving Size
Altoids, peppermint	2 g	10	3 pieces (2 g)
Ginger snaps	9 g	120	4 cookies (28 g)
Critic's Choice Tomato Ketchup	3 g	15	1 tbsp (13 g)
Del Monte Pineapple Cup	13 g	50	Individual cup (113 g)
Dr Pepper soft drink	40 g	150	1.5 cups
French Vanilla Coffee Mate	5 g	40	1 tbsp (15 mL)
Hostess Twinkies	14 g	150	1.5 oz
LifeSavers, WintOGreen	15 g	60	4 mints (16 g)
Tropicana Home Style Orange Juice	22 g	110	8 oz (1 cup)
Snickers bar	29 g	200	2.1 oz
Sunkist orange soda	52 g	190	1.5 cups
Wheatables crackers	4 g	130	13 crackers (29 g)

a. Examine this list. Which item has the highest ratio of grams of sugar to the number of Calories (g sugar/ Cal) in one serving?

b. Does the sugar content of any of these foods surprise you? Explain your response.

c. Do you expect that the specific sugars in Dr Pepper are the same sugars found in Sunkist orange soda? In cranberry juice? In the pineapple cup? Why or why not?

d. The complete label for WintOGreen Lifesavers shows 16 g of total carbohydrates per serving, 15 g of which is sugars. What type of compounds do you think accounts for the other 1 g of carbohydrates?

44. Here is the label information from a popular brand of canned chicken noodle soup.

Serving Size: 1/2 cup (4 oz; 120 g)

Servings per container: about 2.5

Amount per serving

Calories 75   Calories from Fat 25

	Amount	% Daily Value
Total fat	2.5 g	4
Saturated fat	1.5 g	8
Cholesterol	20 mg	7
Sodium	970 mg	40
Total carbohydrates	9 g	3
Dietary fiber	1 g	4
Sugars	1 g	
Protein	4 g	
Vitamin A		15
Vitamin C		2
Calcium		2
Iron		4

a. Analyze this information to see if the soup conforms to *Dietary Guidelines for Americans, 2005*.

b. Is the serving size recommended on the label a reasonable size for you? Explain.

c. What effect would changing the serving size have on your answer to part **a**?

45. When the USDA makes a decision to change dietary recommendations, it also changes the way the information is visually displayed to consumers. The earlier pie chart was replaced by a food pyramid, and now by a more symbolic pyramid and an interactive Web site. What are the advantages and disadvantages of each approach?

46. According to one USDA study, nearly 40% of the food that the average American eats each day consists of milk or dairy products. Would such a diet be possible and still meet the *Dietary Guidelines for Americans, 2005*?

47. Consider this structure for one form of vitamin K. Do you expect it to be water-soluble or lipid-soluble? Explain.

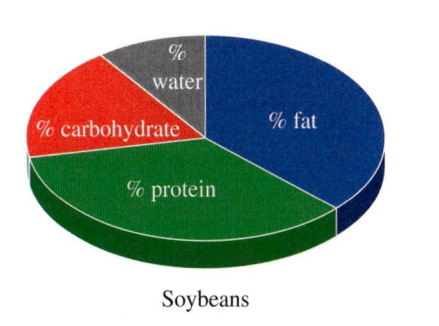

48. Consider the structure for riboflavin, one of the B vitamins found in leafy green vegetables, milk, and eggs. Why is it somewhat safer to take large doses of vitamin B than vitamin D?

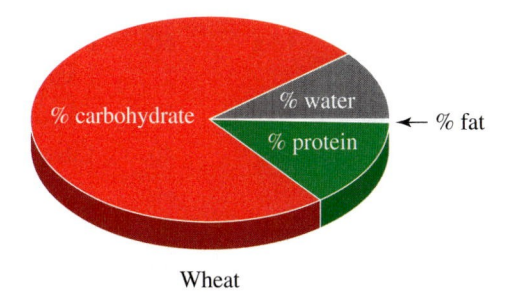

49. American diets depend heavily on bread and other wheat products. A slice of whole wheat bread (36 g) contains approximately 1.5 g of fat (with 0 g saturated fat), 17 g of carbohydrate (with about 1 g of sugar), and 3 g of protein.

  a. Calculate the total Calorie content in a slice of this bread.

  b. Calculate the percent Calories from fat.

  c. Do you consider bread a highly nutritious food? Explain your reasoning.

50. Equation 11.10 and equation 7.10 are the same. Explain the context of each within their respective chapters.

Exploring Extensions

51. The composition of a fast-food meal is given here. Do calculations to determine whether the meal eaten meets the guideline that only 8–10% of total Calories should come from saturated fats.

	Cheeseburger	French Fries	Shake
Calories	330	540	360
Calories from fat	130	230	80
Total fat (g)	14	26	9
Saturated fat (g)	6	4.5	6
Cholesterol (mg)	45	0	40
Sodium (mg)	830	350	250
Carbohydrates (g)	38	68	60
Sugars (g)	7	0	54
Proteins (g)	15	8	11

52. Use the resources of the Web to find the latest scientific studies concerning the effectiveness of some of the so-called fad diets. Is there any new evidence about the ultralow-fat Ornish diet or low-carb Atkins diet? What about the new ultralow-Calorie diets that are reported to increase life span significantly? Report your findings.

53. In September of 2004, the U.S. FDA gave "qualified health claim" status to two omega-3 fatty acids, stating that "supportive but not conclusive research shows that consumption of these omega-3 fatty acids may reduce the risk of coronary heart disease." Find the structure and make a line-angle drawing of any omega-3 fatty acid. What are the food sources for this fatty acid? What is the current thinking about the health benefits of omega-3 fatty acids?

54. How has the proportion of undernourished people in different areas of the world changed over the past 30 years? How have the *total numbers* of undernourished people changed during that time? Focus on any one region of the world and find the necessary information to speak to these two points. Then devise a visual way to represent these data.

55. Compare these two pie charts for the percentage of macronutrients in soybeans and wheat.

Soybeans

Wheat

  a. Use these charts to help explain why the World Health Organization has helped develop several soy-based, rather than wheat-based, food products for distribution in parts of the world where protein deficiency is a major problem.

  b. Suggest some cultural reasons why soy might be preferable to wheat for some areas of the world.

56. The Sceptical Chymist finds the statement that the composition of the human body is ". . . roughly similar to the stuff we stuff into it" an idea hard to believe, but is

willing to try to justify this statement, at least for the macronutrients. Compare the information found in Table 11.1 and Figure 11.2. Does it give you an adequate basis to decide whether the "... roughly similar to the stuff we stuff into it" statement is reasonable, assuming you eat only the foods shown in Table 11.1? Why or why not?

57. How does the elemental composition of the human body compare with the elemental composition of Earth's crust? With the elemental composition of the universe? Table 11.2 gives the values for the human body. Research the composition of Earth's crust and that of the universe, citing your sources. Then comment on the comparative values for the first five elements listed in order of mass abundance in each of the three circumstances—the human body, Earth's crust, and the universe.

58. To use the *Dietary Guidelines for Americans,* 2005, the consumer must know what constitutes a reasonable serving size. Investigate what constitutes reasonable serving sizes for one of the food groups, and then prepare a poster with your results to share with others who are investigating the reasonable serving sizes for other food groups. Were you surprised by any of the serving sizes? Which ones?

59. Not everyone considers milk nature's "perfect food." Compare and contrast the viewpoints of the dairy industry with groups that work against the dairy industry. What are some of the specific benefits attributed to milk, and what are some of the reasons that milk has been called "nature's not so perfect food"?

60. Every month, a certain consumer advocate organization presents an "Unnatural Living Award" to a person, product, or institution that demonstrates an unnatural ability to provide an unnatural product to the American people. What are the criteria by which you would make your nomination for this award? What do you consider would be a good candidate to receive this award? Explain your reasons for suggesting this candidate.

Chapter 12

Genetic Engineering and the Molecules of Life

"No branch of science has created more acute or more subtle and interesting ethical dilemmas than genetics. . . . it is genetics that makes us recall, not simply our responsibilities to the world and to one another, but our responsibilities for how people will be in the future. For the first time we can begin to determine not simply who will live and who will die, but what all those in the future will be like."

Justine Burley and John Harris
Editors, *Companion to Genethics*

A young couple is waiting for their doctor to join them in his office. While they wait they excitedly go over a handwritten list. Boy . . . brown hair . . . blue eyes . . . screened for every disease imaginable. Nine months after the procedure they can expect to deliver a healthy new baby, engineered to their specifications. Is this science fiction or reality? Can genetic technology re-engineer newborn children to the point where "designer babies" will be built rather than born?

The first complete draft of the human genome was completed in 2000, giving us the molecular set of codes that define how we are put together. The media is flooded with stories about stem cell research, recombinant DNA, new drugs and vaccines engineered via DNA, genetic fingerprinting, and more. Genetic material can be mixed between species leading to transgenic or bioengineered foods. And at the top of all this genetic tinkering, we have cloning. Moral and ethical issues are being brought to the forefront as science advances. As citizens we have an obligation to weigh in; to do so capably will require a fundamental understanding of the science involved.

Environmental author Bill McKibben, in *The End of Nature,* opines: "We are on the verge of crossing the line from born to made, from created to built. Sometime in the next few years, a scientist will reprogram a human egg or sperm cell, spawning a genetic change that could be passed down into eternity." To keep society from blindly stumbling toward the future, we present the chemistry of life. Let us take a look at how close we really are to designer babies.

Consider This 12.1	The Process of DNA Discovery

The "discovery" of DNA in 1953 really meant that its helical structure was elucidated. Use the Web to create a timeline that shows how our understanding of the structure and function of DNA developed. The benchmarks are often recognized by awards such as the Nobel Prizes in physiology or medicine, and chemistry.

12.1 The Chemistry of Heredity

Each second the human body hosts millions of chemical reactions. Some compounds are decomposed and others are synthesized; energy is released, transformed, and used; chemical signals are transferred and processed. But, in spite of the dazzling complexity of these processes, the last half-century has seen a phenomenal increase in our knowledge of the chemistry of life. A great deal of current biological research has refocused on molecules rather than on cells or organisms. This "molecular" research has led to an understanding of the very basis of life itself.

In 1900, average life expectancy at birth in the United States was less than 50 years; today it is almost 78 years (Figure 12.1). Reasons for this dramatic increase include better nutrition, improved sanitation, advances in public health, more accurate medical diagnoses, new medical procedures, and numerous new medicines and vaccines. Chemistry has contributed to all these innovations. Biotechnology and molecular engineering are integral parts of the latest revolution in health care. There seems little doubt that genetic engineering will profoundly affect human life in the 21st century.

In 1900, the life expectancies for men and women were only 46 and 48, respectively.

Your Turn 12.2	Increases in Life Expectancy

a. Use Figure 12.1 to determine the percent increase in average human life expectancy at birth for women and for men in the United States between 1950 and 2004.

b. Speculate about why the average life expectancy of women and men differs.

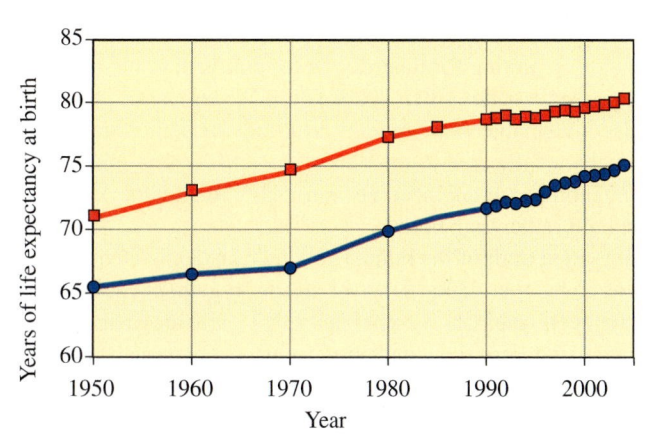

Figure 12.1

Average life expectancy at birth in the United States since 1950 for females (red) and males (blue).

Source: Centers for Disease Control and Prevention, National Center for Health Statistics, http://www.cdc.gov/nchs/fastats/lifexpec.htm

Human red blood cells do not have nuclei.

To begin from a personal perspective, you contain about 100 trillion cells that have a nucleus. Each of these cell nuclei contains a complete set of the genetic instructions that make you what you are—at least biologically. This information is organized into **chromosomes,** 46 self-replicating, rod-shaped strands of DNA and associated proteins found in the nucleus of cells that contain the hereditary information necessary for life. Within the chromosomes, information is also carried by approximately 30,000 **genes,** short pieces of DNA that code for the production of proteins, giving an organism its particular inherited characteristics.

Your special template of life is written in a molecular code on a tightly coiled thread, one invisible to the unaided eye. This thread is **deoxyribonucleic acid (DNA),** the molecule that carries genetic information in all species. Unraveled, the DNA in *each* of your cells is about 2 m long. If all of the DNA in all 100 trillion of your cells were placed end to end, the resulting ribbon would stretch from here to the Sun and back more than 600 times! But as you will soon discover, this astronomical figure is far from the most astounding feature of this amazing molecule.

Sceptical Chymist 12.3   Stretching DNA

Sometimes authors get carried away with their rhetoric. Check the correctness of the claim that the DNA in an adult human being would stretch from Earth to the Sun over 600 times.

Hint: It is approximately 93 million miles from Earth to the Sun. Other necessary information is in the preceding paragraphs; unit conversion factors are in Appendix 1.

The molecular structure of deoxyribonucleic acid dictates how it encodes genetic information. A strand of DNA consists of fundamental chemical units, repeated thousands of times. Each of the units is composed of three parts: a nitrogen-containing base, the sugar deoxyribose, and a phosphate group. All are illustrated in Figure 12.2.

Two of the bases, adenine (symbolized by **A**) and guanine (**G**), are made of six- and five-membered rings fused together. Cytosine (**C**) and thymine (**T**) are each made of a six-membered ring of carbon and nitrogen atoms. Notice that all of these compounds have nitrogen atoms imbedded in their rings, leading to the name "nitrogen-containing bases." A quick review reminds us that bases are defined as substances that either produce

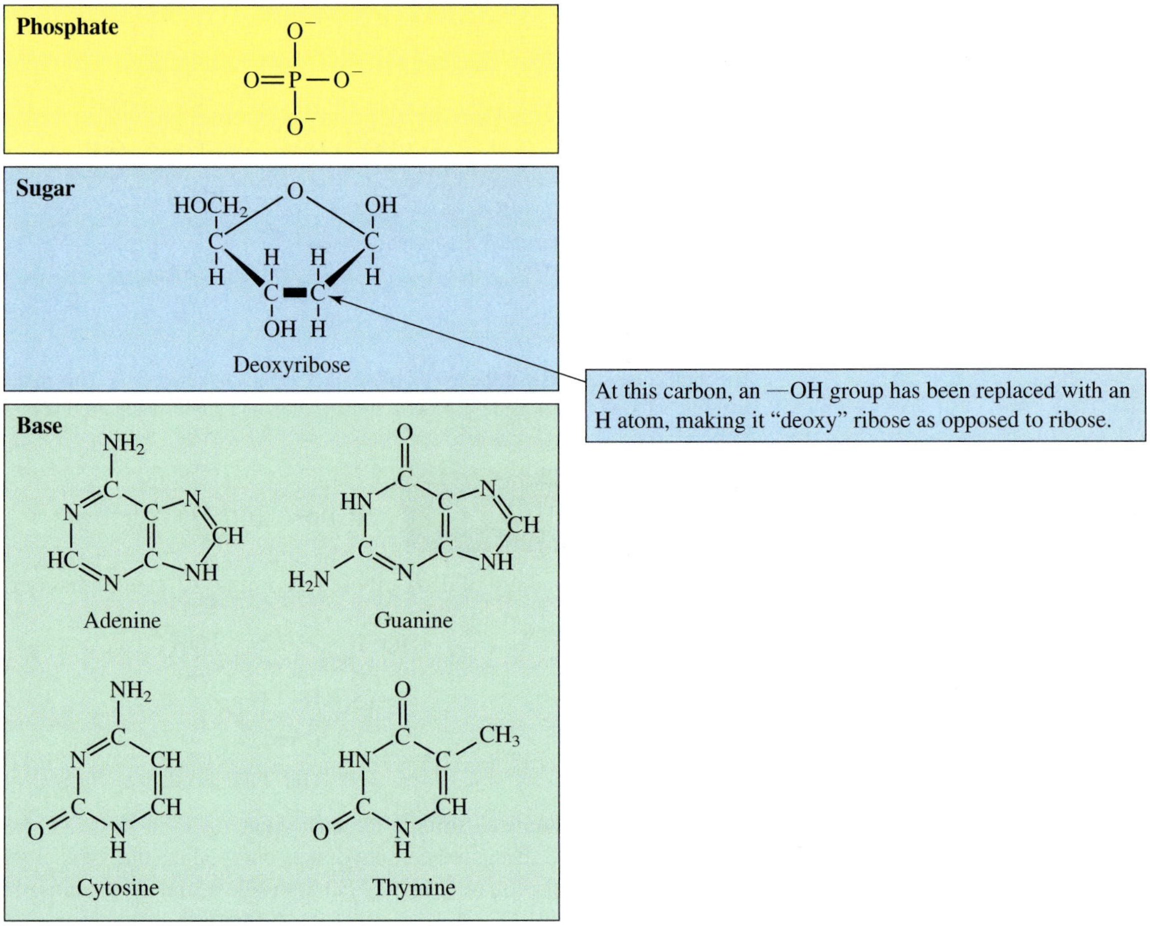

Figure 12.2

The components of deoxyribonucleic acid, DNA.

Figures Alive! Visit the *Online Learning Center* to learn more about the structure of DNA.

OH⁻ ions, or accept H⁺ ions in chemical reactions. Consider what happens when cytosine is treated with an acid (represented as H⁺).

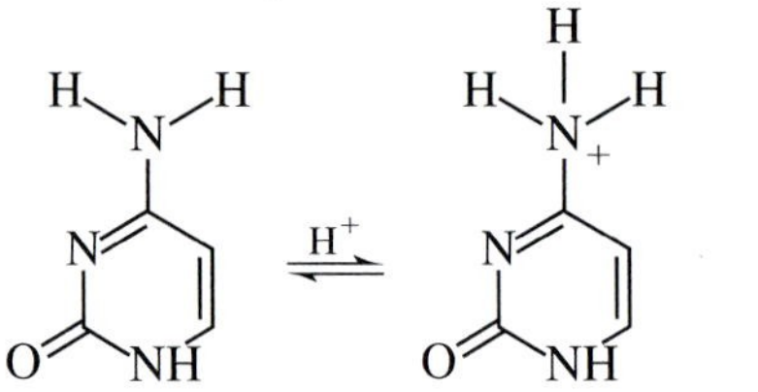

$$[12.1]$$

There is a lone pair of electrons on the nitrogen atom outside of the ring that can form a bond with the hydrogen ion. The nitrogen accepts the H^+, resulting in the ion shown in equation 12.1. The double arrows used in equation 12.1 indicate that the reaction does not go to completion, or that less than 100% of the reactant is converted into product.

Deoxyribose is a monosaccharide (a "single" sugar) with the formula $C_5H_{10}O_4$ (see Figure 12.2). The "deoxy" of deoxyribose means that a hydroxyl (—OH) group in ribose has been replaced by a hydrogen atom. That replacement in the sugar structure occurs at the ring carbon with two hydrogen atoms (indicated on the second structure from the top in Figure 12.2). The third part of a DNA strand is the phosphate group. It is often represented as PO_4^{3-}, but depending on the pH, H^+ can be attached to one or more of the O atoms. The form of phosphate in which each of three oxygen atoms has a proton attached is H_3PO_4, phosphoric acid. The ionizable hydrogen atoms on the phosphate groups are what make nucleic acids acidic.

It is the phosphate groups that make the DNA molecule acidic.

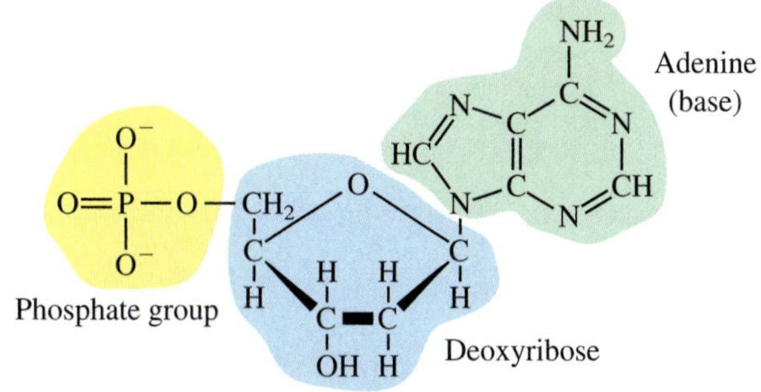

Figure 12.3

The molecular structure of the nucleotide called adenosine phosphate. This molecule is built from adenine, a base (green), deoxyribose, a sugar (blue), and a phosphate group (yellow).

Consider This 12.4   Ribose and Deoxyribose

Compare the structure of deoxyribose (see Figure 12.2) with that of ribose:

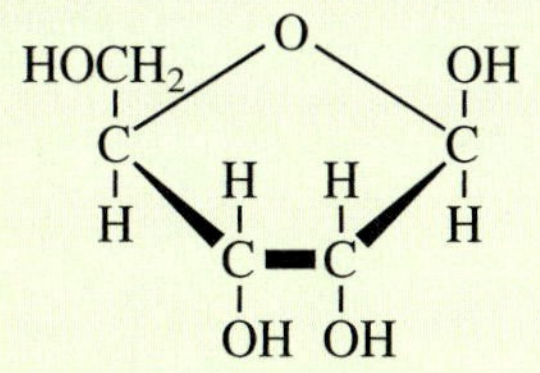

a. Give the chemical formula for each sugar.
b. Section 11.3 discussed carbohydrates, compounds that typically have the chemical formula of $C_nH_{2n}O_n$. Do both ribose and deoxyribose follow this pattern?
c. In these two sugar molecules, which atoms have lone pairs of electrons?
d. Which atoms can be involved in hydrogen bonds?
   Hint: See Section 5.6 for a review of hydrogen bonding.

A **nucleotide** is a combination of a base, a deoxyribose molecule, and a phosphate group. Figure 12.3 indicates how these units are linked in a specific nucleotide called adenosine phosphate. The three units are covalently bonded together; similar nucleotides can be formed using any of the other three bases shown in Figure 12.2.

A typical DNA molecule consists of thousands of nucleotides covalently bonded to form a long chain. Consequently, a single strand of DNA may have a molecular mass in the millions. Figure 12.3 shows that one —OH group on the deoxyribose ring remains unreacted. The phosphate group of another nucleotide can react with this —OH group, forming and eliminating a H_2O molecule, thereby connecting the two nucleotides. This reaction is an example of a condensation polymerization; a polymer is formed when more and more nucleotides are joined, each time splitting out a water molecule. Figure 12.4 shows four nucleotides that have been linked in this manner to form a segment of DNA. The schematic drawing in the inset of Figure 12.4 shows the polymeric nature of DNA where the monomers are the nucleotides.

See Sections 9.5 and 9.6 for more examples of condensation polymerization.

Your Turn 12.5   Nucleotides

Use Figures 12.2 and 12.3 to aid in drawing the structures of nucleotides containing guanine, cytosine, and thymine. Which of the atoms possess nonbonding electron pairs? Use another source to name the nucleotides you have drawn.

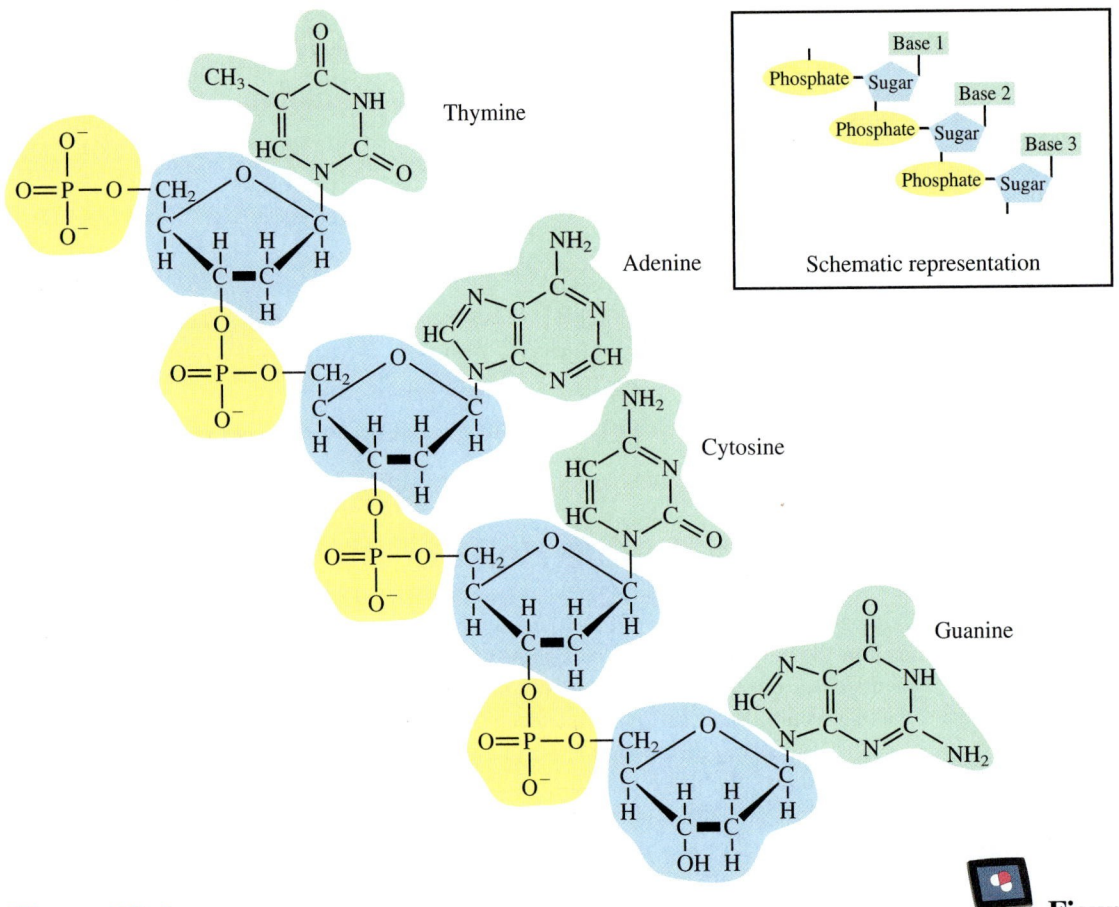

Figure 12.4

A segment of DNA. A phosphate group connects one deoxyribose to an adjacent one. Each of the four bases, thymine (T), adenine (A), cytosine (C), or guanine (G), is attached to a deoxyribose sugar.

Figures Alive! Visit the *Online Learning Center* to learn more about the four bases and the sugar-phosphate backbone.

The specific bases and their sequence in a strand of DNA turn out to have great significance; all living species have the same four bases in their DNA, but the percentages vary. Some of the early clues to the structure of DNA and the mechanism by which it conveys genetic information came as a result of the research of Erwin Chargaff in the 1940s and 1950s. Chargaff and his coworkers were able to determine the percentage of the four bases present in DNA from a variety of species. They found that the relative amounts of the bases in a DNA sample are identical for all members of the same species. Moreover, these percentages are independent of the age, nutritional state, or environment of the organism studied. For example, according to Chargaff's data, the DNA from all members of our species, *Homo sapiens*, contains 31.0% adenine, 31.5% thymine, 19.1% guanine, and 18.4% cytosine. Table 12.1 contains the percent base compositions for the DNA of other species.

A more careful examination of Table 12.1 discloses that the DNA of *Homo sapiens* and of *Escherichia coli* (*E. coli*), a species of bacteria that inhabits our intestines, share an important characteristic. Their base composition obeys a common principle, now called **Chargaff's rules.** In every species, the percent of adenine almost exactly equals the percent of thymine. Similarly, the percent of guanine is essentially identical to the percent of cytosine. Put more simply: %A = %T and %G = %C. Such a correlation can hardly be coincidental. As soon as Chargaff's rules were communicated, the conclusion seemed obvious: The nitrogen-containing DNA bases come in pairs. Adenine always appears to be associated with thymine, and guanine with cytosine.

All living species have the same four bases in their DNA; but the percentage of each base varies from species to species.

Table 12.1	The Percent Base Compositions of DNA for Various Species				
Species	**Common Name**	**Adenine**	**Thymine**	**Guanine**	**Cytosine**
Homo sapiens	human	31.0	31.5	19.1	18.4
Drosophila melanogaster	fruit fly	27.3	27.6	22.5	22.5
Zea mays	corn	25.6	25.3	24.5	24.6
Neurospora crassa	mold	23.0	23.3	27.1	26.6
Escherichia coli	bacterium	24.6	24.3	25.5	25.6
Bacillus subtilis	bacterium	28.4	29.0	21.0	21.6

Note that the percentages of adenine and thymine are consistently similar, as are the percentages of cytosine and guanine.

Source: From I. Edward Alcamo, *DNA Technology: The Awesome Skill,* Second Edition. © 2000 The McGraw-Hill Companies, Inc. All rights reserved. Reprinted with permission.

12.2   The Double Helix of DNA

Although Chargaff's rules of base pairs represented a key breakthrough, it was not obvious how the paired bases were part of the overall molecular structure of DNA. Therefore, scientists set out to determine the way in which nucleotides were incorporated into the DNA molecule. X-ray diffraction, a technique known since early in the 20th century was highly useful. **X-ray diffraction** is a crystallographic technique that generates a pattern of deflected X-rays that have passed through a crystal in order to reveal the nature of the crystal lattice. The X-ray photons interact with the electrons of the atoms in the crystal and are diffracted, or scattered. The crucial point is that the X-rays are only scattered at certain angles that are related to the distance between atoms, and that information can be used to determine the structures of a wide variety of crystalline materials. The X-ray diffraction pattern of a DNA fiber was obtained in late 1952 by the British crystallographer Rosalind Franklin (Figure 12.5).

Two scientists, James D. Watson, a young American, and Francis H. C. Crick, a Cambridge University biophysicist, combined a collection of data, including Franklin's X-ray information. Watson and Crick (Figure 12.6) concluded that a pattern in Franklin's

(a)

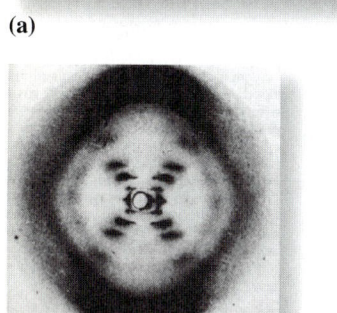

(b)

Figure 12.5

(**a**) Rosalind Franklin, whose work contributed significantly to elucidating the DNA structure.
(**b**) Franklin's X-ray diffraction photo of a hydrated DNA fiber. The cross in the center is indicative of a helical structure, and the darkened arcs at the top and bottom are due to the stack of base pairs.

Figure 12.6

James Watson and Francis H. C. Crick, awarded the 1962 Nobel Prize in physiology or medicine for the discovery of the double-helical structure of DNA. A third scientist, Maurice Wilkins, also shared in the 1962 Prize for his contributions made while working at King's College, London.

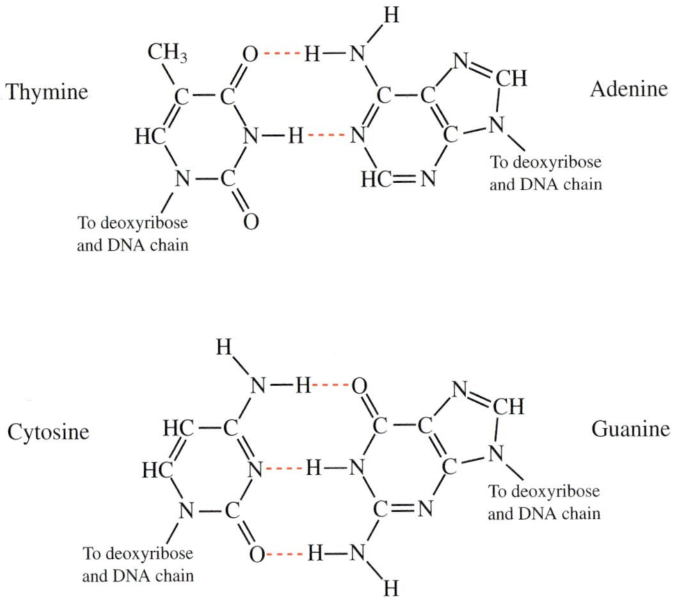

Figure 12.7

Base pairing of adenine with thymine and cytosine with guanine in DNA. Chemical bonds are solid lines, and the hydrogen bonds are dashed red lines.

$1 \text{ nm} = 1 \times 10^{-9} \text{m}$

diffraction photograph was consistent with a repeating helical arrangement of atoms, similar to a loosely coiled spring. Moreover, the X-ray photographs contained evidence of a regular repeat distance of 0.34 nm within a DNA molecule.

Crick and Watson created a structural model and found that the adenine and thymine portions of the molecule fit together almost perfectly, like pieces in a jigsaw puzzle. Moreover, these two bases can be linked by two hydrogen bonds (Figure 12.7). Similarly, cytosine and guanine link by forming three hydrogen bonds. Adenine and thymine are said to be complementary bases, as are cytosine and guanine, since they are capable of forming a hydrogen-bonded base pair. This base-pairing is the molecular basis underlying Chargaff's rules: A pairs with T and C pairs with G.

In the model of DNA developed by Watson and Crick, the hydrogen bonds (along with other weaker forces) between the complementary bases hold together two poly-nucleotide strands. The strands adopt the configuration of a **double helix**, a spiral consisting of two strands that coil around a central axis (Figure 12.8). Watson and Crick concluded that the base pairs are parallel to each other, perpendicular to the axis of the DNA molecule, and separated by 0.34 nm, the repeat distance calculated from the diffraction pattern. In addition, Franklin's results also suggested another repeat distance of 3.4 nm. Watson and Crick took this to be the length of a complete helical turn consisting of 10 base pairs.

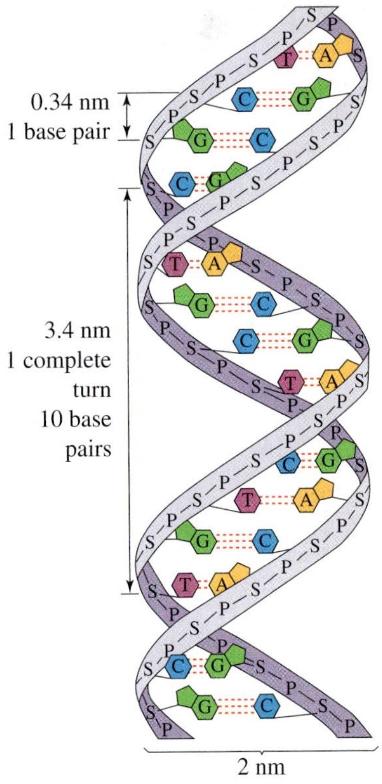

0.34 nm
1 base pair

3.4 nm
1 complete turn
10 base pairs

2 nm

Figure 12.8

A model of DNA with P = phosphate group; S = the sugar, deoxyribose; A = adenine; T = thymine; C = cytosine; G = guanine. The sugar and phosphate groups alternate on the two twisting ribbons. The four bases attach to this backbone and are paired A to T, C to G.

Your Turn 12.6   Complementary Base Sequences

Identify the base sequences that are complementary to each of these.

a. ATACCTGC   **b.** GATCCTA

Answers
a. TATGGACG   **b.** CTAGGAT

Your Turn 12.7   Is Your DNA Doin' the Twist?

The distance between base pairs in a molecule of DNA is 0.34 nm.

a. Calculate the length (in centimeters) of the shortest human chromosome, which consists of 50,000,000 base pairs.

b. Mark off that length on your paper. If this is the length of the unstretched DNA molecule, what does this imply about the organization of DNA in the chromosome?

Answer

a. $\dfrac{0.34 \text{ nm}}{1 \text{ base pair}} \times \dfrac{1 \text{ m}}{1 \times 10^9 \text{ nm}} \times \dfrac{1 \times 10^2 \text{ cm}}{1 \text{ m}} \times \dfrac{5 \times 10^7 \text{ base pairs}}{1 \text{ chromosome}}$

$$= \dfrac{1.7 \text{ cm}}{1 \text{ chromosome}}$$

b. This is a small distance; 1.7 cm is about two thirds of an inch. The best way that 50 million base pairs could fit into such a small space would be to have them tightly packed in a spiral or folded.

Your Turn 12.8   The Length of DNA

The DNA in each human cell consists of 3 billion base pairs. Calculate the length of this DNA. Does this length agree with that given in Section 12.1? Explain.

Watson and Crick's research paper, "Molecular Structure of Nucleic Acids: A Structure for Deoxyribose Nucleic Acid," appeared in the scientific journal *Nature* on April 24, 1953. It is only one page long and written with the customary passionless detachment of contemporary scientific prose. Even the most significant statement in the communication is delivered with typical British understatement: "It has not escaped our notice that the specific pairing we have postulated immediately suggests a possible copying mechanism for the genetic material."

The history of science is full of examples of how a single discovery can release a flood of related research. So it was with the discovery of the structure of DNA. Scientists immediately set out to discover the molecular details of how DNA is replicated, how it encodes genetic information, and how that information is translated into physiological characteristics. **Replication** is the process of cell reproduction where the cell must copy and transmit its genetic information to its progeny. The process is well understood and is diagrammed in Figure 12.9.

Consider This 12.9   DNA Replication

Figures Alive! (at the *Online Learning Center*) offers you interactive activities to explore the features of the DNA molecule. The last of these is an animation of DNA replication. We invite you to critique it.

a. Examine the drawing used to represent DNA, one that shows no atoms. List its strengths and weaknesses.

b. Play and replay the animation of DNA replication to see what it shows and what it fails to show. Make a list of both.

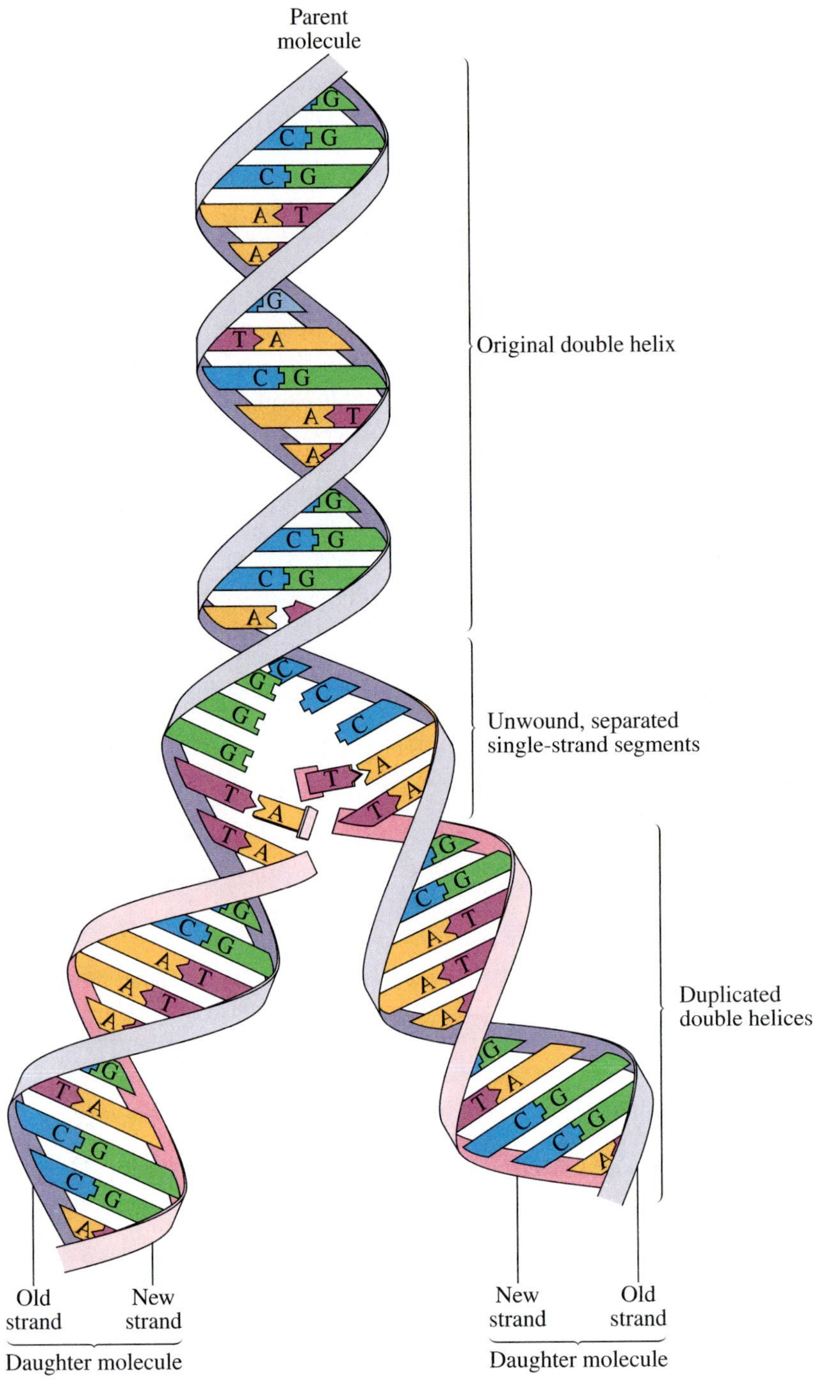

Parent
molecule

Original double helix

Unwound, separated
single-strand segments

Duplicated
double helices

Old	New		New	Old
strand	strand		strand	strand
Daughter molecule			Daughter molecule	

Figure 12.9

Diagram of DNA replication. The original DNA double helix (*top portion of figure*) partially unwinds, and the two complementary portions separate (*middle*). Each of the strands serves as a template for the synthesis of a complementary strand (*bottom*). The result is two complete and identical DNA molecules.

Before a cell divides, the double helix rapidly but only partially unwinds. This results in a region of separated strands of DNA, as pictured in the middle portion of Figure 12.9. Individual nucleotides in the cell are selectively hydrogen-bonded to these two single strands that serve as templates for a new DNA molecule: A to T, T to A, C to G, and G to C. Held in these positions, the nucleotides are bonded together by the action of an enzyme. Every minute, about 90,000 nucleotides are added to the growing chain. By this mechanism, each strand of the original DNA generates a complementary copy of itself. The original template strand and its newly synthesized complement coil about each other

to form a new double helix—a daughter molecule identical to the first. Similarly, the other separated strand of the original molecule twines around its new partner, forming another new daughter molecule. Thus, where there was originally only one double helix, there are now two (see Figure 12.9). As the nucleus splits and the cell divides into two daughter cells, one complete set of chromosomes is incorporated into each. This process is repeated again and again, so that each of the trillions of cells in a newborn baby contains all the genetic information first assembled from parental DNA when the sperm combined with the ovum.

12.3   Cracking the Chemical Code

The discovery of the molecular code for genetic information arguably is history's most amazing example of cryptography, the science of writing in secret code. Key to the code is the sequence of bases in DNA and the order of amino acids in a protein. The 3 billion base pairs repeated in every human cell provide the blueprint for producing one human being. Although these specifications are carried in DNA, they are expressed in proteins. Proteins are everywhere in the body: in skin, muscle, hair, blood, and the thousands of enzymes that regulate the chemistry of life. It follows that, by directing the synthesis of proteins, DNA can dictate the characteristics of the organism.

Proteins are large molecules formed by the combination of amino acids. The 20 amino acids that commonly occur in proteins can be represented by this general structural formula.

For more information about amino acids see Figure 11.12 and Section 11.4.

The amine or amino group is $-NH_2$, the acid group is $-COOH$, and R represents a side chain that is different for each of the 20 amino acids. In a condensation reaction, the $-COOH$ group of one amino acid reacts with the $-NH_2$ group of another. In this process a peptide bond is formed and a molecule of H_2O is formed and eliminated, or "split out." When many amino acids are connected, the result is a **protein, a long chain of amino acid residues.** *Amino acid residue* is the term used for amino acids that have lost an $-H$ and an $-OH$ in the process of joining together.

The biochemists who set out to decipher the genetic code correctly assumed that somehow the order of bases in DNA determines the order of amino acids in a protein. Furthermore, it seemed obvious that the code could not be a simple one-to-one correlation between bases and amino acids. There are only four bases in DNA. If each base corresponded to an individual amino acid, DNA could encode for only four amino acids. But 20 amino acids appear in our proteins. Therefore, the DNA code must consist of at least 20 distinct code "words," each word representing a different amino acid. And the words must be made up of only four letters—A, T, C, and G—or, more accurately, the bases corresponding to those letters.

Some simple statistics can help us determine the minimum length of these code words. To find out how many words of a given length can be made from an alphabet of known size, one raises the number of letters available to a power n, corresponding to the number of letters per word.

$$words = (letters)^n$$

Thus, using four letters to make two-letter words generates 4^2, or 16, different two-letter words. Similarly, DNA bases taken in pairs (akin to two letters per word) could code for only 16 amino acids. This vocabulary is too limited to provide a unique representation for each of the 20 amino acids. So we repeat the calculation, this time assuming that the code is based on three sequential base pairs or, if you prefer, three-letter words. Now the number of different triplet-base combinations is 4^3, or $4 \times 4 \times 4 = 64$. This system provides more than enough capacity to do the job.

Your Turn 12.10 Quadruplet-Base Code

Suppose that the DNA code used four sequential base pairs instead of a triplet-base code. How many different four-base sequences would result?

Answer

$4 \times 4 \times 4 \times 4 = 4^4 = 256$ different four-base sequences

Obviously, more than mathematical reasoning was required to prove the molecular basis of genetics. Once again, Francis Crick was a leader in this research. His work clearly established that the genetic code is written in groupings of three DNA nucleotides. Called **codons,** these sequences of three adjacent nucleotides determine the insertion of a specific amino acid during protein synthesis or they signal to start and stop protein synthesis. And today, thanks to other scientists, this trinucleotide code has been cracked, and specific amino acids have been related to particular codons.

No Rosetta Stone was available to aid these scientists in their efforts at translation. Instead, they relied on elegant and imaginative experiments that ultimately yielded a genetic dictionary. If you were to use the letters A, T, C, and G in a game of Scrabble, you could generate 64 different three-letter combinations. A few, CAT, TAG, and ACT, for example, make sense. Most are like AGC, TCT, and GGG and are meaningless—at least in English. Nature does far better than that; 61 of the 64 possible triplet codons specify amino acids. Thus, the codon sequence GTA in a DNA molecule signals that a molecule of the amino acid histidine should be incorporated into the protein, AAA codes for phenylalanine, and GGC stands for proline. The three-base sequences that do not correspond to amino acids are signals to start or stop the synthesis of the protein chain. An example of a nine-base nucleic acid segment and how it codes for three amino acids is shown in Figure 12.10.

The markings on the Rosetta Stone were in several languages and helped to decipher Egyptian hieroglyphics.

Because there are more codons than amino acids, the code has redundancy. Some amino acids have more than one codon. For example, leucine, serine, and arginine have six codons each. On the other hand, tryptophan and methionine each are represented by only a single codon. Significantly, the code is identical in all living things. The instructions to make people, bacteria, and trees are written in the same molecular language.

The amount of information carried by your deoxyribonucleic acid is truly phenomenal. The DNA in each of your cell nuclei consists of approximately 1 billion (1×10^9) triplet codons. You have just read that each triplet is at least potentially capable of encoding 1 of the 20 amino acids found in human protein. If each codon could be assigned a letter of the English alphabet rather than an amino acid, your DNA could encode 1×10^9 letters or about 2×10^8 five-letter words. These words would fill 1000 volumes of 400 pages each.

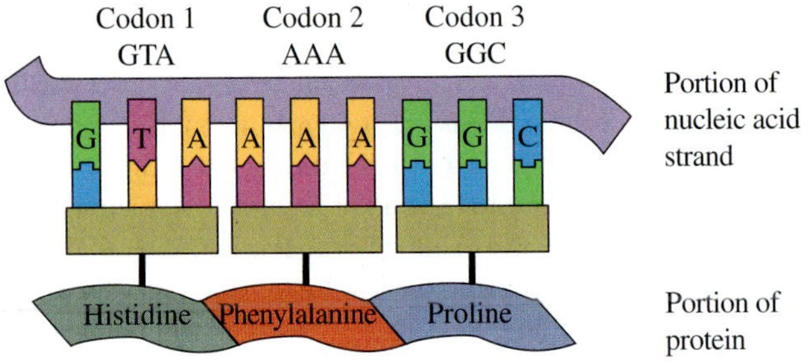

Figure 12.10

A nine-base nucleic acid sequence showing three codons.

Your Turn **12.11**   **Duplicate Codons**

Suggest some advantages of a genetic code in which several codons represent the same amino acid.

Your Turn **12.12**   **DNA Unique to Humans**

The human genome contains about 3 billion base pairs, but only about 2% of this DNA consists of unique genes. The number of genes is estimated at 30,000. Use this information to calculate the average number of base pairs per gene.

Answer

$$\frac{3 \times 10^9 \text{ base pairs}}{1 \text{ human genome}} \times \frac{2 \text{ unique genes}}{100 \text{ base pairs}} \times \frac{1 \text{ human genome}}{3 \times 10^4 \text{ genes}} \times \frac{1 \text{ base pair}}{1 \text{ unique gene}}$$

$$= \frac{2 \times 10^3 \text{ base pairs}}{1 \text{ gene}}$$

12.4   Protein Structure and Activity: Form and Function

The mechanism by which DNA directs protein synthesis is known in great detail, but this discussion lies beyond the scope of this text. There are many good books and Web sites available for the student who wishes to further understand this elegantly complex process. Our discussion will now turn to the functions of proteins and how they are determined largely by their conformation, or three-dimensional shape.

In Chapter 11 the discussion of protein structure was limited to their **primary structure**, the unique identity and sequence of the amino acids that make up each protein (Figure 12.11). But the ordering of amino acid residues and the chemical interactions between them then gives rise to a characteristic conformation of the polypeptide chain. Interactions such as hydrogen bonding cause the chain to form regular, repeating structures. This intermediate level of organization is referred to as the **secondary structure**, or the periodic, localized arrangement of the backbone segments of a protein chain. Sometimes the particular bond angles along the polypeptide chain and intramolecular attractions give rise to an alpha helix. Other sequences and interactions lead to what is known as a beta pleated sheet. These primary and secondary structures are shown in Figure 12.11.

An interesting point about the secondary structure of a protein is the geometry around the peptide bond. Experiments show that this part of the molecule is planar and does not have the free rotation around the C-to-N bond as you would expect for a single bond. However, this can be explained when you consider the Lewis structures and their resonance forms. For example, consider the two resonance forms for the dipeptide formed from glycine and alanine (Figure 12.12).

There is no free rotation around the C-to-N bond because it has a partial double bond character, and we know that double bonds are rigid and do not rotate freely. Experiments that corroborate this theory show that the C-to-N bond length of the peptide bond in a protein is 10% shorter than the C-to-N single bond of an amine; the length is in between the length of a C-to-N double and a C-to-N single bond. Furthermore, the C-to-N-to-H bond angles in this region of the molecule are close to 120°, which is what you would predict for a trigonal planar geometry around the nitrogen atom. All of these factors: planar geometry, 120° bond angles, and intermolecular hydrogen bonding lead to the characteristic secondary structure of a protein. Environmental variables such as pH and temperature also influence the shape.

See Section 2.3 for a discussion on resonance forms.

C-to-N Bond Lengths

Single bond
147 pm

Peptide bond
133 pm

Double bond
127 pm

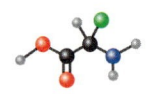

Amino acid monomer

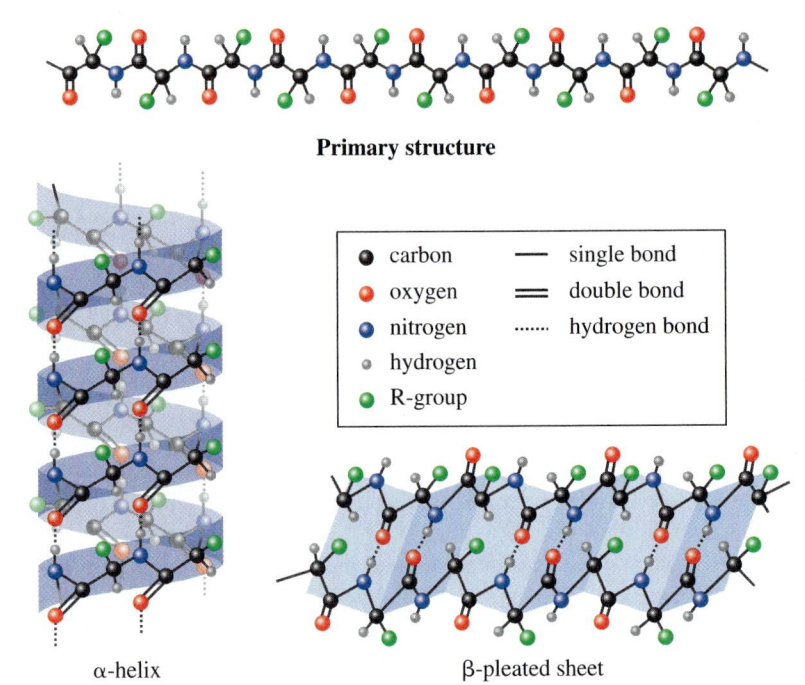

Primary structure

α-helix   β-pleated sheet

Secondary structures

Figure 12.11

Representations of primary and secondary structures of proteins. The two major types of secondary structures are the alpha (α)-helix and the beta (β)-pleated sheet. The beta-pleated sheet shown is a single polypeptide strand that has crossed back along itself, not two individual parallel strands.

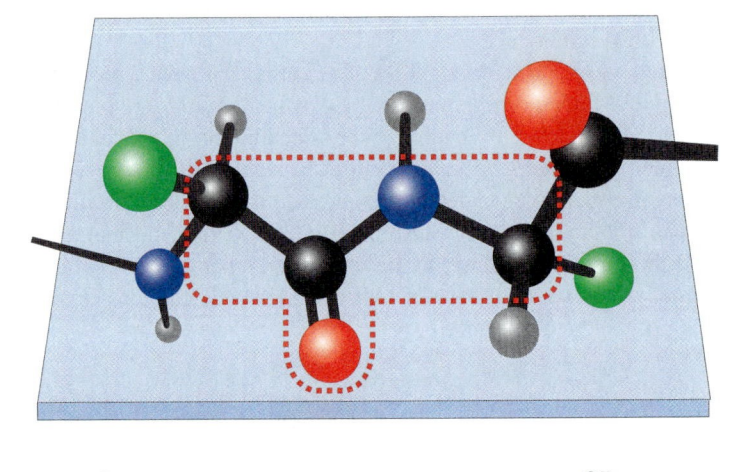

resonance forms

Figure 12.12

Interesting geometric features around the peptide bond contribute to the overall shape of a protein. The concept of resonance can be used to help explain the planar shape of this portion of a protein. Atom colors are the same as in Figure 12.11.

Your Turn   12.13   **Match the Bond Lengths**

Bond lengths for four different C-to-O bonds in carbon monoxide, methanol (CH$_3$OH), carbon dioxide, and the dipeptide formed from glycine and alanine are 143, 111, 121, and 123 pm. Use the information provided in this section in order to match each bond with its correct bond length.

Hint: You may want to draw the Lewis structures for some of these molecules.

Finally, proteins are typically very large molecules, and we would like to have a more "global" description of their shapes. To understand the **tertiary structure,** or the overall shape or conformation of the protein, consider a telephone cord as representing the secondary structure. It has a helical structure. But how often does a typical telephone cord look nice and orderly as the one in Figure 12.13a? Hardly ever! The typical phone cord is all folded up on itself in loops and twists, as shown in Figure 12.13b. The carefully ordered backbone of the polypeptide chain with its secondary folds has an overall topology that is the tertiary structure.

The overall fold is the result of a net increase in stability, and the shape is maintained through hydrogen bonds, intermolecular ionic and covalent bonds, and interactions of the amino acid residues with water.

Most protein enzymes act as catalysts, and their function is related to the overall structure and placement of different functional groups. For an enzyme to carry out its chemistry, functional groups on certain amino acid residues must come close enough to form an active site. The **active site** is the region of the enzyme molecule where its catalytic effect occurs (Figure 12.14). Sometimes, the amino acids involved are adjacent; in other cases, they are widely separated in the protein chain, but close together in the tertiary structure. Figure 12.14 shows the active site of chymotrypsin. Its active site consists of three amino acids that would be far apart if the protein were unwound. These groups help to latch onto the **substrate,** the molecule (or molecules) that is being acted upon by the catalytic enzyme. Some enzymes catalyze the breaking of bonds in the substrate, and other enzymes will have an active site that promotes the formation of chemical bonds. In all cases, the orientation of the active site and the conformation of the rest of the enzyme molecule are of critical importance.

The lock-and-key model described in Section 10.5 can explain the mode of action of many enzymes.

A subtle change in the primary structure of a protein can have a profound effect on its properties. A much-studied example is provided by hemoglobin, the blood protein that transports oxygen. Hemoglobin is involved with a condition called sickle-cell anemia. When an individual with a genetic tendency toward sickle-cell disease is subjected to conditions that involve high oxygen demand, some red blood cells distort into rigid sickle or crescent shapes (Figure 12.15). Because these cells lose their normal deformability, they cannot pass through tiny openings in the spleen and other organs. Some of the sickled cells are destroyed and anemia results. Other sickled cells can clog organs so badly that the blood supply to them is reduced.

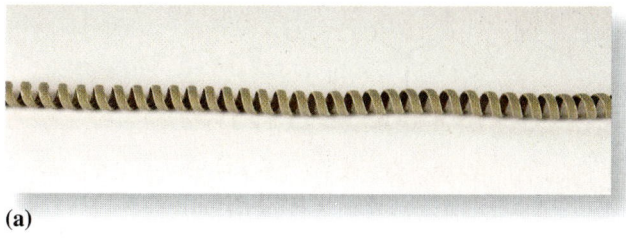
(a)

(b)

Figure 12.13

Using a telephone cord to model the secondary (**a**) and tertiary (**b**) structure of proteins.

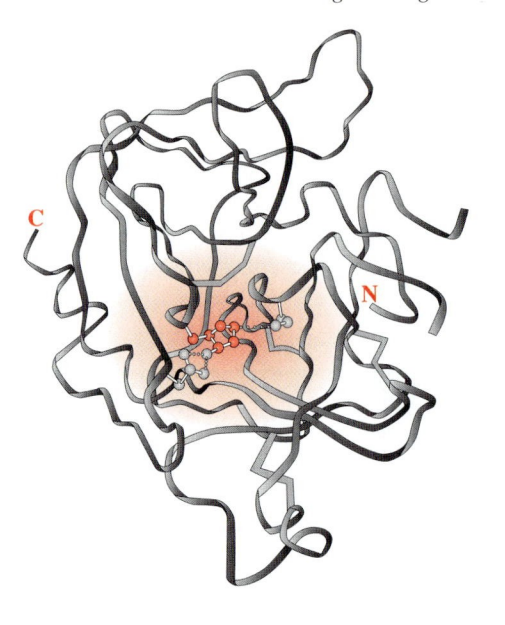

Figure 12.14

Tertiary structure of the enzyme, chymotrypsin. The "ribbon" portion represents the amino acid chain; the central colored portion is the active site at which the enzymatic chemistry takes place. One end is marked C for the carboxylic acid group; another N for the amine group.

The property of sickling has been traced to a minor change in the amino acid composition of human hemoglobin. A hemoglobin molecule has 574 amino acid residues. The only difference between normal hemoglobin and hemoglobin S (sickle-cell disease affected) in persons with the sickle-cell trait is in two of these amino acids. In hemoglobin S, two of the residues that should be glutamic acid are replaced with valine. Apparently this substitution causes the hemoglobin to convert to the abnormal form at low oxygen concentration.

Sickle-cell anemia is hereditary; the error in the amino acid sequence reflects a corresponding error in a DNA codon. Normally, mutations detrimental to a species are eliminated by natural selection. Perhaps the sickle-cell trait has survived because it may also convey some benefit. A clue to what the benefit might be comes from studying the carriers of the gene for hemoglobin S. The gene is most common in people native to Africa and other tropical and subtropical regions and in their descendants. The fact that these are also areas with the highest incidence of malaria has led to speculation that an individual whose hemoglobin has a tendency to sickle may be protected against malaria. Specific mechanisms have been proposed to account for this protection. If the hypothesis is correct, it is an interesting example of how a genetic trait that originally had survival advantage can become a detriment in a different environment. Of course, the fact that sickle-cell anemia is a genetic disease at least raises the possibility that genetic engineering may some day eliminate it.

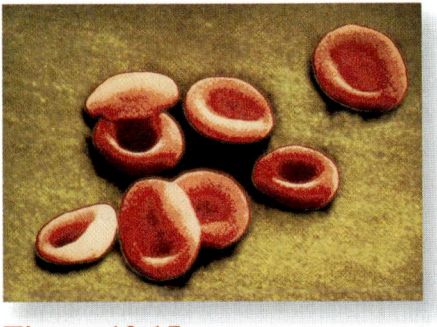 

Figure 12.15

Scanning electron micrographs of normal red blood cells (*left*) and red blood cells showing the effect of sickle-cell disease (*right*).

Scientists who want to know the order of amino acids in proteins such as hemoglobin and hemoglobin S have two alternatives. They can take samples of each protein and painstakingly determine the order of the amino acids. This process often involves using enzymes to clip the whole protein into smaller pieces and then removing one amino acid at a time from one end of each piece and identifying it. The second means of determining amino acid order, another brute force approach, is to determine the sequence of the bases in DNA that code for that protein. Although not intended to address the sequence in a specific protein, the following section describes how the sequence of bases for all genetic material has been mapped.

Consider This 12.14   **Function Follows Form**

As we just mentioned, two of the hemoglobin residues that should be glutamic acid in healthy cells are replaced with valine in sickled cells.

valine   glutamic acid

a. Describe how the molecular structures of valine and glutamic acid differ.
b. Predict the different solubilities for these molecules.
c. Explain how these differences could give rise to the deformed cells typified by sickle cell anemia.

12.5   The Human Genome Project

The **Human Genome Project** was an international effort to map all the genes in the human organism. After more than a decade of research, on June 26, 2000, scientists announced that a rough draft of the project to decode the genetic makeup of humans had been completed. The goal, to determine the sequence of all 3 billion base pairs in the entire genome, was completed for the approximately 30,000 genes found on the 46 human chromosomes. The Human Genome Project was supported by the U.S. National Institutes of Health (NIH), the Wellcome Trust (a philanthropic organization based in London), and by Celera Genomics, a private company in Maryland. On the completion of this first phase, this massive enterprise was described as ". . . the most important, wondrous map ever produced by humankind . . ." by then-President Bill Clinton and as ". . . the outstanding achievement not only of our lifetime but perhaps in the history of mankind . . . " by Dr. Michael Dexter of the Wellcome Trust. Nonetheless, it will take many years to discover the traits the genome conveys.

The Human Genome Project began in 1989, with James Watson of DNA fame as its first director. At an estimated $3 billion, the project cost about $1 per base pair. Instrumental in accomplishing so much so quickly was the high level of cooperation among international teams of researchers, and the competition to be the first to complete the project between publicly funded efforts (NIH) and corporate research (Celera).

The DNA analyzed in the Human Genome Project came from members of over 60 multigenerational French families whose lineage is well documented. But it does not really matter; any one of us could have served as a DNA donor and a representative of *Homo sapiens*. In spite of our apparent differences and our long history of disputes based

on those differences, the DNA of all humans is remarkably similar. It differs from individual to individual by about 0.1% of the base sequences. Within that tiny fraction resides our genetic uniqueness. Biology and chemistry provide irrefutable evidence of a lesson we as nations and as individuals have been slow to learn, a lesson stated by former President Clinton: "The most important fact of life on this Earth is our common humanity."

Determining the sequence of the DNA base pairs was difficult and time-consuming. The smallest human chromosome contains 50,000,000 base pairs. Given those numbers, researchers had to develop automated base sequencers that were both fast and accurate. The accuracy is very important, because a single missed base will throw off all the subsequent base assignments, just as a skipped buttonhole is transmitted down the length of a shirt. By spring 2003, the "finished" draft covered 99% of the genome and left only 300 gaps of the 300,000 base pairs in the rough draft.

Why bother solving this monumental genetic puzzle? The more we know about our genetic makeup, the more likely we will be to diagnose and cure disease, understand human development, trace our evolutionary roots, and recreate our family tree. The information obtained from the project has already helped scientists identify genes that are responsible for leukemia and eczema. There are over 3000 disorders that result from single altered genes, stripping healthy lives from millions of people. Often, little can be done to treat, let alone cure, the majority of these diseases. When scientists know the structure of a gene and are allowed to study the way it causes mutations or alterations that result in disease, they are peering at the very essence of the molecular basis for illnesses. The future might see treatment of genetic disorders by repairing the defective gene itself. And one hopes these discoveries will be utilized for the benefit of our species and others.

Despite the many potential benefits of the Human Genome Project, the enterprise was not without its critics. Some point out that a genetic map represents the ultimate invasion of privacy ("being caught with your genes down"). Information about an individual's genetic makeup might be used by insurance companies to discriminate against those with a hereditary tendency toward certain diseases. The information might be used by businesses to refuse to hire people who may be genetically at risk, or by ruthless governments to identify the "genetically inferior." Research continues to refine the initial data and to produce the genetic map for other species. Comparison of such data across species is likely to generate information about fundamental, biochemical functions in living organisms. The even longer task remains—turning the base pair sequences into useful information regarding the specific details of their cellular functions.

Consider This 12.15   The Human Genome Project

In May 2006, the human DNA sequence, down to the very last chromosome, was published by the Human Genome Project.

a. Explain how the terms *genome* and *chromosome* are related.
Hint: The glossary at the Human Genome Project can help you to construct your answer. A direct link is provided at the *Online Learning Center.*

b. This finished sequence still has minor errors, on the order of 1 per every 10,000 DNA base pairs. Assuming that your DNA contains about 3 billion base pairs, calculate the approximate number of these errors.

c. What lies ahead? List three ways in which the Human Genome Project is expected to shape our lives in the next few decades.

12.6   Genetically Engineered Medical Treatments

Mythology is full of fanciful creatures: the sphinx with the head of a woman and the body of a lion; the griffin, which is half-lion and half-eagle; and the chimera, which combines a lion, a goat, and a serpent (Figure 12.16). In 1973, two American scientists, Herbert

Figure 12.16

Image of a chimera: part lion, goat, and serpent. Recombinant DNA is sometimes referred to as a chimera.

Boyer and Stanley Cohen, created another hybrid. They introduced a gene for manufacturing a protein from the African clawed toad into a common bacterium, *E. coli.* On replication, the *E. coli* bacteria produced the toad's protein. This is an example of the use of **recombinant DNA,** DNA that has incorporated DNA from another organism. Humans have been manipulating the gene pool for thousands of years. We have created mules by crossbreeding horses and donkeys, dogs as diverse as Chihuahuas and Saint Bernards, and fruits and vegetables that never existed in nature. All these were done by selective breeding. Boyer and Cohen created their chemical fantasy in laboratory glassware through the manipulations of genetic engineering.

To illustrate the technique of recombinant DNA, consider a real response to a very real need. Insulin is a small protein consisting of 51 amino acids. It is produced by the pancreas and influences many metabolic processes. Most familiar is its role in reducing the level of glucose in blood by promoting the entry of that sugar into muscle and fat cells. People who suffer from a common type of diabetes have an insufficient supply of insulin, and hence elevated levels of blood sugar. Left untreated, the disease can result in poor blood circulation, especially to the arms and legs, amputations, blindness, kidney failure, and early death. But, diabetes can be controlled by diet, exercise, and insulin injections.

Before 1982, all insulin used by diabetics was isolated from the pancreas glands of cows and pigs, collected in slaughterhouses. It turns out that the insulin produced by cattle and hogs is not identical to human insulin. Bovine (cow) insulin differs from the human hormone in three out of 51 amino acids; porcine (pig) and human insulins differ in only one. These differences are slight, but sufficient to undermine the effectiveness of bovine and porcine insulin in some human diabetics. For many years, there seemed to be no hope of obtaining enough human insulin to meet the need. Although insulin has been synthesized in the laboratory, the process is far too complex for large-scale production. However, since 1982 the lowly bacterium, *E. coli,* has been tricked into making human insulin.

This unlikely bit of interspecies cooperation is a consequence of using recombinant DNA techniques to introduce the gene for human insulin into this simple organism (Figure 12.17). Bacteria contain **plasmids,** or rings of DNA. These rings can be removed and cut open by the action of special enzymes, called restriction enzymes. Meanwhile, the gene for human insulin is either prepared synthetically or isolated from human tissue. This human DNA is inserted into the plasmid ring by other enzymes, resulting in interspecies recombinant DNA. **Vectors** are modified plasmids that are used to carry DNA back into the bacterial "host." Once inside the cell, the biochemistry of the bacterium takes over. Every 20 minutes, the *E. coli* population doubles, soon producing millions of copies or clones of the "guest" (human) DNA.

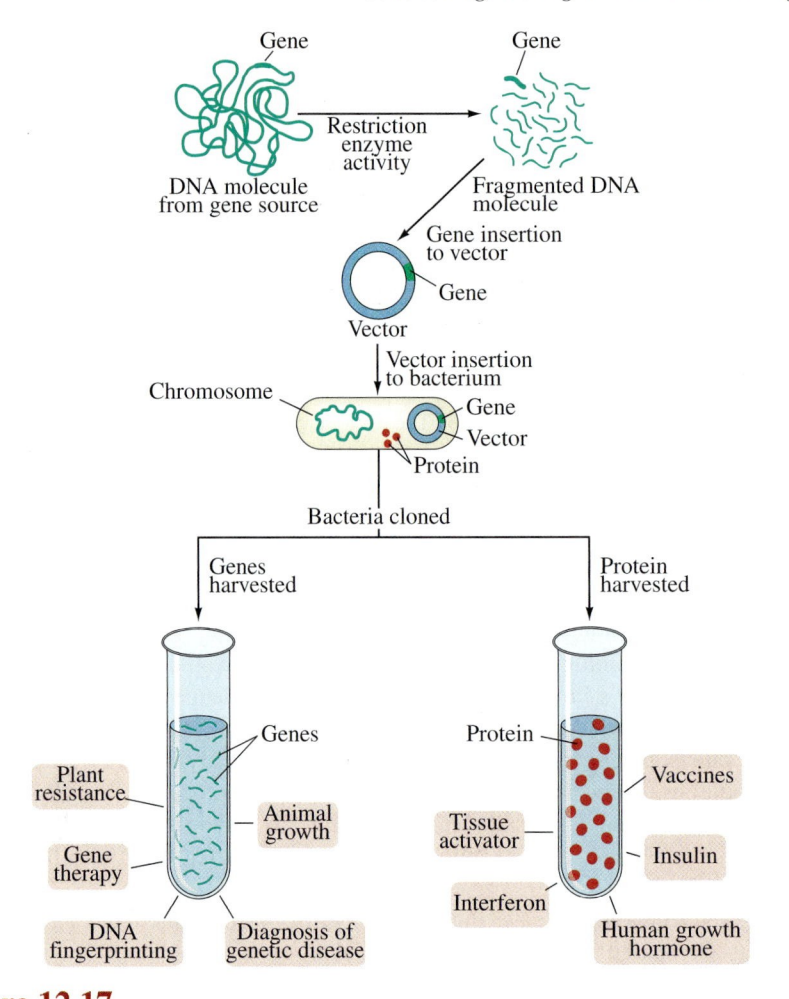

Figure 12.17

A representation of genetic engineering.

Clones are a collection of cells or molecules identical to an original cell or molecule. It is possible to harvest the cloned DNA, but in the insulin example we are interested in a supply of the protein, not its gene. Therefore, the bacteria are allowed to synthesize the proteins encoded in the recombinant DNA. Although the recombinant *E. coli* has no use for human insulin, it generates it in sufficient quantities to harvest, purify, and distribute to diabetics. Today, the cost of human insulin produced by bacteria is less than that of insulin isolated from animal pancreas. Currently, more than 3 million Americans use genetically engineered insulin to treat their diabetes.

Figure 12.17 is a representation of the recombinant DNA techniques just described. The actual operations are a good deal more complicated than the figure suggests, and many details have been omitted. A variety of vectors and host organisms have been used in molecular engineering. Other bacterial species are sometimes used instead of *E. coli,* and yeasts and fungi are often employed because all of these species reproduce very quickly. DNA with specific properties is fragmented and the appropriate portion is introduced into a vector that carries it into a host organism. Cloning of the host, a bacterium in this case, gives rise to two categories of end products: cloned genes and proteins.

Another success for genetic engineering has been the synthesis of human growth hormone (HGH). This protein, produced by the pituitary gland, stimulates body growth by promoting protein synthesis and the use of fat as an energy source. Children with insufficient HGH fail to reach normal size. If the condition is diagnosed early, injections of the hormone over 8–10 years can prevent this form of dwarfism. Formerly, a year's HGH therapy for one person required the pituitary glands from about 80 human cadavers. That source is no longer used, thanks to the production of HGH in bacteria. However, the cost of treatment, even with cloned HGH, can be as high as $20,000–$50,000 per year.

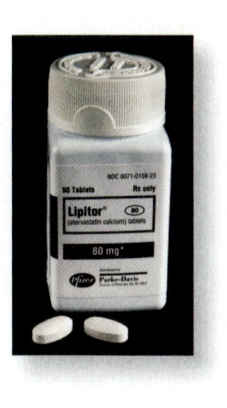

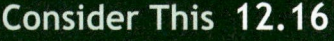

Figure 12.18

Atorvastatin, the active ingredient in Lipitor. The cholesterol-lowering drug has annual sales exceeding $10 billion.

Similar biochemical methods also are being used to create new drugs or larger supplies of already known drugs. The gene coding for the drug is introduced into a host organism, which then synthesizes the desired product. This is currently one of the most rapidly growing applications of recombinant DNA technology.

 In 2006, a biotechnology company called Codexis, Inc. won the U.S. Environmental Protection Agency's Presidential Green Chemistry Challenge Award in the category of "Greener Reaction Conditions." Codexis uses recombinant technology and a "green-by-design" philosophy to create biocatalytic enzymes that catalyze reactions under mild, neutral conditions. Three new enzymes are being used to produce ethyl (R)-4-cyano-3-hydroxybutyrate (commonly called hydroxynitrile, HN), a key chiral building block in the synthesis of atorvastatin, or Lipitor, a cholesterol-lowering drug with blockbuster sales (Figure 12.18).

Chiral drugs were discussed in Section 10.6.

This green technology creates highly pure hydroxynitrile while reducing waste and by-products. Compared with previous commercial processes, it increases the yield of HN with fewer steps and eliminates a labor-intensive purification step. The Codexis process avoids the use of harsh reagents, which is better for the environment and increases overall worker safety.

The biocatalysts were created via a process called directed evolution. It begins with the selection of genes that possess certain DNA sequence diversity. Codexis then uses a proprietary (secret) technique that recombines the DNA, creating a new generation of biocatalysts, which are then screened for desired properties. Codexis repeats the selection and crossing through multiple generations ultimately giving rise to enzymes that are highly selective and stable under various conditions.

Consider This 12.16   **Going for the Green**

The winners of the EPA's Presidential Green Chemistry Award program are selected by experts at the American Chemical Society. Research the past winners of these awards, select one, and write a short report describing the accomplishment. As part of your answer, explain the benefits of the new process or product over existing ones. Be prepared to present your findings to your classmates. A direct link to the winners is provided at the *Online Learning Center*.

12.7   Genetically Engineered Agriculture

One of the more remarkable accomplishments of genetic engineering is **transgenic organisms,** artificially created higher plants and animals that share the genes of another species. Inserting foreign DNA becomes progressively more difficult as one moves up

the evolutionary ladder from bacteria through plants to animals. Therefore, some of the most interesting uses of recombinant DNA technology have involved modifications of agricultural crops. Altering the genetic makeup of plants by genetic engineering is faster and more reliable than relying on traditional crossbreeding.

The National Academy of Sciences of the United States, the Royal Society of the United Kingdom, and equivalent science academies of Brazil, China, India, and Mexico have publicly supported transgenic crop research and development. They have claimed that the research should focus on plants that will (1) improve production stability; (2) give nutritional benefits to the consumer; (3) reduce the environmental effects of intensive and extensive agriculture; and (4) increase the availability of pharmaceuticals and vaccines. Most of the commercially developed crop varieties in the United States and Canada have centered on increasing the shelf life of fruits and vegetables, conferring resistance to pests or viruses, and producing tolerance to specific herbicides.

Although one might look at delayed ripening as only a commercial issue, that is, a source of increased profit for producers, the issue also addresses nutrition and world hunger. Considering that as much as 50% of agricultural commodities in underdeveloped countries spoil before reaching the consumer, extending the access to fresh food has tremendous consequences. The same arguments hold for developing transgenic plants that are resistant to specific pests. About half of all soybeans grown in the United States have a transgenic gene for resistance to a certain herbicide, improving crop yields by controlling weeds. There is no arguing that reduced amounts of pesticides or herbicides are positive for the environment. But the use of herbicides may actually increase as more and more crops become immune to their poison.

Transgenic rice plants have been developed for use in Africa where yellow mottle virus has destroyed a majority of the annual production of rice (Figure 12.19). Researchers are investigating ways of incorporating the DNA of nitrogen-fixing bacteria into wheat, rice, and corn. These bacteria are present in the root systems of soybeans, alfalfa, and other legumes. Thanks to the bacteria, these plants can absorb N_2 directly from the atmosphere and use it in biochemical reactions.

Figure 12.19
Virus-resistant transgenic rice.

The removal of N_2 from the air by certain plants is an important part of the nitrogen cycle, as discussed in Section 6.12.

Consider This 12.17   **Concern About Corn**

Starlink corn, a bioengineered plant, was approved for animal use only. Human consumption did not receive a green light because it was feared that people who ate the corn might develop an allergy.

a. For what purpose was Starlink genetically modified?

b. Starlink corn accidentally found its way into the human food chain via several brands of taco shells. Subsequently, several dozen people who ate the corn shells reported allergic reactions. What are the symptoms, sometimes life-threatening, of an allergic reaction?
Hint: A stomachache, although uncomfortable, is not an allergic reaction.

c. A true allergic reaction produces antibodies in the bloodstream. In 2001, the U.S. Centers for Disease Control and Prevention ran carefully controlled experiments searching for antibodies in the people who claimed an adverse reaction from Starlink corn. What were the findings of the study?
Hint: The *Online Learning Center* provides helpful Web links.

Consider This 12.18   Potatoes and You

The New Leaf Superior potato, first marketed in 1995, was genetically engineered to produce its own insecticide. More recently, the New Leaf Superior potato was replaced with the New Leaf Plus potato.

a. According to the USDA, the average person eats over 100 pounds of potatoes per year. List at least five foods that you eat that contain potatoes in some form.

b. Potatoes are a challenging crop to grow. Learn about these challenges and then draw up a list of at least five challenges. Which items on your list might genetically engineered potatoes be able to address?

c. As noted in Consider This 12.17, one of the concerns of genetically modified food is that people may develop new allergies. But this is not the only point of concern. Name some other points that merit careful consideration.

Hint: The *Online Learning Center* provides helpful links for parts **b** and **c**.

Plant production of medicines and drug precursors suggest that transgenic plants could be developed to produce a host of pharmaceuticals and vaccines. Most vaccines require refrigeration or other special handling and trained professionals to administer them; some developing countries can't even afford the needles to inoculate their populations. Vaccines against infectious diseases of the intestinal tract have been produced in potatoes and bananas. Anticancer antibodies have been expressed and introduced into wheat recently. Considering that petroleum serves as the raw feedstock for many pharmaceutical drugs, the development of this technology should be embraced by any country concerned about dwindling oil supplies. The development of transgenic plants that produce therapeutic agents for addressing problems of disease is just beginning to realize its own potential.

So why isn't the whole world enthusiastic about genetically modified (GM) or transgenic food? The creature from Mary Shelley's *Frankenstein* evokes images of an entity composed of "parts" from several sources but who collectively produce a being that is out of control. Those opposed to GM food often refer to it as "Frankenfood." In Europe, Japan, and other parts of the world, legislation has been introduced to ban or control it.

From an environmental standpoint, many Europeans are worried about the contamination of non-genetically modified crops with transgenic organisms. There is also concern about genetically engineered foods accidentally finding their way into the human food chain. Chris Wietrzny, a representative with the International Coalition to Protect the Polish Countryside, explains that Poland has over 2 million small family farms with rich biodiversity that people want preserved. In his opinion, Polish people do not believe that trans- and non-transgenic crops can coexist.

A 1998 report from a research institute in the United Kingdom attacking the safety of genetically modified potatoes may be largely responsible for the "Frankenfood frenzy" in Europe. That document reported that GM potatoes have a specific and negative effect on organ development and the immune system in rats. In June 1999, the Royal Society of the United Kingdom conducted a thorough, independent review and concluded that the report and the experiments on which it was based were flawed. They cited poor experimental design, possibly exacerbated by the lack of "blind measurements," uncertainty from the small number of samples, the application of inappropriate statistical techniques, possible dietary differences due to nonsystematic dietary enrichment of protein in the rats, and a lack of consistency of findings within and between experiments. Critics, however, remain undaunted and vocal in spite of few reproducible experiments to support any health concerns.

Consider This 12.19   The Latest Lingo

Have you heard of *lexpionage*? This term refers to the sleuthing of new words that start to appear regularly in newspapers, magazines, and Web sites. *Frankenfood* is one example. *Googling* and *phishing* are two others. Actually so is the word *lexpionage*!

a. Look back through this chapter. What words do you spy that you suspect are relatively new?

b. WordSpy.com not only lists *Frankenfood*, but also other new words related to genetically modified plants and animals such as *immunocows* and *plantibody*. At WordSpy, select any five terms and explain them. A direct link is provided at the *Online Learning Center*.

Genetic engineering, however, is more than simply a collection of techniques. It has a human face among those who have benefited from it, whether medically or as a result of the GM food items for the underdeveloped world. Before recombinant DNA, there was no hepatitis B vaccine, insufficient erythropoietin (a protein used to stimulate red blood cell growth in dialysis patients), a lack of tissue plasminogen activator (TPA) to dissolve blood clots in cardiac and stroke patients, and a form of insulin that didn't quite match the human molecule.

Have the benefits of genetically modified foods been overstated while the problems swept under the table by those who benefit the most economically? Should we be worried about consuming these products? Much of the opposition to Frankenfoods is based on the concern that "natural" plants could be changed forever and perhaps irreversibly on exposure to pollen from altered crops. Opponents also fear possible adverse economic effects as well as unanticipated health problems that could arise in people and animals that consume GM food.

Consider This 12.20   The Frankenfood Frenzy

As Figure 12.20 shows, opposition to GM foods was a major theme of protesters at the 2005 World Trade Organization meeting in Hong Kong, China (Figure 12.20). What are people saying about GM foods in the European Union today? For example, you might find a recent editorial or thought piece that conveys a particular point of view. Another possibility is to find a political cartoon. Still another is to research the status of laws in the EU that require foods to carry labels if genetically modified. Whatever approach you take, clearly state the evidence that you used.

Figure 12.20
Protesters at the 2005 WTO meeting in China.

12.8 Cloning Mammals and Humans

The announcement of the successful cloning of human embryos in February 2004 sparked fresh debate and raised concern around the world. But this was not a sudden discovery nor was it unexpected. Since first envisioned in 1938 by Hans Spemann, slow but steady progress has been made to realize the goal. Through the 1960s and 1970s, British molecular biologist John B. Gurdon made several breakthroughs, eventually cloning frogs. In 1996, embryologist Ian Wilmut and colleagues at the Roslin Institute in Scotland made headlines with the birth of Dolly, the first mammal cloned from an adult cell (Figure 12.21a). The birth of Dolly, who died an early death of progressive lung disease in 2003, placed her in the center of worldwide controversy that still exists today. The controversial heat was turned up further when in 2005, Wilmut was granted a license to clone human embryos for medical research.

In August 2005, scientists at the Seoul National University in South Korea reported in *Nature* magazine that they had produced "Snuppy," the first cloned dog (**S**eoul **N**ational **U**niversity pu**ppy** Figure 12.21b,c). The pup was made from a cell taken from a 3-year-old Afghan hound's ear.

Both Dolly and Snuppy were cloned by a technique called nuclear transfer (Figure 12.22). **Nuclear transfer** is a laboratory procedure in which a cell's nucleus is removed and placed into an egg cell that has had its own nucleus removed (Figure 12.23). The genetic information from the donor nucleus controls the resulting cell, which can be induced to form embryos. Growth was then initiated by an electrical "jump start" to create the growing embryo. In Dolly's case, the egg was then implanted into another sheep's uterus, carried full term, and then she was born. Dolly carried DNA identical to the sheep who was the cell donor.

The excitement emanating from Hwang's labs reached its peak in 2004 when it was announced that the Hwang group had successfully cloned a human embryo, and then culled from it master stem cells. Hwang stated "Our research team has successfully culled stem cells from a cloned human embryo through mature growing process in a test tube." Immediately, ethical concerns were raised around the world. New calls for a ban on all forms of human cloning in the United States were issued, and President George W. Bush described the research as "reckless experiments," and called for a "comprehensive and effective ban."

The Hwang group claimed that the embryo had been created via the nuclear transfer technique—genetic material was supposedly added to human donor eggs from which the native nuclear material had been removed. It was reported that after an electrical jolt, the embryos started growing. However, in four to five days at the stage of about 100 cells,

> Dolly's cells contain the DNA of only one parent, unlike identical twins who share DNA from both parents.

(a)   (b)   (c)

Figure 12.21

(a) Dolly the cloned sheep. (b) Snuppy (*right*), the cloned dog next to his "father," an Afghan hound. (c) Snuppy and his surrogate mother.

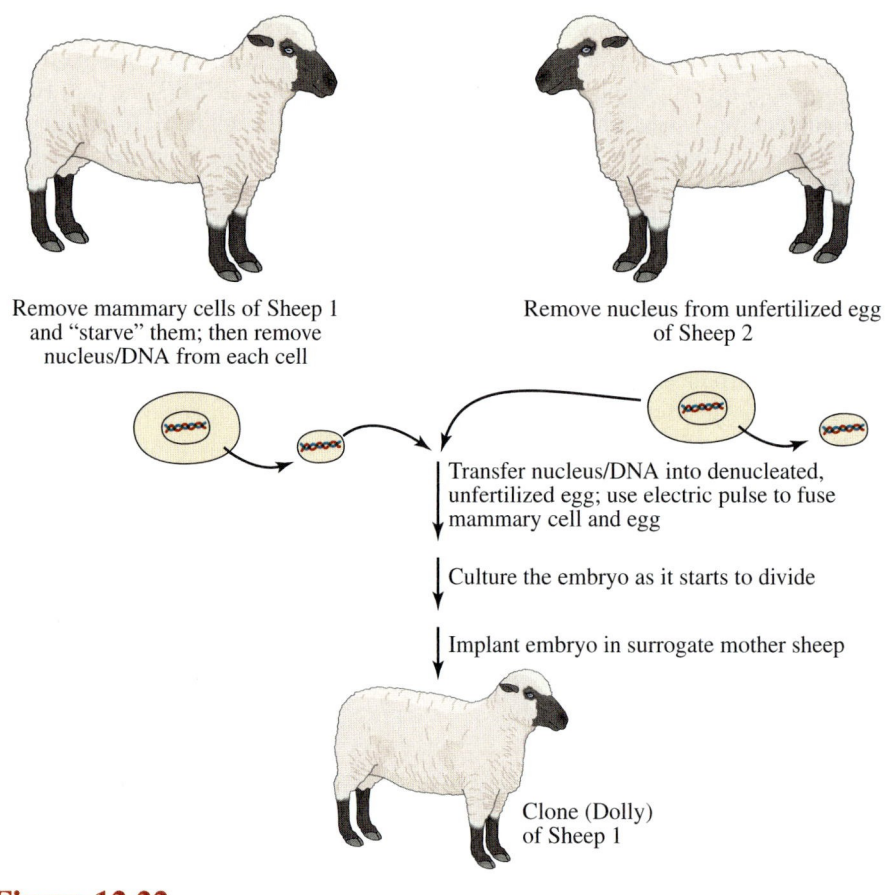

Figure 12.22

The cloning of Dolly from a mature animal.

these blastocysts, as they are called, were not implanted. Taking them to the stage of an infant, called reproductive cloning, was never their intention. At the blastocyst stage, the researchers claimed to have harvested a set of **stem cells**, identical, undifferentiated cells that, by successive divisions, can give rise to specialized ones like blood cells. They are the basic building blocks or master cells that can be cultured or reproduced and theoretically divide without limit to replenish other cells. When a stem cell divides, the new cells can remain stem cells or develop into specialized cells with particular functions such as heart or nerve cells. If the harvested stem cells are artificially coaxed into forming specific cells like heart, brain, or nerve cells, the technique is called therapeutic cloning since the goal is to create a new line of stem cells. In principle a person with a spinal cord injury could donate DNA from her cells, have it converted into stem cells with her genetic material, and then have nerve cells formed to repair the spinal cord injury. Advances in stem cell technology have been touted as holding potential cures for many crippling illnesses and injuries.

The promise of therapeutic advances was short-lived however. In January 2006 Hwang admitted the embryonic stem cells documented in the prestigious journal *Science* in February 2004 were faked. The scientific community fell into shock. An investigative panel at the Seoul National University announced that the 2004 *Science* article and a 2005 paper on tailor-made stem cells (also published in *Science*) were "outright falsified." Follow-up investigations into the validity of Snuppy proved that the pup was indeed a clone, but the already embattled field of stem cell research was dealt a terrific blow. Editors at *Science* said that both articles would be withdrawn, but the peer review process was called into question. Peer-reviewed research—the cornerstone of modern science—is research that has been published in a scientific journal only after it has been reviewed for accuracy and validity by one or more independent, qualified experts in the same field. Although not foolproof, the process is the best we have at this time. As this text went to press, Hwang had resigned from his post and was facing possible jail time for misuse of government funding.

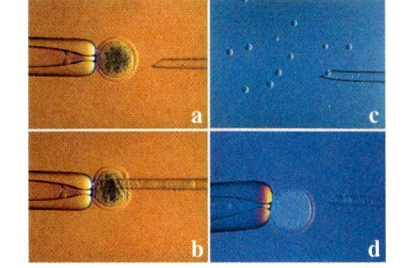

Figure 12.23

(**a, b**) Nucleus is sucked out of cell.
(**c**) New nucleus is picked up.
(**d**) Nucleus/DNA is inserted into egg.

But several ethical and moral issues remain. Some people believe that an embryo, even just several minutes old, is a human life and that harvesting stem cells from the blastocyst, thereby destroying its viability, is equivalent to ending that life. In 2003, the U.S. House of Representatives passed a ban on human cloning although the bill failed to get approval in the Senate. So there is no federal law prohibiting human cloning altogether. The Food and Drug Administration has claimed authority over the regulation of human cloning technology as an investigational new drug (IND) and stated that at this time, they would not approve any projects involving human cloning for safety reasons, but Congress has not passed legislation confirming the FDA's authority to prohibit cloning.

As President, George W. Bush forbade any federally funded research on stem cell lines created after August 2001. In July 2006 Mr. Bush vetoed a bill that would have overturned this policy. State laws vary widely on the issue. California, Connecticut, Maryland, Massachusetts, and New Jersey have set their own statutes, and Illinois has set an Executive Order encouraging embryonic stem cell research. Missouri law forbids the use of state funds for reproductive cloning but not for cloning for research. South Dakota law forbids research on embryos regardless of the source. Arkansas, Indiana, Iowa, Michigan, and North Dakota prohibit research on cloned embryos. Virginia's law calls for a ban on research using cloned embryos, but because human being is not defined, the statute may leave room for interpretation. There is sure to be disagreement about whether the definition of human being includes blastocysts, embryos, or fetuses.

> In September 2004, California approved legislation making it legal to derive embryonic stem cells from human embryos and to conduct research using stem cells from "any source."

12.9   The New Prometheus?

Nature is an indispensable aid and ally in medicinal and biological chemistry. Much of our success has come from understanding and imitating natural processes. For centuries, animal breeders, agricultural researchers, and observant farmers have brought about genetic transformations in animals and plants by selective breeding. Recent applications of chemical methods to biological systems have greatly increased our capacity to effect such changes. Molecular engineering has made it possible to create nucleic acids, proteins, enzymes, hormones, drugs, and other biologically important molecules that do not exist in nature.

The new molecules can be designed to be more efficient catalysts than their naturally occurring counterparts, more effective and less toxic drugs for treating a wide range of diseases, or modified hormones that actually work better than the original. It is not at all fanciful to imagine a whole range of enzymes, engineered to consume environmentally hazardous wastes that are impervious to naturally occurring enzymes. Likewise, it might be possible to create an enzyme that far surpasses the one that catalyzes photosynthesis, one of the most inefficient of all natural enzymes. As Mark Twain suggested: "Predictions are extremely difficult, especially those about the future." Yet, it is likely that because of genetic engineering, HIV-AIDS and at least some forms of cancer may some day become as infrequent as polio or smallpox are now. Even more tantalizing is the possibility of eradicating certain genetic defects. Our growing knowledge of the human genome, coupled with our understanding of the chemistry of genetics, holds the promise of altering our inherited gene pool. Prospects include the elimination of sickle cell anemia, diabetes, hemophilia, phenylketonuria, and dozens of other debilitating hereditary traits.

The next logical step would seem to be the creation of new organisms. Scientists have already cloned "new and improved" animals and vegetables. Mammals have been cloned from adult cells. Human cloning has been dealt with in fiction such as Huxley's *Brave New World*—using eugenics to create different classes of people, and in films like Woody Allen's *Sleeper*—attempts to clone a dead dictator, and *The Boys from Brazil*—cloning Nazis. How should we view the possibility of cloning human beings, perhaps even "new and improved" ones? The question goes straight to our identity as a species and as individuals; our very humanity is at stake. Are we at a point in history where "to build or not to build" is the question?

One could reason that the best way to treat a genetic disease or disability would be to remove the defective gene from the gene pool by intentionally altering the suspect DNA

in the sperm or ova. But one could also argue that *Homo sapiens* was not yet sufficiently wise to assume such god-like power. Our species has had some tragic experiences in the past century with political leaders who used less subtle methods of "genetic cleansing."

Consider This 12.21   Cloning Clinics

Plans to develop human cloning clinics have been announced in the last several years. What risks and benefits are associated with such clinics? Has such a clinic been completed at this time, and are we closer to the scenario presented in the chapter opener? Write a short report detailing your answers to these questions.

Cloning through genetic engineering raises not only the possibility for humans to duplicate themselves by unconventional means but also the chilling potential to design a master race or to subjugate or eliminate "defectives" through genetic manipulation. Hence, there is an intentional irony in the title of this final section. Prometheus was the demigod who stole fire and the flame of learning from the gods and brought these incomparable gifts to humanity. "The New Prometheus" is the subtitle of *Frankenstein*, Mary Shelley's classic study of scientific knowledge run amok. Using our ever-growing knowledge of the chemistry of heredity wisely and well will surely be one of the greatest challenges of the 21st century.

Consider This 12.22   Send in the Clones

The late Isaac Asimov, noted science fiction writer and biochemist, coauthored this verse, called "The Misunderstood Clone."

> *Oh, give me a clone*
> *Of my own flesh and bone*
> *With its Y chromosome changed to an X.*
> *And when it has grown*
> *Then my own little clone*
> *Will be of the opposite sex.*

Source: *The Sun Shines Bright* by Isaac Asimov. Copyright © 1981 Nightfall, Inc. Used by permission of Doubleday, a division of Random House, Inc. and Ralph Vicinanca Agency.

Is the verse describing gene therapy or cloning? Support your answer by explaining the two different processes.

Consider This 12.23   Cloning Humans

James D. Watson said this about manipulating germ cells to create superpersons, "When they are finally attempted, germ-line genetic manipulations will probably be done to change a death sentence into a life verdict—by creating children who are resistant to a deadly virus, for example, much the same way we can already protect plants from viruses by inserting antiviral DNA segments into their genomes."

Source: *Time,* "All for the Good," January 11, 1999, Vol. 153, No. 1.

Draft a statement supporting or opposing Watson's position on this issue. Your argument should be grounded in scientific principles.

Conclusion

The last chapter of this book, like almost all that preceded it, ends with a dilemma: How can we balance the great benefits of modern chemical sciences and technology and the risks that seem inevitably to accompany them? Throughout this text, the authors have occasionally looked, with myopic professorial vision, into the cloudy crystal ball of the future. It is in the nature of science that we cannot confidently predict what new discoveries will be made by tomorrow's chemists. Nor can we know the applications of those discoveries, good or bad. Such uncertainty is one of the delights of our discipline. A chemist must learn to live with ambiguity, indeed, to thrive on it, in the search to better understand the nature of atoms and their intricate combinations, in all their various guises.

But all citizens of this planet must at least develop a tolerance for ambiguity and a willingness to take reasonable risks, especially considering that life itself is a biological, intellectual, and emotional risk. Of course, we all seek to maximize benefits, but we must recognize that individual gain must sometimes be sacrificed for the benefit of society. We live in multiple contexts—the context of our families and friends, our towns and cities, our states, our countries, our special planet. We have responsibilities to all. *You,* the readers of this book, will help create the context of the future. We wish you well.

Chapter Summary

Having studied this chapter, you should be able to:

- Understand the chemical composition of deoxyribonucleic acid (DNA), a polymer of nitrogen-containing bases, deoxyribose, and phosphate groups (12.1)

- Recognize the utility of Chargaff's rules: %A = %T, %G = %C (12.1)

- Interpret evidence for the double-helical structure of DNA and its base pairing (12.2)

- Understand DNA replication (12.2)

- Explain how the genetic code is written in groupings of three DNA bases called codons (12.3)

- Relate to the amount of information encoded in the human genome (12.3)

- Discuss the primary, secondary, and tertiary structure of proteins (12.4)

- Describe the molecular basis of sickle cell anemia (12.4)

- Relate to the Human Genome Project: its aims, significance, and ethical implications (12.5)

- Understand the production of human proteins (insulin and human growth hormone) in bacteria (12.6)

- Understand the essential steps in carrying out recombinant DNA techniques (12.6)

- Describe what is meant by transgenic organisms and give examples (12.7)

- Discuss ethical issues associated with transgenic organisms and the fears about Frankenfood (12.7)

- Describe the role of nuclear transfer in the cloning of mammalian cells (12.8)

- Debate issues associated with the prudent and ethical applications of mammalian cloning and genetic engineering (12.9)

Questions

Emphasizing Essentials

1. The letters DNA have been called three of the most important letters of the late 20th century. For what is DNA an acronym?

2. **a.** Use Figure 12.1 to determine the percent increase in human life expectancy at birth in the United States between 1940 and 1980 for males and females.

   b. Are these percentages the same as those calculated in Your Turn 12.2? Why or why not?

3. Consider the structural formulas in Figure 12.2.

   a. What functional groups are in adenine?

   b. What functional groups are in deoxyribose?

   c. What is the function of the phosphate group?

4. Consider the structural formula of deoxyribose given in Figure 12.2.

   a. Why is deoxyribose classified as a monosaccharide?

   b. What is the chemical formula for deoxyribose?

   c. Why isn't deoxyribose an acid in aqueous solution?

5. Using the structural formula of thymine in Figure 12.2, write an equation showing how thymine could react with water to generate hydroxide ions. *Hint:* See Sections 6.2 and 6.3.

6. **a.** What three units must be present in a nucleotide?

   b. What type of bonding holds these units together?

7. Table 12.1 lists the base composition of DNA for various species. The four bases are adenine, cytosine, guanine, and thymine. What relationships exist among these bases, no matter what the species?

8. **a.** What happens experimentally during X-ray diffraction?

   b. The first X-ray diffraction patterns were of simple salts, such as sodium chloride. The X-ray diffraction studies of nucleic acids and proteins did not come until much later. Suggest reasons why.

9. Figure 12.7 shows the pairing of nucleotide bases in DNA.

   a. What type of *intramolecular* bonding occurs within each base?

   b. What type of *intermolecular* bonding holds the base pairs together?

10. Identify the base sequence that is complementary to each of these sequences.

    a. ATGGCAT   **b.** TATCTAG

11. Given that the distance between adjacent bases is 0.34 nm, how many base pairs are present in a chromosome that is 3.0 cm long? *Hint:* 1 m = 10^2 cm; 1 m = 10^9 nm.

12. During cell division, as many as 90,000 nucleotides per minute can be added to the growing DNA chain.

    a. The shortest human chromosome contains 50 million bases. What is the minimum time required to form a strand of this chromosome?

    b. Determine the length of this chromosome (in centimeters) that would be formed in 1 minute if the distance between bases were 0.34 nm. *Hint:* 1 m = 10^2 cm; 1 m = 10^9 nm.

13. Amino acids are the monomers used to build proteins.

    a. What is the *general* structural formula for an amino acid?

    b. What functional groups are present in all amino acids?

14. The text states that if you were to use the letters A, T, C, and G in a game of Scrabble, you could generate 64 different three-letter combinations. (Your Scrabble opponent would surely challenge some of these combinations!) Nature pairs the four bases A, T, C, and G, and uses the pairs to encode for amino acids. Write down all the possible paired combinations of A, T, C, and G to find the maximum number of amino acids that can be encoded from these four bases. *Hint:* In nature, unlike Scrabble, a letter can be used more than once in forming a pair.

15. What is a codon and what is its role in the genetic code?

16. Only 61 of the possible 64 triplet codons specify certain amino acids. What is the function of the other three codons?

17. Describe what is meant by the primary, secondary, and tertiary structure of a protein. Is one of these more important than the others? Explain.

18. Explain how an error in the primary structure of a protein in hemoglobin causes sickle cell anemia.

19. What, in relation to DNA, is a gene?

20. What is meant by the genetic code?

21. Which set of bases makes up DNA nucleotides?

    a. adenine, thymine, uracil, and guanine

    b. adenine, uracil, guanine, and cytosine

    c. adenine, thymine, cytosine, and guanine

    d. adenine, thymine, cytosine, and uracil

22. There are many DNA-damaging agents that effectively kill bacterial cells in culture dishes. Why are DNA-damaging agents not typically used as antibacterial drugs?

23. Circle and name the functional groups present in this nucleotide.

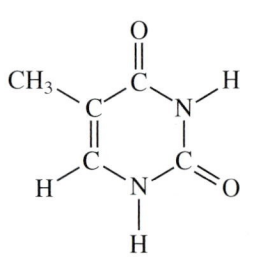

24. The structure for thymine is shown. Use the concept of resonance to explain why neither of the nitrogen atoms is likely to accept a hydrogen ion (H^+). In other words, use resonance forms to explain why thymine is not a very strong base.

25. **a.** What is the Human Genome Project?

    b. Why is it a significant step in understanding the genetic basis of humans?

26. **a.** What are stem cells?

    b. How are stem cells harvested?

    c. What aspect of stem cell research makes it controversial?

Concentrating on Concepts

27. What is meant by the term *cloning?*

28. Life expectancy in the United States increased dramatically during the 20th century, as shown in Figure 12.1 and surrounding text. Is a similar increase possible during the 21st century? Explain your answer.

29. Compare the two representations of a segment of DNA in Figure 12.4. Discuss the strengths and weakness of each representation.

30. Consider Chargaff's discovery that there are equal percentages of adenine and thymine and of cytosine and guanine in DNA. Was his discovery as important to understanding the nature of DNA as Crick and Watson's discovery of the double helix? Explain.

31. Use Figure 12.7 to help explain why stable base pairing does *not* occur between adenosine and cytosine, thymine and guanine, adenine and guanine, and thymine and cytosine. Use molecular structures in your explanation.

32. One of the mechanisms by which DNA sustains damage from UV light is via the formation of covalent bonds between two thymine bases.

    a. What are the potential consequences of cross linking between two thymines on a single strand of DNA?

    b. What possible problems do you foresee if one thymine from each of two complementary strands undergoes a cross-linking reaction?

33. Errors sometimes occur in the base sequence of a strand of DNA. But not all of these errors result in the incorporation of an incorrect amino acid in a protein for which the DNA codes. Explain how this happens and why it is advantageous.

34. Human insulin and human growth hormone have both been made through the use of recombinant DNA technology. Which do you believe is a more significant use of this technology and why? Discuss what factors have influenced your opinion.

35. The unusual C-to-N bond length of a peptide bond was discussed in this chapter. Using similar reasoning, describe the relative bond length of the C-to-O bond of the carbonyl group in a peptide bond compared to that of a ketone.

36. What role did Chargaff's rules play in the discovery of the DNA double helix?

37. Lou Gehrig's disease is caused by the alteration of a single amino acid in the enzyme superoxide dismutase. How many base pairs are responsible for specifying this amino acid in the gene that codes for the protein? What is the minimum number of base pairs that would have to be changed to produce the disease?

38. Do you favor the patenting of genes? What are the advantages and disadvantages of this approach?

39. Use the Web to research the concept of eugenics. How did the term come into use and how has its meaning changed over time?

40. How widely available is the New Leaf potato? If you wanted to plant this in your garden, would you be able to obtain these potato starter sets? For what reasons would you want to obtain this plant?

41. Use the Web to gather data and create a graph showing the relative amounts of GM or transgenic crops grown by six different countries. Include with your graph the source of your data.

42. Consider the idea of mixing genes as an improvement on nature.

a. What are transgenic organisms?

b. Why is the alteration of the genetic makeup of plants by genetic engineering preferred to traditional cross-breeding methods?

43. Make a list of the advantages and disadvantages of transgenic crops.

44. The cloning of Dolly in 1996 and the *would-be* human embryos in 2004 are remarkable scientific events.

    a. In what ways were the techniques used similar?

    b. In what ways were they different?

Exploring Extensions

45. Dolly's birth surprised genetic engineering experts and shocked members of the media responsible for reporting the birth and the method. Locate one or two early reports about the cloning of Dolly. Evaluate these reports for their scientific accuracy and what they reveal about the opinion of experts at that time.

46. Use this information to act as a Sceptical Chymist in checking the correctness of the claim that the DNA in an adult human would stretch from Earth to the Moon and back more than a million times.
    Distance from Earth to the Moon = 3.8×10^5 km
    Number of DNA-nucleated cells in an adult
    human = 1×10^{13}
    Length of stretched human DNA strand = 2 m

47. Consider the structural formula of deoxyribose shown in Figure 12.2. The prefix *deoxy-* means without oxygen; the —OH group is replaced by a hydrogen atom. In the specific case of deoxyribose, the —OH group replaced by a hydrogen was bonded to the only carbon in the ring that bonds to two hydrogen atoms in deoxyribose. The structural formula of ribose appears in Consider This 12.4. Draw another isomer of deoxyribose, one that would still form a nucleoside with a base and with phosphate. Compare your isomer side-by-side with the structural formula of deoxyribose.

48. The text states that the more closely species are related, the more similar the DNA base compositions are in those species. The Sceptical Chymist is having trouble believing this, particularly if it means a close relationship between a fruit fly and a bacterium. Use the information in Table 12.1 to determine whether it supports the generalization about similar base pairs in similar species.

49. Perhaps you have learned some memory aids (mnemonics) when taking music lessons (Every Good Boy Does Fine), memorizing the names of the Great Lakes (HOMES), or learning about oxidation and reduction (OIL RIG). One of the authors learned "All-Together, Go-California" as the mnemonic to remember the correct base pairings in DNA.

    a. What is the relationship in this mnemonic to DNA base pairing?

    b. Design a different mnemonic that will help you remember such base pairings.

50. Of the major players in the discovery of the structure of DNA, only Rosalind Franklin had a degree in chemistry. What was her background and experience that enabled her to make significant contributions? Did her contributions receive adequate credit and recognition? Write a short report, citing your sources.

51. Gene therapy involves the use of recombinant DNA techniques. Use the Web to gather information on this medical tool that has met with mixed results. Write a one- to two-page report on gene therapy including specific examples of diseases that are being treated and how the patients are faring.

52. Transgenic plants have not been widely accepted in all countries. Give reasons for their rejection in some European markets.

53. A recent focus of participants in the Human Genome Project has been to determine the base sequence for many different microbes (that is, bacteria and viruses). Use the Web to document recent efforts and progress. Suggest reasons why these sequences would generate so much interest.

54. Genetic diseases are also called inborn errors of metabolism. You may be familiar with some of these diseases, such as hemophilia, PKU, Tay–Sachs disease, or sickle-cell anemia. One that does not get much attention is a condition known as Niemann–Pick disease. Find out what inborn metabolic error causes this condition, how many children are born with this disease in the United States each year, what treatments are available, and whether a cure is possible. Write a report to be discussed with your classmates.

55. One reason why science fiction is successful is that it starts with a known scientific principle and extends, elaborates, and sometimes embroiders it. Consider how *Jurassic Park* began with the scientific principle of cloning DNA and was extended into a successful story. Now it is your turn. Take a scientific principle from this or one of the other 11 chapters in this text. Then write a one- or two-page outline for a story based on that principle. Be sure to identify the chemical concepts that you plan to include and any pseudoscience that you might employ.

56. Use the Web to create a time line with events to account for the rapid increase and subsequent leveling of the adoption of transgenic crops in the United States.

57. The cloning of human embryos to produce infants is a very controversial possibility. Try to find reports of human cloning of this type. What evidence can you find about whether these reports are credible? What are the U.S. regulations about cloning to produce infants?

58. The creation of "Snuppy" the cloned canine must be of significant interest to professional dog breeders and their organizations. Phil Buckley, spokesman for the Kennel Club in England was quoted on the BBC News Web site as saying "Canine cloning runs contrary to the Kennel Club's objective to promote in every way the general improvement of dogs," and "Cloning cannot be used to make improvements because the technique simply produces genetic replicas of existing dogs." Use the Web to find out the impact cloning has had on similar U.S. dog organizations.

59. Advances in pet cloning has not escaped the interest of entrepreneurs, both illegal and legit. Use the Web to find:

a. Instances of fraud concerning the cloning of pets.

b. Companies that offer cloning services legally. How much do they charge? What kinds of guarantees are made? What locations do these companies operate?

60. Ian Wilmut and colleagues at the Roslin Institute in Scotland were granted a license to clone human embryos for medical research. Has this group made any progress to date?

Measure for Measure
Conversion Factors and Constants

Metric Prefixes

deci (d)	$1/10 = 10^{-1}$	deka (da)	$10 = 10^1$
centi (c)	$1/100 = 10^{-2}$	hecto (h)	$100 = 10^2$
milli (m)	$1/1000 = 10^{-3}$	kilo (k)	$1000 = 10^3$
micro (μ)	$1/10^6 = 10^{-6}$	mega (M)	$= 10^6$
nano (n)	$1/10^9 = 10^{-9}$	giga (G)	$= 10^9$

Length

1 centimeter (cm) = 0.394 inch (in.)

1 meter (m) = 39.4 in. = 3.28 feet (ft) = 1.08 yard (yd)

1 kilometer (km) = 0.621 miles (mi)

1 in. = 2.54 cm = 0.0833 ft

1 ft = 30.5 cm = 0.305 m = 12 in.

1 yd = 91.44 cm = 0.9144 m = 3 ft = 36 in.

1 mi = 1.61 km

Volume

1 cubic centimeter (cm^3) = 1 milliliter (mL)

1 liter (L) = 1000 mL = 1000 cm^3 = 1.057 quarts (qt)

1 qt = 0.946 L

1 gallon (gal) = 4 qt = 3.78 L

Mass

1 gram (g) = 0.0352 ounce (oz) = 0.00220 pound (lb)

1 kilogram (kg) = 1000 g = 2.20 lb

1 lb = 454 g = 0.454 kg

1 metric ton (t) = 1 tonne = 1 long ton = 1000 kg = 2200 lb
$\qquad$ = 1.10 ton

1 ton = 1 short ton = 2000 lb = 909 kg = 0.909 t

Time

1 year (yr) = 365.24 days (d)

1 day = 24 hours (hr or h)

1 hr = 60 minutes (min)

1 min = 60 seconds (s)

Energy

1 joule (J) = 0.239 calorie (cal)

1 cal = 4.184 joule (J)

1 exajoule (EJ) = 10^{18} J

1 kilocalorie (kcal) = 1 dietary Calorie (Cal)
$\qquad$ = 4184 J = 4.184 kilojoule (kJ)

1 kilowatt-hour (kWh) = 3,600,000 J = 3.60×10^6 J

Constants

Speed of light (c) = 3.00×10^8 m/s

Planck's constant (h) = 6.63×10^{-34} J $\cdot$ s

Avogadro's number (N_A) = 6.02×10^{23} objects per mole

Atomic mass unit (m) = 1.66×10^{-24} g

The Power of Exponents

Scientific (or exponential) notation provides a compact and convenient way of writing very large and very small numbers. The idea is to use positive and negative powers of 10. Positive exponents are used to represent large numbers. The exponent, written as a superscript, indicates how many times 10 is multiplied by itself. For example,

$$10^1 = 10$$
$$10^2 = 10 \times 10 = 100$$
$$10^3 = 10 \times 10 \times 10 = 1000$$

Note that the positive exponent is equal to the number of zeros between the 1 and the decimal point. Thus, 10^6 corresponds to 1 followed by six zeros or 1,000,000. This same rule applies to 10^0, which equals 1. One billion, 1,000,000,000, can be written as 10^9.

When 10 is raised to a negative exponent, the number being represented is always less than 1. This is because a negative exponent implies a reciprocal, that is, 1 over 10 raised to the corresponding positive exponent. For example,

$$10^{-1} = 1/10^1 = 1/10 = 0.1$$
$$10^{-2} = 1/10^2 = 1/100 = 0.01$$
$$10^{-3} = 1/10^3 = 1/1000 = 0.001$$

It follows that the larger the negative exponent, the smaller the number. The negative exponent is always one more than the number of zeros between the decimal point and the 1. Thus, 1×10^{-4} is equal to 0.0001. Conversely, 0.000001 in scientific notation is 1×10^{-6}.

Of course, most of the quantities and constants used in chemistry are not simple whole-number powers of 10. For example, Avogadro's number is 6.02×10^{23}, or 6.02 multiplied by a number equal to 1 followed by 23 zeros. Written out, this corresponds to $6.02 \times 100{,}000{,}000{,}000{,}000{,}000{,}000{,}000$, or 602,000,000,000,000,000,000,000. Switching to very small numbers, a wavelength at which carbon dioxide absorbs infrared radiation is 4.257×10^{-6} m. This number is the same as 4.257×0.000001, or 0.000004257 m.

Your Turn
Appendix 2.1

Express these numbers in scientific notation.

a. 10,000   **b.** 430   **c.** 9876.54

d. 0.000001   **e.** 0.007   **f.** 0.05339

Answers

a. 1×10^4 **b.** 4.3×10^2 **c.** 9.87654×10^3

d. 1×10^{-6} **e.** 7×10^{-3} **f.** 5.339×10^{-2}

Your Turn
Appendix 2.2

Express these numbers in conventional decimal notation.

a. 1×10^6 **b.** 3.123×10^6 **c.** 25×10^5

d. 1×10^{-5} **e.** 6.023×10^{-7} **f.** 1.723×10^{-16}

Answers

a. 1,000,000   **b.** 3,123,000

c. 2,500,000   **d.** 0.00001

e. 0.0000006023   **f.** 0.0000000000000001723

Appendix

3

Clearing the Logjam

You may have encountered logarithms in mathematics courses but wondered if you would ever use them. In fact, logarithms (or "logs" for short) are extremely useful in many areas of science. The essential idea is that they make it much easier to deal with very large *ranges* of numbers, for example, moving by powers of 10 from 0.0001 to 1,000,000.

It is likely that you have met logarithmic scales without necessarily knowing it. The Richter scale for expressing magnitudes of earthquakes is one example. On this scale, an earthquake of magnitude 6 is 10 times more powerful than one of magnitude 5. An earthquake of magnitude 8 would be 100 times more powerful than one of magnitude 6. Another example is the decibel (dB) scale. Each increase of 10 units represents a 10-fold increase in sound level. Therefore, a normal conversation between two people 1 m apart (60 dB) is 10 times louder than quiet music (50 dB) at the same distance. Loud music (70 dB) and extremely loud music (80 dB) are 10 times and 100 times as loud, respectively, as a normal conversation.

A simple exercise using a pocket calculator can be a good way to learn about logs. You will need a calculator that "does" logs and preferably has a "scientific notation" option. Start by finding the logarithm of 10. Simply enter 10 and press the "log" button. The answer should be 1. Next find the log of 100 and then the log of 1000. Write down the answers. What pattern do you see? (The pattern may be more obvious if you recall that 100 can be written as 10^2 and 1000 is the same as 10^3.) Predict the log of 10,000 and then check it out. Then try the log of 0.1 or 10^{-1} and log of 0.01 (10^{-2}). Predict the log of 0.0001 and check it out.

So far so good, but we have only been considering whole-number powers of 10. It would be helpful to be able to obtain the logarithm of any number. Once again, your handy little calculator comes to the rescue. Try calculating the logs of 20 and 200, then 50 (5×10^1) and 500 (5×10^2). Predict the log of 5×10^3, or 5000. Now for something slightly trickier: the log of 0.05. Finally, try the log of 2473 and the log of 0.000404. In each of the three cases, does the answer seem to be in the right ballpark? Remember that

your calculator will happily provide you with many more digits than have any meaning, so you will need to do some reasonable rounding.

In Chapter 6, the concept of pH is introduced as a quantitative way to describe the acidity of a substance. A pH value is simply a special case of a logarithmic relationship. It is defined as the negative of the logarithm of the H^+ concentration, expressed in units of molarity (M). Square brackets are used to indicate molar concentrations. The mathematical relationship is given by the equation $pH = -\log [H^+]$. The negative sign indicates an inverse relationship; as the H^+ concentration diminishes, the pH increases. Let us apply the equation by using it to calculate the pH of a beverage with a hydrogen ion concentration of 0.000546 M. We first set up the mathematical equation and substitute the hydrogen ion concentration into it.

$$pH = -\log [H^+] = -\log (5.46 \times 10^{-4} \text{ M})$$

Next, we take the negative logarithm of the H^+ concentration by entering it into a calculator and pressing the log button, then the "plus/minus" key to change the sign. This gives 3.26 as the pH of the beverage. (It may display 3.262807357 if you have not preset the number of digits, but common sense prompts you to round the displayed value.) Apply the same procedure to calculate the pH of milk with a hydrogen ion concentration of 2.20×10^{-7} M.

If we can convert hydrogen ion concentration into pH, how do we go in the reverse direction, that is, how to convert pH into a hydrogen ion concentration? Your calculator can do this for you if it has a button labeled "10^x." Alternatively, it may use two buttons: first "Inv" and then "log." To demonstrate the procedure, suppose you wish to find the hydrogen ion concentration of human blood with a pH of 7.40. Proceed as follows: Enter 7.40, use the "plus/minus" key to change the sign to negative, and then hit 10^x (or follow whatever steps are appropriate for your calculator). The display should give the hydrogen ion concentration as 3.98×10^{-8} M. Now apply the same procedure to calculate the H^+ concentration of an acid rain sample with a pH of 3.6.

Your Turn
Appendix 3.1

Find the pH concentration in each sample.

a. tap water, $[H^+] = 1.0 \times 10^{-6}$ M

b. milk of magnesia, $[H^+] = 3.2 \times 10^{-11}$ M

c. lemon juice, $[H^+] = 5.0 \times 10^{-3}$ M

d. saliva, $[H^+] = 2.0 \times 10^{-7}$ M

Answers

a. 6.0 b. 10.5

c. 2.3 d. 6.7

Your Turn
Appendix 3.2

Find the H^+ concentration in each sample.

a. tomato juice, pH = 4.5

b. acid fog, pH = 3.3

c. vinegar, pH = 2.5

d. blood, pH = 7.6

Answers

a. 3.2×10^{-5} M b. 5.0×10^{-4} M

c. 3.2×10^{-3} M d. 2.5×10^{-8} M

Answers to Your Turn Questions
Not Answered in the Text

Chapter 1

1.9 Both ozone and sulfur dioxide show the same pattern of higher limits of exposure for shorter time periods. For ozone, the 1-hr standard is 0.12 ppm, higher than the 8-hr standard of 0.08 ppm. For sulfur dioxide, the 3-hr standard of 0.50 ppm is higher than the 24-hr standard of 0.14 ppm, which in turn, is higher than the 0.03 ppm standard for annual exposure.

1.10 Yes. Judging by Table 1.5, $PM_{2.5}$ must have more serious health consequences than PM_{10} because the concentrations are set at lower limits.

1.11 **a.** $\dfrac{1050 \ \mu g \ SO_2}{15 \ m^3} = \dfrac{70 \ \mu g \ SO_2}{1 \ m^3}$

b. The woman's exposure does not exceed either the 24-hr standard of 365 $\mu g/m^3$ or the annual standard of 80 $\mu g/m^3$.

1.13 In general, concentrations of pollutants are likely to be higher nearer Earth's surface, where the atmosphere is denser and the pollutants are generated. Temperature inversion layers may trap the pollutants in layers, causing some variation in concentration as a function of altitude. Tall smokestacks may deliver pollutants higher into the atmosphere, helping to distribute them more widely.

1.15 **d.** element **e.** compound **f.** mixture

1.16 **b.** carbon, chlorine (compound)

c. hydrogen, oxygen (compound)

d. carbon, hydrogen, oxygen (compound)

f. nitrogen, oxygen (compound)

1.18 **b.** SO_2 is sulfur dioxide; SO_3 is sulfur trioxide.

1.19 **a.** The "eth" in ethanol indicates two carbon atoms in the chemical formula.

c. The "prop" in propanol indicates three carbon atoms in the chemical formula.

1.20 **b.**

Balanced equation: $N_2 + 2\,O_2 \longrightarrow 2\,NO_2$

1.22 Equation 1.7 contains 16 C, 36 H, and 34 O on each side.

1.24 Your list should include: O_2, N_2, CO_2, CO, H_2O, NO, soot (particulate matter), and VOCs. The exhaust also contains tiny amounts of Ar and even tinier amounts of He, but we usually omit these as they are inert gases at low concentration.

1.25 **b.** $CuS + O_2 \longrightarrow Cu + SO_2$

1.26 Small engines can be big polluters. For example, emissions from gasoline-powered lawn mowers, leaf vacuums, chain saws, or snow blowers can include significant amounts of CO, NO_x, hydrocarbons, and particulate matter.

1.31 Indoor activities that generate pollutants include burning incense, painting, smoking, cooking, using a fireplace or woodstove fires, operating a faulty furnace or space heater, using aerosol products for cleaning or personal grooming, installing new carpet, tracking in pollen and mold from yard work, and many others.

1.34 **b.** No, it wouldn't be more valid. Considering the margin of error expected for a handheld meter, it is unlikely that the additional digits would provide any real information. Compare this to dividing $20 evenly among seven people. Would each person receive $2.857142857 just because your calculator displayed this result?

Chapter 2

2.2 **c.** The maximum is 12,000 ozone molecules per billion molecules and atoms of all gases that make up the stratosphere.

    d. The EPA limit is 0.08 ppm for an 8-hr average, equivalent to 80 ppb or 80 ozone molecules per billion molecules and atoms of all gases that make up the troposphere.

2.4 **c.** 17 protons, 17 electrons

    d. 24 protons, 24 electrons

2.5 **c.** Group 5A; 5 outer electrons

    d. Group 8A; 8 outer electrons

2.6 **b.** Beryllium (Be), magnesium (Mg), calcium (Ca), strontium (Sr), barium (Ba), and radium (Ra) all have two outer electrons and are members of Group 2A.

2.7 **c.** 53 protons, 53 electrons, 78 neutrons

2.8 **b.** :B̈r·   2 Br atoms × 7 outer electrons per atom = 14 outer electrons. These are the Lewis structures for Br_2.

$$:\ddot{B}r\!:\!\ddot{B}r: \ \text{or} \ :\ddot{B}r\!-\!\ddot{B}r:$$

2.9 **b.** ·Ċ·   1 C atom × 4 outer electrons per atom = 4 outer electrons

       :C̈l·   2 Cl atoms × 7 outer electrons per atom = 14 outer electrons

       :F̈·   2 F atoms × 7 outer electrons per atom = 14 outer electrons

    Total = 32 outer electrons

These are the Lewis structures for CCl_2F_2.

$$:\ddot{C}l\!:\!\ddot{C}\!:\!\ddot{F}: \ \text{or} \ :\ddot{C}l\!-\!\ddot{C}\!-\!\ddot{F}:$$

2.10 **b.** ·S̈·   1 S atom × 7 outer electrons per atom = 14 outer electrons

       ·Ö·   2 O atoms × 6 outer electrons per atom = 12 outer electrons

    Total = 26 outer electrons

These are the Lewis structures for SO_2.

$$\ddot{O}\!=\!\overset{\cdot\cdot}{S}\!-\!\ddot{O}: \ \longleftrightarrow \ :\ddot{O}\!-\!\overset{\cdot\cdot}{S}\!=\!\ddot{O}$$

2.12 **b.** The average wavelength for a radio wave is about 10^1 m. The average wavelength for an X-ray is about 10^{-9} m, making a radio wave approximately 10^{10} times longer.

2.16 **a.** O_3 forms if O and O_2 combine.

$$O + O_2 \longrightarrow O_3$$

    b. O_3 is removed by two mechanisms. It can be decomposed into O_2 and O or can combine with atomic oxygen to form 2 molecules of O_2.

$$O_3 \longrightarrow O_2 + O$$
$$O_3 + O \longrightarrow 2\,O_2$$

    c. Adding the three forward equations reveals no net change has taken place. The Chapman cycle is a set of natural steady-state reactions in which ozone is both formed and destroyed in the stratosphere.

$$O_2 \longrightarrow 2\,O$$
$$O + O_2 \longrightarrow O_3$$
$$\underline{O_3 + O \longrightarrow 2\,O_2}$$
$$2\,O_2 + 2\,O + O_3 \longrightarrow 2\,O_2 + 2\,O + O_3$$

2.22 **a.** H·   1 H atom × 1 outer electron per atom = 1 outer electron

       ·Ö·   1 O atom × 6 outer electrons per atom = 6 outer electrons

    Total = 7 outer electrons

These are the Lewis structures for ·OH.

$$·\ddot{O}\!:\!H \ \text{or} \ ·\ddot{O}\!-\!H$$

    b. No. The concentration of water vapor in the stratosphere is too low to cause the observed decrease in ozone concentration.

2.30 **b.** There are several possibilities. This is the structure for one possible halon with two carbon atoms, halon 2402.

$$:\ddot{B}r\!:\!\ddot{C}\!:\!\ddot{C}\!:\!\ddot{B}r: \ \text{or} \ :\ddot{B}r\!-\!\ddot{C}\!-\!\ddot{C}\!-\!\ddot{B}r:$$

Chapter 3

3.2 **c.** Higher energy, shorter wavelength UV radiation passes through the windows of the car, and its energy is absorbed. The energy is reradiated as heat, but the lower energy, longer wavelength IR radiation is unable to pass back out through the windows and so it is trapped.

3.3 **a.** Twenty-five percent of energy from the Sun is reflected from the atmosphere, 6% reflected from the surface, 9% is emitted from the surface, and 60% is emitted from the atmosphere. These processes sum to 100%.

    b. Twenty-three percent is directly absorbed in the atmosphere, and 37% is absorbed in the atmosphere when Earth radiates longer wavelength heat energy. These values sum to 60%, the energy emitted from the atmosphere.

    c. Yellow represents a mix of incoming wavelengths, blue the higher energy, shorter wavelength UV radiation, and red the lower energy, longer wavelength IR radiation.

3.10 **b.** Total the outer electrons: 4 + 2(7) + 2(7) = 32. Eight of these go around the central C atom to form four single bonds, one to each Cl atom and one to each F atom. The other 24 outer electrons are nonbonding

pairs on the Cl or F atoms. The CCl_2F_2 molecule is tetrahedral, the same shape as CH_4 and CCl_2F_2.

$$:\overset{..}{\underset{..}{F}}: \quad\quad :\overset{..}{F}: \quad\quad F$$
$$:\overset{..}{\underset{..}{Cl}}:\overset{..}{\underset{..}{C}}:\overset{..}{\underset{..}{Cl}}: \text{ or } :\overset{..}{Cl}-\overset{|}{\underset{|}{C}}-\overset{..}{\underset{..}{Cl}}: \text{ or } \overset{}{\underset{Cl}{C}}\cdots Cl$$
$$:\overset{..}{\underset{..}{F}}: \quad\quad :\overset{..}{\underset{..}{F}}: \quad\quad 109.5° \; F$$

c. Total the outer electrons: $2(1) + 6 = 8$. Eight of these go around the central S atom to form two single bonds, one to each H atom. There also are two nonbonded pairs. The H_2S molecule is bent.

$$H:\overset{..}{\underset{..}{S}}:H \text{ or } H-\overset{..}{\underset{..}{S}}-H \text{ or } H\overset{S}{\underset{<109°}{\diagdown\diagup}}H$$

3.11 a. Total the outer electrons: $6 + 2(6) = 18$. Eight of these go around the central S atom to form one single bond to an O atom and one double bond to an O atom. The other 10 outer electrons are nonbonding pairs on either the S or O atoms. The SO_2 molecule is bent, just as the O_3 molecule is bent. Only one resonance form is shown.

$$\overset{..}{\underset{..}{O}}::\overset{..}{\underset{..}{S}}:\overset{..}{\underset{..}{O}}: \text{ or } \overset{..}{\underset{..}{O}}=\overset{..}{\underset{..}{S}}-\overset{..}{\underset{..}{O}}: \text{ or } \overset{S}{\underset{120°}{\cdot\overset{..}{O}\diagdown\diagup\overset{..}{O}\cdot}}$$

b. Total the outer electrons: $6 + 3(6) = 24$. Eight of these go around the central S atom to form one double bond and two single bonds to an O atom. The other 18 outer electrons are nonbonding pairs on O atoms. The SO_3 molecule is triangular planar. Only one resonance form is shown.

$$:\overset{..}{\underset{..}{O}}: \quad\quad :\overset{..}{\underset{..}{O}}: \quad\quad :\overset{..}{\underset{..}{O}}:$$
$$\overset{..}{\underset{..}{O}}::\overset{..}{\underset{..}{S}}:\overset{..}{\underset{..}{O}}: \text{ or } \overset{..}{\underset{..}{O}}=\overset{}{\underset{}{S}}-\overset{..}{\underset{..}{O}}: \text{ or } \overset{S}{\underset{120°}{\cdot\overset{..}{O}\diagdown\diagup\overset{..}{O}\cdot}}$$

3.13 b. $\dfrac{10,000}{15.00 \; \mu m} = 666.7 \; cm^{-1}$

c. $2350 \; cm^{-1}$ corresponds to stretching vibrations and $666.7 \; cm^{-1}$ corresponds to bending vibrations, both causing low transmittance and high absorbance of IR radiation.

3.17 a. Transportation-related CO_2 emissions are the largest sector. They overtook industrial emissions in 1999 and remain the largest source of energy-related CO_2.

b. The commercial sector grew from about 750 million metric tons per year in 1990 to about 1000 million metric tons per year in 2005. This is the calculation to verify the percent growth per year over the 15-year period.

Percent growth per year
$$= \dfrac{(1000 \; MMT - 750 \; MMT)}{(750 \; MMT)(15 \; yr)} \times 100 = 2\% \text{ per year}$$

c. Long-term growth in CO_2 emissions can be influenced by economic growth and policies regarding choices of energy sources. Short-term variations can be affected by the weather, economic fluctuations, and the relative prices of fossil fuels.

3.18 a. The atomic number of N is 7 and the atomic mass is 14.01.

b. N-14 has 7 protons, 7 neutrons, and 7 electrons.

c. N-15 has 7 protons, 8 neutrons, and 7 electrons. Only the number of neutrons differs.

d. N-14 is the most abundant natural isotope, given that the atomic mass is 14.01.

3.20 b. $5 \times 10^9 \text{ N atoms} \times \dfrac{2.34 \times 10^{-23} \text{ g N}}{\text{N atom}}$
$$= 1.17 \times 10^{-13} \text{ g N}$$

c. $6 \times 10^{15} \text{ N atoms} \times \dfrac{2.34 \times 10^{-23} \text{ g N}}{\text{N atom}}$
$$= 1.40 \times 10^{-7} \text{ g N}$$

3.21 b. 44.0 g/mol N_2O **c.** 137.5 g/mol CCl_3F

3.22 c. The mass ratio compares the molar mass of N to the molar mass of N_2O.

$$\dfrac{14.0 \text{ g N}}{44.0 \text{ g } N_2O} = \dfrac{0.318 \text{ g N}}{1.00 \text{ g } N_2O}$$

To find the mass percent of N in N_2O, multiply the mass ratio by 100.

$$\dfrac{0.318 \text{ g N}}{1.00 \text{ g } N_2O} \times 100 = 31.8\% \text{ N in } N_2O$$

3.23 b. The mass ratio of S to SO_2 is known from Your Turn 3.22.

$$142 \times 10^6 \text{ t } SO_2 \times \dfrac{32.1 \times 10^6 \text{ t S}}{64.1 \times 10^6 \text{ t } SO_2} = 71.1 \times 10^6 \text{ t S}$$

3.25 a. The GWP of HFC-134a is 1300 times that of CO_2. However, HFC-134a has a far lower tropospheric abundance, 7.5 ppt compared with 385,000 ppt (385 ppm) for CO_2.

b. Despite being banned by the Montreal Protocol, Freon-12 remains in the atmosphere and will be for years to come. The comparison shows that Freon-12, CCl_2F_2, has the highest GWP of these gases and is more abundant that HFC-134a in the troposphere. However, Freon-12 is far less abundant than CO_2, the gas most closely identified with global warming.

	GWP	Tropospheric Abundance (ppt)
Freon-12, CCl_2F_2	10,600	553
HFC-134a	1,300	7.5
CO_2	1	385,000

c. Global atmospheric lifetimes are important because more direct and indirect effects on global warming can take place if a greenhouse gas lasts a long time in the atmosphere. Freon-12 has a higher GWP, a higher abundance, and a longer atmospheric lifetime than HFC-134a. Therefore, it is able to have

more effects on global warming over a longer period of time, a major reason it has been banned.

3.35 **a.** The total tonnes of world CO_2 emissions are projected to nearly double between 1995 and 2035. In 1995, the total was 6.46×10^9 t. In 2035, the total is projected to be 11.71×10^9 t.

b. The percentage for the developed world is expected to drop from 73% to 50% from 1995 to 2035.

c. In 1995, the developed world contributed 73% of 6.46×10^9 t, or 4.72×10^9 t. By 2035, the developed world is expected to contribute 50% of 11.71×10^9 t, or 5.86×10^9 t. Thus, although the percentage of emissions from the developed world is dropping, the total amount of CO_2 emissions is increasing.

Chapter 4

4.2 **c.** $92 \text{ kcal} \times \dfrac{4.184 \text{ kJ}}{1 \text{ kcal}} = 380 \text{ kJ}$

d. 2.9×10^3 22-lb concrete blocks

4.10 Breaking the double bond in O_2 requires 498 kJ/mol. The resonance structures for O_3 include both a double and a single bond. A single O-to-O bond only requires 146 kJ, and the true bond energy between O atoms in O_3 will be intermediate between the single and double bond values. Energy is inversely proportional to wavelength. Therefore, the *higher* bond energy of O_2 requires radiation of *shorter* wavelength to break its bonds.

4.17 **a.** $\dfrac{16.0 \text{ g O}}{88.2 \text{ g } C_5H_{12}O} \times 100 = 18\%$ O in MTBE

b. $\dfrac{16.0 \text{ g O}}{46.1 \text{ g } C_2H_6O} \times 100 = 35\%$ O in ethanol

Chapter 5

5.6 **a.** soluble   **b.** very soluble
c. insoluble   **d.** soluble
e. insoluble   **f.** partially soluble (pure aspirin)

5.7 **a.** 975 mg/day

b. No. Calcium requirements vary considerably depending on age and sex, among other factors.

c. 25 500-mL bottles

5.9 **a.** 16 ppb; 1.6×10^{-2} ppm

b. No. The lead concentration is over the standard of 15 ppb.

5.10 **a.** 7.5×10^{-1} mol NaCl in 500 mL of a 1.5 M NaCl solution

7.5×10^{-2} mol NaCl in 500 mL of 0.15 M NaCl solution

b. To calculate molarity, divide moles of solute by liters of solution. The second solution is 3.0 M; this is greater than the first solution which is 2.0 M.

c. No. The precise concentration will be less than 2.0 M. The 80.0 g of NaOH has volume, and when the 1.0 L of water is added, the resulting solution will be slightly more than 1.0 L. To prepare a solution of exactly 2.0 mol/L, add the 80.0 g NaOH to a volumetric flask, then add enough water to make 1.0 L of solution.

d. The resulting concentration is 2.5 M.

5.11 **b.** An O-to-H bond is more polar than a N-to-H one; the electron pair forming the bond will be more strongly attracted by the oxygen atom.

c. A S-to-O bond is more polar than a N-to-O one; the electron pair forming the bond will be more strongly attracted by the oxygen atom.

5.13 **a.** Four hydrogen bonds surround the central H_2O molecule.

b. Hydrogen bonds are intermolecular forces because they are between, not within, water molecules.

5.16 **b.** Mg^{2+}, the Lewis structures are $\cdot Mg \cdot$ and Mg^{2+}

c. O^{2-}, the Lewis structures are $\cdot \ddot{O} \cdot$ and $\left[:\ddot{O}: \right]^{2-}$

d. Al^{3+}, the Lewis structures are $\cdot \dot{Al} \cdot$ and Al^{3+}

5.17 **b.** F forms F^- ions and K forms K^+ ions. A neutral compound will have one F^- for every one K^+ ion; the formula is KF.

c. Mn forms Mn^{2+} ions, and O forms O^{2-} ions. The formula is MnO.

d. Cl forms Cl^- ions, and Al forms Al^{3+} ions. There must be 3 Cl^- ions for every one Al^{3+} ion; the formula is $AlCl_3$.

e. Co forms two different ions: Co^{2+} and Co^{3+}. S forms only 2^- ions. So the two possible formulas are CoS and Co_2S_3.

5.18 **c.** $Al(C_2H_3O_2)_3$ **d.** NH_4OH

5.19 **c.** sodium hydrogen carbonate or sodium bicarbonate

d. calcium carbonate

e. magnesium phosphate

5.20 **b.** Li_2CO_3 **c.** KNO_3 **d.** $BaSO_4$

5.22 **b.** Soluble. All sodium salts are soluble.

c. Insoluble. Most sulfides are insoluble (except those with Group 1A or NH_4^+ cations).

d. Insoluble. Most hydroxides are insoluble (except those with Group 1A or NH_4^+ cations).

5.23

----- indicates a hydrogen bond

5.31 a. 20 ppb = 20 μg/L; higher concentration than 0.003 mg/L = 3 μg/L = 3 ppb

b. 20 ppb is greater than the 15 ppb standard; 3 ppb is less than the 15 ppb standard.

5.33 b. Approximately 2–3 ppb Pb^{2+}

c. Approximately 18–19 ppb Pb^{2+}

Chapter 6

6.1

Name	Formula	Lewis Structure and Points of Interest
nitrogen ("nitrogen gas")	N_2	$:N \equiv N:$ The major component of our atmosphere and much of what you breathe. (Section 1.1)
nitrogen monoxide (nitric oxide)	NO	$\dot{N} = \ddot{O}$ A primary air pollutant produced from N_2 and O_2 in the presence of a high temperature. (Section 1.9)
nitrogen dioxide	NO_2	(two resonance structures shown) An air pollutant produced in the atmosphere from NO. (Section 1.11)
nitrate ion	NO_3^-	(resonance structures shown in brackets) A common polyatomic ion whose salts are soluble in water. Used in fertilizers. Can be toxic to infants. (Sections 5.8, 5.9, 5.11)
ammonium ion	NH_4^+	(structure shown in brackets with + charge) A common polyatomic ion whose salts are soluble in water. Used in fertilizers. (Sections 5.8, 5.9)
ammonia	NH_3	(structure shown) A compound formerly used as a refrigerant gas, later replaced by CFCs. (Section 2.9)

6.2 a. $HI(aq) \longrightarrow H^+(aq) + I^-(aq)$

b. $HNO_3(aq) \longrightarrow H^+(aq) + NO_3^-(aq)$

6.4 a. $KOH(s) \longrightarrow K^+(aq) + OH^-(aq)$

b. $LiOH(s) \longrightarrow Li^+(aq) + OH^-(aq)$

6.5 a.

$$HNO_3(aq) + KOH(aq) \longrightarrow KNO_3(aq) + H_2O(l)$$

$$H^+(aq) + NO_3^-(aq) + K^+(aq) + OH^-(aq) \longrightarrow K^+(aq) + NO_3^-(aq) + H_2O(l)$$

$$H^+(aq) + OH^-(aq) \longrightarrow H_2O(l)$$

b.

$$H_2SO_4(aq) + 2\,NH_4OH(aq) \longrightarrow (NH_4)_2SO_4(aq) + 2\,H_2O(l)$$

$$2\,H^+(aq) + SO_4^{2-}(aq) + 2\,NH_4^+(aq) + 2\,OH^-(aq) \longrightarrow 2\,NH_4^+(aq) + SO_4^{2-}(aq) + 2\,H_2O(l)$$

$$2\,H^+(aq) + 2\,OH^-(aq) \longrightarrow 2\,H_2O(l)$$

$$H^+(aq) + OH^-(aq) \longrightarrow H_2O(l)$$

6.6 b. The solution is basic because $[OH^-] > [H^+]$.

$$[H^+] = \frac{1 \times 10^{-14}}{1 \times 10^{-6}} = 1 \times 10^{-8}\ M$$

c. The solution is basic because $[OH^-] > [H^+]$.

6.7 a. The solution is basic because $[OH^-] > [H^+]$.
$OH^-(aq) = K^+(aq) > H^+(aq)$; basic

b. The solution is acidic because $[H^+] > [OH^-]$.
$H^+(aq) = NO_2^-(aq) > OH^-(aq)$; acidic

c. The solution is acidic because $[H^+] > [OH^-]$.
$H^+(aq) = HSO_3^- > SO_3^{2-}(aq) = OH^-(aq)$; acidic

6.9 a. The sample of lake water with pH = 4 is 10 times more acidic and has 10 times more H^+ than the sample of rainwater with pH = 5.

6.12 $H_2SO_3(aq) \longrightarrow H^+(aq) + HSO_3^-(aq)$
$HSO_3^-(aq) \longrightarrow H^+(aq) + SO_3^{2-}(aq)$
$H_2SO_3(aq) \longrightarrow 2\,H^+(aq) + SO_3^{2-}(aq)$

6.13 c. 3.36×10^4 tons SO_2

d. The SO_2 is carried by the wind into the surrounding areas. It eventually dissolves in the water droplets of clouds, fog, and rain to form acidic precipitation.

6.14 Between 1940 and 1995, the NO_x emissions from industry did not change much. In contrast, the emissions from combustion engines used for transportation and from fossil-fuel combustion increased. For SO_2, industrial emissions led to the large increase in the 1970s.

6.15 a. Iron is a metal with chemical symbol Fe.

b. Aluminum is a metal with chemical symbol Al.

c. Fluorine is a nonmetal with chemical symbol F.

d. Calcium is a metal with chemical symbol Ca.

e. Zinc is a metal with chemical symbol Zn.

f. Oxygen is a nonmetal with chemical symbol O.

6.16 Add equations 6.22 and 6.23 together, and cancel like terms.

$$4\,Fe(s) + 2\,O_2(g) + 8\,H^+(aq) \longrightarrow 4\,Fe^{2+}(aq) + 4\,H_2O(l)$$
$$4\,Fe^{2+}(aq) + O_2(g) + 4\,H_2O(l) \longrightarrow 2\,Fe_2O_3(s) + 8\,H^+(aq)$$

$$4\,Fe(s) + 3\,O_2(g) \longrightarrow 2\,Fe_2O_3(s)$$

6.17 a. Fe

　c. With respect to Fe, Fe^{2+} has lost two valence electrons and Fe^{3+} has lost three.

6.18 a. $MgCO_3(s) + 2\,H^+(aq) \longrightarrow$
$$Mg^{2+}(aq) + CO_2(g) + H_2O(l)$$

　b. Sodium bicarbonate is soluble in water.

6.19 $CaCO_3(s) + H_2SO_4(aq) \longrightarrow$
$$CaSO_4(s) + CO_2(g) + H_2O(l)$$

6.22 $S(s) + O_2(g) \longrightarrow SO_2(g)$
$SO_2(g) + \frac{1}{2}\,O_2(g) \longrightarrow SO_3(g)$
$SO_3(g) + H_2O(l) \longrightarrow H_2SO_4(aq)$

6.23 a. $H_2SO_4(aq) + NH_4OH(aq) \longrightarrow$
$$NH_4HSO_4(aq) + H_2O(l)$$

　b. $H_2SO_4(aq) + 2\,NH_4OH(aq) \longrightarrow$
$$(NH_4)_2SO_4(aq) + 2\,H_2O(l)$$

6.25 a. 78%

　b. $:N \equiv N:$

　c. The N-to-N triple bond is one of the most difficult bonds to break, requiring 946 kJ/mol. In comparison, the C-to-C triple bond requires 813 kJ/mol. See Table 4.2.

6.26 a. Using NH_4^+ as an example:
$H_2SO_4(aq) + 2\,NH_4OH(aq) \longrightarrow$
$$(NH_4)_2SO_4(aq) + 2\,H_2O(l)$$

　b. Lewis structure for NH_4^+. See Your Turn 6.1 for other structures.

$$\left[\begin{array}{c} H \\ | \\ H-N-H \\ | \\ H \end{array}\right]^+$$

6.27 b. The hydrogen carbonate ion is accepting a hydrogen ion, so it is acting as a base.

6.33 a. $N_2 + O_2 \xrightarrow{\text{high temperature}} 2\,NO$

　$2\,NO + O_2 \longrightarrow 2\,NO_2$ (net reaction)

　$NO_2 \xrightarrow{\text{sunlight}} NO + O$

　$O + O_2 \longrightarrow O_3$

　b. Nitrogen saturation is observed in many lakes.

Chapter 7

7.4 U-234 has 92 protons and 142 neutrons.

7.5 b. $_0^1n + {}_{92}^{235}U \longrightarrow {}_{52}^{137}Te + {}_{40}^{97}Zr + 2\,_0^1n$

7.6 For anthracite coal:

$$9.0 \times 10^{10}\,\text{kJ} \times \frac{1.0\,\text{g anthracite coal}}{30.5\,\text{kJ}} \times \frac{1\,\text{kg}}{1000\,\text{g}}$$
$$= 3.0 \times 10^6\,\text{kg anthracite coal}$$

The equivalent masses for the other grades are calculated in the same way. Bituminous: 2.9×10^6 kg; Subbituminous: 3.8×10^6 kg; Lignite (brown coal): 5.6×10^6 kg.

7.7 $_{95}^{241}Am \longrightarrow {}_{93}^{237}Np + {}_2^4He$

$_4^9Be + {}_2^4He \longrightarrow {}_6^{12}C + {}_0^1n + {}_0^0\gamma$

7.8 In an emergency such as an earthquake, the fuel rods should be inserted to slow down the fission process in the reactor core in preparation for shutdown.

7.9 The cloud is condensed water vapor. People sometimes call this "steam," although technically this is not correct. Steam is water vapor and is invisible until it condenses. The cloud does not contain any nuclear fission products.

7.12 c. UV rays refer to UV radiation, a part of the electromagnetic spectrum. The word *rays* is used to connect UV radiation with the Sun's rays.

　d. In this case, radiation refers to the particles and rays spontaneously emitted by radioactive uranium nuclei.

7.13 b. $_{94}^{239}Pu \longrightarrow {}_{92}^{235}U + {}_2^4He$

7.14 a.

　atom　　molecule　　ion

　b. The iodine atom is the most reactive of the three because it has one unpaired electron (it is a free radical).

7.20 a. Answers will vary because of cosmic ray exposure (elevation) and medical exposures.

　b. The results from the two links are not identical, particularly for smokers. The Los Alamos National Laboratory calculator does not include smoking as a variable.

7.21 $\dfrac{3.0 \times 10^{14}\,\text{C-14 atoms}}{3.5 \times 10^{26}\,\text{C atoms}} \times 100 = 8.57 \times 10^{-11}\%\,\text{C-14}$

7.23 After 36.9 years (3 half-lives), only 12.5% of the original tritium will remain.

7.24 a. Radon-222 is part of the natural decay series of uranium-238, the isotope of uranium that is most common in nature. Therefore if uranium is present, radon will be as well.

　b. Four half-lives (4×3.8 days = 15.2 days) are required for the level of radioactivity to drop from 16 pCi to 1 pCi.

　c. Radon will continue to enter your basement because the uranium in the soils and rocks underneath your home is continuously producing it. See part a.

7.25 $_0^1n + {}_{92}^{235}U \longrightarrow {}_{54}^{143}Xe + {}_{38}^{90}Sr + 3\,_0^1n$

7.30 a. The other isotope is U-238, which is not fissionable under the conditions in a nuclear reactor. A trace amount of U-234 is present as well.

　b. The origin of these other radioisotopes is the process of fission itself. The splitting of U-235 does not occur the same way each time, so many different product nuclei are formed.

Chapter 8

8.2 Equations **a, c**, and **e** are reduction half-reactions because electrons are gained, appearing on the reactant side of the equations. Equations **b** and **d** are oxidation

half-reactions because electrons are lost, appearing on the product side.

8.3 **a.** Each equation is balanced from the standpoint of number and type of atoms.

b. Each equation is balanced from the standpoint of charge, but the total charge does not have to be zero on each side, just the same.

c. Electrons do not appear in the overall cell reaction, but are shown in half-reactions.

8.4 **a.** $Zn \longrightarrow Zn^{2+} + 2\,e^-$

b. $Cu^{2+} + 2\,e^- \longrightarrow Cu$

c. In this galvanic cell, $Zn(s)$ is oxidized to $Zn^{2+}(aq)$. The mobile zinc ions cannot easily reform into a solid zinc electrode when reduced, which would be essential during a recharging step.

8.7 **a.** $Pb(s)$ is being oxidized and $PbO_2(s)$ is being reduced.

b. $PbSO_4(s)$ is both oxidized and reduced during recharging.

8.17 **a.** For equation 8.22, bonds broken: 4 mol (C-to-H single bonds) + 4 mol (O-to-H single bonds)

$$= 4 \text{ mol } (416 \text{ kJ/mol}) + 4 \text{ mol } (467 \text{ kJ/mol})$$
$$= 1664 \text{ kJ} + 1868 \text{ kJ}$$
$$= 3532 \text{ kJ (endothermic step)}$$

For equation 8.22, bonds formed: 4 mol (H-to-H single bonds) + 2 mol (C-to-O double bonds)

$$= 4 \text{ mol } (436 \text{ kJ/mol}) + 2 \text{ mol } (803 \text{ kJ/mol})$$
$$= 1744 \text{ kJ} + 1606 \text{ kJ}$$
$$= 3350 \text{ kJ (exothermic step)}$$

For equation 8.22, overall reaction: $(+3532 \text{ kJ}) + (-3350 \text{ kJ}) = +182 \text{ kJ}$. A similar calculation for equation 8.23 gives $+252$ kJ.

b. Both reactions are endothermic. More energy is required to break bonds than is released in bond formation.

c. Although there is general agreement, remember that Table 4.2 gives *average* bond energies, not specific energies associated with the bonds in these compounds (with the exception of CO_2).

8.21 **a.** $M(s) + H_2(g) \longrightarrow MH_2(s)$ (charging)
$MH_2(s) \longrightarrow M(s) + H_2(g)$ (discharging)

b. The metals used in NiMH batteries are quite similar to those used for hydrogen storage. Maximizing the amount of hydrogen contained in the metal is an important consideration and both applications require the incorporation of hydrogen to be fast and reversible. A difference is that there is no gaseous H_2 in the battery application. Instead, protons are combined directly with OH^- ions to produce water.

8.24 **a.** Doping with phosphorus forms an *n*-type semiconductor. Phosphorus is in Group VA and has an additional electron per atom than a silicon atom.

b. B-doping forms a *p*-type semiconductor. Boron is in Group IIIA and has one fewer electron per atom than a silicon atom.

8.25 Ga is in Group 3A and has 3 outer electrons. As is in Group 5A and has 5 outer electrons. When they bond in a similar array to Ge or Si, the Group 4A element shown in Figure 8.20, each will have a share in a stable octet of electrons. Cd is in Group IIB (a Group that behaves much like the IIA Group) and has 2 outer electrons. Se is in Group VIA and has 6 outer electrons, allowing each to have a share in a stable octet of electrons when bonded.

Chapter 9

9.4 **a.**

b.

9.5 Pearly shampoo bottles, brightly colored detergent bottles, translucent milk jugs, squeeze glue bottles—these and many other plastic containers that hold soaps and nonoily foods typically are made of HDPE. This plastic is usually opaque and sometimes brightly colored (because it is mixed with a pigment). In contrast, you are more likely to find LDPE as a plastic packing sheet material or plastic baggie. It is usually flexible and often clear rather than translucent.

9.11 **a.**

b.

9.13

The head-to-tail arrangement is favored because the bulky benzene rings are spaced as far apart as possible, minimizing repulsions between them. In the head-to-head, tail-to-tail arrangement, the rings would fall on adjacent carbons.

9.14 a.

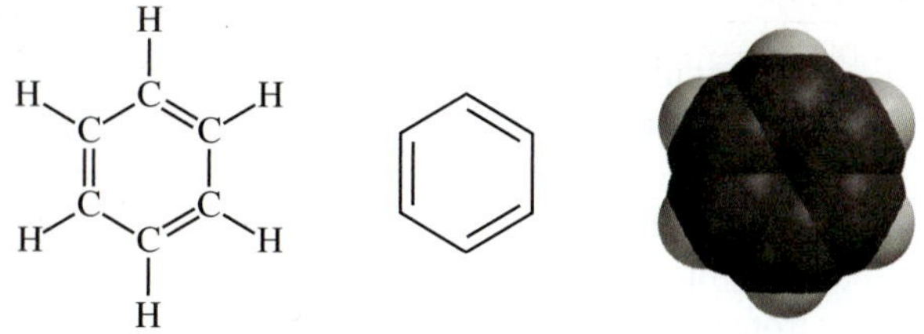

b.

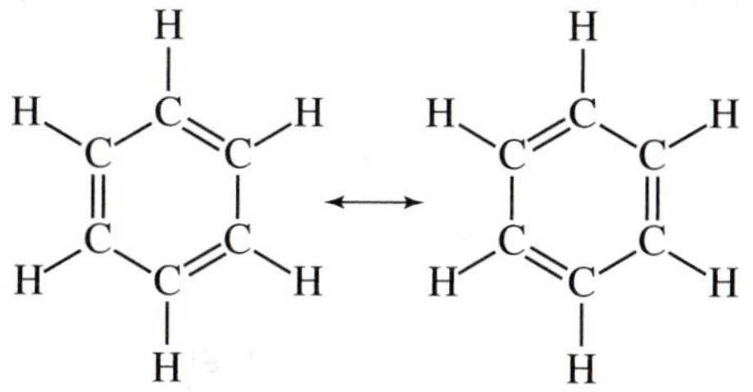

benzene

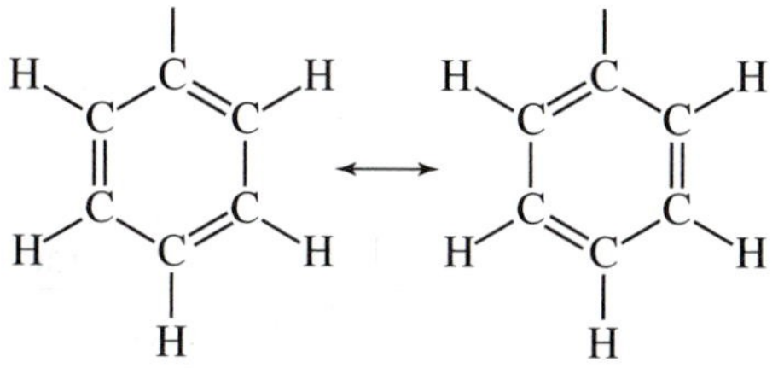

phenyl group

9.17

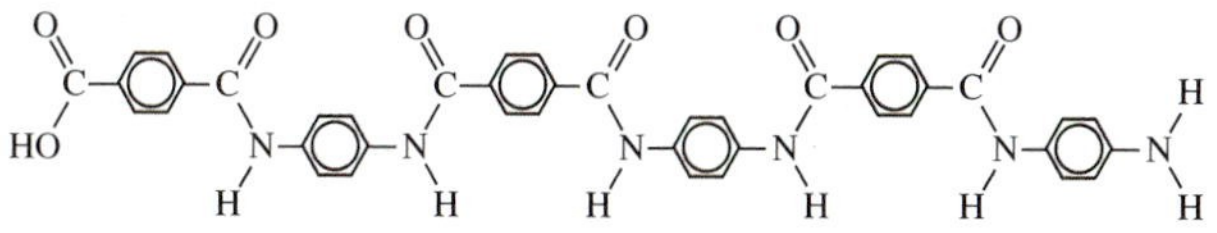

9.22 Propylene is $H_2C=CHCH_3$ or C_3H_6.

$$2500\ C_3H_6 + 11250\ O_2 \longrightarrow 7500\ CO_2 + 75000\ H_2O$$

9.25 When returned for a deposit, most glass bottles are cleaned, refilled, and reused. This is the most preferred scenario (at the top of the solid waste management hierarchy). Deposits on plastic containers are uncommon. However, some food stores, particularly food cooperatives, sell items in bulk. These stores may charge for plastic containers, thus encouraging customers to bring them back and refill them, time after time. This, too, would be a preferred scenario.

9.27 Recyclable items that you might purchase include many plastic containers and jugs, beverages in aluminum cans, and newspapers. Examples of recycled-content items include paper products (some paper napkins, towels, toilet paper), and plastic items (some outdoor picnic tables, park benches, lumber, and even railroad ties). In theory, anything that contains postconsumer waste can be recycled again. In practice, however, this may not occur because municipalities do not offer recycling services for all potentially recyclable products.

Chapter 10

10.4 Each carbon atom in Figure 10.2 is surrounded by eight electrons (four bonds), so the octet rule is followed.

10.5 c.

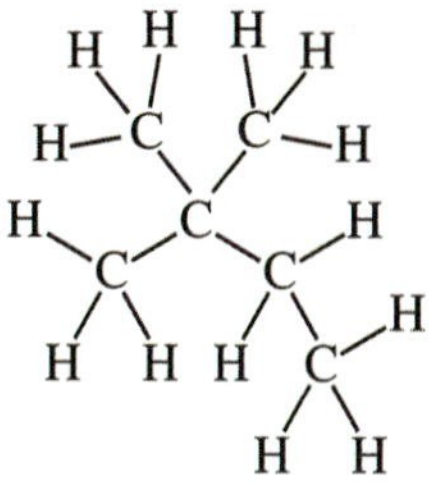

d.

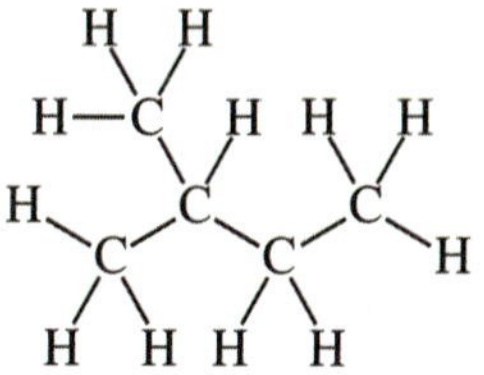

10.6 a. Yes, *n*-butane and *iso*butane are isomers. They have the same formula, C_4H_{10}, but different structures.

b. No, *n*-hexane (C_6H_{14}) and cyclohexane (C_6H_{12}) are not isomers. They have different formulas as well as different structures.

10.7

Structural Formula	Condensed Structural Formula, Line-Angle Drawing
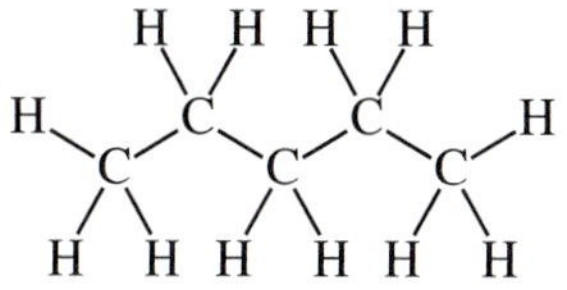	$CH_3CH_2CH_2CH_2CH_3$ 
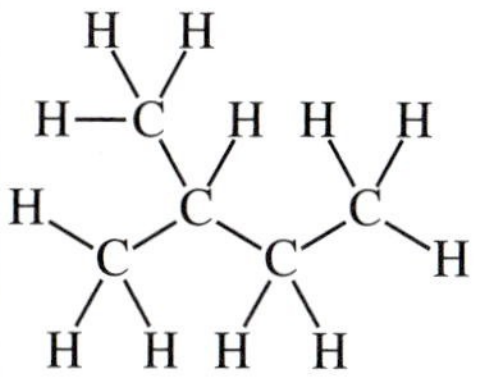	$CH_3CH(CH_3)CH_2CH_3$ 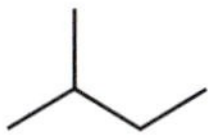
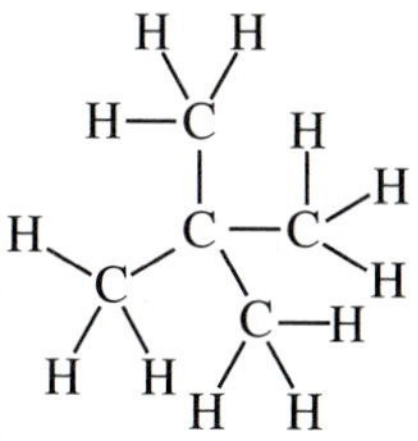	$C(CH_3)_4$ 

10.8 c.

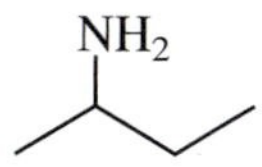

amine

d.

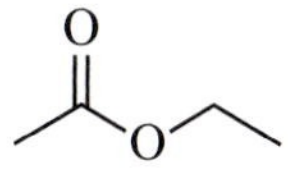

ester

e.

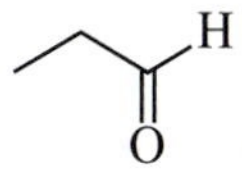

aldehyde

10.9 a.

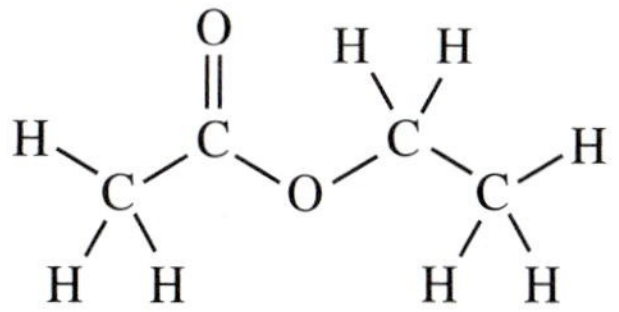

b.

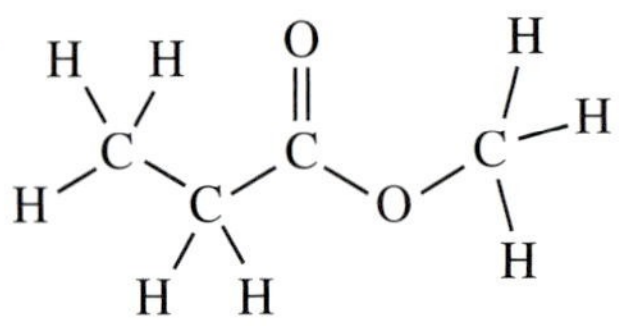

10.10 All three molecules have a benzene ring with attached functional groups that can form hydrogen bonds with water. Both aspirin and ibuprofen have a carboxylic acid group attached.

10.16 b.   OH

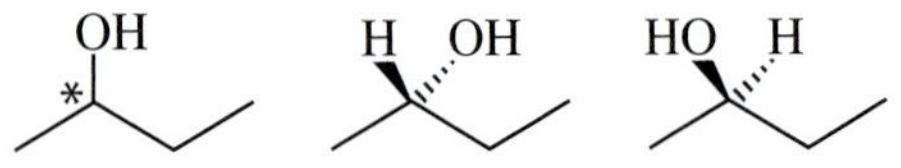

c. *CHClFCH$_3$

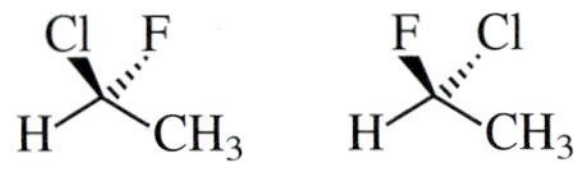

d.

    HNCH$_3$

    CH$_3$NH H

    H HNCH$_3$

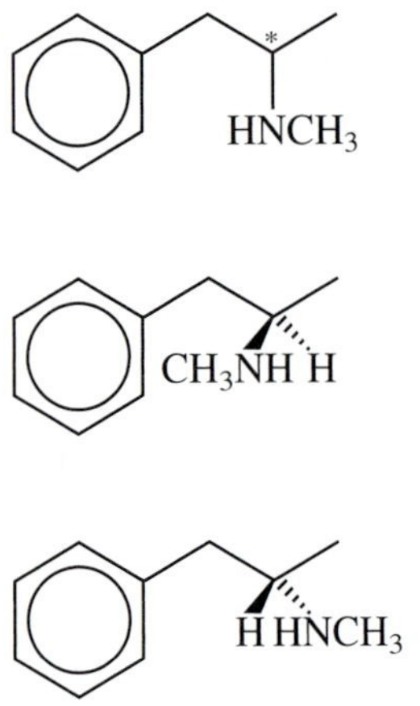

10.17 a. The chiral carbons are indicated with an asterisk in these structures.

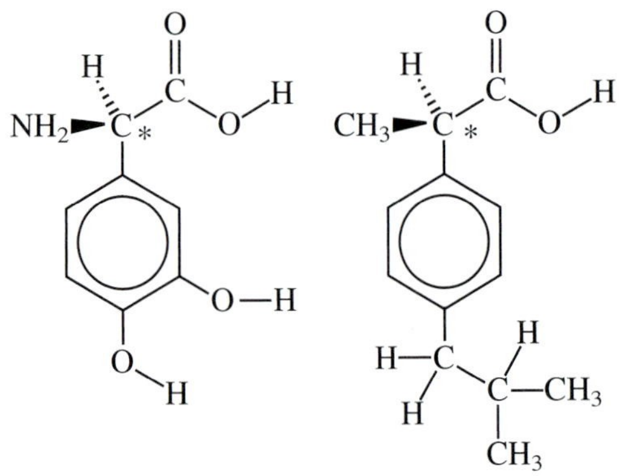

(−)-ibuprofen   (+)-dopa

b. Ibuprofen has a phenyl ring and a carboxylic acid group. The dopa molecule has a phenyl ring, an amine, and a carboxylic acid group.

c. Drawings are given in part **a.**

10.19 a.

estradiol	progesterone
—OH group on the D ring	C=O group off the D ring
—CH$_3$ group on CD ring intersection	C=O and C=C on the A ring
	—CH$_3$ groups on CD and AB ring intersections

estradiol	testosterone
—OH group on the D ring	—OH group on the D ring
—CH$_3$ group on CD ring intersection	—CH$_3$ groups on CD and AD ring intersections
	C=O and C=C on the A ring

cholic acid	cholesterol
—OH groups on the A, B, and C rings	—OH group on the A ring
carboxylic acid group on D ring side chain	double bond in A ring
—CH$_3$ groups on CD and AD ring intersections	—CH$_3$ groups on CD and AD ring intersections

corticosterone	cortisone
C=O and C=C on the A ring	C=O and C=C on the A ring
—OH group on C ring	C=O group on C ring, and also on D ring side chain
—OH group on D ring side chain	—OH groups on D ring and D ring side chain
—CH$_3$ groups on CD and AD ring intersections	—CH$_3$ groups on CD and AD ring intersections

prednisone	cortisone
C=O and 2 C=C bonds on the A ring	C=O and C=C on the A ring
—OH group on C ring	C=O group on C ring, and also on D ring side chain
—OH group on D ring side chain	—OH groups on D ring and D ring side chain
—CH$_3$ groups on CD and AD ring intersections	—CH$_3$ groups on CD and AD ring intersections

b. estradiol $C_{18}H_{24}O_2$ progesterone $C_{21}H_{30}O_2$
corticosterone $C_{21}H_{30}O_4$ testosterone $C_{19}H_{28}O_2$
cholic acid $C_{24}H_{40}O_5$ prednisone $C_{21}H_{26}O_5$
cortisone $C_{21}H_{28}O_5$ cholesterol $C_{27}H_{46}O$

10.31 Hydrochloric acid, HCl(*aq*), is used to form the salt of oxycodone. This is the structure for the salt. The N atom no longer has a lone pair of electrons, which is what characterizes the freebase form.

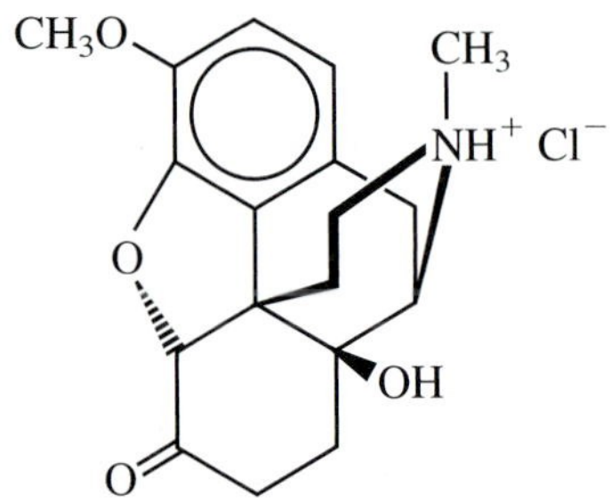

Chapter 11

11.4 There are 64.6 g of oxygen and 10.0 g of hydrogen per 100 g of body mass. However, the number of atoms is dependent on the number of *moles* of each, not the mass.

$$\frac{64.6 \text{ g O}}{100 \text{ g body mass}} \times \frac{1 \text{ mol O}}{16.0 \text{ g O}} = \frac{4.0 \text{ mol O}}{100 \text{ g body mass}}$$

$$\frac{10.0 \text{ g H}}{100 \text{ g body mass}} \times \frac{1 \text{ mol H}}{1.01 \text{ g H}} = \frac{9.9 \text{ mol H}}{100 \text{ g body mass}}$$

$$\frac{9.9 \text{ mol H}}{4.0 \text{ mol O}} = \frac{2.5 \text{ mol H}}{1.0 \text{ mol O}}$$

This predicts that there will be about 2.5 times as many H atoms as there are O atoms, a fact reflected in their relative abundance per million atoms in the body.

$$\frac{630,000 \text{ atoms H}}{255,000 \text{ atoms O}} = \frac{2.5 \text{ atoms H}}{1 \text{ atom O}}$$

11.5 c.

d. The three ester groups are shown in red.

11.6 c.

d. The three ester groups are shown in red.

11.10 Salivary enzymes break down the complex carbohydrates in unsweetened crackers into simple sugars, which are responsible for the sweet taste.

11.12 GlyGlyGly, GlyGlyAla, GlyAlaAla, GlyAlaGly, AlaAlaAla, AlaAlaGly, AlaGlyGly, AlaGlyAla

11.17 Answers will vary depending on the food items chosen.

11.26 Answers will vary depending on the cereals chosen.

Chapter 12

12.2 a. There was a 14% increase (from 66 to 75) for males and an 11% increase for females (from 72 to 80) between 1950 and 2004.

b. Factors that have been proposed are that women have better diets and have hormonal differences that offer some protection against heart attacks before menopause. Another suggested factor is that the percentage of women who smoke is lower. Job-related stress was once suggested as a factor in lowering life expectancy for men more than for women, but that may not be making much difference in today's society.

12.5

All of the nitrogen and oxygen atoms in nucleotides have nonbonding electron pairs. The names are cytidine phosphate, guanosine phosphate, and thymidine

phosphate. The suffix of each name represents the type of base it is; cytosine and thymine are pyrimidines, and adenine and guanine are purines.

12.8 3×10^9 base pairs $\times \dfrac{0.34 \text{ nm}}{\text{base pair}} \times \dfrac{1 \text{ m}}{1 \times 10^9 \text{ nm}} = 1 \text{ m}$

Section 12.1 notes that the length is 2 m, but this is for the combined length of both DNA strands.

12.11 Because multiple codons can represent a certain amino acid, the possibility of mistakes during protein synthesis is reduced. A mutation can occur in several ways. For example, the codon CAA may have had a mutation where a G was inserted between the two adenines forming CAG. Although the codon is now different, it still codes for the same amino acid as CAA. Thus there will be no major problem in protein synthesis.

12.13 Bond length decreases as multiple bonding increases. Methanol has a C-to-O single bond (143 pm), the length of the dipeptide bond is somewhere between a single and a double bond (123 pm), CO_2 contains C-to-O double bonds (121 pm), and CO contains a C-to-O triple bond (111 pm).

Answers to Selected End-of-Chapter Questions Indicated in Color in the Text

Chapter 1

1. $\dfrac{0.5\ \text{L}}{1\ \text{breath}} \times \dfrac{15\ \text{breaths}}{1\ \text{minute}} \times \dfrac{60\ \text{minutes}}{1\ \text{hour}} \times \dfrac{8\ \text{hours}}{1\ \text{working day}}$
$= 3600\ \text{L}$

5. **a.** $N_2 > O_2 > Ar > CO_2 > CO > Rn$

   b. It is more convenient to express the concentrations of CO_2 and CO in ppm.

6. **a.** $9000\ \text{ppm} \times \dfrac{100\ \text{parts per hundred}}{1{,}000{,}000\ \text{ppm}}$
$= 0.9\ \text{parts per hundred or } 0.9\%$

10. **a.** 85,000 g

    b. 10,000,000 gallons

11. **a.** $2.2 \times 10^{-4}\ \text{g/m}^3$

    b. No, because CO is odorless.

13. **a.** Group 1A and Group 7A

    b. 1A: hydrogen, lithium, sodium, potassium, rubidium, cesium, francium
    7A: fluorine, chlorine, bromine, iodine, astatine

16. **a.** A molecule of CH_4 consists of 1 carbon atom and 4 hydrogen atoms.
    A molecule of SO_2 consists of 1 sulfur atom and 2 oxygen atoms.
    A molecule of O_3 consists of 3 oxygen atoms.

    b. CH_4 (methane), SO_2 (sulfur dioxide), O_3 (ozone)

18. **a.** $N_2(g) + O_2(g) \longrightarrow 2\ NO(g)$

    b. $O_3(g) \longrightarrow O_2(g) + O(g)$

    c. $2\ S(s) + 3\ O_2(g) \longrightarrow 2\ SO_3(g)$

24. **a.** platinum (Pt), palladium (Pd), rhodium (Rh)

    b. All three metals are in Group 8B on the periodic table. Platinum is directly under palladium, and rhodium is just to the left of palladium.

    c. These metals are solids at the temperature of the exhaust gases, so they must have relatively high melting points. Also, they do not undergo permanent chemical change when catalyzing the reaction of CO to CO_2 in the exhaust stream.

28. **a.** $5.0 \times 10^{-3}\ \text{m} < 1\ \text{m} < 3.0 \times 10^2\ \text{m}$

30. **a.** Compound (2 molecules of one compound made up of two different elements).

b. Mixture (2 atoms of one element plus 2 atoms of another).

c. Mixture (three different substances, two elements and one compound).

d. Element (4 atoms of the same element).

33. CO is termed the "silent killer" because your senses cannot detect this colorless, tasteless, and odorless gas. The same term cannot be applied to pollutants such as O_3, SO_2, or NO_2 because each has a distinctive odor that can be detected at concentrations below the level of toxicity.

36. Ozone up high is "good" because it protects us by absorbing incoming ultraviolet radiation from the Sun. Ozone nearby is "bad" because it can be dangerous to breathe and damages vegetation and some materials.

37. **a.** The elderly, the young, and people with respiratory problems such as asthma and emphysema are most affected by ozone.

    b. 7 days

    c. Ozone is highly reactive, thus does not persist long in the atmosphere. Since no ozone is produced at night, its concentration falls.

    d. Answers will vary. Possibilities include: overcast skies, rain, or high winds. It could be a day when fewer people are driving or that industries are shut down.

    e. Ozone levels in London, Ontario, are lower in December because there is less daylight in the winter months.

44. Jogging outdoors, as opposed to sitting outdoors, increases your exposure to air pollutants because you will be breathing harder and exchanging more air during your exercise.

47. Formaldehyde can be released from cigarette smoke and from synthetic materials such as foam insulation, and from the adhesives used in dying and gluing carpet pads, carpets, and laminated building materials. The air indoors is often not well circulated, leading to an accumulation of formaldehyde and other pollutants. Efforts to make homes air tight, leading to

greater energy efficiency, have led in some cases to making problems of indoor air pollution worse, rather than better.

Chapter 2

1. a. Yes; 0.118 is equivalent to 118 ppb, above the detection minimum of 10 ppb.

$$\frac{0.118 \text{ parts } O_3}{1,000,000 \text{ parts air}} = \frac{118 \text{ parts } O_3}{1,000,000,000 \text{ parts air}} \text{ or } 118 \text{ ppb}$$

   b. Yes; 25 ppm is equivalent to 25,000 ppb, above the detection minimum of 10 ppb.

$$\frac{25 \text{ parts } O_3}{1,000,000 \text{ parts air}} = \frac{25,000 \text{ parts } O_3}{1,000,000,000 \text{ parts air}}$$
   or 25,000 ppb

4. a. Diamond and graphite are allotropes of carbon. They are two different forms of the same element, carbon.

   b. Water and hydrogen peroxide are not allotropes. They are different compounds of hydrogen and oxygen.

7. a. O; 8 p, 8 e   **b.** N; 7 p, 7 e

9. a. helium, He

   b. potassium, K

   c. copper, Cu

10. c. 92 protons, 146 neutrons, 92 electrons

   f. 88 protons, 138 neutrons, 88 electrons

11. c. The element is radon. The symbol for this isotope is $^{222}_{88}$Rn.

12. a. ·Ca·   **b.** ·N̈·   **c.** :C̈l·

13. b. 2(1) + 2(6) = 14 outer electrons
   H:Ö:Ö:H and H—Ö—Ö—H

   c. 2(1) + 6 = 8 outer electrons
   H:S̈:H and H—S̈—H

15. a. Wave **1** has a longer wavelength than wave **2**.

   b. Wave **1** has a lower frequency than wave **2**.

   c. Wave **1** and wave **2** travel forward at the same speed.

18. This is the order of increasing energy per photon.
   radio waves < infrared radiation < visible light < gamma rays

20. a. UV-C < UV-B < UV-A

   b. UV-A < UV-B < UV-C

   c. UV-A < UV-B < UV-C

22. a. Cl, 7 outer electrons, :C̈l·

   NO_2, 17 outer electrons, Ö::N̈:Ö:

   ClO, 13 outer electrons, ·C̈l:Ö:

   HO, 7 outer electrons, ·Ö:H

   b. Each of these species has less than a full octet of outer electrons on one atom. Their reactivities are based on their inclination to attain a full octet of outer electrons for each atom capable of holding an octet.

27. a. Methane, CH_4, has 4 + 4(1) = 8 outer electrons.

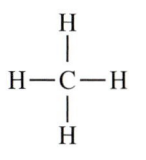

   Ethane, C_2H_6, has 2(4) + 6(1) = 14 outer electrons.

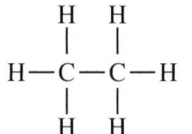

   b. Fourteen different CFCs can be formed from methane.

30. Ozone in the stratosphere is necessary for absorbing harmful UV radiation, but down in the troposphere it is harmful to living beings.

31. *Allotropes* of oxygen, O_2 and O_3, have different molecular forms.

   Isotopes of oxygen differ in the number of neutrons in the nucleus. For example, $^{18}_{8}$O has 8 protons, 10 neutrons, and 8 electrons if electrically neutral. An atom of $^{16}_{8}$O has 8 protons, 8 neutrons, and 8 electrons when electrically neutral.

39. UV-C radiation is extremely dangerous, but it is almost completely absorbed by stratospheric oxygen, O_2, as well as by ozone, O_3, before it can reach the surface of the Earth.

42. This graph describes the amount of solar energy in the UV-A and UV-B regions that reaches the top of the atmosphere and the Earth. The red curve is the amount of energy above the atmosphere while the blue curve represents the amount of energy that reaches the surface of the Earth. The graph shows that the stratospheric ozone layer absorbs light energy in the UV-B region (280–300 nm). However, wavelengths in the UV-A region (320–360 nm) are only slightly absorbed. The y-axis is a log scale. This means that at 330 nm, for example, the amount of energy reaching the surface of the Earth is 10 times less than that above the atmosphere.

   The data in this graph has strong biological implications. First, it shows that a large amount of solar energy in the UV-A region, the region responsible for sunburn, reaches the Earth. Secondly, it suggests that depletion of the ozone layer will allow more UV-B energy to reach the Earth's surface. Exposure to UV-B is strongly associated with skin cancers. Thus, if more solar energy in the UV-B region of the spectrum reaches the Earth, the incidence of skin cancers and other afflictions associated with exposure to UV-radiation will increase.

43. ClO· takes part in a catalytic cycle so a few ClO· species can destroy many O_3 molecules.

53. O_2, O_3, and N_2 all have even numbers of electrons in their Lewis structures. N_3 would have 15 electrons, an odd number. Molecules with odd numbers of electrons are generally more reactive than those in which all electrons are present in pairs.

58. a. 90 + 12 = 102. The compound contains one carbon atom, no hydrogen atoms, and two fluorine atoms. The chemical formula for CFC-12 is CF_2Cl_2.

b. CCl_4 contains one carbon atom, no hydrogen atoms, and no fluorine atoms. Therefore, the code number for CCl_4 is CFC-10.

c. Yes the "90" method will work for HCFCs. 90 + 22 = 112, so HCFC-22 would be composed of one carbon, one hydrogen, and two fluorine atoms and its chemical formula would be CHF_2Cl.

d. No, this method will not work for halons. There are no guidelines for handling bromine.

Chapter 3

1. a. Yes. Earth is warmer than predicted based on distance from the Sun and the amount of radiation reaching the Earth. Without the greenhouse effect, our planet would be too cold to be hospitable to life as we know it.

b. Yes. Most scientists now conclude that observed increases in Earth's average temperature are evidence that enhanced greenhouse effect, or global warming, is taking place. There is also more than 90% likelihood that man's activities are affecting that increase.

4. a. The number of atoms of each element on either side is the same. C = 6, O = 18, H = 12

b. The number of molecules is not the same on either side of the equation. There are 12 on the left, but only 7 on the right. The large molecule glucose has formed on the product side of the equation, using 24 atoms per molecule.

6. a. The rest of the Sun's energy is absorbed in or reflected from the atmosphere.

7. b. The mean atmospheric temperature at present is somewhat above the 1950–1980 mean atmospheric temperature. Twenty thousand years ago, the mean atmospheric temperature was lower by about 9 °C. However, 120,000 years ago the mean atmospheric temperature was lower by only about 1 °C than it is at present.

c. Although there appears to be a *correlation* between mean atmospheric temperature and CO_2 concentration, this figure does not prove *causation* of either factor by the other.

10. These are the two Lewis structures.

$$H-H \quad \text{and} \quad H-\overset{\cdot\cdot}{\underset{\cdot\cdot}{O}}-H$$

The two atoms of H_2 can only be arranged in a straight line. For H_2O, even though the Lewis structure shows atoms in a straight line, this does not mean that the molecule is linear. In fact, the bent structure of water is so well known that the Lewis structure is often written in this manner.

$$H-\overset{\cdot\cdot}{\underset{|}{\underset{H}{O}}}:$$

13. a. 3(1) + 4 + 6 + 1 = 14 outer electrons. This is the Lewis structure.

$$H-\overset{\displaystyle H}{\underset{\displaystyle H}{\overset{|}{\underset{|}{C}}}}-\overset{\cdot\cdot}{\underset{\cdot\cdot}{O}}-H$$

b. The geometry around the C atom is tetrahedral and there are no lone pairs. A H-to-C-to-H bond angle of 109.5° is predicted.

c. There are four pairs of electrons around the O atom, two of which are bonding pairs and two are non-bonded pairs. Repulsion between the two nonbonded electron pairs and their repulsion of the bonding pairs is predicted to cause the H-to-O-to-C bond angle to be slightly less than 109.5°.

16. All of them can contribute to the greenhouse effect. In every case, as the bond stretches the atoms move and therefore the charge distribution changes. Unlike linear CO_2, water molecules are bent and the polarity of the molecule changes with each of these modes of vibration.

17. a. $E = \dfrac{hc}{\lambda}$ can be used to calculate the energies.

$$E = \frac{6.63 \times 10^{-34} \text{ J·s} \times 3.00 \times 10^8 \text{ m/s}}{4.26 \ \mu m \times \dfrac{1 \text{ m}}{10^6 \ \mu m}} = 4.67 \times 10^{-20} \text{ J}$$

$$E = \frac{6.63 \times 10^{-34} \text{ J·s} \times 3.00 \times 10^8 \text{ m/s}}{1500 \ \mu m \times \dfrac{1 \text{ m}}{10^6 \ \mu m}} = 1.33 \times 10^{-22} \text{ J}$$

20. $C_6H_{12}O_6(aq) \xrightarrow{\text{yeast}} 2\ C_2H_5OH(aq) + 2\ CO_2(g)$

22. a. A neutral atom of Ag-107 has 47 protons, 60 neutrons, and 47 electrons.

b. A neutral atom of Ag-109 has 47 protons, 62 neutrons, and 47 electrons. Only the number of neutrons has changed.

25. a. 2(1.0) + 16.0 = 18.0 g/mol

b. 12.0 + 2(19.0) + 2(35.5) = 121.0 g/mol

29. a. CO_2

b. Although CO_2 accounts for the largest percentage of greenhouse gas contributions, the case could be made that N_2O has the greatest effect. It is almost 300 times more effective than CO_2, but its concentration is far lower than that of either CO_2 or CH_4. However, the net effectiveness of N_2O is the largest of these three gases.

Gas	Percent Contribution (graph)	GWP (Table 3.5)	Net Effectiveness (Product)
CO_2	55	1	0.6
CH_4	15	23	3.5
N_2O	5	296	14.8

33. Drilled ocean cores can be analyzed for the number of type of microorganisms present. Another correlating piece of evidence is the changing alignment of magnetic field in particles in the sediment over time. Another possibility is to analyze the deuterium-to-hydrogen ratio in ice cores.

36. Lewis dot structures are very good at showing which atom is joined with another, but the representation on paper does not necessarily indicate the molecular geometry. That must be predicted by looking at the number of bonded electron pairs and lone electron pairs around the central atom. Using H_2O as an example, it has $2(1) + 6 = 8$ electrons. This is the Lewis structure.

$$H-\ddot{\underset{\cdot\cdot}{O}}-H$$

The Lewis structure was drawn to show that a H atom is linked to an O atom, which is linked to another H atom. There are two lone electron pairs. Far from being linear, which might have been incorrectly predicted looking at this two-dimensional Lewis structure, the H_2O molecule is bent with a bond angle somewhat less than $109.5°$ because of electron pair repulsion.

40. **a.** $C_2H_5OH + 3\,O_2 \longrightarrow 3\,H_2O + 2\,CO_2$

    b. $2\ mol\ CO_2$ **c.** $30\ mol\ O_2$

45. 73×10^6 metric tons $CH_4 \times \dfrac{12\ \text{metric tons C}}{16\ \text{metric tons }CH_4}$

$$= 5.5 \times 10^7\ \text{metric tons C}$$

Chapter 4

3. $70\ Cal \times \dfrac{4.184\ kJ}{1\ Cal} \times \dfrac{1000\ J}{1\ kJ} \times \dfrac{1\ beat}{1\ J} \times \dfrac{1\ min}{80\ beats}$

$$= 3700\ min$$

5. **a.** $2\,C_2H_6 + 7\,O_2 \longrightarrow 4\,CO_2 + 6\,H_2O$

b.
$$2\ H-\overset{\displaystyle H}{\underset{\displaystyle H}{\overset{|}{\underset{|}{C}}}}-\overset{\displaystyle H}{\underset{\displaystyle H}{\overset{|}{\underset{|}{C}}}}-H\ +\ 7\ \ddot{\underset{\cdot\cdot}{O}}=\ddot{\underset{\cdot\cdot}{O}} \longrightarrow$$

$$4\ \ddot{\underset{\cdot\cdot}{O}}=C=\ddot{\underset{\cdot\cdot}{O}}\ +\ 6\ \underset{H}{\overset{\ddot{\overset{\cdot\cdot}{O}}}{\diagup\ \diagdown}}H$$

6. $\dfrac{52.0\ kJ}{1\ g\ C_2H_6} \times \dfrac{30.1\ g\ C_2H_6}{1\ mol\ C_2H_6} = \dfrac{1570\ kJ}{1\ mol\ C_2H_6}$

10. **a.** Exothermic; a charcoal briquette releases heat as it burns.

    b. Endothermic; liquid water gains the necessary heat for evaporation from your skin, and your skin feels cool.

12. **a.** Bonds broken in the reactants

    1 mol N-to-N triple bonds $= 1(946\ kJ) = 946\ kJ$

    3 mol H-to-H single bonds $= 3(436\ kJ) = 1308\ kJ$

    Total energy *absorbed* in breaking bonds $= 2254\ kJ$

    Bonds formed in the products

    6 mol N-to-H single bonds $= 6(391\ kJ) = 2346\ kJ$

    Total energy *released* in forming bonds $= 2346\ kJ$

    Net energy change is $(+2254\ kJ) + (-2346\ kJ)$

$$= -92\ kJ$$

Notice that the overall energy change has a negative sign, characteristic of an exothermic reaction.

b. Bonds broken in the reactants

    12 mol C-to-H single bonds $= 12(416\ kJ) = 4992\ kJ$

    4 C-to-C single bonds $= 4(356\ kJ) = 1424\ kJ$

    11 mol O-to-O double bonds $= 11(498\ kJ) = 5478\ kJ$

    Total energy *absorbed* in breaking bonds $= 11,894\ kJ$

    Bonds formed in the products

    10 mol C-to-O triple bonds $= 10(1073\ kJ)$

$$= 10,730\ kJ$$

    24 mol O-to-H single bonds $= 24(467\ kJ)$

$$= 11,208\ kJ$$

    Total energy *released* in forming bonds $= 21,938\ kJ$

    Net energy change is $(+11,894\ kJ) + (-21,938\ kJ)$

$$= -8927\ kJ$$

Notice that the overall energy change has a negative sign, characteristic of an exothermic reaction.

c. Bonds broken in the reactants

    1 mol H-to-H single bonds $= 1(436\ kJ) = 436\ kJ$

    1 mol Cl-to-Cl single bonds $= 1(242\ kJ) = 242\ kJ$

    Total energy *absorbed* in breaking bonds $= 678\ kJ$

    Bonds formed in the products

    2 mol H-to-Cl single bonds $= 2(431\ kJ) = 862\ kJ$

    Total energy *released* in forming bonds $= 862\ kJ$

    Net energy change is $(+678\ kJ) + (-862\ kJ)$

$$= -184\ kJ$$

Notice that the overall energy change has a negative sign, characteristic of an exothermic reaction.

17. $\dfrac{650,000\ kcal}{1\ day} \times \dfrac{365\ days}{1\ yr} = \dfrac{2.4 \times 10^8\ kcal}{1\ yr}$

This value can be related to each of the energy sources.

b. $\dfrac{2.4 \times 10^8\ kcal}{1\ yr} \times \dfrac{1\ yr}{65\ barrels\ oil} \times \dfrac{1\ barrel\ oil}{42\ gal}$

$$= \dfrac{8.8 \times 10^4\ kJ}{1\ gal}$$

c. $\dfrac{2.4 \times 10^8\ kcal}{1\ yr} \times \dfrac{1\ yr}{16\ tons\ coal} = \dfrac{1.5 \times 10^7\ kJ}{1\ ton\ coal}$

19. Pentane is a liquid at room temperature. Triacontane is a solid at room temperature. Octane is a liquid at room temperature.

22. **a.**
$$H-\overset{\displaystyle H}{\underset{\displaystyle H}{\overset{|}{\underset{|}{C}}}}-\overset{\displaystyle H}{\underset{\displaystyle H}{\overset{|}{\underset{|}{C}}}}-\overset{\displaystyle H}{\underset{\displaystyle H}{\overset{|}{\underset{|}{C}}}}-\overset{\displaystyle H}{\underset{\displaystyle H}{\overset{|}{\underset{|}{C}}}}-H$$

b.
$$\begin{array}{c}H\\|\\H-C-H\\H\ \ \ |\ \ \ H\\|\ \ \ \ |\ \ \ \ |\\H-C-C-C-H\\|\ \ \ \ |\ \ \ \ |\\H\ \ \ H\ \ \ H\end{array}$$

c. There are only the two shown in **a** and **b.**

26. a. The physical state of normal hydrocarbons changes from gas to liquid to solid as the number of carbon atoms increases.

b. $C_{50}H_{102}$

28. a. The C-to-F single bond requires 485 kJ/mol, the C-to-Cl single bond requires 327 kJ/mol, and the C-to-Br bond requires 285 kJ/mol to break the bond. The C-to-Br bond is the weakest, and the bromine free radical can interact with ozone even more effectively than the chlorine free radical.

b.

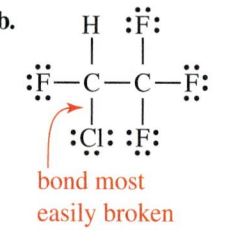

bond most
easily broken

Because the C-to-Cl bond is broken most easily, chlorine free radicals can be released. Such free radicals can catalyze the destruction of ozone.

29. a. When $n=1$, this is the balanced equation.

$$CO + 3 H_2 \longrightarrow CH_4 + H_2O$$

To calculate the heat evolved we use the same method as in Question 4.13.

Bonds broken in the reactants

1 mol C-to-O triple bonds = 1(1073 kJ) = 1073 kJ

3 mol H-to-H single bonds = 3(436 kJ) = 1308 kJ

Total energy *absorbed* in breaking bonds = 2381 kJ

Bonds formed in the products

4 mol C-to-H single bonds = 4(416 kJ) = 1664 kJ

2 mol O-to-H single bonds = 2(467 kJ) = 934 kJ

Total energy *released* in forming bonds = 2598 kJ

Net energy change is (+2361 kJ) + (−2598 kJ)

= −237 kJ

b. Reactions with n greater than 1 will release more energy as n becomes larger.

33. Section 1.11 contained a discussion of catalytic converters in automobiles and the catalytic destruction of ozone by chlorine free radicals was described in Section 2.8.

41. Typically oxygenated fuels have a lower energy content than hydrocarbons. To determine the heat of combustion of MTBE, we first need the balanced combustion reaction.

$$C_5H_{12}O(l) + {}^{15}\!/_2\,O_2(g) \longrightarrow 5\,CO_2(g) + 6\,H_2O(g)$$

We determine the heat of combustion in the usual way.

Bonds broken in the reactants

3 mol C-to-C single bonds = 3(356 kJ) = 1068 kJ

12 mol C-to-H single bonds = 12(416 kJ) = 4992 kJ

2 mol C-to-O single bonds = 2(336 kJ) = 672 kJ

$^{15}\!/_2$ mol O-to-O double bonds = ($^{15}\!/_2$)(498 kJ)

= 3735 kJ

Total energy *absorbed* in breaking bonds = 10,467 kJ

Bonds formed in the products

10 mol C-to-O double bonds = 10(803 kJ)

= 8030 kJ

12 mol O-to-H single bonds = 12(467 kJ) = 5604 kJ

Total energy *released* in forming bonds = 13,634 kJ

Net energy change is (+10,467 kJ) + (−13,634 kJ)

= −3167 kJ

Chapter 5

1. a. 75–98 lb   **b.** 9–12 gal

4. a. Partially soluble. Orange juice concentrate needs to be stirred to prevent some of the suspended pulp from settling out of the mixture.

b. Very soluble. The detergent and water form a solution that spreads evenly throughout the clothes being cleaned.

c. Very soluble. Ammonia and water form a clear solution useful for many purposes.

6. a. 55 mg of Ca^{2+} per liter of bottled water is the same as 55 ppm.

$$\frac{55\ \text{mg Ca}^{2+}}{1\ \text{L H}_2\text{O}} \times \frac{1\ \text{L}}{10^3\ \text{mL}} \times \frac{1\ \text{mL H}_2\text{O}}{1\ \text{g H}_2\text{O}} \times \frac{1\ \text{g}}{10^3\ \text{mg}}$$

$$= \frac{55\ \text{mg Ca}^{2+}}{10^6\ \text{mg H}_2\text{O}} \text{ or 55 ppm Ca}^{2+}$$

b. Evian, with 78 mg Ca^{2+}/L, has a higher Ca^{2+} concentration than this bottled water in which the Ca^{2+} concentration is 55 ppm, or 55 mg Ca^{2+}/L.

9. a. The number of moles of H_2SO_4 in 100 mL of 12 M H_2SO_4 is the same as the number of moles of HCl in 100 mL of 12 M HCl.

$$\frac{12\ \text{mol H}_2\text{SO}_4}{1\ \text{L solution}} \times \frac{1\ \text{L}}{10^3\ \text{mL}} \times 100\ \text{mL solution}$$

$$= 1.2\ \text{mol H}_2\text{SO}_4$$

$$\frac{12\ \text{mol HCl}}{1\ \text{L solution}} \times \frac{1\ \text{L}}{10^3\ \text{mL}} \times 100\ \text{mL solution}$$

$$= 1.2\ \text{mol HCl}$$

b. The number of grams is *not* the same because the molar mass of H_2SO_4 is greater than that of HCl.

$$1.2\ \text{mol H}_2\text{SO}_4 \times \frac{98.1\ \text{g H}_2\text{SO}_4}{1\ \text{mol H}_2\text{SO}_4} = 120\ \text{g H}_2\text{SO}_4$$

$$1.2\ \text{mol HCl} \times \frac{36.5\ \text{g HCl}}{1\ \text{mol HCl}} = 44\ \text{g HCl}$$

11. Although there is slight polarity to each C-to-H bond in CH_4, overall the molecule is symmetrical and nonpolar. The forces among nonpolar molecules are relatively weak, allowing the molecules to escape from the liquid to the gas phase at room temperature and pressure. In contrast to CH_4, the O-to-H bonds in H_2O are very polar and so is the molecule, thanks to its bent shape. The forces among polar water molecules, the hydrogen bonds, are strong. Far more energy is required to change liquid water to a gas than is the case with CH_4.

13. **a.** N and C; $3.0 - 2.5 = 0.5$
   S and O; $3.5 - 2.5 = 1.0$
   N and H; $3.0 - 2.1 = 0.9$
   S and F; $4.0 - 2.5 = 1.5$

   b. N will attract the bonded electron pair more strongly than C.
   O will attract the bonded electron pair more strongly than S.
   N will attract the bonded electron pair more strongly than H.
   F will attract the bonded electron pair more strongly than S.

15. **a.** This is the Lewis structure for ammonia, NH_3.

$$H—\overset{\cdot\cdot}{N}—H$$
$$|$$
$$H$$

   b. Each N-to-H bond is polar. The difference in electronegativity is $3.0 - 2.1$, or 0.9.

   c. The molecule is polar. The molecular shape is triangular pyramidal.

20. **a.** A chlorine atom satisfies the octet rule by gaining one electron to form a chloride ion, with a charge of $1-$.

$$:\overset{\cdot\cdot}{\underset{\cdot\cdot}{Cl}}\cdot \quad \text{and} \quad \left[:\overset{\cdot\cdot}{\underset{\cdot\cdot}{Cl}}:\right]^-$$

   b. A barium atom satisfies the octet rule by losing two electrons to form a barium ion, with a charge of $2+$.

$$\cdot Ba\cdot \quad \text{and} \quad \left[Ba\right]^{2+}$$

21. **a.** Na_2S sodium sulfide
   b. Al_2O_3 aluminum oxide

22. **a.** $Ca(HCO_3)_2$
   b. $CaCO_3$

23. **a.** potassium acetate
   b. calcium hypochlorite

25. **a.** The lightbulb will shine. $CaCl_2$ is an electrolyte. It dissolves in water to form a solution that conducts electricity.

   b. The lightbulb will not shine. C, H, and O atoms are covalently bonded in C_2H_5OH. Although C_2H_5OH dissolves in water to form a solution, the solution will not conduct electricity.

26. **a.** Ca^{2+} and OCl^- ions are present.

   b. No ions are present. C_2H_5OH (ethanol) is not an electrolyte.

28. The concentration of Mg^{2+} is 2.5 M and the concentration of NO_3^- is 5.0 M.

29. **a.** To prepare 2.0 L of 1.5 M KOH, measure out 138 g of KOH and place it into a 2.0 L graduated cylinder. Add water to the 2.0 L mark.

   b. To prepare 1.0 L of .05 M NaBr, measure out 5.15 g of NaBr and place it into a 1.0 L graduated cylinder. Add water to the 1.0 L mark.

35. **a.** The electronegativities of the elements generally increase from left to right across a period (until the 8A Group is reached) and from bottom to top within any group. This means that of the positions indicated, the element in position 2 is predicted to have the highest electronegativity.

   b. Ranking the other elements is not straightforward. Element 1 is expected to be more electronegative than element 3, based on their relative positions in the same group. Element 4 will likely be more electronegative than the 1 and 3 and less electronegative than 2. However, because element 4 is not in the same period with any other element, this prediction cannot be made with certainty. Here are the values found in references; they do not appear in Table 5.4. EN 1 = 0.8; EN 2 = 2.4; EN 3 = 0.7; EN 4 = 1.9. These values confirm the relative order: $3 < 1 < 4 < 2$.

38. NH_3, like water, is a polar molecule. Therefore, despite its low molar mass, considerable energy must be added to liquid NH_3 to break the intermolecular forces among NH_3 molecules.

39. **a.** A single covalent bond holds two hydrogen atoms together in H_2. It is an example of an *intra*molecular force.

   b. Hydrogen bonding is a type of *inter*molecular force, a force between molecules, not within the molecule.

43. With the exception of contaminants that are known carcinogens, MCLG and MCL values are usually very close to being the same.

Chapter 6

1. **a.** Possibilities include nitric acid (HNO_3), hydrochloric acid (HCl), sulfuric acid (H_2SO_4), sulfurous acid (H_2SO_3), phosphoric acid (H_3PO_4), carbonic acid (H_2CO_3), and hydrobromic acid (HBr).

   b. In general, acids taste sour, turn litmus paper red (and have characteristic color changes with other indicators), are corrosive to metals such as iron and aluminum, and release carbon dioxide ("fizz") from a carbonate. These properties may not be observed if the acid is not sufficiently concentrated.

2. **a.** $HBr(aq) \longrightarrow H^+(aq) + Br^-(aq)$

   b. $H_2SO_3(aq) \longrightarrow H^+(aq) + HSO_3^-(aq)$

3. **a.** Possibilities include sodium hydroxide (NaOH), potassium hydroxide (KOH), ammonium hydroxide (NH_4OH), magnesium hydroxide ($Mg(OH)_2$), and calcium hydroxide ($Ca(OH)_2$).

   b. In general, bases taste bitter, turn litmus paper blue (and have characteristic color changes with other indicators), have a slippery feel in water, and are caustic to your skin.

4. **a.** $KOH(s) \longrightarrow K^+(aq) + OH^-(aq)$

6. a.

$$KOH(aq) + HNO_3(aq) \longrightarrow KNO_3(aq) + H_2O(l)$$

$$K^+(aq) + OH^-(aq) + H^+(aq) + NO_3^-(aq) \longrightarrow$$
$$K^+(aq) + NO_3^-(aq) + H_2O(l)$$

$$OH^-(aq) + H^+(aq) \longrightarrow H_2O(l)$$

10. a. The solution of pH = 6 has 100 times more [H$^+$] than the solution of pH = 8.

d. The solution with [OH$^-$] = 1 $\times$ 10^{-2} M has 10 times more [OH$^-$] than the solution with [OH$^-$] = 1 $\times$ 10^{-3} M.

13. a. If we compare Coca-Cola (pH = 2.7) with rainwater (average pH = 5.5), we find that the beverage is about 1000 times more acidic.

14. $S(s) + O_2(g) \longrightarrow SO_2(g)$

21. a. Combustion engines, such as those associated with jet aircraft, directly emit CO, CO$_2$, and NO. If small amounts of sulfur are present in the fuel, SO$_2$ and SO$_3$ will be emitted as well.

22. Fuel combustion contributes greatly to the emissions of SO$_2$, with the major source being the burning of coal. While fuel combustion contributes less to the emissions of NO$_x$, the percentage still is substantial. In contrast, transportation makes a small contribution to the emissions of SO$_2$ whereas a much larger one to the emissions of NO$_x$. Nitrogen monoxide is produced from N$_2$ and O$_2$ in the air wherever there is a high temperature. Thus, both automobile engines and power plants are big contributors of NO$_x$.

25. The molar masses of SO$_2$ and CaCO$_3$ are needed to solve this. Note that it is not necessary to change tons to grams. The ratio of the number of grams per mole is the same as the ratio of the number of kg per kilomole or the ratio of the number of tons per ton·mole.

$$1.00 \text{ ton SO}_2 \times \frac{1 \text{ ton} \cdot \text{mol SO}_2}{64.1 \text{ tons SO}_2} \times \frac{2 \text{ ton} \cdot \text{mol CaCO}_3}{2 \text{ ton} \cdot \text{mol SO}_2}$$

$$\times \frac{100 \cdot \text{tons CaCO}_3}{1 \text{ ton} \cdot \text{mol CaCO}_3} = 1.56 \text{ tons CaCO}_3$$

30. c. Using HC$_2$H$_3$O$_2$ has the advantage of being written in the same format we have used for other acids; that is, with the H atom(s) that have the potential for dissociation being written first. The advantage of using CH$_3$COOH is that it gives a better indication of the order of the linkage of the atoms in the molecule.

32. a. H$_2$O > Na$^+$(aq) = OH$^-$(aq) > H$^+$(aq)

35. a. The slightly higher pH values in the lab indicate that the acidity decreased slightly.

b. Answers will vary. One possibility is that the field samples contained naturally occurring acids that were not stable over time. They decomposed, thus making the solution less acidic. Another possibility is that some of the acids present in the sample reacted with other molecules in the sample or with the sample container itself between the time the sample was collected and analyzed.

Chapter 7

1. All carbon atoms have six protons, but carbon atoms can differ from each other in the number of neutrons (carbon has isotopes). All carbon atoms differ from all uranium atoms in their numbers of protons, neutrons, and electrons.

3. a. 94 protons

b. Np (neptunium), Pu (plutonium)

4. a. C-14 has six protons and eight neutrons.

10. Neutrons are needed to initiate the process of nuclear fission of U-235.

$$_{0}^{1}n + {}_{92}^{235}U \longrightarrow [{}_{92}^{236}U] \longrightarrow {}_{56}^{141}Ba + {}_{36}^{92}Kr + 3\,_{0}^{1}n$$

The fission products include 2 or 3 neutrons that can initiate more fission reactions. In this manner, a self-sustaining chain reaction can be established in which the products of one reaction initiate another.

12. **A** = control rod assembly, **B** = cooling water out of the core, **C** = control rods, **D** = cooling water into the core, **E** = fuel rods.

15. a. $_{0}^{1}n + {}_{5}^{10}B \longrightarrow [{}_{5}^{11}B] \longrightarrow {}_{2}^{4}He + {}_{3}^{7}Li$

b. Boron, like cadmium, can be used to make fuel rods because it is a good neutron absorber.

17. a. ${}_{94}^{239}Pu \longrightarrow {}_{92}^{235}U + {}_{2}^{4}He$

b. As a particulate, plutonium can be inhaled and become lodged in the lungs. If so, the ionizing radiation it produces (alpha particles) can cause damage to lung cells. The product U-235 also is radioactive and similarly can damage tissue.

c. The half-life of Pu-131 is 8.5 minutes, and so 10 half-lives would be 85 minutes. Thus the radioactivity of a sample of Pu-131 would be negligible in a matter of hours.

18. a. ${}_{53}^{131}I \longrightarrow {}_{54}^{131}Xe + {}_{-1}^{0}e$

22. To see what is happening, it may be helpful to construct a chart.

Number of Half-lives	Percent Remaining	Percent Decayed
0	100	0
1	50	50
2	25	75
3	12.5	87.5
4	6.25	93.75
5	3.12	97.88
6	1.56	98.44

26. The natural abundances of U-238 and U-235 are 99.3% and 0.7%, respectively. This would not have any particular significance except for the fact that U-235 can be induced to undergo nuclear fission and thus is suitable both for nuclear power plants and nuclear weapons. Those who would use U-235 for either of

these purposes are not able to readily procure it either in large amount or in pure form.

34. After 7 half-lives, 0.78% of the radioisotope remains (see question #22). Although more than 99% has decayed (reasonably close to being "gone"), the radioactivity actually is *not* gone, as 0.78% of the sample still remains. Thus, if you start with a large amount of a radioactive substance (for example, 2000 lb), after 7 half-lives have passed you have close to 10 lb left. This could be considered a sizeable amount.

50. a. The sum of the masses of the reactants is 5.02838 g, and the sum for the products is 5.00878 g. This means that the mass difference, 0.0196 g, is equivalent to energy following Einstein's equation, $E = mc^2$.

   b. $E = mc^2$

   $$E = 0.0196 \text{ g} \times \frac{1 \text{ kg}}{10^3 \text{ g}} \times \left[\frac{3.00 \times 10^8 \text{ m}}{\text{s}}\right]^2$$

   $$E = 1.76 \times 10^{12} \text{ J}$$

Chapter 8

1. a. Oxidation is a process in which an atom, ion, or molecule *loses* one or more electrons. Reduction is a process in which an atom, ion, or molecule *gains* one or more electrons.

   b. Electrons must be transferred from the species losing electrons to the species gaining electrons.

3. Parts **a** and **c** are redox reactions. Electrons must be transferred to change an element (Zn in part **a**, O_2 in part **c**) into its combined form. Part **b** is a neutralization reaction in which ions combine to form a soluble salt as well as covalently bonded water molecules. Electron transfer does not occur in this case.

4. This is a true statement. We defined combustion as the rapid combination of a fuel and oxygen to make products (usually CO_2 and H_2O), accompanied by the release of energy. The H_2O is formed from O_2 through a reduction reaction that is the reverse reaction of that shown in Your Turn 8.2d.

$$O_2(g) + 4 H^+(aq) + 4 e^- \longrightarrow 2 H_2O(l)$$

The carbon (in whatever compound is used for the fuel) is oxidized to CO_2. Even when substances that do not contain carbon are burned, the reactions can still be described in terms of oxidation and reduction.

6. a. The anode is Zn(s) and the oxidation half-reaction is:

   $$Zn(s) \longrightarrow Zn^{2+}(aq) + 2 e^-$$

   b. The cathode is Ag(s) and the reduction half-reaction is:

   $$2 Ag^+(aq) + 2 e^- \longrightarrow 2 Ag(s)$$

11. a. The electrolyte provides a medium for transfer of ions but blocks the passage of electrons.

   b. KOH paste   **c.** $H_2SO_4(aq)$

13. In current usage, the term *hybrid car* refers to the combination of a gasoline engine with a nickel-metal hydride battery, an electric motor, and an electric generator. Other hybrids using fuel cells are either available or under development.

18. Oxidation half-reaction:
   $$CH_4 + 8 OH^- \longrightarrow CO_2 + 6 H_2O + 8 e^-$$

   Reduction half-reaction:
   $$2 O_2 + 4 H_2O + 8 e^- \longrightarrow 8 OH^-$$

   Overall reaction:
   $$CH_4 + 2 O_2 \longrightarrow CO_2 + 2 H_2O$$

23. $370 \text{ kg H}_2 \times \dfrac{1000 \text{ g}}{1 \text{ kg}} \times \dfrac{1 \text{ mol H}_2}{2.0 \text{ g H}_2} \times \dfrac{286 \text{ kJ}}{1 \text{ mol H}_2}$
   $$= 5.3 \times 10^7 \text{ kJ}$$

24. a. First write the chemical equation for the reaction.

   $$H_2(g) + \tfrac{1}{2} O_2(g) \longrightarrow H_2O(l) + \text{energy}$$

   Then consider the Lewis structures for the reaction.

   $$H\text{—}H + \tfrac{1}{2} \; \overset{..}{\underset{..}{O}} = \overset{..}{\underset{..}{O}} \longrightarrow H^{\overset{..}{O}} {}_{H}$$

   Energy needed to break bonds:
   436 kJ + 1/2 (498 kJ) = 685 kJ
   Energy released as new bonds form:
   2(467 kJ) = −934 kJ
   Net energy change is (685 kJ) + (−934 kJ)
   $$= -249 \text{ kJ}.$$

   b. Average bond energies are based on bonds within molecules in the gaseous state. In the given chemical equation, the H_2O formed is present as a liquid rather than as a gas. Additional energy is released when gaseous water condenses to the liquid state, so the stated value of 286 kJ is greater than the 249 kJ calculated in part **a**.

27. Note that there are eight electrons around each silicon atom, but nine electrons surround the central atom. Each silicon atom has four outer electrons, so the central atom in the figure must have five outer electrons. This is consistent with arsenic, which is in Group 5A. The additional electron forms an *n*-type silicon semiconductor.

29. In every electrochemical process described in this chapter, energy is produced through electron transfer. Such transfer takes place because of a chemical reaction, such as takes place in galvanic cells, batteries, and fuel cells. The transfer may be initiated when light strikes a photovoltaic cell, resulting in the movement of electrons.

33. A storage battery converts chemical energy into electrical energy by means of a reversible reaction. No reactants or products leave the "storage" battery, and the reactants can be reformed during the recharging cycle. A fuel cell also converts chemical energy into electrical energy, but the reaction is not reversible. A fuel cell continues to operate only if fuel and oxidant

are continuously added, which is why it is classed as a "flow" battery.

43. a. Conversion of fuel:

$$C_8H_{18}(l) + 4\ O_2(g) \longrightarrow 9\ H_2(g) + 8\ CO(g)$$

Fuel cell reaction:

$$CO(g) + H_2O(g) \xrightarrow{\text{catalyst}} CO_2(g) + H_2(g)$$

b. This type of fuel cell is convenient because it runs on a liquid fuel, gasoline, rather than using gaseous hydrogen. Such fuel cells would generate electricity more efficiently than possible with current gasoline engines. However, the liquid fuel is still petroleum-based and therefore nonrenewable. Also, this fuel cell still emits CO_2, an undesirable greenhouse gas. Therefore, although such fuel cells may find specialty applications in the near future, their long-term prospects are not superior.

Chapter 9

1. Cotton, silk, rubber, wool, and DNA are examples of natural polymers. Synthetic polymers include Kevlar, polyvinyl chloride (PVC), Dacron, polyethylene, polypropylene, and polyethylene terephthalate.

3. The *n* on the left side of the equation gives the number of monomers that react to form the polymer. Thus, it is a coefficient. The *n* on the right side is a subscript; it represents the number of repeating units in the polymer.

6. The bottle on the left is likely made of low-density polyethylene; the one on the right likely is high-density polyethylene. The molecular structures of LDPE and HDPE can help explain this difference in properties at a molecular level. LDPE is more branched, lessening molecular attractions between the chains and causing the plastic to be softer and more easily deformed. HDPE molecules, with fewer branches, can more closely approach each other, creating greater opportunity for interactions.

9. Each ethylene monomer has a molar mass of 28.052 g. To determine the number of monomers in the polymer, divide 40,000 (the molar mass of the polymer) by 28.052 (the molar mass of the monomer) to get 1426 monomers. To determine the number of carbon atoms present in the polymer, note that each monomer contains two carbon atoms ($H_2C{=}CH_2$). Accordingly, the polymer contains 2×1426 carbon atoms, or 2852 carbon atoms.

11. This is the head-to-head, tail-to-tail arrangement of PVC formed from three monomer units.

22. a. A blowing agent is either a gas or a substance capable of producing a gas to manufacture a foamed plastic. For example, a blowing agent produces Styrofoam from PVC.

b. The CO_2 replaces CFCs/HCFCs that were formerly used as blowing agents. Although CO_2 is a greenhouse gas, the replacement is environmentally beneficial because CFCs and HCFCs deplete the ozone layer.

23. a. In 1997, 8.9×10^{10} lb of plastic was produced in the United States. In contrast, 1.07×10^{11} lb of plastic was produced in 2003.

b. $\dfrac{1.07 \times 10^{11}\ \text{lb plastic}}{2.90 \times 10^8\ \text{people}} = 370$ lb/person in 2003

$\dfrac{8.9 \times 10^{10}\ \text{lb plastic}}{2.69 \times 10^8\ \text{people}} = 330$ lb/person in 1997

c. $\dfrac{370\ \text{lb/person} - 330\ \text{lb/person}}{330\ \text{lb/person}} \times 100$

$= 12\%$ change

28. In addition to the chemical composition of the monomers, other factors influence the properties of the polymer. These include length of the chain (the number of monomer units), three-dimensional arrangement of the chains, branching of the chain, strength/types of intermolecular forces between the chains, and orientation of monomer units within the chain.

30. For addition polymerization, the monomer must have a C-to-C double bond. Although some monomers have benzene rings as part of their structures (styrene, for example), the double bond involved in addition polymerization must not be in the ring. An example is the formation of PP from propylene.

For condensation polymerization, two possibilities for monomers exist. In either case, each monomer must have two functional groups that can react and eliminate a small molecule such as water. For example, an alcohol and a carboxylic acid can react to eliminate water. In this case, (1) the monomers must each have an alcohol and a carboxylic acid group, or (2) there are two monomers, one with two alcohol groups and the other with two carboxylic acid groups. An example of (2) is the formation of PET from ethylene glycol and terephthalic acid.

31. In vinyl chloride, there are three bonds (two single and a double) around each carbon. Thus the geometry is trigonal planar and the Cl-to-C-to-H bond angle is $120°$. In the polymer, each carbon atom is connected to other atoms by four single bonds and the geometry is tetrahedral, with a bond angle of $109.5°$.

32. a. This is the Lewis structure.

b. When Acrilan fibers burn, one of the combustion products is the poisonous gas hydrogen cyanide, HCN.

34. It requires 598 kJ/mol to break C-to-C double bonds. The formation of C-to-C single bonds releases 356 kJ/mol. If we consider the reaction of two ethylene monomers, two double bonds are broken and replaced with four single bonds (two bonds between the C atoms of the monomers, one between the first monomer and the second, and a bond extending to what would be the third ethylene monomer). The calculation is

$(2 \times 598 \text{ kJ/mol}) - (4 \times 356 \text{ kJ/mol})$
$= -228 \text{ kJ/mol}$. Thus, the reaction is exothermic.

38. a. The approximate increase in plastic production for any five-year period is between 8 and 12 billion pounds.

b. Plastics production in 1977 was 34 billion pounds. A doubling of this production would require an output of 68 billion pounds, a level that, according to the graph, occurred around 1992. Thus 15 years were required to double the plastics output of 1977.

43. a. Branched LDPE cannot be used in this application because it would not be strong enough. The low level of protection it would offer against accidental cuts or punctures would not be adequate to protect the surgeon from blood-borne diseases.

b. The linear HDPE polymer Spectra is very resistant to being cut or punctured. With an extremely thin liner, the surgeon can be protected from cuts or punctures while retaining the flexibility needed.

45. The "Big Six" polymers are generally nonpolar molecules and therefore do not dissolve in polar solvents such as water. The generalization, developed in Chapter 5, is that "like dissolves like." Some of the "Big Six" dissolve or soften in hydrocarbons or chlorinated hydrocarbons because these nonpolar solvents interact with the nonpolar polymeric chains.

Chapter 10

1. a. An antipyretic drug is intended to reduce fever.

b. An analgesic drug is intended to reduce pain.

c. An anti-inflammatory drug is intended to reduce inflammation, which is redness, heat, swelling, and pain caused by irritation, injury, or infection.

2. Organic chemists study the chemistry of carbon compounds.

3. The condensed formulas are $CH_3CH_2CH_2CH_2CH_3$ (or $CH_3(CH_2)_3CH_3$), $CH_3CH_2CH(CH_3)CH_3$, and $CH_3C(CH_3)_2CH_3$. These are the line-angle drawings.

5. Four possible isomers have the formula C_4H_9OH. Here is a structural formula for each isomer. The hydrogen atoms bonded to the carbons have been omitted for clarity.

6. a. contains the functional group and is an ether.

b. contains the functional group and is a carboxylic acid.

c. contains the functional group and is a ketone.

d. contains the functional group and is an amide.

e. contains the functional group and is an ester.

7. There are one-carbon examples of alcohols, aldehydes, and acids. The other classes of compounds require more than one carbon atom by the nature of their functional groups.

a. Alcohol. The only example is methanol, CH_3OH.

b. Aldehyde. The only example is methanal (commonly called formaldehyde), CH_2O.

d. Ester. There is no example of a one-carbon ester. To be an ester, a compound must have a carbon-containing group bonded to the oxygen atom that is singly bonded to the carbonyl carbon atom.

8. a. As written, the compound is an alcohol; an isomer is an ether. The structure of the ether is:

b. As written, the compound is an aldehyde; an isomer is a ketone. The structure of the ketone is:

c. As written, the compound is an ester; an isomer is a carboxylic acid. One possible structure of an acid with the same molecular formula is:

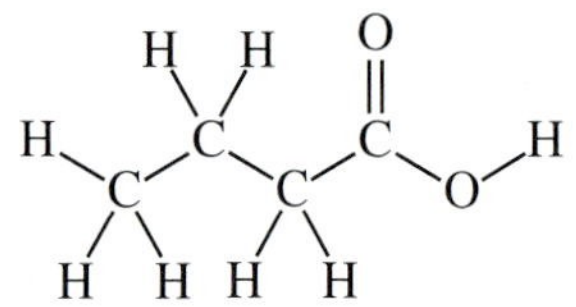

10. a. This is the structural formula for acetaminophen.

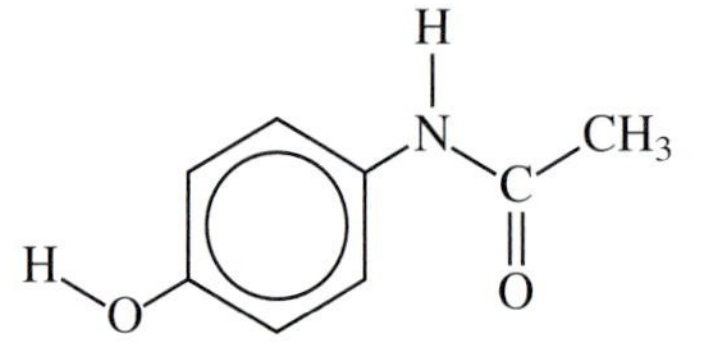

b. The molecular formula is $C_8H_9NO_2$.

11. a. There are two amide groups.

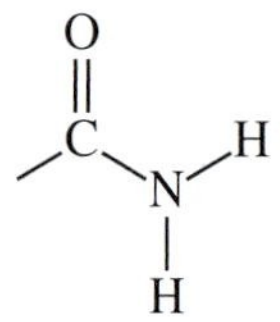

14. a. *n*-propanol, $CH_3CH_2CH_2OH$

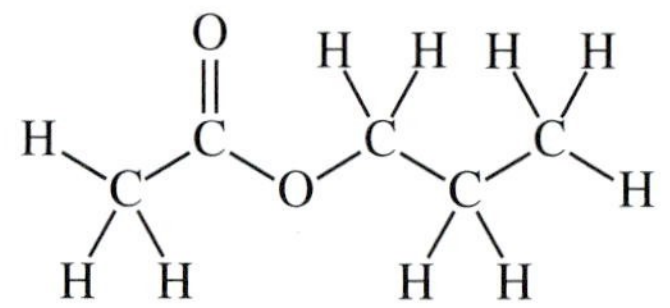

b. *iso*propanol, $(CH_3)_2CHOH$

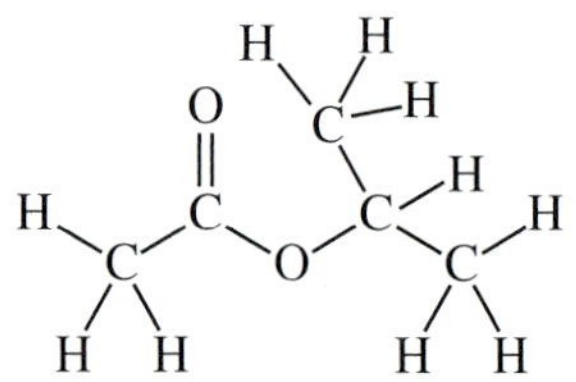

c. *t*-butanol, $(CH_3)_3COH$

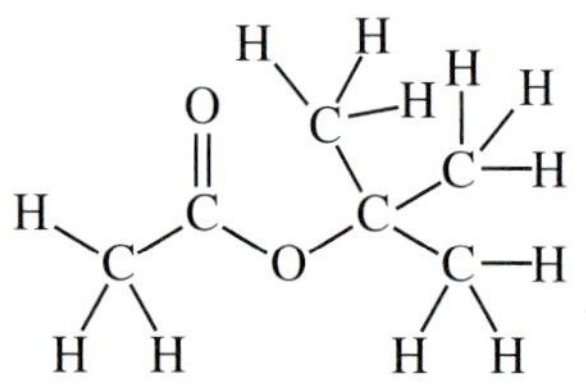

16. a. $H-C\equiv N\!:$

17. No, aspirin would not be more active if it were to interact with prostaglandins directly. To accomplish such blockage would require a direct correspondence between the number of molecules of aspirin and prostaglandin. When aspirin blocks a COX enzyme, it is preventing synthesis of many prostaglandin molecules, since one enzyme is responsible for increasing the rate of synthesis of the prostaglandins.

18. If you started with two active functional groups, you would have four different products after two synthetic steps. If the reagent used in the first step had two reactive groups itself, you would produce eight different products after the two synthetic steps (assuming the second step had a reagent with only one reactive group).

21. A pharmacophore is the three-dimensional arrangement of atoms, or groups of atoms, responsible for the biological activity of a drug molecule.

22. Sulfanilamide has the same basic shape and contains similar functional groups in the same regions as *para*-aminobenzoic acid, so it replaces the nutrient in some biologically important process. Without the key nutrient, the bacteria die.

23. a. This compound cannot exist in chiral forms. The central carbon atom is bonded to two equivalent —CH_3 groups.

b. This compound can exist in chiral forms. The four groups attached to the central carbon atom are all different.

c. This compound can exist in chiral forms. The four groups attached to the central carbon atom are all different.

d. This compound cannot exist in chiral forms. The central carbon atom is bonded to two equivalent —CH_3 groups.

25. A freebase is a nitrogen-containing molecule in which the nitrogen is in possession of its lone pair of electrons. Treating methamphetamine hydrochloride with a base (such as hydroxide ions, OH^-) will strip off one of the hydrogen atoms attached to the nitrogen atom, freeing up its lone pair.

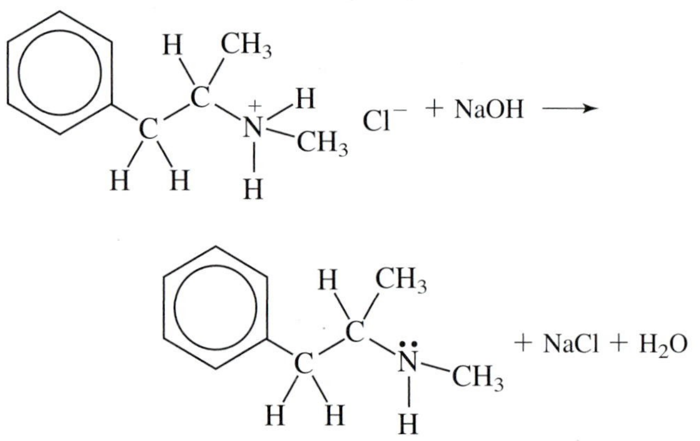

27.

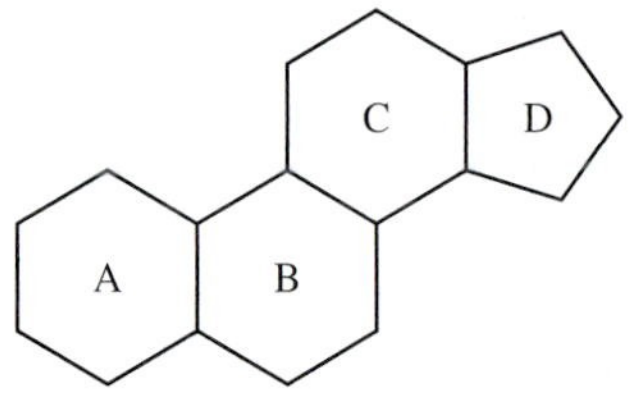

29. a. Four single bonds, one triple bond

b. Six single bonds, one double bond

31. Just three distinct isomers are shown here, because some of the structures are duplicates. Numbers 1 and 5 are different paper-and-pencil representations of the *same* isomer. Numbers 2, 3, and 4 are all different paper-and-pencil representations of the *same* isomer. Number 6 is an isomer *different* from numbers 1 and 5, and from numbers 2–4.

32.

35. a. Aspirin produces a physiological response in the body.

b. Morphine produces a physiological response in the body.

c. Antibiotics kill or inhibit the growth of bacteria that cause infections.

e. Amphetamine produces a physiological response in the body.

37. The drug (−)-dopa is effective because the molecule fits in the receptor site, but the nonsuperimposable mirror image form, (+)-dopa, does not. This is the structure of (−)-dopa, with the chiral carbon atom marked in red. Note that four different groups are attached to the starred carbon atom.

39. A chiral molecule and its asymmetric binding site fit together in a very specific way. It must be that the (−)-methorphan has a much better fit and is able to act as a narcotic, but (+)-methorphan does not fit as well and is therefore not as potent a drug. Because (+)-methorphan has less activity, it can be added to over-the-counter medicines with some safety, assuming that consumers follow the label directions on the over-the-counter drugs containing this compound.

45. a. There are 6(4) + 6(1) or 30 electrons available. This is a possible linear isomer of benzene.

Structural formula:

b. This is the condensed formula:
$$CH_2{=}C{=}CH{-}CH{=}C{=}CH_2$$

c. First check to see if all of the structures correctly represent C_6H_6 and that each carbon has four bonds. If these conditions are met, the structures with double bonds should differ only in the placement of the lone C—C single bond. However, structures including carbon-carbon triple bonds can also be drawn, and they would be distinctly different.

Chapter 11

1. The four fundamental types of materials provided by food are water, energy sources, raw materials, and metabolic regulators.

3. a. Macronutrients are materials that are consumed in relatively large amounts from the foods we eat. They provide essentially all of the energy and most of the raw material for repair and synthesis.

b. The three major classes of macronutrients are fats, carbohydrates, and proteins.

5. The chart indicates more carbohydrate is present than would be found in steak, and more protein than would be found in chocolate chip cookies. The chart is likely to be a representation of peanut butter. (See Table 11.1 for confirmation.)

9. a. There are 580 potassium atoms to every 300 sodium atoms in every million atoms in the body. The ratio is 580 to 300, or 1.9 to 1.

b. There are 0.36 g of potassium and 0.11 g of sodium in every 100 g of body weight. The ratio is 0.36 to 0.11, or 3.3 to 1.

c. These elements, although important, are not listed individually in Figure 11.2 because they represent a relatively small proportion of the types of atoms present in the human body. These elements are included in the category listed as "vitamins and other minerals (less than 1%).

11. Similarities: Fats and oils are both characterized by the presence of long, nonpolar hydrocarbon chains. Edible fats and oils both contain some oxygen. Most fats and oils are triglycerides, which are esters of three fatty acid molecules and one glycerol molecule. Both oils and fats feel greasy and are insoluble in water.

Differences: Oils tend to contain more polyunsaturated fatty acids and smaller fatty acids than fats. Triglycerides that are solids at room temperature are called fats. Triglycerides that are liquids at room temperature are called oils.

12. a. Here is a structural formula for lactic acid.

The carbon atom on the top in this diagram has three regions of bonding electrons. The geometry is trigonal planar with bond angles of approximately 120°. Coming down the chain, the second carbon and third carbon atoms are each surrounded by four single bonds. The geometry is tetrahedral with bond angles of approximately 109.5° for each of those carbon atoms.

b. Lactic acid is saturated because the hydrocarbon chain contains only single bonds between the carbon atoms.

c. Yes, lactic acid is a fatty acid. It has a carboxylic acid group and a hydrocarbon chain, the two necessary components for a fatty acid.

16. According to Table 11.1, peanut butter is 25% protein, which makes it a very good protein source indeed. However, it is also 50% fat, which is quite high if one needs to limit fat intake. On the positive side, much of the fat in peanut butter is unsaturated, unless the peanut butter has been hydrogenated.

18. See Figure 11.9 for structures of fructose and glucose. Observe that the structure of fructose is based on a five-membered ring composed of four C atoms and one O atom. The structure of glucose is based on a six-member ring composed of five C atoms and one O atom. Glucose has one $-CH_2OH$ side chain, and fructose has two.

26. a. Region A. 6–11 servings of bread, cereal, grain, and pasta

Region B. 3–5 servings of vegetables

Region C. 2–4 servings of fruits

Region D. 2–3 servings of milk, yogurt, and cheese

Region E. 2–3 servings of meat, poultry, fish, dry beans, eggs, and nuts

Region F. fats, oils, and sweets used sparingly

28. a. Food shortages are being experienced in the countries of sub-Saharan Africa, in many countries of the CIS (Commonwealth of Independent States, formerly known as the Soviet Union), Afghanistan, and Cuba.

b. Areas of North America (United States, Canada, and Mexico) are not included because, although there are many malnourished people living in North America, the overall educational and economic levels of the population are sufficient to enable individuals to select and pay for nourishing foods. Food production in North America, and, where necessary, food imports are sufficient to provide reasonably priced food resources for the population. In fact, rather than starving, many individuals in this part of the world suffer from the consequences of overeating.

30. A gamma ray is a high-energy photon. Gamma rays are used to preserve food because they kill insects, fungi, or bacteria that can contribute to food spoilage.

32. a. Very few elements form the macronutrients. The major ones are hydrogen, oxygen, carbon, and nitrogen. At this time, elements 1-111 have been accepted by IUPAC. The percentage is:

$$\frac{4 \text{ elements used to form macronutrients}}{111 \text{ elements accepted by IUPAC}} \times 100 = 3.6\%$$

b. Hydrogen and oxygen atoms form covalent bonds with each other in water molecules. Carbon atoms can covalently bond with one another in a grand array of different biochemical compounds. Nitrogen, hydrogen, and oxygen are able to form covalent bonds with carbon and with one another.

36. a. Saturated fats, with their long "straight" hydrocarbon chains, can fit together closely. There is free rotation of the atoms involved in single C-to-C bonds in saturated fats. This allows for twisting and turning until the molecules approach each other as closely as possible. The presence of C-to-C double bonds in unsaturated fats prevent these molecules from such close packing because double bonds introduce "kinks" in the hydrocarbon chain. C-to-C double bonds do not allow for free rotation but rather fix the relative position of the bonded atoms. This is easier to understand with the help of models.

b. The melting points decrease as the number of C-to-C double bonds increases. With no C-to-C double bonds, stearic acid molecules can bend and fold so they can fit closely together, increasing the extent of intermolecular attractions between molecules and therefore increasing the melting point. As the number of C-to-C double bonds increases, the molecules fit together less closely, resulting in reduced intermolecular interactions and, therefore, lower melting points. Less energy is required to separate the molecules from one another in the case of the polyunsaturated fatty acids because the intermolecular forces are lower.

c. Trans fats are created by the full or partial hydrogenation of fatty acids. When the H atoms are on the opposite side of the double bond, such as in elaidic acid shown in Figure 11.8, they straighten the hydrocarbon chain of the fatty acid. This trans fatty acid has a higher melting temperature than its counterpart, which has the H atoms are on the same side of the double bond. Trans fatty acids more closely resemble the packing of saturated acids in fats and lead to a rise in triglycerides and "bad" cholesterol in the blood.

46. *The Dietary Guidelines for Americans 2005* include a recommendation that 30% or less of total Calories should come from fat. For a typical 2000 Cal/day diet, that would mean only 600 Cal from fat. Meeting this recommendation, while also maintaining a diet low in harmful saturated fats, might be possible if all of the milk or dairy products consumed were low-fat or even no-fat. For example, 1 oz of regular cheddar cheese has 6.0 g of saturated fat and 114 Cal. However, 1 oz of low-fat cheddar cheese has only 1.2 g of saturated fat and 49 Cal. One cup of whole milk has 4.6 g of saturated fat and 146 Cal, but switching to 1% milk means consuming 1.5 g saturated fat and 102 Calories per cup.

Chapter 12

1. DNA stands for **d**eoxyribo**n**ucleic **a**cid.
3. **a.** The base adenine has the amine group, $-NH_2$.
   b. The sugar deoxyribose has several hydroxyl groups, $-OH$.
   c. There are no functional groups *in* the phosphate, but the phosphate itself is a functional group. Do not mistake the doubly bonded oxygen atom for a ketone, for there is no bond to a carbon atom.
6. **a.** A nucleotide must contain a base, a deoxyribose molecule, and a phosphate group linked together.
   b. Covalent bonding holds the units together.
8. **a.** A beam of X-rays, which have relatively high energy and relatively short wavelengths, is directed at a target. The X-rays are then diffracted at certain angles, which are related to the distance between atoms. The process of diffraction makes it appear like the X-ray waves are deflected or bent by the atoms in the structure.
   b. Ions in a salt like sodium chloride have a very regular structure that is easily determined by X-ray studies. Atoms in nucleic acids and proteins do not show the same well-known patterns of crystalline regularity, making the interpretation of the X-ray diffraction pattern more difficult.
13. **a.** This is the general formula for an amino acid, where R represents a side chain that is different in each of the 20 amino acids.

   b. The functional groups are $-COOH$, which is the carboxylic acid group, and the $-NH_2$ group, the amine group.
15. A codon is a grouping of three RNA bases. An appropriate RNA molecule transfers the order of bases in DNA into a specific amino acid that should appear in a protein sequence. Therefore, codons are used to signal that a molecule of a certain amino acid should be incorporated into a protein.
18. Only a minor change in the amino acid composition of human hemoglobin leads to sickle cell anemia. In hemoglobin S, two of the residues that should be glutamic acid are replaced with valine. This seemingly innocuous change has rather drastic results for the person with this genetic disease.
21. The correct answer is c. The bases adenine, thymine, cytosine, and guanine make up DNA nucleotides.
30. The discovery that %A = %T and that %C = %G provided the basis for asking *why* this pattern was observed. Chargaff's contribution was his finding that the bases were paired. Crick and Watson took this information a step further to discover both *how* and *why* they were paired, and the influence the pairing had on the structure of DNA.
46. The statement is false. Calculations show that the DNA in an adult would stretch just about 26,000 times to the Moon and back.

$$\frac{2 \text{ m}}{1 \text{ DNA thread}} \times \frac{1 \times 10^{13} \text{ cells}}{1 \text{ adult}}$$
$$= 2 \times 10^{13} \text{ m of DNA in an adult}$$

$$3.8 \times 10^5 \text{ km} \times \frac{1 \times 10^3 \text{ m}}{1 \text{ km}} \times 2 = 7.6 \times 10^8 \text{ m},$$
the distance to the Moon and back

The ratio of these two distances is:

$$\frac{2 \times 10^{13} \text{ m}}{7.6 \times 10^8 \text{ m}} = \frac{26,000}{1}$$

Glossary

The numbers at the end of each definition indicate the page(s) where the term is defined and explained in the text.

A

acid anhydride literally "an acid without water" *250*

acid deposition deposition of either wet forms or dry forms such as rain, snow, fog, and cloud-like suspensions of microscopic water droplets often more acidic and damaging than acid rain *246*

acid rain rain that is more acidic than "normal" rain and that has a lower pH *245*

acid a compound that releases hydrogen ions, H^+, in aqueous solution *240*

acid-neutralizing capacity (ANC) capacity of a lake or other body of water to resist a decrease in pH *269*

activation energy energy necessary to initiate a chemical reaction *179*

active site region of an enzyme molecule where its catalytic effect occurs *510*

addition polymerization type of polymerization in which the monomers add to the growing chain in such a way that the polymer contains all the atoms of the monomer. No other products are formed. *372*

aerosols particles, both liquid and solid, that remain suspended in the air rather than settling out *35*

albedo ratio of electromagnetic radiation *reflected* relative to the amount of radiation *incident* on a surface *131*

alkane a hydrocarbon with only single bonds between the carbon atoms *171*

allotropes two or more forms of the same element that differ in their chemical structure and therefore in their properties *57*

alpha particle (α) positively charged (2^+) particle consisting of two protons and two neutrons (the nucleus of a helium atom) *296*

ambient air the outside air, that is, the air surrounding or encircling us *18*

amino acid monomer from which our body builds proteins. Each amino acid molecule contains two functional groups: an amine group ($—NH_2$) and a carboxylic acid group ($—COOH$). *385*

amino acid residue amino acid that was once incorporated into a peptide chain *468*

amorphous region in a polymer, a region in which the long polymer molecules are in a random, disordered arrangement *377*

anaerobic bacteria bacteria that can function without the use of molecular oxygen *127*

anion negatively charged ion *209*

anode electrode where oxidation takes place *333*

antioxidant compound added to foods, drugs, and cosmetics to minimize the oxidation of unsaturated oils and fats that can cause rancidity, color loss, and flavor changes *486*

aqueous solution solution in which water is the solvent *199*

aquifer great pool of water trapped in sand and gravel 50–500 ft below the surface *197*

atom smallest unit of an element that can exist as a stable, independent entity *25*

atomic mass average mass of an atom of an element compared with an atomic mass of exactly 12 amu for carbon-12 *121*; mass (in grams) of the same number of atoms found in exactly 12 g of carbon-12 *122*

atomic number number of protons in an atom of that element *60*

Avogadro's number number of atoms in exactly 12 g of carbon-12 *122*

B

background radiation the radiation, on average, that exists at a particular location, usually due to natural sources *304*

basal metabolism rate (BMR) minimum amount of energy required daily to support basic body functions *479*

base compound that produces hydroxide ions, OH^-, in aqueous solution *242*

battery device consisting of one or more cells that can produce a direct current by converting chemical energy to electrical energy *332*

beta particle (β) high-speed electron emitted from a nucleus *296*

biomass general term for plant matter such as trees, grasses, agricultural crops or other biological material *179*

blowing agent either a gas or a substance capable of producing a gas used to manufacture a foamed plastic *381*

bond energy amount of energy that must be absorbed to break a specific chemical bond *160*

breeder reactor a nuclear reactor that can produce more fissionable fuel (usually Pu-239) than it consumes (usually U-235) *312*

C

calibration graph graph made by carefully measuring the absorbencies of several solutions of known concentration for the species being analyzed *226*

calorie formerly defined as the amount of heat necessary to raise

the temperature of exactly 1 g of water by 1 °C. Now redefined as exactly 4.184 J. *152*

calorimeter device with which the quantity of heat energy released in a combustion reaction can be determined experimentally *158*

carbohydrates compounds containing carbon, hydrogen, and oxygen, the last two elements in the same 2:1 atomic ratio as found in water *465*

carbon nanotubes nano-sized tubes of pure carbon with wall thicknesses as thin as a single atom *430*

carbon sink natural reservoir that removes CO_2 from the atmosphere *121*

carcinogen compound capable of causing cancer *219*

carcinogenic capable of causing cancer *43*

catalyst chemical substance that participates in a chemical reaction and influences its rate or speed without undergoing permanent change *36, 84*

catalytic converter device installed in the exhaust stream of an engine to reduce emissions *36*

catalytic cracking catalysts used to promote molecular breakdown at lower temperatures than thermal cracking *175*

cathode electrode where reduction takes place. The cathode receives the electrons sent from the anode through the external circuit. *333*

cation positively charged ion *209*

chain reaction term that generally refers to any reaction in which one of the products becomes a reactant *288*

Chapman cycle set of natural steady-state reactions for stratospheric ozone *74*

Chargaff's rules observation that in every species, the percent of adenine almost exactly equals the percent of thymine. Similarly, the percent of guanine is essentially identical to the percent of cytosine. Put more simply: %A = %T and %G = %C. *501*

chemical equation representation of a chemical reaction using chemical formulas *29*

chemical formula symbolic way to represent the elementary composition of a substance, indicating the kinds and numbers of atoms present in a molecule *26, 408*

chemical reaction process whereby reactants are transformed into products *29*

chemical symbol one- or two-letter abbreviation for an element. Also called atomic symbol *21*

chiral (optical) isomers compounds with the same chemical formula but different three-dimensional molecular structures and different interaction with plane polarized light *423*

chlorofluorocarbons (CFCs) compounds composed only of the elements chlorine, fluorine, and carbon *82*

chromosomes the 46 self-replicating, rod-shaped strands of deoxyribonucleic acid (DNA) and associated proteins found in the nucleus of cells that contain the hereditary information necessary for life *498*

clone collection of cells or molecules identical to an original cell or molecule *515*

codon sequence of three adjacent nucleotides that determines the insertion of a specific amino acid during protein synthesis or that signals the starting and stopping of protein synthesis *507*

coenzyme molecule that works in conjunction with an enzyme to enhance the enzyme's activity *474*

combinatorial chemistry systematic creation of large numbers of molecules in "libraries" that can be rapidly screened in the lab for biological activity and the potential for becoming new drugs *422*

combustion chemical process in which a fuel combines rapidly with oxygen to release energy and form products *29, 158*

complex carbohydrates polysaccharides such as starch, cellulose, and glycogen *466*

compound pure substance made up of two or more elements in a fixed, characteristic chemical combination *24*

concentration ratio of amount of solute to amount of solution *201*

condensation polymerization a type of polymerization in which a small molecule such as water is split out (eliminated) when the monomers join to form a polymer *383*

condensed structural formula chemical formula in which bonds are not drawn out explicitly, but simply understood to contain an appropriate number of bonds *408*

conductivity meter an apparatus that produces a signal to indicate that electricity is being conducted *208*

control rods rods composed primarily of an excellent neutron absorber such as cadmium or boron that can be positioned in a nuclear reactor to absorb fewer or more neutrons, thereby regulating the rate of fission *292*

copolymer polymer built from two or more different monomers *383*

covalent bond a chemical bond in which two electrons are shared by the atoms involved *63*

cracking chemical process by which large molecules are broken into smaller ones, such as those suitable for use in gasoline *174*

criteria pollutants air pollutants for which EPA has set permissible levels based on their effects on human health and on the environment *10*

critical mass amount of fissionable fuel required to sustain an atomic chain reaction *288*

crystalline region in a polymer, a region in which the long polymer molecules are arranged neatly and tightly in a regular pattern *377*

curie (Ci) unit of radioactivity, equal to 3.7×10^{10} disintegrations/s and roughly equivalent to the level of radioactivity from 1 g of radium *305*

current rate of electron flow *335*

D

denitrification process of converting nitrate ions, typically in soil, to nitrogen gas *266*

density the ratio of mass per unit volume *207*

deoxyribonucleic acid (DNA) molecule that carries genetic information in all species *498*

depleted uranium contains almost entirely U-238 (99.8%) and has been depleted of most of the U-235 that it once naturally contained *317*

desalination any process that removes ions from salty water *230*

diatomic molecule molecule that contains two atoms *26*

dietary supplement vitamins, minerals, amino acids, enzymes, herbs, and other botanicals *440*

dipeptide a compound formed from two amino acids *468*

"dirty bomb" device that employs a conventional explosive to disperse a radioactive substance *319*

*di***saccharide** "double sugar" formed by joining two monosaccharide units *465*

dispersion forces attractions between molecules that result from a distortion of the electron cloud that causes an uneven distribution of the negative charge *375*

distillation separation process in which a solution is heated to its boiling point and the vapors of the various components are condensed and collected *171, 230*

distributed generation placing power-generating modules of 30 megawatts or less near the end user *345*

"doping" process of intentionally adding small amounts of other elements to pure silicon *355*

double bond covalent bond consisting of two pairs of shared electrons *66*

double helix spiral consisting of two strands that coil around a central axis *503*

E

effective stratospheric chlorine chlorine- and bromine-containing gases in the stratosphere *89*

electricity flow of electrons from one region to another that is driven by a difference in potential energy *332*

electrode electrical conductor placed in the cell as sites for chemical reactions *332*

electrolysis process of passing a direct current of electricity of sufficient voltage through water to decompose it into H_2 and O_2 *348*

electrolyte conducting solute in solution *209*

electrolytic cell device in which electrical energy is converted to chemical energy *332*

electromagnetic spectrum continuum of waves ranging from very long and low-energy radio waves to very short and high-energy X-rays and gamma rays *69*

electron subatomic particle with a much smaller mass than a proton or neutron and a negative electrical charge equal in magnitude to that of a proton, but opposite in sign *60*

electronegativity (EN) measure of an atom's attraction for the electrons it shares in a covalent bond *204*

element pure substance that cannot be broken down into simpler ones by any *chemical* means *21*

endothermic term applied to any chemical or physical change that absorbs energy *160*

energy the capacity to do work or supply heat *152*

"enhanced greenhouse effect" process in which atmospheric gases trap and return *more than* 80% of the heat energy radiated by the Earth *105*

enriched uranium uranium that has a higher percent of U-235 than its natural abundance of about 0.7% *316*

entropy randomness in position or energy level *157*

enzyme protein that acts as a biochemical catalyst, influencing the rate of a chemical reaction *417*

essential amino acid an amino acid that is required for protein synthesis but that must be obtained from the diet because the body cannot synthesize it *470*

exothermic term applied to any chemical or physical change accompanied by the release of heat *158*

exposure amount of a substance encountered, generally in reference to human contact with a toxic substance or a disease-causing organism *17*

F

fat triglyceride that is solid at room temperature *459*

fatty acid molecule molecule with two structural features: a nonpolar long hydrocarbon chain generally containing an even number of carbon atoms (typically 12 to 24) and a polar carboxylic acid group at the end of the chain *457*

first law of thermodynamics the statement that energy is neither created nor destroyed, also called the law of conservation of energy *153*

food irradiation process of subjecting food to high-energy ionizing radiation to kill or reduce the levels of undesirable contaminants such as bacteria, spores, and insects *487*

forcings factors that affect the annual global mean surface temperature *131*

free radical unstable chemical species with one or more unpaired electrons *80*

freebase nitrogen-containing molecule in which the nitrogen is in possession of its lone pair of electrons *414*

frequency number of waves passing a fixed point in one second *68*

fuel cell galvanic cell that produces electricity by converting the chemical energy of a fuel directly into electricity without burning the fuel *340*

functional group distinctive arrangement of groups of atoms that impart characteristic physical and chemical properties to the molecules that contain them *382, 411*

G

galvanic cell device that converts the energy released in a spontaneous chemical reaction into electrical energy *332*

galvanized iron iron coated with zinc *260*

gamma ray (γ) high-energy, short-wavelength photon emitted from the nucleus with no charge or mass *296*

gaseous diffusion a process used to separate gases with different molecular weights by forcing them

through a series of permeable membranes *316*

gene short piece of DNA that codes for the production of proteins, giving an organism its particular inherited characteristics *498*

generic drug medication that is the chemical equivalent of a pioneer drug, but that cannot be marketed until the patent protection on the pioneer drug has run out after 20 years *435*

global atmospheric lifetime time required for a gas added to the atmosphere to be removed. Also referred to as the "turnover time." *126*

global warming popular term used to describe the increase in average global temperatures *101*

global warming potential (GWP) number that represents the relative contribution of a molecule of an atmospheric gas to global warming *128*

green chemistry the designing of chemical products and processes that reduce or eliminate the use or generation of hazardous substances *38*

greenhouse effect process by which atmospheric gases trap and return a major portion of the heat (infrared radiation) radiated by the Earth *103*

greenhouse gases those gases capable of absorbing and re-emitting infrared radiation *102*

groundwater water pumped from wells that have been drilled into underground aquifers *197*

group vertical column in the periodic table *23*

H

half-life ($t_{1/2}$) time required for half the nuclei in a sample of a radioisotope to undergo radioactive decay *309*

half-reaction type of chemical equation that shows the electrons either lost or gained *332*

halons compounds similar to CFCs, in which bromine or fluorine atoms replace some or all of the chlorine atoms *82*

HDL (high-density lipoprotein) called the "good" lipoproteins because it is more effective than LDL in transporting cholesterol through the blood *464*

heat energy that flows from a hotter to a colder object *152*

heat of combustion quantity of heat energy given off when a specified amount of a substance burns in oxygen *159*

high-level radioactive waste (HLW) products of nuclear reactions that have high levels of radioactivity and, because of the long half-lives of the radioisotopes involved, require essentially permanent isolation from the biosphere *311*

hormesis concept that low doses of a harmful substance (such as radiation) may actually be beneficial *308*

hormone chemical messengers produced by the body's endocrine glands *416*

Human Genome Project international effort to map all the genes in the human organism *512*

hybrid vehicle vehicle that combines conventional gasoline engines with battery technology *338*

hydrocarbon compound that contains only the elements hydrogen and carbon *28, 171*

hydrochlorofluorocarbons (HCFCs) compounds of hydrogen, chlorine, fluorine, and carbon *91*

hydrofluorocarbons (HFCs) compounds of hydrogen, fluorine, and carbon *92*

hydrogen bond electrostatic attraction between a hydrogen atom bearing a partial positive charge in one molecule and an O, N, or F atom bearing a partial negative charge in a neighboring molecule *206*

hydrogenation process in which hydrogen gas, in the presence of a metal catalyst, is added to a double bond and converts it to a single bond *462*

hydronium ion (H_3O^+) water molecule plus a proton *343*

hygroscopic describes a substance that readily absorbs water from the atmosphere and retains it *262*

I

infrared (IR) heat radiation; the region of the electromagnetic spectrum with wavelengths longer than those of red visible light *69*

intermolecular attractive force attraction between two molecules resulting from the interactions of their electron clouds and nuclei *374*

intermolecular force force that occurs *between* molecules *206*

intramolecular force force that exists *within* a molecule *204*

ion atom or group of atoms that has acquired a net electrical charge as a result of gaining or losing one or more electrons *209*

ionic bond chemical bond formed by the attraction between oppositely charged ions *209*

ionic compound compound composed of electrically charged ions that are present in fixed proportions and are arranged in a regular, geometric pattern *209*

isomers molecules with the same chemical formula (same number and kinds of atoms), but with different structures and properties *175, 408*

isotopes two or more atoms of the same element. Isotopes have the same number of protons but differ in the number of neutrons, and hence in mass. *62*

K

kinetic energy energy of motion *153*

L

law of conservation of matter and mass in a chemical reaction, matter and mass are conserved *30*

LDL (low-density lipoprotein) called "bad" lipoprotein because it is less effective than HDL in transporting cholesterol through the blood *464*

lead compound drug (or a modified version of that drug) that shows high promise for becoming an approved drug *422*

Lewis structure representation of an atom or molecule that shows its outer electrons *63*

line-angle drawing simplified version of a structural formula that is most useful for representing larger molecules *410*

linear, nonthreshold model model that assumes that the adverse effects of radiation increase linearly with dose, with radiation being harmful at all doses, even low ones *307*

lipid class of compounds that includes not only the edible fats and oils but also diverse materials such as cholesterol and other steroids *457*

liter (L) volume occupied by 1000 g of water at 4 °C *201*

low-level radioactive waste (LLW) waste that is contaminated with smaller quantities of radioactive materials than HLW and specifically excludes spent nuclear fuel *311*

M

macrominerals seven elements (Ca, P, Cl, K, S, Na, and Mg) that are necessary for life but are not nearly as abundant as O, C, H, or N *475*

macromolecules molecules of high molecular mass that have characteristic properties because of their large size *370*

macronutrients fats, carbohydrates, and proteins that provide essentially all of the energy and most of the raw material for repair and synthesis *455*

malnutrition condition caused by a diet lacking in the proper mix of nutrients, even though the energy content of the food eaten may be adequate *454*

mass number sum of the number of protons and neutrons in the nucleus of an atom *62*

maximum contaminant level (MCL) legal limit for the concentration of a contaminant *220*

maximum contaminant level goal (MCLG) maximum level of a contaminant in drinking water at which no known or anticipated adverse effect on a person's health would occur *219*

mesosphere region of the atmosphere above the stratosphere; found at an altitude starting about 50 km *20*

metallic bonding "electron sea" model, in which outermost (valence) electrons are shared among all the atoms in the substance *353*

metalloids elements between metals and nonmetals on the periodic table that do not fall cleanly into either group. Sometimes called semi-metals. *23*

metals elements that are shiny and conduct electricity and heat well. They tend to lose their valence electrons to form cations. *23, 210*

microcell very tiny fuel cell *345*

microgram (μg) a millionth (10^{-6}) of a gram *18*

micrometer (μm) a millionth (10^{-6}) of a meter. Sometimes simply referred to as a micron. *10*

microminerals quantities of Fe, Cu, and Zn that the body requires in lesser amounts *475*

micronutrients substances needed only in miniscule amounts, but essential for the body to produce enzymes, hormones, and other substances needed for proper growth and development *472*

microtubule nano-sized hollow cylinder with outer diameter between 20 nm and 30 nm *429*

mineral naturally occurring element or compound that usually has a definite chemical composition, a crystalline structure, and is formed as a result of geological processes *199*

mixture physical combination of two or more substances present in variable amounts *11*

moderator material that slows the neutrons in a nuclear reactor, making them more effective in producing fission *293*

molar mass mass of one Avogadro's number, or "mole," of whatever particles are specified *124*

molarity (M) number of moles of solute present in 1 L of solution *202*

mole an Avogadro's number of objects *123*

molecule two or more atoms held together by chemical bonds in a certain spatial arrangement *26*

monomer (from *mono* meaning "one" and *meros* meaning "unit"). Small molecules used to synthesize polymers. *370*

monosaccharide single sugar *465*

monounsaturated property of fats in which only one double bond exists between carbon atoms per molecule. Oleic acid is an example. *459*

municipal solid waste (MSW) garbage, that is, everything you discard or throw into your trash, including food scraps, grass clippings, and old appliances. MSW does not include industrial waste or waste from construction sites. *388*

N

nanomedicine the union of nanoscale technology and medical treatment *429*

nanometer (nm) a billionth (10^{-9}) of a meter (m) *69*

nanotechnology technology at the atomic and molecular (nanometer) scale: 1 nanometer (nm) = 1×10^{-9} m *25*

nanotubes (nanocapsules) thin, single-walled tubes that may be synthetic or partially synthetic (bio-nanotubes) *429*

neutral solution a solution that is neither acidic nor basic; that is, one that has equal concentrations of H^+ and OH^- *244*

neutralization chemical reaction in which the hydrogen ions from an acid combine with the hydroxide ions from a base to form molecules of water *243*

neutron electrically neutral subatomic particle with the same mass as a proton *60*

nitrification process of converting ammonia, typically in soil, to nitrate ions *266*

nitrogen cycle set of chemical pathways whereby nitrogen moves through the biosphere *267*

nitrogen saturation process by which an area is overloaded with "nitrogen"; that is, when the reactive forms of nitrogen entering an ecosystem exceed the system's capacity to absorb the nitrogen *270*

nitrogen-fixing bacteria bacteria that remove nitrogen from the air and convert it to ammonia *266*

noble gases elements that are inert and do not readily undergo chemical reactions *24*

nonelectrolyte nonconducting solute in solutions *209*

nonmetals elements with varied appearances that do not conduct electricity or heat well. Nonmetals tend to gain electrons to form anions. *23, 210*

***n*-type semiconductor** in which there are freely-moving negative charges, the electrons *355*

nuclear fission the splitting of a large nucleus into smaller ones with the release of energy and neutrons *287*

nuclear transfer laboratory procedure in which a cell's nucleus is removed and placed into an egg cell that has had its own nucleus removed. The genetic information from the donor nucleus controls the resulting cell, which can be induced to form embryos. *520*

nucleotide combination of a base, a deoxyribose molecule, and a phosphate group *500*

nucleus minuscule but highly dense region at the center of an atom that is composed of protons and neutrons *60*

O

octet rule the generalization that electrons in many molecules are arranged so that every atom (except hydrogen) shares in eight electrons *64*

oils triglycerides that are liquid at room temperature *459*

organic chemistry the branch of chemistry devoted to the study of carbon compounds *407*

organic compound compound that contains mainly carbon and hydrogen *36*

osmosis natural tendency for a solvent to move through a membrane from a region of higher solvent concentration to a region of lower solvent concentration *230*

outer (valence) electrons electrons that help account for many of the observed trends in chemical properties *61*

oxidation half-reaction type of chemical equation that shows the reactant that loses electrons *332*

oxygenated gasolines blends of petroleum-derived hydrocarbons with added oxygen-containing compounds such as MTBE, ethanol, or methanol (CH_3OH) *177*

ozone layer region of the stratosphere with the maximum ozone concentration *58*

P

parts per billion (ppb) 1 part out of a billion parts, unit of concentration. 1 ppb is 1000 times less concentrated than 1 ppm. *19, 202*

parts per million (ppm) 1 part out of a million parts, unit of concentration. 1 ppm is 10,000 times less concentrated than 1 part per hundred (pph). *13, 201*

peptide bond covalent bond that forms when the —COOH group of one amino acid reacts with the —NH₂ group of another, thus joining the two amino acids *386*

percent parts per hundred; sometimes abbreviated as pph *11, 201*

periodic properties regular recurrence of certain chemical aspects of atoms that is demonstrated with increasing atomic number. These attributes are repeated at regular intervals in the periodic table. *61*

periodic table an orderly arrangement of all the elements based on similarities in their properties *22*

pH a number, usually between 0 and 14, that indicates the acidity of a solution *245*

pharmacophore three-dimensional arrangement of atoms or groups of atoms responsible for the biological activity of a drug molecule *421*

photons individual bundles of energy *71*

photovoltaic cell (solar cell) a device that converts radiant energy directly to electrical energy *353*

plasmid ring of DNA *514*

plasticizer compound added in small amounts to polymers to make them softer and more pliable *380*

PM₁₀ particulate matter with an average diameter of 10 μm or less (on the order of 0.0004 in) *10*

PM₂.₅ particulate matter with an average diameter less than 2.5 μm, also called fine particles *10*

polar covalent bond covalent bond in which the electrons are not equally shared, but rather displaced toward the more electronegative atom *204*

polar stratospheric clouds (PSCs) thin stratospheric clouds composed of a small amount of frozen water vapor *86*

polyamide condensation polymer that contains the amide functional group *386*

polyatomic ion ion that is made up of two or more atoms covalently bound together *213*

polyatomic molecule molecule consisting of three or more atoms *65*

polymer large molecule built from monomers consisting of a long chain or chains of atoms covalently bonded together *370*

polysaccharide polymer made up of thousands of glucose units *466*

polyunsaturated hydrocarbon fatty acids that contain more than one double bond between carbon atoms per molecule *459*

postconsumer content used material that would otherwise have been discarded as waste *396*

potable water water that is fit for human consumption *195*

potential energy energy that is stored. Also called the energy of position. *153*

power density energy capacity per unit of fuel cell mass *343*

preconsumer content waste left over from the manufacturing process itself, such as scraps and clipping *396*

primary coolant liquid that comes in direct contact with the nuclear reactor core to carry away heat *293*

primary structure the unique identity and sequence of the amino acids that make up each protein *508*

product substance formed from reactants as a result of a chemical reaction *29*

protein polyamides (polypeptides) built from a long chain of amino acids *468, 506*

protein complementarity combining foods that complement essential amino acid content so that the total diet provides a complete supply of amino acids *471*

proton positively charged subatomic particle having the same mass as a neutron *60*

***p*-type semiconductor** semiconductor that contains freely moving positive charges, or "holes" *355*

Q

quantized noncontinuous energy distribution that consists of many individual steps *71, 118*

R

racemic mixture ($\pm$) mixture consisting of equal amounts of each optical isomer of a compound *425*

rad (radiation absorbed dose) unit of radiation that indicates absorption of 0.01 J of radiant energy per kilogram of tissue *305*

radiant energy the entire collection of different wavelengths, each with its own energy *69*

radiation sickness illness characterized by early symptoms of anemia, nausea, malaise, and susceptibility to infection that results from a large exposure to radiation *304*

radioactive decay series characteristic pathway of radioactive decay that begins with a radioisotope and progresses through a series of steps to eventually produce a stable isotope *299*

radioactivity spontaneous emission of radiation by certain elements *296*

reactant starting material that is transformed into a product during a chemical reaction *29*

reactive nitrogen compounds of nitrogen that are biologically active, chemically active, or active with light in our atmosphere *265*

recombinant DNA DNA that has incorporated DNA from another organism *514*

recyclable product product that can be recycled. They do not necessarily contain any recycled materials. *396*

recycled-content product product made from materials that otherwise would have been in the waste stream *396*

reduction half-reaction type of chemical equation that shows the reactant that gains electrons *332*

reforming process using heat, pressure, and catalysts to rearrange the atoms within molecules *344*

reformulated gasolines (RFGs) oxygenated gasolines that also contain a lower percentage of certain more volatile hydrocarbons such as benzene found in nonoxygenated conventional gasoline *177*

rem (roentgen equivalent man) unit of equivalent dose that indicates the damage done to human tissue by a particular dose of radiation. A rem is the number of rads multiplied by the quality factor Q. *305*

replication process of cell reproduction in which the cell copies and transmits its genetic information to its progeny *504*

resonance forms Lewis structures that represent hypothetical extremes of electron arrangements in a molecule *67*

respiration process by which humans and animals exchange the oxygen necessary for metabolism with the carbon dioxide produced by it *166*

reverse osmosis purification process that uses pressure to force the movement of a solvent through a semipermeable membrane from a region of high solute concentration to a region of lower solute concentration *231*

risk assessment organized evaluation of scientific data to predict the probability of an occurrence *17*

S

saturated hydrocarbon hydrocarbon chain containing only single bonds between the carbon atoms *459*

scientific notation system for writing numbers as the product of a number and 10 raised to the appropriate power *18*

second law of thermodynamics the statement that the entropy of the universe is constantly increasing *157*

secondary coolant water in the steam generators of a nuclear reactor core that does not come in contact with the core *293*

secondary pollutant pollutant produced from chemical reactions among two or more other pollutants *40*

secondary structure periodic, localized arrangement of the backbone segments of a protein chain *508*

semiconductor material that does not normally conduct electricity or heat well, but that can do so under certain conditions, such as exposure to sunlight *353*

sequestration process of keeping some things apart. Chemically this is accomplished by forming stable bonds between the sequestering agent and the substance "trapped." *136*

sievert (Sv) international unit equal to 100 rem *305*

significant figure a number that correctly represents the accuracy with which an experimental quantity is known *46*

single covalent bond a bond formed when only one pair of shared electrons forms the linkage between atoms *64*

solute substance that dissolves in a solvent *199*

solution homogeneous mixture of uniform composition *199*

solvent substance capable of dissolving other substances *199*

specific heat quantity of heat energy that must be absorbed to increase the temperature of 1 g of a substance by 1 °C *208*

spent nuclear fuel (SNF) radioactive material remaining in fuel rods after they have been used to generate power in a nuclear reactor. SNF is regulated as high-level radioactive waste (HLW). *312*

steady state condition in which a dynamic system is in balance so that no net change occurs in the concentration of the major species involved *74*

stem cells identical, undifferentiated cells that, by successive divisions, can give rise to specialized ones like blood cells *521*

steroids class of naturally occurring or synthetic fat-soluble organic compounds that share a common carbon skeleton arranged in four rings *427*

storage battery battery that is capable of storing electrical energy *336*

stratosphere region of the atmosphere above the troposphere; includes the ozone layer *20*

structural formula chemical representation that shows the atoms and their arrangement with respect to one another in a molecule. A structural formula replaces each bonded electron pair in a Lewis structure with a line. *64, 408*

structure–activity relationship (SAR) study systematic changes made to a drug molecule and assessment of the resulting changes in activity *420*

substituent atom or functional group substituted for a hydrogen atom *415*

substrate molecule (or molecules) being acted on, often catalytically by an enzyme *421, 510*

surface water water from lakes, rivers, and reservoirs *197*

T

temperature property of matter that determines the direction of heat flow *152*

tertiary structure overall shape or conformation of a protein molecule *510*

tetrahedron four-cornered figure with four equal triangular sides *111*

thermal cracking heating of starting materials to a high temperature *175*

thermoplastic polymer plastics that can be melted and reshaped over and over again *377*

toxicity intrinsic health hazard of a substance *17*

trace minerals minerals in the diet that are usually required in micrograms *475*

trans fats fats that have been transformed by the addition of hydrogen to unsaturated vegetable oils *462*

transgenic organisms artificially created higher plants and animals that share the genes of another species *516*

triglyceride an ester of three fatty acid molecules and one glycerol molecule *458*

triple bond covalent linkage made up of three pairs of shared electrons *67*

troposphere region of the atmosphere that lies directly above the surface of the Earth *20*

U

ultraviolet (UV) region portion of the electromagnetic spectrum that includes wavelengths shorter than those of the visible color of violet *69*

undernourishment condition in which a person's daily caloric intake is insufficient to meet metabolic needs *454*

unsaturated hydrocarbon hydrocarbon molecule that contains one or more double bonds between the carbon atoms *459*

V

vector modified plasmid used to carry DNA back into the bacterial "host" *514*

vitamin organic molecule with a wide range of physiological functions. Although only small amounts are needed in the diet, vitamins are essential for good health, proper metabolic functioning, and disease prevention. *473*

vitrification process in which the spent fuel elements or other mixed waste from a nuclear reactor are encased in ceramic or glass *313*

volatile refers to a substance that readily passes into the vapor phase *36*

volatile organic compounds (VOCs) vapors of incompletely burned gasoline molecules or fragments of these molecules *37*

voltage difference in electrochemical potential between the two electrodes *333*

volumetric flask type of glassware that contains a precise amount of solution when filled to the mark on its neck *202*

W

wavelength distance between successive peaks of waves in the electromagnetic spectrum *68*

wavenumbers numbers often expressed in units of cm^{-1} and used on the x-axis of an infrared spectrum; inversely proportional to wavelength *116*

work form of energy describing movement against a restraining force. Mathematically, work is equal to the force multiplied by the distance over which the motion occurs. *152*

X

X-ray diffraction crystallography technique that generates a pattern of deflected X-rays passing through a crystal to reveal the nature of the crystal lattice *502*

Credits

Photographs

CHAPTER 0

OPENER: © R. Ian Lloyd/Masterfile; **PAGE 5:** © The McGraw-Hill Companies, Inc./Jill Braaten, photographer; **PAGE 6:** © Michael Barnes/University of California; **PAGE 7:** Lucy Pryde Eubanks.

CHAPTER 1

OPENER: Image by Reto Stockli, NASA, Goddard Space Flight Center. Enhancements by Robert Simmon; **FIGURE 1.1:** Image provided by ORBIMAGE. © Orbital Imaging Corporation and processing by NASA Goddard Space Flight Center; **FIGURE 1.2:** Cathy Middlecamp; **FIGURE 1.4:** © Lon C. Diehl/PhotoEdit; **FIGURE 1.5:** EPA; **FIGURE 1.7:** © Galen Rowell/Corbis; **FIGURE 1.8:** Cathy Middlecamp; **FIGURE 1.11:** Image reproduced by permission of IBM Research, Almaden Research Center. Unauthorized use not permitted; **FIGURE 1.12:** © The McGraw-Hill Companies, Inc./Photo by Bob Coyle; **FIGURE 1.14:** Courtesy Corning Incorporated; **FIGURE 1.15a:** © The McGraw-Hill Companies, Inc./Photo by Eric Misko, Elite Images Photography; **FIGURE 1.15b:** © Chinch Gryniewicz, Ecoscene/Corbis; **FIGURE 1.17:** EPA; **FIGURE 1.18a:** © Image Source/ Corbis RF website; **FIGURE 1.18b:** © Digital Vision Vol. DV384/Getty; **FIGURE 1.19:** © The McGraw-Hill Companies, Inc./Photo by Ken Karp; **FIGURE 1.20:** © David M. Grossman/ Photo Researchers, Inc.; **FIGURE 1.21:** © Sheila Terry/Photo Researchers, Inc.; **PAGE 50:** Cathy Middlecamp; **PAGE 53 (left):** Courtesy EPA; **(right):** © Associated Press/AP; **PAGE 54 (all):** National Science Foundation.

CHAPTER 2

OPENER: © KNMI/ESA; **FIGURE 2.2 (all)** © The McGraw-Hill Companies, Inc./Photo by Stephen Frisch; **FIGURE 2.5:** © Philip Schermeister/National Geographic Image Collection; **FIGURE 2.12:** Courtesy Blue Lizard Products; **FIGURE 2.14 (all):** © The McGraw-Hill Companies, Inc./Photo by Stephen Frisch; **PAGE 85:** Courtesy Carlye Calvin; **FIGURE 2.17:** © David Hay Jones/Photo Researchers, Inc.; **FIGURE 2.20:** © Courtesy Pyrocool Technologies, Inc.

CHAPTER 3

OPENER: © Raymond Geham/ National Geographic Society; **FIGURE 3.1:** NASA; **FIGURE 3.3:** © Maria Stenzel/ National Geographic Image Collection; **FIGURE 3.7:** © Michael Newman/ PhotoEdit, Inc.; **FIGURE 3.19:** Courtesy Conrad Stanitski; **FIGURE 3.20 (all):** Courtesy Ocean Drilling Program; **FIGURE 3.21:** Courtesy of the Oak Ridge National Laboratory, managed by the U.S. Department of Energy by UT-Battelle, LLC; **PAGE 134:** www. wisconsinbutterflies.org; **FIGURE 3.25:** Illustration: Graphics, Statoil.

CHAPTER 4

OPENER (clockwise from upper left): © Craig Aurness/Corbis; Courtesy Jasper Environmental Association; © Associated Press/AP; © Vol. 31/PhotoDisc/Getty; Courtesy Dave Warren, USDA; Courtesy National Renewable Energy Laboratory; Courtesy National Renewable Energy Laboratory; **FIGURE 4.1:** © Digital Vision, Vol. DV418/Getty; **FIGURE 4.2:** © Charles D. Winters/Photo Researchers, Inc.; **FIGURE 4.10a:** © Martin Shields/ Photo Researchers, Inc.; **FIGURE 4.10b:** © Mark A. Schneider/Photo Researchers, Inc.; **FIGURE 4.13:** © Claudius/zefa/Corbis; **FIGURE 4.17:** © Justin Sullivan/Getty; **FIGURE 4.18:** Courtesy of the Reynolds Group; **FIGURE 4.20:** © Bob Daemmrich/Stock Boston; **FIGURE 4.21 (all):** © AP/Wide World Photos; **FIGURE 4.22:** Courtesy A. Truman Schwartz; **FIGURE 4.24 (all):** © Associated Press/AP.

CHAPTER 5

OPENER (clockwise from upper left): © Babu/Reuters/Corbis; © Jacques Langevin/Corbis Sygma; © Brand X Pictures/PunchStock; © Piyal Adhikary/ epa/Corbis; © AFP/Getty Images; © Vladimir Pirogov/Reuters/Corbis; **FIGURE 5.1a:** © Norbert Schaefer/ Corbis; **FIGURE 5.1b:** © LWA-Stephen Welstead/Corbis; **PAGE 195:** Courtesy Morgantown Utility Board, Morgantown, WV; **FIGURE 5.3:** © PunchStock RF; **PAGE 200:** © The McGraw-Hill Companies, Inc./Jill Braaten, photographer; **FIGURE 5.12 (all):** © Tom Pantages; **FIGURE 5.20:** © Robert Landau/Corbis; **FIGURE 5.25:** © 2006 Compare Infobase Limited; **FIGURE 5.28:** Courtesy of Katadyn.

CHAPTER 6

OPENER: © PressNet/Topham/The Image Works; **FIGURE 6.1:** © Ted Spiegel/Corbis; **FIGURE 6.2:** © PhotoDisc Vol. 77/Getty; **FIGURE 6.3:** © The McGraw-Hill Companies, Inc./Photo by Eric Misko, Elite Images Photography; **FIGURE 6.4:** © The McGraw-Hill Companies, Inc./Photo by Eric Misko, Elite Images Photography; **FIGURE 6.5:** © The McGraw-Hill Companies, Inc./Photo by C. P. Hammond; **FIGURE 6.8:** © Charles D. Winters/Photo Researchers, Inc.; **FIGURE 6.9a:** Cathy Middlecamp; **FIGURE 6.10 (all):** Cathy Middlecamp; **FIGURE 6.13:** © E. R. Degginger/ Color Pic; **FIGURE 6.15:** © Kennon Cooke/Valan Photos; **FIGURE 6.18a:** © NYC Parks Photo Archive/Fundamental Photographs; **FIGURE 6.18b:** © Kristen Brochmann/

Fundamental Photographs; **FIGURE 6.19:** © A. J. Copley/Visuals Unlimited; **FIGURE 6.20 (all):** U.S. National Park Service; **FIGURE 6.21b:** © Pittsburgh Post Gazette Archives, 2004. All rights reserved. Reprinted with permission; **FIGURE 6.22:** © M. Kaleb/Custom Medical Stock Photo; **FIGURE 6.26:** © Phil McCarten/PhotoEdit, Inc.; **FIGURE 6.27:** Clean School Bus, USA, USEPA Office of Air Transportation and Air Quality; **FIGURE 6.28:** National Energy Technology Laboratory/ U.S. Department of Energy; **PAGE 277:** © Stock Portfolio/Stock Connection/Picture Quest; **PAGE 278:** Cathy Middlecamp; **PAGE 279:** Cathy Middlecamp.

CHAPTER 7

OPENER (top): © Associated Press/ AP; **(bottom):** © Brand X Pictures/ PunchStock; **FIGURE 7.1:** © The McGraw-Hill Companies, Inc./Photo by Eric Misko, Elite Images Photography; **PAGE 285:** © Rob Crandall/The Image Works; **FIGURE 7.3:** © Bettmann/ Corbis; **FIGURE 7.5:** Department of Energy; **FIGURE 7.7:** © The McGraw-Hill Companies, Inc./ Photo by C. P. Hammond; **FIGURE 7.8:** © Associated Press/AP; **FIGURE 7.9:** © Associated Press/AP; **PAGE 294:** © Larry Lee Photography/Corbis; **FIGURE 7.12:** © Hulton-Deutsch Collection/ Corbis; **FIGURE 7.15:** © Associated Press/AP; **FIGURE 7.16:** © Chuck Nacke/Time Life Pictures/Getty Images; **FIGURE 7.17:** © James Hill/ Contact Press Images; **FIGURE 7.18:** © Southern Illinois University/Photo Researchers, Inc.; **FIGURE 7.22:** © Peter Essick/ Aurora Photos; **FIGURE 7.24:** © Science Source/ Photo Researchers, Inc.; **FIGURE 7.25b:** U.S. Department of Energy; **FIGURE 7.26:** © Dan Lamont/Corbis; **FIGURE 7.27:** Courtesy USEG, Inc.; **FIGURE 7.28:** Copyright by Space Imaging/NASA; **FIGURE 7.29:** © AP/ Wide World Photos; **FIGURE 7.30:** © 2006 Westinghouse Electric Company, LLC. All rights reserved.

CHAPTER 8

OPENER (clockwise from upper left): © Digital Vision Vol. DV673/ Getty; © Getty Images; © AFP/Getty Images; Courtesy Steven W. Keller; © AFP/Getty Images; **FIGURE 8.1:** © The McGraw-Hill Companies, Inc./ Jill Braaten, photographer; **FIGURE 8.3:** © The McGraw-Hill Companies, Inc./Photo by Eric Misko, Elite Images Photography; **FIGURE 8.6 (all):** © AP/Wide World Photos; **FIGURE 8.10:** © Getty Images; **FIGURE 8.12:** Pacific Northwest National Laboratory; **FIGURE 8.13:** Siemens Press Picture; **FIGURE 8.17:** © Michael Barnes, University of California; **FIGURE 8.18:** © J. Karl Johnson, University of Pittsburgh and National Energy Technology Laboratory; **PAGE 353:** © Arctic Images/Corbis; **FIGURE 8.20:** Warren Gretz/DOE/NREL; **FIGURE 8.24:** Figure courtesy of www.qahill. com; **FIGURE 8.27:** © Daniel Karmann/ DPA/Corbis; **FIGURE 8.28:** Image courtesy of www. cleanenergy.org; **FIGURE 8.29:** Courtesy DOE/NREL; **FIGURE 8.30:** Photo by Stefano Paltera/North American Solar Challenge.

CHAPTER 9

OPENER: © Mark Richards/ PhotoEdit, Inc.; **PAGE 369:** © Dynamic Graphics/JupiterImages; **FIGURE 9.2 (all), FIGURE 9.3a:** Cathy Middlecamp; **FIGURE 9.5 (all):** © The McGraw-Hill Companies, Inc./Jill Braaten, photographer; **FIGURE 9.6a:** © Bill Aaron/PhotoEdit; **FIGURES 9.9, 9.11:** © The McGraw-Hill Companies, Inc./Jill Braaten, photographer; **FIGURES 9.12, 9.13:** Courtesy DuPont; **FIGURE 9.16:** © The Garbage Project, University of Arizona; **FIGURE 9.23:** Courtesy A. N. Wyeth; **FIGURE 9.24:** © Gayna Hoffman/Stock Boston.

CHAPTER 10

OPENER (clockwise from upper left): © Bushnell/Soifer/Getty; © The McGraw-Hill Companies, Inc./Erica Simone, photographer; © Comstock Images/Getty R-F; © Quill/Getty Images; **FIGURE 10.1:** © Terry Wild Studio; **FIGURE 10.7 (all):** © The McGraw-Hill Companies, Inc./Jill Braaten,

photographer; **FIGURE 10.10:** Courtesy of the Alexander Fleming Laboratory Museum, St. Mary's Hospital, Paddington, London; **FIGURE 10.11:** Library of Congress; **FIGURE 10.15:** Photo courtesy John M. Rimoldi, University of Mississippi; **FIGURE 10.27a:** © Odd Andersen/AFP/Getty Images; **FIGURE 10.27b:** © Rick Nederstigt/ AFP/Getty Images; **FIGURE 10.29a:** © Bill Aron/PhotoEdit, Inc.; **FIGURE 10.29b:** © The McGraw-Hill Companies, Inc./Jill Braaten, photographer; **FIGURE 10.30(all):** © Michael P. Gadomski/Photo Researchers, Inc.; **FIGURE 10.31a:** © Gerald & Buff Corsi/Visuals Unlimited; **FIGURE 10.31b:** © James Leynse/Corbis; **FIGURE 10.33:** © Chris Knapton/Photo Researchers, Inc.; **FIGURE 10.35:** © 2004, Publishers Group www.streetdrugs.org.

CHAPTER 11

FIGURE 11.1: Federal Citizen Information Center; **FIGURE 11.15:** © Len Lessin/Peter Arnold; **PAGE 482:** © Onoky/SuperStock; **FIGURE 11.21:** Courtesy FAO; **FIGURE 11.24:** © Tony Freeman/PhotoEdit.

CHAPTER 12

OPENER: © George B. Diebold/ Corbis; **FIGURE 12.5a:** © Science Source/Photo Researchers, Inc.; **FIGURE 12.5b:** King's College London Archives; **FIGURE 12.6:** © Bettmann/Corbis; **FIGURE 12.13 (all):** © The McGraw-Hill Companies, Inc./Jill Braaten, photographer; **FIGURE 12.15 (all):** © Bill Longcore/ Photo Researchers, Inc.; **FIGURE 12.16:** © Sandro Castelli; **FIGURE 12.18:** © The McGraw-Hill Companies, Inc./Jill Braaten, photographer; **FIGURE 12.19:** © Dung Vo Trung/ Corbis; **PAGE 517:** © Cone 6 Productions/Brand X/Corbis; **PAGE 518:** © Bettmann/ Corbis; **FIGURE 12.20:** © CLARO CORTES IV/ Reuters/Corbis; **FIGURE 12.21a:** © Reuters/Corbis; **FIGURE 12.21b, c:** © Seoul National University/Handout/ Reuters/Corbis; **FIGURE 12.23 (all)** Courtesy of the Roslin Institute.

Index

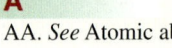